Spring 5 高级编程

(第 5 版)

[美] 尤莉安娜·科斯米纳(Iuliana Cosmina)
罗布·哈罗普(Rob Harrop)
克里斯·舍弗(Chris Schaefer)
克拉伦斯·厚(Clarence Ho) 著

王 净 译

清华大学出版社

北 京

Pro Spring 5, Fifth Edition

Iuliana Cosmina, Rob Harrop, Chris Schaefer, Clarence Ho

EISBN：978-1-4842-2807-4

北京市版权局著作权合同登记号　图字：01-2018-2643

图书在版编目(CIP)数据

Spring 5高级编程：第5版 / (美)尤莉安娜•科斯米纳(Iuliana Cosmina) 等著；王净 译. —北京：清华大学出版社，2019

书名原文：Pro Spring 5, Fifth Edition

ISBN 978-7-302-51644-6

Ⅰ. ①S…　Ⅱ. ①尤… ②王…　Ⅲ. JAVA 语言—程序设计—教材　Ⅳ. ①TP312.8

中国版本图书馆 CIP 数据核字(2018)第 257354 号

责任编辑： 王　军
装帧设计： 孔祥峰
责任校对： 牛艳敏
责任印制： 李红英

出版发行： 清华大学出版社
网　址：http://www.tup.com.cn，http://www.wqbook.com
地　址：北京清华大学学研大厦 A 座　　邮　编：100084
社 总 机：010-62770175　　邮　购：010-62786544
投稿与读者服务：010-62776969，c-service@tup.tsinghua.edu.cn
质 量 反 馈：010-62772015，zhiliang@tup.tsinghua.edu.cn

印 装 者： 三河市铭诚印务有限公司

经　销： 全国新华书店
开　本： 190mm×260mm　　**印　张：** 34.75　　**字　数：** 1279 千字
版　次： 2019 年 1 月第 1 版　　**印　次：** 2019 年 1 月第 1 次印刷
定　价： 128.00 元

产品编号：079785-01

译 者 序

Spring 是一种多层的 J2EE 应用程序框架，以 Rod Johnson 撰写的 *Expert One-on-One J2EE Design and Development* 一书的代码为基础发展而来。Spring 的核心是提供一种新的机制来管理业务对象及其依赖关系。例如，可以指定一个 DAO(数据访问对象)类依赖一个 DataSource 类。此外，还允许开发人员通过接口编程，使用 XML 文件来简单地定义其实现。Spring 拥有许多类来支持其他的框架(如 Hibernate 和 Struts)，这使得集成变得易如反掌。

Spring 5 是自 2013 年 12 月的版本 4 发布之后，Spring 的第一个主发行版。Spring 项目的负责人 Juergen Hoeller 于 2016 年 7 月 28 日发布了第一个 Spring 5 里程碑版本。整个 Spring 5 代码库运行在 Java 8 上，因此 Spring 5 对环境的最低要求就是 Java 8。作为开发者的我们而言，已经能够借此享受到现代 Java 发行版中的所有新特性了。

《Spring 5 高级编程(第 5 版)》一书共 18 章，深入浅出地介绍了 Spring 5 所有的相关知识，其中重点介绍如何使用 Java 配置类、lambda 表达式、SpringBoot 以及反应式编程，此外，还与读者分享了一些实际的开发经验，包括远程处理、事务、Web 和表示层等。按照内容，大致可分为四大部分。第一部分(包括第 1~5 章)主要介绍 Spring 5 相关的基础知识，包括 Spring 框架的演变、翻转控制和依赖注入、Spring 配置以及 AOP。第二部分(第 6~12 章)主要介绍 Spring 的一些特殊功能，包括 JDBC 支持、使用 Hibernate、数据访问、事务管理、任务调度以及远程控制。第三部分(第 13~15 章)主要介绍 Spring 测试、脚本支持以及应用程序监控。第四部分(第 16~18 章)主要介绍 Spring 在网络方面的应用以及对 Spring 项目组合中的一些项目进行概述，特别是 Spring Batch、Spring Integration、Spring XD 以及 Spring 5 中增加的一些新功能。

本书适合有一定 Java 编程基础的读者阅读，适用于在 Java 平台下进行各类软件开发的开发人员、测试人员，尤其适用于企业级 Java 开发人员。既可以被刚开始学习 Spring 的读者当作学习指南，也可以被那些想深入了解 Spring 某方面功能的资深用户作为参考用书。

参与本书翻译的人有王净、田洪、范园芳、范桢、胡训强、纪红、彭洁、晏峰、余佳隽、张洁以及赵翊含，最终由王净负责统稿。此外，还要感谢我的家人，他们总是无怨无悔地支持我的一切工作，我为有这样的家庭而感到幸福。

译者在翻译过程中，尽量保持原书的特色，并对书中出现的术语和难词、难句进行了仔细推敲和研究。但毕竟有少量技术是译者在自己的研究领域中不曾遇到的，所以疏漏和争议之处在所难免，望广大读者提出宝贵意见。

最后，希望广大读者能多花些时间细细品味这本凝聚作者和译者大量心血的书籍，为将来的职业生涯奠定良好的基础。

王 净

作于广州

作 者 简 介

Iuliana Cosmina 是一名 Spring 认证的 Web 应用程序开发人员，也是 Spring 认证的 Spring 专家(由 Pivotal 定义，Pivotal 是 Spring 框架、Spring Boot 以及其他工具的制造商)。她曾与 Apress 出版社合作出版了多本关于核心 Spring 认证和 Spring 认证 Web 开发的书籍。她是来自 Bearing Point Software 的一名软件架构师，也是 GitHub、Stack Overflow 等平台上活跃的编码者和软件贡献者。

Rob Harrop 是一位软件顾问，致力于提供高性能、高度可扩展的企业级应用程序。他是一位经验丰富的架构师，对于理解和解决复杂的设计问题具有极高天赋。凭借扎实的 Java 和.NET 开发功力，Harrop 已经成功在两种平台上部署不少项目。此外，他还在其他行业拥有丰富的经验，尤其是零售和政府领域。Harrop 共独自撰写或参与撰写了 5 本书，其中就包括本书(当然不是第 5 版)，该书广受好评，包含大量关于 Spring 框架的资源。

Chris Schaefer 是 Pivotal Spring 项目的主要软件开发人员，Pivotal 是 Spring 框架、Spring Boot 以及其他 Spring 工具的制造商。

Clarence Ho 是中国香港软件咨询公司 SkywideSoft Technology Limited 的 Java 高级架构师。Clarence 在 IT 领域工作了 20 多年，一直担任许多内部应用程序开发项目的团队负责人，并为客户提供有关企业解决方案的咨询服务。

技术审校者简介

Massimo Nardone 在安全、Web/移动开发、云计算和 IT 架构方面拥有超过 23 年的经验，但他真正感兴趣的是安全和 Android。

他目前担任 Cargotec Oyj 的首席信息安全官(CISO)，并且是 ISACA 芬兰分会董事会成员。在他长期的职业生涯中，曾担任过以下职位：项目经理、软件工程师、研究工程师、首席安全架构师、信息安全经理、PCI/SCADA 审计师以及高级 IT 安全/云/ SCADA 架构师。此外，他还是赫尔辛基技术大学网络实验室的客座讲师和主管。

Massimo 拥有意大利萨勒诺大学的计算科学硕士学位，拥有 4 项国际专利(PKI、SIP、SAML 和代理领域)。除审校本书外，Massimo 还为多家出版公司审校了 40 多本 IT 书籍，并且是 *Pro Android Games*(Apress 出版社于 2015 年出版)一书的合著者。

致　谢

作为本书第 5 版的主要作者，我感到非常荣幸。一开始，我以为是要审校本书，直到后来，才意识到自己要主笔，激动不已。

在求学甚至以后的工作过程中，我读过 Apress 出版的许多书，不断提高自己的专业水平。本书是我与 Apress 合作出版的第三本书，能为培养下一代开发人员做出贡献是一件好事。

我十分感谢那些耐心倾听我抱怨失眠、工作太多、写作思路不畅的朋友。谢谢大家的支持，是你们让我在编写这本书时感到一些乐趣。

另外，我还要感谢我最喜爱的所有歌手，是他们的美妙音乐让我更轻松愉快地完成工作，尤其是 John Mayer；我之所以决心按时完成这本书，是为了可以去美国参加他的一场音乐会。这也是为什么我将书中的示例主题更改为这些歌手及其音乐的原因，这是对他们艺术和才能的致敬。

—Iuliana Cosmina

前　言

本书涵盖 Spring 5 的所有内容，如果想要充分利用这一领先的企业级 Java 应用程序开发框架的强大功能，本书是最全面的 Spring 参考和实用指南。

本书第 5 版涵盖核心的 Spring 及其与其他领先的 Java 技术(比如 Hibernate、JPA 2、Tiles、Thymeleaf 和 WebSocket)的集成。本书的重点是介绍如何使用 Java 配置类、lambda 表达式、Spring Boot 以及反应式编程。同时，将与企业级应用程序开发人员分享一些见解和实际经验，包括远程处理、事务、Web 和表示层，等等。

通过本书，你可以学习如何完成以下事情：

- 使用控制反转(IoC)和依赖注入(DI)。
- 了解 Spring 5 中的新功能。
- 使用 Spring MVC 和 WebSocket 构建基于 Spring 的 Web 应用程序。
- 使用 Spring WebFlux 构建 Spring Web 反应式应用程序。
- 使用 JUnit 5 测试 Spring 应用程序。
- 使用新的 Java 8 lambda 语法。
- 使用 Spring Boot 达到更高的水平，以获取任何类型的 Spring 应用程序并立即运行。
- 在 Spring 应用程序中使用 Java 9 的新功能。

由于 Java 9 的发布日期不断推迟，因此 Spring 5 是基于 Java 8 发布的。书中介绍的与 Java 9 的互操作性都是基于早期的可访问版本完成的。

有一个与本书相关的多模块项目，使用 Gradle 4 进行配置。该项目位于 Apress 的官方存储库中，参见 https://github.com/Apress/pro-spring-5。只要在本机上安装了 Gradle，根据 README.adoc 文件中的说明即可在克隆后立即构建该项目。如果尚未安装 Gradle，那么可以使用 IntelliJ IDEA 下载并使用 Gradle Wrapper 来构建自己的项目(https://docs.gradle.org/current/userguide/gradle_wrapper.html)。本书末尾的附录 A 介绍了项目结构、配置以及可用于开发和运行本书代码示例的开发工具的其他详细信息，这些工具可在 GitHub 上找到。

在编写本书的过程中，Spring 5 新的候选版本发布了，还发布了新版本的 IntelliJ IDEA，同时书中使用的 Gradle 以及其他技术的版本也更新了。因此，我们及时升级到新版本，以便提供最新信息，并使本书与官方文档保持同步。多位专家已经对本书的技术准确性进行了审查，但是如果发现任何不一致的地方，请发送电子邮件至 editorial@apress.com 并创建勘误项。

可以通过 www.apress.com/9781484228074 网页上的 Download Source Code 按钮访问本书的示例源代码。我们将对这些代码进行及时维护，以便与新技术保持同步，同时根据开发人员在学习 Spring 的过程中提出的建议不断进行丰富。

真心希望你喜欢用本书来学习 Spring，就像我们喜欢编写本书一样。

目　录

第1章

Spring 介绍

提到Java开发人员社区，我们会不自觉地想起19世纪40年代末的淘金者，他们在北美的河流上疯狂地淘金，寻找黄金碎片。作为Java开发人员，我们的“河流”充斥着开源项目，但像淘金者一样，找到一个有用的项目可能会既费时又费力。

对许多开源Java项目的常见抱怨是，它们仅仅是为了填补与最新技术或模式的差距而创建的。话虽如此，但许多高质量、可用的项目满足并解决了真实应用程序的实际需求，在阅读本书的过程中，就会遇到部分此类项目，从而更好地了解Spring。Spring的第一个版本于2002年10月发布，由一个带有易于配置和使用的控制反转(IoC)容器的小型内核组成。多年来，Spring已经成为Java Enterprise Edition(JEE)服务器的主要替代品，并且发展成为一个由许多不同项目组成的成熟技术，每个项目都有自己的目的，因此无论是想要构建微服务、应用程序还是经典的ERP，Spring都有一个项目可以满足需求。

在本书中，你将会看到许多使用了不同开源技术的应用程序，所有这些应用程序都整合在Spring框架下。在使用Spring时，应用程序开发人员可以使用各种开源工具，而无须编写大量代码，也不需要将应用程序与任何特定工具紧密耦合。

如章名所示，本章的主要内容是介绍Spring框架，而不是提供任何可靠的示例或说明。如果你已经熟悉Spring，那么可以跳过本章并直接进入第2章。

1.1 什么是Spring

如果想要解释Spring，那么最难的部分就是对其进行分类。通常情况下，Spring被描述为构建Java应用程序的轻量级框架，但这种描述带来了两个有趣的观点。

首先，与许多其他框架(比如仅限于Web应用程序的Apache Struts)不同，可以使用Spring构建Java中的任何应用程序(例如，独立的应用程序、Web应用程序或JEE应用程序)。

其次，该描述中的*轻量级*一词真正指的并不是类的数量或发布的大小，而是整体性定义Spring原则：最轻的影响。从某种意义上讲，Spring是轻量级的，因为只需要对应用程序代码进行很少的更改(如果有的话)就可以获得Spring Core所带来的好处。如果想要在任何时候停止使用Spring，那么你会发现可以很容易做到。

请注意，上述描述仅针对Spring Core——许多额外的Spring组件(例如数据访问)需要更紧密地与Spring框架耦合。然而，这种耦合的好处是非常明显的，在本书中将介绍相关的技术来最大限度地减少这种耦合对应用程序的影响。

1.1.1 Spring框架的演变

Spring框架源自Rod Johnson编写的Expert One-on-One：J2EE Design and Development一书(Wrox出版社，2002年出版)。在过去十年中，Spring框架在核心功能、相关项目以及社区支持方面发展迅猛。随着Spring框架的新主要版本的推出，有必要快速回顾一下Spring的每个里程碑版本所带来的重要特性，并最终发展到Spring Framework 5.0。

- Spring 0.9：这是该框架第一个公开发布的版本，以 *Expert One-on-One：J2EE Design and Development* 一书为基础，提供了 bean 配置基础、AOP 支持、JDBC 抽象框架、抽象事务支持等。该版本没有官方参考文档，但可以在 SourceForge 上找到现有的源代码和文档[1]。
- Spring 1.x：这是发布的第一个带有官方参考文档的版本。它由图 1-1 所示的七个模块组成。
  - Spring Core：bean 容器以及支持的实用程序。
  - Spring Context：ApplicationContext、UI、验证、JNDI、Enterprise JavaBean(EJB)、远程处理和邮件支持。
  - Spring DAO：事务基础结构、Java Database Connectivity(JDBC)和数据访问对象(DAO)支持。
  - Spring ORM：Hibernate、iBATIS 和 Java Data Object(JDO)支持。
  - Spring AOP：符合 AOP 联盟的面向方面编程(AOP)实现。
  - Spring Web：基本集成功能，比如多部分功能、通过 servlet 侦听器进行上下文初始化以及面向 Web 的应用程序上下文。
  - Spring Web MVC：基于 Web 的 Model-View-Controller(MVC)框架。
- Spring 2.x：该版本由图 1-2 所示的六个模块组成。现在，Spring Context 模块包含在 Spring Core 中，而在 Spring 2.x 版本中，所有的 Spring Web 组件都由单个项目表示。

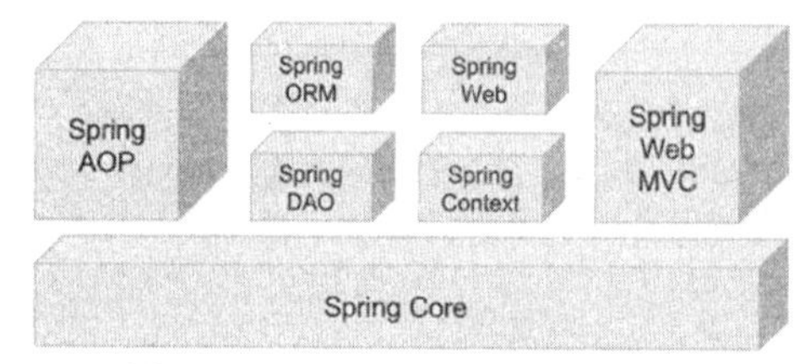

图 1-1　Spring Framework 1.x 版本概述

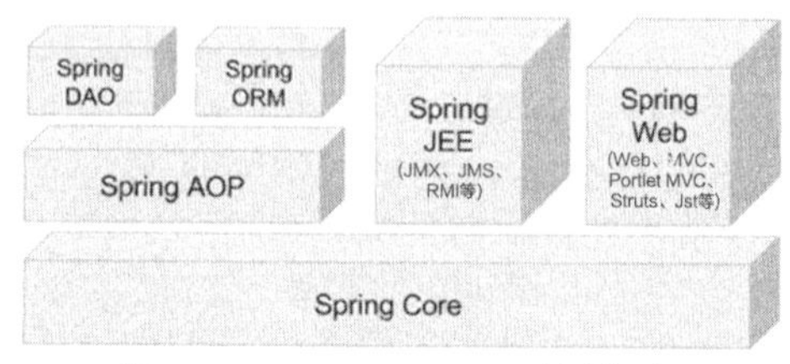

图 1-2　Spring Framework 2.x 版本概述

  - 通过使用新的基于 XML Schema 的配置而不是 DTD 格式来简化 XML 配置。值得注意的改进方面包括 bean 定义、AOP 以及声明式事务。
  - 用于 Web 和门户的新 bean 作用域(请求、会话和全局会话)。
  - 支持 AOP 开发的@AspectJ 注解。
  - Java Persistence API(JPA)抽象层。
  - 完全支持异步 JMS 消息驱动的 POJO(用于普通的旧 Java 对象)。
  - JDBC 简化包括在使用 Java 5+时的 SimpleJdbcTemplate。
  - JDBC 命名参数支持(NamedParameterJdbcTemplate)。
  - 针对 Spring MVC 的表单标签库。
  - 对 Portlet MVC 框架的介绍。
  - 动态语言支持。可以使用 JRuby、Groovy 以及 BeanShell 来编写 bean。
  - JMX 中的通知支持以及可控的 MBean 注册。
  - 为调度任务而引入的 TaskExecutor 抽象。
  - Java 5 注解支持，特别针对@Transactional、@Required 和@AspectJ。
- Spring 2.5.x：该版本包含以下功能。
  - 名为@Autowired 的新配置注解以及对 JSR-250 注解(@Resource、@PostConstruct 和@PreDestroy)的支持。
  - 新的构造型注解：@Component、@ Repository、@Service 和@Controller。
  - 自动类路径扫描支持，可以检测和连接带有构造型注解的类。
  - AOP 更新，包括一个新的 bean 切入点元素以及 AspectJ 加载时织入(weaving)。
  - 完整的 WebSphere 事务管理支持。
  - 除了 Spring MVC @Controller 注解，还添加了@RequestMapping、@RequestParam 和@ModelAttribute 注解，从而支持通过注解配置进行请求处理。
  - 支持 Tiles 2。

1　可以从 SourceForge 站点下载 Spring 的早期版本(包括 0.9 版本)：https://sourceforge.net/projects/springframework/files/ springframework/。

  - 支持 JSF 1.2。
  - 支持 JAX-WS 2.0/2.1。
  - 引入了 Spring TestContext Framework，提供注解驱动和集成测试支持，不受所用测试框架的影响。
  - 能够将 Spring 应用程序上下文部署为 JCA 适配器。
- Spring 3.0.x：这是基于 Java 5 的 Spring 的第一个版本，旨在充分利用 Java 5 的功能，如泛型、可变参数和其他语言改进。该版本引入了基于 Java 的@Configuration 模型。目前已经对框架模块进行了修改，分别针对每个模块 JAR 使用一棵源代码树进行管理。图 1-3 中对此进行了抽象描述。

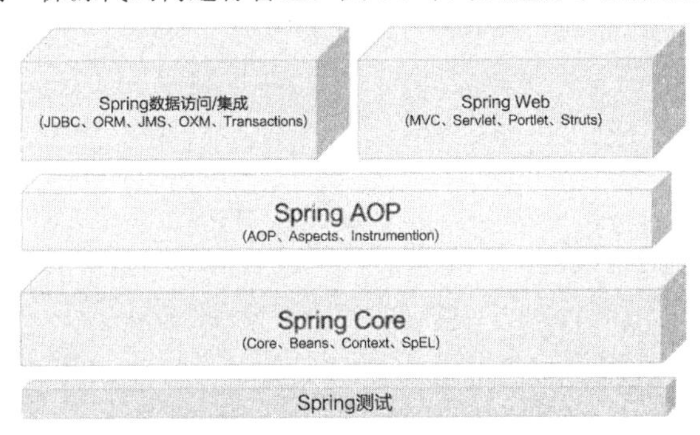

图 1-3　Spring Framework 3.0.x 版本概述

  - 支持 Java 5 功能，例如泛型、可变参数以及其他改进。
  - 对 Callables、Futures、ExecutorService 适配器和 ThreadFactory 集成提供很好的支持。
  - 框架模块目前针对每个模块 JAR 都使用一棵源代码树进行分别管理。
  - Spring Expression Language(SpEL)的引入。
  - 核心 Java Config 功能和注解的集成。
  - 通用型转换系统和字段格式化系统。
  - 全面支持 REST。
  - 新的 MVC XML 名称空间和其他注解，例如 Spring MVC 中的@CookieValue 和@RequestHeaders。
  - 验证增强功能和 JSR-303(bean 验证)支持。
  - 对 Java EE 6 的早期支持，包括@ Async/@Asynchronous 注解、JSR-303、JSF 2.0、JPA 2.0 等。
  - 支持嵌入式数据库，例如 HSQL、H2 和 Derby。
- Spring 3.1.x：该版本包含以下功能。
  - 新的缓冲抽象。
  - 可以用 XML 定义 bean 定义配置文件，同时也支持@Profile 注解。
  - 针对统一属性管理的环境抽象。
  - 与常见 Spring XML 名称空间元素等价的注解，如@ComponentScan、@EnableTransactionManagement、@EnableCaching、@EnableWebMvc、@EnableScheduling、@EnableAsync、@EnableAspectJAutoProxy、@EnableLoadTimeWeaving 和@EnableSpringConfigured。
  - 支持 Hibernate 4。
  - Spring TestContext Framework 对@Configuration 类和 bean 定义配置文件的支持。
  - 名称空间 c:简化了构造函数注入。
  - 支持 Servlet 3 中 Servlet 容器的基于代码的配置。
  - 能够在不使用 persistence.xml 的情况下启动 JPA EntityManagerFactory。
  - 将 Flash 和 RedirectAttributes 添加到 Spring MVC 中，从而允许通过使用 HTTP 会话重定向属性
  - URI 模板变量增强功能。
  - 能够使用@Valid 来注解 Spring MVC @RequestBody 控制器方法参数。
  - 能够使用@RequestPart 来注解 Spring MVC 控制器方法参数。

- Spring 3.2.x：该版本包含以下功能。
  - ◆ 支持基于 Servlet 3 的异步请求处理。
  - ◆ 新的 Spring MVC 测试框架。
  - ◆ 新的 Spring MVC 注解@ControllerAdvice 和@MatrixVariable。
  - ◆ 支持 RestTemplate 和@RequestBody 参数中的泛型类型。
  - ◆ 支持 Jackson JSON 2。
  - ◆ 支持 Tiles 3。
  - ◆ 现在，@RequestBody或@RequestPart参数的后面可以跟一个Errors参数，从而可以对验证错误进行处理。
  - ◆ 能够通过使用 MVC 名称空间和 Java Config 配置选项来排除 URL 模式。
  - ◆ 支持没有 Joda Time 的@DateTimeFormat。
  - ◆ 全局日期和时间格式化。
  - ◆ 跨框架的并发优化，从而最小化锁定，并改进了作用域/原型 bean 的并发创建。
  - ◆ 新的基于 Gradle 的构建系统。
  - ◆ 迁移到 GitHub(https://github.com/SpringSource/spring-framework)。
  - ◆ 在框架和第三方依赖中支持精简的 Java SE 7/OpenJDK 7。现在，CGLIB 和 ASM 已经成为 Spring 的一部分。除了 AspectJ 1.6，其他版本都支持 AspectJ 1.7。
- Spring 4.0x：这是一个重要的 Spring 版本，也是第一个完全支持 Java 8 的版本。虽然仍然可以使用较旧版本的 Java，但 Java SE6 已经提出了最低版本要求。弃用的类和方法已被删除，但模块组织几乎相同，如图 1-4 所示。

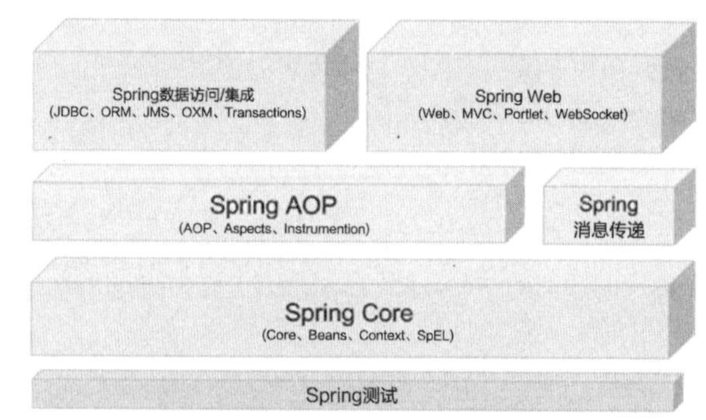

图 1-4　Spring Framework 4.0.x 版本概述

  - ◆ 通过新的 www.spring.io/guides 网站上的一系列入门指南提高了入门体验。
  - ◆ 从先前的 Spring 3 版本中删除了弃用的软件包和方法。
  - ◆ 支持 Java 8，将最低 Java 版本提高到 6 update 18。
  - ◆ Java EE 6 及以上版本现在被认为是 Spring Framework 4.0 的基准。
  - ◆ Groovy bean 定义 DSL，允许通过 Groovy 语法配置 bean 定义。
  - ◆ 核心容器、测试和一般 Web 改进。
  - ◆ WebSocket、SockJS 和 STOMP 消息。
- Spring 4.2.x：该版本包含以下功能。
  - ◆ 核心改进(例如，引入@AliasFor，并修改现有的注解以使用它)。
  - ◆ 全面支持 Hibernate ORM 5.0。
  - ◆ JMS 和 Web 改进。
  - ◆ 对 WebSocket 消息传递的改进。
  - ◆ 测试改进，最引人注目的是引入了@Commit 来替换@Rollback(false)，并引入了 AopTestUtils 实用工具类，允许访问隐藏在 Spring 代理后面的底层对象。
- Spring 4.3.x：该版本包含以下功能。
  - ◆ 完善了编程模型。

    - 在核心容器(包含 ASM 5.1、CGLIB 3.2.4 以及 spring-core.jar 中的 Objenesis 2.4)和 MVC 方面有了相当大的改进。
    - 添加了组合注解。
    - Spring TestContext Framework 需要 JUnit 4.12 或更高版本。
    - 支持新的库，包括 Hibernate ORM 5.2、Hibernate Validator 5.3、Tomcat 8.5 和 9.0、Jackson 2.8 等。
- Spring 5.x：这是一个主要版本。整个框架代码库都基于 Java 8，并且自 2016 年 7 月起与 Java 9 完全兼容[1]。
    - 支持 Portlet、Velocity、JasperReports、XMLBeans、JDO、Guava、Tiles 2 和 Hibernate 3。
    - 现在，XML 配置名称空间被流式传输到未版本化的模式；虽然特定版本的声明仍然被支持，但要针对最新的 XSD 架构进行验证。
    - 充分利用 Java 8 的强大功能，从而在性能上得到极大的改进。
    - Resource 抽象为防御 getFile 访问提供了 isFile 指示符。
    - Spring 提供的 Filter 实现完全支持 Servlet 3.1 签名。
    - 支持 Protobuf 3.0。
    - 支持 JMS 2.0+和 JPA 2.1+。
    - 引入了 Spring Web Flow，这是一个用于替代 Spring MVC 的项目，构建在反应式基础之上，这意味着它完全是异步和非阻塞的，主要用于事件循环执行模型，而非传统的每个请求执行模型都带有一个线程的大型线程池(基于 Project Reactor 构建[2])。
    - Web 和核心模块适用于反应式编程模型[3]。
    - Spring 测试模块有了很大的改进。现在支持 JUnit 5，引入了新的注解来支持 Jupiter 编程和扩展模型，例如@SpringJUnitConfig、@SpringJUnitWebConfig、@EnabledIf 和@DisabledIf。
    - 支持在 Spring TestContext Framework 中实现并行测试执行。

1.1.2 翻转控制或依赖注入

Spring 框架的核心是基于控制反转(Inversion of Control，IoC)的原理。IoC 是一种将组件依赖项的创建和管理外部化的技术。请参考一个示例，比如类 Foo 依赖于 Bar 类的一个实例来执行某种处理。传统上，Foo 通过使用 new 操作符来创建 Bar 实例或从某个工厂类中获取 Bar 实例。而如果使用 IoC 方法，Bar 的一个实例(或一个子类)在运行时由某个外部进程提供给 Foo。这种在运行时注入依赖项的行为促使 Martin Fowler 将 IoC 重命名为更具描述性的依赖注入(Dependency Injection，DI)。第 3 章将讨论 DI 所管理的依赖项的具体特性。

⚠ 正如将在第 3 章中看到的那样，当提到控制反转时，使用术语依赖注入始终是正确的。在 Spring 的上下文中，可以互换使用这些术语，而不会有任何意义上的损失。

Spring 的 DI 实现基于两个核心的 Java 概念：JavaBeans 和接口。当使用 Spring 作为 DI 提供程序时，可以通过不同的方式(例如，XML 文件、Java 配置类、代码中的注解或新的 Groovy bean 定义方法)在应用程序中灵活定义依赖项配置。JavaBean(POJO)提供了一种创建 Java 资源的标准机制，这些资源可以通过多种方式进行配置，例如构造函数和 setter 方法。在第 3 章中，将学习 Spring 如何使用 JavaBean 规范来形成其 DI 配置模型的核心；实际上，任何 Spring 管理的资源都被称为 *bean*。如果你对 JavaBean 不熟悉，请参阅第 3 章开头介绍的快速入门。

接口和 DI 是互利的技术。显而易见，将应用程序设计和编写为接口可以让应用程序更加灵活，但是将使用接口设计的应用程序连接在一起是非常复杂的，并且会给开发人员带来额外的编码负担。通过使用 DI，可以将基于接口的设计所需的代码量减少到几乎为零。同样，通过使用接口，可以充分利用 DI，因为 bean 可以利用任何接口实现来满足它们的依赖项。接口的使用还允许 Spring 利用 JDK 动态代理(Proxy 模式)为横切关注点提供诸如 AOP 的强大概念。

---

1 请记住，根据 http://openjdk.java.net/projects/jdk9/上提供的 Oracle 时间表，Java 9 已于 2017 年 9 月正式向公众发布。

2 Project Reactor 实现了 Reactive Streams API 规范，请参阅 https://projectreactor.io/。

3 反应式编程是一种涉及智能路由和事件消耗的微型架构，会导致非阻塞应用程序异步和事件驱动，并且需要少量线程在 JVM 中垂直扩展，而不是通过集群进行水平扩展。

在 DI 的上下文中，Spring 更像是容器，而不是框架——提供应用程序类的实例以及它们所需的所有依赖项，但它是以一种不受干扰的方式实现的。如果使用 Spring 实现 DI，那么仅依赖于类中的 JavaBeans 命名约定——没有特殊的类可供继承或者没有专门的命名方案可供遵循。如果有，在使用 DI 的应用程序中所做的唯一更改是在 JavaBean 上公开更多的属性，从而允许在运行时注入更多的依赖项。

1.1.3 依赖注入的演变

在过去的几年中，由于 Spring 和其他 DI 框架的普及，DI 在 Java 开发人员社区中获得了广泛的认可。同时，开发人员确信使用 DI 是应用程序开发的最佳实践，使用 DI 的好处也很好理解。

当 Java Community Process(JCP)在 2009 年采用 JSR-330(针对 Java 的依赖注入)时，DI 的流行得到了承认。SRR-330 已经成为正式的 Java 规范请求，正如所预料的那样，规范来源之一是 Rod Johnson——Spring 框架的创始人。在 JEE 6 中，JSR-330 成为整个技术栈的规范之一。与此同时，EJB 架构(从 3.0 版开始)也大幅改进；它采用 DI 模型以简化各种 Enterprise JavaBeans 应用程序的开发。

虽然在第 3 章将会对 DI 进行全面讨论，但关注一下使用 DI 的好处(相比于更传统的方法)还是非常值得的。

- **减少粘合代码**：DI 最大的优点之一是能够用最少的代码将应用程序的组件粘合在一起。通常所编写的代码是非常少的，所以创建依赖项时只需要创建一个对象的新实例即可。但是，当需要在 JNDI 存储库中查找依赖项时，或者无法直接调用依赖项时，粘合代码可能会变得非常复杂，就像调用远程资源一样。在这些情况下，DI 可以通过提供自动 JNDI 查找和远程资源的自动代理来真正简化粘合代码。
- **简化应用程序配置**：通过采用 DI，可以大大简化配置应用程序的过程。可以使用各种选项来配置那些可注入其他类的类。可以使用相同的技术向“注入器”表达依赖项需求，以注入适当的 bean 实例或属性。另外，DI 可以更容易地将一个依赖项的实现变换为另一个依赖项的实现。假设有一个对 PostgreSQL 数据库执行数据操作的 DAO 组件，并且想要将数据库升级到 Oracle。如果使用 DI，只需要简单地重新配置业务对象的适当依赖项，以便使用 Oracle 实现而不是 PostgreSQL 实现。
- **能够在单个存储库中管理常见依赖项**：如果使用传统方法对公共服务的依赖项(例如，数据源连接、事务和远程服务)进行管理，那么可以在需要的地方(依赖类中)创建依赖项的实例(或从某些工厂类中查找依赖项实例)。这样一来，就会导致依赖项遍布应用程序的不同类中，一旦更改这些类，可能就会出现问题。当使用 DI 时，关于这些公共依赖项的所有信息都包含在一个存储库中，从而使得依赖项的管理变得更加简单并且不易出错。
- **改进的可测试性**：当为 DI 设计类时，可以轻松替换依赖项。当测试应用程序时，这特别方便。假设有一个需要执行一些复杂处理的业务对象；其中使用 DAO 访问存储在关系数据库中的数据。当进行测试时，你可能对测试 DAO 并不感兴趣；而只是想要使用各种数据集来测试一下业务对象。在传统的方法中，业务对象负责获取 DAO 本身的实例，因此很难进行测试，因为无法轻松地将 DAO 实现替换为返回测试数据集的模拟实现。相反，需要确保测试数据库包含正确的数据，并使用完整的 DAO 实现进行测试。而如果使用 DI，可以创建返回测试数据集的 DAO 对象的模拟实现，然后将其传递给业务对象进行测试。此机制可以扩展用于测试应用程序的任何层，特别适用于测试可以创建 HttpServletRequest 和 HttpServletResponse 的模拟实现的 Web 组件。
- **培养良好的应用程序设计**：一般而言，DI 的设计意味着针对接口进行设计。在一个典型的面向注入的应用程序的设计中，所有主要组件都被定义为接口，然后使用 DI 容器创建这些接口的具体实现并连接在一起。在 DI 和基于 DI 的容器(例如 Spring)出现之前，这种设计是可能的，但通过使用 Spring，可以免费获得大量的 DI 功能，并且能够专注于构建自己的应用程序逻辑，而不是支持它的框架。

正如前面所介绍的那样，DI 为应用程序提供了很多好处，但它并非没有缺点。特别是对于那些不熟悉代码的人来说，DI 很难让他们知道某个特定依赖项的哪个实现与哪个对象连接。通常，只有当开发人员对 DI 没有任何经验时，才会出现上述问题。一旦经验丰富了并且遵循良好的 DI 编码实践(例如，将每个应用程序层中的所有可注入类放入同一个包)之后，开发人员就能轻松地知道哪个实现与哪个对象相连接。虽然在大多数情况下，优点是远远多于缺点的，但是在规划应用程序时应该考虑这一点。

1.1.4 除了依赖注入

虽然单独使用 Spring Core 及其先进的 DI 功能是非常有价值的，但 Spring 真正的价值在于其数不胜数的附加功能，所有这些功能都是使用 DI 原则进行精心设计和构建的。Spring 为应用程序的所有层提供功能，从用于数据访问权限的辅助应用程序编程接口(API)到高级 MVC 功能。Spring 中这些功能的优点在于，虽然 Spring 通常提供自己的方法，但也可以轻松地将它们与 Spring 中的其他工具集成在一起，从而使这些工具成为 Spring 家族的一流成员。

支持 Java 9

Java 8 带来了 Spring Framework 5 支持的许多令人兴奋的功能，其中最引人注目的是 lambda 表达式以及 Spring 的回调接口的方法引用。Spring 5 发布计划与 JDK 9 的最初发布计划保持一致，尽管 JDK 9 的发布期限已经推迟，但 Spring 5 已按计划发布。据估计，Spring 5.1 将完全支持 JDK 9。Spring 5 将利用 JDK 9 功能，比如紧凑字符串、ALPN 堆栈和新的 HTTP 客户端实现。尽管 Spring Framework 4.0 支持 Java 8，但兼容性仍然保持在 JDK 6 update 18。对于新开发的项目，建议使用更新版本的 Java，如 Java 7 或 Java 8。Spring 5 需要 Java 8+，因为 Spring 开发团队已经将 Java 8 语言级别应用于整个框架代码库，但从一开始，Spring 5 就是基于 JDK 9 构建的，以便为 JDK 9 的功能提供全面的支持。

使用 Spring 实现面向方面编程(AOP)

AOP 提供了在单个地方实现横切逻辑(即适用于应用程序多个部分的逻辑)的能力，并且可以自动在应用程序中应用该逻辑。Spring 对 AOP 的支持是为目标对象创建动态代理，并用配置好的通知来创建对象以执行横切逻辑。鉴于 JDK 动态代理的本质，目标对象必须实现一个接口来声明将应用 AOP 建议的方法。另一个流行的 AOP 库是 Eclipse AspectJ 项目[1]，它提供了更强大的功能，包括对象构建、类加载以及更强大的横切功能。然而，对于 Spring 和 AOP 开发人员来说真正的好消息是，从 2.0 版开始，Spring 提供了与 AspectJ 更紧密的集成。主要有以下几个亮点：

- 支持 AspectJ 风格的切入点表达式。
- 支持@AspectJ 注解样式，同时仍然使用 Spring AOP 进行织入(weaving)。
- 在 AspectJ 中实现 DI 方面的支持。
- 支持 Spring ApplicationContext 中的加载时织入。

从 Spring Framework 3.2 开始，可以使用 Java 配置启用@AspectJ 注解支持。

这两种 AOP 都有各自的位置，大多数情况下，Spring AOP 足以满足应用程序的横切需求。但是，对于更复杂的需求，AspectJ、Spring AOP 和 AspectJ 可以在同一个 Spring 应用程序中混合使用。

AOP 有许多应用程序。许多传统 AOP 示例中给出的典型功能大都是执行某种日志记录，但 AOP 的应用远不止日志记录应用程序。实际上，在 Spring 框架中，AOP 被用于多种用途，特别是在事务管理中。第 5 章将详细介绍 Spring AOP，其中将演示在 Spring 框架和自己的应用程序中 AOP 的典型用法，同时讨论 AOP 性能以及与传统技术相比 AOP 更适合的领域。

Spring 表达式语言

表达式语言(EL)是一种允许应用程序在运行时操纵 Java 对象的技术。然而，EL 的问题在于不同的技术提供了自己的 EL 实现和语法。例如，Java Server Pages(JSP)和 Java Server Faces(JSF)都有自己的 EL，它们的语法不同。为了解决这个问题，创建了统一表达式语言(UEL)。

由于 Spring 框架发展非常迅速，因此需要一种标准表达式语言，它可以在所有 Spring 框架模块以及其他 Spring 项目之间共享。所以，从 3.0 版开始，Spring 引入了 *Spring 表达式语言(Spring Expression Language，SpEL)*。SpEL 为评估表达式以及在运行时访问 Java 对象和 Spring bean 提供了强大功能。结果可以在应用程序中使用或注入其他 JavaBean 中。

Spring 中的验证

验证是任何类型的应用程序中都存在的另一个大问题。在理想情况下，无论数据操作请求来自前端、批处理作

---

1 www.eclipse.org/aspectj。

业还是从远程启动(例如，通过 Web 服务、RESTful Web 服务或远程过程调用[RPC])，包含业务数据的 JavaBeans 中的属性验证规则都应该以一致的方式应用。

为解决这些问题，Spring 通过 Validator 接口提供了一个内置的验证 API。该接口提供了一种简洁的机制，允许将验证逻辑封装到负责验证目标对象的类中。除目标对象外，validate 方法接收一个 Errors 对象，该对象用于收集可能发生的任何验证错误。

Spring 还提供了一个方便的实用程序类 ValidationUtils，它提供了便捷的方法来调用其他验证程序、检查常见问题(如空字符串)，并将错误报告给 Errors 对象。

在需求驱动下，JCP 还开发了 JSR-303(Bean Validation)，它提供了定义 bean 验证规则的标准方法。 例如，当将@NotNull 注解应用于 bean 的属性时，它将强制该属性在能够保存到数据库中之前不包含空值。

从 3.0 版开始，Spring 为 JSR-303 提供了开箱即用支持。如果要使用其 API，只需要声明一个 LocalValidatorFactoryBean并将Validator接口注入到Spring管理的任何bean中即可。Spring负责解析底层实现。默认情况下，Spring将首先查找Hibernate Validator(hibernate.org/subprojects/validator)，这是一个流行的JSR-303实现。包括Spring MVC在内的许多前端技术(例如JSF 2和Google Web Toolkit)也支持在用户界面中应用JSR-303验证。开发人员需要在用户界面和后端编写相同的验证逻辑的时代已经一去不复返了。更多内容，请参阅第10章。

 从 Spring Framework 4.0 版本开始，JSR-349(Bean Validation)的 1.1 版本得到支持。

在 Spring 中访问数据

数据访问和持久化似乎是 Java 世界中讨论最多的话题。Spring 为这些数据访问工具的选择提供了极好的集成。另外，Spring 使得普通的 JDBC 成为许多项目的一个可行的选择，其根据标准的 API 简化了封装的 API。Spring 的数据访问模块对 JDBC、Hibernate、JDO 和 JPA 提供了开箱即用支持。

从 Spring Framework 4.0 版本开始，就已经删除了对 iBATIS 的支持。MyBatis-Spring 项目提供了与 Spring 的集成，读者可以在 http://mybatis.github.io/spring/上找到更多信息。

然而，在过去几年中，由于互联网和云计算的爆炸性增长，除了关系数据库，还开发了许多其他的“专用”数据库。本书的示例使用了基于键值对的用来处理大量数据的数据库(通常称为 NoSQL)、图形数据库以及文档数据库。为了帮助开发人员支持这些数据库并避免 Spring 数据访问模块复杂化，人们创建了一个名为 Spring Data[1]的单独项目。该项目被进一步划分为不同的类别，以便支持更具体的数据库访问需求。

本书不包含 Spring 对非关系数据库的支持。如果对此主题感兴趣，则可以好好地研究一下前面提到的 Spring Data 项目。项目页面详细说明了它所支持的非关系数据库，并链接到这些数据库的主页。

Spring 对 JDBC 的支持使得在 JDBC 之上构建应用程序成为可能，即使创建更复杂的应用程序也是如此。对 Hibernate、JDO 和 JPA 的支持使得原本简单的 API 变得更加简单，从而减轻了开发人员的负担。当使用 Spring API 通过任何工具访问数据时，可以充分利用 Spring 出色的事务支持，第 9 章将对此进行全面讨论。

Spring 中最好的功能之一就是能够轻松地在应用程序中混合和匹配数据访问技术。例如，可以使用 Oracle 运行一个应用程序，同时使用 Hibernate 处理大部分数据访问逻辑。但如果想要使用某些 Oracle 特有的功能，那么可以非常容易地使用 Spring 的 JDBC API 实现数据访问层中的这一部分逻辑。

Spring 中的对象/XML 映射

大多数应用程序需要整合或向其他应用程序提供服务。一个常见的需求是定期或实时地与其他系统交换数据。在数据格式方面，XML 最常用。因此，经常需要将 JavaBean 转换为 XML 格式，反之亦然。Spring 支持许多常见的 Java-to-XML 映射框架，并且仍然不需要直接耦合到任何特定的实现。Spring 提供了用于编组(将 JavaBean 转换为 XML)以及解组(将 XML 转换为 Java 对象)到任何 Spring bean 的通用接口。Spring 支持 Java Architecture for XML Binding(JAXB)、Castor、XStream、JiBX 以及 XMLBeans 等通用库。在第 12 章中，当讨论远程访问 XML 格式的商业数据的 Spring 应用程序时，将介绍如何在自己的应用程序中使用 Spring 的对象/XML 映射(OXM)支持。

1 http://projects.spring.io/spring-data。

管理事务

Spring 为事务管理提供了一个优秀的抽象层，允许编程式和声明式事务控制。通过对事务使用 Spring 抽象层，可以更加简单地更改底层事务协议和资源管理器。可以从简单的、本地的且特定于资源的事务开始，逐步转向使用全局的多资源事务，而无须更改代码。第 9 章将详细介绍事务。

简化以及与 JEE 集成

随着诸如 Spring 之类的 DI 框架逐步被接受，许多开发人员选择通过使用 DI 框架来构建应用程序，以便支持 JEE 的 EJB 方法。因此，JCP 社区也意识到 EJB 的复杂性。从 EJB 规范的 3.0 版开始，API 被简化了，所以它现在包含 DI 的许多概念。

但是，对于构建在 EJB 上的应用程序，或者需要在 JEE 容器中部署基于 Spring 的应用程序并利用应用程序服务器的企业服务(例如，Java Transaction API 的事务管理器、数据源连接池和 JMS 连接工厂)的应用程序，Spring 还为这些技术提供简化的支持。针对 EJB，Spring 提供了一个简单的声明以执行 JNDI 查找并注入 Spring bean。反之，Spring 还提供了将 Spring bean 注入 EJB 中的简单注解。

对于存储在 JNDI 可访问位置的任何资源，Spring 允许删除复杂的查找代码，并将 JNDI 管理的资源作为依赖项在运行时注入其他对象中。这样做的另一个作用是将应用程序与 JNDI 分离，从而可以在未来更多地重用代码。

Web 层中的 MVC

几乎可以在任何环境中使用 Spring，包括从桌面到 Web，Spring 都提供了一组丰富的类来支持创建基于 Web 的应用程序。当使用 Spring 时，在选择如何实现 Web 前端时就有了最大的灵活性。为了开发 Web 应用程序，MVC 模式是最流行的做法。在最近的版本中，Spring 已经从简单的 Web 框架逐渐演变为完整的 MVC 实现。首先，Spring MVC 中的视图支持非常广泛。除了提供对 JSP 以及 Java Standard Tag Library(JSTL)的标准支持(主要由 Spring 标签库提供支持)，还可以充分利用对 Apache Velocity、FreeMarker、Apache Tiles、Thymeleaf 和 XSLT 的完全集成支持。另外，可以找到一组基本视图类，以便将 Microsoft Excel、PDF 和 JasperReports 输出添加到应用程序中。

大多数情况下，Spring MVC 足以满足 Web 应用程序开发需求。不过，Spring 还可以与其他流行的 Web 框架(如 Struts、JSF、Atmosphere、Google Web Toolkit(GWT)等)集成。

在过去几年里，Web 框架的技术发展迅速，用户需要更多的响应式和交互式体验，从而导致 Ajax 的崛起，成为开发富 Internet 应用程序(RIA)时广泛采用的技术。另一方面，用户也希望能够通过任何设备访问他们的应用程序，包括智能手机和平板电脑。这就需要支持 HTML5、JavaScript 和 CSS3 的 Web 框架。第 16 章将讨论如何使用 Spring MVC 开发 Web 应用程序。

WebSocket 支持

从 Spring Framework 4.0 开始就已经支持 JSR-356(用于 WebSocket 的 Java API)了。WebSocket 定义了一个用于在客户端和服务器之间创建持久连接的 API，通常在 Web 浏览器和服务器间实现。WebSocket 式的开发为高效的全双工通信打开了大门，为高度响应式应用程序实现了实时消息交换。对 WebSocket 支持的使用详见第 17 章。

远程支持

在 Java 中访问或公开远程组件从来就不是一件最简单的工作。通过使用 Spring，可以充分利用对各种远程技术的广泛支持来快速公开和访问远程服务。Spring 对各种远程访问机制提供支持，包括 Java Remote Method Invocation(RMI)、JAX-WS、Caucho Hessian and Burlap、JMS、Advanced Message Queuing Protocol(AMQP)和 REST。除了这些远程协议，Spring 还提供了基于标准 Java 序列化的基于 HTTP 的调用器。通过应用 Spring 的动态代理功能，可以将远程资源的代理作为依赖项注入某个类中，从而避免将应用程序与特定远程实现耦合，并且减少了为应用程序所需编写的代码量。第 12 章将讨论 Spring 的远程支持。

邮件支持

发送电子邮件是许多类型应用程序的典型需求，Spring 框架内对此进行了很好的处理。Spring 为发送电子邮件消息提供了一个简化的 API，可以很好地与 Spring DI 功能相匹配。Spring 支持标准的 JavaMail API。Spring 提供了在 DI 容器中创建原型消息的能力，并将其用作从应用程序发送的所有消息的基础。这样一来就可以轻松自定义邮件参数，例如主题和发件人地址。另外，为了自定义消息体，Spring 集成了模板引擎，如 Apache Velocity；从而允许邮件内容在 Java 代码中外部化。

作业调度支持

大多数重要的应用程序都需要某种调度功能。无论是将更新发送给客户还是执行内务处理任务，能够在预定义时间运行代码的能力对于开发人员来说都是非常宝贵的工具。Spring 提供了可以满足大多数常见场景的调度支持。任务可以按固定时间间隔或通过使用 UNIX cron 表达式进行调度。另一方面，对于任务执行和调度，Spring 也与其他调度库进行了集成。例如，在应用程序服务器环境中，Spring 可以将执行委托给许多应用程序服务器所使用的 CommonJ 库。而对于作业调度，Spring 还支持包括 JDK Timer API 和 Quartz(一种常用的开源调度库)在内的库。第 11 章将详细介绍 Spring 中的调度支持。

动态脚本支持

从 JDK 6 开始，Java 就引入了动态语言支持，可以在 JVM 环境中执行用其他语言编写的脚本，比如 Groovy、JRuby 和 JavaScript。Spring 还支持在 Spring 驱动的应用程序中执行动态脚本，或者定义一个用动态脚本语言编写并注入其他 JavaBean 中的 Spring bean。Spring 支持的动态脚本语言包括 Groovy、JRuby 和 BeanShell。在第 14 章，将详细讨论 Spring 对于动态脚本的支持。

简化的异常处理

Spring 中真正有助于减少重复、样板代码量的一项功能是异常处理。在异常处理方面，Spring 思想核心认为在 Java 中检查型异常(checked exception)被过度使用，并且框架不应该强制捕捉那些不可能恢复的异常——这个观点完全正确。实际上，许多框架的设计目的是避免编写代码来处理检查型异常。但是，这些框架中有许多仍然坚持采用检查型异常的方法，从而人为减少了异常类层次结构的粒度。在使用 Spring 时你会注意到，由于使用未检查型异常(unchecked exception)给开发人员带来了极大便利，异常层次显著细化。在本书中，你将会看到一些例子，其中 Spring 异常处理机制可以减少必须编写的代码量，同时提高识别、分类和诊断应用程序内错误的能力。

1.2 Spring 项目

关于 Spring 项目最值得关注的事情之一是社区活动的水平以及 Spring 与其他项目(比如 CGLIB、Apache Geronimo 和 AspectJ)之间的交流。开源代码最受欢迎的优点之一是，即使项目明天结束，也会留下代码；但是当面对这些代码时，你可能并不希望使用一个与 Spring 大小相当的代码库来完成支持和改进。因此，了解 Spring 社区的建立和活跃程度是很有必要的。

1.2.1 Spring 的起源

如本章前面所述，Spring 的起源可以追溯到 *Expert One-to-One：J2EE Design and Development* 一书。在该书中，Rod Johnson 提出了自己的名为 Interface 21 Framework 的框架，该框架主要用于他自己的应用程序。正如你在今天所看到的，这个框架已经被发布到开源世界，形成 Spring 框架的基础。在经过早期测试版以及发布候选版阶段之后，Spring 得到了快速发展。第一个官方版本 1.0 于 2004 年 3 月发布。从那时起，Spring 发生了翻天覆地的变化，在编写本书时，Spring 框架的最新主要版本是 5.0。

1.2.2 Spring 社区

如果想要找到所需的任何开源项目，Spring 社区是最好的地方之一。邮件列表和论坛总是很活跃，不断有新的功能被开发出来。开发团队真正致力于使 Spring 成为所有 Java 应用程序框架中最成功的一个，主要体现在所复制代码的质量上。正如前面已经提到的，Spring 的发展也受益于其他的开源项目，这一事实在考虑 Spring 发行版与其他框架的依赖关系时非常重要。从用户的角度看，也许 Spring 的最佳特性之一就是随着版本一起发布的优秀文档和测试套件。文档提供了 Spring 的几乎所有功能，使得新用户可以轻松地选择使用该框架。Spring 提供的测试套件非常全面——开发团队对所有内容都进行了测试。如果他们发现一个 bug，那么首先会编写一个突出该 bug 的测试，然后让测试通过，从而修复这个 bug。当然，修复 bug 和创建新功能并不仅限于开发团队！如果想要提供自己的代码，可通过官方 GitHub 存储库(http://github.com/spring-projects)提交针对任何 Spring 项目组合的请求。此外，可通过官方的 Spring JIRA(https://jira.spring.io/secure/Dashboard.jspa)创建和跟踪问题。这一切意味着什么呢？简言之，这意味着应该对 Spring 框架的质量充满信心，并且相信在可预见的未来，Spring 开发团队将继续改进这个已经非常优秀的框架。

1.2.3 Spring 工具套件

为简化在 Eclipse 中开发基于 Spring 的应用程序，Spring 创建了 Spring IDE 项目。不久之后，Rod Johnson 创立的 SpringSource 公司创建了一个名为 Spring Tool Suite(STS)的集成工具，该工具可从 https://spring.io/tools 下载。虽然该工具以前需要付费，但现在可免费使用。它将 Eclipse IDE、Spring IDE、Mylyn(基于任务的 Eclipse 开发环境)、Maven for Eclipse、AspectJ Development Tools 以及许多其他有用的 Eclipse 插件集成到一个包中。每个新版本都添加了更多功能，如 Groovy 脚本语言支持、图形化 Spring 配置编辑器、用于项目(比如 Spring Batch 和 Spring Integration)的可视化开发工具，以及对 Pivotal tc Server 应用程序服务器的支持。

SpringSource 已被 VMware 收购并且被并入 Pivotal Software。

除基于 Java 的套件外，Groovy/Grails 工具套件也具有相似的功能，但主要针对 Groovy 和 Grails 开发(http://spring.io/tools)。

1.2.4 Spring Security 项目

Spring Security 项目(http://projects.spring.io/spring-security)，以前被称为 Spring 的 Acegi Security System，是 Spring 产品组合中的另一个重要项目。Spring Security 为 Web 应用程序和方法级安全性提供全面的支持。它与 Spring 框架以及其他常用的身份验证机制(比如 HTTP 基本身份验证、基于表单的登录、X.509 证书以及单点登录(SSO)产品(例如 CA SiteMinder))紧密集成。它为应用程序资源提供基于角色的访问控制，在具有更复杂安全需求的应用程序(例如数据分隔)中，支持使用访问控制列表(ACL)。但 Spring Security 主要用于保护 Web 应用程序，相关内容将在第 16 章详细讨论。

1.2.5 Spring Boot

建立应用程序的基础是一项烦琐的工作。必须创建项目的配置文件，并且安装和配置其他工具(如应用程序服务器)。Spring Boot(http://projects.spring.io/spring-boot/)是一个 Spring 项目，可以轻松创建可运行的、独立的、生产级的、基于 Spring 的应用程序。Spring Boot 针对不同类型的 Spring 应用程序提供开箱即用的配置，这些配置都封装在*启动器(starter)*包中。例如，web-starter 包提供了一个预配置且易于自定义的 Web 应用程序上下文，并支持 Tomcat 7+、Jetty 8+和 Undertow 1.3 嵌入式 servlet 容器。

Spring Boot 还包含 Spring 应用程序所需的所有依赖项，同时考虑到版本之间的兼容性。在编写本书时，Spring Boot 的当前版本是 2.0.0.RELEASE。

第 4 章将详述 Spring Boot，作为 Spring 项目的替代配置，后续章节中的大多数项目都将使用 Spring Boot 运行，因为它使开发和测试更加实用和快速。

1.2.6 Spring 批处理和集成

不用多说，批处理作业的执行和集成是应用程序中的常见用例。为了满足该需求并为这些领域的开发人员提供便利，Spring 创建了 Spring Batch 和 Spring Integration 项目。Spring Batch 为批处理作业的实现提供了一个通用框架以及各种策略，减少了大量的样板代码。通过实现 Enterprise Integration Patterns(EIP)，Spring Integration 可使 Spring 应用程序与外部系统轻松集成。相关内容将在第 20 章详细讨论。

1.2.7 许多其他项目

前面已经介绍了 Spring 的核心模块以及 Spring 组合中的一些主要项目，但还有很多其他项目是为了满足社区的不同需求而创建的，比如 Spring Boot、Spring XD、Android for Spring、Spring Mobile、Spring Social 和 Spring AMQP。其中一些项目将在第 20 章中进一步讨论。更多详细信息，请参阅 Spring by Pivotal 网站(www.spring.io/projects)。

1.3 Spring 的替代品

回顾前面关于开源项目数量的讨论，如果说 Spring 并不是唯一提供依赖注入功能或用于构建应用程序的完整端到端解决方案的框架，那么我想你不应该感到惊讶。实际上，相关的项目实在太多了。本着开放的精神，这里简要讨论一下其中的几个框架，但这些平台都不能提供与 Spring 完全一样的解决方案。

1.3.1 JBoss Seam 框架

Seam 框架(http://seamframework.org)由 Gavin King(Hibernate ORM 库的创建者)创建，是另一个完整的基于 DI 的框架。它支持 Web 应用程序前端开发(JSF)、业务逻辑层(EJB 3)和 JPA 持久化。正如你看到的，Seam 和 Spring 之间的主要区别在于 Seam 框架完全基于 JEE 标准构建。JBoss 还将 Seam 框架中的想法提供给 JCP，并成为 JSR-299(Java EE 平台的上下文和依赖注入)。

1.3.2 Google Guice

Google Guice(http://code.google.com/p/google-guice)是另一个流行的 DI 框架。在搜索引擎巨头 Google 的领导下，Guice 成了一个轻量级框架，专注于为应用程序配置管理提供 DI。它也是 JSR-330(用于 Java 的依赖注入)的参考实现。

1.3.3 PicoContainer

PicoContainer(http://picocontainer.com)是一个非常小的 DI 容器，允许在不引入除 PicoContainer 外的任何依赖项的情况下为应用程序使用 DI。由于 PicoContainer 只是一个 DI 容器，因此如果随着应用程序的增长需要引入另一个框架，比如 Spring，那么在这种情况下最好从一开始就使用 Spring。但是，如果只需要一个小型的 DI 容器，那么 PicoContainer 是一个不错的选择。由于 Spring 将 DI 容器与框架的其余部分分开打包，因此可以轻松使用 PicoContainer 并保持未来的灵活性。

1.3.4 JEE 7 容器[1]

如前所述，DI 的概念被 JCP 广泛采用并实现。当为应用程序服务器开发符合 JEE 7(JSR-342)的应用程序时，可以在所有层中使用标准 DI 技术。

1.4 小结

本章概述 Spring 框架，并讨论了所有主要功能，同时详细介绍了本书的相关章节。阅读本章后，你应该了解 Spring 能做些什么；接下来将学习它是如何做的。在下一章中，将讨论你需要知道的有关启动和运行基本 Spring 应用程序的所有信息，同时会展示如何获得 Spring 框架并讨论打包选项、测试套件和文档。另外，第 2 章将介绍一些基本的 Spring 代码，其中包括历史悠久的 Hello World 示例。

---

1 JEE 8 的发布日期被推迟至 2017 年底，请参阅 https://jcp.org/en/jsr/detail?id=366。

第 2 章

入　门

学习使用任何新的开发工具时，最困难的部分往往是搞清楚从哪里开始。通常，当新工具与 Spring 一样提供了多种选择时，这个问题会变得更糟糕。但幸运的是，如果知道首先看什么，那么 Spring 入门并不难。在本章中，将提供开始学习 Spring 所需的所有基本知识。具体来说，会介绍以下内容。

- 获取 Spring：第一个逻辑步骤是获取或构建 Spring JAR 文件。如果想快速启动并运行，只需要根据 http://projects.spring.io/spring-framework 上提供的示例，在构建系统中使用依赖项管理片段即可。但是，如果想要始终处于 Spring 开发的前沿，则需要查看 Spring 的 GitHub 存储库中的最新版源代码[1]。
- Spring 打包选项：Spring 打包是模块化的；允许选择想要在应用程序中使用哪些组件，并在分发应用程序时只包含这些组件。虽然 Spring 有许多模块，但你可能只需要这些模块的一部分，具体取决于应用程序的需求。每个模块都将其编译的二进制代码与相应的 Javadoc 和源 JAR 一起放入 JAR 文件中。
- Spring 指南：新的 Spring 网站包含一个 Guides 部分，位于 http: //spring.io/guides。这些指南旨在为使用 Spring 构建任何开发任务的 Hello World 版本提供快速的实践指导。这些指南还反映了最新的 Spring 项目发布和技术，提供最新的示例。
- 测试套件和文档：Spring 社区成员最引以为傲的一件事就是他们的综合测试套件和文档集。测试是团队工作的重要组成部分，当然随标准分发提供的文档集也非常优秀。
- 将 Spring 放入 Hello World 中：先将所有烦琐的事情放在一边，开始学习使用任何新编程工具的最好方法是直接编写一些代码。接下来将给出一个简单示例，它是每个人最喜欢的完全基于 DI 实现的 Hello World 应用程序。即使无法立即理解所有的代码，也不要惊慌；本书后面会详细讨论。

如果你已经熟悉 Spring 框架的基础知识，那么可以直接学习第 3 章，以深入了解 Spring 中的 IoC 和 DI。然而，即使熟悉 Spring 的基础知识，也会发现本章中的一些讨论很有趣，尤其是那些关于打包和依赖项的讨论。

2.1　获取 Spring 框架

在开始任何 Spring 开发之前，首先需要获取 Spring 库。有两个选择来检索这些库：可以使用构建系统来引入想要使用的模块，也可以在 Spring GitHub 存储库中检索并构建代码。使用诸如 Maven 或 Gradle 之类的依赖项管理工具通常是最直接的方法；只需要在配置文件中声明依赖项并让该工具获取所需的库即可。

> 如果有 Internet 连接并将 Maven 或 Gradle 等构建工具与 Eclipse 或 IntelliJ IDEA 等智能 IDE 结合使用，则可以自动下载 Javadoc 和库，以便在开发时访问它们。在生成项目时如果升级构建配置文件中的版本，库和 Javadoc 也会更新。

2.1.1　快速入门

访问 Spring 框架的项目页面[2]，为构建系统获取依赖项管理代码片段，从而在项目中包含 Spring 最新发布的 RELEASE 版本。还可以为即将发布的版本或以前的版本使用快照。

使用 Spring Boot 时，不需要指定要使用的 Spring 版本，因为 Spring Boot 提供了自定义的“starter”项目对象模型(POM)文件，以简化 Maven 配置和默认的 Gradle 入门配置。请记住，版本 2.0.0.RELEASE 之前的 Spring Boot 版

1　GitHub 存储库位于 http://github.com/spring-projects/spring-framework。
2　http://projects.spring.io/spring-framework。

本使用 Spring 4.x 版本。

2.1.2 在 GitHub 中查找 Spring

如果想要在进入快照之前了解一下新功能，可以直接查看 Pivotal 的 GitHub 存储库中的源代码。要查看最新版本的 Spring 代码，请先安装 Git，可以从 http://git-scm.com 下载。然后打开一个终端 shell 并运行以下命令：

git clone git://github.com/spring-projects/spring-framework.git

请参阅项目根目录中的 README.md 文件以获取有关如何从源代码构建的完整详细信息和要求。

2.1.3 使用正确的 JDK

Spring 框架使用 Java 构建，这意味着需要能够在计算机上执行 Java 应用程序。为此，需要安装 Java。当谈到 Java 应用程序开发时，有三种广泛使用的 Java 缩写词。

- Java 虚拟机(JVM)是一台抽象机器。它是一个规范，提供了可以执行 Java 字节码的运行时环境。
- Java Runtime Environment(JRE)用于提供运行时环境。它是实际存在的 JVM 的实现，包含一组 JVM 在运行时使用的库和其他文件。Oracle 在 2010 年收购了 Sun Microsystems；此后，就不断提供新版本和补丁。诸如 IBM 的其他公司也提供了自己的 JVM 实现。
- Java Development Kit(JDK)包含JRE、文档和Java工具。Java开发人员在自己的机器上安装的就是JDK。IntelliJ IDEA 或 Eclipse 等智能编辑器要求提供 JDK 的位置，以便在开发期间加载和使用类和文档。

如果使用的是诸如 Maven 或 Gradle 之类的构建工具(本书附带的源代码是在 Gradle 多模块项目中组织的)，那么也需要一个 JVM；Maven 和 Gradle 都是基于 Java 的项目。

Java 最新的稳定版本是 Java 8，Java 9 已于 2017 年 9 月 21 日发布。可以从 https://www.oracle.com/下载 JDK。默认情况下，它将安装在计算机上的某个默认位置，具体位置取决于所使用的操作系统。如果想通过命令行使用 Maven 或 Gradle，需要为 JDK 和 Maven/Gradle 定义环境变量，并将其可执行文件的路径添加到系统路径中。读者可以在本书的附录中找到有关如何在每个产品的官方网站上执行此操作的说明。

第 1 章介绍了 Spring 版本和所需 JDK 版本的列表。本书所讨论的 Spring 版本是 5.0.x。书中介绍的源代码是使用 Java 8 语法编写的，因此至少需要 JDK 8 才能编译和运行示例。

2.2 了解 Spring 打包

*Spring 模块*只是简单的 JAR 文件，它打包了该模块所需的代码。在了解每个模块的用途之后，可以选择项目中所需的模块并将其包含在代码中。从 Spring 5 RELEASE 版本开始，Spring 提供了 21 个模块，打包成 21 个 JAR 文件。表 2-1 描述了这些 JAR 文件及其相应的模块。例如，实际的 JAR 文件名称是 spring-aop-5.0.0.RELEASE.jar，但为简单起见，只包含特定的模块部分(例如，在 aop 中)。

表 2-1 Spring 模块

模块	描述
aop	该模块包含在应用程序中使用 Spring 的 AOP 功能时所需的所有类。如果打算在 Spring 中使用其他使用了 AOP 的功能，例如声明式事务管理，则需要在应用程序中包含此 JAR 文件。此外，支持与 AspectJ 集成的类也封装在此模块中
aspects	该模块包含与 AspectJ AOP 库进行高级集成的所有类。例如，如果为完成 Spring 配置而使用 Java 类，并且需要 AspectJ 风格的注解驱动的事务管理，则需要使用此模块
beans	该模块包含所有支持 Spring 对 Spring bean 进行操作的类。该模块中的大多数类都支持 Spring 的 bean 工厂实现。例如，处理 Spring XML 配置文件和 Java 注解所需的类被封装到此模块中
beans-groovy	此模块包含用于支持 Spring 对 Spring bean 进行操作的 Groovy 类
context	该模块包含为 Spring Core 提供许多扩展的类。你会发现所有类都需要使用 Spring 的 ApplicationContext 功能(将在第 5 章介绍)以及 Enterprise JavaBeans(EJB)、Java Naming and Directory Interface(JNDI)和 Java Management Extensions(JMX)集成的类。此模块中还包含 Spring 远程处理类，与动态脚本语言(例如 JRuby、Groovy 和 BeanShell)、JSR-303(Bean Validation)、调度和任务执行等集成的类

(续表)

模块	描述
context-indexer	该模块包含一个索引器实现，它提供对 META-INF/spring.components 中定义的候选项的访问功能。但核心类 CandidateComponentsIndex 并不能在外部使用
context-support	该模块包含对 spring-context 模块的进一步扩展。在用户界面方面，有一些用于支持邮件并与模板引擎(例如 Velocity、FreeMarker 和 JasperReports)集成的类。此外，还包括与各种任务执行和调度库(包括 CommonJ 和 Quartz)的集成
core	这是每个 Spring 应用程序都需要的主要模块。在该 JAR 文件中，可以找到所有其他 Spring 模块(例如，用于访问配置文件的类)所共享的所有类。另外，在该 JAR 文件中，会发现在整个 Spring 代码库中都使用的非常有用的实用程序类，可以在自己的应用程序中使用它们
expression	该模块包含 Spring Expression Language(SpEL)的所有支持类
instrument	该模块包含用于 JVM 启动的 Spring 工具代理。如果在 Spring 应用程序中使用 AspectJ 实现加载时织入，那么该模块是必需的
dbc	该模块包含所有的 JDBC 支持类。对于需要数据库访问的所有应用程序，都需要此模块。支持数据源、JDBC 数据类型、JDBC 模板、本地 JDBC 连接等的类都被打包在此模块中
jms	该模块包含 JMS 支持的所有类
orm	该模块扩展了 Spring 的标准 JDBC 功能集，支持流行的 ORM 工具，包括 Hibernate、JDO、JPA 和数据映射器 iBATIS。该 JAR 文件中的许多类都依赖于 spring-jdbc JAR 文件中所包含的类，因此也需要把它包含在应用程序中
oxm	该模块为 Object/XML 映射(OXM)提供支持。用于抽象 XML 编组和解组以及支持 Castor、JAXB、XMLBeans 和 XStream 等常用工具的类都包含在此模块中
test	Spring 提供一组模拟类来帮助测试应用程序，并且许多模拟类都在 Spring 测试套件中使用，所以它们都经过了很好的测试，从而使测试应用程序变得更简单。在对 Web 应用程序进行单元测试时会发现模拟 HttpServletRequest 和 HttpServletResponse 类所带来的好处。另一方面，Spring 提供了与 JUnit 单元测试框架的紧密集成，并且在该模块中提供了许多支持 JUnit 测试用例开发的类；例如，SpringJUnit4ClassRunner 提供了一种在单元测试环境中引导 Spring ApplicationContext 的简单方法
tx	该模块提供支持 Spring 事务基础架构的所有类。 可以从事务抽象层找到相应的类来支持 Java Transaction API(JTA)以及与主要供应商的应用程序服务器的集成
web	此模块包含在 Web 应用程序中使用 Spring 所需的核心类，包括用于自动加载 ApplicationContext 功能的类、文件上传支持类以及一些用于执行重复任务(比如从查询字符串中解析整数值)的有用类
web-reactive	该模块包含 Spring Web Reactive 模型的核心接口和类
web-mvc	该模块包含 Spring 自己的 MVC 框架的所有类。如果想要为应用程序使用单独的 MVC 框架，则不需要此 JAR 文件中的任何类。Spring MVC 在第 16 章中有更详细的介绍
webfsocket	该模块提供对 JSR-356(WebSocket 的 Java API)的支持

2.2.1　为自己的应用程序选择模块

如果没有像 Maven 或 Gradle 这样的依赖项管理工具，那么在应用程序中选择所需使用的模块可能有点棘手。例如，如果需要 Spring 的 bean 工厂和 DI 支持，那么可能还需要其他几个模块，包括 spring-core、spring-beans、spring-context 和 spring-aop。如果需要 Spring 的 Web 应用程序支持，则需要进一步添加 spring-web 等。借助于诸如 Maven 传递依赖项支持等工具功能，所有必需的第三方库将自动包含在应用程序中。

2.2.2　在 Maven 存储库上访问 Spring 模块

从开源到企业环境，由 Apache Software Foundation 创建的 Maven[1]已经成为管理 Java 应用程序依赖项的最受欢迎的工具之一。Maven 是一个功能强大的应用程序构建、打包和依赖项管理工具。它管理应用程序的整个构建周期，

1　http://maven.apache.org。

从资源处理和编译，再到测试和打包。针对各种任务，还存在大量的 Maven 插件，比如，更新数据库和将打包应用程序部署到特定服务器(例如 Tomcat、JBoss 或 WebSphere)。在编写本书时，Maven 的最新版本是 3.3.9。

几乎所有的开源项目都支持通过 Maven 存储库进行库的分发。其中最流行的是在 Apache 上托管的 Maven Central 存储库，可以在 Maven Central 网站上访问和搜索存在哪些工件以及相关信息[1]。如果将 Maven 下载并安装到自己的开发计算机中，就会自动获得对 Maven Central 存储库的访问权限。其他一些开源社区(例如，Pivotal 的 JBoss 和 Spring)也为其用户提供了自己的 Maven 存储库。但是，如果想要能够访问这些存储库，需要将存储库添加到 Maven 的设置文件或项目的 POM 文件中。

关于 Maven 的详细信息不在本书的讨论范围之内，可以随时参阅联机文档或书籍，以便详细了解 Maven。但是，由于 Maven 被广泛采用，重点提一下 Maven 存储库中项目打包的典型结构还是很有必要的。

每个 Maven 工件由组 ID、工件 ID、打包类型和版本标识。例如，对于工件 log4j 来说，组 ID 是 log4j，工件 ID 是 log4j，打包类型是 jar。此外，还定义了不同的版本。例如，对于版本 1.2.12，工件的文件名变为组 ID、工件 ID 和版本文件夹下的 log4j-1.2.17.jar。Maven 配置文件是用 XML 编写的，所以必须遵守 http://maven.apache.org/maven-v4_0_0.xsd 模式定义的 Maven 标准语法。项目的 Maven 配置文件的默认名称是 om.xml，接下来显示一个示例文件：

```
<project xmlns="http://maven.apache.org/POM/4.0.0"
   xmlns:xsi="http://www.w3.org/2001/XMLSchema-instance"
  xsi:schemaLocation="http://maven.apache.org/POM/4.0.0
   http://maven.apache.org/maven-v4_0_0.xsd">
  <modelVersion>4.0.0</modelVersion>
  <groupId>com.apress.prospring5.ch02</groupId>
  <artifactId>hello-world</artifactId>
  <packaging>jar</packaging>
  <version>5.0-SNAPSHOT</version>
  <name>hello-world</name>
<properties>
  <project.build.sourceEncoding>UTF-8</project.build.sourceEncoding>
  <spring.version>5.0.0.RELEASE</spring.version>
</properties>
<dependencies>
    <!-- https://mvnrepository.com/artifact/log4j/log4j -->
      <dependency>
            <groupId>log4j</groupId>
            <artifactId>log4j</artifactId>
            <version>1.2.17</version>
      </dependency>
  </dependencies>
  <build>
    <plugins>
      <plugin>
       ...
      </plugin>
    </plugins>
    </build>
</project>
```

Maven 还定义了一种标准的项目结构，如图 2-1 所示。

main 目录包含应用程序的类(java 目录)和配置文件(resources 目录)。

test 目录包含用来对 main 目录中的应用程序进行测试的类(java 目录)和配置文件(resources 目录)。

```
▼ hello-world
  ▼ src
    ▼ main
      ► java
      ► resources
    ▼ test
      ► java
      ► resources
    pom.xml
```

图 2-1　一种标准的 Maven 项目结构

2.2.3　使用 Gradle 访问 Spring 模块

Maven 项目的标准结构以及工件分类和组织非常重要，因为 Gradle 遵循相同的规则，甚至使用 Maven 中央存储库来检索工件。Gradle 是一个功能强大的构建工具，它放弃了用于配置的臃肿 XML，并转而利用 Groovy 的简单性和灵活性。在编写本书时，Gradle 的当前版本是 4.0[2]。从版本 4.x 开始，Spring 团队已经转向使用 Gradle 来配置每个 Spring 产品。这就是为什么本书的源代码可以使用 Gradle 构建和执行的原因。项目的 Gradle 配置文件的默认名称是 build.gradle。下面所示的代码与前面所述的 pom.xml 文件(其中的一个版本)等价：

1　http://search.maven.org。

2　在官方项目网站上，可以找到有关如何下载、安装和配置 Gradle 以进行开发的详细说明：https://gradle.org/install。

```
group 'com.apress.prospring5.ch02'
version '5.0-SNAPSHOT'

apply plugin: 'java'

repositories {
   mavenCentral()
}

ext{
      springVersion = '5.0.0.RELEASE'
}

tasks.withType(JavaCompile) {
   options.encoding = "UTF-8"
}

dependencies {
    compile group: 'log4j', name: 'log4j', version: '1.2.17'
    ...
}
```

上面的代码是不是更具有可读性？正如你看到的那样，工件是使用先前在Maven中引入的组、工件和版本来标识的，但属性名称不同。由于Gradle不在本书的讨论范围之内，因此对它的介绍也就到此结束。

2.2.4 使用Spring文档

对于那些正在构建实际应用程序的开发人员来说，Spring之所以是一个非常有用的框架，是因为它提供了丰富且准确的文档。在每个发行版中，Spring框架的文档团队都努力确保所有文档都由开发团队完成并完善。这意味着Spring的每个功能不仅在Javadoc中被完整记录，而且还包含在每个发行版的Spring参考手册中。如果还不熟悉Spring Javadoc和参考手册，现在就请熟悉一下吧！本书并不能完全取代这些资源；相反，它是一份补充参考，演示了如何从头开始构建基于Spring的应用程序。

2.2.5 将Spring放入Hello World中

到目前为止，本书希望可以让你明白Spring是一个可靠的、支持良好的项目，它具有应用程序开发工具所应具备的所有条件。然而，却遗漏了一件事——还没有展示任何代码。也许你现在非常想了解实际应用中的Spring是什么样子的，同时如果不列举一些代码就无法完成后面的学习，因此接下来展示相关的代码。如果无法完全理解本节中的所有代码，请不要担心；在通读本书时会详细讨论所有主题。

2.2.6 构建示例Hello World应用程序

现在，我们确信你应该已经非常熟悉传统的Hello World示例，但如果在过去30年里你一直在月球上生活，那么以下代码片段展示了Java版本的所有荣耀：

```
package com.apress.prospring5.ch2;

public class HelloWorld {
   public static void main(String... args) {
      System.out.println("Hello World!");
   }
}
```

该例非常简单，它可以完成相关工作，但不具有可扩展性。如果想改变消息，应该怎么做呢？如果想以不同的方式输出消息(可能是标准错误而不是标准输出，或者用HTML标签而不是纯文本)，应该怎么做呢？需要重新定义示例应用程序的需求，并声明它必须支持用于更改消息的简单且灵活的机制，还必须易于更改渲染行为。在基本的Hello World示例中，只需要根据需求更改代码即可快速轻松地完成这两项更改。但是，在更大的应用程序中，重新编译需要时间，并且需要对应用程序进行完全测试。更好的解决方案是将消息内容外部化并在运行时读取它，可以使用以下代码片段中所示的命令行参数：

```
package com.apress.prospring5.ch2;

public class HelloWorldWithCommandLine {
```

```
    public static void main(String... args) {
        if (args.length > 0) {
            System.out.println(args[0]);
        } else {
            System.out.println("Hello World!");
        }
    }
}
```

该例完成了所需的功能，现在可以在不更改代码的情况下更改消息。但是，这个应用程序仍然存在问题：负责渲染消息的组件同时也负责获取消息。更改消息的获取方式意味着需要更改渲染器中的代码。但事实却是，改变渲染器是非常不容易实现的；这样做意味着更改启动应用程序的类。

如果对该应用程序进行进一步扩展(远离 Hello World 的基础)，更好的解决方案是将渲染和消息检索逻辑重构为单独的组件。另外，如果真的想让应用程序更加灵活，应该让这些组件实现接口并定义组件和使用这些接口的启动程序之间的相互依赖关系。通过重构消息检索逻辑，可以定义一个带有单个方法 getMessage()的简单的 MessageProvider 接口，如以下代码片段所示：

```
package com.apress.prospring5.ch2.decoupled;

public interface MessageProvider {
    String getMessage();
}
```

MessageRenderer 接口由可渲染消息的所有组件实现，并且对此类组件在以下代码片段中进行了描述：

```
package com.apress.prospring5.ch2.decoupled;

public interface MessageRenderer {
    void render();
    void setMessageProvider(MessageProvider provider);
    MessageProvider getMessageProvider();
}
```

如你所看到的，MessageRenderer 接口声明了一个方法 render()以及一个 JavaBean 样式的方法 setMessageProvider()。任何 MessageRenderer 实现都与消息检索分离并将该责任委托给它们所提供的 MessageProvider 实例。此时，MessageProvider 是 MessageRenderer 的依赖项。创建这些接口的简单实现非常简单，如以下代码片段所示：

```
package com.apress.prospring5.ch2.decoupled;

public class HelloWorldMessageProvider implements MessageProvider {
    @Override
    public String getMessage() {
        return "Hello World!";
    }
}
```

可以看到已经创建了一个简单的 MessageProvider，它始终返回“Hello World!”作为消息。下面所示的 StandardOutMessageRenderer 类也非常简单：

```
package com.apress.prospring5.ch2.decoupled;

public class StandardOutMessageRenderer implements MessageRenderer {
    private MessageProvider messageProvider;

    @Override
    public void render() {
        if (messageProvider == null) {
            throw new RuntimeException(
                "You must set the property messageProvider of class:"
                + StandardOutMessageRenderer.class.getName());
        }
        System.out.println(messageProvider.getMessage());
    }

    @Override
    public void setMessageProvider(MessageProvider provider) {
        this.messageProvider = provider;
    }

    @Override
    public MessageProvider getMessageProvider() {
        return this.messageProvider;
```

```
    }
}
```

现在剩下要做的就是重写入口类的 main()方法：

```
package com.apress.prospring5.ch2.decoupled;

public class HelloWorldDecoupled {
    public static void main(String... args) {
        MessageRenderer mr = new StandardOutMessageRenderer();
        MessageProvider mp = new HelloWorldMessageProvider();
        mr.setMessageProvider(mp);
        mr.render();
    }
}
```

图 2-2 描述了到目前为止所构建的应用程序的抽象模式。

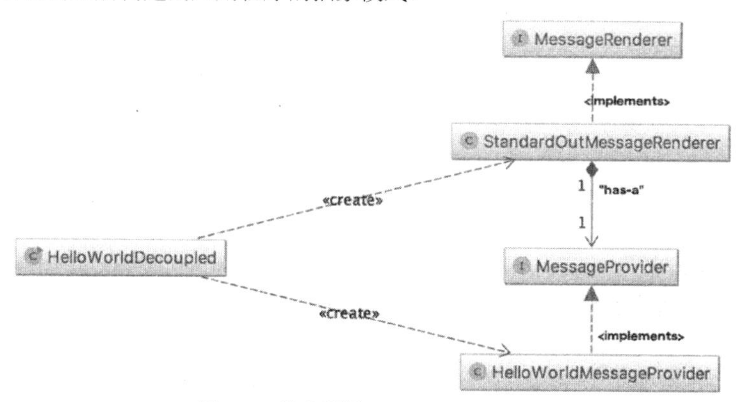

图 2-2 更加解耦的 Hello World 应用程序

此时的代码非常简单。我们实例化了 HelloWorldMessageProvider 和 StandardOutMessageRenderer 的实例，虽然声明的类型分别是 MessageProvider 和 MessageRenderer。这是因为只需要在编程逻辑中与接口提供的方法进行交互，并且 HelloWorldMessageProvider 和 StandardOutMessageRenderer 已经分别实现了这些接口。然后，将 MessageProvider 传递给 MessageRenderer 并调用 MessageRenderer.render()。如果编译并运行该程序，将会得到预期的“Hello World!”输出。现在，该例更能满足我们的要求，但存在一个小问题。更改 MessageRenderer 或 MessageProvider 接口的实现意味着对代码进行更改。为了解决这个问题，可以创建一个简单的工厂类，它从属性文件中读取实现类的名称，并代表应用程序对它们进行实例化，如下所示：

```
package com.apress.prospring5.ch2.decoupled;
import java.util.Properties;

public class MessageSupportFactory {
    private static MessageSupportFactory instance;

    private Properties props;
    private MessageRenderer renderer;
    private MessageProvider provider;

    private MessageSupportFactory() {
        props = new Properties();

        try {
                props.load(this.getClass().getResourceAsStream("/msf.properties"));
                String rendererClass = props.getProperty("renderer.class");
                String providerClass = props.getProperty("provider.class");

                renderer = (MessageRenderer) Class.forName(rendererClass).newInstance();
                provider = (MessageProvider) Class.forName(providerClass).newInstance();
        } catch (Exception ex) {
                ex.printStackTrace();
        }
    }

    static {
        instance = new MessageSupportFactory();
    }
```

```
        public static MessageSupportFactory getInstance() {
            return instance;
        }
    public MessageRenderer getMessageRenderer() {
            return renderer;
        }

        public MessageProvider getMessageProvider() {
            return provider;
        }
    }
```

虽然上述实现过程没有太多价值且非常简单，错误处理也非常简单，并且配置文件的名称是硬编码的，但是已经有了一定数量的代码。该类的配置文件非常简单。

```
renderer.class=
    com.apress.prospring5.ch2.decoupled.StandardOutMessageRenderer
provider.class=
    com.apress.prospring5.ch2.decoupled.HelloWorldMessageProvider
```

要使用前面的实现，必须再次修改 main 方法。

```
package com.apress.prospring5.ch2.decoupled;

public class HelloWorldDecoupledWithFactory {
    public static void main(String... args) {
        MessageRenderer mr =
            MessageSupportFactory.getInstance().getMessageRenderer();
        MessageProvider mp =
            MessageSupportFactory.getInstance().getMessageProvider();
        mr.setMessageProvider(mp);
        mr.render();
    }
}
```

在继续介绍如何将 Spring 引入此应用程序之前，先快速回顾一下前面已经完成的事情。从简单的 Hello World 应用程序开始，定义了应用程序必须满足的两个附加要求。第一个要求是改变消息应该更加简单，第二个要求是改变渲染机制也应该很简单。为了满足这些要求，使用了两个接口：MessageProvider 和 MessageRenderer。MessageRenderer 接口依赖于 MessageProvider 接口的实现，以便能够检索要渲染的消息。最后，添加了一个简单的工厂类来检索实现类的名称并根据需要对它们进行实例化。

2.2.7 用 Spring 重构

虽然前面所示的最后一个示例满足了示例应用程序的要求，但仍然存在问题。第一个问题是必须编写大量的粘合代码将应用程序拼凑在一起，同时保持组件松耦合。第二个问题是仍然需要手动提供 MessageProvider 实例的 MessageRenderer 实现。可以通过使用 Spring 来解决这两个问题。为了解决粘合代码太多的问题，可以从应用程序中完全删除 MessageSupportFactory 类，并用 Spring 接口 ApplicationContext 替换它。不要太担心该接口，目前只需要知道该接口用于存储 Spring 所管理的有关应用程序的所有环境信息。该接口扩展了另一个接口 ListableBeanFactory，而后者充当 Spring 管理的任何 bean 实例的提供程序。

```
package com.apress.prospring5.ch2;

import org.springframework.context.ApplicationContext;
import org.springframework.context.support.ClassPathXmlApplicationContext;

public class HelloWorldSpringDI {
    public static void main(String args) {
        ApplicationContext ctx = new ClassPathXmlApplicationContext
            ("spring/app-context.xml");

        MessageRenderer mr = ctx.getBean("renderer", MessageRenderer.class);
        mr.render();
    }
}
```

在前面的代码片段中，可以看到 main()方法获取了 ClassPathXmlApplicationContext 的实例(从项目的类路径中的文件 spring/app-context.xml 加载应用程序配置信息)，类型为 ApplicationContext，并通过使用 ApplicationContext.getBean()方法获得 MessageRenderer 实例。现在不必太担心 getBean()方法；只需要知道该方法读取应用程序配置(此时为一个

XML文件)，并初始化Spring的ApplicationContext环境，然后返回配置好的bean[1]实例。该XML文件(app-context.xml)的作用与MessageSupportFactory所使用文件的作用相同。

```
<?xml version="1.0" encoding="UTF-8"?>
<beans xmlns="http://www.springframework.org/schema/beans"
   xmlns:xsi="http://www.w3.org/2001/XMLSchema-instance"
   xmlns:p="http://www.springframework.org/schema/p"
   xsi:schemaLocation="http://www.springframework.org/schema/beans
      http://www.springframework.org/schema/beans/spring-beans.xsd">

<bean id="provider"
      class="com.apress.prospring5.ch2.decoupled.HelloWorldMessageProvider"/>

<bean id="renderer"
      class="com.apress.prospring5.ch2.decoupled.StandardOutMessageRenderer"
      p:messageProvider-ref="provider"/>
</beans>
```

上面的文件显示了典型的Spring ApplicationContext配置。首先，声明Spring的名称空间，默认的名称空间是beans。名称空间beans用于声明需要由Spring管理的bean，并声明它们的依赖项需求(对于前面的示例，渲染器bean的messageProvider属性引用了提供者bean)。Spring将解析并注入这些依赖项。

之后，用ID provider和相应的实现类来声明bean。当Spring在ApplicationContext初始化过程中看到这个bean定义时，就会实例化该类并将其存储为指定的ID。

然后声明渲染器bean以及相应的实现类。请记住，该bean依赖于MessageProvider接口来获取需要渲染的消息。为了将DI需求通知给Spring，可以使用名称空间属性p。标记属性p: messageProviderref="provider"告诉Spring，bean的属性messageProvider应该与另一个bean一起注入。要注入属性中的bean应该使用ID provider引用另一个bean。当Spring看到该定义时，将会实例化类，然后查找名为messageProvider的bean属性，并使用ID provider将其注入bean实例中。

正如你所看到的，在初始化Spring的ApplicationContext时，main()方法通过使用类型安全的getBean()方法(传入ID和期望的返回类型，即MessageRenderer接口)获得MessageRenderer bean，并调用render()；Spring创建了MessageProvider实现并将其注入MessageRenderer实现中。请注意，不必对使用Spring连线在一起的类进行任何更改。事实上，这些类没有引用Spring，而是完全忽略它的存在。但是，情况并非总是如此。你自己的类可以实现Spring指定的接口，从而以各种方式与DI容器进行交互。

现在已经完成新的Spring配置以及修改过的main()方法，接下来看看实际运行情况如何。使用Gradle，并在终端中输入以下命令来构建项目和源代码的根目录：

```
gradle clean build copyDependencies
```

Spring配置文件中唯一需要声明的Spring模块是spring-context。Gradle将自动引入此模块所需的任何传递依赖项。在图2-3中，可以看到spring-context.jar的传递依赖项。

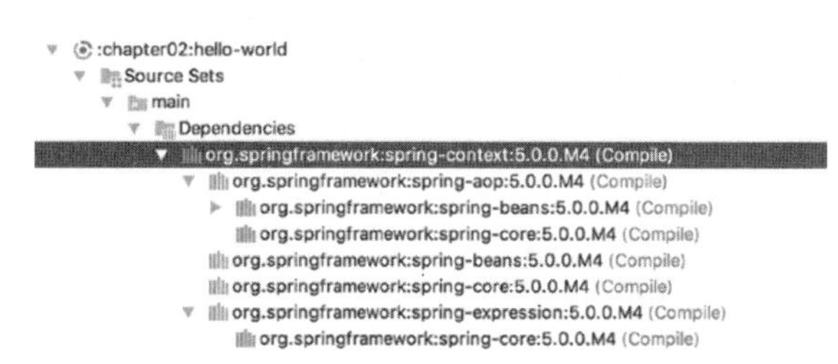

图2-3 在IntelliJ IDEA中描述了spring-context及其传递依赖项

上面的命令将从头开始构建项目，删除以前生成的文件，并在build/libs下将所有需要的依赖项复制到放置所生成工件的相同位置。此路径值也将用作构建JAR时添加到MANIFEST.MF的库文件的附加前缀。如果对Gradle JAR构建配置和流程不太熟悉，可以参阅第2章的源代码(可在Apress网站上获得)，特别是Gradle文件hellorworld/build.properties以获取更多信息。最后，请输入以下命令，以便运行Spring DI示例：

```
cd build/libs; java -jar hello-world-5.0-SNAPSHOT.jar
```

此时，你应该首先看到由Spring容器的启动过程生成的一些日志语句，随后是预期的Hello World输出。

使用注解的Spring配置

从Spring 3.0开始，开发Spring应用程序时不再需要XML配置文件。可以将它们替换为注解和配置类。配置类是用@Configuration注解的Java类，它们包含了bean定义(用@Bean注解的方法)，或者通过使用

1 bean是Spring中类的实例。

@ComponentScanning 对 bean 定义进行注解，从而识别应用程序中的 bean 定义。下面所示的代码与前面提供的 app-context.xml 文件等价：

```
package com.apress.prospring5.ch2.annotated;

import com.apress.prospring5.ch2.decoupled.HelloWorldMessageProvider;
import com.apress.prospring5.ch2.decoupled.MessageProvider;
import com.apress.prospring5.ch2.decoupled.MessageRenderer;
import com.apress.prospring5.ch2.decoupled.StandardOutMessageRenderer;
import org.springframework.context.annotation.Bean;
import org.springframework.context.annotation.Configuration;

@Configuration
public class HelloWorldConfiguration {

   // equivalent to <bean id="provider" class=".."/>
   @Bean
   public MessageProvider provider() {
      return new HelloWorldMessageProvider();
   }

   // equivalent to <bean id="renderer" class=".."/>
   @Bean
   public MessageRenderer renderer(){
      MessageRenderer renderer = new StandardOutMessageRenderer();
      renderer.setMessageProvider(provider());
      return renderer;
   }
}
```

必须修改 main()方法以将 ClassPathXmlApplicationContext 替换为另一个知道如何从配置类读取 bean 定义的 ApplicationContext 实现，该类为 AnnotationConfigApplicationContext。

```
package com.apress.prospring5.ch2.annotated;

import com.apress.prospring5.ch2.decoupled.MessageRenderer;
import org.springframework.context.ApplicationContext;
import org.springframework.context.annotation.AnnotationConfigApplicationContext;

public class HelloWorldSpringAnnotated {

   public static void main(String... args) {
      ApplicationContext ctx = new AnnotationConfigApplicationContext
            (HelloWorldConfiguration.class);
      MessageRenderer mr = ctx.getBean("renderer", MessageRenderer.class);
      mr.render();
   }
}
```

以上只是使用注解和配置类的其中一个配置版本。如果不使用 XML，那么在 Spring 配置方面事情会变得非常灵活。本书后面将详细介绍这一点，但关于配置的重点在于 Java 配置和注解。

⚠ 在 Hello World 示例中定义的一些接口和类可能会在后面的章节中使用。虽然在本示例中展示了完整的源代码，但在后续章节中可能还会出现该代码的精简版本(不会那么冗长)，尤其是在修改增量代码的情况下。该代码的组织有点简单，所有可以在 Spring 未来示例中使用的类都放在 com.apress.prospring5.ch2.decoupled 和 com.apress.prospring5.ch2.annotated 包中，但是请记住，在实际的应用程序中，还是应该适当地对代码进行分层。

2.3 小结

在本章中，主要介绍了启动并运行 Spring 所需的所有背景信息；演示了如何通过依赖项管理系统使用 Spring 以及目前可以直接从 GitHub 获取的 Spring 的开发版本；描述了 Spring 是如何打包的，以及 Spring 的每项功能所需的依赖项。根据这些信息，可以决定应用程序需要哪些 Spring JAR 文件，以及需要在应用程序中分发哪些依赖项。Spring 的文档、指南和测试套件为 Spring 用户提供了开始 Spring 开发的理想基础，因此值得花一些时间来研究 Spring 所提供的内容。本章最后给出了一个示例，演示了如何使用 Spring DI 使传统的 Hello World 成为松耦合的、可扩展的消息渲染应用程序。重要的是，本章只是触碰到了 Spring DI 的皮毛，而没有将 Spring 作为一个整体来学习。在下一章中，将介绍 Spring 中的 IoC 和 DI。

第 3 章

在 Spring 中引入 IoC 和 DI

在第 2 章中，我们已经介绍了控制反转的基本原理。实际上，依赖注入是 IoC 的一种特殊形式，尽管你会经常发现这两个术语可以互换使用。本章将更详细地介绍 IoC 和 DI，将这两个概念之间的关系正式化，并详细探讨 Spring 如何融入其中。

在定义 IoC 和 DI 并了解 Spring 与两者之间的关系之后，将了解一下 Spring 实现 DI 所必需的相关概念。本章仅介绍 Spring 的 DI 实现的基础知识；更多高级的 DI 功能将在第 4 章讨论。具体来说，本章包含以下主题：

- **控制反转的概念**：主要讨论各种类型的 IoC，包括依赖注入和依赖查找，还介绍各种 IoC 方法之间的差异以及每种方法的优缺点。
- **Spring 中的控制反转**：介绍 Spring 中可用的 IoC 功能以及它们的实现方式，特别是介绍 Spring 所提供的依赖注入服务，包括 setter、构造函数和方法注入(Method Injection)。
- **Spring 中的依赖注入**：介绍 Spring 对 IoC 容器的实现。对于 bean 定义和 DI 需求来说，BeanFactory 是应用程序与之交互的主要接口。然而，除前几部分外，本章提供的示例代码将着重使用 Spring 的 ApplicationContext 接口，该接口是 BeanFactory 的一个扩展，并提供了更强大的功能。在后续章节中将介绍 BeanFactory 和 ApplicationContext 之间的区别。
- **配置 Spring 应用程序上下文**：本章最后重点介绍如何使用 XML 以及注解方法来完成 ApplicationContext 配置。Groovy 和 Java 配置将在第 4 章中进一步讨论。这部分首先讨论 DI 配置，然后介绍由 BeanFactory 提供的其他服务，如 bean 继承、生命周期管理和自动装配。

3.1 控制反转和依赖注入

IoC 的核心是 DI，旨在提供一种更简单的机制来设置组件依赖项(通常称为对象的*协作者*)，并在整个生命周期中管理这些依赖项。需要某些依赖项的组件通常被称为*依赖对象*，或者在 IoC 的情况下被称为*目标对象*。通常，IoC 可以分解为两种子类型：依赖注入和依赖查找。这些子类型被进一步分解为 IoC 服务的具体实现。通过这个定义可以清楚地看到，当谈论 DI 时，通常是在谈论 IoC，而当谈论 IoC 时，则并不总是在谈论 DI(例如，依赖查找也是 IoC 的一种形式)。

3.2 控制反转的类型

你可能想知道为什么有两种类型的 IoC 以及为什么这些类型被进一步分解为不同的实现。这个问题似乎没有明确的答案。当然，不同类型都提供了一定程度的灵活性，但对我们来说，IoC 似乎更多是新旧观念的混合体。两种类型的 IoC 就印证了这一点。依赖查找是一种更传统的方法，乍一看，Java 程序员似乎更熟悉它。虽然初看起来依赖注入有悖常理，但实际上比依赖查找更灵活且更有用。使用依赖查找时，组件必须获取对依赖项的引用，而使用依赖注入时，依赖项将通过 IoC 容器注入组件。依赖查找有两种类型：依赖拉取(dependency pull，DL)和上下文依赖查找(contextualized dependency lookup，CDL)。依赖注入也有两种常见的风格：构造函数和 setter 依赖注入。

在本节的讨论过程中，我们并不关心虚构的 IoC 容器如何了解所有不同的依赖项，它只是在某些时候执行每种机制所描述的操作。

3.2.1 依赖拉取

对于 Java 开发人员来说，*依赖拉取*是一种最常见的 IoC 类型。在依赖拉取中，根据需要从注册表中提取依赖项。任何曾经编写代码访问过 EJB(2.1 或更早版本)的开发人员都使用过依赖拉取(即通过 JNDI API 查找 EJB 组件)。图 3-1 显示了通过查找机制进行依赖拉取的场景。

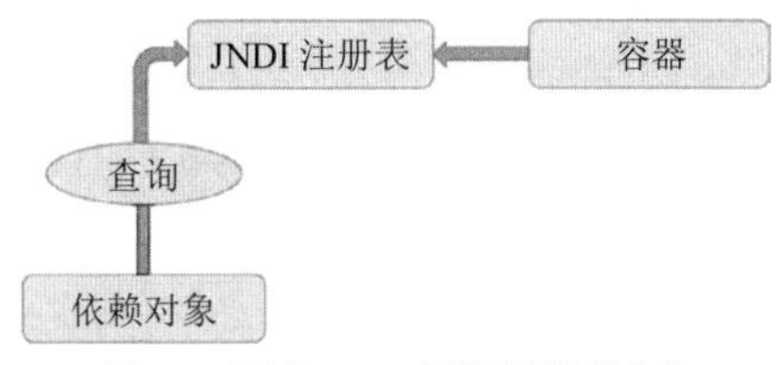

图 3-1 通过 JNDI 查找进行依赖拉取

Spring 还提供依赖拉取作为一种检索框架所管理组件的机制；在第 2 章中曾经介绍过相关内容。以下代码示例显示了基于 Spring 的应用程序中典型的依赖拉取查询：

```
package com.apress.prospring5.ch3;

import org.springframework.context.ApplicationContext;
import org.springframework.context.support.ClassPathXmlApplicationContext;

public class DependencyPull {
    public static void main(String... args) {
       ApplicationContext ctx = new ClassPathXmlApplicationContext
          ("spring/app-context.xml");

       MessageRenderer mr = ctx.getBean("renderer", MessageRenderer.class);
       mr.render();
    }
}
```

这种类型的 IoC 不仅在基于 JEE 的应用程序(使用 EJB 2.1 或以前的版本)中非常流行，它广泛使用 JNDI 查找来获得注册表的依赖项，而且在许多环境中也是使用 Spring 的关键。

3.2.2 上下文依赖查找

上下文依赖查找(CDL)在某些方面与依赖拉取类似，但在 CDL 中，查找是针对管理资源的容器执行的，而不是来自某个中央注册表，并且通常在某个设定点执行。图 3-2 显示了 CDL 机制。

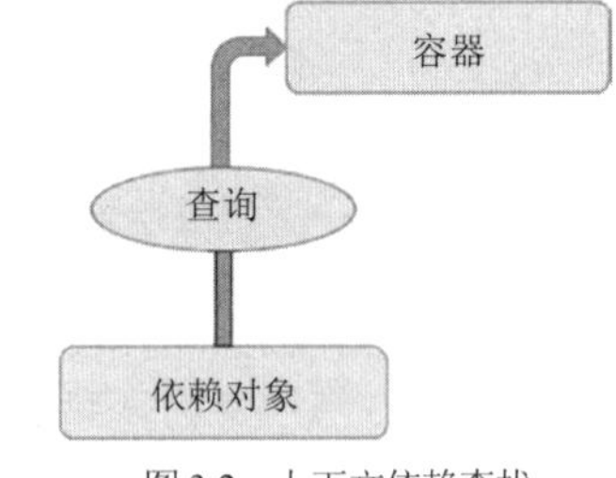

图 3-2 上下文依赖查找

CDL 通过让组件实现类似于以下代码片段的接口来进行工作：

```
package com.apress.prospring5.ch3;

public interface ManagedComponent {
    void performLookup(Container container);
}
```

通过实现该接口，一个组件可以向容器发送它想要获得依赖项的信号。容器通常由底层应用程序服务器或框架(例如 Tomcat、JBoss 或 Spring)提供。以下代码片段显示了一个提供依赖查找服务的简单 Container 接口：

```
package com.apress.prospring5.ch3;

public interface Container {
    Object getDependency(String key);
}
```

当容器准备将依赖项传递给组件时，会依次调用每个组件的 performLookup()方法。然后，组件可以使用 Container 接口查找所需的依赖项，如以下代码片段所示：

```
package com.apress.prospring5.ch3;

public class ContextualizedDependencyLookup
        implements ManagedComponent {
    private Dependency dependency;

    @Override
    public void performLookup(Container container) {
       this.dependency = (Dependency) container.getDependency("myDependency");
```

```
    }

    @Override
    public String toString() {
        return dependency.toString();
    }
}
```

3.2.3 构造函数依赖注入

当在组件的构造函数中提供依赖项时，就会发生*构造函数依赖注入*。首先，组件声明一个或一组构造函数，并将其依赖项作为参数，然后在组件实例化时由 IoC 容器将依赖项传递给组件，如以下代码片段所示：

```
package com.apress.prospring5.ch3;

public class ConstructorInjection {
    private Dependency dependency;

    public ConstructorInjection(Dependency dependency) {
        this.dependency = dependency;
    }

    @Override
    public String toString() {
        return dependency.toString();
    }
}
```

使用构造函数注入的一个显而易见的结果是，如果没有依赖项，就不能创建对象；因此，必须有依赖项。

3.2.4 setter 依赖注入

在 setter 依赖注入中，IoC 容器通过 JavaBean 样式的 setter 方法注入组件的依赖项。组件的 setter 方法公开了 IoC 容器可以管理的依赖项。以下代码示例显示了一个典型的基于 setter 依赖注入的组件：

```
package com.apress.prospring5.ch3;

public class SetterInjection {
    private Dependency dependency;

    public void setDependency(Dependency dependency) {
        this.dependency = dependency;
    }

    @Override
    public String toString() {
        return dependency.toString();
    }
}
```

使用 setter 注入的一个显而易见的后果是，可以在没有依赖项的情况下创建对象，然后可以通过调用 setter 来提供依赖项。

在容器中，setDependency()方法所公开的依赖需求由 JavaBeans 风格的名称 dependency 引用。实际上，setter 注入是使用最广泛的注入机制，也是最简单的 IoC 机制之一。

> 在 Spring 中，还支持另一种被称为字段注入(field injection)的注入类型，在本章后面学习如何使用@Autowire 注解进行自动装配时将介绍该注入类型。

3.2.5 注入与查找

选择使用哪种类型的 IoC(注入或查找)通常不难作决定。在许多情况下，所使用的 IoC 类型由使用的容器确定。例如，如果使用的是 EJB 2.1 或更早版本，则必须使用查找式 IoC(通过 JNDI)从 JEE 容器中获取 EJB。在 Spring 中，除了初始的 bean 查找，组件及其依赖项始终使用注入式 IoC 连接在一起。

> 当使用 Spring 时，可以访问 EJB 资源而无须执行显式查找。Spring 可以充当查找式和注入式 IoC 系统之间的适配器，因此可以使用注入来管理所有资源。

真正的问题是：如果可以选择，那么应该使用哪种方法，注入还是查找？答案绝对是注入。如果查看前面示例中的代码，可以清楚地看到使用注入对组件的代码没有任何影响。另一方面，依赖拉取代码必须主动获得对注册表的引用并与其交互以获取依赖项，而如果使用 CDL，则需要你的类实现特定的接口并手动查找所有依赖项。当使用注入时，你的类最需要做的就是允许使用构造函数或 setter 来注入依赖项。

如果使用注入，则可以自由地使用与 IoC 容器完全分离的类，而 IoC 容器通过手动为它们的协作者提供依赖对象；而如果使用查找，那么你的类总是依赖于容器定义的类和接口。查找的另一个缺点是难以独立于容器来测试类。而使用注入则可以非常容易地测试自己的组件，因为可以通过使用适当的构造函数或 setter 来简单地提供依赖项。

有关使用依赖注入和 Spring 进行测试的更多讨论，请参阅第 13 章。

基于查找的解决方案必然比基于注入的解决方案更加复杂。虽然复杂性并没有什么可害怕的，但应该质疑的是，向作为依赖管理的应用程序添加不必要的复杂性是否正确。

抛开以上原因，之所以选择注入而不是查找的最大理由是，它能够让开发更加轻松。在使用注入时，所编写的代码非常少，并且代码也很简单，通常可以通过良好的 IDE 自动完成。你会发现，注入示例中的所有代码都是被动的，因为它们并不积极尝试完成任务。在注入代码中看到的最令人兴奋的事情是，对象仅存储在字段中；没有任何其他代码可以从任何注册表或容器中提取依赖项。因此，注入代码非常简单并且不易出错。被动代码比主动代码更容易维护，因为很少出错。请思考下面从 CDL 示例中获得的代码：

```
public void performLookup(Container container) {
    this.dependency = (Dependency) container.getDependency("myDependency");
}
```

在这段代码中，可能会出现很多错误：依赖项键可能会改变，容器实例可能为空，或者返回的依赖项可能是不正确的类型。通常将此类代码称为具有多个移动部分的代码，因为很多部分都可能会被破坏。虽然使用依赖查找可能解耦应用程序的组件，但会增加将这些组件连接在一起以执行任何有用任务时所需的附加代码的复杂性。

3.2.6 setter 注入与构造函数注入

现在虽然已经确定了哪种 IoC 方法更可取，但仍然需要选择是使用 setter 注入还是使用构造函数注入。当在使用组件之前必须拥有一个依赖类的实例时，*构造函数注入*就特别有用了。包括 Spring 在内的许多容器都提供了一种机制，可确保在使用 setter 注入时定义所有依赖项；但如果使用构造函数注入，则可以通过一种容器无关(container-agnostic)的方式声明对依赖项的需求。此外，构造函数注入也有助于实现不可变对象的使用。

*setter 注入*在各种情况下都很有用。如果组件向容器公开了它的依赖项，并乐于提供自己的默认值，那么 setter 注入通常是实现此目的的最佳方法。setter 注入的另一个好处是，它允许在接口上声明依赖项，尽管这种方法并没有前面的方法有用。假设有一个带有业务方法 defineMeaningOfLife()的典型业务接口。如果除了该方法还定义了一个用于注入的 setter 方法(例如 setEncylopedia())，就强制要求所有的实现都必须使用或者至少知道依赖项。其实并不需要在业务接口中定义 setEncylopedia()。相反，可以在实现业务接口的类中定义方法。当以这种方式进行编程时，包括 Spring 在内的所有最新 IoC 容器都可以以业务接口的形式使用组件，同时仍然提供了实现类的依赖项。提供一个示例可以更加清楚地说明这一点，请思考下面代码片段中的业务接口：

```
package com.apress.prospring5.ch3;

public interface Oracle {
    String defineMeaningOfLife();
}
```

请注意，业务接口没有为依赖注入定义任何 setter 方法。该接口可以按照以下代码片段中所示的方式实现：

```
package com.apress.prospring5.ch3;

public class BookwormOracle implements Oracle {
    private Encyclopedia encyclopedia;

    public void setEncyclopedia(Encyclopedia encyclopedia) {
       this.encyclopedia = encyclopedia;
    }

    @Override
    public String defineMeaningOfLife() {
```

```
        return "Encyclopedias are a waste of money - go see the world instead";
    }
}
```

正如你所看到的，BookwormOracle 类不仅实现了 Oracle 接口，还定义了用于依赖注入的 setter 方法。虽然 Spring 处理类似于这样的结构会更加轻松，但没有必要在业务接口上定义依赖项。使用接口来定义依赖项的能力是 setter 注入的一个经常被提及的好处，但实际上，应该尽量保持 setter 仅用于注入接口。除非完全确定特定业务接口的所有实现都需要一个特定的依赖项，否则让每个实现类定义自己的依赖项并为业务方法保留业务接口。

虽然不应该在业务接口中放置用于依赖项的 setter 方法，但在业务接口中放置用来获取配置参数的 setter 和 getter 方法却是一个好主意，并且使 setter 注入成为一个有价值的工具。可以将配置参数视为依赖项的特例。当然，虽然组件依赖于配置数据，但配置数据与前面看到的依赖项类型明显不同。稍后会讨论这些差异，但现在先考虑一下下面的代码片段所显示的业务接口：

```
package com.apress.prospring5.ch3;

public interface NewsletterSender {
    void setSmtpServer(String smtpServer);
    String getSmtpServer();
    void setFromAddress(String fromAddress);
    String getFromAddress();

    void send();
}
```

通过电子邮件发送一组新闻简报的类都实现了 NewsletterSender 接口。send()方法是唯一的业务方法，但请注意，在接口上还定义了两个 JavaBean 属性。前面讲了不应该在业务接口中定义依赖项，那么现在为什么还要这样做呢？原因是，从实际意义上来说，这些值(即 SMTP 服务器地址和电子邮件地址)并不是依赖项；相反，它们是影响 NewsletterSender 接口函数的所有实现的配置细节信息。但问题是：配置参数和任何其他类型的依赖项之间有什么区别呢？在大多数情况下，可以明确地确定是否应该将依赖项归类为配置参数，但如果不确定，请查看是否具备以下三个配置参数的特征：

- 配置参数是被动的。在前面代码片段所描述的 NewsletterSender 示例中，SMTP 服务器参数是被动依赖项的一个示例。被动依赖项不直接用于执行操作，而在内部使用或由其他依赖项使用来完成相关的操作。在第 2 章的 MessageRenderer 示例中，MessageProvider 依赖项不是被动的；它执行了一个 MessageRenderer 以完成其任务所必需的函数。
- 配置参数通常是信息，而不是其他组件。这意味着配置参数通常是某一组件完成其工作所需的一些信息。显然，SMTP 服务器是 NewsletterSender 所需的一部分信息，而 MessageProvider 实际上是 MessageRenderer 正确运行所需的另一个组件。
- 配置参数通常是简单值或简单值的集合。实际上，这是根据前两点得到的，配置参数通常是简单值。在 Java 中，这意味着它们是基本数据类型(或相应的包装类)、字符串或这些值的集合。简单值通常是被动的。这意味着除了处理它所表示的数据，你不能用 String 来完成更多的事情；并且这些值始终作为信息目的来使用，例如，表示网络套接字应侦听的端口号的 int 值，或者表示电子邮件程序应通过其发送邮件的 SMTP 服务器的 String 值。

在考虑是否在业务接口中定义配置选项时，还应该考虑配置参数是适用于业务接口的所有实现，还是仅适用于特定业务接口。例如，在实现 NewsletterSender 接口时，很显然，所有实现都需要知道发送电子邮件时要使用哪个 SMTP 服务器。但也可以选择保留配置选项，以标记是否将安全电子邮件从业务接口发送出去，因为并非所有电子邮件 API 都可以实现此目的，并且假设许多实现没有考虑到安全性是一种正确的做法。

回想一下，在第 2 章中，配置选项被用来定义业务目的中的依赖项。这里只是出于说明目的，并不是最佳做法。

---

setter 注入还允许即时交换针对不同实现的依赖项，而无须创建父组件的新实例。Spring 的 JMX 支持使这项功能成为可能。也许 setter 注入最大的好处是，它是注入机制中侵入性最小的。

一般来说，应根据使用情况选择注入类型。基于 setter 的注入允许在不创建新对象的情况下交换依赖项，并且还可以让类选择适当的默认值，而无须显式注入对象。当想要确保将依赖项传递给组件和设计不可变对象时，构造函数注入是一个不错的选择。请记住，虽然构造函数注入可确保向组件提供所有的依赖项，但大多数容器也都提供一种机制来确保这一点，不过付出的代价是使代码与框架产生耦合。

3.3 Spring 中的控制反转

如前所述，控制反转是 Spring 的重要组成部分。Spring 实现的核心是基于依赖注入，尽管也提供了依赖查找功能。当 Spring 自动将协作者提供给依赖对象时，是使用依赖注入实现的。在基于 Spring 的应用程序中，始终优先使用依赖注入将协作者传递给依赖对象，而不是让依赖对象通过查找获取协作者。图 3-3 显示了 Spring 的依赖注入机制。虽然依赖注入是将协作者和依赖对象连接在一起的首选机制，但有时仍然需要使用依赖查找来访问依赖对象。在许多环境中，Spring 无法通过使用依赖注入来自动连接所有应用程序组件，并且必须使用依赖查找来访问初始组件组。例如，在独立的 Java 应用程序中，需要在 main()方法中启动 Spring 容器，并获取依赖项(通过 ApplicationContext 接口)，以便以编程方式进行处理。但是，当使用 Spring 的 MVC 支持构建 Web 应用程序时，Spring 可以通过将整个应用程序自动粘贴在一起来避免这种情况。只要可以使用 Spring 的依赖注入，就应该尽量使用；否则，可以使用依赖查找功能。在本章的学习过程中会看到这两种情况的示例，当它们第一次出现时会指出来。

图 3-3　Spring 的依赖注入机制

Spring 的 IoC 容器有一个非常有趣的功能，可以在自己的依赖注入容器和外部依赖查询容器之间充当适配器。在本章后面将会讨论该功能。

Spring 支持构造函数注入和 setter 注入，并支持标准的 IoC 功能集，同时提供大量有用的附加功能，从而让开发工作更轻松。

本章其余部分将介绍 Spring 的 DI 容器的基础知识，并附有大量示例。

3.4 Spring 中的依赖注入

Spring 对依赖注入的支持是全面的，正如在第 4 章中看到的，超出了到目前为止所讨论的标准 IoC 功能集。本章其余部分将介绍 Spring 的依赖注入容器的基础知识，学习一下 setter 注入、构造函数注入和方法注入，并详细介绍 Spring 如何配置依赖注入。

3.4.1 bean 和 BeanFactory

Spring 的依赖注入容器的核心是 BeanFactory 接口。BeanFactory 负责管理组件，包括依赖项以及它们的生命周期。在 Spring 中，术语 *bean* 用于引用由容器管理的任何组件。通常，bean 在某种程度上遵守 JavaBean 规范，但这并不是必需的，尤其是当想要使用构造函数注入来将 bean 连接在一起的。

如果应用程序只需要 DI 支持，则可以通过 BeanFactory 接口与 Spring DI 容器进行交互。这种情况下，应用程序必须创建一个实现了 BeanFactory 接口的类的实例，并使用 bean 和依赖信息对其进行配置。完成后，应用程序就可以通过 BeanFactory 访问 bean 并进行处理。

某些情况下，所有这些设置都是自动处理的(例如，在 Web 应用程序中，Web 容器将在应用程序启动期间通过 web.xml 描述符文件中声明的 ContextLoaderListener 类启动 Spring 的 ApplicationContext)。但在很多情况下，需要自己编写设置代码。本章中的所有示例都需要手动设置 BeanFactory 实现。

虽然 BeanFactory 可以通过编程方式配置，但更常见的做法是使用某种配置文件在外部对其进行配置。在内部，bean 配置由实现 BeanDefinition 接口的类的实例表示。bean 配置不仅存储有关 bean 本身的信息，还存储有关它所依赖的 bean 的信息。对于任何实现了 BeanDefinitionReader 接口的 BeanFactory 实现类来说，都可以通过使用 PropertiesBeanDefinitionReader 或 XmlBeanDefinitionReader 从配置文件中读取 BeanDefinition 数据。PropertiesBeanDefinitionReader 从属性文件读取 bean 定义，而 XmlBeanDefinitionReader 则从 XML 文件中读取相关信息。

所以，可以在 BeanFactory 中识别自己的 bean；为每个 bean 分配一个 ID、一个名称或两者兼具。一个 bean 也可以在没有任何 ID 或名称(称为*匿名 bean*)的情况下被实例化，或者作为另一个 bean 的内部 bean 被实例化。每个 bean 至少有一个名称，但也可以有任意数量的名称(名称用逗号分隔)。第一个名称后面的任何名称都被认为是同一个 bean 的别名。可以使用 bean ID 或名称从 BeanFactory 检索一个 bean，并建立依赖关系(也就是说，bean X 依赖于 bean Y)。

3.4.2 BeanFactory 实现

对 BeanFactory 接口的描述可能看起来过于复杂，但实际上并非如此。接下来看一个简单的示例。假设有一个模仿神谕的实现，可以告诉你生命的意义。

```
//interface
package com.apress.prospring5.ch3;

public interface Oracle {
    String defineMeaningOfLife();
}
//implementation
package com.apress.prospring5.ch3;

public class BookwormOracle implements Oracle {
    private Encyclopedia encyclopedia;

    public void setEncyclopedia(Encyclopedia encyclopedia) {
        this.encyclopedia = encyclopedia;
    }

    @Override
    public String defineMeaningOfLife() {
        return "Encyclopedias are a waste of money - go see the world instead";
    }
}
```

接下来让我们看一下在独立的 Java 程序中，Spring 的 BeanFactory 如何被初始化并获得用于处理的 oracle bean。代码如下：

```
package com.apress.prospring5.ch3;

import org.springframework.beans.factory.support.DefaultListableBeanFactory;
import org.springframework.beans.factory.xml.XmlBeanDefinitionReader;
import org.springframework.core.io.ClassPathResource;

public class XmlConfigWithBeanFactory {

  public static void main(String... args) {
     DefaultListableBeanFactory factory = new DefaultListableBeanFactory();
     XmlBeanDefinitionReader rdr = new XmlBeanDefinitionReader(factory);
     rdr.loadBeanDefinitions(new
       ClassPathResource("spring/xml-bean-factory-config.xml"));
     Oracle oracle = (Oracle) factory.getBean("oracle");
     System.out.println(oracle.defineMeaningOfLife());
  }
}
```

在上面的代码示例中，可以看到所使用的是 DefaultListableBeanFactory，它是 Spring 提供的两个主要 BeanFactory 实现之一，并且使用 XmlBeanDefinitionReader 从 XML 文件读取 BeanDefinition 信息。一旦创建并配置了 BeanFactory 实现，就可以通过使用在 XML 配置文件中配置的名称 oracle 来检索 oracle bean。

```
<?xml version="1.0" encoding="UTF-8"?>

<beans xmlns="http://www.springframework.org/schema/beans"
    xmlns:xsi="http://www.w3.org/2001/XMLSchema-instance"
    xsi:schemaLocation="http://www.springframework.org/schema/beans
       http://www.springframework.org/schema/beans/spring-beans.xsd">
    <bean id="oracle"
       name="wiseworm"
       class="com.apress.prospring5.ch3.BookwormOracle"/>
</beans>
```

在声明 Spring XSD 位置时，最好不要包含版本号。这个问题已经由 Spring 代为处理，因为版本化的 XSD 文件是通过 spring.schemas 文件中的指针配置的。该文件驻留在定义为项目依赖项的 spring-beans 模块中。这样做也可以防止升级到新版本的 Spring 时对所有的 bean 文件进行修改。

前面的文件声明了一个 Spring bean，并将其 ID 和名称分别设置为 oracle 和 wiseworm，同时告诉 Spring 基础实现类是 com.apress.prospring4.ch3.BookwormOracle。目前不需要太担心配置的含义，在后面的章节中将会更详细地加以讨论。

定义配置后，运行前面的代码示例中所示的程序；将会在控制台输出中看到由 defineMeaningOfLife()方法返回的短语。

除了 XmlBeanDefinitionReader，Spring 还提供了 PropertiesBeanDefinitionReader，它允许通过使用属性而不是 XML 来管理 bean 配置。尽管属性对于小型简单应用程序来说非常理想，但当处理大量的 bean 时，就可能会很快变得非常麻烦。出于这个原因，最好的做法是除最简单的应用程序外都使用 XML 配置格式。

当然，也可以自由定义自己的 BeanFactory 实现，但要注意这样做是相当复杂的；需要实现比 BeanFactory 更多的接口才能获得与 BeanFactory 实现相同水平的功能。如果只想定义一种新的配置机制，那么可以通过开发一个扩展了 DefaultListableBeanFactory 类的类来创建定义读取器，该类实现了 BeanFactory 接口。

3.4.3 ApplicationContext

在 Spring 中，ApplicationContext 接口是 BeanFactory 的一个扩展。除了 DI 服务，ApplicationContext 还提供了其他服务，例如事务和 AOP 服务、国际化(i18n)的消息源以及应用程序事件处理等。在开发基于 Spring 的应用程序时，建议通过 ApplicationContext 接口与 Spring 进行交互。Spring 通过手动编码(通过手动实例化并加载适当的配置)或在 Web 容器环境中通过 ContextLoaderListener 来支持 ApplicationContext 的启动。从现在开始，本书中的所有示例代码都使用 ApplicationContext 及其实现。

3.5 配置 ApplicationContext

在讨论了 IoC 和 DI 的基本概念并学习了 Spring 的 BeanFactory 接口的简单示例之后，接下来深入介绍如何配置 Spring 应用程序。下面将介绍配置 Spring 应用程序的各方面内容。具体来说，应该将注意力集中在 ApplicationContext 接口上，该接口提供了比传统的 BeanFactory 接口更多的配置选项。

3.5.1 设置 Spring 配置选项

在深入学习配置 Spring 的 ApplicationContext 的细节之前，先来看看在 Spring 中定义应用程序的配置时可用的选项。最初，Spring 支持通过属性或 XML 文件定义 bean。但自从 JDK 5 发布以及 Spring 支持 Java 注解以来，Spring(从 Spring 2.5 开始)在配置 ApplicationContext 时也支持使用 Java 注解。那么，哪一种方法更好呢？XML 还是注释？关于这个问题已经有很多争论，可以在 Internet 上找到许多讨论[1]。但至今没有明确的答案，每种方法都有其优缺点。使用 XML 文件可以从 Java 代码中外部化所有配置，而注解允许开发人员从代码中定义和查看 DI 设置。Spring 还支持在单个 ApplicationContext 中混合使用这两种方法。一种常见的方法是在 XML 文件中定义应用程序基础结构(例如数据源、事务管理器、JMS 连接工厂或 JMX)，同时在注解中定义 DI 配置(可注入 bean 和 bean 的依赖项)。但是，无论选择哪种方法，都要坚持使用并在整个开发团队中清楚地传达相关信息。如果在整个开发团队中就所使用方法达成一致意见并在整个应用程序中保持一致，就会使开发和维护变得更容易。

为便于理解 XML 和注解配置，本书在适当的时候会提供 XML 和注解的示例代码，但重点将放在注解和 Java 配置上，因为 XML 已在前几版的书中介绍过。

3.5.2 基本配置概述

对于 XML 配置，需要声明应用程序需要的由 Spring 提供的名称空间基础信息。下面所示的配置示例显示了最

1 例如，Spring 社区论坛 http://forum.spring.io。

基本的示例，该例仅声明了用于定义 Spring bean 的 bean 的名称空间。在整个示例中，将配置文件 app-contextxml.xml 用于 XML 样式的配置。

```
<?xml version="1.0" encoding="UTF-8"?>

<beans xmlns="http://www.springframework.org/schema/beans"
    xmlns:xsi="http://www.w3.org/2001/XMLSchema-instance"
    xmlns:c="http://www.springframework.org/schema/c"
    xsi:schemaLocation="http://www.springframework.org/schema/beans
    http://www.springframework.org/schema/beans/spring-beans.xsd">
</beans>
```

除了 bean，Spring 还为不同目的提供了大量其他名称空间，比如针对 ApplicationContext 配置的上下文、用于 AOP 支持的 aop 以及用于事务支持的 tx。本书将在适当的章节介绍这些名称空间。

要想在应用程序中使用 Spring 的注解支持，需要在 XML 配置中声明以下配置示例中的标记。可以将此配置文件称为 app-context-annotation.xml，用于在整个示例中支持带有注解的 XML 配置。

```
<?xml version="1.0" encoding="UTF-8"?>

<beans xmlns="http://www.springframework.org/schema/beans"
       xmlns:xsi="http://www.w3.org/2001/XMLSchema-instance"
       xmlns:context="http://www.springframework.org/schema/context"
       xmlns:c="http://www.springframework.org/schema/c"
       xsi:schemaLocation="http://www.springframework.org/schema/beans
          http://www.springframework.org/schema/beans/spring-beans.xsd
          http://www.springframework.org/schema/context
          http://www.springframework.org/schema/context/spring-context.xsd">

    <context:component-scan
       base-package="com.apress.prospring5.ch3.annotation"/>

</beans>
```

<context:component-scan>标记告诉 Spring 扫描代码，从而找到使用@Component、@Controller、@Repository 和@Service 注解的注入 bean 以及支持在指定包(及其所有子包)下使用@Autowired、@Inject 和@Resource 注解的 bean。在<context:component-scan>标记中，可以使用逗号、分号或空格作为分隔符来定义多个包。此外，该标记支持组件扫描的包含和排除，从而实现更细粒度的控制。例如，思考以下配置示例：

```
<?xml version="1.0" encoding="UTF-8"?>

<beans xmlns="http://www.springframework.org/schema/beans"
       xmlns:xsi="http://www.w3.org/2001/XMLSchema-instance"
       xmlns:context="http://www.springframework.org/schema/context"
       xmlns:c="http://www.springframework.org/schema/c"
       xsi:schemaLocation="http://www.springframework.org/schema/beans
          http://www.springframework.org/schema/beans/spring-beans.xsd
          http://www.springframework.org/schema/context
          http://www.springframework.org/schema/context/spring-context.xsd">

    <context:component-scan
          base-package="com.apress.prospring5.ch3.annotation">
          <context:exclude-filter type="assignable"
            expression="com.example.NotAService"/>
    </context:component-scan>
</beans>
```

上面所示的标记告诉 Spring 扫描指定的包，同时忽略那些在表达式中指定了类型的类(可以是类或接口)。除了排除过滤器，还可以使用包含过滤器。对于类型，可以使用注解、正则表达式、赋值、AspectJ 或自定义类(使用实现了 org.springframework.core.type.filter.TypeFilter 的过滤器类)作为过滤条件。表达式的格式取决于所指定的类型。

3.5.3 声明 Spring 组件

在开发完某种服务类并想在基于 Spring 的应用程序中使用它时，需要告诉 Spring：这些 bean 有资格注入其他 bean 并让 Spring 代为管理。请思考第 2 章中的示例，其中 MessageRender 输出消息，并依赖于 MessageProvider 来提供要渲染的消息。以下代码示例描述了这两种服务的接口和实现：

```
package com.apress.prospring5.ch2.decoupled;

//renderer interface
public interface MessageRenderer {
```

```
    void render();
    void setMessageProvider(MessageProvider provider);
    MessageProvider getMessageProvider();
}
// rendered implementation
public class StandardOutMessageRenderer
    implements MessageRenderer {

        private MessageProvider messageProvider;

        @Override

        public void render() {
                if (messageProvider == null) {
                        throw new RuntimeException(
                          "You must set the property messageProvider of class:"
                          + StandardOutMessageRenderer.class.getName());
        }
                System.out.println(messageProvider.getMessage());
                }

        @Override
        public void setMessageProvider(MessageProvider provider) {
                this.messageProvider = provider;
        }

        @Override
                public MessageProvider getMessageProvider() {
        return this.messageProvider;
        }
}

//provider interface
public interface MessageProvider {
    String getMessage();
}

//provider implementation
public class HelloWorldMessageProvider implements MessageProvider {

    @Override
    public String getMessage() {
        return "Hello World!";
    }
}
```

上面所示的类是 com.apress.prospring5.ch2.decoupled 包的一部分。它们也被用于本章特定的项目中，因为在实际的生产应用程序中，开发人员会尝试重复使用代码而不是复制它们。这就是为什么第 2 章的项目被定义为第 3 章中一些项目的依赖(在获取资源时你会看到这一点)的原因。

如果想要在 XML 文件中声明 bean 定义，可以使用<bean ../>标记，最终的 app-context-xml.xml 文件如下所示：

```
<?xml version="1.0" encoding="UTF-8"?>

<beans xmlns="http://www.springframework.org/schema/beans"
    xmlns:xsi="http://www.w3.org/2001/XMLSchema-instance"
    xmlns:p="http://www.springframework.org/schema/p"
    xsi:schemaLocation="http://www.springframework.org/schema/beans
        http://www.springframework.org/schema/beans/spring-beans.xsd">

    <bean id="provider"
        class="com.apress.prospring5.ch2.decoupled.HelloWorldMessageProvider"/>

    <bean id="renderer"
        class="com.apress.prospring5.ch2.decoupled.StandardOutMessageRenderer"
        p:messageProvider-ref="provider"/>
</beans>
```

前面所示的标记声明了两个 bean，一个是具有 HelloWorldMessageProvider 实现且 ID 为 provider 的 bean，另一个是具有 StandardOutMessageRenderer 实现且 ID 为 renderer 的 bean。

从该例开始，相关的名称空间将不再添加到配置示例中，除非引入新的名称空间，因为这将使 bean 定义更加清晰可见。

要想使用注解创建 bean 定义，必须使用适当的构造型注解来注解 bean 类[1]，并且必须使用@Autowired 注解方法或构造函数，以便告诉 Spring IoC 容器：查找该类型的 bean 并将其用作调用该方法时的参数。在下面的代码片段中，用于创建 bean 定义的注解以下画线标出。构造型注解可以将结果 bean 的名称作为参数。

```
package com.apress.prospring5.ch3.annotation;

import com.apress.prospring5.ch2.decoupled.MessageProvider;
import org.springframework.stereotype.Component;
//simple bean
@Component("provider")
public class HelloWorldMessageProvider implements MessageProvider {
    @Override
    public String getMessage() {
        return "Hello World!";
    }
}

import com.apress.prospring5.ch2.decoupled.MessageProvider;
import com.apress.prospring5.ch2.decoupled.MessageRenderer;
import org.springframework.stereotype.Service;
import org.springframework.beans.factory.annotation.Autowired;

//complex, service bean
@Service("renderer")
public class StandardOutMessageRenderer
   implements MessageRenderer {
    private MessageProvider messageProvider;

    @Override
    public void render() {
        if (messageProvider == null) {
           throw new RuntimeException(
           "You must set the property messageProvider of class:"
           + StandardOutMessageRenderer.class.getName());
        }

        System.out.println(messageProvider.getMessage());
    }

    @Override
    @Autowired
    public void setMessageProvider(MessageProvider provider) {
        this.messageProvider = provider;
    }

    @Override
    public MessageProvider getMessageProvider() {
        return this.messageProvider;
    }
}
```

当使用下面所示的 XML(在 app-context-annotation.xml 文件中)配置来启动 Spring 的 ApplicationContext 时，Spring 将查找这些组件并使用指定名称实例化 bean：

```
<?xml version="1.0" encoding="UTF-8"?>

<beans xmlns="http://www.springframework.org/schema/beans"
       xmlns:xsi="http://www.w3.org/2001/XMLSchema-instance"
       xmlns:context="http://www.springframework.org/schema/context"
       xsi:schemaLocation="http://www.springframework.org/schema/beans
       http://www.springframework.org/schema/beans/spring-beans.xsd
       http://www.springframework.org/schema/context
       http://www.springframework.org/schema/context/spring-context.xsd">

<context:component-scan
       base-package="com.apress.prospring5.ch3.annotation"/>
</beans>
```

使用这两种方法不会影响从 ApplicationContext 获取 bean 的方式。

```
package com.apress.prospring5.ch3;
```

1 这些注解被称为构造型(stereotype)注解，因为它们是 org.springframework.stereotype 包的一部分。该包将用于定义 bean 的所有注解组合在一起。这些注解也与 bean 的角色相关。例如，@Service 用于定义一个服务 bean，它是一个更复杂的功能 bean，提供其他 bean 可能需要的服务，而@Repository 用于定义一个完成从数据库中检索/保存数据等操作的 bean。

```
import com.apress.prospring5.ch2.decoupled.MessageRenderer;
import org.springframework.context.support.GenericXmlApplicationContext;

public class DeclareSpringComponents {

    public static void main(String... args) {
       GenericXmlApplicationContext ctx = new GenericXmlApplicationContext();
       ctx.load("classpath:spring/app-context-xml.xml");
       ctx.refresh();
       MessageRenderer messageRenderer = ctx.getBean("renderer",
       MessageRenderer.class);
          messageRenderer.render();
       ctx.close();
    }
}
```

GenericXmlApplicationContext而不是DefaultListableBeanFactory的一个实例被实例化。GenericXmlApplicationContext类实现了ApplicationContext接口，并能够通过XML文件中定义的配置启动Spring的ApplicationContext。

读者可以在本章提供的源代码中交换 app-context-xml.xml 与 app-context-annotation.xml，并且发现这两种情况都会产生相同的结果：打印出“Hello World！”。唯一的区别是在交换之后，提供相关功能的 bean 是使用 com.apress.prospring5.ch3.annotation 包中的注解定义的 bean。

使用 Java 配置

在第 1 章曾经讲过，可以将 app-context-xml.xml 替换为配置类，而无须修改表示正在创建的 bean 类型的类。当应用程序所需要的 bean 类型是不能修改的第三方库时，这样做是很有用的。配置类使用@Configuration 注解，并包含用@Bean 注解的方法，这些方法由 Spring IoC 容器直接调用来实例化 bean。bean 名称与用于创建它的方法的名称相同。下面的代码示例显示了一个配置类，同时在方法名称中使用了下划线，从而清楚地说明所创建的 bean 是如何被命名的：

```
package com.apress.prospring5.ch2.annotated;

import com.apress.prospring5.ch2.decoupled.HelloWorldMessageProvider;
import com.apress.prospring5.ch2.decoupled.MessageProvider;
import com.apress.prospring5.ch2.decoupled.MessageRenderer;
import com.apress.prospring5.ch2.decoupled.StandardOutMessageRenderer;
import org.springframework.context.annotation.Bean;
import org.springframework.context.annotation.Configuration;

@Configuration
public class HelloWorldConfiguration {

    @Bean
    public MessageProvider provider() {
       return new HelloWorldMessageProvider();
    }

    @Bean
    public MessageRenderer renderer(){
       MessageRenderer renderer = new StandardOutMessageRenderer();
       renderer.setMessageProvider(provider());
       return renderer;
    }
}
```

如果想要从该类中读取配置信息，还需要一个不同的 ApplicationContext 实现。

```
package com.apress.prospring5.ch2.annotated;

import com.apress.prospring5.ch2.decoupled.MessageRenderer;
import org.springframework.context.ApplicationContext;
import org.springframework.context.annotation.AnnotationConfigApplicationContext;
public class HelloWorldSpringAnnotated {

    public static void main(String... args) {
       ApplicationContext ctx = new AnnotationConfigApplicationContext
             (HelloWorldConfiguration.class);
       MessageRenderer mr = ctx.getBean("renderer", MessageRenderer.class);
       mr.render();
    }
}
```

此时，实例化的是 AnnotationConfigApplicationContext 而不是 DefaultListableBeanFactory 的一个实例。

AnnotationConfigApplicationContext 类实现了 ApplicationContext 接口，并且能够根据 HelloWorldConfiguration 类定义的配置信息启动 Spring 的 ApplicationContext。

也可以使用配置类读取带注解的 bean 定义。在这种情况下，因为 bean 的定义配置是 bean 类的一部分，所以类将不再需要任何@Bean 注解的方法。但是，为了能够在 Java 类中查找 bean 定义，必须启用组件扫描。可以使用与<context:component-scanning ../>元素等效的注解来注解配置类，从而完成启动。此注解为@ComponentScanning，并具有与 XML 类似元素相同的参数。

```
package com.apress.prospring5.ch3.annotation;

import org.springframework.context.annotation.ComponentScan;
import org.springframework.context.annotation.Configuration;

@ComponentScan(basePackages = {"com.apress.prospring5.ch3.annotation"})
@Configuration
public class HelloWorldConfiguration {
}
```

使用 AnnotationConfigApplicationContext 启动 Spring 环境的代码也可以在这个类中使用，无须进行额外的修改。

在现实的生产应用程序中，可能存在旧版代码，这些代码是使用旧版本的 Spring 开发的，或者需要使用 XML 和配置类。幸运的是，XML 和 Java 配置可以以多种方式混合使用。例如，配置类可以使用@ImportResource 从一个或多个 XML 文件中导入 bean 定义，在这种情况下也可以使用 AnnotationConfigApplicationContext 完成相同的启动过程。

```
package com.apress.prospring5.ch3.mixed;

import org.springframework.context.annotation.ComponentScan;
import org.springframework.context.annotation.Configuration;
import org.springframework.context.annotation.ImportResource;

@ImportResource(locations = {"classpath:spring/app-context-xml.xml"})
@Configuration
public class HelloWorldConfiguration {
}
```

所以，Spring 能够让 bean 的创建更具有创造性；第 4 章将会更多地介绍相关的内容，其中着重介绍 Spring 应用程序配置。

使用 setter 注入

要想使用 XML 配置来配置 setter 注入，需要在<bean>标记下指定<property>标记(为每一个<property>标记注入一个依赖项)。例如，如果想要将消息提供程序 bean 分配给 messageRenderer bean 的 messageProvider 属性，那么只需要更改 renderer bean 的<bean>标记即可，如下面的代码片段所示：

```
<beans ...>
    <bean id="renderer"
       class="com.apress.prospring5.ch2.decoupled.StandardOutMessageRenderer">
       <property name="messageProvider" ref="provider"/>
    </bean>

    <bean id="provider"
       class="com.apress.prospring5.ch2.decoupled.HelloWorldMessageProvider"/>
</beans>
```

从上述代码中可以看到，provider bean 被分配给了 messageProvider 属性。可以使用 ref 为属性分配一个 bean 引用(稍后会更详细地讨论)。

如果使用的是 Spring 2.5 或更新版本，并且在 XML 配置文件中声明了 p 名称空间，则可以声明注入，如以下代码片段所示：

```
<?xml version="1.0" encoding="UTF-8"?>

<beans xmlns="http://www.springframework.org/schema/beans"
    xmlns:xsi="http://www.w3.org/2001/XMLSchema-instance"
    xmlns:p="http://www.springframework.org/schema/p"
    xsi:schemaLocation="http://www.springframework.org/schema/beans
       http://www.springframework.org/schema/beans/spring-beans.xsd">

    <bean id="renderer"
       class="com.apress.prospring5.ch2.decoupled.StandardOutMessageRenderer"
       p:messageProvider-ref="provider"/>
```

```
    <bean id="provider"
        class="com.apress.prospring5.ch2.decoupled.HelloWorldMessageProvider"/>
</beans>
```

> p 名称空间没有在 XSD 文件中定义，而只存在于 Spring Core 中；因此，在 schemaLocation 属性中没有声明 XSD。

如果使用注解，则更加简单。只需要向 setter 方法添加一个@Autowired 注解即可，如下面的代码片段所示：

```
package com.apress.prospring5.ch3.annotation;
...
import org.springframework.beans.factory.annotation.Autowired;

@Service("renderer")
public class StandardOutMessageRenderer implements MessageRenderer {
...
    @Override
    @Autowired
    public void setMessageProvider(MessageProvider provider) {
        this.messageProvider = provider;
    }
}
```

由于在 XML 配置文件中声明了<context:component-scan>标记，因此在 Spring 的 ApplicationContext 初始化期间，Spring 会发现这些@Autowired 注解并根据需要注入依赖项。

> 除了使用@Autowired，还可以使用@Resource(name ="messageProvider")取得相同的结果。@Resource 是 JSR-250 标准中的注解之一，它定义了在 JSE 和 JEE 平台上使用的一组通用的 Java 注解。与@Autowired 不同，@Resource 注解支持 name 参数以获得更精细的 DI 要求。此外，Spring 支持使用作为 JSR-299(用于 Java EE 平台的上下文和依赖注入)一部分而引入的@Inject 注解。@Inject 在行为上与 Spring 的@Autowired 注解等价。

要验证结果，可以使用前面介绍的 DeclareSpringComponents。如前一节所述，可以在本章提供的源代码中将 app-context-xml.xml 与 app-context-annotation.xml 交换，并且会发现这两种情况会产生相同的结果：打印出“Hello World!”。

使用构造函数注入

在上例中，MessageProviderd 的实现 HelloWorldMessageProvider 为 getMessage()方法的每次调用都返回相同的硬编码消息。在 Spring 配置文件中，可以轻松创建一个可配置的 MessageProvider，从而允许从外部定义消息，如以下代码片段所示：

```
package com.apress.prospring5.ch3.xml;

import com.apress.prospring5.ch2.decoupled.MessageProvider;

public class ConfigurableMessageProvider
    implements MessageProvider {
        private String message;

        public ConfigurableMessageProvider(String message) {
            this.message = message;
        }

        @Override
        public String getMessage() {
            return message;
        }
}
```

正如你所看到的，如果没有为消息提供一个值(除非提供了 null)，就不可能创建 ConfigurableMessageProvider 实例。而这也正是我们想要的，该类非常适合与*构造函数注入一起使用*。以下代码片段显示了如何重新定义 provider bean 以创建 ConfigurableMessageProvider 实例，并使用构造函数注入来注入消息：

```
<beans xmlns="http://www.springframework.org/schema/beans"
       xmlns:xsi="http://www.w3.org/2001/XMLSchema-instance"
       xsi:schemaLocation="http://www.springframework.org/schema/beans
           http://www.springframework.org/schema/beans/spring-beans.xsd">
    <bean id="messageProvider"
          class="com.apress.prospring5.ch3.xml.ConfigurableMessageProvider">
```

```
        <constructor-arg value="I hope that someone gets my message in a bottle"/>
    </bean>
</beans>
```

在上面这段代码中，使用了一个<constructor-arg>标记而不是<property>标记。因为这次没有传入另一个 bean，只是传入一个 String 文本，所以使用 value 属性而不是 ref 来指定构造函数参数的值。当有多个构造函数参数或者你的类有多个构造函数时，需要为每个<constructor-arg>标记指定一个 index 属性，以指定构造函数签名中参数的索引，从 0 开始。对于具有多个参数的构造函数，通常最好使用 index 属性，以避免混淆参数并确保 Spring 选择正确的构造函数。

除了 p 名称空间，从 Spring 3.1 开始，还可以使用 c 名称空间，如下所示：

```
<beans xmlns="http://www.springframework.org/schema/beans"
    xmlns:xsi="http://www.w3.org/2001/XMLSchema-instance"
    xmlns:c="http://www.springframework.org/schema/c"
    xsi:schemaLocation="http://www.springframework.org/schema/beans
    http://www.springframework.org/schema/beans/spring-beans.xsd">

    <bean id="provider"
       class="com.apress.prospring5.ch3.xml.ConfigurableMessageProvider"
       c:message="I hope that someone gets my message in a bottle"/>
</beans>
```

c 名称空间没有在 XSD 文件中定义，并且只存在于 Spring Core 中；因此，没有在 schemaLocation 属性中声明 XSD。

要为构造函数注入使用注解，还需要在目标 bean 的构造函数方法中使用@Autowired 注解，这是 setter 注入方法的替代方法，如以下代码片段所示：

```
package com.apress.prospring5.ch3.annotated;

import com.apress.prospring5.ch2.decoupled.MessageProvider;
import org.springframework.beans.factory.annotation.Autowired;
import org.springframework.stereotype.Service;

@Service("provider")
public class ConfigurableMessageProvider implements MessageProvider {

    private String message;

    @Autowired
    public ConfigurableMessageProvider(
    (@Value("Configurable message") String message) {
       this.message = message;
    }

    @Override
    public String getMessage() {
       return this.message;
    }
}
```

从上面的代码中可以看到，使用了另一个注解@Value 来定义要注入构造函数的值。这是 Spring 中将值注入 bean 的方式。除了简单的字符串，还可以使用强大的 SpEL 进行动态值注入(本章后面会详细介绍)。

但是，对代码中的值进行硬编码并不是一个好主意；如果想要改变值，需要重新编译程序。既然选择了注解式 DI，一种好的做法是对这些值进行外部注入。为了使消息外部化，需要在注解配置文件中将消息定义为一个 Spring bean，如以下代码片段所示：

```
<beans ...>
    <context:component-scan
    base-package="com.apress.prospring5.ch3.annotated"/>

    <bean id="message" class="java.lang.String"
       c:_0="I hope that someone gets my message in a bottle"/>
</beans>
```

此时定义了一个 ID 为 message 且类型为 java.lang.String 的 bean。请注意，还使用了 c 名称空间作为构造函数注入来设置字符串值，_0 表示构造函数参数的索引。在声明了 bean 之后，就可以从目标 bean 中提取@Value 注解，如以下代码片段所示：

```
package com.apress.prospring5.ch3.annotated;
```

```
import com.apress.prospring5.ch2.decoupled.MessageProvider;
import org.springframework.beans.factory.annotation.Autowired;
import org.springframework.stereotype.Service;

@Service("provider")
public class ConfigurableMessageProvider implements MessageProvider {

    private String message;

    @Autowired
    public ConfigurableMessageProvider(String message) {
       this.message = message;
    }

    @Override
    public String getMessage() {
       return this.message;
    }
}
```

由于声明的消息 bean 及其 ID 与构造函数中指定参数的名称相同，因此 Spring 将检测注解并将该值注入构造函数方法中。现在针对 XML(app-context.xml.xml)和注解配置(app-context-annotation.xml)使用以下代码运行测试，会发现在这两种情况下都会显示所配置的消息：

```
package com.apress.prospring5.ch3;

import com.apress.prospring5.ch2.decoupled.MessageProvider;
import org.springframework.context.support.GenericXmlApplicationContext;

public class DeclareSpringComponents {
    public static void main(String... args) {
       GenericXmlApplicationContext ctx = new GenericXmlApplicationContext();
       ctx.load("classpath:spring/app-context-annotation.xml");
       ctx.refresh();

       MessageProvider messageProvider = ctx.getBean("provider",
          MessageProvider.class);

       System.out.println(messageProvider.getMessage());
    }
}
```

在某些情况下，Spring 会发现无法确定将哪个构造函数用于构造函数注入。当使用了两个具有相同参数数量的构造函数并且参数所使用的类型以相同的方式表示时，就会出现这种情况。请考虑下面的代码：

```
package com.apress.prospring5.ch3.xml;

import org.springframework.context.support.GenericXmlApplicationContext;

public class ConstructorConfusion {
    private String someValue;

    public ConstructorConfusion(String someValue) {
       System.out.println("ConstructorConfusion(String) called");
       this.someValue = someValue;
    }

    public ConstructorConfusion(int someValue) {
       System.out.println("ConstructorConfusion(int) called");
       this.someValue = "Number: " + Integer.toString(someValue);
    }

    public String toString() {
       return someValue;
    }

       public static void main(String... args) {
          GenericXmlApplicationContext ctx = new GenericXmlApplicationContext();
          ctx.load("classpath:spring/app-context-xml.xml");
          ctx.refresh();
       ConstructorConfusion cc = (ConstructorConfusion)
          ctx.getBean("constructorConfusion");
       System.out.println(cc);
       ctx.close
    }
}
```

上面的代码从 ApplicationContext 中检索一个 ConstructorConfusion 类型的 bean，并将该值写入控制台输出。现在看看下面的配置代码：

```
<beans ...>
    <bean id="provider"
        class="com.apress.prospring5.ch3.xml.ConfigurableMessageProvider"
        c:message="I hope that someone gets my message in a bottle"/>

    <bean id="constructorConfusion"
        class="com.apress.prospring5.ch3.xml.ConstructorConfusion">
        <constructor-arg>
            <value>90</value>
        </constructor-arg>
    </bean>

</beans>
```

在这种情况下调用哪个构造函数呢？运行该例将产生以下输出：

```
ConstructorConfusion(String) called
```

该输出表明带有 String 参数的构造函数被调用。但这并不是想要的结果，因为我们想要通过使用构造函数注入为传入的每个任何整数值添加前缀 Number:，如 int 构造函数中所示。为了解决这个问题，需要对配置做一些小的修改，如下面的代码片段所示：

```
<beans ...>
    <bean id="provider"
        class="com.apress.prospring5.ch3.xml.ConfigurableMessageProvider"
        c:message="I hope that someone gets my message in a bottle"/>

    <bean id="constructorConfusion"
        class="com.apress.prospring5.ch3.xml.ConstructorConfusion">
        <constructor-arg type="int">
            <value>90</value>
        </constructor-arg>
    </bean>

</beans>
```

现在，注意<constructor-arg>标记有一个额外的属性 type，它指定了 Spring 应该查找的参数的类型。再次使用修改后的配置运行该例会生成正确的输出。

```
ConstructorConfusion(int) called
Number: 90
```

对于注解式的构造函数注入，可以通过将注解直接应用于目标构造函数方法来避免混淆，如以下代码片段所示：

```
package com.apress.prospring5.ch3.annotated;

import org.springframework.beans.factory.annotation.Autowired;
import org.springframework.beans.factory.annotation.Value;
import org.springframework.context.support.GenericXmlApplicationContext;
import org.springframework.stereotype.Service;

@Service("constructorConfusion")
public class ConstructorConfusion {

    private String someValue;

    public ConstructorConfusion(String someValue) {
        System.out.println("ConstructorConfusion(String) called");
        this.someValue = someValue;
    }

    @Autowired
    public ConstructorConfusion(@Value("90") int someValue) {
        System.out.println("ConstructorConfusion(int) called");
        this.someValue = "Number: " + Integer.toString(someValue);
    }

    public String toString() {
        return someValue;
    }

    public static void main(String... args) {
        GenericXmlApplicationContext ctx = new GenericXmlApplicationContext();
        ctx.load("classpath:spring/app-context-annotation.xml");
```

```
        ctx.refresh();

        ConstructorConfusion cc = (ConstructorConfusion)
            ctx.getBean("constructorConfusion");
        System.out.println(cc);
        ctx.close();
    }
}
```

通过将@Autowired 注解应用于所需的构造函数方法，Spring 将使用该方法来实例化 bean 并按照指定注入值。和前面一样，应该将配置的值外部化。

@Autowired 注解只能应用于其中一个构造函数方法。如果将该注解应用于多个构造函数方法，Spring 会在启动 ApplicationContext 时产生错误。

使用字段注入

Spring 中支持的第三种依赖注入被称为*字段注入(field injection)*。顾名思义，依赖项直接注入字段中，不需要构造函数或 setter。字段注入是通过使用 Autowired 注解来注解类成员完成的。这看起来确实可行，因为当在所属的对象之外不再需要依赖项时，这样做可以避免开发人员编写一些在初始创建 bean 后不再使用的代码。在下面的代码片段中，Singer 类型的 bean 有一个 Inspiration 类型的字段：

```
package com.apress.prospring5.ch3.annotated;

import org.springframework.beans.factory.annotation.Autowired;
import org.springframework.stereotype.Service;

@Service("singer")
public class Singer {

    @Autowired
    private Inspiration inspirationBean;

    public void sing() {
        System.out.println("... " + inspirationBean.getLyric());
    }
}
```

该字段是私有的，但 Spring IoC 容器并不关心这个问题。它使用反射来填充所需的依赖项。Inspiration 类代码如下所示，它是一个带有 String 成员的简单 bean。

```
package com.apress.prospring5.ch3.annotated;

import org.springframework.beans.factory.annotation.Value;
import org.springframework.stereotype.Component;

@Component
public class Inspiration {

    private String lyric =
        "I can keep the door cracked open, to let light through";

    public Inspiration(
        @Value("For all my running, I can understand") String lyric) {
        this.lyric = lyric;
    }

    public String getLyric() {
        return lyric;
    }

    public void setLyric(String lyric) {
        this.lyric = lyric;
    }
}
```

以下配置使用组件扫描来发现将由 Spring IoC 容器创建的 bean 定义：

```
<beans ...>
    <context:component-scan
        base-package="com.apress.prospring5.ch3.annotated"/>
</beans>
```

当找到一个 Inspiration 类型的 bean 时，Spring IoC 容器会将该 bean 注入 singer bean 的 inspirationBean 成员中。

这就是为什么在运行下面代码片段中的示例时，在控制台中会打印出“For all my running, I can understand”的原因。

```
package com.apress.prospring5.ch3.annotated;

import org.springframework.context.support.GenericXmlApplicationContext;

public class FieldInjection {

    public static void main(String... args) {

        GenericXmlApplicationContext ctx =
            new GenericXmlApplicationContext();
        ctx.load("classpath:spring/app-context.xml");
        ctx.refresh();

        Singer singerBean = ctx.getBean(Singer.class);
        singerBean.sing();

        ctx.close();
    }
}
```

但字段注入也存在缺点，这就是为什么通常要避免使用字段注入的原因。

- 虽然使用字段注入可以很容易地添加依赖项，但必须格外小心，不要违反单一责任原则。拥有更多的依赖项意味着担负更多的责任，这可能会在重构时出现难以分离的问题。当使用构造函数或 setter 设置依赖项时，可以很容易看出类变得臃肿，但是当使用字段注入时却很难看出来。
- 虽然注入依赖项的责任在 Spring 中由容器承担，但类应该通过方法或构造函数清楚地传达使用公共接口所需的依赖项类型。而如果使用字段注入，则可能很难搞清楚什么类型的依赖项是真正需要的，以及依赖项是不是强制性的。
- 字段注入引入了 Spring 容器的依赖项，因为@Autowired 注解是 Spring 组件；因此，这个 bean 不再是一个 POJO，并且不能独立实例化。
- 字段注入不能用于 final 字段。这种类型的字段只能使用构造函数注入来初始化。
- 由于必须手动注入依赖项，因此在编写测试时，字段注入会带来困难。

使用注入参数

在前面的三个示例中，学习了如何使用 setter 注入和构造函数注入将其他组件和值注入 bean 中。Spring 支持多种注入参数选项，不仅可以注入其他组件和简单值，还可以注入 Java 集合、外部定义的属性以及其他工厂中的 bean。通过分别使用<property>和<constructor-args>标记下的相应标记，可以将所有这些注入参数类型用于 setter 注入和构造函数注入。

注入简单值

将简单值注入 bean 中是很容易的。只需要在配置标记中指定值，并将其包含在<value>标记内即可。默认情况下，<value>标记不仅可以读取 String 值，还可以将这些值转换为任何原始值或原始值包装类。下面的代码片段显示了一个简单的 bean，它具有多种用于注入的属性：

```
package com.apress.prospring5.ch3.xml;

import org.springframework.context.support.GenericXmlApplicationContext;

public class InjectSimple {

    private String name;
    private int age;
    private float height;
    private boolean programmer;
    private Long ageInSeconds;

    public static void main(String... args) {
        GenericXmlApplicationContext ctx =
            new GenericXmlApplicationContext();
        ctx.load("classpath:spring/app-context-xml.xml");
        ctx.refresh();

        InjectSimple simple = (InjectSimple) ctx.getBean("injectSimple");
        System.out.println(simple);
```

```
        ctx.close();
    }

    public void setAgeInSeconds(Long ageInSeconds) {
        this.ageInSeconds = ageInSeconds;
    }

    public void setProgrammer(boolean programmer) {
        this.programmer = programmer;
    }

    public void setAge(int age) {
        this.age = age;
    }

    public void setHeight(float height) {
        this.height = height;
    }

    public void setName(String name) {
        this.name = name;
    }

    public String toString() {
        return "Name: "+ name + "\n"
                    + "Age: " + age + "\n"
                    + "Age in Seconds: " + ageInSeconds + "\n"
                    + "Height: " + height + "\n"
                    + "Is Programmer?: " + programmer;
    }
}
```

除了这些属性，InjectSimple 类还定义了 main()方法，该方法首先创建 ApplicationContext，然后从 Spring 中检索 InjectSimple bean，最后将这个 bean 的属性值写入控制台输出。下面的代码片段描述了该 bean 的 app-context-xml.xml 中所包含的配置：

```
<beans ...>

    <bean id="injectSimpleConfig"
        class="com.apress.prospring5.ch3.xml.InjectSimpleConfig"/>

    <bean id="injectSimpleSpel"
          class="com.apress.prospring5.ch3.xml.InjectSimpleSpel"
        p:name="John Mayer"
        p:age="39"
        p:height="1.92"
        p:programmer="false"
        p:ageInSeconds="1241401112"/>
</beans>
```

从上面的两个代码片段中看到，可以在 bean 上定义接收 String 值、原始值或原始包装器值的属性，然后通过使用<value>标记为这些属性注入值。运行示例，会生成如下所示的输出：

```
Name: John Mayer
Age: 39
Age in Seconds: 1241401112
Height: 1.92
Is Programmer?: false
```

对于注解式的简单值注入，可以将@Value 注解应用于 bean 属性。这一次是将注解应用于属性声明语句而不是 setter 方法，正如你在下面的代码片段中看到的那样(Spring 支持 setter 方法或属性中的注解)：

```
package com.apress.prospring5.ch3.annotated;

import org.springframework.beans.factory.annotation.Value;
import org.springframework.context.support.GenericXmlApplicationContext;
import org.springframework.stereotype.Service;

@Service("injectSimple")
public class InjectSimple {

    @Value("John Mayer")
    private String name;
    @Value("39")
    private int age;
    @Value("1.92")
    private float height;
```

```
    @Value("false")
    private boolean programmer;
    @Value("1241401112")
    private Long ageInSeconds;

    public static void main(String... args) {
        GenericXmlApplicationContext ctx =
            new GenericXmlApplicationContext();
        ctx.load("classpath:spring/app-context-annotation.xml");
        ctx.refresh();

        InjectSimple simple = (InjectSimple) ctx.getBean("injectSimple");
        System.out.println(simple);

        ctx.close();
    }

    public String toString() {
        return "Name: "+ name + "\n"
                    + "Age: " + age + "\n"
                    + "Age in Seconds: " + ageInSeconds + "\n"
                    + "Height: " + height + "\n"
                    + "Is Programmer?: " + programmer;
    }
}
```

上述代码实现了与 XML 配置相同的结果。

通过使用 SpEL 注入值

Spring 3 中引入的一个强大功能是 Spring Expression Language(SpEL)。SpEL 能够动态评估表达式，然后在 Spring 的 ApplicationContext 中使用它。可以将评估结果注入 Spring bean。下面将通过前面那个示例学习如何使用 SpEL 从其他 bean 注入属性。

假设现在想要在配置类中将注入 Spring bean 的值外部化，如以下代码片段所示：

```
package com.apress.prospring5.ch3.annotated;

import org.springframework.stereotype.Component;

@Component("injectSimpleConfig")
public class InjectSimpleConfig {

    private String name = "John Mayer";
    private int age = 39;
    private float height = 1.92f;
    private boolean programmer = false;
    private Long ageInSeconds = 1_241_401_112L;

    public String getName() {
        return name;
    }

    public int getAge() {
        return age;
    }

    public float getHeight() {
        return height;
    }

    public boolean isProgrammer() {
        return programmer;
    }

    public Long getAgeInSeconds() {
        return ageInSeconds;
    }
}
```

然后，可以在 XML 配置中定义 bean，并使用 SpEL 将 bean 的属性注入依赖 bean，如以下配置代码片段所示：

```
<beans ...>

    <bean id="injectSimpleConfig"
        class="com.apress.prospring5.ch3.xml.InjectSimpleConfig"/>
    <bean id="injectSimpleSpel"
        class="com.apress.prospring5.ch3.xml.InjectSimpleSpel"
```

```
        p:name="#{injectSimpleConfig.name}"
        p:age="#{injectSimpleConfig.age + 1}"
        p:height="#{injectSimpleConfig.height}"
        p:programmer="#{injectSimpleConfig.programmer}"
        p:ageInSeconds="#{injectSimpleConfig.ageInSeconds}"/>
</beans>
```

请注意，在引用另一个bean的属性时使用了SpEL # {injectSimpleConfig.name}。针对年龄值，在bean的值上加1，表示可以使用SpEL来操纵认为合适的属性并将其注入依赖bean。现在可以使用以下代码片段所示的程序来测试配置：

```java
package com.apress.prospring5.ch3.xml;

import org.springframework.context.support.GenericXmlApplicationContext;

public class InjectSimpleSpel {
    private String name;
    private int age;
    private float height;
    private boolean programmer;
    private Long ageInSeconds;

    public String getName() {
        return this.name;
    }

    public void setName(String name) {
        this.name = name;
    }

    public int getAge() {
        return this.age;
    }

    public void setAge(int age) {
        this.age = age;
    }

    public float getHeight() {
        return this.height;
    }

    public void setHeight(float height) {
        this.height = height;
    }

    public boolean isProgrammer() {
        return this.programmer;
    }

    public void setProgrammer(boolean programmer) {
        this.programmer = programmer;
    }

    public Long getAgeInSeconds() {
        return this.ageInSeconds;
    }

    public void setAgeInSeconds(Long ageInSeconds) {
        this.ageInSeconds = ageInSeconds;
    }

    public String toString() {
        return "Name: "+ name + "\n"
                        + "Age: " + age + "\n"
                        + "Age in Seconds: " + ageInSeconds + "\n"
                        + "Height: " + height + "\n"
                        + "Is Programmer?: " + programmer;
    }

    public static void main(String... args) {
        GenericXmlApplicationContext ctx = new GenericXmlApplicationContext();
        ctx.load("classpath:spring/app-context-xml.xml");
        ctx.refresh();

        InjectSimpleSpel simple = (InjectSimpleSpel)ctx.getBean("injectSimpleSpel");
        System.out.println(simple);
```

```
        ctx.close();
    }
}
```

以下是该程序的输出：

```
Name: John Mayer
Age: 40
Age in Seconds: 1241401112
Height: 1.92
Is Programmer?: false
```

当使用注解式的值注入时，只需要用 SpEL 表达式替换值注解即可(参见下面的代码片段)：

```
package com.apress.prospring5.ch3.annotated;

import org.springframework.beans.factory.annotation.Value;
import org.springframework.context.support.GenericXmlApplicationContext;
import org.springframework.stereotype.Service;

@Service("injectSimpleSpel")
public class InjectSimpleSpel {

    @Value("#{injectSimpleConfig.name}")
    private String name;

    @Value("#{injectSimpleConfig.age + 1}")
    private int age;

    @Value("#{injectSimpleConfig.height}")
    private float height;

    @Value("#{injectSimpleConfig.programmer}")
    private boolean programmer;

    @Value("#{injectSimpleConfig.ageInSeconds}")
    private Long ageInSeconds;

    public String toString() {
        return "Name: " + name + "\n"
            + "Age: " + age + "\n"
            + "Age in Seconds: " + ageInSeconds + "\n"
            + "Height: " + height + "\n"
            + "Is Programmer?: " + programmer;
    }

    public static void main(String... args) {
        GenericXmlApplicationContext ctx = new GenericXmlApplicationContext();
        ctx.load("classpath:spring/app-context-annotation.xml");
        ctx.refresh();

        InjectSimpleSpel simple = (InjectSimpleSpel)ctx.getBean("injectSimpleSpel");
        System.out.println(simple);

        ctx.close();
    }
}
```

InjectSimpleConfig 的版本如下所示：

```
package com.apress.prospring5.ch3.annotated;

import org.springframework.stereotype.Component;

@Component("injectSimpleConfig")
public class InjectSimpleConfig {
    private String name = "John Mayer";
    private int age = 39;
    private float height = 1.92f;
    private boolean programmer = false;
    private Long ageInSeconds = 1_241_401_112L;

    // getters here ...
}
```

在前面的代码片段中，使用了@Component 而不是@Service 注解。基本上，使用@Component 与使用@Service 具有相同的效果。这两个注解都告诉 Spring，注解的类是使用基于注解的配置和类路径扫描进行自动检测时的候选对象。但是，由于 InjectSimpleConfig 类存储了应用程序配置，而不是提供业务服务，因此使用@Component 更有意义。实际上，@Service 是@Component 的一个特例，它表明注解的类正在向应用程序中的其他层提供业务服务。

测试程序将产生相同的结果。通过使用 SpEL，Spring 可以访问任何 Spring 管理的 bean 和属性，并通过 Spring 对复杂语言功能和语法的支持来操作它们，从而供应用程序使用。

在相同的 XML 单元中注入 bean

正如你在前面看到的，通过使用 ref 标记可以将一个 bean 注入另一个 bean。接下来的代码片段显示了一个公开了 setter 方法的类，从而允许注入一个 bean：

```
package com.apress.prospring5.ch3.xml;

import org.springframework.context.support.GenericXmlApplicationContext;
import com.apress.prospring5.ch3.Oracle;

public class InjectRef {
   private Oracle oracle;

   public void setOracle(Oracle oracle) {
      this.oracle = oracle;
   }

   public static void main(String... args) {
      GenericXmlApplicationContext ctx = new GenericXmlApplicationContext();
      ctx.load("classpath:spring/app-context-xml.xml");
      ctx.refresh();

      InjectRef injectRef = (InjectRef) ctx.getBean("injectRef");
      System.out.println(injectRef);

      ctx.close();
   }

   public String toString() {
      return oracle.defineMeaningOfLife();
   }
}
```

要想配置 Spring，从而将一个 bean 注入另一个 bean，首先需要配置两个 bean：一个是要注入的 bean，另一个是要作为注入目标的 bean。然后，只需要使用目标 bean 上的<ref>标记配置注入即可。下面的代码片段显示了此配置一个的示例(app-context-xml.xml 文件)：

```
<beans ...>

   <bean id="oracle" name="wiseworm"
      class="com.apress.prospring5.ch3.BookwormOracle"/>
   <bean id="injectRef"
      class="com.apress.prospring5.ch3.xml.InjectRef">
      <property name="oracle">
         <ref bean="oracle"/>
      </propcrty>
   </bean>
</beans>
```

运行 InjectRef 类会产生以下输出：

```
Encyclopedias are a waste of money - go see the world instead
```

需要注意的一点是，被注入的类型不一定是目标 bean 上所定义的确切类型；类型只需要兼容即可。兼容意味着如果目标 bean 上声明的类型是一个接口，那么注入类型必须实现此接口。如果声明的类型是一个类，那么注入类型必须是相同类型或子类型。在示例中，InjectRef 类定义了 setOracle()方法来接收 Oracle 的一个实例，其中 Oracle 是一个接口，而注入类型是 BookwormOracle，它是一个实现了 Oracle 接口的类。虽然这一点会让一些开发人员感到困惑，但它确实非常简单。注入与任何 Java 代码一样都要遵守相同的输入规则，如果熟悉 Java 输入的工作原理，理解注入中的输入将会是很容易的。

在前面的示例中，使用了<ref>标记的 local 属性指定了要注入的 bean 的 ID。稍后你将会看到，在 3.5.5 节“了解 bean 命名”中，可以给一个 bean 指定多个名称，以便可以使用各种别名来引用它。当使用 local 属性时，这意味着<ref>标记只查看 bean 的 ID，而不会查看其别名。此外，bean 定义应该位于同一个 XML 配置文件中。如果想要以任何名称注入 bean 或从其他 XML 配置文件导入 bean，请使用<ref>标记的 bean 属性而不是 local 属性。下面的代码片段显示了前一个示例的替代配置，这里使用了注入的 bean 的替代名称：

```
<beans ...>

   <bean id="oracle" name="wiseworm"
      class="com.apress.prospring5.ch3.BookwormOracle"/>

   <bean id="injectRef"
      class="com.apress.prospring5.ch3.xml.InjectRef">
      <property name="oracle">
         <ref bean="wiseworm"/>
      </property>
   </bean>
</beans>
```

在该例中，通过使用 name 属性为 oracle bean 赋予一个别名，然后通过使用该别名以及<ref>标记的 bean 属性，将 oracle bean 注入 injectRef bean。现在不要太担心命名语义。我们将在本章后面更详细地讨论这一点。再次运行 InjectRef 类，将产生与前一个示例相同的结果。

注入和 ApplicationContext 嵌套

到目前为止，注入的 bean 与被注入的 bean 位于相同的 ApplicationContext 中(因此也是相同的 BeanFactory)。但是，Spring 支持 ApplicationContext 的层次结构，因此一个上下文(以及相关的 BeanFactory)被认为是另一个上下文的父级。通过 ApplicationContexts 嵌套，能够将配置分割成不同的文件，这对于包含许多 bean 的大型项目来说简直就是天赐之物。

在嵌套 ApplicationContext 实例时，Spring 允许子上下文中的 bean 引用父上下文中的 bean。使用 GenericXmlApplicationContextis 实现的 ApplicationContext 嵌套简单易懂。要想将一个 GenericXmlApplicationContext 嵌套在另一个 GenericXmlApplicationContext 中，只需要调用子 ApplicationContext 的 setParent()方法即可，代码示例如下所示：

```
package com.apress.prospring5.ch3;

import org.springframework.context.support.GenericXmlApplicationContext;

public class HierarchicalAppContextUsage {

   public static void main(String... args) {
      GenericXmlApplicationContext parent = new GenericXmlApplicationContext();
      parent.load("classpath:spring/parent-context.xml");
      parent.refresh();

      GenericXmlApplicationContext child = new GenericXmlApplicationContext();
      child.load("classpath:spring/child-context.xml");
      child.setParent(parent);
      child.refresh();

      Song song1 = (Song) child.getBean("song1");
      Song song2 = (Song) child.getBean("song2");
      Song song3 = (Song) child.getBean("song3");

      System.out.println("from parent ctx: " + song1.getTitle());
      System.out.println("from child ctx: " + song2.getTitle());
      System.out.println("from parent ctx: " + song3.getTitle());

      child.close();
      parent.close();
   }
}
```

Song 类非常简单，如下所示：

```
package com.apress.prospring5.ch3;

public class Song {
   private String title;

   public void setTitle(String title) {
      this.title = title;
   }
   public String getTitle() {
      return title;
   }
}
```

在子 ApplicationContext 的配置文件的内部，在父 ApplicationContext 中引用一个 bean 的工作原理与在子 ApplicationContext 中引用一个 bean 的工作原理是完全相同的，除非在子 ApplicationContext 中有一个共享了相同名称的 bean。在这种情况下，只需要使用 parent 替换 ref 元素的 bean 属性，然后就可以继续使用了。以下配置片段描述了父 BeanFactory 的配置文件(名为 parent-context.xml)的内容：

```
<beans ...>
   <bean id="childTitle" class="java.lang.String" c:_0="Daughters"/>

   <bean id="parentTitle" class="java.lang.String" c:_0="Gravity"/>
</beans>
```

如你所见，该配置只是定义了两个 bean：childTitle 和 parentTitle。两者都是带有 Daughters 和 Gravity 值的 String 对象。以下配置片段描述了 child-context.xml 中包含的子 ApplicationContext 的配置：

```
<beans ...>

   <bean id="song1" class="com.apress.prospring5.ch3.Song"
      p:title-ref="parentTitle"/>

   <bean id="song2" class="com.apress.prospring5.ch3.Song"
      p:title-ref="childTitle"/>

   <bean id="song3" class="com.apress.prospring5.ch3.Song">
      <property name="title">
         <ref parent="childTitle"/>
      </property>
   </bean>

   <bean id="childTitle" class="java.lang.String" c:_0="No Such Thing"/>
</beans>
```

请注意，此时定义了四个 bean。该代码中的 childTitle 类似于父上下文中的 childTitle，不同之处在于它表示的 String 具有不同的值，以表明它位于子 ApplicationContext 中。

song1 bean 使用 bean ref 属性来引用名为 parentTitle 的 bean。因为这个 bean 只存在于父 BeanFactory 中，所以 song1 接收对该 bean 的引用。这里有两点值得注意。首先，可以使用 bean 属性引用子 ApplicationContext 和父 ApplicationContext 中的 bean。这样一来，便可以更容易地引用 bean，从而允许在应用程序增长时在配置文件之间移动 bean。其次，不能使用 local 属性来引用父 ApplicationContext 中的 bean。XML 解析器检查本地属性的值是否作为同一文件中的有效元素存在，从而防止它被用于引用父上下文中的 bean。

song2 bean 使用 bean ref 属性来引用 childTitle。由于该 bean 是在两个 ApplicationContext 中定义的，因此 song2 bean 接收自己的 ApplicationContext 中 childTitle 的引用。

song3 bean 使用<ref>标记直接引用父 ApplicationContext 中的 childTitle。因为 song3 使用了<ref>标记的 parent 属性，所以在子 ApplicationContext 中声明的 childTitle 实例将被完全忽略。

⚠ 你可能已经注意到，与 song1 和 song2 不同，song3 bean 没有使用 p 名称空间。虽然 p 名称空间提供了方便的快捷方式，但它不提供使用属性标记时的所有功能，比如引用父 bean。虽然此时将它作为示例展示，但最好选择 p 名称空间或属性标记来定义 bean，而不是混合使用(除非有绝对必要)。

以下是运行 HierarchicalAppContextUsage 类的输出：

```
from parent ctx: Gravity
from child ctx: No Such Thing
from parent ctx: Daughters
```

正如预期的那样，song1 和 song3 bean 都引用了父 ApplicationContext 中的 bean，而 song2 bean 则引用了子 ApplicationContext 中的 bean。

注入集合

通常 bean 需要访问对象的集合，而不仅仅是单独的 bean 或值。因此，Spring 允许将一组对象注入 bean 中，对此你应该不会感到意外。使用集合很简单：可以选择<list>、<map>、<set>或<props>来表示 List、Map、Set 或 Properties 实例，然后像其他的注入一样在各个项中传递。<props>标记只允许将 String 作为值传入，因为 Properties 类只允许属性值为 String。当使用<list>、<map>或<set>时，可以在注入属性时使用任何标记，甚至可以是另一个集合标记。这样一来，就可以传入 List of Maps、Map of Sets 甚至 List of Maps of Sets of Lists！下面的代码片段显示了一个可以注入所有四种集合类型的类：

```
package com.apress.prospring5.ch3.xml;

import org.springframework.context.support.GenericXmlApplicationContext;

import java.util.List;
import java.util.Map;
import java.util.Properties;
import java.util.Set;

public class CollectionInjection {

    private Map<String, Object> map;
    private Properties props;
    private Set set;
    private List list;
    public static void main(String... args) {
        GenericXmlApplicationContext ctx =
            new GenericXmlApplicationContext();
        ctx.load("classpath:spring/app-context-xml.xml");
        ctx.refresh();

        CollectionInjection instance =
             (CollectionInjection) ctx.getBean("injectCollection");
        instance.displayInfo();
        ctx.close();
    }

    public void displayInfo() {
        System.out.println("Map contents:\n");
        map.entrySet().stream().forEach(e -> System.out.println(
            "Key: " + e.getKey() + " - Value: " + e.getValue()));

        System.out.println("\nProperties contents:\n");
        props.entrySet().stream().forEach(e -> System.out.println(
            "Key: " + e.getKey() + " - Value: " + e.getValue()));

        System.out.println("\nSet contents:\n");
        set.forEach(obj -> System.out.println("Value: " + obj));

        System.out.println("\nList contents:\n");
        list.forEach(obj -> System.out.println("Value: " + obj));
    }

    public void setList(List list) {
        this.list = list;
    }

    public void setSet(Set set) {
        this.set = set;
    }

    public void setMap(Map<String, Object> map) {
        this.map = map;
    }

    public void setProps(Properties props) {
        this.props = props;
    }
}
```

上述代码看似很多，但实际上完成的工作却非常少。main()方法从 Spring 中检索一个 CollectionInjection bean，然后调用 displayInfo()方法。此方法只输出从 Spring 注入的 Map、Properties、Set 和 List 实例的内容。下面所示的代码描述了为 CollectionInjection 类的每个属性注入值所需的配置，并且该配置文件被命名为 app-context-xml.xml。

另外，请注意 Map <String, Object>属性的声明。对于 JDK 5 之后的版本，Spring 还支持强类型的 Collection 声明，并执行从 XML 配置到相应指定类型的转换。

```
<beans ...>

    <bean id="lyricHolder"
        lass="com.apress.prospring5.ch3.xml.LyricHolder"/>

    <bean id="injectCollection"
        class="com.apress.prospring5.ch3.xml.CollectionInjection">
```

```
        <property name="map">
            <map>
                <entry key="someValue">
                    <value>It's a Friday, we finally made it</value>
                </entry>
                <entry key="someBean">
                    <ref bean="lyricHolder"/>
                </entry>
            </map>
        </property>
        <property name="props">
            <props>
                <prop key="firstName">John</prop>
                <prop key="secondName">Mayer</prop>
            </props>
        </property>
        <property name="set">
            <set>
                <value>I can't believe I get to see your face</value>
                <ref bean="lyricHolder"/>
            </set>
        </property>
        <property name="list">
            <list>
                <value>You've been working and I've been waiting</value>
                <ref bean="lyricHolder"/>
            </list>
        </property>
    </bean>
</beans>
```

在上面这段代码中，可以看到已经将值注入 CollectionInjection 类所公开的四个 setter 中。对于 map 属性，通过使用<map>标记注入了一个 Map 实例。请注意，每个条目都使用<entry>标记来指定，并且每个条目都有一个 String 键以及一个条目值。条目值是可以单独注入属性的任何值；该例演示了如何使用<value>和<ref>标记为 Map 添加一个 String 值和一个 bean 引用。下面所示的 LyricHolder 类是前面配置中注入 Map 的 lyricHolder bean 的类型：

```
package com.apress.prospring5.ch3.xml;

import com.apress.prospring5.ch3.ContentHolder;

public class LyricHolder implements ContentHolder{
    private String value = "'You be the DJ, I'll be the driver'";

    @Override public String toString() {
        return "LyricHolder: { " + value + "}";
    }
}
```

对于 props 属性，可以使用<props>标记创建一个 java.util.Properties 实例，并使用<prop>标记填充它。请注意，尽管<prop>标记的键入方式与<entry>标记类似，但只能为 Properties 实例中的每个属性指定 String 值。

另外，对于<map>元素可以使用一种替代的、更紧凑的配置，即使用 value 和 value-ref 属性，而不是使用<value>和<ref>元素。此时声明的 map 与前面配置中的 map 相同：

```
<property name="map">
    <map>
        <entry key="someValue" value="It's a Friday, we finally made it"/>
        <entry key="someBean" value-ref="lyricHolder"/>
    </map>
</property>
```

<list>和<set>标记都以相同的方式工作：通过使用任何单个值标记(例如用来将单个值注入单个属性的<value>和<ref>)来指定每个元素。在之前的配置中，可以看到已经为 List 和 Set 实例添加了一个 String 值和一个 bean 引用。

以下是类 CollectionInjection 中 main()方法生成的输出。如你所料，它只是列出了被添加到配置文件的集合中的元素。

```
Map contents:

Key: someValue - Value: It's a Friday, we finally made it
Key: someBean - Value: LyricHolder: { 'You be the DJ, I'll be the driver'}

Properties contents:

Key: secondName - Value: Mayer
```

```
Key: firstName - Value: John

Set contents:

Value: I can't believe I get to see your face
Value: LyricHolder: { 'You be the DJ, I'll be the driver'}

List contents:

Value: You've been working and I've been waiting
Value: LyricHolder: { 'You be the DJ, I'll be the driver'}
```

请记住，通过使用<list>、<map>和<set>元素，可以使用任何用于设置非集合属性值的标记来指定集合中某个条目的值。这是一个相当有用的概念，因为这意味着不仅限于注入原始值的集合，还可以注入 bean 的集合或其他集合。

通过使用该功能，可以更加容易地模块化应用程序和为应用程序逻辑的关键部分提供不同的、用户可选的实现。假设有一个允许企业员工在线创建、校对和订购个性化商务信纸的系统。在此系统中，每个订单的完成图稿在准备生产时会被发送到适当的打印机。唯一的麻烦是，一些打印机希望通过电子邮件接收图稿，而另一些打印机希望通过 FTP 以及其他使用 Secure Copy Protocol(SCP)的协议接收图稿。通过使用 Spring 的集合注入，可以为此功能创建一个标准接口，如以下代码片段所示：

```
package com.apress.prospring5.ch3;

public interface ArtworkSender {
    void sendArtwork(String artworkPath, Recipient recipient);
    String getFriendlyName();
    String getShortName();
}
```

在上面的示例中，Recipient 是一个空类。通过该接口可以创建多个实现，而每个实现都能够将自己描述为一个人，如下所示：

```
package com.apress.prospring5.ch3;

public class FtpArtworkSender
        implements ArtworkSender {

    @Override
    public void sendArtwork(String artworkPath, Recipient recipient) {
        // ftp logic here...
    }

    @Override
    public String getFriendlyName() {
        return "File Transfer Protocol";
    }

    @Override
    public String getShortName() {
        return "ftp";
    }
}
```

假设开发一个类 ArtworkManager，它支持 ArtworkSender 接口的所有可用的实现。完成所有实现后，只需要将 List 传递给 ArtworkManager 类就可以了。通过使用 getFriendlyName()方法，可以显示系统管理员在配置每个信纸模板时可供选择的交付选项列表。另外，如果只是编写 ArtworkSender 接口，那么应用程序可以与各个实现保持完全分离。可以将 ArtworkManager 类的实现作为练习留给读者完成。

除 XML 配置外，还可以使用注解进行集合注入。此外，可以将集合的值外部化到配置文件中以便于维护。以下代码片段是四个不同 Spring bean 的配置，它们模拟了前面示例的相同集合属性(配置文件 app-context-annotation.xml)：

```
<?xml version="1.0" encoding="UTF-8"?>
<beans xmlns="http://www.springframework.org/schema/beans"
    xmlns:xsi="http://www.w3.org/2001/XMLSchema-instance"
    xmlns:context="http://www.springframework.org/schema/context"
    xmlns:util="http://www.springframework.org/schema/util"
    xsi:schemaLocation="http://www.springframework.org/schema/beans
        http://www.springframework.org/schema/beans/spring-beans.xsd
        http://www.springframework.org/schema/context
        http://www.springframework.org/schema/context/spring-context.xsd
        http://www.springframework.org/schema/util
        http://www.springframework.org/schema/util/spring-util.xsd">
```

```
    <context:component-scan
        base-package="com.apress.prospring5.ch3.annotated"/>

    <util:map id="map" map-class="java.util.HashMap">
        <entry key="someValue" value="It's a Friday, we finally made it"/>
        <entry key="someBean" value-ref="lyricHolder"/>
    </util:map>

    <util:properties id="props">
        <prop key="firstName">John</prop>
        <prop key="secondName">Mayer</prop>
    </util:properties>

    <util:set id="set" set-class="java.util.HashSet">
        <value>I can't believe I get to see your face</value>
        <ref bean="lyricHolder"/>
    </util:set>

    <util:list id="list" list-class="java.util.ArrayList">
        <value>You've been working and I've been waiting</value>
        <ref bean="lyricHolder"/>
    </util:list>
</beans>
```

开发 LyricHolder 类的一个注解版本。类的内容如下所示：

```
package com.apress.prospring5.ch3.annotated;

import com.apress.prospring5.ch3.ContentHolder;
import org.springframework.stereotype.Service;

@Service("lyricHolder")
public class LyricHolder implements ContentHolder{
    private String value = "'You be the DJ, I'll be the driver'";

    @Override public String toString() {
        return "LyricHolder: { " + value + "}";
    }
}
```

在前面所示的配置中，利用 Spring 提供的 util 名称空间来声明用来存储集合属性的 bean。与前面的 Spring 版本相比，它极大地简化了配置。在用来测试配置的类中，注入前面的 bean，并使用 JSR-250 @Resource 注解以及指定的名称作为一个参数来正确标识 bean。displayInfo()方法与之前相同，因此不在此处显示。

```
@Service("injectCollection")
public class CollectionInjection {
    @Resource(name="map")
    private Map<String, Object> map;

    @Resource(name="props")
    private Properties props;

    @Resource(name="set")
    private Set set;

    @Resource(name="list")
    private List list;

    public static void main(String... args) {
        GenericXmlApplicationContext ctx =
            new GenericXmlApplicationContext();
        ctx.load("classpath:spring/app-context-annotation.xml");
        ctx.refresh();

        CollectionInjection instance = (CollectionInjection)
            ctx.getBean("injectCollection");
        instance.displayInfo();

        ctx.close();
    }
    ...
}
```

运行测试程序，将获得与使用 XML 配置的示例相同的结果。

你可能想知道为什么使用注解@Resource 而不是@Autowired。这是因为@Autowired 注解的语义定义方式是，它始终将数组、集合和映射视为相应 bean 的集合，而目标 bean 类型从声明的集合值类型派生。例如，如果一个类具有 List <ContentHolder>类型的属性并且定义了@Autowired 注解，那么 Spring 会尝试将当前 ApplicationContext 中所有 ContentHolder 类型的 bean 注入该属性(而不是配置文件中声明的<util:list>)，这将导致意外的依赖项被注入；或者如果没有 ContentHolder 类型的 bean，Spring 将抛出一个异常。因此，对于集合类型注入，必须明确指示 Spring 通过指定@Resource 注解支持的 bean 名称来执行注入。

虽然@Autowired 和@Qualifier 的组合可以用来实现相同的目的，但最好使用其中一个注解而不是两个都使用。在下面的代码片段中，可以看到，通过使用@Autowired 和@Qualifier，可以获得使用 bean 名称注入集合的等效配置。

```
package com.apress.prospring5.ch3.annotated;

import org.springframework.beans.factory.annotation.Autowired;
import org.springframework.beans.factory.annotation.Qualifier;
@Service("injectCollection")
public class CollectionInjection {

    @Autowired
    @Qualifier("map")
    private Map<String, Object> map;
...
}
```

3.5.4 使用方法注入

除了构造函数注入和 setter 注入，Spring 提供的另一个不常用的 DI 功能是方法注入。Spring 的方法注入功能有两种形式：查找方法注入(Lookup Method Injection)和方法替换(Method Replacement)。查找方法注入提供了另一种机制，通过该机制，bean 可以获得它的一个依赖项；而方法替换允许随意替换 bean 上任何方法的实现，而无须更改原始源代码。为了提供这两项功能，Spring 使用了 CGLIB 的动态字节码增强功能[1]。

查找方法注入

查找方法注入在 1.1 版本中被添加到 Spring 中，以解决当 bean 依赖于另一个具有不同生命周期的 bean 时所遇到的问题，特别是当单例(singleton)依赖于非单例(nosingleton)时。在这种情况下，setter 注入和构造函数注入都会导致单例 bean 维护一个应该是非单例 bean 的实例。在某些情况下，会希望单例 bean 在每次需要相关的 bean 时获得一个非单例 bean 的新实例。

假设 LockOpener 类提供打开任何储物柜的服务。LockOpener 类依赖于 KeyHelper 类来打开储物柜(KeyHelper 已被注入 LockOpener 中)。但是 KeyHelper 类的设计涉及一些内部状态，使其不适合重用。每次调用 openLock()方法时，都需要一个新的 KeyHelper 实例。这种情况下，LockOpener 就是一个单例。然而，如果通过使用正常机制注入 KeyHelper 类，将重用 KeyHelper 类的相同实例(在 Spring 首次执行注入时实例化)。为了确保每次调用时都将 KeyHelper 的新实例都传递给 openLock()方法，需要使用查找方法注入。

通常情况下，可以通过让单例 bean 实现 ApplicationContextAware 接口来实现这一点(我们将在下一章中讨论该接口)。然后通过使用 ApplicationContext 实例，单例 bean 就可以在每次需要时查找非单例依赖项的新实例。查找方法注入允许单例 bean 声明它需要一个非单例的依赖项，并且每次需要与非单例 bean 进行交互时都会收到一个新的非单例 bean 实例，而不需要实现任何 Spring 特定的接口。

查找方法注入的工作方式是让单例声明一个查找方法，该方法返回非单例 bean 的一个实例。当在应用程序中获取对单例的引用时，实际上正在接收一个对动态创建的子类的引用，而 Spring 已经在该类上实现了查找方法。典型的实现都涉及定义查找方法，因此 bean 类是抽象的。当忘记配置方法注入并且直接使用带有空方法实现的 bean 类而不是 Spring 增强的子类时，就会出现一些奇怪的错误，而使用查找方法注入可以防止此类错误出现。该内容非常复杂，最好通过示例来演示。

1 cglib 是一个功能强大、高性能、高质量的代码生成库。它可以扩展 Java 类并在运行时实现接口。它是开源的，可以在 https://github.com/cglib 上找到官方库。

在这个示例中，创建了一个非单例 bean 和两个实现了相同接口的单例 bean。其中一个单例 bean 通过使用“传统”setter 注入来获得非单例 bean 的一个实例，而另一个使用方法注入。以下代码示例描述了 Singer 类，在本例中它是非单例 bean 的类型：

```
package com.apress.prospring5.ch3;

public class Singer {
   private String lyric = "I played a quick game of chess with the salt
            and pepper shaker";
   public void sing() {
         //commented because it pollutes the output
      //System.out.println(lyric);
   }
}
```

虽然该类没有什么特别的地方，但它却完全适用于本例。接下来，可以看到 DemoBean 接口，它由两个单例 bean 类实现。

```
package com.apress.prospring5.ch3;
public interface DemoBean {
   Singer getMySinger();
   void doSomething();
}
```

该 bean 有两个方法：getMySinger()和 doSomething()。示例应用程序使用 getMySinger()方法获取对 Singer 实例的引用，并且执行实际的方法查找。DoSomething()是一个依赖于 Singer 类来完成处理的简单方法。以下代码片段显示了 StandardLookupDemoBean 类，该类使用 setter 注入来获取 Singer 类的实例：

```
package com.apress.prospring5.ch3;

public class StandardLookupDemoBean
       implements DemoBean {

   private Singer mySinger;

   public void setMySinger(Singer mySinger) {
         this.mySinger = mySinger;
   }

   @Override
   public Singer getMySinger() {
         return this.mySinger;
   }

   @Override
   public void doSomething() {
         mySinger.sing();
   }
}
```

上面这段代码应该看起来很熟悉，但请注意，doSomething()方法使用存储的 Singer 实例来完成其处理。在下面的代码片段中，可以看到 AbstractLookupDemoBean 类使用方法注入来获取 Singer 类的实例。

```
package com.apress.prospring5.ch3;

public abstract class AbstractLookupDemoBean
   implements DemoBean {
  public abstract Singer getMySinger();

   @Override
   public void doSomething() {
      getMySinger().sing();
   }
}
```

请注意，getMySinger()方法被声明为抽象的，并且该方法由 doSomething()方法调用以获取 Singer 实例。该例的 Spring XML 配置包含在名为 app-context-xml.xml 的文件中，如下所示：

```
<beans ...>
   <bean id="singer" class="com.apress.prospring5.ch3.Singer"
      scope="prototype"/>
   <bean id="abstractLookupBean"
      class="com.apress.prospring5.ch3.AbstractLookupDemoBean">
      <lookup-method name="getMySinger" bean="singer"/>
   </bean>
```

```
    <bean id="standardLookupBean"
        class="com.apress.prospring5.ch3.StandardLookupDemoBean">
        <property name="mySinger" ref="singer"/>
    </bean>
</beans>
```

现在，singer和standardLookupBean bean的配置应该看起来很熟悉。对于abstractLookupBean，需要使用<lookup-method>标记来配置查找方法。<lookup-method>标记的name属性告诉Spring应该覆盖的bean方法的名称。该方法不接收任何参数，并且所返回的类型应该是希望从该方法返回的bean的类型。此时，该方法应返回一个类型为Singer的类或其子类。bean属性告诉Spring查找方法应该返回哪个bean。下面的代码片段显示了该例的最后一段代码，包含用于运行该例的main()方法的类：

```
package com.apress.prospring5.ch3;

import org.springframework.context.support.GenericXmlApplicationContext;
import org.springframework.util.StopWatch;

public class LookupDemo {
    public static void main(String... args) {
        GenericXmlApplicationContext ctx = new GenericXmlApplicationContext();
        ctx.load("classpath:spring/app-context-xml.xml");
        ctx.refresh();

        DemoBean abstractBean = ctx.getBean("abstractLookupBean",
                DemoBean.class);
        DemoBean standardBean = ctx.getBean("standardLookupBean",
                DemoBean.class);

        displayInfo("abstractLookupBean", abstractBean);
        displayInfo("standardLookupBean", standardBean);

        ctx.close();
    }

    public static void displayInfo(String beanName, DemoBean bean) {
        Singer singer1 = bean.getMySinger();
        Singer singer2 = bean.getMySinger();

        System.out.println("" + beanName + ": Singer Instances the Same? "
                + (singer1 == singer2));
                StopWatch stopWatch = new StopWatch();
        stopWatch.start("lookupDemo");
        for (int x = 0; x < 100000; x++) {
            Singer singer = bean.getMySinger();
        singer.sing();
        }

        stopWatch.stop();

        System.out.println("100000 gets took "
                + stopWatch.getTotalTimeMillis() + " ms");
    }
}
```

在上面这段代码中，可以看到来自 GenericXmlApplicationContext 的 abstractLookupBean 和 standardLookupBean 被检索到，并且每个引用都被传递给 displayInfo()方法。只有当使用查找方法注入时才支持抽象类的实例化，其中 Spring 将使用 CGLIB 来生成 AbstractLookupDemoBean 类的一个子类，该类将动态地重写该方法。displayInfo()方法的第一部分创建了两个 Singer 类型的局部变量，并通过调用所传递的 bean 上的 getMySinger()来为这些变量分配一个值。通过使用这两个变量，可以向控制台写入一条消息，指示这两个引用是否指向相同的对象。

对于 abstractLookupBean bean，每次调用 getMySinger()都应该检索一个新的 Singer 实例，因此引用不应该相同。

而对于 standardLookupBean，一个单一的 Singer 实例通过 setter 注入被传递给 bean，并且该实例被存储且在每次调用 getMySinger()时返回，所以这两个引用应该是相同的。

前面的例子中使用的 StopWatch 类是 Spring 提供的一个实用类。当需要执行简单的性能测试以及测试应用程序时，你会发现 StopWatch 非常有用。

---

displayInfo()方法的最后一部分运行了一个简单的性能测试，以查看哪个 bean 更快。显然，standardLookupBean 应该更快，因为它每次都返回相同的实例，但看到两者之间的差异也是很有趣的。现在可以运行 LookupDemo 类来

进行测试。下面是从该例中得到的输出：

```
[abstractLookupBean]: Singer Instances the Same? false
100000 gets took 431 ms

[standardLookupBean]: Singer Instances the Same? true
100000 gets took 1 ms
```

正如你所看到的，当使用 standardLookupBean 时，Singer 实例和预期的一样；而当使用 abstractLookupBean 时，则有所不同。当使用 standardLookupBean 时，会存在明显的性能差异，但这是可以预料的。

当然，还有一种使用注解来配置前面介绍的 bean 的等效方法。此时，Singer bean 必须有额外的注解来指定 prototype 作用域。

```
package com.apress.prospring5.ch3.annotated;
import org.springframework.context.annotation.Scope;
import org.springframework.stereotype.Component;

@Component("singer)
@Scope("prototype")
public class Singer {
    private String lyric = "I played a quick game of chess
            with the salt and pepper shaker";

    public void sing() {
        // commented to avoid console pollution
        //System.out.println(lyric);
    }
}
```

AbstractLookupDemoBean 不再是一个抽象类，方法 getMySinger()有一个空的主体，并用@Lookup 注解，它接收 Singer bean 的名称作为参数。方法的主体将在动态生成的子类中被重写。

```
package com.apress.prospring5.ch3.annotated;

import org.springframework.beans.factory.annotation.Lookup;
import org.springframework.stereotype.Component;

@Component("abstractLookupBean")
public class AbstractLookupDemoBean implements DemoBean {
    @Lookup("singer")
    public Singer getMySinger() {
        return null; // overriden dynamically
    }

    @Override
    public void doSomething() {
        getMySinger().sing();
    }
}
```

StandardLookupDemoBean 类只能用@Component 注解，而 setMySinger 必须用@Autowired 和@Qualifier 注解来注入 Singer bean。

```
package com.apress.prospring5.ch3.annotated;

import org.springframework.beans.factory.annotation.Autowired;
import org.springframework.beans.factory.annotation.Qualifier;
import org.springframework.stereotype.Component;

@Component("standardLookupBean")
public class StandardLookupDemoBean implements DemoBean {

    private Singer mySinger;
    @Autowired
    @Qualifier("singer")
    public void setMySinger(Singer mySinger) {
        this.mySinger = mySinger;
    }

    @Override
    public Singer getMySinger() {
        return this.mySinger;
    }

    @Override
    public void doSomething() {
```

```
        mySinger.sing();
    }
}
```

名为 app-context-annotated.xml 的配置文件只能对包含注解的类的包进行组件扫描。

```
<beans ...>

    <context:component-scan
        base-package="com.apress.prospring5.ch3.annotated"/>
</beans>
```

用于执行代码的类与 LookupDemo 类是相同的；唯一的区别是使用 XML 文件作为创建 GenericXmlApplicationContext 对象的参数。

如果想要完全摆脱 XML 文件，可以通过一个配置类在 com.apress.prospring5.ch3.annotated 包上启用组件扫描。该类可以在需要的地方声明，此时意味着运行该类来测试 bean，如下所示：

```
package com.apress.prospring5.ch3.config;

import com.apress.prospring5.ch3.annotated.DemoBean;
import com.apress.prospring5.ch3.annotated.Singer;
import org.springframework.context.annotation.AnnotationConfigApplicationContext;
import org.springframework.context.annotation.ComponentScan;
import org.springframework.context.annotation.Configuration;
import org.springframework.context.support.GenericApplicationContext;
import org.springframework.util.StopWatch;

import java.util.Arrays;

public class LookupConfigDemo {

    @Configuration
    @ComponentScan(basePackages = {"com.apress.prospring5.ch3.annotated"})
    public static class LookupConfig {}
    public static void main(String... args) {
        GenericApplicationContext ctx =
            new AnnotationConfigApplicationContext(LookupConfig.class);

        DemoBean abstractBean = ctx.getBean("abstractLookupBean",
        DemoBean.class);
        DemoBean standardBean = ctx.getBean("standardLookupBean",
        DemoBean.class);

        displayInfo("abstractLookupBean", abstractBean);
        displayInfo("standardLookupBean", standardBean);

        ctx.close();
    }

    public static void displayInfo(String beanName, DemoBean bean) {
        // same implementation as before
        ...
    }
}
```

第 4 章将详细介绍使用了注解和 Java 配置的替代配置。

查找方法注入的注意事项

当想要使用两个具有不同生命周期的 bean 时，可以使用查找方法注入。当 bean 共享相同的生命周期时，应该避免使用查找方法注入，尤其是当这些 bean 都是单例时。运行上面示例后的输出显示了使用方法注入所获取的依赖项的新实例与使用标准 DI 所获取的单个依赖项实例之间的性能差异。此外，只有在必要的时候才使用查找方法注入，即使当拥有不同生命周期的 bean 时。

假设有三个共享相同依赖项的单例。如果想要每个单例都有自己的依赖项实例，那么可以创建依赖项作为一个单例，同时又希望每个单例在其生命周期中使用相同的协作者实例。在这种情况下，setter 注入是理想的解决方案；查找方法注入只会增加不必要的开销。

当使用查找方法注入时，有几个设计准则在构建类时应该记住。在前面的示例中，都是在接口中声明了查找方法。这样做的唯一原因是，不需要针对两种不同的 bean 类型重复 displayInfo()方法两次。如前所述，一般来说，不需要使用专门用于 IoC 目的的不必要的定义来“污染”业务接口。另一点是，虽然没必要一定让查找方法抽象化，但这样做可以防止忘记配置查找方法，所以有时也可以使用空白的实现。当然，这只适用于 XML 配置。基于注解的配置强制方法的空白实现；否则，就不会创建 bean。

方法替换

虽然 Spring 文档将方法替换归类为一种注入形式，但它与前面介绍的注入类型不同。到目前为止，已经使用注入向 bean 提供协作者。通过使用方法替换，可以任意替换任何 bean 的任何方法的实现，而无须更改所修改 bean 的源代码。例如，有一个在 Spring 应用程序中使用的第三方库，并且需要更改某个方法的逻辑。此时，无法更改源代码，因为它是由第三方提供的，所以一种解决方案是使用方法替换，使用自己的实现来替换该方法的逻辑。

在内部，可以通过动态创建 bean 类的子类来实现此目的。使用 CGLIB 并将想要替换的方法的调用重定向为另一个实现了 MethodReplacer 接口的 bean。在下面的代码示例中，可以看到一个简单的 bean，它声明了 formatMessage()方法的两个重载版本：

```
package com.apress.prospring5.ch3;

public class ReplacementTarget {
    public String formatMessage(String msg) {
        return "<h1>" + msg + "</h1>";
    }

    public String formatMessage(Object msg) {
        return "<h1>" + msg + "</h1>";
    }
}
```

可以使用 Spring 的方法替换功能替换 ReplacementTarget 类的任何方法。在本例中，演示了如何替换 formatMessage(String)方法，并且还将替换方法的性能与原始方法的性能做了比较。

如果想要替换一个方法，首先需要创建 MethodReplacer 接口的一个实现，如下面的代码示例所示：

```
package com.apress.prospring5.ch3;

import org.springframework.beans.factory.support.MethodReplacer;

import java.lang.reflect.Method;

public class FormatMessageReplacer
        implements MethodReplacer {

    @Override
    public Object reimplement(Object arg0, Method method, Object... args)
                throws Throwable {
            if (isFormatMessageMethod(method)) {
                String msg = (String) args0;
                return "<h2>" + msg + "</h2>";
            } else {
                throw new IllegalArgumentException("Unable to reimplement method "
                    + method.getName());
            }
    }
    private boolean isFormatMessageMethod(Method method) {
            if (method.getParameterTypes().length != 1) {
                return false;
            }

            if (!("formatMessage".equals(method.getName()))) {
                return false;
            }

            if (method.getReturnType() != String.class) {
                return false;
            }

            if (method.getParameterTypes()[0] != String.class) {
                return false;
            }

            return true;
    }
}
```

MethodReplacer 接口有一个必须实现的方法 reimplement()。分别将三个参数传递给 reimplement()：被调用的原始方法的 bean，表示正在被重写的方法的 Method 实例以及传递给该方法的参数数组。reimplement()方法应该返回重新实现的逻辑的结果，显然，返回值的类型应该与要替换的方法所返回的类型兼容。在前面的代码示例中，FormatMessageReplacer 首先检查被重写的方法是否是 formatMessage(String)方法；如果是，就执行替换逻辑(此时，用<h2>和</ h2>包含的消息替换)，并将格式化的消息返回给调用者。一般来说，没有必要检查消息是否正确，但如果使用几个具有类似参数的 MethodReplacer，那么检查消息则可能很有用。通过检查有助于防止出现不小心使用兼容参数和返回类型的不同 MethodReplacer 的情况发生。

在下面的配置示例中，可以看到一个 ApplicationContext 实例，该实例定义了两个类型为 ReplacementTarget 的 bean；一个 bean 替换了 formatMessage(String)方法，而另一个 bean 则没有(该文件被命名为 app-context-xml.xml)：

```
<beans ...>

   <bean id="methodReplacer"
      class="com.apress.prospring5.ch3.FormatMessageReplacer"/>

   <bean id="replacementTarget"
      class="com.apress.prospring5.ch3.ReplacementTarget">
      <replaced-method name="formatMessage" replacer="methodReplacer">
         <arg-type>String</arg-type>
      </replaced-method>
   </bean>

   <bean id="standardTarget"
      class="com.apress.prospring5.ch3.ReplacementTarget"/>
</beans>
```

如你所见，在 ApplicationContext 中，MethodReplacer 实现被声明为一个 bean。然后使用<replaced-method>标记替换 replacementTargetBean 上的 formatMessage(String)方法。<replaced-method>标记的 name 属性指定了要替换方法的名称，而 replacer 属性用于指定想要替换方法实现的 MethodReplacer bean 的名称。在有重载方法(参考 ReplacementTarget 类)的情况下，可以使用<arg-type>标记指定要匹配的方法签名。<arg-type>标记支持模式匹配，因此 String 可以与 java.lang.String 以及 java.lang.StringBuffer 匹配。

以下代码片段显示了一个简单的演示应用程序，它从 ApplicationContext 中检索 standardTarget 和 replacement-Target bean，并执行它们的 formatMessage(String)方法，然后运行简单的性能测试以查看哪个方法更快。

```
package com.apress.prospring5.ch3;

import org.springframework.context.support.GenericXmlApplicationContext;
import org.springframework.util.StopWatch;

public class MethodReplacementDemo {
   public static void main(String... args) {
      GenericXmlApplicationContext ctx =
         new GenericXmlApplicationContext();
      ctx.load("classpath:spring/app-context-xml.xml");
      ctx.refresh();

      ReplacementTarget replacementTarget = (ReplacementTarget) ctx
            .getBean("replacementTarget");
      ReplacementTarget standardTarget = (ReplacementTarget) ctx
            .getBean("standardTarget");

      displayInfo(replacementTarget);
      displayInfo(standardTarget);

      ctx.close();
   }

   private static void displayInfo(ReplacementTarget target) {
      System.out.println(target.formatMessage("Thanks for playing, try again!"));

      StopWatch stopWatch = new StopWatch();
      stopWatch.start("perfTest");

      for (int x = 0; x < 1000000; x++) {
         String out = target.formatMessage("No filter in my head");
         //commented to not pollute the console
         //System.out.println(out);
      }

      stopWatch.stop();

      System.out.println("1000000 invocations took: "
```

```
                + stopWatch.getTotalTimeMillis() + " ms");
    }
}
```

现在，你应该对上述代码比较熟悉了，所以在此不做详细讨论。运行该例将产生以下输出：

```
<h2>Thanks for playing, try again!</h2>
1000000 invocations took: 188 ms

<h1>Thanks for playing, try again!</h1>
1000000 invocations took: 24 ms
```

正如预期的那样，replacementTarget bean 的输出反映了 MethodReplacer 所提供重写的实现。有趣的是，动态替换的方法比静态定义的方法慢很多倍。在 MethodReplacer 中删除对有效方法的检查，这在许多执行过程中产生的影响微不足道，所以可以得出结论，大部分开销都在 CGLIB 子类中。

何时使用方法替换

在不同情况下，可以证明方法替换是非常有用的，尤其是当只想覆盖单个 bean 的特定方法而不是相同类型的所有 bean 时。即使有这种说法，很多人也仍然倾向于使用标准的 Java 机制来重写方法，而不是依赖于运行时字节码的增强。

如果想要使用方法替换作为应用程序的一部分，建议为每个方法或一组重载的方法使用一个 MethodReplacer。应该避免针对多个不相关的方法使用单个 MethodReplacer；这样会导致在代码查找应该重新实现的方法时执行不必要的字符串比较。我们发现执行简单的检查以确保 MethodReplacer 使用正确的方法是很有用的，并且不会为代码增加太多开销。如果你真的关心性能，那么可以简单地在 MethodReplacer 中添加一个 Boolean 属性，从而确定在使用依赖注入时是否进行检查。

3.5.5 了解 bean 命名

Spring 支持相当复杂的 bean 命名结构，从而允许灵活地处理多种情况。每个 bean 在其所在的 ApplicationContext 中至少包含一个唯一的名称。Spring 遵循一个简单的解析过程来确定 bean 使用的是什么名称。如果为<bean>标记赋予一个 id 属性，那么该属性的值将用作名称。如果没有指定 id 属性，Spring 会查找 name 属性，如果定义了 name 属性，那么将使用 name 属性中定义的第一个名称(之所以是第一个名称，是因为可以在 name 属性中定义多个名称；稍后会更详细地介绍这一点)。如果既没有指定 id 属性，也没有指定 name 属性，那么 Spring 使用该 bean 的类名作为名称，当然，前提是没有其他 bean 使用相同的类名。如果声明了多个没有 ID 或名称的相同类型的 bean，那么 Spring 将在 ApplicationContext 初始化期间，在注入时抛出异常(类型为 org.springframework.beans.factory.NoSuchBeanDefinitionException)。以下配置示例描述了这三种命名方案(app-context-01.xml)：

```
<beans ...>
    <bean id="string1" class="java.lang.String"/>
    <bean name="string2" class="java.lang.String"/>
    <bean class="java.lang.String"/>
</beans>
```

从技术角度看，每种方法都是同样有效的，但哪种方法是应用程序的最佳选择呢？首先，应该避免按照类行为使用自动名称，否则就无法灵活地定义多个相同类型的 bean，所以最好定义自己的名称。这样一来，即使 Spring 将来更改了默认行为，应用程序也会继续工作。如果想查看 Spring 如何使用先前的配置来命名 bean，运行以下示例：

```
package com.apress.prospring5.ch3.xml;

import org.springframework.context.support.GenericXmlApplicationContext;

public class BeanNamingTest {
    public static void main(String... args) {
        GenericXmlApplicationContext ctx = new GenericXmlApplicationContext();
        ctx.load("classpath:spring/app-context-01.xml");
        ctx.refresh();

        Map<String,String> beans = ctx.getBeansOfType(String.class);

        beans.entrySet().stream().forEach(b -> System.out.println(b.getKey()));

        ctx.close();
    }
}
```

ctx.getBeansOfType(String.class)用于获取 ApplicationContext 中存在的所有 String 类型的 bean 及其 ID 的映射。映射的键就是前面的代码中使用 lambda 表达式打印的 bean ID。以上配置的输出如下所示：

```
string1
string2
java.lang.String#0
```

输出示例中的最后一行是 Spring 提供给未在配置中明确指定的 String 类型的 bean 的 ID。如果想要修改配置以添加另一个未命名的 String bean，可以使用以下代码::

```
<beans ...>
   <bean id="string1" class="java.lang.String"/>
   <bean name="string2" class="java.lang.String"/>
   <bean class="java.lang.String"/>
   <bean class="java.lang.String"/>
</beans>
```

输出将变为以下内容:

```
string1
string2
java.lang.String#0
java.lang.String#1
```

在 Spring 3.1 之前，id 属性与 XML 标识(即 xsd:ID)相同，这样一来，就会限制可用的字符。从 Spring 3.1 开始，Spring 使用 xsd:String 作为 id 属性，因此对可用字符的限制也就消失了。但是，Spring 继续确保 id 在整个 ApplicationContext 中是唯一的。作为一种惯例，应该使用 id 属性作为 bean 的名称，然后使用名称别名将 bean 与其他名称相关联，如下一节所述。

bean 名称别名

Spring 允许一个 bean 拥有多个名称。可以通过在 bean 的<bean>标记的 name 属性中指定以空格、逗号或分号分隔的名称列表来实现多个名称。name 属性可以代替 id 属性，或者组合使用这两个属性。除了使用 name 属性，还可以使用<alias>标记为 Spring bean 名称定义别名。以下配置示例显示了一个简单的<bean>配置，该配置为单个 bean(app-context-02.xml)定义了多个名称：

```
<beans ...>
   <bean id="john" name="john johnny,jonathan;jim" class="java.lang.String"/>
   <alias name="john" alias="ion"/>
</beans>
```

如你所见，此时已经定义了六个名称：一个名称使用 id 属性定义，另外四个名称使用 name 属性中分隔的 bean 名称列表定义(该例仅用于演示目的，不建议用于实际开发)。在实际开发中，建议对分隔符进行标准化，以便在应用程序中分离 bean 名称的声明。最后，使用<alias>标记定义了一个别名。下面的代码示例描述了一个 Java 例程，它使用不同的名称从 ApplicationContext 实例中获取同一个 bean 六次，并验证它们是否是相同的 bean。此外，利用先前介绍的 ctx.getBeansOfType(..)方法确保上下文中只有一个 String bean。

```
package com.apress.prospring5.ch3.xml;

import org.springframework.context.support.GenericXmlApplicationContext;

import java.util.Map;

public class BeanNameAliasing {
   public static void main(String... args) {
      GenericXmlApplicationContext ctx = new GenericXmlApplicationContext();
      ctx.load("classpath:spring/app-context-02.xml");
      ctx.refresh();

      String s1 = (String) ctx.getBean("john");
      String s2 = (String) ctx.getBean("jon");
      String s3 = (String) ctx.getBean("johnny");
      String s4 = (String) ctx.getBean("jonathan");
      String s5 = (String) ctx.getBean("jim");
      String s6 = (String) ctx.getBean("ion");
      System.out.println((s1 == s2));

      System.out.println((s2 == s3));
      System.out.println((s3 == s4));
      System.out.println((s4 == s5));
      System.out.println((s5 == s6));
```

```
        Map<String,String> beans = ctx.getBeansOfType(String.class);

        if(beans.size() == 1) {
           System.out.println("There is only one String bean.");
        }

        ctx.close();
     }
  }
```

执行前面的代码将打印 5 次 true，同时“There is only one String bean”文本证实了使用不同名称访问的 bean 实际上是相同的 bean。

可以通过调用 ApplicationContext.getAliases(String)并传入任何 bean 的名称或 ID 来获取 bean 的别名列表。除指定别名外，别名列表将作为字符串数组返回。

前面曾经讲过，在 Spring 3.1 之前，id 属性与 XML 标识(即 xsd:ID)是相同的，这意味着 bean ID 不能包含空格、逗号或分号之类的特殊字符。从 Spring 3.1 开始，xsd:String 用于 id 属性，因此以前对可用字符的限制消失了。但是，这并不意味着可以使用以下代码：

```
<bean name="jon johnny,jonathan;jim" class="java.lang.String"/>
```

而是应该使用以下代码：

```
<bean id="jon johnny,jonathan;jim" class="java.lang.String"/>
```

Spring IoC 对 name 和 id 属性值的处理方式不同。可以通过调用 ApplicationContext.getAliases(String)并传入任何一个 bean 的名称或 ID 来获取 bean 的别名列表。除指定别名外，别名列表将作为字符串数组返回。这意味着在第一种情况下，jon 将变成 id，而其余的值将变成别名。

在第二种情况下，当使用相同的字符串作为 id 属性的值时，完整字符串将成为 bean 的唯一标识符。可以通过下面所示的配置(可在文件 app-context-03.xml 中找到)进行测试：

```
<beans ...>
   <bean name="jon johnny,jonathan;jim" class="java.lang.String"/>

   <bean id="jon johnny,jonathan;jim" class="java.lang.String"/>
</beans>
```

以及如下代码示例中所示的主类：

```
package com.apress.prospring5.ch3.xml;

import org.springframework.context.support.GenericXmlApplicationContext;

import java.util.Arrays;
import java.util.Map;

public class BeanCrazyNaming {
   public static void main(String... args) {
      GenericXmlApplicationContext ctx = new GenericXmlApplicationContext();
      ctx.load("classpath:spring/app-context-03.xml");
      ctx.refresh();

      Map<String,String> beans = ctx.getBeansOfType(String.class);

      beans.entrySet().stream().forEach(b ->
      {
         System.out.println("id: " + b.getKey() +
            "\n aliases: " + Arrays.toString(ctx.getAliases(b.getKey())) +"\n");
      });
      ctx.close();
   }
}
```

当运行上述示例时，会产生以下输出：

```
id: jon
aliases: jonathan, jim, johnny

id: jon johnny,jonathan;jim
aliases:
```

正如你所看到的，带有 String bean 的映射包含两个 bean：一个 bean 具有唯一标识符 jon 和三个别名；另一个 bena 具有唯一标识符“jon johnny, jonathan; jim”，没有别名。

bean 名称别名是一项非常奇怪的功能，因为在构建新的应用程序时往往不会使用该功能。即使将许多其他的 bean 注入另一个 bean，也可以使用相同的名称来访问该 bean。但是，随着应用程序进入使用和维护环节，可能会进行修改等工作，此时 bean 名称别名就变得更加有用了。

假设有一个应用程序，其中有 50 个使用 Spring 配置的 bean 都需要实现 Foo 接口。25 个 bean 使用 StandardFoo 实现，bean 名称为 standardFoo，另外 25 个 bean 使用 SuperFoo 实现，bean 名称为 superFoo。在使用该应用程序 6 个月后，决定将前 25 个 bean 移至 SuperFoo 实现。要做到这一点，有三个选择。

- 第一种方法是将 standardFoo bean 的实现类更改为 SuperFoo。这种方法的缺点是，当只需要 SuperFoo 类的一个实例时，却有两个。另外，当配置更改时，需要对两个 bean 进行更改。
- 第二种方法是针对正在更改的 25 个 bean 更新注入配置，这样一来就可以将 bean 的名称从 standardFoo 更改为 superFoo。但这并不是最优雅的方式。虽然可以执行查找和替换，但是当公司管理层对所做的更改不满意而需要回滚更改时，意味着需要从版本控制系统中检索旧版本的配置。
- 第三种方法是最理想的方法，就是删除(或注释掉)standardFoo bean 的定义，并将 standardFoo 作为 superFoo 的别名。这种更改非常容易实现，并且将系统恢复到之前的配置也同样简单。

使用注解配置的 bean 命名

当使用注解声明 bean 定义时，bean 命名与 XML 稍有不同，并且可以完成更多有趣的事情。然而，让我们从基础开始学起：使用构造型注解(@Component 及其所有特例，如 Service、Repository 和 Controller)声明 bean 定义。

请思考下面的 Singer 类：

```java
package com.apress.prospring5.ch3.annotated;

import org.springframework.stereotype.Component;

@Component
public class Singer {

   private String lyric = "We found a message in a bottle we were drinking";

   public void sing() {
      System.out.println(lyric);
   }
}
```

该类包含了使用@Component 注解编写的 Singer 类型的单例 bean 的声明。@Component 注解没有任何参数，因此 Spring IoC 容器决定了该 bean 的唯一标识符。在这种情况下遵循的惯例是将 bean 命名为类本身，但是将首字母小写，这意味着该 bean 将被命名为 singer。其他的构造型注解也遵循该惯例。为了便于测试，可以使用以下类：

```java
package com.apress.prospring5.ch3.annotated;

import org.springframework.context.support.GenericXmlApplicationContext;

import java.util.Arrays;

import java.util.Map;

public class AnnotatedBeanNaming {

   public static void main(String... args) {
      GenericXmlApplicationContext ctx =
         new GenericXmlApplicationContext();
      ctx.load("classpath:spring/app-context-annotated.xml");
      ctx.refresh();

      Map<String,Singer> beans =
            ctx.getBeansOfType(Singer.class);
      beans.entrySet().stream().forEach(b ->
         System.out.println("id: " + b.getKey()));
      ctx.close();
   }
}
```

app-context-annotated.xml 配置文件仅包含 com.apress.prospring5.ch3.annotated 的组件扫描声明，因此不会再显示。当运行上面的类时，以下输出将被打印在控制台中：

```
id: singer
```

因此，使用@Component("singer")相当于使用@Component 注解 Singer 类。如果想以不同的方式命名 bean，@Component 注解必须接收 bean 名称作为参数。

```
package com.apress.prospring5.ch3.annotated;

import org.springframework.stereotype.Component;

@Component("johnMayer")
public class Singer {

    private String lyric = "Down there below us, under the clouds";

    public void sing() {
        System.out.println(lyric);
    }
}
```

正如所料，如果运行 AnnotatedBeanNaming，则会生成以下输出：

```
id: johnMayer
```

但是，别名呢？由于@Component 注解的参数成为 bean 的唯一标识符，因此以这种方式声明 bean 时不可能使用 bean 别名。此时就需要使用 Java 配置。请思考一下下面的类，它包含一个静态配置类(是的，Spring 允许这样做，并且是非常有用的，可以将所有逻辑保存在同一个文件中)：

```
package com.apress.prospring5.ch3.config;

import com.apress.prospring5.ch3.annotated.Singer;
import org.springframework.context.annotation.AnnotationConfigApplicationContext;
import org.springframework.context.annotation.Bean;
import org.springframework.context.annotation.Configuration;
import org.springframework.context.support.GenericApplicationContext;
import org.springframework.context.support.GenericXmlApplicationContext;

import java.util.Arrays;
import java.util.Map;

public class AliasConfigDemo {

    @Configuration
    public static class AliasBeanConfig {
        @Bean
        public Singer singer(){
            return new Singer();
        }
    }

    public static void main(String... args) {
        GenericApplicationContext ctx =
          new AnnotationConfigApplicationContext(AliasBeanConfig.class);

        Map<String,Singer> beans = ctx.getBeansOfType(Singer.class);
        beans.entrySet().stream().forEach(b ->
            System.out.println("id: " + b.getKey()
            + "\n aliases: "
                + Arrays.toString(ctx.getAliases(b.getKey())) + "\n")
        );

        ctx.close();
    }
}
```

该类包含一个 Singer 类型的 bean 定义，而 Singer 类型是通过使用@Bean 注解 singer()方法声明的。当没有为@Bean 注解提供任何参数时，bean 的唯一标识符(即 bean 的 id)将成为方法名称。因此，当运行上述类时，可以得到以下输出：

```
id: singer
 aliases:
```

要声明别名，可以使用@Bean 注解的 name 属性。该属性是@Bean 注解的默认属性，也就是说，如果通过使用@Bean、@Bean("singer")或@Bean(name ="singer")注解 singer()方法来声明 bean，将导致相同的结果。Spring IoC 容器将创建一个 Singer 类型的带有 singer ID 的 bean。

如果该属性的值是一个包含别名特定分隔符(空格、逗号和分号)的字符串，那么该字符串将成为该 bean 的 ID。但如果该属性的值是一个字符串数组，则第一个数组成为 id，而其他数组成为别名。修改 bean 配置，如下所示：

```
@Configuration
   public static class AliasBeanConfig {
      @Bean(name={"johnMayer","john","jonathan","johnny"})
      public Singer singer(){
         return new Singer();
      }
   }
```

当运行 AliasBeanConfig 类时，输出将变为如下所示的内容：

```
id: johnMayer
 aliases: jonathan, johnny, john
```

当涉及别名时，在 Spring 4.2 中引入了@AliasFor 注解。该注解用于为注解属性声明别名，并且大多数 Spring 注解都会使用它。例如，@Bean 注解有两个属性——name 和 value，它们被声明为彼此的别名。当使用@AliasFor 注解时，它们都是显式别名。下面所示的代码片段是@Bean 注解代码的快照，来自官方的 Spring GitHub 存储库。此时不相关的代码和文档被省略：[1]

```
package org.springframework.context.annotation;

import java.lang.annotation.Documented;
import java.lang.annotation.ElementType;
import java.lang.annotation.Retention;
import java.lang.annotation.RetentionPolicy;
import java.lang.annotation.Target;

import org.springframework.core.annotation.AliasFor;
...
@Target({ElementType.METHOD, ElementType.ANNOTATION_TYPE})
@Retention(RetentionPolicy.RUNTIME)
@Documented
public @interface Bean {
   @AliasFor("name")
   String value() default {};

   @AliasFor("value")
   String name() default {};

   ...
}
```

在下面的示例中，声明了一个可以在 Singer 实例上使用的名为@Award 的注解。

```
package com.apress.prospring5.ch3.annotated;

import org.springframework.core.annotation.AliasFor;

public @interface Award {

   @AliasFor("prize")
   String value() default {};

   @AliasFor("value")
   String prize() default {};
}
```

通过使用该注解，可以修改 Singer 类，如下所示：

```
package com.apress.prospring5.ch3.annotated;

import org.springframework.beans.factory.annotation.Qualifier;
import org.springframework.stereotype.Component;

@Component("johnMayer")
@Award(prize = {"grammy", "platinum disk"})
public class Singer {

   private String lyric = "We found a message in a bottle we were drinking";

   public void sing() {
```

1 可以通过以下链接看到完整的实现过程：https://github.com/spring-projects/spring-framework/blob/master/spring-core/src/main/java/org/springframework/core/annotation/AliasFor.java。

```
        System.out.println(lyric);
    }
}
```

上面所示的注释等同于@Award(value = {"grammy", "platinum disk"})和@Award({"grammy", "platinum disk"})。

此外，还可以使用@AliasFor 注解来完成一些更有趣的事情：可以声明元注解属性的别名。在下面的代码片段中，声明了@Award 注解的一个特例以及一个名为 name 的属性，该属性是@Award 注解的 value 属性的别名。之所以这样做，是为了想明确表示参数是唯一的 bean 标识符。

```
package com.apress.prospring5.ch3.annotated;

import org.springframework.core.annotation.AliasFor;

@Award
public @interface Trophy {

    @AliasFor(annotation = Award.class, attribute = "value")
    String name() default {};
}
```

因此，无须像下面所示的那样编写 Singer 类：

```
package com.apress.prospring5.ch3.annotated;

import org.springframework.stereotype.Component;

@Component("johnMayer")
@Award(value={"grammy", "platinum disk"})
public class Singer {

    private String lyric = "We found a message in a bottle we were drinking";

    public void sing() {
        System.out.println(lyric);
    }
}
```

可以编写如下所示的 Singer 类：

```
package com.apress.prospring5.ch3.annotated;

@Component("johnMayer")
@Trophy(name={"grammy", "platinum disk"})
public class Singer {
    private String lyric = "We found a message in a bottle we were drinking";

    public void sing() {
        System.out.println(lyric);
    }
}
```

使用另一个注解@AliasFor 为注解的属性创建别名是有一定局限性的。@AliasFor 不能用于任何构造型注解(@Component 及其特例)。原因是，对这些 value 属性的特殊处理在@AliasFor 出现的前几年就已经存在了。因此，由于向后兼容性的问题，根本不可能使用具有此类值属性的 @AliasFor。即使在编写代码时这么做了(即在构造型注解中为 value 属性起别名)，也不会出现编译错误，代码甚至可能会运行，但为别名提供的任何参数都将被忽略。@Qualifier 注解也是如此。

3.5.6 了解 bean 实例化模式

默认情况下，Spring 中的所有 bean 都是单例的。这意味着 Spring 维护一个 bean 的实例，所有依赖对象都使用同一个实例，对 ApplicationContext.getBean()的所有调用都会返回同一个实例。我们在前一节中演示了这一点，可以使用标识比较(==)而不是 equals()比较来检查 bean 是否相同。

术语*单例(singleton)*在 Java 中可互换使用，分别指两个不同的概念：应用程序中具有单个实例的对象以及 Singleton 设计模式。通常将第一个概念称为*单例*，而将 Singleton 设计模式简称为 Singleton。在 Erich Gamma 等人编写的 *Design Patterns: Elements of Reusable Object-Oriented Software* 一书(Addison-Wesley 出版社，1994 年出版)中，Singleton 设计模式得到了推广。如果把使用单例实例与使用 Singleton 设计模式的需求搞混淆，就会出现问题。以下代码片段显示了 Java 中 Singleton 设计模式的典型实现：

```
package com.apress.prospring5.ch3;

public class Singleton {
   private static Singleton instance;
   static {
      instance = new Singleton();
   }

   public static Singleton getInstance() {
      return instance;
   }
}
```

虽然上述设计模式实现了自己的目标，即允许在整个应用程序中维护和访问单个类的实例，但这样做的代价是增加了耦合。应用程序代码必须对 Singleton 类有明确的了解才能获得实例，从而完全清除了对接口进行编码的能力。

实际上，Singleton 设计模式是两种模式。第一种模式和预期的一样，主要用来维护对象的单个实例。第二种模式是一种用于查找对象的模式，该模式不太常用，因为它完全消除了使用接口的可能性。使用 Singleton 设计模式也使得任意替换的实现变得更困难，因为大多数需要 Singleton 实例的对象都直接访问 Singleton 对象。当试图对应用程序进行单元测试时，这可能会导致各种令人头疼的问题，因为无法用模拟替代 Singleton 以进行测试。

幸运的是，如果使用 Spring，则可以利用单例实例化模型，而无须使用 Singleton 设计模式。Spring 中的所有 bean 默认情况下都创建为 Singleton 实例，Spring 使用相同的实例来完成对 bean 的所有请求。当然，Spring 不仅限于使用 Singleton 实例；仍然可以创建新的 bean 实例，以满足每个依赖项以及对 getBean()的调用。Spring 在完成这些工作的同时不会对应用程序代码产生任何影响，因此，通常将 Spring 称为实例化模式不可知的。这是一个强大的概念。如果正在使用一个单例对象，但发现它并不适合多线程访问，那么可以将它改为非单例(原型)，同时不会影响任何应用程序代码。

虽然改变 bean 的实例化模式不会影响应用程序代码，但如果使用了 Spring 的生命周期接口，则会引起一些问题。我们将在第 4 章中更详细地介绍这一点。

将实例化模式从单例更改为非单例是非常简单的。以下配置片段展示了如何在 XML 中使用注解完成此操作：

```
<!-- app-context-xml.xml -->
<beans ...>
   <bean id="nonSingleton" class="com.apress.prospring5.ch3.annotated.Singer"
         scope="prototype" c:_0="John Mayer"/>
   </beans>

\\Singer.java
package com.apress.prospring5.ch3.annotated;

import org.springframework.beans.factory.annotation.Value;
import org.springframework.context.annotation.Scope;
import org.springframework.stereotype.Component;

@Component("nonSingleton")
@Scope("prototype")
public class Singer {
   private String name = "unknown";

   public Singer(@Value("John Mayer") String name) {
      this.name = name;
   }
}
```

在 XML 配置中，Singer 类被用作 XML 中所声明 bean 的类型。如果组件扫描未启用，类中的注解将被忽略。

正如你所看到的，该 bean 的声明和你在前面看到的任何声明之间的唯一区别是添加了 scope 属性并将其值设置为 prototype。Spring 默认作用域为 singleton 值。原型作用域指示 Spring 在每次应用程序请求 bean 实例时实例化一个新的 bean 实例。以下代码片段显示了此设置对应用程序的影响：

```
package com.apress.prospring5.ch3;

import com.apress.prospring5.ch3.annotated.Singer;
import org.springframework.context.support.GenericXmlApplicationContext;
public class NonSingletonDemo {
   public static void main(String... args) {
```

```
        GenericXmlApplicationContext ctx =
           new GenericXmlApplicationContext();
        ctx.load("classpath:spring/app-context-xml.xml");
        ctx.refresh();

        Singer singer1 = ctx.getBean("nonSingleton", Singer.class);
        Singer singer2 = ctx.getBean("nonSingleton", Singer.class);

        System.out.println("Identity Equal?: " + (singer1 ==singer2));
        System.out.println("Value Equal:? " + singer1.equals(singer2));
        System.out.println(singer1);
        System.out.println(singer2);

        ctx.close();
    }
}
```

运行该例，得到以下输出：

```
Identity Equal?: false
Value Equal:? false
John Mayer
John Mayer
```

由此可以看出，虽然两个 String 对象的值明显相等，但身份不同，即使两个实例使用相同的 bean 名称进行检索。

选择实例化模式

大多数情况下，很容易看出哪种实例化模式是合适的。通常单例是 bean 的默认模式。一般来说，单例应该在下列情况下使用：

- **没有状态的共享对象**：假设有一个不保持状态且有许多依赖对象的对象。因为如果没有状态就不需要同步，所以每次依赖对象需要使用该对象进行一些处理时，不需要创建 bean 的新实例。
- **具有只读状态的共享对象**：这与前一种情况类似，但有一些只读状态。在这种情况下，仍然不需要同步，所以创建一个实例来满足 bean 的每个请求只会增加开销。
- **具有共享状态的共享对象**：如果有一个必须共享状态的 bean，那么单例是最理想的选择。在这种情况下，确保状态写入同步尽可能细化。
- **具有可写状态的高通量(high-throughput)对象**：如果有一个 bean 在应用程序中被大量使用，那么你会发现相对于不断创建数百个 bean 实例，使用一个单例并保持对该 bean 状态的所有写访问同步会获得更好的性能。当使用这种方法时，保持同步要尽可能精细，同时不要牺牲一致性。当应用程序长时间创建大量实例时，当共享对象只有少量可写状态时，以及当实例化新实例的代价非常高时，你会发现这种方法特别有用。

在以下情况下，应该考虑使用非单例：

- **具有可写状态的对象**：如果有一个拥有大量可写状态的 bean，那么你会发现保持同步所需的成本大于创建新实例以处理来自依赖对象的每个请求的成本。
- **具有私有状态的对象**：某些依赖对象需要具有私有状态的 bean，以便它们可以与依赖于该 bean 的其他对象分开进行处理。在这种情况下，单例显然不适合，应该使用非单例。

从 Spring 的实例化管理中获得的主要好处是，只需要付出很少的努力，应用程序就可以从与单例相关的较低内存使用率中受益。如果发现单例模式无法满足应用程序的需要，那么修改配置以使用非单例模式也是一项很容易完成的任务。

实现 bean 作用域

除单例和原型作用域外，当需要为更特定的目的而定义 Spring bean 时，还可以使用其他作用域。也可以实现自己的自定义作用域，并在 Spring 的 ApplicationContext 中注册它。Spring 从版本 4 起支持以下 bean 作用域：

- **单例作用域**：默认为单例作用域。每个 Spring IoC 容器只会创建一个对象。
- **原型作用域**：当应用程序请求时，Spring 将创建一个新实例。
- **请求作用域**：用于 Web 应用程序。当为 Web 应用程序使用 Spring MVC 时，首先针对每个 HTTP 请求实例化带有请求作用域的 bean，然后在请求完成时销毁。
- **会话作用域**：用于 Web 应用程序。当为 Web 应用程序使用 Spring MVC 时，首先针对每个 HTTP 会话实例化带有会话作用域的 bean，然后在会话结束时销毁。

- **全局会话作用域**：用于基于 Portlet 的 Web 应用程序。带有全局会话作用域的 bean 可以在同一个 Spring MVC 驱动的门户应用程序的所有 Portlet 之间共享。
- **线程作用域**：当一个新线程请求 bean 实例时，Spring 将创建一个新的 bean 实例，而对于同一个线程，返回相同的 bean 实例。请注意，线程作用域默认情况下未注册。
- **自定义作用域**：可以通过实现org.springframework.beans.factory.config.Scope接口创建自定义作用域，并在Spring配置中注册自定义作用域(对于XML，请使用org.springframework.bean.factory.config.CustomScopeConfigurer类)。

3.6 解析依赖项

在正常的操作过程中，Spring 可以通过查看配置文件或类中的注解来解析依赖项。通过这种方式，Spring 可以确保每个 bean 都按正确的顺序进行配置，以便每个 bean 都具有正确配置的依赖项。如果 Spring 没有执行此操作，而只是创建 bean 并以任意顺序配置它们，那么也可以在依赖项之前创建并配置一个 bean，这显然不是我们想要的，因为这会导致应用程序产生各种问题。

但遗憾的是，Spring 并不知道代码中 bean 之间存在的任何依赖项，这些依赖项在配置中没有指定。例如，有一个名为 johnMayer 的 Singer 类型的 bean，它通过使用 ctx.getBean()获得另一个名为 gopher 的 Guitar 类型的 bean 实例，然后在调用 johnMayer.sing()方法时使用该实例。在该方法中，通过调用 ctx.getBean("gopher")来获得 Guitar 类型的实例，而不是请求 Spring 注入依赖项。在这种情况下，Spring 并不知道 johnMayer 依赖于 gopher，因此 Spring 可能会在 gopher 之前实例化 johnMayer。此时，可以使用<bean>标记的 depends-on 属性为 Spring 提供有关 bean 的依赖项的附加信息。以下配置代码片段(包含在名为 app-context-01.xml 的文件中)显示了如何配置 johnMayer 和 gopher 的方案：

```
<beans ...">
   <bean id="johnMayer" class="com.apress.prospring5.ch3.xml.Singer"
      depends-on="gopher"/>
   <bean id="gopher" class="com.apress.prospring5.ch3.xml.Guitar"/>
</beans>
```

在该配置中，声明 johnMayer bean 依赖于 gopher bean。用 Spring 实例化 bean 时应考虑到这一点，并确保在 johnMayer 之前创建 gopher。为此，johnMayer 需要访问 ApplicationContext。还必须告诉 Spring 注入该引用，所以当调用 johnMayer.sing()方法时，可以使用该引用来获取 gopher bean，而该引用是通过让 Singer bean 实现 ApplicationContextAware 接口来完成的。ApplicationContextAware 接口是一个特定于 Spring 的接口，它为 ApplicationContext 对象强制实现一个 setter。Spring IoC 容器会自动进行检测，并且注入 bean 段创建时所在的 ApplicationContext。这一切都是在调用 bean 的构造函数之后完成的，因此在构造函数中使用 ApplicationContext 将导致 NullPointerException 异常。可以看一下如下所示的 Singer 类的代码：

```
package com.apress.prospring5.ch3.xml;

import org.springframework.beans.BeansException;
import org.springframework.context.ApplicationContext;
import org.springframework.context.ApplicationContextAware;
public class Singer implements ApplicationContextAware {

   ApplicationContext ctx;

   @Override
   public void setApplicationContext(
      ApplicationContext applicationContext) throws BeansException {
         this.ctx = applicationContext;
   }

   private Guitar guitar;

   public Singer(){
   }

   public void sing() {
      guitar = ctx.getBean("gopher", Guitar.class);
      guitar.sing();
   }
}
```

Guitar 类很简单，它只包含 sing 方法，如下所示：

```
package com.apress.prospring5.ch3.xml;

public class Guitar {

    public void sing(){
        System.out.println("Cm Eb Fm Ab Bb");
    }
}
```

为了测试该例，可以使用下面所示的类：

```
package com.apress.prospring5.ch3.xml;

import org.springframework.context.support.GenericXmlApplicationContext;

public class DependsOnDemo {

    public static void main(String... args) {
        GenericXmlApplicationContext
            ctx = new GenericXmlApplicationContext();
        ctx.load("classpath:spring/app-context-01.xml");
        ctx.refresh();

        Singer johnMayer = ctx.getBean("johnMayer", Singer.class);
        johnMayer.sing();

        ctx.close();
    }
}
```

当然，还有一个与前面的 XML 配置相同的注解配置。必须使用一种原型注解(此时使用的是@Component)将 Singer 和 Guitar 声明为 bean。此时的新奇之处在于@DependsOn 注解，它被放置在 Singer 类中。该注解等同于 XML 配置中的 depends-on 属性。

```
package com.apress.prospring5.ch3.annotated;

import org.springframework.beans.BeansException;
import org.springframework.context.ApplicationContext;
import org.springframework.context.ApplicationContextAware;
import org.springframework.context.annotation.DependsOn;
import org.springframework.stereotype.Component;

@Component("johnMayer")
@DependsOn("gopher")
public class Singer implements ApplicationContextAware{

    ApplicationContext applicationContext;

    @Override public void setApplicationContext(ApplicationContext
        applicationContext) throws BeansException {
         this.applicationContext = applicationContext;
    }

    private Guitar guitar;

    public Singer(){
    }

    public void sing() {
        guitar = applicationContext.getBean("gopher", Guitar.class);
        guitar.sing();
    }
}
```

现在要做的就是启用组件扫描，然后在DependsOnDemo类中使用application-context-02.xml创建ApplicationContext。

```
<!-- application-context-02.xml -->
<beans...>
    <context:component-scan
        base-package="com.apress.prospring5.ch3.annotated"/>
</beans>
```

运行该例，输出"Cm Eb Fm Ab Bb"。

当开发应用程序时，应该避免应用程序使用此功能；相反，应该通过 setter 和构造函数注入协定来定义依赖项。但是，如果将 Spring 与遗留代码集成在一起，可能会发现代码中定义的依赖项要求向 Spring 框架提供额外的信息。

3.7 自动装配 bean

Spring 支持五种自动装配模式。

- **byName 模式**：当使用 byName 模式进行自动装配时，Spring 会尝试将每个属性连接到同名的 bean。因此，如果目标 bean 具有名为 foo 的属性并且在 ApplicationContext 中定义了 foo bean，那么 foo bean 将被分配给目标 bean 的 foo 属性。
- **byType 模式**：当使用 byType 进行自动装配时，Spring 通过在 ApplicationContext 中自动使用相同类型的 bean 来尝试连接目标 bean 模式的每个属性。
- **构造函数模式**：该模式与 byType 模式在功能上是相同的，只不过使用的是构造函数而不是 setter 来执行注入。Spring 试图匹配构造函数中最大数量的参数。所以，如果 bean 有两个构造函数，一个接收一个 String，另一个接收一个 String 和一个 Integer，并且 ApplicationContext 中有一个 String 和一个 Integer bean，那么 Spring 将使用带有两个参数的构造函数。
- **默认模式**：Spring 将自动在构造函数模式和 byType 模式之间进行选择。如果 bean 有一个默认的(无参数)构造函数，那么 Spring 使用 byType 模式；否则，就使用构造函数模式。
- **无**：这是默认设置。

因此，如果在目标 bean 上有一个 String 类型的属性，而在 ApplicationContext 中有一个 String 类型的 bean，那么 Spring 会将该 String bean 连接到目标 bean 的 String 属性。如果在同一个 ApplicationContext 实例中有多个相同类型的 bean(此时为 String 类型)，那么 Spring 将无法决定哪一个用于自动装配，并引发异常(类型为 org.springframework.beans.factory.NoSuchBeanDefinitionException)。

以下配置代码片段显示了一个简单的配置，它通过使用每种模式来自动连接相同类型的三个 bean(app-context-03.xml)：

```
<beans ...>

   <bean id="fooOne" class="com.apress.prospring5.ch3.xml.Foo"/>
   <bean id="barOne" class="com.apress.prospring5.ch3.xml.Bar"/>

   <bean id="targetByName" autowire="byName"
      class="com.apress.prospring5.ch3.xml.Target" lazy-init="true"/>

   <bean id="targetByType" autowire="byType"
      class="com.apress.prospring5.ch3.xml.Target" lazy-init="true"/>

   <bean id="targetConstructor" autowire="constructor"
      class="com.apress.prospring5.ch3.xml.Target" lazy-init="true"/>
</beans>
```

上述配置现在看起来应该很熟悉。Foo 和 Bar 是空类。请注意，针对 autowire 属性，每个 Target bean 都有不同的值。此外，lazy-init 属性被设置为 true，从而告诉 Spring 仅在第一次请求时才实例化 bean，而不是在启动时实例化，以便可以将结果输出到测试程序中的正确位置。以下代码示例显示了一个简单的 Java 应用程序，它从 ApplicationContext 中检索每个 Target bean：

```
package com.apress.prospring5.ch3.xml;

import org.springframework.context.support.GenericXmlApplicationContext;

public class Target {
   private Foo fooOne;
   private Foo fooTwo;
   private Bar bar;

   public Target() {
   }

   public Target(Foo foo) {
      System.out.println("Target(Foo) called");
   }

   public Target(Foo foo, Bar bar) {
      System.out.println("Target(Foo, Bar) called");
```

```
    }

    public void setFooOne(Foo fooOne) {
        this.fooOne = fooOne;
        System.out.println("Property fooOne set");
    }

    public void setFooTwo(Foo foo) {
        this.fooTwo = foo;
        System.out.println("Property fooTwo set");
    }

    public void setBar(Bar bar) {
        this.bar = bar;
        System.out.println("Property bar set");
    }

    public static void main(String... args) {
        GenericXmlApplicationContext ctx = new GenericXmlApplicationContext();
        ctx.load("classpath:spring/app-context-03.xml");
        ctx.refresh();

        Target t = null;

        System.out.println("Using byName:\n");
        t = (Target) ctx.getBean("targetByName");

        System.out.println("\nUsing byType:\n");
        t = (Target) ctx.getBean("targetByType");

        System.out.println("\nUsing constructor:\n");
        t = (Target) ctx.getBean("targetConstructor");

        ctx.close();
    }
}
```

在上面这段代码中，可以看到 Target 类有三个构造函数：无参数构造函数、接收 Foo 实例的构造函数以及接收 Foo 和 Bar 实例的构造函数。除了这些构造函数，Target bean 还有三个属性：两种 Foo 类型和一种 Bar 类型。当调用这些属性和构造函数时，每个属性和构造函数都会向控制台输出写入一条消息。main()方法简单地检索 ApplicationContext 中声明的每个 Target bean，从而触发自动装配过程。以下是运行此例的输出：

```
Using byName:

Property fooOne set

Using byType:

Property bar set
Property fooOne set
Property fooTwo set

Using constructor:

Target(Foo, Bar) called
```

从输出结果可以看出，当 Spring 使用 byName 模式时，所设置的唯一属性是 foo，因为这是配置文件中唯一具有相应 bean 条目的属性。当使用 byType 模式时，Spring 设置所有三个属性的值。fooOne 和 fooTwo 属性由 fooOne bean 设置，而 bar 属性由 barOne bean 设置。当使用构造函数时，Spring 使用双参数构造函数，因为 Spring 可以为两个参数提供 bean，并且不需要回退到另一个构造函数。

当按类型进行自动装配时，如果 bean 类型彼此相关，就会让事情变得复杂。同时，当有多个类实现了相同的接口并且需要自动装配的属性将该接口指定为类型时，就会抛出异常，因为 Spring 不知道注入哪个 bean。为模拟出这么一种场景，可将 Foo 转换为一个接口并声明两种实现了该接口的 bean 类型，每个 bean 都有自己的 bean 声明。保持默认配置，不要额外命名。

```
package com.apress.prospring5.ch3.xml.complicated;

public interface Foo {
    // empty interface, used as a marker interface
```

```
}

public class FooImplOne implements Foo {
}

public class FooImplOne implements Foo {
}
```

如果要添加一个新的名为 app-context-04.xml 的配置文件，那么它将包含以下配置：

```
<beans ...>

    <bean id="fooOne"
        class="com.apress.prospring5.ch3.xml.complicated.FooImplOne"/>

    <bean id="fooTwo"
        class="com.apress.prospring5.ch3.xml.complicated.FooImplOne"/>

    <bean id="bar" class="com.apress.prospring5.ch3.xml.Bar"/>

    <bean id="targetByType" autowire="byType"
        class="com.apress.prospring5.ch3.xml.complicated.CTarget"
        lazy-init="true"/>

</beans>
```

对于这个更简单的示例，仅介绍一下 CTarget 类。该类与前面介绍的 Target 类相同；只是 main()方法不同。代码片段如下所示：

```
package com.apress.prospring5.ch3.xml.complicated;

import com.apress.prospring5.ch3.xml.*;
import org.springframework.context.support.GenericXmlApplicationContext;

public class CTarget {
    ...

    public static void main(String... args) {
        GenericXmlApplicationContext
            ctx = new GenericXmlApplicationContext();
        ctx.load("classpath:spring/app-context-04.xml");
        ctx.refresh();
        System.out.println("\nUsing byType:\n");
        CTarget t = (CTarget) ctx.getBean("targetByType");
        ctx.close();
    }
```

运行上述类将产生以下输出：

```
Using byType:

Exception in thread "main"
    org.springframework.beans.factory.UnsatisfiedDependencyException:
    Error creating bean with name 'targetByType' defined in class path
        resource spring/app-context-04.xml:
        Unsatisfied dependency expressed through bean property 'foo';
        nested exception is
        org.springframework.beans.factory.NoUniqueBeanDefinitionException:
    No qualifying bean of type
        'com.apress.prospring5.ch3.xml.complicated.Foo' available:
        expected single matching bean but found 2: fooOne,fooTwo
        ...
```

虽然此时输出的内容更多，但输出中的第一行以一种可读性更强的方式显示了问题。当 Spring 不知道自动装配什么 bean 时，会抛出 UnsatisfiedDependencyException 异常并提供明确的消息。它会告知发现了哪些 bean，但是无法选择在哪里使用哪个 bean。此时有两种方法可以解决这个问题。第一种方法是，在希望 Spring 首先考虑自动装配的 bean 定义中使用 primary 属性，并将其设置为 true。

```
<beans ...>

    <bean id="fooOne"
        class="com.apress.prospring5.ch3.xml.complicated.FooImpl1"
        primary="true"/>
    <bean id="fooTwo"
```

```
        class="com.apress.prospring5.ch3.xml.complicated.FooImpl2"/>

    <bean id="bar" class="com.apress.prospring5.ch3.xml.Bar"/>

    <bean id="targetByType" autowire="byType"
        class="com.apress.prospring5.ch3.xml.complicated.CTarget"
        lazy-init="true"/>

</beans>
```

因此，如果按照前面所述修改配置，那么在运行该例时，将输出以下内容：

```
Using byType:

Property bar set
Property fooOne set
Property fooTwo set
```

此时，一切都恢复正常。但是，只有当仅有两种 bean 相关的类型时，primary 属性才是解决方案。如果有更多的 bean 相关的类型，则会抛出 UnsatisfiedDependencyException 异常。另一种解决方案是命名 bean 并通过 XML 将它们配置到要注入的地方(这种方法可以完全控制将哪些 bean 注入到哪里)。前面的例子是一种相当复杂和令人讨厌的实现，它只是为了证明每种自动装配类型如何在 XML 中进行配置。当使用注解时，情况会有所变化。有一个与 lazy-init 属性等价的注解；可以在类级别使用@Lazy 注解来声明在第一次访问时将被实例化的 bean。通过使用构造型注解，可以只为 bean 创建一个配置，这看起来似乎非常合乎逻辑，bean 的名称并不重要，因为每种类型只有一个 bean。因此，通过注解使用配置时的默认自动装配模式是 byType。当存在 bean 相关的类型时，如果能够指定自动装配应按名称完成，那将是非常有用的。想要完成上述操作，需要使用@Qualifier 和@Autowired 注解，并提供被注入的 bean 的名称作为参数。

考虑下面的代码：

```java
package com.apress.prospring5.ch3.sandbox;

import org.springframework.beans.factory.annotation.Autowired;
import org.springframework.beans.factory.annotation.Qualifier;
import org.springframework.context.annotation.Lazy;
import org.springframework.context.support.GenericXmlApplicationContext;
import org.springframework.stereotype.Component;
@Component
@Lazy
public class TrickyTarget {

    Foo fooOne;
    Foo fooTwo;
    Bar bar;
    public TrickyTarget() {
        System.out.println("Target.constructor()");
    }

    public TrickyTarget(Foo fooOne) {
        System.out.println("Target(Foo) called");
    }

    public TrickyTarget(Foo fooOne, Bar bar) {
        System.out.println("Target(Foo, Bar) called");
    }

    @Autowired
    public void setFooOne(Foo fooOne) {
        this.fooOne = fooOne;
        System.out.println("Property fooOne set");
    }

    @Autowired
    public void setFooTwo(Foo foo) {
        this.fooTwo = foo;
        System.out.println("Property fooTwo set");
    }

    @Autowired
    public void setBar(Bar bar) {
        this.bar = bar;
        System.out.println("Property bar set");
    }
```

```
    public static void main(String... args) {
        GenericXmlApplicationContext ctx =
            new GenericXmlApplicationContext();
        ctx.load("classpath:spring/app-context-04.xml");
        ctx.refresh();

        TrickyTarget t = ctx.getBean(TrickyTarget.class);

        ctx.close();
    }
}
```

如果 Foo 是一个类，如下所示：

```
package com.apress.prospring5.ch3.sandbox;

@Component
public class Foo {

}
```

那么当运行 TrickyTarget 类时，将产生以下输出：

```
Property fooOne set
Property fooTwo set
Property bar set
```

Bar 类非常简单：

```
package com.apress.prospring5.ch3.sandbox;

import org.springframework.stereotype.Component;

@Component
public class Bar {

}
```

如果要修改 TrickyTarget 类并对该 bean 命名，可以如下所示：

```
@Component("gigi")
@Lazy
public class TrickyTarget {
...
}
```

那么当运行该类时，会产生相同的输出，因为只有一个 Target 类型的 bean，并且当使用 ctx.getBean(TrickyTarget.class) 从上下文中请求时，上下文将返回此类的唯一 bean，而不考虑 bean 的名称。另外，如果想为 Bar 类型的 bean 提供一个名称，可以如下所示：

```
package com.apress.prospring5.ch3.sandbox;

import org.springframework.stereotype.Component;

@Component("kitchen")
public class Bar {
}
```

当再次运行该例时，会看到相同的输出。这意味着默认的自动装配模式是 byType。

如前所述，当 bean 类型相关时，情况就会变得复杂。接下来将 Foo 转换成一个接口并声明两种实现了该接口的 bean 类型，每个 bean 都有自己的 bean 声明。保持默认配置，不需要额外命名。

```
package com.apress.prospring5.ch3.sandbox;

//Foo.java
public interface Foo {
    // empty interface, used as a marker interface
}

//FooImplOne.java
@Component
public class FooImplOne implements Foo {
}

//FooImplTwo.java
```

```
@Component
public class FooImplTwo implements Foo{
}
```

TrickyTarget 类保持不变，当运行它时，会看到如下所示的输出：

```
Property bar set
Exception in thread "main"
   org.springframework.beans.factory.UnsatisfiedDependencyException:
Error creating bean with name 'gigi':
   Unsatisfied dependency expressed through method 'setFoo' parameter 0;
nested exception is
   org.springframework.beans.factory.NoUniqueBeanDefinitionException:
No qualifying bean of type 'com.apress.prospring5.ch3.sandbox.Foo' available:
   expected single matching bean but found 2: fooImplOne,fooImplTwo
...
```

此时的输出会比较多，所以只显示了前几行，正如你所看到的，Spring 给出的信息是非常明确的。它告知你不知道通过 setFoo 方法自动加载哪个 bean，并且还告知你它拥有哪些可以选择的 bean。这些 bean 的名称是由 Spring 根据类名决定的，是通过小写类名的第一个字母确定的。通过使用这些信息，可以修复 TrickyTarget。有两种方法可以做到这一点。第一种方法是在定义 bean 的类上使用@Primary 注解(与之前介绍的 primary 属性等价)，从而告诉 Spring 在按类型自动装配时优先考虑这个 bean。此时注解了 FooImplOne。

```
package com.apress.prospring5.ch3.sandbox;

import org.springframework.context.annotation.Primary;
import org.springframework.stereotype.Component;

@Component
@Primary
public class FooImplOne implements Foo {
}
```

@Primary 注解是一个标记接口；它没有属性。当需要使用 byType 模式自动装配这种类型的 bean 时，它在 bean 配置中存在标志着该 bean 拥有优先权。如果运行 TrickyTarget 类，则会再次打印预期的输出。

```
Property fooOne set
Property fooTwo set
Property bar set
```

与 primary 属性的情况一样，@Primary 注解仅在有两种相关的 bean 类型时才有用。如果想要处理更多相关的 bean 类型，@Qualifier 注解更合适。它通常紧跟 setter(setFooOne()和 setFooTwo())的@Autowired(其余代码保持不变，因此不再显示)。

```
@Component("gigi")
@Lazy
public class TrickyTarget {
    ...
    @Autowired
    @Qualifier("fooImplOne")
    public void setFoo(Foo foo) {
        this.foo = foo;
        System.out.println("Property fooOne set");
    }

    @Autowired
    @Qualifier("fooImplTwo")
    public void setFooTwo(Foo fooTwo) {
        this.fooTwo = fooTwo;
        System.out.println("Property fooTeo set");
    }
    ...
}
```

现在，如果运行该例，则会再次打印预期的输出。

```
Property fooOne set
Property fooTwo set
Property bar set
```

当使用 Java 配置时，唯一改变的就是 bean 的定义方式。此时，在配置类的 bean 声明方法中使用的是@Bean 而不是@Component 注解，如下面的示例所示：

```
package com.apress.prospring5.ch3.config;

import com.apress.prospring5.ch3.sandbox.*;
```

```
import org.springframework.context.annotation.AnnotationConfigApplicationContext;
import org.springframework.context.annotation.Bean;
import org.springframework.context.annotation.Configuration;
import org.springframework.context.support.GenericApplicationContext;

public class TargetDemo {

    @Configuration
    static class TargetConfig {

        @Bean
        public Foo fooImplOne() {
            return new FooImplOne();
        }

        @Bean
        public Foo fooImplTwo() {
            return new FooImplTwo();
        }

        @Bean
        public Bar bar() {
            return new Bar();
        }

        @Bean
        public TrickyTarget trickyTarget() {
            return new TrickyTarget();
        }
    }

    public static void main(String args) {
        GenericApplicationContext ctx =
            new AnnotationConfigApplicationContext(TargetConfig.class);
        TrickyTarget t = ctx.getBean(TrickyTarget.class);
        ctx.close();
    }
}
```

此时，重用 com.apress.prospring5.ch3.sandbox 包中的现有类以避免代码重复，由于未启用组件扫描，因此使用构造型注解的任何 bean 声明都将被忽略。如果运行上面的类，你会注意到与前面示例的输出是相同的。如前所述，当使用带有@Bean 的 bean 声明时，约定是方法的名称变成 bean 的名称，所以使用@Qualifier 注解的 TrickyTarget 仍然可以正常工作。

何时使用自动装配

在大多数情况下，对于是否应该使用自动装配的问题，答案绝对是不应该！虽然在编写小型应用程序时自动装配可以节省一定的时间，但在许多情况下，这样做会导致不好的实践，尤其是在大型应用程序中更加不灵活。使用 byName 模式似乎是一个好主意，但它可能会导致为了利用自动装配功能而为类指定虚假的属性名称。其实，Spring 的主要思想是，可按自己的喜好创建自己的类，并且让 Spring 为你工作，而不是反过来。你可能会试图使用 byType 模式，但是很快你就会意识到在 ApplicationContext 中，每种类型只能有一个 bean。当需要使用相同类型的不同配置维护 bean 时，这种限制是有问题的。同样的理由适用于使用构造函数自动装配。

在某些情况下，自动装配可以节省时间，但实际上并不需要付出额外的努力来显式定义布线，并且可以从属性命名的明确语义、灵活性以及可管理的相同类型的多个实例中受益。对于任何简单的应用程序来说，应该尽量避免自动装配。

3.8 设置 bean 继承

某些情况下，许多人需要定义多个相同类型或实现共享接口的 bean。如果想让这些 bean 共享某些配置设置而不是全部配置，就可能会产生问题。保持共享的配置设置同步的过程相当容易出错，而在大型项目中，这样做可能非常耗时。为了解决这个问题，Spring 允许在同一个 ApplicationContext 实例中提供一个继承了另一个 bean 的属性设置的<bean>定义。可以根据需要覆盖子 bean 中任何属性的值，也就是说，对子 bean 拥有完全控制权，但父 bean 可以为每个 bean 提供基本配置。以下代码示例显示了具有两个 bean 的简单配置，其中一个是另一个

(app-context-xml.xml)的子节点：

```
<beans ...>

   <bean id="parent" class="com.apress.prospring5.ch3.xml.Singer"
      p:name="John Mayer" p:age="39"/>

   <bean id="child" class="com.apress.prospring5.ch3.xml.Singer"
      parent="parent" p:age="0"/>
</beans>
```

在上面这段代码中，可以看到 child bean 的<bean>标记有一个额外的属性 parent，它表明 Spring 应该把 parent bean 当作该 bean 的父 bean，并继承它的配置。如果不想从 ApplicationContext 中查找父 bean，可以在声明父 bean 时，在<bean>标记中添加属性 abstract ="true"。因为 child bean 对 age 属性有自己的值，所以 Spring 将这个值传递给该 bean。但是，对于 name 属性 child bean 没有值，所以 Spring 使用了赋给 inheritParent bean 的值。

Singer bean 非常简单。

```
package com.apress.prospring5.ch3.xml;

public class Singer {
   private String name;
   private int age;

   public void setName(String name) {
      this.name = name;
   }

   public void setAge(int age) {
      this.age = age;
   }

   public String toString() {
      return "\tName: " + name + "\n\t" + "Age: " + age;
   }
}
```

为进行测试，可编写如下所示的简单类：

```
package com.apress.prospring5.ch3.xml;

import org.springframework.context.support.GenericXmlApplicationContext;

public class InheritanceDemo {

   public static void main(String... args) {
      GenericXmlApplicationContext ctx =
         new GenericXmlApplicationContext();
      ctx.load("classpath:spring/app-context-xml.xml");
      ctx.refresh();

      Singer parent = (Singer) ctx.getBean("parent");
      Singer child = (Singer) ctx.getBean("child");

      System.out.println("Parent:\n" + parent);
      System.out.println("Child:\n" + child);
   }
}
```

正如你所看到的，Singer 类的 main()方法从 ApplicationContext 获取 child bean 和 parent bean，并将其属性的内容写入 stdout。该例的输出如下所示：

```
Parent:
      Name: John Mayer
      Age: 39
Child:
      Name: John Mayer
      Age: 0
```

正如预期的那样，inheritChild bean 继承了 inheritParent bean 的 name 属性值，同时还为 age 属性提供了自己的值。

由于子 bean 继承父 bean 的构造函数参数和属性值，因此可将两种注入方式与 bean 继承一起使用。这种灵活性使得 bean 继承成为一个强大工具，可用来构建具有多个 bean 定义的应用程序。如果使用共享的属性值声明大量具有相同值的 bean，则应该避免通过复制和粘贴来共享值，而是应该在配置中设置继承层次结构。

在使用继承时，请记住，bean 继承不必与 Java 继承层次结构相匹配。在相同类型的五个 bean 上使用 bean 继承是完全可以接受的。与其说 bean 继承是一项继承功能，还不如说更像是模板功能。但是请注意，如果要更改子 bean 的类型，那么该类型必须扩展父 bean 的类型。

3.9 小结

在本章中，我们介绍了一下 Spring Core 和 IoC 的大体情况，演示了 IoC 类型的示例，并介绍了在应用程序中使用每种机制的优缺点，研究了 Spring 提供的 IoC 机制以及在应用程序中何时使用每种机制。在学习 IoC 时，首先介绍了 Spring BeanFactory(它是 Spring IoC 功能的核心组件)，然后介绍了 ApplicationContext，它扩展了 BeanFactory 并提供额外的功能。对于 ApplicationContext，重点介绍了 GenericXmlApplicationContext，它允许使用 XML 对 Spring 进行外部配置。此外，还讨论了另一种为 ApplicationContext 声明 DI 要求的方法，即使用 Java 注解。我们提供了一些包含了 AnnotationConfigApplicationContext 和 Java 配置的示例，从而更详细地介绍这种配置 bean 的方式。

本章还介绍了 Spring 的 IoC 功能集的基础知识，包括 setter 注入、构造函数注入、方法注入、自动装配和 bean 继承。在讨论配置时，演示了如何使用各种值来配置 bean 属性，包括其他 bean、使用 XML 和注解类型配置以及 GenericXmlApplicationContext。

本章只是介绍了 Spring 和 Spring 的 IoC 容器的一些基本知识。在下一章中，你将看到 Spring 特有的一些与 IoC 相关的功能，并且更详细地了解 Spring Core 中可用的其他功能。

第 4 章

详述 Spring 配置和 Spring Boot

在前一章中，详细介绍了控制反转(IoC)的概念及其如何适用于 Spring 框架，但只是简要介绍了一下 Spring Core 可以完成的事情。Spring 提供了大量的服务来补充和扩展其基本的 IoC 功能。在本章中，将详细探讨这些内容。具体来说，主要包括以下内容：

- **管理 bean 的生命周期**：到目前为止，你所看到的 bean 都相当简单，并且完全与 Spring 容器分离。在本章中，将介绍一些策略，通过这些策略可以让 bean 在其整个生命周期的各个阶段接收来自 Spring 容器的通知。可以使用三种方法实现该功能：通过实现 Spring 指定的特定接口，指定 Spring 可通过反射调用的方法，以及使用 JSR-250 JavaBean 生命周期注解。
- **让"Spring 感知" bean**：在某些情况下，可能希望 bean 能够与配置它的 ApplicationContext 实例进行交互。基于该原因，Spring 提供了两个接口 BeanNameAware 和 ApplicationContextAware(在第 3 章结束时介绍过)，它们允许 bean 分别获取所分配的名称以及引用它的 ApplicationContext。本章将介绍如何实现这些接口，并提供在应用程序中使用这些接口时所需考虑的一些事项。
- **使用 FactoryBean 接口**：顾名思义，任何充当其他 bean 的工厂的 bean 都需要实现 FactoryBean 接口。FactoryBean 接口提供了一种机制，通过该机制，可以轻松地将自己的工厂与 Spring BeanFactory 接口集成。
- **使用 JavaBean PropertyEditor 接口**：PropertyEditor 接口是 java.beans 包中提供的标准接口。PropertyEditor 接口用于属性值与 String 表示形式之间的相互转换。Spring 广泛使用 PropertyEditor，主要用于读取 BeanFactory 配置中指定的值并将它们转换为正确的类型。在本章中，将讨论 Spring 提供的 PropertyEditor 集合以及如何在应用程序中使用它们。同时，还会学习如何实现定制的 PropertyEditor。
- **学习更多关于 Spring ApplicationContext 的知识**：我们知道，ApplicationContext 是 BeanFactory 的一个扩展，主要用在完整的应用程序中。ApplicationContext 接口提供了一组有用的附加功能，包括国际化消息支持、资源加载和事件发布等。在本章中，将详细介绍 ApplicationContext 提供的除 IoC 外的其他功能。此外，还会演示 ApplicationContext 如何在构建 Web 应用程序时简化对 Spring 的使用。
- **使用 Java 类进行配置**：在 3.0 版本之前，Spring 仅支持带有 bean 注解和依赖项配置的 XML 基本配置。从 3.0 版本开始，Spring 为开发人员提供了使用 Java 类配置 Spring ApplicationContext 接口的另一种选择。本章将详细讨论 Spring 应用程序配置中的这个新选项。
- **使用 Spring Boot**：通过使用 Spring Boot，Spring 应用程序配置变得更加实用。通过 Spring Boot 项目可以轻松创建独立的、生产级的、基于 Spring 的应用程序，而我们所需要做的只是"运行"。
- **使用配置增强功能**：Spring 提供了使应用程序配置更容易的相关功能，例如配置文件管理、环境和属性源抽象等。在这部分内容中，将介绍这些功能并演示如何使用它们来解决特定的配置需求。
- **使用 Groovy 进行配置**：Spring 4.0 的新功能是能够使用 Groovy 语言配置 bean 定义，bean 定义可用作现有 XML 和 Java 配置方法的替代或补充。

4.1 Spring 对应用程序可移植性的影响

本章讨论的大部分功能都是 Spring 特有的，在很多情况下，这些功能在其他 IoC 容器中都不可用。尽管许多 IoC 容器提供了生命周期管理功能，但它们可能是通过与 Spring 不同的一组接口来实现的。如果应用程序需要在不同的 IoC 容器之间进行移植，就应该避免使用将应用程序耦合到 Spring 的一些功能。

但请记住，如果设置了一个约束(意味着应用程序可以在 IoC 容器之间移植)，将失去 Spring 提供的丰富功能。如果正在做是否使用 Spring 的战略决策，那么尽可能使用 Spring 是有道理的。

注意不要凭空产生对可移植性的需求。在很多情况下，应用程序的最终用户不关心应用程序是否可以在三个不同的 IoC 容器上运行；他们只是想让应用程序运行即可。根据我们的经验，选择使用可用功能最少的技术来构建应用程序往往是错误的。这样做经常会使应用程序从一开始就处于劣势。但是，如果应用程序确实需要 IoC 容器具有可移植性，就不要将其视为缺点，而是视为真正的需求，因此应用程序应该满足这一需求。在 *Expert One-on-One:J2EE Development without EJB*(Wrox 出版社，2004 年出版)一书中，Rod Johnson 和 JürgenHöller 将这些类型的需求描述为“幻像”需求，并详细讨论了这些需求以及它们如何影响项目。

虽然使用这些功能可能会将应用程序耦合到 Spring 框架，但事实上，这样做在更广泛的范围内提升了应用程序的可移植性。假设正在使用一个免费的开源框架，它没有特定的供应商隶属关系。使用 Spring 的 IoC 容器构建的应用程序可在 Java 运行的任何位置运行。对于 Java 企业级应用程序，Spring 为可移植性开创了新的可能性。Spring 提供了许多与 JEE 相同的功能，同时提供了一些类来抽象和简化 JEE 的许多其他方面。在很多情况下，可以使用 Spring 构建一个 Web 应用程序，该 Web 应用程序运行在一个简单的 servlet 容器中，但具有与针对全面 JEE 应用程序服务器的应用程序相同的复杂程度。通过耦合到 Spring，就可以使用 Spring 中的等效功能来替换许多功能(比如特定于供应商或依赖特定供应商配置的功能)，从而提升应用程序的可移植性。

4.2 管理 bean 生命周期

任何 IoC 容器(包括 Spring 在内)的一个重要部分是，可以构建 bean，使得它们可以在生命周期中的某些点接收通知。这样一来，就能够在 bean 的整个生命周期中的某些点上完成相关处理。通常，有两个生命周期事件与 bean 特别相关：post-initialization 和 pre-destruction。

在 Spring 的上下文中，一旦 Spring 完成 bean 的所有属性值设置以及所配置的依赖项检查，就会触发 post-initialization 事件。在 Spring 销毁 bean 实例之前，pre-destruction 事件被触发。但是，对于具有原型作用域的 bean 来说，Spring 并不会触发 pre-destruction 事件。Spring 的设计思想是，不管 bean 的作用域如何，初始化生命周期回调方法将在对象上被调用，而对于具有原型作用域的 bean，销毁生命周期回调方法则不会被调用。bean 可以使用 Spring 提供的三种机制挂钩到每个事件中并执行一些额外的处理，分别是基于接口的、基于方法的和基于注解的机制。

如果使用基于接口的机制，那么 bean 需要实现一个特定于想要接收的通知类型的接口，并且 Spring 通过接口中定义的回调方法来通知 bean。如果使用基于方法的机制，Spring 允许在 ApplicationContext 配置中指定在初始化 bean 时要调用的方法名以及在销毁 bean 时要调用的方法名。如果使用基于注解的机制，则可以使用 JSR-250 注解来指定 Spring 在构建之后或销毁之前要调用的方法。

对于这两种事件，这些机制实现了完全相同的目标。接口机制在 Spring 中被广泛使用，以便每次使用 Spring 的组件时都不必指定初始化或销毁。但是，在你自己的 bean 中，使用基于方法或基于注解的机制可能会更好，因为 bean 不需要实现任何 Spring 特定的接口。虽然我们说可移植性通常不像许多书籍中所讲的那样重要，但这并不意味着当存在完美的替代方案时，应该牺牲可移植性。也就是说，如果以其他方式将应用程序耦合到 Spring，那么可以使用接口方法指定回调一次，然后忘掉它。如果定义了大量需要利用生命周期通知的相同类型的 bean，那么使用接口机制可以避免为 XML 配置文件中的每个 bean 指定生命周期回调方法。使用 JSR-250 注解也是一种可行的选择，因为它是由 JCP 定义的标准，并且也不与 Spring 特定的注解相耦合。只要确保运行应用程序的 IoC 容器支持 JSR-250 标准即可。

总体而言，选择使用哪种机制来接收生命周期通知取决于应用程序的需求。如果关心可移植性，或者只需要定义一个或两个需要回调的特定类型的 bean，那么请使用基于方法的机制。如果使用注解类型的配置并确定正在使用支持 JSR-250 的 IoC 容器，那么请使用基于注解机制。如果不太在意可移植性，或者正在定义需要生命周期通知的相同类型的多个 bean，那么使用基于接口的机制是确保 bean 始终能够收到所期望通知的最佳方式。如果打算在多个不同的 Spring 项目中使用一个 bean，你肯定希望这个 bean 的功能尽可能独立，所以应该使用基于接口的机制。

图 4-1 高度概括了 Spring 如何管理其容器中 bean 的生命周期。

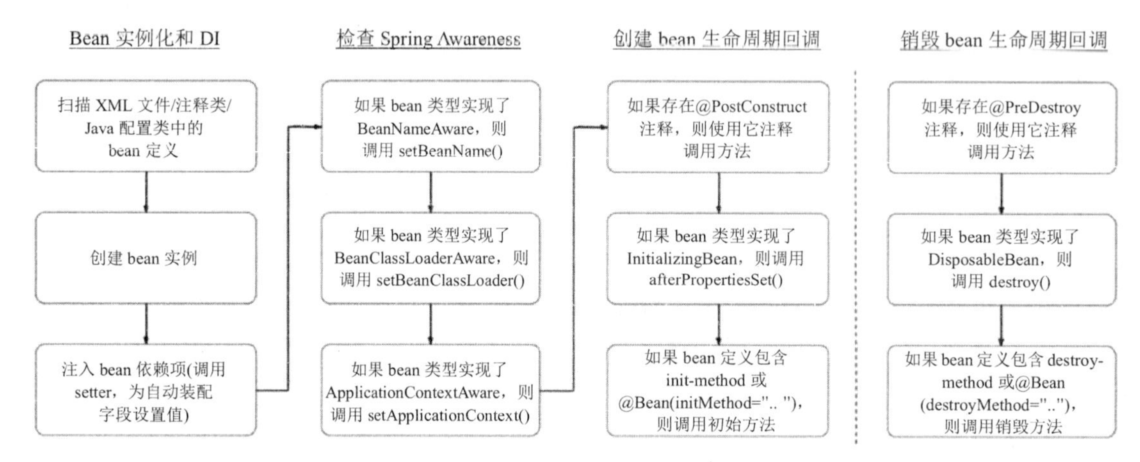

图 4-1　Spring bean 的生命周期

4.3　挂钩到 bean 的创建

通过了解初始化的时间，bean 可以检查是否满足其所需的所有依赖项。尽管 Spring 可以帮助我们检查依赖项，但它几乎是一种全有或全无的方法，并且不会提供任何机会来将其他逻辑应用于依赖项的解析过程中。假设一个 bean 有四个被声明为 setter 的依赖项，其中两个是必需的，还有一个 bean 在没有提供依赖项的情况下提供合适的默认值。通过使用初始化回调，bean 可以检查它所需的依赖项，触发异常或根据需要提供默认值。

一个 bean 不能在其构造函数中执行这些检查，因为 Spring 无法为所需的依赖项提供值。Spring 的初始化回调函数在 Spring 完成提供依赖项之后调用，并执行你所要求的依赖项检查。

不限于使用初始化回调来检查依赖项；可以在回调中做任何想要做的事情，但是它对于我们描述的目的来说非常有用。在许多情况下，可以在初始化回调中触发为响应其配置而必须自动执行的任何操作。例如，如果构建一个 bean 来运行计划任务，则初始化回调为启动调度程序提供了理想的位置。毕竟，配置数据是在 bean 上设置。

> 无须编写 bean 来运行计划任务，因为 Spring 可以通过其内置的调度功能或通过与 Quartz 计划程序集成来自动执行相关操作。第 11 章将会详细介绍该内容。

4.3.1　在创建 bean 时执行方法

正如前面提到的，接收初始化回调的一种方法是在 bean 上指定一个方法作为初始化方法，并告诉 Spring 将此方法用作初始化方法。正如所讨论的，当只有几个相同类型的 bean，或者希望应用程序与 Spring 分离时，这种回调机制很有用。使用这种机制的另一个原因，是为了能够让 Spring 应用程序使用先前构建的或由第三方供应商提供的 bean。指定回调方法只是在 bean 的<bean>标记的 init-method 属性中指定名称的一种情况。下面的代码示例显示了一个具有两个依赖项的基本 bean：

```
package com.apress.prospring5.ch4;

import org.springframework.beans.factory.BeanCreationException;
import org.springframework.context.ApplicationContext;
import org.springframework.context.support.GenericXmlApplicationContext;

public class Singer {
    private static final String DEFAULT_NAME = "Eric Clapton";

    private String name;
    private int age = Integer.MIN_VALUE;

    public void setName(String name) {
        this.name = name;
    }

    public void setAge(int age) {
        this.age = age;
```

```
    }

    public void init() {
        System.out.println("Initializing bean");

        if (name == null) {
            System.out.println("Using default name");
            name = DEFAULT_NAME;
        }

        if (age == Integer.MIN_VALUE) {
            throw new IllegalArgumentException(
            "You must set the age property of any beans of type " + Singer.class);
        }
    }

    public String toString() {
        return "\tName: " + name + "\n\tAge: " + age;
    }

    public static void main(String... args) {
        GenericXmlApplicationContext ctx =
            new GenericXmlApplicationContext();
        ctx.load("classpath:spring/app-context-xml.xml");
        ctx.refresh();

        getBean("singerOne", ctx);
        getBean("singerTwo", ctx);
        getBean("singerThree", ctx);
        ctx.close();
    }

    public static Singer getBean(String beanName,
        ApplicationContext ctx) {
        try {
            Singer bean = (Singer) ctx.getBean(beanName);
            System.out.println(bean);
            return bean;
        } catch (BeanCreationException ex) {
            System.out.println("An error occured in bean configuration: "
                    + ex.getMessage());
            return null;
        }
    }
}
```

请注意，前面已经定义了方法 init()作为初始化回调函数。init()方法检查 name 属性是否已设置，如果未设置，则使用存储在 DEFAULT_NAME 常量中的默认值。init()方法还会检查 age 属性是否已设置，如果未设置，则会抛出 IllegalArgumentException 异常。

SimpleBean 类的 main()方法尝试使用自己的 getBean()方法从 GenericXmlApplicationContext 获取三个 bean(类型都为 Singer)。请注意，在 getBean()方法中，如果成功获取 bean，就把其详细信息写入控制台输出。如果在 init()方法中抛出一个异常，则可能是因为未设置 age 属性，Spring 会将该异常封装到 BeanCreationException 中。getBean()方法捕获这些异常，并向控制台输出写入消息，从而告诉我们错误，并返回 null 值。

以下配置片段显示了一个 ApplicationContext 配置，该配置定义了在前面的代码片段(app-context-xml.xml)中使用的 bean：

```
<?xml version="1.0" encoding="UTF-8"?>

<beans xmlns="http://www.springframework.org/schema/beans"
    xmlns:xsi="http://www.w3.org/2001/XMLSchema-instance"
    xmlns:p="http://www.springframework.org/schema/p"
    xsi:schemaLocation="http://www.springframework.org/schema/beans
        http://www.springframework.org/schema/beans/spring-beans.xsd"
    default-lazy-init="true">

    <bean id="singerOne"
        class="com.apress.prospring5.ch4.Singer"
        init-method="init" p:name="John Mayer" p:age="39"/>

    <bean id="singerTwo"
        class="com.apress.prospring5.ch4.Singer"
        init-method="init" p:age="72"/>
```

```
    <bean id="singerThree"
       class="com.apress.prospring5.ch4.Singer"
       init-method="init" p:name="John Butler"/>
</beans>
```

正如你所看到的，每个 bean 的<bean>标记都有一个 init-method 属性，它告诉 Spring，只要完成 bean 的配置，就应该调用 init()方法。singerOne bean 具有 name 和 age 属性的值，因此通过 init()方法没有进行任何更改。singerTwo bean 针对 name 属性没有任何值，这意味着在 init()方法中，name 属性将被赋予默认值。最后，singerThreee bean 针对 age 属性没有任何值。init()方法中定义的逻辑将其视为错误，因此抛出 IllegalArgumentException 异常。还要注意，在<beans>标记中，添加属性 default-lazy-init="true"来指示 Spring 仅在应用程序请求 bean 时才实例化配置文件中定义的 bean。如果没有指定该属性，那么 Spring 将尝试在启动 ApplicationContext 的过程中初始化所有的 bean，并且在 singerThree 初始化期间将失败。

当配置文件中的所有 bean 具有相同的 init-method 配置时，可以通过在<beans>元素上设置 default-init-method 属性来简化该文件。bean 可以是不同的类型；唯一的条件是拥有一个名为 default-init-method 属性值的方法。所以，前面的配置也可以如下所示：

```
<beans ...
    default-lazy-init="true" default-init-method="init">

    <bean id="singerOne"
       class="com.apress.prospring5.ch4.Singer"
         p:name="John Mayer" p:age="39"/>

    <bean id="singerTwo"
       class="com.apress.prospring5.ch4.Singer"
         p:age="72"/>

    <bean id="singerThree"
       class="com.apress.prospring5.ch4.Singer"
        p:name="John Butler"/>
</beans>
```

运行上面的示例，将产生如下输出：

```
Initializing bean
        Name: John Mayer
        Age: 39
Initializing bean
Using default name
        Name: Eric Clapton
        Age: 72
Initializing bean
An error occured in bean configuration: Error creating bean
with name 'singerThree' defined in class path
resource spring/app-context-xml.xml: Invocation of init method failed;
nested exception is java.lang.IllegalArgumentException:
You must set the age property of any beans of type class
com.apress.prospring5.ch4.Singer
```

从这个输出中可以看到，我们使用配置文件中指定的值配置了 singerOne。对于 singerTwo 来说，因为没有在配置中指定值，所以使用 name 属性的默认值。最后，对于 singerThree，没有创建 bean 实例，因为 init()方法由于缺少 age 属性的值而引发了错误。

如你所见，使用初始化方法是确保 bean 正确配置的理想方法。通过使用这种机制，便可以充分利用 IoC 的优势，而不会失去通过手动定义依赖项获得的任何控制权。初始化方法的唯一限制是不能接收任何参数。可以定义任何返回类型，虽然返回值会被 Spring 忽略，甚至可以使用静态方法，但也不能接收任何参数。

当使用静态初始化方法时，这种机制带来的好处就被否定了，因为无法访问任何 bean 的状态来加以验证。如果 bean 正在使用静态状态作为节约内存的机制，同时正在使用静态初始化方法来验证状态，那么应考虑将静态状态移至实例状态并使用非静态初始化方法。虽然也可以使用 Spring 的单例管理功能来实现相同的效果，但前一种方法可以得到一个更加容易测试的 bean，并且可以在必要时使用自己的状态创建该 bean 的多个实例。当然，在某些情况下，需要使用在多个 bean 实例间共享的静态状态，此时可以使用静态初始化方法。

4.3.2 实现 InitializingBean 接口

Spring 中定义的 InitializingBean 接口允许在 bean 代码中定义希望 bean 接收的 Spring 已经完成配置的通知。与使用初始化方法的方式相同，通过实现 InitializingBean 接口可以创造机会来检查 bean 配置以确保它是有效的，并提供任何默认值。InitializingBean 接口定义了一个方法，即 afterPropertiesSet()，它的作用与前一节介绍的 init()方法的作用相同。以下代码片段重新实现了前面的示例，但此时使用 InitializingBean 接口替换了初始化方法：

```
package com.apress.prospring5.ch4;

import org.springframework.beans.factory.BeanCreationException;
import org.springframework.beans.factory.InitializingBean;
import org.springframework.context.ApplicationContext;
import org.springframework.context.support.GenericXmlApplicationContext;

public class SingerWithInterface implements InitializingBean {
    private static final String DEFAULT_NAME = "Eric Clapton";

    private String name;
    private int age = Integer.MIN_VALUE;

    public void setName(String name) {
        this.name = name;
    }

    public void setAge(int age) {
        this.age = age;
    }

    public void afterPropertiesSet() throws Exception {
        System.out.println("Initializing bean");

        if (name == null) {
            System.out.println("Using default name");
            name = DEFAULT_NAME;
        }

        if (age == Integer.MIN_VALUE) {
            throw new IllegalArgumentException(
                    "You must set the age property of any beans of type "
                    + SingerWithInterface.class);
        }
    }

    public String toString() {
        return "\tName: " + name + "\n\tAge: " + age;
    }

    public static void main(String... args) {
        GenericXmlApplicationContext ctx =
            new GenericXmlApplicationContext();
        ctx.load("classpath:spring/app-context-xml.xml");
        ctx.refresh();

        getBean("singerOne", ctx);
        getBean("singerTwo", ctx);
        getBean("singerThree", ctx);

        ctx.close();
    }

    private static SingerWithInterface getBean(String beanName,
            ApplicationContext ctx) {
        try {
            SingerWithInterface bean =
                (SingerWithInterface) ctx.getBean(beanName);
            System.out.println(bean);
            return bean;
        } catch (BeanCreationException ex) {
            System.out.println("An error occured in bean configuration: "
                    + ex.getMessage());
            return null;
        }
    }
}
```

正如你所看到的，该例没有太多改变。除了类名发生了明显的更改，唯一的区别是该类实现了 InitializingBean 接口，并且将初始化逻辑放入 afterPropertiesSet()方法中。在下面的代码片段中，可以看到该例的配置(app-context-xml.xml)：

```
<beans ... default-lazy-init="true">

    <bean id="singerOne"
       class="com.apress.prospring5.ch4.SingerWithInterface"
       p:name="John Mayer" p:age="39"/>

    <bean id="singerTwo"
       class="com.apress.prospring5.ch4.SingerWithInterface"
       p:age="72"/>
    <bean id="singerThree"
       class="com.apress.prospring5.ch4.SingerWithInterface"
       p:name="John Butler"/>
</beans>
```

同样，此时介绍的配置代码与上一节中的配置代码相比没有多大区别。明显的区别在于省略了 init-method 属性。由于 SimpleBeanWithInterface 类实现了 InitializingBean 接口，因此 Spring 知道调用哪个方法作为初始化回调，不需要任何其他配置。该例的输出如下所示：

```
Initializing bean
        Name: John Mayer
        Age: 39
Initializing bean
Using default name
        Name: Eric Clapton
        Age: 72
Initializing bean
An error occured in bean configuration: Error creating bean with name 'singerThree'
  defined in class path resource spring/app-context-xml.xml: Invocation of
  init method failed; nested exception is java.lang.IllegalArgumentException:
  You must set the age property of any beans of type class
  com.apress.prospring5.ch4.SingerWithInterface
```

4.3.3 使用 JSR-250 @PostConstruct 注解

另一种可以实现相同目的的方法是使用 JSR-250 生命周期注解@PostConstruct。从 Spring 2.5 开始，可以使用 JSR-250 注解来指定在类中存在与 bean 的生命周期相关的相应注解时 Spring 应该调用的方法。以下代码示例显示了应用@PostConstruct 注解的程序：

```
package com.apress.prospring5.ch4;

import javax.annotation.PostConstruct;
import org.springframework.beans.factory.BeanCreationException;
import org.springframework.context.ApplicationContext;
import org.springframework.context.support.GenericXmlApplicationContext;

public class SingerWithJSR250 {
    private static final String DEFAULT_NAME = "Eric Clapton";

    private String name;
    private int age = Integer.MIN_VALUE;

    public void setName(String name) {
       this.name = name;
    }

    public void setAge(int age) {
       this.age = age;
    }
    @PostConstruct
    public void init() throws Exception {
       System.out.println("Initializing bean");

       if (name == null) {
          System.out.println("Using default name");
          name = DEFAULT_NAME;
       }

       if (age == Integer.MIN_VALUE) {
          throw new IllegalArgumentException(
                "You must set the age property of any beans of type " +
                SingerWithJSR250.class);
       }
    }
```

```
    public String toString() {
        return "\tName: " + name + "\n\tAge: " + age;
    }

    public static void main(String... args) {
        GenericXmlApplicationContext ctx =
            new GenericXmlApplicationContext();
        ctx.load("classpath:spring/app-context-annotation.xml");
        ctx.refresh();

        getBean("singerOne", ctx);
        getBean("singerTwo", ctx);
        getBean("singerThree", ctx);

        ctx.close();
    }

    public static SingerWithJSR250 getBean(String beanName,
          ApplicationContext ctx) {
        try {
           SingerWithJSR250 bean =
                (SingerWithJSR250) ctx.getBean(beanName);
            System.out.println(bean);
            return bean;
        } catch (BeanCreationException ex) {
            System.out.println("An error occured in bean configuration: "
                    + ex.getMessage());
            return null;
        }
    }
}
```

该程序与使用 init-method 方法相同；只是在 init()方法前应用了@PostConstruct 注解。请注意，可以为该方法分配任何名称。就配置而言，由于使用了注解，因此需要将上下文名称空间中的<context:annotation-driven>标记添加到配置文件中。

```
<?xml version="1.0" encoding="UTF-8"?>

<beans xmlns="http://www.springframework.org/schema/beans"
       xmlns:xsi="http://www.w3.org/2001/XMLSchema-instance"
       xmlns:context="http://www.springframework.org/schema/context"
       xmlns:p="http://www.springframework.org/schema/p"
       xsi:schemaLocation="http://www.springframework.org/schema/beans
           http://www.springframework.org/schema/beans/spring-beans.xsd
           http://www.springframework.org/schema/context
           http://www.springframework.org/schema/context/spring-context.xsd"
      default-lazy-init="true">

    <context:annotation-config/>
    <bean id="singerOne"
        class="com.apress.prospring5.ch4.SingerWithJSR250"
        p:name="John Mayer" p:age="39"/>

    <bean id="singerTwo"
        class="com.apress.prospring5.ch4.SingerWithJSR250"
        p:age="72"/>

    <bean id="singerThree"
        class="com.apress.prospring5.ch4.SingerWithJSR250"
            p:name="John Butler"/>
</beans>
```

运行上述程序，将会看到与其他机制相同的输出。

```
Initializing bean
        Name: John Mayer
        Age: 39
Initializing bean
Using default name
        Name: Eric Clapton
        Age: 72
Initializing bean
An error occurred in bean configuration: Error creating bean with name 'singerThree':
  Invocation of init method failed; nested exception is
  java.lang.IllegalArgumentException: You must set the age property of any beans
  of type class com.apress.prospring5.ch4.SingerWithJSR250
```

所有这三种方法都有各自的优缺点。使用初始化方法时，可以使应用程序与 Spring 分离，但必须记住为每个需要初始化方法的 bean 配置相应的初始化方法。如果使用 InitializingBean 接口，则可以一次性为 bean 类的所有实例指定初始化回调，但必须结合应用程序才能完成此操作。如果使用注解，需要将注解应用于方法，并确保 IoC 容器支持 JSR-250。总之，应该根据应用程序的需求决定使用哪种方法。如果可移植性是一个需要关注的问题，那么请使用初始化或注解方法；否则，请使用 InitializingBean 接口，以减少应用程序所需的配置数量，以及由于配置错误而导致出现错误的可能性。

当使用 init-method 或@PostConstruct 配置初始化时，声明带有不同访问权限的初始化方法是有一定好处的。初始化方法应该在 bean 创建时由 Spring IoC 调用一次，而随后的调用将导致意外的结果甚至失败。可以通过将初始化私有化来禁止外部调用。Spring IoC 能够通过反射来调用它，但是代码中的任何其他调用都不被允许。

4.4 使用@Bean 声明一个初始化方法

为 bean 声明初始化方法的另一种方法是指定@Bean 注解的 initMethod 属性，并将初始化方法的名称设置为它的值。该注解用于在 Java 配置类中声明 bean。尽管本章后面会详细介绍 Java 配置，但对 bean 初始化部分的介绍在本节完成。在下面的示例中，之所以使用初始 Singer 类，是因为配置是外部的，就像使用 init-method 属性一样。只需要编写一个配置类和一个新的 main()方法来测试它。此外，在每个 bean 声明中，default-lazy-init ="true"将被替换为@Lazy 注解。

```
package com.apress.prospring5.ch4.config;

import com.apress.prospring5.ch4.Singer;
import org.springframework.context.annotation.AnnotationConfigApplicationContext;
import org.springframework.context.annotation.Bean;
import org.springframework.context.annotation.Configuration;
import org.springframework.context.annotation.Lazy;
import org.springframework.context.support.GenericApplicationContext;

import static com.apress.prospring5.ch4.Singer.getBean;

public class SingerConfigDemo {

    @Configuration
    static class SingerConfig{

        @Lazy
        @Bean(initMethod = "init")
        Singer singerOne() {
            Singer singerOne = new Singer();
            singerOne.setName("John Mayer");
            singerOne.setAge(39);
            return singerOne;
        }

        @Lazy
        @Bean(initMethod = "init")
        Singer singerTwo() {
            Singer singerTwo = new Singer();
            singerTwo.setAge(72);
            return singerTwo;
        }
        @Lazy
        @Bean(initMethod = "init")
        Singer singerThree() {
            Singer singerThree = new Singer();
            singerThree.setName("John Butler");
            return singerThree;
        }
    }

    public static void main(String args) {
        GenericApplicationContext ctx =
            new AnnotationConfigApplicationContext(SingerConfig.class);

        getBean("singerOne", ctx);
        getBean("singerTwo", ctx);
        getBean("singerThree", ctx);
```

```
        ctx.close();
    }
}
```

运行上述代码，将产生与前面相同的结果，如下所示：

```
Initializing bean
        Name: John Mayer
        Age: 39
Initializing bean
Using default name
        Name: Eric Clapton
        Age: 72
Initializing bean
An error occurred in bean configuration: Error creating bean with name 'singerThree'
defined in com.apress.prospring5.ch4.config.SingerConfigDemo$SingerConfig:
Invocation of init method failed; nested exception is
java.lang.IllegalArgumentException: You must set the age property of any beans
of type class com.apress.prospring5.ch4.Singer
```

了解解析顺序

所有初始化机制都可以在同一个 bean 实例上使用。在这种情况下，Spring 首先调用使用了@PostConstruct 注解的方法，然后调用 afterPropertiesSet()，最后调用配置文件中指定的初始化方法。该顺序是由一个技术原因决定的，通过了解前面的图 4-1 中的路径，可以注意到在 bean 创建过程中主要完成以下步骤：

(1) 首先调用构造函数来创建 bean。

(2) 注入依赖项(调用 setter)。

(3) 现在 bean 已经存在并且提供了依赖项，预初始化的 BeanPostProcessor 基础结构 bean 将被查询，以查看它们是否想从创建的 bean 中调用任何东西。这些都是特定于 Spring 的基础架构 bean，它们在创建后执行 bean 修改操作。@PostConstruct 注解由 CommonAnnotationBeanPostProcessor 注册，所以该 bean 将调用使用了@PostConstruct 注解的方法。该方法在 bean 被构建之后，在类被投入使用之前[1]且在 bean 的实际初始化之前(即在 afterPropertiesSet() 和 init-method 之前)执行。

(4) InitializingBean 的 afterPropertiesSet()方法在注入依赖项后立即执行。如果 BeanFactory 设置了提供的所有 Bean 属性并且满足 BeanFactoryAware 和 ApplicationContextAware，将会调用 afterPropertiesSet()方法。

(5) 最后执行 init-method 属性，这是因为它是 bean 的实际初始化方法。

如果你有一个在特定方法中执行某些初始化操作的 bean，同时在使用 Spring 时需要添加更多的初始化代码，那么理解不同类型 bean 的初始化顺序是非常有用的。

4.5 挂钩到 bean 的销毁

当使用封装了DefaultListableBeanFactory接口的ApplicationContext实现(例如，通过getDefaultListableBeanFactory()方法获取的GenericXmlApplicationContext)时，可以通过调用ConfigurableBeanFactory.destroySingletons()向BeanFactory发出信号，告知销毁所有单例实例。通常，在应用程序关闭时执行此操作，并允许清理bean可能保持打开的任何资源，从而使应用程序可以正常关闭。此外，在该回调中还可以将存储在内存中的任何数据刷新到持久存储库中，并允许bean关闭可能已启动的长时间运行的任何进程。

为了让 bean 接收到 destroySingletons()被调用的通知，存在三种选择，这些选择类似于用来接收初始化回调的机制。销毁回调通常与初始化回调一起使用。在许多情况下，在初始化回调中创建并配置资源，然后在销毁回调中释放资源。

4.5.1 在 bean 被销毁时执行一个方法

如果想要指定一个在 bean 被销毁时调用的方法，只需要在 bean 的<bean>标记的 destroy-method 属性中指定该方法的名称即可。Spring 在销毁 bean 的单例实例之前会调用该方法(对于那些有原型作用域的 bean，Spring 不会调用此方法)。以下代码片段提供了使用 destroy-method 回调的示例：

---

1 可从 JEE 官方 Javadoc 中检查此代码段：http://docs.oracle.com/javaee/7/api/javax/annotation/ PostConstruct.html。

```java
package com.apress.prospring5.ch4;

import java.io.File;
import org.springframework.beans.factory.InitializingBean;
import org.springframework.context.support.GenericXmlApplicationContext;

public class DestructiveBean implements InitializingBean {
    private File file;
    private String filePath;

    public void afterPropertiesSet() throws Exception {
        System.out.println("Initializing Bean");
        if (filePath == null) {
            throw new IllegalArgumentException(
                    "You must specify the filePath property of"
                        + DestructiveBean.class);
        }

        this.file = new File(filePath);
        this.file.createNewFile();

        System.out.println("File exists: " + file.exists());
    }

    public void destroy() {
        System.out.println("Destroying Bean");

        if(!file.delete()) {

            System.err.println("ERROR: failed to delete file.");
        }

        System.out.println("File exists: " + file.exists());
    }

    public void setFilePath(String filePath) {
        this.filePath = filePath;
    }

    public static void main(String... args) throws Exception {
        GenericXmlApplicationContext ctx =
            new GenericXmlApplicationContext();
        ctx.load("classpath:spring/app-context-xml.xml");
        ctx.refresh();

        DestructiveBean bean = (DestructiveBean) ctx.getBean("destructiveBean");

        System.out.println("Calling destroy()");
        ctx.destroy();
        System.out.println("Called destroy()");
    }
}
```

上述代码定义了一个 destroy()方法，用来删除所创建的文件。main()方法从 GenericXmlApplicationContext 中检索一个类型为 DestructiveBean 的 bean，然后调用它的 destroy()方法(并依次调用由 ApplicationContext 封装的 ConfigurableBeanFactory.destroySingletons())，指示 Spring 销毁它所管理的所有单例。初始化回调和销毁回调都会向控制台输出写入一条消息，告知它们已被调用。在下面的代码片段中，可以看到 destructiveBean bean(app-context-xml.xml)的配置：

```xml
<?xml version="1.0" encoding="UTF-8"?>

<beans xmlns="http://www.springframework.org/schema/beans"
    xmlns:xsi="http://www.w3.org/2001/XMLSchema-instance"
    xmlns:p="http://www.springframework.org/schema/p"
    xsi:schemaLocation="http://www.springframework.org/schema/beans
        http://www.springframework.org/schema/beans/spring-beans.xsd">

    <bean id="destructiveBean"
        class="com.apress.prospring5.ch4.DestructiveBean"
        destroy-method="destroy"
        p:filePath="#{systemProperties'java.io.tmpdir'}#{systemProperties'file.separator'}test.txt"/>
</beans>
```

请注意，这里通过使用 destroymethod 属性指定 destroy()方法作为销毁回调。filePath 属性值是使用 SpEL 表达式构建的，在文件名 test.txt 之前将系统属性 java.io.tmpdir 和 file.separator 连接起来以确保跨平台兼容性。运行该例将产生以下输出：

```
Initializing Bean
File exists: true
Calling destroy()
Destroying Bean
File exists: false
Called destroy()
```

如你所见，Spring 首先调用初始化回调，并且使用 DestructiveBean 实例创建并存储了 File 实例。接下来，在调用 destroy()期间，Spring 遍历它所管理的一组单例(在本例中，只有一个单例)，并调用指定的销毁回调。此外，在回调中，使用 DestructiveBean 实例删除所创建的文件并将消息记录到屏幕中，以指示这些文件不再存在。

4.5.2 实现 DisposableBean 接口

与初始化回调一样，Spring 也提供了一个接口(即 DisposableBean)，你的 bean 可以实现该接口，作为接收销毁回调的机制。DisposableBean 接口定义了一个方法 destroy()，该方法在 bean 被销毁之前被调用。使用该机制与使用 InitializingBean 接口接收初始化回调是正交的。以下代码片段显示了实现 DisposableBean 接口的 DestructiveBean 类的修改后的实现：

```java
package com.apress.prospring5.ch4;

import java.io.File;
import org.springframework.beans.factory.DisposableBean;
import org.springframework.beans.factory.InitializingBean;
import org.springframework.context.support.GenericXmlApplicationContext;

public class DestructiveBeanWithInterface implements InitializingBean, DisposableBean {
    private File file;
    private String filePath;

    @Override
    public void afterPropertiesSet() throws Exception {
       System.out.println("Initializing Bean");
       if (filePath == null) {
          throw new IllegalArgumentException(
              "You must specify the filePath property of " +
              DestructiveBeanWithInterface.class);
       }
       this.file = new File(filePath);
       this.file.createNewFile();

       System.out.println("File exists: " + file.exists());
    }

    @Override
    public void destroy() {
       System.out.println("Destroying Bean");

       if(!file.delete()) {
          System.err.println("ERROR: failed to delete file.");
       }

       System.out.println("File exists: " + file.exists());
    }
    public void setFilePath(String filePath) {
       this.filePath = filePath;
    }
    public static void main(String... args) throws Exception {
       GenericXmlApplicationContext ctx =
          new GenericXmlApplicationContext();
       ctx.load("classpath:spring/app-context-xml.xml");
       ctx.refresh();

       DestructiveBeanWithInterface bean =
          (DestructiveBeanWithInterface) ctx.getBean("destructiveBean");

       System.out.println("Calling destroy()");
       ctx.destroy();
```

```
        System.out.println("Called destroy()");
    }
}
```

使用回调方法机制的代码与使用回调接口机制的代码之间没有太大区别。甚至可以使用相同的方法名称。下面描述了该例的配置(app-context-xml.xml)：

```
<beans ...>

    <bean id="destructiveBean"
       class="com.apress.prospring5.ch4.DestructiveBeanWithInterface"
       p:filePath=
    "#{systemProperties'java.io.tmpdir'}#{systemProperties'file.separator'}test.txt"/>
</beans>
```

除不同的类名外，唯一的区别就是省略了 destroy-method 属性。运行该例将产生以下输出：

```
Initializing Bean
File exists: true
Calling destroy()
Destroying Bean
File exists: false
Called destroy()
```

4.5.3 使用 JSR-250 @PreDestroy 注解

定义在bean销毁之前所调用方法的第三种方式是使用JSR-250 生命周期注解@PreDestroy，它与@PostConstruct注解相反。以下代码片段是DestructiveBean的一个版本，它在同一个类中同时使用@PostConstruct和@PreDestroy来执行程序的初始化和销毁操作：

```
package com.apress.prospring5.ch4;

import java.io.File;
import javax.annotation.PostConstruct;
import javax.annotation.PreDestroy;
import org.springframework.context.support.GenericXmlApplicationContext;

public class DestructiveBeanWithJSR250 {
    private File file;
    private String filePath;

    @PostConstruct
    public void afterPropertiesSet() throws Exception {
       System.out.println("Initializing Bean");

       if (filePath == null) {
          throw new IllegalArgumentException(
                "You must specify the filePath property of " +
                DestructiveBeanWithJSR250.class);
       }

       this.file = new File(filePath);
       this.file.createNewFile();

       System.out.println("File exists: " + file.exists());
    }

    @PreDestroy
    public void destroy() {
       System.out.println("Destroying Bean");

       if(!file.delete()) {
          System.err.println("ERROR: failed to delete file.");
       }
       System.out.println("File exists: " + file.exists());
    }

    public void setFilePath(String filePath) {
       this.filePath = filePath;
    }

    public static void main(String... args) throws Exception {
       GenericXmlApplicationContext ctx =
          new GenericXmlApplicationContext();
       ctx.load("classpath:spring/app-context-annotation.xml");
       ctx.refresh();
```

```
        DestructiveBeanWithJSR250 bean =

            (DestructiveBeanWithJSR250) ctx.getBean("destructiveBean");
        System.out.println("Calling destroy()");
        ctx.destroy();
        System.out.println("Called destroy()");
    }
}
```

在下面的代码片段中，可以看到该 bean 的配置文件，它使用<context：annotation-config>标记(app-context-annotation.xml)。

```
<beans ...>

    <context:annotation-config/>

    <bean id="destructiveBean"
       class="com.apress.prospring5.ch4.DestructiveBeanWithJSR250"
       p:filePath="#{systemProperties'java.io.tmpdir'}#{systemProperties'file.separator'}test.txt"/>
</beans>
```

4.6 使用@Bean 声明销毁方法

声明 bean 的销毁方法的另一种方式是为@Bean 注解指定 destroyMethod 属性，并将销毁方法的名称设置为该属性的值。该注解用于在 Java 配置类中声明 bean。尽管本章稍后会详细介绍 Java 配置，但是对 bean 销毁部分的介绍在本节完成。在这个示例中，之所以使用初始的 DestructiveBeanWithJSR250 类，是因为配置是外部的，就像使用 destroy-method 属性一样。只需要编写一个配置类和一个新的 main()方法来测试它。此外，在每个 bean 声明中，default-lazy-init ="true"将被替换为@Lazy 注解。

```
package com.apress.prospring5.ch4.config;

import com.apress.prospring5.ch4.DestructiveBeanWithJSR250;
import org.springframework.context.annotation.AnnotationConfigApplicationContext;
import org.springframework.context.annotation.Bean;
import org.springframework.context.annotation.Configuration;
import org.springframework.context.annotation.Lazy;
import org.springframework.context.support.GenericApplicationContext;

/**
 * Created by iuliana.cosmina on 2/27/17.
 */
public class DestructiveBeanConfigDemo {

    @Configuration
    static class DestructiveBeanConfig {

        @Lazy
        @Bean(initMethod = "afterPropertiesSet", destroyMethod = "destroy")
         DestructiveBeanWithJSR250 destructiveBean() {
            DestructiveBeanWithJSR250 destructiveBean =
                    new DestructiveBeanWithJSR250();
            destructiveBean.setFilePath(System.getProperty("java.io.tmpdir") +
                   System.getProperty("file.separator") + "test.txt");
             return destructiveBean;
            }

    }

    public static void main(String... args) {
            GenericApplicationContext ctx =
              new AnnotationConfigApplicationContext(DestructiveBeanConfig.class);

            ctx.getBean(DestructiveBeanWithJSR250.class);
            System.out.println("Calling destroy()");
            ctx.destroy();
            System.out.println("Called destroy()");
    }
}
```

@PostConstruct 注解也在 bean 配置中使用；因此，运行上述代码将产生与前面相同的结果。

```
Initializing Bean
File exists: true
```

```
Calling dcstroy()
Destroying Bean
File exists: false
Called destroy()
```

销毁回调是确保应用程序正常关闭并且不使资源处于打开或不一致状态的理想机制。但是，需要决定是使用销毁方法回调、DisposableBean 接口、@PreDestroy 注解、XML destroy-attribute 属性还是 destroyMethod。再次强调一下，要根据应用程序的需求做出合适的决定；当需要考虑可移植性问题时使用方法回调，同时可以使用 DisposableBean 接口或 JSR-250 注解来减少所需的配置量。

4.7 了解解析的顺序

与创建 bean 的情况一样，可以在同一 bean 实例上使用所有机制来进行 bean 销毁。这种情况下，Spring 首先调用用@PreDestroy 注解的方法，然后调用 DisposableBean.destroy()，最后调用 XML 定义中配置的 destroy()方法。

使用关闭钩子

在 Spring 中销毁回调函数的唯一缺点是它们不会自动触发；需要记住在应用程序关闭之前调用 AbstractApplicationContext.destroy()。当应用程序作为 servlet 运行时，可以简单地在 servlet 的 destroy()方法中调用 destroy()。但是，在独立的应用程序中，事情并不那么简单，尤其是在应用程序中存在多个退出点时。幸运的是，有一个可行的解决方案。Java 允许创建一个*关闭钩子(shutdown hook)*，它是在应用程序关闭之前执行的一个线程。这是调用 AbstractApplicationContext(所有具体的 ApplicationContext 实现都扩展了 AbstractApplicationContext)的 destroy()方法的一种理想方式。利用此机制的最简单方法是使用 AbstractApplicationContext 的 registerShutdownHook()方法。该方法自动指示 Spring 注册底层 JVM 运行时的关闭钩子。bean 的声明和配置和之前一样；唯一改变的是 main()方法：添加对 ctx. registerShutdownHook 的调用，同时删除对 ctx.destroy()或 close()的调用。

```
...
public class DestructiveBeanWithHook {

    public static void main(String... args) {
        GenericApplicationContext ctx =
            new AnnotationConfigApplicationContext(
                DestructiveBeanConfig.class);

            ctx.getBean(DestructiveBeanWithJSR250.class);
            ctx.registerShutdownHook();
    }
}
```

运行上述代码，将产生与前面相同的结果。

```
Initializing Bean
File exists: true
Destroying Bean
File exists: false
```

正如你所看到的那样，即使没有编写任何代码以在应用程序关闭时进行显式调用，也会调用 destroy()方法。

4.8 让 Spring 感知 bean

相对于依赖查找，作为实现控制反转机制的依赖注入的最大亮点之一是，bean 不需要知道正在管理它们的容器是如何实现的。对于使用构造函数注入或 setter 注入的 bean 而言，Spring 容器与 Google Guice 或 PicoContainer 所提供的容器是相同的。但是，在某些情况下，可能需要一个使用依赖注入来获取其依赖项的 bean，以便出于某种其他原因而与容器进行交互。比如一个用来自动配置关闭钩子的 bean，它需要访问 ApplicationContext。在其他情况下，bean 可能想知道它的名称是什么(即在当前 ApplicationContext 中分配的 bean 名称)，以便可以根据名称进行一些额外的处理。

也就是说，此功能真正用于 Spring 内部使用。为 bean 名称提供某种业务含义通常是一个糟糕的主意，并且可能导致配置问题，因为必须人为地操纵 bean 名称以支持其业务含义。但是，我们发现能够让 bean 在运行时找到它的名称，对于日志记录来说是非常有用的。假设有许多在不同配置下运行的相同类型的 bean。此时，在日志消息中可以包含 bean 名称，以便当出现错误时帮助区分哪些 bean 发生了错误，而哪些 bean 正常工作。

4.8.1 使用 BeanNameAware 接口

想要获取自己名称的 bean 可以实现 BeanNameAware 接口，它有一个方法 setBeanName(String)。在完成 bean 的配置之后且在调用任何生命周期回调(初始化回调或销毁回调)之前，Spring 会调用 setBeanName()方法(参见图 4-1)。大多数情况下，setBeanName()的实现仅仅是一行代码，它将容器传入的值存储在字段中供以后使用。下面的代码片段显示了一个简单的 bean，它通过使用 BeanNameAware 获取自己的名称，然后将名称打印到控制台：

```
package com.apress.prospring5.ch4;

import org.springframework.beans.factory.BeanNameAware;

public class NamedSinger implements BeanNameAware {
    private String name;

    /** @Implements {@link BeanNameAware#setBeanName(String)} */
    public void setBeanName(String beanName) {
       this.name = beanName;
    }

    public void sing() {
       System.out.println("Singer " + name + " - sing()");
    }
}
```

这个实现相当简单。请记住，在通过调用 ApplicationContext.getBean()将 bean 的第一个实例返回给应用程序之前调用 BeanNameAware.setBeanName()，因此不需要检查 bean 名称是否在 sing()方法中可用。下面显示了本例中使用的 app-context-xml.xml 文件中包含的配置：

```
<beans ...>
    <bean id="johnMayer"
       class="com.apress.prospring5.ch4.NamedSinger"/>
</beans>
```

如你所见，不需要特殊配置即可使用 BeanNameAware 接口。在下面的代码片段中，可以看到一个简单的示例应用程序，它从 ApplicationContext 中检索 Singer 实例，然后调用 sing()方法：

```
package com.apress.prospring5.ch4;

import org.springframework.context.support.GenericXmlApplicationContext;

public class NamedSingerDemo {
    public static void main(String... args) {
       GenericXmlApplicationContext ctx =
          new GenericXmlApplicationContext();
       ctx.load("classpath:spring/app-context-xml.xml");
       ctx.refresh();

       NamedSinger bean = (NamedSinger) ctx.getBean("johnMayer");
       bean.sing();

       ctx.close();
    }
}
```

本例生成以下日志输出；注意在调用 sing()的日志消息中包含了 bean 名称：

```
Singer johnMayer - sing()
```

使用 BeanNameAware 接口非常简单，当需要提高日志消息的质量时可以很好地使用它。之说以要避免将业务含义赋予 bean 名称，是因为可以访问它们；如果赋予了业务含义，就会将自己的类与 Spring 耦合起来，但由此带来的好处确是微不足道的。如果 bean 在内部需要某种名称，可以让它们实现一个带有 setName()方法的接口，比如 Nameable，而该接口是特定于应用程序的，然后通过依赖注入为每个 bean 命名。通过这种方式，既可以保留用于配置的名称，又无须完成不必要的配置以为 bean 名称赋予业务含义。

4.8.2 使用 ApplicationContextAware 接口

第 3 章末尾介绍过 ApplicationContextAware 接口，并演示了如何使用 Spring 来处理需要其他 bean 来完成相关操作的 bean，这些 bean 不会在配置中使用构造函数注入或 setter 注入(参见 depends-on 示例)。

通过使用 ApplicationContextAware 接口，bean 可以获得对配置它们的 ApplicationContext 实例的引用。创建此接口的主要原因，是为了允许 bean 在应用程序中访问 Spring 的 ApplicationContext，例如，使用 getBean()以编程方式获取其他 Spring bean。但是，应该避免这种做法，并使用依赖注入为 bean 提供协作者。如果在可以使用依赖注入时使用了基于查找的 getBean()方法来获得依赖项，那么将会为 bean 添加不必要的复杂性，并且会在没有任何充分理由的情况下将它们耦合到 Spring 框架。

当然，ApplicationContext 并不仅仅用于查找 bean；它可以执行许多其他任务。正如你在前面看到的，其中一项任务是销毁所有单例，但在销毁之前会依次通知每个单例。在前面的章节中，已经介绍了如何创建一个关闭钩子来确保在应用程序关闭之前指示 ApplicationContext 销毁所有单例。可以使用 ApplicationContextAware 接口创建一个 bean，该 bean 可以在 ApplicationContext 中配置，并自动创建和配置关闭钩子 bean。以下配置显示了该 bean 的代码：

```
package com.apress.prospring5.ch4;

import org.springframework.beans.BeansException;
import org.springframework.context.ApplicationContext;
import org.springframework.context.ApplicationContextAware;
import org.springframework.context.support.GenericApplicationContext;

public class ShutdownHookBean implements ApplicationContextAware {
    private ApplicationContext ctx;

        /** @Implements {@link ApplicationContextAware#s
           etApplicationContext(ApplicationContext)} }*/
    public void setApplicationContext(ApplicationContext ctx)
        throws BeansException {

        if (ctx instanceof GenericApplicationContext) {
            ((GenericApplicationContext) ctx).registerShutdownHook();
        }
    }
}
```

现在，大部分代码对你来说应该是很熟悉的。ApplicationContextAware 接口定义了一个方法 setApplicationContext(ApplicationContext)，Spring 调用它，从而将 ApplicationContext 的引用传递给 bean。在前面的代码片段中，ShutdownHookBean 类检查 ApplicationContext 是否属于 GenericApplicationContext 类型，这意味着它支持 registerShutdownHook()方法；如果属于，则会向 ApplicationContext 注册一个关闭钩子。以下配置代码片段显示了如何配置此 bean 以使用 DestructiveBeanWithInterface bean(app-context-annotation.xml)：

```
<beans ...">

    <context:annotation-config/>

    <bean id="destructiveBean"
       class="com.apress.prospring5.ch4.DestructiveBeanWithInterface"
       p:filePath="#{systemProperties'java.io.tmpdir'}#{systemProperties'file.separator'}test.txt"/>

    <bean id="shutdownHook"
       class="com.apress.prospring5.ch4.ShutdownHookBean"/>
</beans>
```

请注意，此时不需要特殊配置。以下代码片段显示了一个简单的示例应用程序，它使用 ShutdownHookBean 来管理单例 bean 的销毁：

```
package com.apress.prospring5.ch4;

import javax.annotation.PostConstruct;
import javax.annotation.PreDestroy;
import java.io.File;
import org.springframework.context.support.GenericXmlApplicationContext;

public class DestructiveBeanWithInterface {
    private File file;
    private String filePath;
```

```
    @PostConstruct
    public void afterPropertiesSet() throws Exception {
        System.out.println("Initializing Bean");

        if (filePath == null) {
        throw new IllegalArgumentException(
              "You must specify the filePath property of " +
              DestructiveBeanWithInterface.class);
        }

        this.file = new File(filePath);
        this.file.createNewFile();

        System.out.println("File exists: " + file.exists());
    }

    @PreDestroy
    public void destroy() {
        System.out.println("Destroying Bean");

        if(!file.delete()) {
        System.err.println("ERROR: failed to delete file.");
        }

        System.out.println("File exists: " + file.exists());
    }

    public void setFilePath(String filePath) {
        this.filePath = filePath;
    }

    public static void main(String... args) throws Exception {
        GenericXmlApplicationContext ctx =
            new GenericXmlApplicationContext();
        ctx.load("classpath:spring/app-context-annotation.xml");
        ctx.registerShutdownHook();
        ctx.refresh();
        ctx.getBean("destructiveBean",
            DestructiveBeanWithInterface.class);
    }
}
```

上述代码对你来说应该也很熟悉。当 Spring 启动 ApplicationContext 并且在配置中定义了 destructiveBean 时，Spring 将 ApplicationContext 的引用传递给 shutdownHook bean，以便注册关闭钩子。运行该例，将产生预期的输出，如下所示：

```
Initializing Bean
File exists: true
Destroying Bean
File exists: false
```

正如你所看到的，即使主应用程序中没有调用 destroy()，由于 ShutdownHookBean 被注册为关闭钩子，它也会在应用程序关闭之前调用 destroy()。

4.9 使用 FactoryBean

在使用 Spring 时会面临一个问题：如何创建并注入不能简单地使用 new 运算符创建的依赖项。为了解决这个问题，Spring 提供了 FactoryBean 接口，该接口充当不能使用标准 Spring 语义创建和管理的对象的适配器。通常，使用 FactoryBean 创建不能通过使用 new 运算符创建的 bean，例如通过静态工厂方法访问的 bean，尽管情况并非总是如此。简而言之，FactoryBean 是一个 bean，可以作为其他 bean 的工厂。FactoryBean 像任何普通 bean 一样在 ApplicationContext 中配置，但是当 Spring 使用 FactoryBean 接口来满足依赖或查找请求时，它并不返回 FactoryBean；相反，它调用 FactoryBean.getObject()方法并返回调用的结果。

FactoryBean 在 Spring 中的使用效果很好；最明显的用途是创建事务代理(第 9 章将会详细介绍代理的相关内容)以及从 JNDI 上下文中自动获取资源。但是，FactoryBean 不仅仅用于构建 Spring 的内部组件，当构建自己的应用程序时，你会发现它们也非常有用，因为它们允许通过使用 IoC 来管理更多的资源。

FactoryBean 示例：MessageDigestFactoryBean

一般来说，所开发的项目需要进行某种密码处理；通常包括生成存储在数据库中的消息摘要或者用户密码的哈希值。在Java中，MessageDigest类提供了创建任意数据摘要的功能。MessageDigest本身是抽象的，通过调用MessageDigest.getInstance()并传入想要使用的摘要算法的名称来获得具体的实现。例如，如果想使用MD5 算法创建一个摘要，可以使用下面的代码来创建MessageDigest实例：

```
MessageDigest md5 = MessageDigest.getInstance("MD5");
```

如果想用 Spring 来管理 MessageDigest 对象的创建，那么在没有 FactoryBean 的情况下可以采用的最好方法是使用 bean 上的属性 algorithmName，然后使用初始化回调来调用 MessageDigest.getInstance()；而如果使用 FactoryBean，则可以将相关逻辑封装在一个 bean 中。然后需要 MessageDigest 实例的任何 bean 都可以简单地声明属性 messageDigest，并使用 FactoryBean 来获取实例。下面的代码片段中的 FactoryBean 实现了上述功能：

```
package com.apress.prospring5.ch4;

import java.security.MessageDigest;
import org.springframework.beans.factory.FactoryBean;
import org.springframework.beans.factory.InitializingBean;

public class MessageDigestFactoryBean implements
      FactoryBean<MessageDigest>, InitializingBean {
   private String algorithmName = "MD5";

   private MessageDigest messageDigest = null;

   public MessageDigest getObject() throws Exception {
      return messageDigest;
   }

   public Class<MessageDigest> getObjectType() {
      return MessageDigest.class;
   }

   public boolean isSingleton() {
      return true;
   }

   public void afterPropertiesSet() throws Exception {
      messageDigest = MessageDigest.getInstance(algorithmName);
   }

   public void setAlgorithmName(String algorithmName) {
      this.algorithmName = algorithmName;
   }
}
```

Spring 调用 getObject()方法来检索由 FactoryBean 创建的对象。该对象将被传递给使用 FactoryBean 作为协作者的其他 bean。在上述代码片段中，可以看到MessageDigestFactoryBean 传递了一个在 InitializingBean.afterPropertiesSet()回调中创建的用于存储 MessageDigest 实例的副本。

getObjectType()方法允许告诉 Spring FactoryBean 所返回对象的类型。如果事先不知道返回类型(例如，FactoryBean 根据配置创建不同类型的对象，具体的类型只有在 FactoryBean 初始化后才能确定)，那么对象类型可以为 null，但如果指定了类型，那么 Spring 可以使用该类型实现自动装配。在这个示例中，将 MessageDigest 作为类型返回(此时返回的是一个类，但也可以尝试返回一个接口并让 FactoryBean 实例化具体的实现类，不过只有在必要的情况下才会这么做)，因为我们不知道将返回什么样的具体类型(但这并不重要，因为所有的 bean 都会使用 MessageDigest 来定义它们的依赖项)。

通过使用 isSingleton()属性可以告知 Spring FactoryBean 是否正在管理一个单例实例。请记住，通过设置 FactoryBean 的<bean>标记的 singleton 属性，可以告诉 Spring FactoryBean 的单例状态，而不是所返回的对象。接下来看看在应用程序中如何使用 FactoryBean。在下面的代码片段中，可以看到一个简单的 bean，它维护了两个 MessageDigest 实例，然后显示传递给 digest()方法的消息的摘要：

```
package com.apress.prospring5.ch4;

import java.security.MessageDigest;
public class MessageDigester {
```

```
    private MessageDigest digest1;
    private MessageDigest digest2;

    public void setDigest1(MessageDigest digest1) {
        this.digest1 = digest1;
    }

    public void setDigest2(MessageDigest digest2) {
        this.digest2 = digest2;
    }

    public void digest(String msg) {
        System.out.println("Using digest1");
        digest(msg, digest1);

        System.out.println("Using digest2");
        digest(msg, digest2);
    }

    private void digest(String msg, MessageDigest digest) {
        System.out.println("Using alogrithm: " + digest.getAlgorithm());
        digest.reset();
        byte[] bytes = msg.getBytes();
        byte[] out = digest.digest(bytes);
        System.out.println(out);
    }
}
```

以下配置代码片段显示了两个 MessageDigestFactoryBean 类的示例配置，一个用于 SHA1 算法，另一个使用默认的(MD5)算法(app-context-xml.xml)：

```
<beans ...>

    <bean id="shaDigest"
        class="com.apress.prospring5.ch4.MessageDigestFactoryBean"
        p:algorithmName="SHA1"/>

    <bean id="defaultDigest"
        class="com.apress.prospring5.ch4.MessageDigestFactoryBean"/>

    <bean id="digester"
        class="com.apress.prospring5.ch4.MessageDigester"
        p:digest1-ref="shaDigest"
        p:digest2-ref="defaultDigest"/>
</beans>
```

如你所见，不仅配置了两个 MessageDigestFactoryBean 类，而且还使用两个 MessageDigestFactoryBean 类配置了 MessageDigester，以提供 digest1 和 digest2 属性的值。对于 defaultDigest bean，由于没有指定 algorithmName 属性，因此不会发生注入，并且将使用类中编写的默认算法(MD5)。在下面的代码示例中，将会看到一个基本示例类，它从 BeanFactory 中检索 MessageDigester bean 并创建简单消息的摘要：

```
package com.apress.prospring5.ch4;

import org.springframework.context.support.GenericXmlApplicationContext;

public class MessageDigestDemo {
    public static void main(String... args) {
        GenericXmlApplicationContext ctx =
            new GenericXmlApplicationContext();
        ctx.load("classpath:spring/app-context-xml.xml");
        ctx.refresh();

        MessageDigester digester = ctx.getBean("digester",
            MessageDigester.class);
        digester.digest("Hello World!");

        ctx.close();
    }
}
```

运行该例，将产生如下所示的输出：

```
Using digest1
Using alogrithm: SHA1
[B@130f889
Using digest2
Using alogrithm: MD5
[B@1188e820
```

正如你所看到的，虽然在 BeanFactory 中没有配置 MessageDigest bean，但 MessageDigest bean 提供了两个 MessageDigest 实现，即 SHA1 和 MD5，而这正是 FactoryBean 要完成的工作。

当使用的类无法通过 new 操作符创建时，FactoryBean 是完美的解决方案。如果使用通过工厂方法创建的对象，并且希望在 Spring 应用程序中使用这些类，那么可以创建 FactoryBean 以充当适配器，从而使类可以充分利用 Spring 的 IoC 功能。

当使用的配置是通过 Java 配置时，使用 FactoryBean 是不同的，因为在这种情况下，编译器对设置属性的类型有限制；因此，必须显式调用 getObject()方法。在下面的代码片段中，可以看到一个配置与上例相同的 bean 的示例，但此时使用的是 Java 配置：

```
package com.apress.prospring5.ch4.config;

import com.apress.prospring5.ch4.MessageDigestFactoryBean;
import com.apress.prospring5.ch4.MessageDigester;
import org.springframework.context.annotation.AnnotationConfigApplicationContext;
import org.springframework.context.annotation.Bean;
import org.springframework.context.annotation.Configuration;
import org.springframework.context.support.GenericApplicationContext;

public class MessageDigesterConfigDemo {
   @Configuration
   static class MessageDigesterConfig {

      @Bean
      public MessageDigestFactoryBean shaDigest() {
         MessageDigestFactoryBean factoryOne =
            new MessageDigestFactoryBean();
            factoryOne.setAlgorithmName("SHA1");
            return factoryOne;
      }

      @Bean
      public MessageDigestFactoryBean defaultDigest() {
         return new MessageDigestFactoryBean();
      }

      @Bean
      MessageDigester digester() throws Exception {
         MessageDigester messageDigester = new MessageDigester();
         messageDigester.setDigest1(shaDigest().getObject());
         messageDigester.setDigest2(defaultDigest().getObject());
         return messageDigester;
      }
   }

   public static void main(String... args) {
      GenericApplicationContext ctx =
         new AnnotationConfigApplicationContext(MessageDigesterConfig.class);
      MessageDigester digester = (MessageDigester) ctx.getBean("digester");
      digester.digest("Hello World!");
      ctx.close();
   }
}
```

如果运行该类，将生成相同的输出。

4.10 直接访问 FactoryBean

假设 Spring 可以自动满足由 FactoryBean 生成的对象对 FactoryBean 的引用，那么你可能会问是否可以直接访问 FactoryBean。答案是肯定的。

访问 FactoryBean 很简单，在调用 getBean()时用“&”符号作为 bean 名称的前缀即可，如以下代码示例所示：

```
package com.apress.prospring5.ch4;

import org.springframework.context.support.GenericXmlApplicationContext;

import java.security.MessageDigest;

public class AccessingFactoryBeans {
```

```
    public static void main(String... args) {
        GenericXmlApplicationContext ctx =
            new GenericXmlApplicationContext();
        ctx.load("classpath:spring/app-context-xml.xml");
        ctx.refresh();
        ctx.getBean("shaDigest", MessageDigest.class);

        MessageDigestFactoryBean factoryBean =
                    (MessageDigestFactoryBean) ctx.getBean("&shaDigest");
        try {
                MessageDigest shaDigest = factoryBean.getObject();
                System.out.println(shaDigest.digest("Hello world".getBytes()));
        } catch (Exception ex) {
                ex.printStackTrace();
        }
        ctx.close();
    }
}
```

运行上述程序将生成如下输出：

```
[B@130f889
```

虽然该功能可以在一些 Spring 代码中使用，但是你的应用程序应该没有理由使用它。FactoryBean 的用途是作为支持基础结构的一部分，从而可以在 IoC 设置中使用更多的应用程序类。应该避免直接访问 FactoryBean 和手动调用其 getObject()方法，而是让 Spring 完成该操作；如果手动执行此操作，就是在做一些无用的工作，并且会将应用程序与未来可能随时改变的特定实现细节耦合在一起。

4.11 使用 factory-bean 和 factory-method 属性

有时需要实例化由非 Spring 的第三方应用程序提供的 JavaBean。此时，你可能不知道如何实例化该类，但知道第三方应用程序提供了一个可用于获取 Spring 应用程序所需的 JavaBean 实例的类。在这种情况下，可以使用<bean>标记中的 factory-bean 和 factory-method 属性。

为了解上述方法的具体工作过程，下面的代码片段显示了 MessageDigestFactory 的另一个版本，该版本提供了一个返回 MessageDigest bean 的方法：

```
package com.apress.prospring5.ch4;

import java.security.MessageDigest;

public class MessageDigestFactory {
    private String algorithmName = "MD5";

    public MessageDigest createInstance() throws Exception {
        return MessageDigest.getInstance(algorithmName);
    }

    public void setAlgorithmName(String algorithmName) {
        this.algorithmName = algorithmName;
    }
}
```

以下配置代码片段显示了如何配置用于获取相应 MessageDigest bean 实例的工厂方法(app-context-xml.xml)：

```
<beans...>

    <bean id="shaDigestFactory"
        class="com.apress.prospring5.ch4.MessageDigestFactory"
        p:algorithmName="SHA1"/>

    <bean id="defaultDigestFactory"
        class="com.apress.prospring5.ch4.MessageDigestFactory"/>

    <bean id="shaDigest"
            factory-bean="shaDigestFactory"
            factory-method="createInstance">

    </bean>

    <bean id="defaultDigest"
            factory-bean="defaultDigestFactory"
            factory-method="createInstance"/>
```

```
        <bean id="digester"
            class="com.apress.prospring5.ch4.MessageDigester"
            p:digest1-ref="shaDigest"
            p:digest2-ref="defaultDigest"/>
    </beans>
```

请注意，此时定义了两个摘要工厂 bean，一个使用 SHA1，另一个使用默认算法。然后，针对 bean shaDigest 和 defaultDigest，指示 Spring 通过 factory-bean 属性使用相应的消息摘要工厂 bean 来实例化 bean，并且通过 factory-method 属性指定用于获取 bean 实例的方法。以下代码片段描述了测试类：

```
package com.apress.prospring5.ch4;

import org.springframework.context.support.GenericXmlApplicationContext;

public class MessageDigestFactoryDemo {
    public static void main(String... args) {
        GenericXmlApplicationContext ctx =
            new GenericXmlApplicationContext();
        ctx.load("classpath:spring/app-context-xml.xml");
        ctx.refresh();

        MessageDigester digester = ctx.getBean("digester",
            MessageDigester.class);
        digester.digest("Hello World!");

        ctx.close();
    }
}
```

运行程序，将生成如下输出：

```
Using digest1
Using alogrithm: SHA1
[B@77a57272
Using digest2
Using alogrithm: MD5
[B@7181ae3f
```

4.12 JavaBean PropertyEditor

如果还不完全熟悉 JavaBean 概念，可以看看 PropertyEditor，它是一个接口，它将属性值从其本机类型表示形式转换为字符串。最初，该接口的设计目的是允许将属性值作为字符串值输入到编辑器中，并将它们转换为正确的类型。但是，由于 PropertyEditor 本身就是轻量级的类，因此在许多设置中都可以找到它们，包括 Spring。

因为基于 Spring 的应用程序中，很大一部分属性值都在 BeanFactory 配置文件中开始生命周期，所以它们基本上都是 Strings。但是，用这些值设置的属性可能并不是字符串类型的。因此，为了避免人为创建 String 类型的属性，Spring 允许定义 PropertyEditor 以实现基于字符串的属性值到正确的类型的转换。图 4-2 显示了属于 spring-beans 包的 PropertyEditor 的完整列表；可以使用任何智能 Java 编辑器查看此列表。

它们都扩展了 java.beans.PropertyEditorSupport，并且可用来将字符串文字隐式转换为要注入 bean 中的属性值；因此，这里使用 BeanFactory 预先注册了它们。

图 4-2 Spring PropertyEditor 列表

4.12.1 使用内置的 PropertyEditor

以下代码片段显示了一个简单的 bean，它声明了 14 个属性，每个属性的类型对应于内置 PropertyEditor 实现所支持的每种类型：

```
package com.apress.prospring5.ch4;

import java.io.File;
import java.io.InputStream;
import java.net.URL;
import java.util.Date;
import java.util.List;
import java.util.Locale;
import java.util.Properties;
import java util.regex.Pattern;
import java.text.SimpleDateFormat;
import org.springframework.beans.PropertyEditorRegistrar;
import org.springframework.beans.PropertyEditorRegistry;
import org.springframework.beans.propertyeditors.CustomDateEditor;
import org.springframework.beans.propertyeditors.StringTrimmerEditor;

import org.springframework.context.support.GenericXmlApplicationContext;

public class PropertyEditorBean {

)

    private byte[] bytes;                 // ByteArrayPropertyEditor
)

    private Character character;          // CharacterEditor
    private Class cls;                    // ClassEditor
    private Boolean trueOrFalse;          // CustomBooleanEditor
    private List<String> stringList;      // CustomCollectionEditor
    private Date date;                    // CustomDateEditor
    private Float floatValue;             // CustomNumberEditor
    private File file;                    // FileEditor
    private InputStream stream;           // InputStreamEditor
    private Locale locale;                // LocaleEditor
    private Pattern pattern;              // PatternEditor
    private Properties properties;        // PropertiesEditor
    private String trimString;            // StringTrimmerEditor
    private URL url;                      // URLEditor

    public void setCharacter(Character character) {
        System.out.println("Setting character: " + character);
        this.character = character;
    }

    public void setCls(Class cls) {
        System.out.println("Setting class: " + cls.getName());
        this.cls = cls;
    }

    public void setFile(File file) {
        System.out.println("Setting file: " + file.getName());
        this.file = file;
    }

    public void setLocale(Locale locale) {
        System.out.println("Setting locale: " + locale.getDisplayName());
        this.locale = locale;
    }

    public void setProperties(Properties properties) {
        System.out.println("Loaded " + properties.size() + " properties");
        this.properties = properties;
    }

    public void setUrl(URL url) {
        System.out.println("Setting URL: " + url.toExternalForm());
        this.url = url;
    }

    public void setBytes(byte... bytes) {
        System.out.println("Setting bytes: " + Arrays.toString(bytes));
        this.bytes = bytes;
    }

    public void setTrueOrFalse(Boolean trueOrFalse) {
        System.out.println("Setting Boolean: " + trueOrFalse);
        this.trueOrFalse = trueOrFalse;
    }
```

```
    public void setStringList(List<String> stringList) {
        System.out.println("Setting string list with size: "
            + stringList.size());

        this.stringList = stringList;

        for (String string: stringList) {
            System.out.println("String member: " + string);
        }
    }

    public void setDate(Date date) {
        System.out.println("Setting date: " + date);
        this.date = date;
    }

    public void setFloatValue(Float floatValue) {
        System.out.println("Setting float value: " + floatValue);
        this.floatValue = floatValue;
    }

    public void setStream(InputStream stream) {
        System.out.println("Setting stream: " + stream);
        this.stream = stream;
    }

    public void setPattern(Pattern pattern) {
        System.out.println("Setting pattern: " + pattern);
        this.pattern = pattern;
    }

    public void setTrimString(String trimString) {
        System.out.println("Setting trim string: " + trimString);
        this.trimString = trimString;
    }
    public static class CustomPropertyEditorRegistrar
            implements PropertyEditorRegistrar {
        @Override
        public void registerCustomEditors(PropertyEditorRegistry registry) {
            SimpleDateFormat dateFormatter = new SimpleDateFormat("MM/dd/yyyy");
            registry.registerCustomEditor(Date.class,
                    new CustomDateEditor(dateFormatter, true));

            registry.registerCustomEditor(String.class, new StringTrimmerEditor(true));
        }
    }

    public static void main(String... args) throws Exception {
        File file = File.createTempFile("test", "txt");
        file.deleteOnExit();

        GenericXmlApplicationContext ctx =
            new GenericXmlApplicationContext();
        ctx.load("classpath:spring/app-context-01.xml");
        ctx.refresh();

        PropertyEditorBean bean =
            (PropertyEditorBean) ctx.getBean("builtInSample");
        ctx.close();
    }
}
```

在下面的配置示例中，可以看到用于声明类型为 PropertyEditorBean 的 bean 的配置，它为前面所示的所有属性指定了值(app-config-01.xml)：

```
<?xml version="1.0" encoding="UTF-8"?>

<beans xmlns="http://www.springframework.org/schema/beans"
    xmlns:xsi="http://www.w3.org/2001/XMLSchema-instance"
    xmlns:util="http://www.springframework.org/schema/util"
    xmlns:p="http://www.springframework.org/schema/p"
    xsi:schemaLocation="http://www.springframework.org/schema/beans
        http://www.springframework.org/schema/beans/spring-beans.xsd
        http://www.springframework.org/schema/util
        http://www.springframework.org/schema/util/spring-util.xsd">

    <bean id="customEditorConfigurer"
```

```
        class="org.springframework.beans.factory.config.CustomEditorConfigurer"
        p:propertyEditorRegistrars-ref="propertyEditorRegistrarsList"/>

    <util:list id="propertyEditorRegistrarsList">
    <bean class="com.apress.prospring5.ch4.PropertyEditorBean$
        CustomPropertyEditorRegistrar"/>
    </util:list>
    <bean id="builtInSample"
         class="com.apress.prospring5.ch4.PropertyEditorBean"
        p:character="A"
        p:bytes="John Mayer"
        p:cls="java.lang.String"
        p:trueOrFalse="true"
        p:stringList-ref="stringList"
        p:stream="test.txt"
        p:floatValue="123.45678"
        p:date="05/03/13"
          p:file="#{systemProperties'java.io.tmpdir'}
                  #{systemProperties'file.separator'}test.txt"
        p:locale="en_US"
        p:pattern="a*b"
        p:properties="name=Chris age=32"
        p:trimString=" String need trimming "
        p:url="https://spring.io/"
    />

    <util:list id="stringList">
        <value>String member 1</value>
        <value>String member 2</value>
    </util:list>
</beans>
```

正如你所看到的，虽然 PropertyEditorBean 上的所有属性都不是字符串，但属性的值都被指定为简单的字符串。另外请注意，还注册了 CustomDateEditor 和 StringTrimmerEditor，因为这两个编辑器默认情况下并未在 Spring 中注册。运行该例将产生以下输出：

```
Setting bytes: [74, 111, 104, 110, 32, 77, 97, 121, 101, 114]
Setting character: A
Setting class: java.lang.String
Setting date: Wed May 03 00:00:00 EET 13
Setting file: test.txt
Setting float value: 123.45678
Setting locale: English (United States)
Setting pattern: a*b
Loaded 1 properties
Setting stream: java.io.BufferedInputStream@42e25b0b
Setting string list with size: 2
String member: String member 1
String member: String member 2
Setting trim string: String need trimming
Setting Boolean: true
Setting URL: https://spring.io/
```

如你所见，Spring 使用内置的 PropertyEditor 将各种属性的 String 表示转换为正确的类型。表 4-1 列出了 Spring 中最重要的内置 PropertyEditor。

表 4-1 Spring PropertyEditor

PropertyEditor	描述
ByteArrayPropertyEditor	将字符串值转换为相应的字节表示形式
CharacterEditor	从 String 值填充 Character 或 char 类型的属性
ClassEditor	从完全限定的类名转换为Class实例。当使用此PropertyEditor时，请小心在使用GenericXmlApplicationContext时，不要在类名的任何一侧包含任何多余的空格，因为这会导致 ClassNotFoundException 异常
CustomBooleanEditor	将字符串转换为 Java Boolean 类型
CustomCollectionEditor	将源集合(例如，由 Spring 中的 util 名称空间表示)转换为目标 Collection 类型
CustomDateEditor	将日期的字符串表示形式转换为 java.util.Date 值。需要在 Spring 的 ApplicationContext 中以所需的日期格式注册 CustomDateEditor 实现
FileEditor	将 String 文件路径转换为 File 实例。Spring 不检查文件是否存在
InputStreamEditor	将资源的字符串表示(例如，使用 file:D:/temp/test.txt 或 classpath:test.txt 的文件资源)转换为输入流属性

(续表)

PropertyEditor	描述
LocaleEditor	将语言环境的字符串表示形式(如 en-GB)转换为 java.util.Locale 实例
PatternEditor	将字符串转换为 JDK Pattern 对象或其他方式
PropertiesEditor	以格式 key1 = value1 key2 = value2 keyn = valuen 将字符串转换为配置了相应属性的 java.util.Properties 实例
StringTrimmerEditor	在注入之前对字符串值进行修剪。需要明确注册该编辑器
URLEditor	将 URL 的字符串表示形式转换为 java.net.URL 的实例

这组 PropertyEditor 为使用 Spring 提供了良好的基础，并且可以更加简单地使用常用组件(如文件和 URL)配置应用程序。

4.12.2 创建自定义 PropertyEditor

虽然内置的 PropertyEditor 涵盖了属性类型转换的一些标准情况，但有时可能需要创建自己的 PropertyEditor 来支持应用程序中所使用的一个类或一组类。Spring 完全支持注册自定义 PropertyEditor；唯一的缺点是 java.beans.PropertyEditor 接口有很多方法，其中很多方法与当前的任务(即当前转换属性类型的任务)无关。值得庆幸的是，JDK 5 或更新的版本提供了其 PropertyEditor 可以扩展的 PropertyEditorSupport 类，该类只有一个方法 setAsText()。接下来用一个简单的示例演示如何实现一个自定义属性编辑器。假设有一个 FullName 类，它只有两个属性 firstName 和 lastName，定义如下所示：

```
package com.apress.prospring5.ch4.custom;

public class FullName {
    private String firstName;
    private String lastName;

    public FullName(String firstName, String lastName) {
        this.firstName = firstName;
        this.lastName = lastName;
    }

    public String getFirstName() {
        return firstName;
    }

    public void setFirstName(String firstName) {
        this.firstName = firstName;
    }

    public String getLastName() {
        return lastName;
    }

    public void setLastName(String lastName) {
        this.lastName = lastName;
    }

    public String toString() {
        return "First name: " + firstName + " - Last name: " + lastName;
    }
}
```

为了简化应用程序配置，开发一个自定义编辑器，将带有空格分隔符的字符串分别转换为 FullName 类的名字和姓氏。以下代码片段描述了自定义属性编辑器的实现：

```
package com.apress.prospring5.ch4.custom;

import java.beans.PropertyEditorSupport;

public class NamePropertyEditor extends PropertyEditorSupport {
    @Override
    public void setAsText(String text) throws IllegalArgumentException {
        String[] name = text.split("\\s");

        setValue(new FullName(name[0], name[1]));
    }
}
```

该编辑器很简单。它扩展了 JDK 的 PropertyEditorSupport 类并实现了 setAsText()方法。在该方法中，简单地将 String 分隔成一个以空格作为分隔符的字符串数组。之后，实例化 FullName 类的一个实例，将空格字符之前的字符串作为第一个名称并将空格字符之后的字符串作为姓氏传递。最后，通过调用带结果的 setValue()方法返回转换后的值。如果想要在应用程序中使用 NamePropertyEditor，需要在 Spring 的 ApplicationContext 中注册该编辑器。以下配置示例显示了 CustomEditorConfigurer 和 NamePropertyEditor 的 ApplicationContext 配置(app-context-02.xml)：

```
<beans ...>

    <bean name="customEditorConfigurer"
  class="org.springframework.beans.factory.config.CustomEditorConfigurer">
        <property name="customEditors">
            <map>
                <entry key="com.apress.prospring5.ch4.custom.FullName"
                  value="com.apress.prospring5.ch4.custom.NamePropertyEditor"/>
            </map>
        </property>
    </bean>

    <bean id="exampleBean"
      class="com.apress.prospring5.ch4.custom.CustomEditorExample"
      p:name="John Mayer"/>
</beans>
```

上述配置中有两件事值得注意。首先，通过使用 Map 类型的 customEditors 属性将自定义 PropertyEditor 注入 CustomEditorConfigurer 类中。其次，Map 中的每个条目表示一个 PropertyEditor，条目的键是 PropertyEditor 所使用的类的名称。正如你所看到的，NamePropertyEditor 的键是 com.apress.prospring4.ch4.FullName，表示这是编辑器应该使用的类。以下代码片段显示了在前面的配置中注册为 bean 的 CustomEditorExample 类的代码：

```
package com.apress.prospring5.ch4.custom;

import org.springframework.context.support.GenericXmlApplicationContext;

public class CustomEditorExample {
    private FullName name;

    public FullName getName() {
        return name;
    }

    public void setName(FullName name) {
        this.name = name;
    }

    public static void main(String... args) {
        GenericXmlApplicationContext ctx =
            new GenericXmlApplicationContext();
        ctx.load("classpath:spring/app-context-02.xml");
        ctx.refresh();

        CustomEditorExample bean =
             (CustomEditorExample) ctx.getBean("exampleBean");

        System.out.println(bean.getName());

        ctx.close();
    }
}
```

上述代码没什么特别之处。运行该例，将看到以下输出：

```
First name: John - Last name: Mayer
```

这是在 FullName 类中实现的 toString()方法的输出，可以看到 Spring 使用配置的 NamePropertyEditor 正确填充了 FullName 对象的名字和姓氏。从版本 3 开始，Spring 引入了类型转换 API(Type Conversion API)和字段格式化 SPI(Field Formatting Service Provider Interface)，它们提供了一个更简单且结构良好的 API 来执行类型转换和字段格式化。这对于 Web 应用程序开发来说尤其有用。第 10 章将详细讨论类型转换 API 和字段格式化 SPI。

4.13 更多的 Spring ApplicationContext 配置

到目前为止，虽然讨论的是 Spring 的 ApplicationContext，但是介绍的大部分功能主要围绕 ApplicationContext 封装的 BeanFactory 接口。在 Spring 中，BeanFactory 接口的各种实现负责 bean 实例化，为 Spring 管理的 bean 提供依赖注入和生命周期支持。然而，如前所述，作为 BeanFactory 接口的扩展，ApplicationContext 还提供其他有用的功能。ApplicationContext 的主要功能是提供一个更丰富的框架来构建应用程序。ApplicationContext 更加了解所配置的 bean(与 BeanFactory 相比)，在使用 Spring 基础结构类和接口(例如 BeanFactoryPostProcessor)的情况下，ApplicationContext 可以代替你与这些类或接口进行交互，从而减少为了使用 Spring 所需编写的代码量。

使用 ApplicationContext 的最大好处是，它允许以完全声明的方式配置和管理 Spring 以及 Spring 所管理的资源。这意味着 Spring 尽可能提供支持类来自动将 ApplicationContext 加载到应用程序中，从而不需要编写任何代码来访问 ApplicationContext。实际上，该功能目前仅在使用 Spring 构建 Web 应用程序时才可用，它允许在 Web 应用程序的部署描述符中初始化 Spring 的 ApplicationContext。当使用独立的应用程序时，也可以通过简单编码来初始化 Spring 的 ApplicationContext。

除了提供更关注声明式配置的模型，ApplicationContext 还支持以下功能：

- 国际化
- 事件发布
- 资源管理和访问
- 额外的生命周期接口
- 改进了基础设施组件的自动配置

在下面的章节中，将讨论除 DI 外的 ApplicationContext 中一些最重要的功能。

4.13.1 使用 MessageSource 进行国际化

Spring 真正擅长的领域是支持国际化(i18n)。通过使用 MessageSource 接口，应用程序可以访问以各种语言存储的字符串资源(称为*消息*)。对于希望在应用程序中得到支持的每种语言，都会维护一个与其他语言中的消息相对应的消息列表。例如，如果想用英语和德语显示“The quick brown fox jumped over the lazy dog”，那么可以创建两条消息，这两条消息的键为 msg；英国人会说“The quick brown fox jumped over the lazy dog”，而德国人会说“Der schnelle braune Fuchs sprang über den faulen Hund”。

虽然不需要通过 ApplicationContext 来使用 MessageSource，但 ApplicationContext 接口扩展了 MessageSource，并对加载消息以及特定环境下的可用性提供了特别的支持。虽然消息的自动加载在任何环境中都可用，但是只有在某些 Spring 管理的场景中才提供自动访问，例如使用 Spring 的 MVC 框架构建 Web 应用程序时。虽然任何类都可以实现 ApplicationContextAware 并访问自动加载的消息，但本章后面的 4.13.2 节“在独立的应用程序中使用 MessageSource”中提供了一个更好的解决方案。

在继续学习之前，如果不熟悉 Java 中的 i18n 支持，那么建议查看一下 Javadocs(http://download.java.net/jdk8/docs/api/index.html)。

使用 MessageSource 进行国际化

除了 ApplicationContext，Spring 还提供了三个 MessageSource 实现。

- ResourceBundleMessageSource
- ReloadableResourceBundleMessageSource
- StaticMessageSource

不应该在生产应用程序中使用 StaticMessageSource 实现，因为无法在外部配置它。当需要向应用程序添加国际化功能时，从外部进行配置是主要的需求之一。

ResourceBundleMessageSource 通过使用 Java ResourceBundle 加载消息。ReloadableResourceBundleMessageSource 基本上与之相同，只不过它支持对基础源文件进行预定的重新加载。

所有这三个 MessageSource 实现都实现了另一个名为 HierarchicalMessageSource 的接口，它允许嵌套多个 MessageSource 实例。这是 ApplicationContext 使用 MessageSource 实例的关键。

要想充分利用 ApplicationContext 对 MessageSource 的支持，必须在 MessageSource 类型的配置中定义一个 bean，并命名为 messageSource。ApplicationContext 使用此 MessageSource 并将其嵌套在自身中，从而允许使用 ApplicationContext 访问消息。该过程可能很难直观显示，所以请看下面的示例。以下代码示例显示了一个简单的应用程序，用于访问英语和德语语言环境下的一组消息：

```
package com.apress.prospring5.ch4;
import java.util.Locale;

import org.springframework.context.support.GenericXmlApplicationContext;

public class MessageSourceDemo {
   public static void main(String... args) {
      GenericXmlApplicationContext ctx = new GenericXmlApplicationContext();
      ctx.load("classpath:spring/app-context-xml.xml");
      ctx.refresh();
      Locale english = Locale.ENGLISH;
      Locale german = new Locale("de", "DE");

      System.out.println(ctx.getMessage("msg", null, english));
      System.out.println(ctx.getMessage("msg", null, german));

      System.out.println(ctx.getMessage("nameMsg", new Object[]
        { "John", "Mayer" }, english));
      System.out.println(ctx.getMessage("nameMsg", new Object[]
      { "John", "Mayer" }, german));

      ctx.close();
   }
}
```

不要担心对 getMessage()的调用；稍后会介绍该方法。现在，只需要知道它们为指定的语言环境检索了关键信息。在以下配置代码片段中，可以看到此应用程序使用的配置(app-context-xml.xml)：

```
<beans ...>

     <bean id="messageSource"
   class="org.springframework.context.support.ResourceBundleMessageSource"
      p:basenames-ref="basenames"/>

   <util:list id="basenames">
      <value>buttons</value>
      <value>labels</value>
   </util:list>
</beans>
```

此时，根据需要定义了一个名为 messageSource 的 ResourceBundleMessageSource bean，并使用一组名称来配置它，从而形成文件集的基础。ResourceBundleMessageSource 所用的 Java ResourceBundle 处理一组由基本名称标识的属性文件。在查找特定语言环境下的消息时，ResourceBundle 会查找以基本名称和语言环境名称组合命名的文件。例如，如果基本名称是 foo，并且正在 en-GB(英式英语)语言环境下查找消息，那么 ResourceBundle 会查找名为 foo_en_GB.properties 的文件。

对于前面的示例，下面显示了英语(labels_en.properties)和德语(labels_de_DE.properties)的属性文件的内容：

```
#labels_en.properties
msg=My stupid mouth has got me in trouble
nameMsg=My name is {0} {1}
#labels_de_DE.properties
msg=Mein dummer Mund hat mich in Schwierigkeiten gebracht
nameMsg=Mein Name ist {0} {1}
```

针对该例，你可能会提出更多的问题。对 getMessage()的那些调用意味着什么？ 为什么使用 ApplicationContext.getMessage()而不是直接访问 ResourceBundleMessageSource bean？接下来依次回答每个问题。

使用 getMessage()方法

MessageSource 接口为 getMessage()方法定义了三个重载版本，如表 4-2 所示。

表 4-2　MessageSource.getMessage()的重载版本

方法签名	描述
getMessage(String, Object[], Locale)	这是标准的 getMessage()方法。String 参数是与属性文件中的键相对应的消息的键。在前面的代码示例中，第一次调用 getMessage()时使用 msg 作为键，这与 en 语言环境下属性文件中的以下条目相对应：msg=The quick brown fox jumped over the lazy dog。Object []数组参数用于消息中的替换。在第三次调用 getMessage()时，传入一个由两个字符串组成的数组。键为 nameMsg 的消息是 My name is {0} {1}。花括号中的数字是占位符，它们中的每个都用参数数组中的相应条目替换。最后一个参数 Locale 告诉 ResourceBundleMessageSource 要查找哪个属性文件。尽管在示例中对 getMessage()的第一次和第二次调用使用相同的键，但它们返回与传递给 getMessage()的 Locale 设置相对应的不同消息
getMessage(String, Object[], String, Locale)	除了第二个 String 参数，此重载版本与 getMessage(String, Object[], Locale)相同，它允许传入默认值，以防止与所提供键对应的消息不可用于提供的 Locale
getMessage (MessageSourceResolvable, Locale)	该重载版本是一种特殊情况。在 4.13.2 节的“MessageSourceResolvable 接口”部分将会进一步地详细讨论它

为什么使用 ApplicationContext 作为 MessageSource

为了回答这个问题，需要稍微提前讲一下 Spring 中的 Web 应用程序支持。一般来说，该问题的答案是不应该使用 ApplicationContext 作为 MessageSource，因为这样做会不必要地将 bean 耦合到 ApplicationContext(在下一节中会有更详细的讨论)。当使用 Spring 的 MVC 框架构建 Web 应用程序时，应该使用 ApplicationContext。

Spring MVC 中的核心接口是 Controller。与 Struts 等需要通过继承具体类来实现控制器的框架不同，Spring 只需要实现 Controller 接口(或者使用@Controller 注解来标注控制器类)即可。话虽如此，Spring 还是提供了一组可用来实现自己的控制器的有用基类。这些基类中的每一个都是 ApplicationObjectSupport 类的一个子类(直接或间接子类)，对于任何想要了解 ApplicationContext 的应用程序对象来说，该类都是一个方便的超类。请记住，在 Web 应用程序设置中，ApplicationContext 会自动加载。

ApplicationObjectSupport 访问此 ApplicationContext，将其封装在 MessageSourceAccessor 对象中，并通过受保护的 getMessageSourceAccessor()方法将其提供给控制器。MessageSourceAccessor 为使用 MessageSource 实例提供了一系列方便的方法。这种自动注入形式是非常有用的；不需要所有的控制器公开 MessageSource 属性。

但是，以上介绍的并不是在 Web 应用程序中将 ApplicationContext 用作 MessageSource 的最佳理由。使用 ApplicationContext 而不是手动定义的 MessageSource bean 的主要原因是，Spring 可以尽可能地将 ApplicationContext 作为 MessageSource 公开到视图层。这意味着当使用 Spring 的 JSP 标记库时，<spring:message>标记会自动从 ApplicationContext 中读取消息，而当使用 JSTL 时，<fmt:message>标记也会执行相同的操作。

以上这些好处意味着，当构建 Web 应用程序时，最好在 ApplicationContext 中使用 MessageSource 支持，而不是单独管理 MessageSource 实例。当认为利用此功能所需要做的只是用名称 MessageSource 配置 MessageSource bean 时，更是如此。

4.13.2　在独立的应用程序中使用 MessageSource

在独立的应用程序中使用 MessageSource 时，除了将 MessageSource bean 自动嵌套到 ApplicationContext 中，Spring 没有提供其他额外的支持，因此最好通过依赖注入来使用 MessageSource。你可能会选择使用 ApplicationContextAware bean，但这样做就无法在 BeanFactory 上下文中使用该 bean。此外，还会使测试变得复杂且没有任何明显的好处，显然应该坚持使用依赖注入来访问独立设置中的 MessageSource 对象。

MessageSourceResolvable 接口

从 MessageSource 查找消息时，可以使用实现了 MessageSourceResolvable 的对象代替键和一组参数。该接口在 Spring 验证库中被广泛使用，用来将 Error 对象链接到对应的国际化错误消息。

4.13.3 应用程序事件

BeanFactory 中不存在的 ApplicationContext 的另一个功能是通过使用 ApplicationContext 作为代理发布和接收事件的能力。在本节中，将介绍它的用法。

使用应用程序事件

事件是派生自 ApplicationEvent 的类，而 ApplicationEvent 类又派生自 java.util.EventObject。任何 bean 都可以通过实现 ApplicationListener<T>接口来监听事件；当配置时，ApplicationContext 会自动注册实现此接口的任何 bean 作为监听器。事件是通过使用 ApplicationEventPublisher.publishEvent()方法发布的，所以发布类必须可以访问 ApplicationContext(它扩展了 ApplicationEventPublisher 接口)。在 Web 应用程序中，这很简单，因为许多类都是从 Spring 框架类派生的，这些类允许通过受保护的方法访问 ApplicationContext。在独立的应用程序中，可以让发布 bean 实现 ApplicationContextAware，使其可以发布事件。

以下代码示例显示了一个基本事件类的示例：

```
package com.apress.prospring5.ch4;
import org.springframework.context.ApplicationEvent;

public class MessageEvent extends ApplicationEvent {
   private String msg;
   public MessageEvent(Object source, String msg) {
      super(source);
      this.msg = msg;
   }

   public String getMessage() {
      return msg;
   }
}
```

这段代码很简单，唯一值得注意的是，ApplicationEvent 具有一个接收指向事件源的引用的构造函数。这一点反映在 MessageEvent 的构造函数中。监听器的代码如下所示：

```
package com.apress.prospring5.ch4;

import org.springframework.context.ApplicationListener;

public class MessageEventListener
      implements ApplicationListener<MessageEvent> {
   @Override
   public void onApplicationEvent(MessageEvent event) {
      MessageEvent msgEvt = (MessageEvent) event;
      System.out.println("Received: " + msgEvt.getMessage());
   }
}
```

ApplicationListener 接口定义了一个方法 onApplicationEvent，当触发一个事件时由 Spring 调用它。MessageEventListener 通过实现强类型的 ApplicationListener 接口，在 MessageEvent(或其子类)类型的事件中显示其感兴趣的操作。如果接收到 MessageEvent，就将消息写入 stdout。发布事件很简单：只需要创建一个事件类的实例并将其传递给 ApplicationEventPublisher.publishEvent()方法即可，如下所示：

```
package com.apress.prospring5.ch4;

import org.springframework.beans.BeansException;
import org.springframework.context.ApplicationContext;
import org.springframework.context.ApplicationContextAware;
import org.springframework.context.support.ClassPathXmlApplicationContext;

public class Publisher implements ApplicationContextAware {
   private ApplicationContext ctx;

   public void setApplicationContext(ApplicationContext applicationContext)
         throws BeansException {
      this.ctx = applicationContext;
   }

   public void publish(String message) {
      ctx.publishEvent(new MessageEvent(this, message));
   }
```

```
    public static void main(String... args) {
        ApplicationContext ctx = new ClassPathXmlApplicationContext(
            "classpath:spring/app-context-xml.xml");

        Publisher pub = (Publisher) ctx.getBean("publisher");
        pub.publish("I send an SOS to the world... ");
        pub.publish("... I hope that someone gets my...");
        pub.publish("... Message in a bottle");
    }
}
```

在这里可以看到，Publisher 类从 ApplicationContext 中检索自己的实例，然后使用 publish()方法将两个 MessageEvent 实例发布到 ApplicationContext。Publisher bean 实例通过实现 ApplicationContextAware 来访问 ApplicationContext 实例。以下是该例的配置(app-context-xml.xml)：

```
<beans ...>

    <bean id="publisher"
        class="com.apress.prospring5.ch4.Publisher"/>

    <bean id="messageEventListener"
        class="com.apress.prospring5.ch4.MessageEventListener"/>
</beans>
```

请注意，不需要特殊配置就可以使用 ApplicationContext 注册 MessageEventListener；它由 Spring 自动获取。运行此例将生成以下输出：

```
Received: I send an SOS to the world...
Received: ... I hope that someone gets my...
Received: ... Message in a bottle
```

使用事件时的注意事项

在许多情况下，在某个应用程序中，某些组件需要通知某些事件。通常为了做到这一点，需要通过编写代码来明确地通知每个组件，或者使用诸如 JMS 之类的消息传递技术。通过编写代码来依次通知每个组件的缺点是，将这些组件与发布者耦合起来，而在大多数情况下这种耦合是没有必要的。

考虑一种情况，假设在应用程序中缓存产品详细信息以避免访问数据库，而另一个组件允许修改产品详细信息并将其保存到数据库中。为避免使缓存无效，更新组件需要明确通知缓存详细信息已更改。在该例中，更新组件被耦合到一个与其业务无关的组件。此时，更好的解决方案是让更新组件在每次修改产品的详细信息时发布一个事件，然后让感兴趣的组件(比如缓存)监听该事件。这有利于保持组件解耦，并可以在需要时移除缓存或添加另一个有兴趣知道产品详细信息发生更改的监听器。

在这种情况下没有必要使用 JMS，因为在缓存中使产品条目失效的过程是很快的并且不是关键业务。Spring 事件基础结构的使用仅为应用程序增加很少的开销。

通常，使用事件作为快速执行的反应逻辑，而不是作为主要应用逻辑的一部分。在前面的示例中，缓存中产品的失效是由于对产品详细信息更新而产生的，该失效过程会很快(或者说应该很快)执行，而且不是应用程序主要功能的一部分。对于那些长时间运行且构成主要业务逻辑的一部分的流程，建议使用 JMS 或类似的消息传递系统，如 RabbitMQ。使用 JMS 的主要好处是它更适合长时间运行的流程，随着系统的不断发展，如有必要，可以将包含业务信息的消息的 JMS 驱动处理放在单独的机器上。

4.14 访问资源

通常，应用程序需要以不同形式访问各种资源。可能需要访问存储在文件系统中的一些配置数据、存储在类路径上的 JAR 文件中的某些图像数据，以及位于其他位置的服务器上的某些数据。Spring 以独立于协议的方式提供了访问资源的统一机制。这意味着应用程序可以以相同的方式访问文件资源，无论它们存储在文件系统中、类路径中还是远程服务器上。

Spring的资源支持的核心是org.springframework.core.io.Resource接口。Resource接口定义了十个自解释方法：contentLength()、exists()、getDescription()、getFile()、getFileName()、getURI()、getURL()、isOpen()、isReadable()和 lastModified()。除了这十个方法，还有一个方法：createRelative()。createRelative()方法通过使用与被调用的实例相关的路径来创建新的Resource实例。虽然该方法不在本章的讨论范围之内，但通过该方法可以提供自己的Resource实现。

在大多数情况下，使用内置实现之一来访问文件(FileSystemResource类)、类路径(ClassPathResource类)或URL资源(UrlResource类)。在内部，Spring使用另一个接口ResourceLoader和默认实现DefaultResourceLoader来定位和创建资源实例。但是，我们通常不会与DefaultResourceLoader进行交互，而是使用另一个名为ApplicationContext的ResourceLoader实现。以下是使用ApplicationContext访问三个资源的示例应用程序：

```
package com.apress.prospring5.ch4;

import java.io.File;
import org.springframework.context.ApplicationContext;
import org.springframework.context.support.ClassPathXmlApplicationContext;
import org.springframework.core.io.Resource;

public class ResourceDemo {
    public static void main(String... args) throws Exception{
        ApplicationContext ctx = new ClassPathXmlApplicationContext();

        File file = File.createTempFile("test", "txt");
        file.deleteOnExit();

        Resource res1 = ctx.getResource("file://" + file.getPath());
        displayInfo(res1);

        Resource res2 = ctx.getResource("classpath:test.txt");
        displayInfo(res2);

        Resource res3 = ctx.getResource("http://www.google.com");
        displayInfo(res3);

    }
    private static void displayInfo(Resource res) throws Exception{
        System.out.println(res.getClass());
        System.out.println(res.getURL().getContent());
        System.out.println("");
    }
}
```

请注意，在每次调用 getResource()时，都会为每个资源传递一个 URI。可以看到，针对 res1 和 res3 传入了常见的 file:以及 http:协议。此外，用于 res2 的 classpath:协议是特定于 Spring 的，指示 ResourceLoader 应该在类路径中查找资源。运行此例将生成以下输出：

```
class org.springframework.core.io.UrlResource
java.io.BufferedInputStream@3567135c

class org.springframework.core.io.ClassPathResource
sun.net.www.content.text.PlainTextInputStream@90f6bfd

class org.springframework.core.io.UrlResource
sun.net.www.protocol.http.HttpURLConnection$HttpInputStream@735b5592
```

注意，对于 file:和 http:协议，Spring 都会返回一个 UrlResource 实例。Spring 包含一个 FileSystemResource 类，但 DefaultResourceLoader 根本不使用这个类。这是因为 Spring 的默认资源加载策略将 URL 和文件视为具有不同协议(file:和 http:)的相同类型的资源。如果需要 FileSystemResource 的实例，请使用 FileSystemResourceLoader。一旦获得 Resource 实例，就可以使用 getFile()、getInputStream()或 getURL()来自由地访问内容。在某些情况下，比如当使用 http:协议时，对 getFile()的调用会导致 FileNotFoundException 异常。出于这个原因，建议使用 getInputStream()来访问资源内容，因为它可能适用于所有可能的资源类型。

4.15 使用 Java 类进行配置

除 XML 和属性文件配置外，还可以使用 Java 类来配置 Spring 的 ApplicationContext。前面演示的代码示例可能会让你对注解样式的配置感到非常满意。Spring JavaConfig 过去是一个单独的项目，但从 Spring 3.0 开始，其使用 Java 类进行配置的主要功能被合并到核心 Spring 框架中。在本节中，将演示如何使用 Java 类来配置 Spring 的 ApplicationContext 及其等价物(使用 XML 配置时)。

4.15.1 Java 中的 ApplicationContext 配置

接下来看看如何使用 Java 类来配置 Spring 的 ApplicationContext；此时将参考在第 2 章和第 3 章中介绍消息提供

程序和渲染器时使用的相同示例。以下代码回顾了一下消息提供程序接口和可配置消息提供程序：[1]

```
//chapter02/hello-world/src/main/java/com/apress/prospring5/
//    ch2/decoupled/MessageProvider.java
package com.apress.prospring5.ch2.decoupled;
public interface MessageProvider {
   String getMessage();
}

//chapter03/constructor-injection/src/main/java/com/apress/prospring5/
//    ch3/xml/ConfigurableMessageProvider.java
package com.apress.prospring5.ch3.xml;

import com.apress.prospring5.ch2.decoupled.MessageProvider;

public class ConfigurableMessageProvider implements MessageProvider {
   private String message = "Default message";

   public ConfigurableMessageProvider() {

   }

   public ConfigurableMessageProvider(String message) {
         this.message = message;
   }

   public void setMessage(String message) {
         this.message = message;
   }

   public String getMessage() {
         return message;
   }
}
```

以下代码片段显示了 MessageRenderer 接口和 StandardOutMessageRenderer 实现：

```
//chapter02/hello-world/src/main/java/com/apress/prospring5/
//    ch2/decoupled/MessageRenderer.java
package com.apress.prospring5.ch2.decoupled;

public interface MessageRenderer {
   void render();
   void setMessageProvider(MessageProvider provider);
   MessageProvider getMessageProvider();
}

//chapter02/hello-world/src/main/java/com/apress/prospring5/
//    ch2/decoupled/StandardOutMessageRenderer.java
package com.apress.prospring5.ch2.decoupled;

public class StandardOutMessageRenderer
   implements MessageRenderer {
private MessageProvider messageProvider;

public StandardOutMessageRenderer(){
   System.out.println(" -->
      StandardOutMessageRenderer: constructor called");
   }
   @Override
   public void render() {
      if (messageProvider == null) {
         throw new RuntimeException(
            "You must set the property messageProvider of class:"
               + StandardOutMessageRenderer.class.getName());
      } System.out.println(messageProvider.getMessage());
   }

   @Override
   public void setMessageProvider(MessageProvider provider) {
      System.out.println(" -->
         StandardOutMessageRenderer: setting the provider");
      this.messageProvider = provider;
   }

   @Override
```

1 这些类不用在 com.apress.prospring5.ch4 包中再次创建，但定义它们的项目被用作 java-config 项目的依赖项。

```
    public MessageProvider getMessageProvider() {
        return this.messageProvider;
    }
}
```

以下配置代码片段描述了 XML 配置(app-context-xml.xml)：

```
<beans ...>

    <bean id="messageRenderer"
      class="com.apress.prospring5.ch2.decoupled.StandardOutMessageRenderer"
      p:messageProvider-ref="messageProvider"/>

    <bean id="messageProvider"
      class="com.apress.prospring5.ch3.xml.ConfigurableMessageProvider"
      c:message="This is a configurable message"/>
</beans>
```

测试该类的类可能看起来非常熟悉，如下所示：

```
package com.apress.prospring5.ch4;

import com.apress.prospring5.ch2.decoupled.MessageRenderer;
import org.springframework.context.ApplicationContext;
import org.springframework.context.support.ClassPathXmlApplicationContext;

public class JavaConfigXMLExample {
    public static void main(String... args) {
            ApplicationContext ctx = new ClassPathXmlApplicationContext("
                classpath:spring/app-context-xml.xml");

    MessageRenderer renderer =
        ctx.getBean("messageRenderer", MessageRenderer.class);
    renderer.render();
    }
}
```

运行程序，将产生下面的输出：

```
--> StandardOutMessageRenderer: constructor called
--> StandardOutMessageRenderer: setting the provider
This is a configurable message
```

为了删除 XML 配置，必须使用一个被称为*配置类*的特殊类(被@Configuration 注解)来替换 app-context-xml.xml 文件。@Configuration 注解用来告知 Spring 这是一个基于 Java 的配置文件。该类将包含用@Bean 定义注解的方法，它们表示 bean 声明。@Bean 注解用来声明 Spring bean 和 DI 需求。@Bean 注解等同于<bean>标记，方法名称等同于<bean>标记内的 id 属性，并且在实例化 MessageRender bean 时，setter 注入是通过调用相应的方法来获得消息提供程序的，这与在 XML 配置中使用<ref>属性是相同的。虽然这些注解以及这种类型的配置在前面的章节中介绍过，你应该比较熟悉了，但之前并没有详细介绍它们。以下代码片段描述了与之前介绍的 XML 配置等效的 AppConfig 内容：

```
package com.apress.prospring5.ch4;

import com.apress.prospring5.ch2.decoupled.MessageProvider;
import com.apress.prospring5.ch2.decoupled.MessageRenderer;
import com.apress.prospring5.ch2.decoupled.StandardOutMessageRenderer;
import com.apress.prospring5.ch3.xml.ConfigurableMessageProvider;
import org.springframework.context.annotation.Bean;
import org.springframework.context.annotation.Configuration;

@Configuration
public class AppConfig {
    @Bean
    public MessageProvider messageProvider() {
        return new ConfigurableMessageProvider();
    }

    @Bean
    public MessageRenderer messageRenderer() {
        MessageRenderer renderer = new StandardOutMessageRenderer();
        renderer.setMessageProvider(messageProvider());

        return renderer;
    }
}
```

下面所示的代码片段演示了如何通过 Java 配置文件初始化 ApplicationContext 实例：

```
package com.apress.prospring5.ch4;

import com.apress.prospring5.ch2.decoupled.MessageRenderer;
import org.springframework.context.ApplicationContext;
import org.springframework.context.annotation.AnnotationConfigApplicationContext;

public class JavaConfigExampleOne {
   public static void main(String... args) {
      ApplicationContext ctx = new
         AnnotationConfigApplicationContext(AppConfig.class);

      MessageRenderer renderer =
         ctx.getBean("messageRenderer", MessageRenderer.class);

      renderer.render();
   }
}
```

在上面所示的代码清单中，使用了 AnnotationConfigApplicationContext 类，并传入配置类作为构造函数参数(可以通过 JDK varargs 功能将多个配置类传递给构造函数)。随后，可以像往常一样使用返回的 ApplicationContext。有时出于测试目的，可以将配置类声明为静态内部类，如下所示：

```
package com.apress.prospring5.ch4;

import com.apress.prospring5.ch2.decoupled.MessageProvider;
import com.apress.prospring5.ch2.decoupled.MessageRenderer;
import com.apress.prospring5.ch2.decoupled.StandardOutMessageRenderer;
import com.apress.prospring5.ch3.xml.ConfigurableMessageProvider;
import org.springframework.context.ApplicationContext;
import org.springframework.context.annotation.AnnotationConfigApplicationContext;
import org.springframework.context.annotation.Bean;
import org.springframework.context.annotation.Configuration;

public class JavaConfigSimpleExample {

   @Configuration
   static class AppConfigOne {
      @Bean
      public MessageProvider messageProvider() {
         return new ConfigurableMessageProvider();
   }

   @Bean
   public MessageRenderer messageRenderer() {
      MessageRenderer renderer = new StandardOutMessageRenderer();
      renderer.setMessageProvider(messageProvider());
         return renderer;
      }
   }

   public static void main(String... args) {
      ApplicationContext ctx = new
         AnnotationConfigApplicationContext(AppConfig.class);

      MessageRenderer renderer =
         ctx.getBean("messageRenderer", MessageRenderer.class);

      renderer.render();
   }
}
```

返回的 ApplicationContext 实例可以像往常一样使用，并且输出与使用 XML 配置的输出是相同的。

```
  --> StandardOutMessageRenderer: constructor called
  --> StandardOutMessageRenderer: setting the provider
Default message
```

在了解 Java 配置类的基本用法后，让我们继续学习更多配置选项。对于消息提供程序，假设想要将消息外部化为一个属性文件(message.properties)，然后使用构造函数注入将其注入 ConfigurableMessageProvider 中。message.properties 的内容如下所示：

```
message=Only hope can keep me together
```

接下来看看修改后的测试程序，它使用@PropertySource 注解加载属性文件，然后将它们注入消息提供程序的实现中。

```
package com.apress.prospring5.ch4;

import com.apress.prospring5.ch2.decoupled.MessageProvider;
import com.apress.prospring5.ch2.decoupled.MessageRenderer;
import com.apress.prospring5.ch2.decoupled.StandardOutMessageRenderer;
import org.springframework.beans.factory.annotation.Autowired;
import org.springframework.context.annotation.Bean;
import org.springframework.context.annotation.Configuration;
import org.springframework.context.annotation.PropertySource;
import org.springframework.core.env.Environment;

@Configuration
@PropertySource(value = "classpath:message.properties")
public class AppConfigOne {

    @Autowired
    Environment env;
    @Bean
    public MessageProvider messageProvider() {
          return new ConfigurableMessageProvider(env.getProperty("message"));
    }

    @Bean(name = "messageRenderer")
    public MessageRenderer messageRenderer() {
       MessageRenderer renderer = new StandardOutMessageRenderer();
       renderer.setMessageProvider(messageProvider());
       return renderer;
    }
}
```

第 3 章介绍了配置类，并显示了与 XML 元素和属性等价的配置类。可以使用与 bean 作用域、加载类型和依赖项相关的其他注解来注解 bean 声明。在下面的代码片段中，使用了 bean 声明的注解充实了 AppConfigOne 配置类：

```
package com.apress.prospring5.ch4;

import com.apress.prospring5.ch2.decoupled.MessageProvider;
import com.apress.prospring5.ch2.decoupled.MessageRenderer;
import com.apress.prospring5.ch2.decoupled.StandardOutMessageRenderer;
import org.springframework.beans.factory.annotation.Autowired;
import org.springframework.context.annotation.*;
import org.springframework.core.env.Environment;

@Configuration
@PropertySource(value = "classpath:message.properties")
public class AppConfig {

    @Autowired
    Environment env;

    @Bean
    @Lazy
    public MessageProvider messageProvider() {
       return new ConfigurableMessageProvider(env.getProperty("message"));
    }

    @Bean(name = "messageRenderer")
    @Scope(value="prototype")
    @DependsOn(value="messageProvider")
    public MessageRenderer messageRenderer() {
       MessageRenderer renderer = new StandardOutMessageRenderer();
       renderer.setMessageProvider(messageProvider());
       return renderer;
    }
}
```

在上面的代码示例中，引入了一些注解，对这些注解在表 4-3 中进行了解释。使用构造型注解(比如@Component、@Service 等)定义的 bean 可以在 Java 配置类中使用，方法是启用组件扫描并在需要时自动装配它们。在下面的实例中，将 ConfigurableMessageProvider 声明为一个服务 bean。

表 4-3 Java 配置注解表

注解	描述
@PropertySource	该注解用于将属性文件加载到 Spring 的 ApplicationContext 中，它将接收位置作为参数(可以提供多个位置)。而在 XML 中，可以使用<context:property-placeholder>完成相同的功能

(续表)

注解	描述
@Lazy	该注解指示 Spring 仅在请求时才实例化 bean(与 XML 中的 lazy-init ="true"相同)。此注解具有默认值为 true 的 value 属性；因此，使用@Lazy(value = truc)相当于使用@Lazy
@Scope	当期望的作用域不是 singleton 时，可以使用该注解定义 bean 作用域
@DependsOn	该注解告诉 Spring 某个 bean 依赖于其他 bean，所以 Spring 将确保这些 bean 首先被实例化
@Autowired	该注解用于 Environment 类型的 env 变量。这是 Spring 提供的 Environment 抽象功能。本章后面将会详细讨论该功能

```
package com.apress.prospring5.ch4.annotated;

import com.apress.prospring5.ch2.decoupled.MessageProvider;
import org.springframework.beans.factory.annotation.Autowired;
import org.springframework.beans.factory.annotation.Value;
import org.springframework.stereotype.Service;

@Service("provider")
public class ConfigurableMessageProvider implements MessageProvider {

   private String message;

   public ConfigurableMessageProvider(
         @Value("Love on the weekend")String message) {
      this.message = message;
   }

   @Override
   public String getMessage() {
      return this.message;
   }
}
```

在这里可以看到配置类：

```
package com.apress.prospring5.ch4;

import com.apress.prospring5.ch2.decoupled.MessageProvider;
import com.apress.prospring5.ch2.decoupled.MessageRenderer;
import com.apress.prospring5.ch2.decoupled.StandardOutMessageRenderer;
import org.springframework.beans.factory.annotation.Autowired;
import org.springframework.context.annotation.*;

@Configuration
@ComponentScan(basePackages={"com.apress.prospring5.ch4.annotated"})
public class AppConfigTwo {

   @Autowired
   MessageProvider provider;

   @Bean(name = "messageRenderer")
   public MessageRenderer messageRenderer() {
      MessageRenderer renderer =
         new StandardOutMessageRenderer();
      renderer.setMessageProvider(provider);
      return renderer;
   }
}
```

@ComponentScan 告诉 Spring 应该扫描那些带有 bean 定义注解的包。它与 XML 配置中的<context:component-scan>标签相同。执行以下示例中的代码：

```
package com.apress.prospring5.ch4;

import com.apress.prospring5.ch2.decoupled.MessageRenderer;
import org.springframework.context.ApplicationContext;
import org.springframework.context.annotation.AnnotationConfigApplicationContext;

public class JavaConfigExampleTwo {
   public static void main(String... args) {
      ApplicationContext ctx = new
         AnnotationConfigApplicationContext(AppConfigTwo.class);
```

```
        MessageRenderer renderer =
            ctx.getBean("messageRenderer", MessageRenderer.class);

        renderer.render();
    }
}
```

将会得到以下输出结果：

```
 --> StandardOutMessageRenderer: constructor called
 --> StandardOutMessageRenderer: setting the provider
Love on the weekend
```

一个应用程序也可以有多个配置类，可以根据目的分离配置和组织 bean(例如，一个类可以专用于 DAO bean 声明，另一个类用于 Service bean 声明，等等)。接下来使用另一个名为 AppConfigFour 的配置类来定义 provider bean。通过导入该类所定义的 bean，可以从另一个配置类访问该 bean。这是通过使用@Import 注解目标配置类 AppConfigThree 来实现的。

```
//AppConfigFour.java
package com.apress.prospring5.ch4.multiple;

import org.springframework.context.annotation.ComponentScan;
import org.springframework.context.annotation.Configuration;

@Configuration
@ComponentScan(basePackages={"com.apress.prospring5.ch4.annotated"})
public class AppConfigFour { }

package com.apress.prospring5.ch4.multiple;

import com.apress.prospring5.ch2.decoupled.MessageProvider;
import com.apress.prospring5.ch2.decoupled.MessageRenderer;
import com.apress.prospring5.ch2.decoupled.StandardOutMessageRenderer;
import org.springframework.beans.factory.annotation.Autowired;
import org.springframework.context.annotation.Bean;
import org.springframework.context.annotation.Configuration;
import org.springframework.context.annotation.Import;

@Configuration
@Import(AppConfigFour.class)
public class AppConfigThree {
    @Autowired
    MessageProvider provider;

    @Bean(name = "messageRenderer")
    public MessageRenderer messageRenderer() {
        MessageRenderer renderer = new StandardOutMessageRenderer();
        renderer.setMessageProvider(provider);
        return renderer;
    }
}
```

如果在类 JavaConfigExampleTwo 的 main()方法中将 AppConfigTwo 类替换为 AppConfigThree，那么当运行该例时，会打印相同的输出。

4.15.2 Spring 混合配置

Spring 可以做的远不止这些。Spring 允许混合使用 XML 和 Java 配置类。当应用程序因为某种原因无法更改遗留代码时，这种混合配置是非常有用的。如果想从 XML 文件导入 bean 声明，可以使用@ImportResource 注解。在以下配置代码片段中，可以看到在名为 app-context-xml-01.xml 的 XML 文件中声明的 provider bean：

```
<beans ...>

    <bean id="provider"
          class="com.apress.prospring5.ch4.ConfigurableMessageProvider"
        p:message="Love on the weekend" />

</beans>
```

下面的代码示例描述了类 AppConfigFive，其中导入了 XML 文件中声明的 bean。如果在 classJavaConfigExampleTwo 的 main()方法中用 AppConfigFive 替换 AppConfigTwo 类，那么运行该例时将打印相同的输出。

```
package com.apress.prospring5.ch4.mixed;

import com.apress.prospring5.ch2.decoupled.MessageProvider;
import com.apress.prospring5.ch2.decoupled.MessageRenderer;
import com.apress.prospring5.ch2.decoupled.StandardOutMessageRenderer;
import org.springframework.beans.factory.annotation.Autowired;
import org.springframework.context.annotation.Bean;
import org.springframework.context.annotation.Configuration;
import org.springframework.context.annotation.ImportResource;

@Configuration
@ImportResource(value="classpath:spring/app-context-xml-01.xml")
public class AppConfigFive {
   @Autowired
   MessageProvider provider;

   @Bean(name = "messageRenderer")
   public MessageRenderer messageRenderer() {
      MessageRenderer renderer = new StandardOutMessageRenderer();
      renderer.setMessageProvider(provider);
      return renderer;
   }
}
```

另外，也可以执行相反的操作：在 Java 配置类中定义的 bean 可以导入 XML 配置文件。在下面的示例中，messageRenderer bean 在 XML 文件中定义，其依赖项(即 provider bean)在配置类 AppConfigSix 中定义。XML 配置文件 app-context-xml-02.xml 的内容如下所示：

```
<beans ...>

   <context:annotation-config/>

   <bean class="com.apress.prospring5.ch4.mixed.AppConfigSix"/>
   <bean id="messageRenderer"
         class="com.apress.prospring5.ch2.decoupled.StandardOutMessageRenderer"
      p:messageProvider-ref="provider"/>

</beans>
```

必须声明配置类类型的 bean，并且必须使用<context:annotation-config/>启用对注解方法的支持。这使得在类中声明的 bean 可以被配置为在 XML 文件中声明的 bean 的依赖项。配置类 AppConfigSix 非常简单。

```
package com.apress.prospring5.ch4.mixed;

import com.apress.prospring5.ch2.decoupled.MessageProvider;
import com.apress.prospring5.ch4.annotated.ConfigurableMessageProvider;
import org.springframework.context.annotation.Bean;
import org.springframework.context.annotation.Configuration;

@Configuration
public class AppConfigSix {

   @Bean
   public MessageProvider provider() {
      return new ConfigurableMessageProvider("Love on the weekend");
   }
}
```

使用 ClassPathXmlApplicationContext 完成 ApplicationContext 实例的创建，到现在为止这已经多次使用了。

```
package com.apress.prospring5.ch4;

import com.apress.prospring5.ch2.decoupled.MessageRenderer;
import com.apress.prospring5.ch4.mixed.AppConfigFive;
import org.springframework.context.ApplicationContext;
import org.springframework.context.annotation.AnnotationConfigApplicationContext;
import org.springframework.context.support.ClassPathXmlApplicationContext;

public class JavaConfigExampleThree {
   public static void main(String... args) {
      ApplicationContext ctx =
            new ClassPathXmlApplicationContext
             ("classpath:spring/app-context-xml-02.xml");

      MessageRenderer renderer =
```

```
            ctx.getBean("messageRenderer", MessageRenderer.class);
        renderer.render();
    }
}
```

运行前面的代码将产生与之前描述相同的输出。

应用程序基础结构服务也可以在 Java 配置类中定义。例如，@EnableTransactionManagement 定义了所使用的 Spring 的事务管理功能，具体将在第 9 章中进一步讨论，@EnableWebSecurity 和@EnableGlobalMethodSecurity 用于启用 Spring 安全性上下文，它们将在第 16 章中讨论。

4.15.3 Java 或 XML 配置?

正如你所看到的，使用 Java 类可以实现与 XML 同级的 ApplicationContext 配置。那么，应该使用哪一种方法呢？这个问题与到底是使用 XML 还是 Java 注解进行 DI 配置非常相似。每种方法都有各自的优缺点。但是，答案都是一样的；也就是说，当决定使用其中一种方法时，就应该坚持使用并保持配置样式一致，而不是一会儿用 Java 类，一会儿用 XML 文件。使用一种方法会让维护工作更容易。

4.16 配置文件

Spring 提供的另一个有趣功能是配置文件(configuration profile)的概念。一般来说，配置文件指示 Spring 仅配置在指定配置文件处于活动状态时定义的 ApplicationContext 实例。在本节中，将演示如何在简单程序中使用配置文件。

使用 Spring 配置文件功能的示例

假设有一个名为 FoodProviderService 的服务，负责向学校(包括幼儿园和高中)提供食物。FoodProviderService 接口只有一个供学校呼叫使用的名为 provideLunchSet()的方法，它负责为每个学生提供午餐套餐。午餐套餐是 Food 对象的列表，它是一个只有 name 属性的简单类。以下代码片段显示了 Food 类：

```
package com.apress.prospring5.ch4;

public class Food {
    private String name;

    public Food() {
    }

    public Food(String name) {
        this.name = name;
    }

    public String getName() {
        return name;
    }

    public void setName(String name) {
        this.name = name;
    }
}
```

下面所示的是 FoodProviderService 接口：

```
package com.apress.prospring5.ch4;

import java.util.List;

public interface FoodProviderService {
    List<Food> provideLunchSet();
}
```

现在假设午餐套餐有两个供应商，一个供幼儿园使用，另一个供高中使用。由他们制作的午餐套餐是不同的，尽管他们提供的服务是相同的，都是为学生提供午餐。因此，现在创建 FoodProviderService 的两个实现，并使用相同的名称，但将它们放入不同的包以确定目标学校。这两个类如下所示：

```
//chapter04/profiles/src/main/java/com/apress/prospring5/ch4/highschool/FoodProviderServiceImpl.java
```

```
package com.apress.prospring5.ch4.highschool;

import java.util.ArrayList;
import java.util.List;

import com.apress.prospring5.ch4.Food;
import com.apress.prospring5.ch4.FoodProviderService;

public class FoodProviderServiceImpl implements FoodProviderService {
    @Override
    public List<Food> provideLunchSet() {
        List<Food> lunchSet = new ArrayList<>();
        lunchSet.add(new Food("Coke"));
        lunchSet.add(new Food("Hamburger"));
        lunchSet.add(new Food("French Fries"));

        return lunchSet;
    }
}

//chapter04/profiles/src/main/java/com/apress/prospring5/ch4/kindergarten/FoodProviderServiceImpl.java
package com.apress.prospring5.ch4.kindergarten;

import java.util.ArrayList;
import java.util.List;

import com.apress.prospring5.ch4.Food;
import com.apress.prospring5.ch4.FoodProviderService;

public class FoodProviderServiceImpl implements FoodProviderService {
    @Override
    public List<Food> provideLunchSet() {
        List<Food> lunchSet = new ArrayList<>();
        lunchSet.add(new Food("Milk"));
        lunchSet.add(new Food("Biscuits"));

        return lunchSet;
    }
}
```

从上面的代码可以看到，这两个实现提供相同的FoodProviderService接口，但在午餐套餐中生成不同的食物组合。所以，现在假设一家幼儿园希望供应商为他们的学生提供午餐套餐。接下来看看如何使用Spring的配置文件来实现该功能。首先运行XML配置。创建两个XML配置文件，一个用作幼儿园配置文件，另一个用作高中配置文件。以下配置代码片段描述了这两个配置文件：

```
<!-- highschool-config.xml -->
<beans xmlns="http://www.springframework.org/schema/beans"
    xmlns:xsi="http://www.w3.org/2001/XMLSchema-instance"
    xsi:schemaLocation="http://www.springframework.org/schema/beans
        http://www.springframework.org/schema/beans/spring-beans.xsd"
        profile="highschool">

    <bean id="foodProviderService"
      class="com.apress.prospring5.ch4.highschool.FoodProviderServiceImpl"/>
</beans>

<!-- kindergarten-config.xml -->
<beans xmlns="http://www.springframework.org/schema/beans"
    xmlns:xsi="http://www.w3.org/2001/XMLSchema-instance"
    xsi:schemaLocation="http://www.springframework.org/schema/beans
        http://www.springframework.org/schema/beans/spring-beans.xsd"
        profile="kindergarten">

    <bean id="foodProviderService"
      class="com.apress.prospring5.ch4.kindergarten.FoodProviderServiceImpl"/>
</beans>
```

在这两个配置中，请注意<beans>标记中分别使用了profile ="kindergarten"和profile ="highschool"。实际上，它告诉Spring，只有当指定的配置文件处于活动状态时，文件中的这些bean才应该被实例化。现在看看当在独立的应用程序中使用Spring的ApplicationContext时如何激活正确的配置文件。以下代码片段显示了测试程序：

```
package com.apress.prospring5.ch4;

import java.util.List;
import org.springframework.context.support.GenericXmlApplicationContext;
```

```
public class ProfileXmlConfigExample {
    public static void main(String... args) {
        GenericXmlApplicationContext ctx =
              new GenericXmlApplicationContext();
        ctx.load("classpath:spring/*-config.xml");
        ctx.refresh();

        FoodProviderService foodProviderService =
            ctx.getBean("foodProviderService", FoodProviderService.class);

        List<Food> lunchSet = foodProviderService.provideLunchSet();

        for (Food food: lunchSet) {
            System.out.println("Food: " + food.getName());
        }

        ctx.close();
    }
}
```

ctx.load()方法将加载 kindergarten-config.xml 和 highschool-config.xml，因为此时将通配符作为前缀传递给了该方法。在该例中，根据 profile 属性值，只有文件 kindergartenconfig.xml 中的 bean 被 Spring 实例化，而该文件通过传递 JVM 参数-Dspring.profiles.active ="kindergarten"来激活。使用此 JVM 参数运行程序会生成以下输出：

```
Food: Milk
Food: Biscuits
```

以上输出就是幼儿园供应商提供的午餐套餐。如果将配置文件参数更改为高中(-Dspring.profiles.active = "highschool")，输出将变为以下内容：

```
Food: Coke
Food: Hamburger
Food: French Fries
```

也可以通过调用 ctx.getEnvironment().setActiveProfiles("kindergarten")，以编程的方式在代码中设置所使用的配置文件。此外，只需要将@Profile 注解添加到类中，即可使用 Java Config 通过配置文件注册类。

4.17 使用 Java 配置来配置 Spring 配置文件

当然，还有一种使用 Java 配置来配置 Spring 配置文件的方法，对于那些不喜欢 XML 配置的开发人员来说，这种方法应该会比较喜欢。上一节声明的 XML 文件必须用等价的 Java 配置类替换。下面所示的是 kindergarten 配置文件的配置类：

```
package com.apress.prospring5.ch4.config;

import com.apress.prospring5.ch4.FoodProviderService;
import com.apress.prospring5.ch4.kindergarten.FoodProviderServiceImpl;
import org.springframework.context.annotation.Bean;
import org.springframework.context.annotation.Configuration;
import org.springframework.context.annotation.Profile;

@Configuration
@Profile("kindergarten")
public class KindergartenConfig {

    @Bean
    public FoodProviderService foodProviderService(){
        return new FoodProviderServiceImpl();
    }
}
```

如你所见，使用@Bean 注解来定义 foodProviderService bean。该类使用@Profile 注解标记为特定于 kindergarten 配置文件。除 bean 类型、配置文件名称和类名外，特定于 highschool 配置文件的类是相同的。

```
package com.apress.prospring5.ch4.config;

import com.apress.prospring5.ch4.FoodProviderService;
import com.apress.prospring5.ch4.highschool.FoodProviderServiceImpl;
import org.springframework.context.annotation.Bean;
import org.springframework.context.annotation.Configuration;
import org.springframework.context.annotation.Profile;
```

```
@Configuration
@Profile("highschool")
public class HighschoolConfig {

    @Bean
    public FoodProviderService foodProviderService(){
        return new FoodProviderServiceImpl();
    }
}
```

这些类的使用方式与 XML 文件相同。声明一个上下文以使用它们，并且根据-Dspring.profiles.active JVM 选项的值，选择其中一个类来创建 ApplicationContext 实例。

```
package com.apress.prospring5.ch4.config;

import com.apress.prospring5.ch4.Food;
import com.apress.prospring5.ch4.FoodProviderService;
import org.springframework.context.annotation.AnnotationConfigApplicationContext;
import org.springframework.context.support.GenericApplicationContext;

import java.util.List;

public class ProfileJavaConfigExample {

    public static void main(String... args) {
        GenericApplicationContext ctx =
            new AnnotationConfigApplicationContext(
                    KindergartenConfig.class,
                    HighschoolConfig.class);

        FoodProviderService foodProviderService =
            ctx.getBean("foodProviderService",
            FoodProviderService.class);

        List<Food> lunchSet = foodProviderService.provideLunchSet();
        for (Food food : lunchSet) {
            System.out.println("Food: " + food.getName());
        }
        ctx.close();
    }
}
```

如果使用 kindergarten 作为-Dspring.profiles.active JVM 选项的值，那么运行示例将打印预期的输出。

```
Food: Milk
Food: Biscuits
```

还有一个注解用于配置替换-Dspring.profiles.active JVM 选项的已用配置文件，但只能用于测试类。由于第 13 章将会详细介绍如何测试 Spring 应用程序，因此这里不再详细介绍，只给出一些示例代码。

```
package com.apress.prospring5.ch4.config;

import com.apress.prospring5.ch4.FoodProviderService;
import org.junit.Test;
import org.junit.runner.RunWith;
import org.springframework.beans.factory.annotation.Autowired;
import org.springframework.test.context.ActiveProfiles;
import org.springframework.test.context.ContextConfiguration;
import org.springframework.test.context.junit4.SpringJUnit4ClassRunner;

import static org.junit.Assert.assertEquals;
import static org.junit.Assert.assertFalse;
import static org.junit.Assert.assertTrue;

@RunWith(SpringJUnit4ClassRunner.class)
@ContextConfiguration(classes={KindergartenConfig.class,
    HighschoolConfig.class})
@ActiveProfiles("kindergarten")
public class ProfilesJavaConfigTest {

    @Autowired FoodProviderService foodProviderService;

    @Test
    public void testProvider(){
        assertTrue(foodProviderService.provideLunchSet() != null);
        assertFalse(foodProviderService.provideLunchSet().isEmpty());
```

```
        assertEquals(2, foodProviderService.provideLunchSet().size());
    }
}
```

你可能已经明白了，指定用于运行此测试的配置文件的注解是@ActiveProfiles("kindergarten")。在复杂的应用程序中，通常有多个配置文件，其中更多配置文件可用于组成测试的上下文配置。该类可以在任何 Java 智能编辑器中运行，并在执行 gradle clean build 时自动运行。

使用配置文件时的注意事项

Spring 中的配置文件功能为开发人员管理应用程序的运行配置创造了另一种方式，通常运行配置在构建工具中完成(例如，Maven 的配置文件支持)。构建工具首先根据传入的参数将正确的配置/属性文件打包到 Java 归档文件(JAR 或 WAR 文件，取决于应用程序的类型中)，然后部署到目标环境。Spring 的配置文件功能可以让应用程序开发人员自行定义配置文件，并以编程方式或通过传入的 JVM 参数来激活它们。通过使用 Spring 的配置文件支持功能，可以使用相同的应用程序存档文件并部署到所有环境中(在 JVM 启动过程中将正确的配置文件作为参数传入)。例如，可以使用具有不同配置文件的应用程序，比如(dev,hibernate)、(prd,jdbc)等，每种组合代表运行环境(开发或生产)以及数据访问库(Hibernate 或 JDBC)。这将应用程序配置文件管理引入到了编程端。

但是这种方法也有缺点。例如，有些人可能会争辩说，如果不谨慎处理，将不同环境的所有配置放入应用程序配置文件或 Java 类中并捆绑在一起很容易出错(例如，管理员可能会忘记在应用程序中设置正确的 JVM 参数服务器环境)。将所有配置文件打包在一起也会使包比平时大一些。再次强调一下，应该根据应用程序和配置要求选择最适合自己项目的方法。

4.18 Environment 和 PropertySource 抽象

想要设置活动配置文件，需要访问 Environment 接口。Environment 接口是一个抽象层，用于封装正在运行的 Spring 应用程序的环境。

除配置文件外，Environment 接口封装的其他关键信息都是属性。属性用来存储应用程序的底层环境配置，例如应用程序文件夹的位置、数据库连接信息等。

Spring 中的 Environment 和 PropertySource 抽象功能帮助开发人员访问来自运行平台的各种配置信息。在抽象环境中，所有系统属性、环境变量和应用程序属性都由 Environment 接口提供，Spring 启动 ApplicationContext 时将填充该接口。以下代码片段显示了一个简单示例：

```
package com.apress.prospring5.ch4;

import java.util.HashMap;
import java.util.Map;

import org.springframework.context.support.GenericXmlApplicationContext;
import org.springframework.core.env.ConfigurableEnvironment;
import org.springframework.core.env.MapPropertySource;
import org.springframework.core.env.MutablePropertySources;

public class EnvironmentSample {
    public static void main(String... args) {
        GenericXmlApplicationContext ctx =
            new GenericXmlApplicationContext();
        ctx.refresh();

        ConfigurableEnvironment env = ctx.getEnvironment();
        MutablePropertySources propertySources = env.getPropertySources();

        Map<String,Object> appMap = new HashMap<>();
        appMap.put("user.home", "application_home");

        propertySources.addLast(new MapPropertySource("prospring5_MAP", appMap));

        System.out.println("user.home: " + System.getProperty("user.home"));
        System.out.println("JAVA_HOME: " + System.getenv("JAVA_HOME"));

        System.out.println("user.home: " + env.getProperty("user.home"));
        System.out.println("JAVA_HOME: " + env.getProperty("JAVA_HOME"));
```

```
        ctx.close();
    }
}
```

在上面的代码片段中，在 ApplicationContext 初始化之后，获得对 ConfigurableEnvironment 接口的引用。通过该接口，可以获得 MutablePropertySource 的句柄(PropertySource 接口的默认实现，它可以对所包含的属性源进行操作)。之后，构建一个映射，并将应用程序属性放入映射中，然后使用该映射构造 MapPropertySource 类(从 Map 实例读取键和值的 PropertySource 子类)。最后，通过 addLast()方法将 MapPropertySource 类添加到 MutablePropertySources 中。运行该程序，将打印以下内容：

```
user.home: /home/jules
JAVA_HOME: /home/jules/bin/java
user.home: /home/jules
JAVA_HOME: /home/jules/bin/java
application.home: application_home
```

对于前两行，与通过使用 JVM 的 System 类一样，JVM 系统属性 user.home 和环境变量 JAVA_HOME 被检索到。但是，对于最后三行，可以看到所有的系统属性、环境变量以及应用程序属性都可以通过 Environment 接口访问。通过该例可以了解 Environment 抽象是如何帮助开发人员管理和访问应用程序运行环境中的所有各种属性的。

而对于 PropertySource 抽象，Spring 将按照以下默认顺序访问属性：

- 运行 JVM 的系统属性
- 环境变量
- 应用程序定义的属性

例如，假设定义了相同的应用程序属性 user.home，并通过 MutablePropertySources 类将其添加到 Environment 接口。如果运行该程序，仍然会看到 user.home 是从 JVM 属性中检索的，而不是你所创建的 user.home。但是，Spring 允许对 Environment 检索属性的顺序进行控制。以下代码片段显示了修改后的版本：

```
package com.apress.prospring5.ch4;

import java.util.HashMap;
import java.util.Map;

import org.springframework.context.support.GenericXmlApplicationContext;
import org.springframework.core.env.ConfigurableEnvironment;
import org.springframework.core.env.MapPropertySource;
import org.springframework.core.env.MutablePropertySources;

public class EnvironmentSample {
    public static void main(String... args) {
        GenericXmlApplicationContext ctx =
            new GenericXmlApplicationContext();
        ctx.refresh();

        ConfigurableEnvironment env = ctx.getEnvironment();
        MutablePropertySources propertySources = env.getPropertySources();

        Map<String,Object> appMap = new HashMap<>();
        appMap.put("application.home", "application_home");

        propertySources.addFirst(new MapPropertySource("prospring5_MAP", appMap));

        System.out.println("user.home: " + System.getProperty("user.home"));
        System.out.println("JAVA_HOME: " + System.getenv("JAVA_HOME"));

        System.out.println("user.home: " + env.getProperty("user.home"));
        System.out.println("JAVA_HOME: " + env.getProperty("JAVA_HOME"));

        ctx.close();
    }
}
```

在前面的代码示例中，定义了一个名为 user.home 的应用程序属性，并将其添加为通过 MutablePropertySources 类的 addFirst()方法进行搜索的第一个属性。运行该程序时，将看到以下输出：

```
user.home: /home/jules
JAVA_HOME: /home/jules/bin/java
user.home: application_home
JAVA_HOME: /home/jules/bin/java
```

前两行保持不变，因为仍然使用 JVM System 类的 getProperty()和 getenv()方法来检索它们。但是，当使用 Environment 接口时，会看到前面定义的 user.home 属性被优先检索，因为我们将其定义为搜索的第一个属性值。

在现实生活中，很少需要直接与 Environment 接口进行交互，但会以${}的形式使用一个属性占位符(例如${application.home})，并将解析后的值注入 Spring bean 中。接下来看一下该过程是如何完成的。假设有一个类用来存储从属性文件加载的所有应用程序属性。以下所示的是 AppProperty 类：

```
package com.apress.prospring5.ch4;

public class AppProperty {
   private String applicationHome;
   private String userHome;

   public String getApplicationHome() {
      return applicationHome;
   }

   public void setApplicationHome(String applicationHome) {
      this.applicationHome = applicationHome;
   }

   public String getUserHome() {
      return userHome;
   }

   public void setUserHome(String userHome) {
      this.userHome = userHome;
   }
}
```

以下所示的是 application.properties 文件的内容：

```
application.home=application_home
user.home=/home/jules-new
```

请注意，该属性文件还声明了 user.home 属性。接下来看一下 Spring XML 配置，请参阅以下代码(app-context-xml.xml)：

```
<beans xmlns="http://www.springframework.org/schema/beans"
       xmlns:xsi="http://www.w3.org/2001/XMLSchema-instance"
       xmlns:context="http://www.springframework.org/schema/context"
       xmlns:p="http://www.springframework.org/schema/p"
       xsi:schemaLocation="http://www.springframework.org/schema/beans
        http://www.springframework.org/schema/beans/spring-beans.xsd
        http://www.springframework.org/schema/context
        http://www.springframework.org/schema/context/spring-context.xsd">

    <context:property-placeholder
       location="classpath:application.properties"/>

    <bean id="appProperty" class="com.apress.prospring5.ch4.AppProperty"
       p:applicationHome="${application.home}"
       p:userHome="${user.home}"/>
</beans>
```

使用<context:property-placeholder>标记将属性加载到 Spring 的 Environment 中，该 Environment 被封装到 ApplicationContext 接口中。此外，还使用 SpEL 占位符将值注入 AppProperty bean 中。以下代码片段显示了测试程序：

```
package com.apress.prospring5.ch4;

import org.springframework.context.support.GenericXmlApplicationContext;
public class PlaceHolderDemo {
   public static void main(String... args) {
      GenericXmlApplicationContext ctx =
         new GenericXmlApplicationContext();
      ctx.load("classpath:spring/app-context-xml.xml");
      ctx.refresh();

      AppProperty appProperty = ctx.getBean("appProperty",
         AppProperty.class);
      System.out.println("application.home: " +
         appProperty.getApplicationHome());
      System.out.println("user.home: " +
         appProperty.getUserHome());

      ctx.close();
   }
}
```

运行该程序，可以看到以下输出：

```
application.home: application_home
user.home: /Users/jules
```

可以看到 application.home 占位符已被正确解析，而 user.home 属性仍从 JVM 属性中检索，这是正确的，因为它是 PropertySource 抽象的默认行为。如果想要指示 Spring 优先考虑 application.properties 文件中的值，可以将属性 local-override ="true"添加到<context:property-placeholder>标记。

```
<context:property-placeholder local-override="true"
   location="classpath:env/application.properties"/>
```

local-override 属性指示 Spring 使用此占位符中定义的属性覆盖现有属性。运行该程序，将看到 application.properties 文件中的 user.home 属性被检索到。

```
application.home: application_home
user.home: /home/jules-new
```

4.19 使用 JSR-330 注解进行配置

正如第 1 章中讨论的那样，JEE 6 提供了对 JSR-330(Java 的依赖注入)的支持，它是一个注解集合，用来在 JEE 容器或其他兼容的 IoC 框架中表示应用程序的 DI 配置。Spring 也支持并识别这些注解，因此尽管可能没有在 JEE 6 容器中运行应用程序，但仍然可以在 Spring 中使用 JSR-330 注解。使用 JSR-330 注解可帮助我们轻松地从 Spring 迁移到 JEE 6 容器或其他兼容的 IoC 容器(例如，Google Guice)。

接下来，以消息渲染器和消息提供者为例，并使用 JSR-330 注解来实现。要支持 JSR-330 注解，需要将 javax.inject3 依赖项添加到项目中。

以下代码片段显示了 MessageProvider 和 ConfigurableMessageProvider 实现：

```
//chapter04/jsr330/src/main/java/com/apress/prospring5/ch4/MessageProvider.java
package com.apress.prospring5.ch4;

public interface MessageProvider {
   String getMessage();
}
//chapter04/jsr330/src/main/java/com/apress/prospring5/
   ch4/ConfigurableMessageProvider.java
package com.apress.prospring5.ch4;

import javax.inject.Inject;
import javax.inject.Named;

@Named("messageProvider")
public class ConfigurableMessageProvider
   implements MessageProvider {
   private String message = "Default message";

   public ConfigurableMessageProvider() {
   }

   @Inject
   @Named("message")
   public ConfigurableMessageProvider(String message) {
      this.message = message;
   }

   public void setMessage(String message) {
      this.message = message;
   }

   public String getMessage() {
      return message;
   }
}
```

可以看到，所有注解都属于 javax.inject 包，它是 JSR-330 标准。该类在两个地方使用了@Named。首先，@Named 被用来声明一个可注入的 bean(与 Spring 中的@Component 注解或其他构造型注解相同)。在上面的代码清单中，@Named("messageProvider") 注解指定 ConfigurableMessageProvider 是一个注入 bean，并为其提供名称 messageProvider，它与 Spring 的<bean>标记中的 name 属性相同。其次，通过在接收字符串值的构造函数之前使用@Inject 注解来使用构造函数注入。最后，使用@Named 来指定要注入名为 message 的值。下面再看看 MessageRenderer

接口和 StandardOutMessageRenderer 实现。

```
//chapter04/jsr330/src/main/java/com/apress/prospring5/ch4/MessageRenderer.java
package com.apress.prospring5.ch4;

public interface MessageRenderer {
   void render();
   void setMessageProvider(MessageProvider provider);
   MessageProvider getMessageProvider();
}
//chapter04/jsr330/src/main/java/com/apress/prospring5/ch4/StandardOutMessageRenderer.java
package com.apress.prospring5.ch4;

import javax.inject.Inject;
import javax.inject.Named;
import javax.inject.Singleton;

@Named("messageRenderer")
@Singleton
public class StandardOutMessageRenderer
   implements MessageRenderer {

   @Inject
   @Named("messageProvider")
   private MessageProvider messageProvider = null;

   public void render() {
      if (messageProvider == null) {
         throw new RuntimeException(
            "You must set the property messageProvider of class:"
            + StandardOutMessageRenderer.class.getName());
      }

      System.out.println(messageProvider.getMessage());
   }

   public void setMessageProvider(MessageProvider provider) {
      this.messageProvider = provider;
   }
   public MessageProvider getMessageProvider() {
      return this.messageProvider;
   }
}
```

在前面的代码片段中，使用@Named 定义了一个可注入的 bean。请注意@Singleton 注解。值得注意的是，在 JSR-330 标准中，bean 的默认作用域是非单例，就像 Spring 的原型作用域一样。所以，在 JSR-330 环境中，如果想让 bean 成为单例，需要使用@Singleton 注解。然而，在 Spring 中使用该注解实际上没有任何作用，因为 Spring 的 bean 实例化的默认作用域已经是单例。在此使用@Singleton 注解只是为了演示，注意一下 Spring 和其他 JSR-330 兼容容器的区别。

对于 messageProvider 属性，此时使用@Inject 进行 setter 注入，并指定名为 messageProvider 的 bean 用于注入。以下配置代码片段为应用程序定义了一个简单的 Spring XML 配置(app-context-annotation.xml)：

```
<beans ...>

   <context:component-scan
      base-package="com.apress.prospring5.ch4"/>

   <bean id="message" class="java.lang.String">
      <constructor-arg value="Gravity is working against me"/>
   </bean>
</beans>
```

使用 JSR-330 无需任何特殊的标记；只需要像普通的 Spring 应用程序一样配置应用程序即可。使用 <context:component-scan>指示 Spring 扫描与 DI 相关的注解，Spring 将识别这些 JSR-330 注解。同时，还在 ConfigurableMessageProvider 类中声明了一个名为 message 的 Spring bean，主要用于构造函数注入。以下代码片段显示了测试程序：

```
package com.apress.prospring5.ch4;

import org.springframework.context.support.GenericXmlApplicationContext;

public class Jsr330Demo {
```

```
    public static void main(String... args) {
        GenericXmlApplicationContext ctx =
            new GenericXmlApplicationContext();
        ctx.load("classpath:spring/app-context-annotation.xml");
        ctx.refresh();

        MessageRenderer renderer = ctx.getBean("messageRenderer",
            MessageRenderer.class);
        renderer.render();

        ctx.close();
    }
}
```

运行该程序，将产生以下输出：

```
Gravity is working against me
```

通过使用 JSR-330，可以轻松地迁移到其他兼容 JSR-330 的 IoC 容器(比如，JEE 6 兼容的应用程序服务器或其他诸如 Google Guice 之类的 DI 容器)。但是，Spring 的注解相比 JSR-330 注解更加丰富和灵活。接下来重点讲一下两者之间的主要差异：

- 当使用 Spring 的@Autowired 注释时，可以指定 required 属性来表明必须完成 DI(也可以使用 Spring 的@Required 注解来声明这个要求)，而对于 JSR-330 的@Inject 注解，则没有与之等价的属性。此外，Spring 还提供了@Qualifier 注解，它允许 Spring 更精细地控制基于限定符名称的依赖项的自动装配的执行。
- JSR-330 仅支持单例和非单例 bean 作用域，而 Spring 支持更多作用域，这对 Web 应用程序很有用。
- 在 Spring 中，可以使用@Lazy 注解来指示 Spring 仅在应用程序请求时才实例化 bean。在 JSR-330 中没有与之等价的注解。

还可以在同一个应用程序中混用并匹配 Spring 和 JSR-330 注解。但是，建议选择其中一种注解，以便保持应用程序风格的一致性。一种方式是尽可能使用 JSR-330 注解，并在需要时使用 Spring 注解。但是，该方式所带来的好处很少，因为仍然需要完成相当多的工作来迁移到另一个 DI 容器。总之，相比于使用 JSR-330 注解，推荐使用 Spring 的注解，因为 Spring 的注解功能要强大得多，除非应用程序需要独立于 IoC 容器。

4.20 使用 Groovy 进行配置

Spring Framework 4.0 的一项新功能是可以通过使用 Groovy 语言来配置 bean 定义和 ApplicationContext。这为开发人员提供了另一种配置选择，以替换或补充 XML 和/或基于注解的 bean 配置。Spring ApplicationContext 可以直接在 Groovy 脚本中创建，也可以从 Java 中加载，但这两种方法都需要使用 GenericGroovyApplicationContext 类。首先通过演示如何从外部 Groovy 脚本创建 bean 定义并从 Java 中加载它们来深入了解细节。在前面的章节中，已经介绍了各种 bean 类，为了提高代码的可重用性，在本例中将使用第 3 章介绍的 Singer 类。下面的代码片段显示了该类的内容：

```
package com.apress.prospring5.ch3.xml;

public class Singer {
    private String name;
    private int age;

    public void setName(String name) {
        this.name = name;
    }

    public void setAge(int age) {
        this.age = age;
    }

    public String toString() {
        return "\tName: " + name + "\n\t" + "Age: " + age;
    }
}
```

正如你所看到的，这只是一个 Java 类，它使用一些属性来描述一名歌手。之所以在此使用这个简单的 Java 类，是因为在 Groovy 中配置 bean 并不意味着整个代码库都需要在 Groovy 中重写。不仅如此，Java 类可以从依赖项中导入并在 Groovy 脚本中使用。接下来，编写用于创建 bean 定义的 Groovy 脚本(beans.groovy)，如下所示：

```
package com.apress.prospring5.ch4
```

```
import com.apress.prospring5.ch3.xml.Singer

beans {
   singer(Singer, name: 'John Mayer', age: 39)
}
```

该 Groovy 脚本从一个名为 bean 的顶级闭包开始，它向 Spring 提供了 bean 定义。首先指定 bean 名称(singer)，然后以参数的形式提供类类型(Singer)、想要设置的属性名称以及值。接下来，用 Java 创建一个简单的测试驱动程序，从 Groovy 脚本加载 bean 定义，如下所示：

```
package com.apress.prospring5.ch4;

import com.apress.prospring5.ch3.xml.Singer;
import org.springframework.context.ApplicationContext;
import org.springframework.context.support.GenericGroovyApplicationContext;

public class GroovyBeansFromJava {
   public static void main(String... args) {
   ApplicationContext context =
      new GenericGroovyApplicationContext("classpath:beans.groovy");
      Singer singer = context.getBean("singer", Singer.class);
      System.out.println(singer);
   }
}
```

正如你所看到的，虽然 ApplicationContext 的创建是以典型的方式进行的，但它是通过使用 GenericGroovyApplicationContext 类并提供构建 bean 定义的 Groovy 脚本来完成的。

在运行示例之前，需要为该项目添加一个依赖项：groovy-all 库。groovy-config-java 项目的配置文件 build.gradle 的内容如下所示：

```
apply plugin: 'groovy'

dependencies {
      compile misc.groovy
      compile project(':chapter03:bean-inheritance')
}
```

compile project(':chapter03: bean-inheritance')行指定必须编译第 3 章的 bean-inheritance 项目并将其用作本项目的依赖项。bean-inheritance 项目包含了 Singer 类。

Gradle 配置文件使用了 Groovy 语法，而 misc.groovy 引用父项目的 build.gradle 文件中定义的 misc 数组的 groovy 属性。此文件的内容片段(与 Groovy 相关的配置)如下所示：

```
ext {
      springVersion = '5.0.0.M4'
      groovyVersion = '2.4.5'
...

   misc = [
      ...
         groovy: "org.codehaus.groovy:groovy-all:$groovyVersion"
      ]
...
}
```

运行 GroovyBeansFromJava 类，将产生如下输出：

```
Name: John Mayer
Age: 39
```

现在，你已经知道了如何通过外部 Groovy 脚本从 Java 加载 bean 定义，那么如何才能从 Groovy 脚本中单独创建 ApplicationContext 和 bean 定义呢？首先看看下面列出的 Groovy 代码(GroovyConfig.groovy)：

```
package com.apress.prospring5.ch4

import com.apress.prospring5.ch3.xml.Singer
import org.springframework.context.support.GenericApplicationContext
import org.springframework.beans.factory.groovy.GroovyBeanDefinitionReader

def ctx = new GenericApplicationContext()
def reader = new GroovyBeanDefinitionReader(ctx)

reader.beans {
   singer(Singer, name: 'John Mayer', age: 39)
}
```

```
ctx.refresh()

println ctx.getBean("singer")
```

当运行该例时，将得到与前面相同的输出。这一次，创建了一个典型的 GenericApplicationContext 实例，并且使用了 GroovyBeanDefinitionReader(它将被用来传递 bean 定义)。剩下的代码与前面的示例一样，从简单的 POJO 创建一个 bean，然后刷新 ApplicationContext，并打印 Singer bean 的字符串表示形式。该例也比较简单！

正如你所想象的那样，我们只是了解了 Spring 中的 Groovy 支持所能完成的一部分工作。一旦掌握了 Groovy 语言的全部功能，在创建 bean 定义时就可以完成各种有趣的事情。由于可以完全访问 ApplicationContext，因此不仅可以配置 bean，还可以使用配置文件支持、属性文件等。请记住，能力越大，责任越大。

4.21 Spring Boot

到目前为止，你已经学习了多种配置 Spring 应用程序的方法。但无论是使用 XML、注解、Java 配置类、Groovy 脚本还是它们的组合，你现在应该基本了解了如何完成 Spring 应用程序的配置。但是如果我告诉你还有比这些方法更好的方法，你认为会是什么方法呢？

Spring Boot 项目旨在简化使用 Spring 构建应用程序的入门体验。Spring Boot 不再需要手动收集依赖项，而是提供大多数应用程序所需的一些最常见的功能，例如度量和健康检查等。

Spring Boot 采用一种“自以为是”的方式来简化开发人员的工作量，具体方法是为各种类型的应用程序提供起始项目，而这些应用程序已经包含适当的依赖项和版本，这意味着可以用更短的时间开始项目开发。对于那些可能希望彻底摆脱 XML 的人来说，Spring Boot 不需要用 XML 编写任何配置。

在下面的示例中，将创建传统的 Hello World Web 应用程序。与典型的 Java Web 应用程序设置相比，你会惊讶地发现所需的代码量非常少。通常情况下，只需要定义需要添加到项目中的依赖项就可以启动示例。Spring Boot 的部分简化模型为我们准备了所有的依赖项。例如，当使用 Maven 时，开发者可以使用父 POM 来获得该功能。如果使用 Gradle，则事情变得更加简单。除了 Gradle 插件和启动依赖项，不再需要其他依赖项。在下面的示例中，将创建一个列出上下文中所有 bean 并访问 helloWorld bean 的 Spring 应用程序。下面的代码描述了 boot-simple 项目的 Gradle 配置：

```
buildscript {
   repositories {
      mavenLocal()
      mavenCentral()
      maven { url "http://repo.spring.io/release" }
      maven { url "http://repo.spring.io/milestone" }
      maven { url "http://repo.spring.io/snapshot" }
      maven { url "https://repo.spring.io/libs-snapshot" }
   }
   dependencies {
      classpath boot.springBootPlugin
   }
}

apply plugin: 'org.springframework.boot'

dependencies {
   compile boot.starter
}
```

boot.springBootPlugin 行引用了父项目的 build.gradle 文件中定义的 boot 数组的 springBootPlugin 属性。该文件的内容片段(仅与 Spring Boot 相关的配置)如下所示：

```
ext {
   bootVersion = '2.0.0.BUILD-SNAPSHOT'
   ...
   boot = [
      springBootPlugin:
         "org.springframework.boot:spring-boot-gradle-plugin:$bootVersion", starter :
         "org.springframework.boot:spring-boot-starter:$bootVersion", starterWeb :
         "org.springframework.boot:spring-boot-starter-web:$bootVersion"
   ]
   ...
}
```

在编写本书时，Spring Boot 2.0.0 版本尚未发布。这就是为什么该版本是 2.0.0.BUILD-SNAPSHOT，需要在配置中添加 Spring Snapshot 存储库(https://repo.spring.io/libs-snapshot)的原因。本书出版之后，很可能会发布一个正式版本。

Spring Boot 的每个版本都提供了它所支持的一系列依赖项。为了使 API 完美匹配，需要选择所需库的版本，而这一切都是由 Spring Boot 处理的。因此，不需要手动配置依赖项版本。升级 Spring Boot 将确保这些依赖项也被升级。在前面所示的配置中，一组依赖项将被添加到项目中，每个依赖项都有适当的版本，以便它们的 API 相兼容。在诸如 IntelliJ IDEA 之类的智能编辑器中，有一个 Gradle Projects 视图，可以在其中展开每个模块并检查可用的任务和依赖项，如图 4-3 所示。

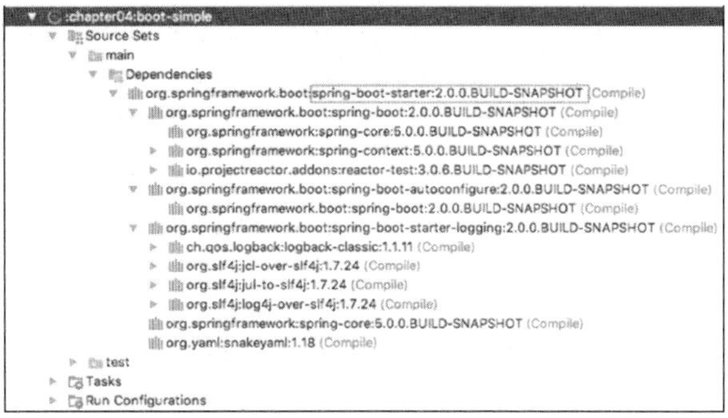

图 4-3 boot-simple 项目的 Gladle Projects 视图

现在设置已经完成，接下来创建相关的类。HelloWorld 类如下所示：

```
package com.apress.prospring5.ch4;

import org.slf4j.Logger;
import org.slf4j.LoggerFactory;
import org.springframework.stereotype.Component;

@Component
public class HelloWorld {

    private static Logger logger =

LoggerFactory.getLogger(HelloWorld.class);

    public void sayHi() {
        logger.info("Hello World!");
    }
}
```

该类没有什么特别或复杂之处；它只是一个带有一个方法和一个 bean 声明注解的类。我们来看看如何使用 Spring Boot 构建一个 Spring 应用程序并创建一个包含该 bean 的 ApplicationContext：

```
package com.apress.prospring5.ch4;

import org.slf4j.Logger;
import org.slf4j.LoggerFactory;
import org.springframework.boot.SpringApplication;
import org.springframework.boot.autoconfigure.SpringBootApplication;
import org.springframework.context.ConfigurableApplicationContext;

import java.util.Arrays;

@SpringBootApplication
public class Application {

    private static Logger logger = LoggerFactory.getLogger(Application.class);

    public static void main(String args) throws Exception {
        ConfigurableApplicationContext ctx =
            SpringApplication.run(Application.class, args);
        assert (ctx != null);
        logger.info("The beans you were looking for:");

        // listing all bean definition names
        Arrays.stream(ctx.getBeanDefinitionNames()).forEach(logger::info);

        HelloWorld hw = ctx.getBean(HelloWorld.class);
        hw.sayHi();

        System.in.read();
        ctx.close();
    }
}
```

以上就是全部代码。的确是这样。本来这个类可以更小，但为了介绍一些其他的内容而特意增加了一些代码。主要增加了以下内容：

- **检查是否有一个上下文**：assert 语句用于测试 ctx 是否为 null。
- **设置日志记录**：Spring Boot 附带一组日志库，因此只需要在 resources 目录下放置想要使用的配置即可。在本例中，选择了 logback。

- **列出上下文中的所有 bean 定义**：通过使用 Java 8 lambda 表达式，可以在一行中列出上下文中的所有 bean 定义。因此，示例中显示了该行，以便可以看到 Spring Boot 为我们自动配置了哪些 bean。其中，在列表中可以找到 helloWorld bean。
- **确认退出**：如果没有使用 System.in.read()方法，那么应用程序将会打印 bean 名称和 HelloWorld，然后退出。此时添加了对该方法的调用，以便应用程序等待开发人员按下某个键后退出。

上述代码的新奇之处在于@SpringBootApplication 注解。此注解是一个顶级注解，仅在类级别使用。它是一个方便的注解，相当于声明以下三个注解：

- @Configuration：将该类标记为可以使用@Bean 声明 bean 的配置类。
- @EnableAutoConfiguration：这是一个来自 org.springframework.boot.autoconfigure 的特定的 Spring Boot 注解，它可以启用 Spring ApplicationContext，并尝试根据指定的依赖项猜测和配置可能需要的 bean。
  虽然@EnableAutoConfiguration 可以很好地使用 Spring 提供的启动依赖项，但它并不直接与它们绑定，因此也可以用于启动之外的其他依赖项。例如，如果在类路径中有特定的嵌入式服务器，则可以使用它，除非项目中有另一个 EmbeddedServletContainerFactory 配置。
- @ComponentScan：可以声明使用构造型注解进行注解的类，这些类将成为某种类型的 bean。可以使用属性 basePackages 列出需要使用@SpringBootApplication 扫描的包。在版本 1.3.0 中，为组件扫描添加了另一个属性：basePackageClasses。此属性为 basePackages 提供了一个类型安全的替代方法，用于指定要扫描注解组件的包。所指定的每个类的包都会被扫描。

如果@SpringBootApplication 注解没有定义组件扫描属性，那么它将仅扫描其注解的类所在的包。这也就是为什么在本例中能够找到 helloWorld bean 定义，并创建 bean 的原因。

前面的Spring Boot应用程序是一个简单的控制台应用程序，其中包含一个开发人员定义的bean以及一个完整的开箱即用环境。但Spring Boot也为Web应用程序提供了起始依赖项。要使用的依赖项是spring-boot-starter-web，从图4-4中可以看到此Boot起始库的传递依赖项。在boot-web项目中，HelloWorld类是一个Spring控制器，它是一种特殊类型的类，用于创建Spring Web bean。这是一个典型的Spring MVC控制器类，相关内容将在第16章中介绍。

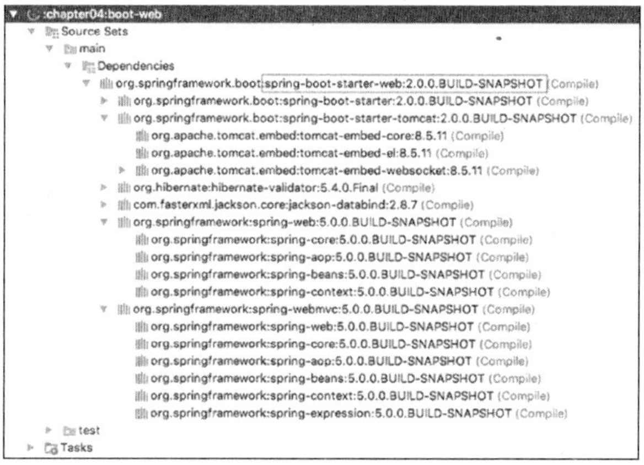

图 4-4 boot-web 项目的 Gradle Projects 视图

```
package com.apress.prospring5.ch4.ctrl;

import org.springframework.web.bind.annotation.RequestMapping;
import org.springframework.web.bind.annotation.RestController;

@RestController
public class HelloWorld {

    @RequestMapping("/")
     public String sayHi() {
       return "Hello World!";
    }
}
```

用于声明 Spring Web bean 的注解是@Component 的特例：@Controller 注解。这些类类型包含用@RequestMapping 注解的方法，而这些方法被映射到某个请求 URL。在示例@RestController 中看到的注解是出于实际原因而使用的。它是用于 REST 服务的@Controller 注解。此时将 helloWorld bean 公开为 REST 服务是非常有用的，因为不必创建具有用户界面和其他 Web 组件的 Web 应用程序，这些 Web 应用程序会干扰对本节内容的学习。第 16 章将会详细介绍上述 Web 组件。

接下来使用一个简单的 main()方法创建引导类，如下所示：

```
package com.apress.prospring5.ch4;

import com.apress.prospring5.ch4.ctrl.HelloWorld;
```

```
import org.slf4j.Logger;
import org.slf4j.LoggerFactory;
import org.springframework.boot.SpringApplication;
import org.springframework.boot.autoconfigure.SpringBootApplication;
import org.springframework.context.ConfigurableApplicationContext;

import java.util.Arrays;

@SpringBootApplication(scanBasePackageClasses = HelloWorld.class)
public class WebApplication {

   private static Logger logger =
      LoggerFactory.getLogger(WebApplication.class);

   public static void main(String args) throws Exception {
         ConfigurableApplicationContext ctx =
            SpringApplication.run(WebApplication.class, args);
         assert (ctx != null);
         logger.info("Application started...");

         System.in.read();
         ctx.close();
   }
}
```

由于 HelloWorld 控制器是在与 WebApplication 类不同的包中声明的，因此我们很好地介绍了一下如何使用 scanBasePackageClasses 属性。

此时，你可能会自问：web.xml 配置文件以及必须为基本 Web 应用程序创建的所有其他组件在哪里呢？实际上，你已经定义了所需的所有内容！不相信吗？编译项目并运行 Application 类，就会看到指示应用程序已启动的日志消息。如果查看一下生成的日志文件，你会看到很多小的代码。最值得注意的是，看起来似乎 Tomcat 正在运行，并且已经定义了各种端点，例如运行状况检查、环境输出信息和指标等。首先导航到 http://localhost:8080，将看到 Hello World 网页按预期显示。接下来查看一些预配置的端点(例如，http://localhost:8080/health，将返回应用程序状态的 JSON 表示)。然后，加载 http://localhost:8080/metrics，以更好地理解正在收集的各种度量，例如堆大小、垃圾回收等。

通过上述示例可以看到，Spring Boot 从根本上简化了创建任何类型应用程序的方式。为了获得简单的 Web 应用程序而必须设置大量文件的时代已经一去不复返了，现在可以使用大量的嵌入式 servlet 容器为 Web 应用程序提供服务，一切都“非常正常”。

前面已经演示了一个简单的示例，但请记住，Spring Boot 不会限制你使用它所选择的内容；它只是采取一种“自以为是”的方法，并为你选择了默认值。如果不想使用嵌入式 Tomcat，而是使用 Jetty，只需要修改配置文件，从 spring-boot-starter-web 依赖项中排除 Tomcat 启动模块即可。使用 Gradle Projects 视图，这是一种帮助你可视化项目中依赖项的方法。Spring Boot 还为其他类型的应用程序提供了许多其他起始依赖项，因此建议阅读文档以获取更多详细信息。

有关 Spring Boot 的更多信息，请参阅项目页面 http://projects.spring.io/spring-boot/。

4.22 小结

在本章中，我们介绍了各种 Spring 特有的功能，它们可以作为核心 IoC 功能的补充。你学习了如何挂钩到 bean 的生命周期，并使其识别 Spring 环境。我们引入了 FactoryBean 作为 IoC 的解决方案，从而支持更广泛的类：还演示了如何使用 PropertyEditor 来简化应用程序配置并消除对人为字符串类型属性的需求；介绍了使用 XML、注解和 Java 配置定义 bean 的多种方法。此外，我们还深入了解了 ApplicationContext 提供的一些附加功能，包括国际化、事件发布和资源访问。

本章还介绍了一些新功能，比如，使用 Java 类和新的 Groovy 语法(而不是 XML 配置)、配置文件支持以及环境和属性源抽象层等。此外，我们讨论了如何在 Spring 中使用 JSR-330 标准注解。

最后，本章锦上添花地介绍了如何使用 Spring Boot 来配置 bean 并尽快且轻松地启动应用程序。

到目前为止，我们已经介绍了 Spring 框架的主要概念及其作为 DI 容器的相关功能，以及核心 Spring 框架所提供的其他服务。在下一章及后续各章中，将讨论如何在特定领域使用 Spring，比如 AOP、数据访问、事务支持和 Web 应用程序支持。

第5章

Spring AOP

除了依赖注入(DI)之外，Spring 框架提供的另一个核心功能是支持面向方面的编程(AOP)。AOP 通常被称为实施横切关注点的工具。术语*横切关注点*(*crosscutting concerns*)是指应用程序中无法从应用程序的其余部分分解并且可能导致代码重复和紧密耦合的逻辑。通过使用 AOP 模块化各个逻辑部分(称为*关注点*(*concerns*))，可以将它们应用于应用程序的多个部分，而无须复制代码或创建硬性依赖关系。日志记录和安全性是许多应用程序中存在的横切关注点的典型示例。假设一个应用程序记录每个方法的开始和结束，以便进行调试。此时，你可能会将日志代码重构为一个特殊的类，但是为了执行日志记录，仍然必须在应用程序中调用两次该类中的方法。而通过使用 AOP，可以简单地指定在应用程序中每个方法调用之前和之后所调用的日志记录类上对应的方法。

了解 AOP 是对面向对象编程(OOP)的补充而非竞争关系是非常重要的。OOP 擅长解决程序员遇到的各种各样的问题。但是，如果再次回到日志记录示例，可以很明显看到在大规模实施横切逻辑时，OOP 就有点力不从心了。鉴于 AOP 功能建立在 OOP 基础之上，因此使用 AOP 来开发整个应用程序实际上是不可能的。同样，尽管使用 OOP 开发整个应用程序当然是可能的，但如果通过使用 AOP 解决涉及横切逻辑的某些问题，则可以更好地工作。

本章包含以下主题：

- **AOP 的基础知识**：在讨论 Spring 的 AOP 实现之前，先介绍一下 AOP 的基础知识。“AOP 概念”部分所涵盖的大部分概念并非特定于 Spring，可以在任何 AOP 实现中找到。如果你已经熟悉其他的 AOP 实现，那么可以跳过“AOP 概念”部分。
- **AOP 的类型**：AOP 有两种截然不同的类型：静态 AOP 和动态 AOP。在静态 AOP 中，比如 AspectJ 编译时织入机制所提供的静态 AOP，横切逻辑将在编译时应用于代码，如果想要修改逻辑，必须修改代码并重新编译。而通过动态 AOP(比如 Spring AOP)，横切逻辑在运行时动态应用,这样一来就可以更改 AOP 配置而无须重新编译应用程序。这些类型的 AOP 是互补的，并且当它们一起使用时，就可以为应用程序提供可使用的强大组合。
- **Spring AOP 架构**：Spring AOP 只是其他实现(如 AspectJ)中完整 AOP 功能集的子集。在本章中，将重点关注 Spring 中提供了哪些功能，它们是如何实现的，以及为什么某些功能被排除在 Spring 实现之外。
- **Spring AOP 中的代理**：代理是 Spring AOP 工作方式的一个重要组成部分，必须了解它们才能充分利用 Spring AOP。在本章中，主要介绍两种代理：JDK 动态代理和 CGLIB 代理。具体来说，了解一下 Spring 使用每个代理的不同场景、两种代理类型的性能以及在应用程序中遵循的一些简单准则，以便从 Spring AOP 中获得最大收益。
- **使用 Spring AOP**：在本章中，将介绍使用 AOP 的一些实例。以一个简单的 Hello World 示例开始，以便更加轻松地进入 Spring 的 AOP 代码，并且继续详细描述 Spring 中提供的 AOP 功能，并附有示例。
- **切入点的高级使用**：将讨论在应用程序中使用切入点时所使用的 ComposablePointcut 和 ControlFlowPointcut 类，以及相关的技巧。
- **AOP 框架服务**：Spring 框架完全支持以透明和声明的方式配置 AOP。本章主要介绍如何使用三种方法(ProxyFactoryBean 类、aop 名称空间和@AspectJ 样式注解)将声明定义的 AOP 代理作为协作者注入到应用程序对象中，从而使应用程序完全不知道自己正在使用的被通知对象。
- **集成 AspectJ**：AspectJ 是一个功能全面的 AOP 实现。AspectJ 和 Spring AOP 之间的主要区别在于，AspectJ 通过织入(编译时或加载时织入)将通知(advice)应用于目标对象，而 Spring AOP 则基于代理。AspectJ 的功能集比 Spring AOP 强大得多，但比 Spring 更复杂。当发现 Spring AOP 缺少所需要的功能时，AspectJ 是一个很好的解决方案。

5.1 AOP 概念

与大多数技术一样，AOP 带有自己特定的一组概念和术语，了解它们的含义非常重要。以下是 AOP 的核心概念。

- **连接点**：连接点是应用程序执行期间明确定义的一个点。连接点的典型示例包括方法调用、方法调用本身、类初始化和对象实例化。连接点是 AOP 的核心概念，并且定义了在应用程序中可以使用 AOP 插入其他逻辑的点。
- **通知**：在特定连接点执行的代码就是通知，它是由类中的方法定义的。有许多类型的通知，比如前置通知(在连接点之前执行)和后置通知(在连接点之后执行)。
- **切入点**：切入点是用于定义何时执行通知的连接点集合。通过创建切入点，可以更细致地控制如何将通知应用于应用程序中的组件。如前所述，典型的连接点是方法调用，或是特定类中所有方法调用的集合。通常情况下，可以在复杂的关系中插入切入点，从而进一步限制执行通知的时间。
- **切面**：切面是封装在类中的通知和切入点的组合。这种组合定义了应该包含在应用程序中的逻辑及其应该执行的位置。
- **织入**：织入是在适当的位置将切面插入到应用程序代码中的过程。对于编译时 AOP 解决方案，织入过程通常在生成时完成。同样，对于运行时 AOP 解决方案，织入过程在运行时动态执行。此外，AspectJ 还支持另一种称为加载时织入(LTW)的织入机制，在该机制中，拦截底层的 JVM 类加载器，并在类加载器加载字节码时向其提供织入功能。
- **目标对象**：执行流由 AOP 进程修改的对象被称为目标对象。通常也会看到目标对象被称为*被通知*(*advised*)对象。
- **引入**：这是通过引入其他方法或字段来修改对象结构的过程。可以通过引入 AOP 来使任何对象实现特定的接口，而无须对象的类显式地实现该接口。

如果你对这些概念感到混乱，请不要担心；只需要看一些示例，就会清楚理解这些概念。另外，请注意，Spring AOP 中的许多概念都被屏蔽了，还有一些概念因为 Spring 的实现选择而没有任何关联。我们将在 Spring 的上下文中讨论每一个功能。

5.2 AOP 的类型

正如前面所提到的，AOP 有两种不同的类型：静态 AOP 和动态 AOP。它们之间的区别就在于织入过程发生的地点以及如何实现这一过程。

5.2.1 使用静态 AOP

在静态 AOP 中，织入过程构成了应用程序生成过程中的另一个步骤。用 Java 术语来说，可以通过修改应用程序的实际字节码并根据需要更改和扩展应用程序代码来实现静态 AOP 实现中的织入过程。这是实现织入过程比较好的方法，因为最终结果只是 Java 字节码，并且在运行时无须使用任何特殊技巧来确定应该何时执行通知。但这种机制的缺点是，对切面所做的任何修改都要求重新编译整个应用程序，即使只是想添加另一个连接点。AspectJ 的编译时织入是静态 AOP 实现的一个很好的例子。

5.2.2 使用动态 AOP

动态 AOP 实现(比如 Spring AOP)与静态 AOP 实现不同，因为织入过程在运行时动态执行。如何实现依赖于具体实现，但正如你所看到的，Spring 采用的方法是为所有被通知对象创建代理，以便根据需要调用通知。动态 AOP 的缺点是，一般来说性能不如静态 AOP，但目前性能在稳步提高。动态 AOP 实现的主要优点是可以轻松修改应用程序的整个切面集，而无须重新编译主应用程序代码。

5.2.3 选择 AOP 类型

选择使用静态 AOP 还是动态 AOP 是相当困难的决定。两者都有各自的优点，并且不限于只使用一种类型。一般来说，静态 AOP 实现的时间更长，倾向于具有更多功能丰富的实现，并且具有更多的可用连接点。通常，如果性能绝对重要，或者需要一项在 Spring 中未实现的 AOP 功能，就需要使用 AspectJ。在大多数情况下，Spring AOP 是理想的选择。请记住，许多基于 AOP 的解决方案(比如事务管理)都已经由 Spring 提供，因此在开发之前应该检查一下框架功能！与往常一样，让应用程序的需求驱动 AOP 实现的选择，并且如果技术组合更适合应用程序，就不要将自己限制为单个实现。一般来说，Spring AOP 并不比 AspectJ 复杂，所以它往往是理想的第一选择。

5.3 Spring 中的 AOP

Spring 的 AOP 实现可以被看作两个逻辑部分。第一部分是 AOP 内核，它提供了完全解耦的纯编程 AOP 功能(也称为 Spring AOP API)。AOP 实现的第二部分是使 AOP 更易于在应用程序中使用的一组框架服务。除此之外，Spring 的其他组件(比如事务管理器和 EJB 助手类)还提供了基于 AOP 的服务，以简化应用程序的开发。

5.3.1 AOP Alliance

AOP Alliance(http://aopalliance.sourceforge.net/)是许多开源 AOP 项目代表共同努力的结果，它为 AOP 实现定义了一组标准接口。只要适用，Spring 就应该使用 AOP Alliance 接口而不是定义自己的接口。这样一来，就可以在支持 AOP Alliance 接口的多个 AOP 实现中重复使用某些通知。

5.3.2 AOP 中的 Hello World 示例

在深入讨论 Spring AOP 实现之前，先来看一个示例。看看如何修改经典的 Hello World 示例，我们将转向电影以获取灵感。首先编写一个名为 Agent 的类，负责打印 Bond。当使用 AOP 时，该类的实例将在运行时转换为打印 James Bond !。以下代码描述了 Agent 类：

```
package com.apress.prospring5.ch5;

public class Agent {
    public void speak() {
        System.out.print("Bond");
    }
}
```

实现名称打印的方法之后，还需要*通知*(在 AOP 术语中意味着添加通知)该方法以使用 speak()打印 James Bond !。

为此，需要在执行方法体之前执行相关代码(写入 James)，并方法体执行之后再执行相关代码(写入！)。用 AOP 术语来讲，需要实现环绕通知(around advice)，即围绕连接点执行哪个通知。在本例中，连接点是 speak()方法的调用。以下代码片段显示了 AgentDecorator 类的代码，该代码实现了环绕通知：

```
package com.apress.prospring5.ch5;

import org.aopalliance.intercept.MethodInterceptor;
import org.aopalliance.intercept.MethodInvocation;
public class AgentDecorator implements MethodInterceptor {
    public Object invoke(MethodInvocation invocation) throws Throwable {
        System.out.print("James ");

        Object retVal = invocation.proceed();

        System.out.println("!");
        return retVal;
    }
}
```

MethodInterceptor 接口是一个标准的 AOP Alliance 接口，用于实现方法调用连接点的环绕通知。MethodInvocation 对象表示正在被通知的方法调用，通过使用此对象，可以控制方法调用何时进行。因为是环绕通知，所以能够在调用方法之前以及之后且返回之前执行相关操作。因此，在前面的代码片段中，只需要将 James 写入到控制台输出，然后调用 invocation.proceed()，最后将!写入到控制台输出。

此例的最后一步是将 AgentDecorator 通知(更具体地说，是 invoke()方法)织入代码中。为此，创建一个 Agent 实例，即目标对象，然后创建此实例的代理，指示代理工厂织入 AgentDecorator 通知，如下所示：

```
package com.apress.prospring5.ch5;

import org.springframework.aop.framework.ProxyFactory;

public class AgentAOPDemo {
    public static void main(String... args) {
        Agent target = new Agent();

        ProxyFactory pf = new ProxyFactory();
        pf.addAdvice(new AgentDecorator());
        pf.setTarget(target);

        Agent proxy = (Agent) pf.getProxy();

        target.speak();
        System.out.println("");
        proxy.speak();
    }
}
```

这里重要的部分是使用 ProxyFactory 类创建目标对象的代理，同时织入通知。通过调用 addAdvice()将 AgentDecorator 通知传递给 ProxyFactory，并通过调用 setTarget()指定织入目标。一旦设置目标并将一些通知添加到 ProxyFactory，就可以通过调用 getProxy()生成代理。最后，调用原始目标对象和代理对象上的 speak()。运行代码会产生以下输出：

```
Bond
James Bond!
```

正如你所看到的，如果调用原始目标对象上的 speak()，将会导致标准的方法调用，并且不会向控制台输出写入额外的内容。但是，代理的调用会导致执行 AgentDecorator 中的代码，从而创建所期望的 James Bond！输出。通过该例可以看到，被通知类不依赖于 Spring 或 AOP Alliance 接口；AOP 的优点在于，即使在不考虑 AOP 的情况下创建了类，也可以为任何类提供通知。在 Spring AOP 中唯一的限制是不能通知最终的类，因为它们不能被覆盖，所有不能被代理。

5.4 Spring AOP 架构

Spring AOP 的核心架构基于代理。当想要创建一个类的被通知实例时，必须使用 ProxyFactory 创建该类的代理实例，首先向 ProxyFactory 提供想要织入到代理的所有切面。使用 ProxyFactory 是创建 AOP 代理的纯程序化方法。大多数情况下，不需要在应用程序中使用它；相反，可以依赖 Spring 所提供的声明式 AOP 配置机制(ProxyFactoryBean 类、aop 名称空间和@AspectJ 样式注解)来完成声明式代理的创建。但是，了解代理创建的工作过程是非常重要的，因此首先演示代理创建的编程方法，然后深入讨论 Spring 的声明式 AOP 配置。

在运行时，Spring 会分析为 ApplicationContext 中的 bean 定义的横切关注点，并动态生成代理 bean(封装了底层目标 bean)。此时，不会直接调用目标 bean，而是将调用者注入代理 bean。然后，代理 bean 分析运行条件(即，连接点、切入点或通知)并相应地织入适当的通知。图 5-1 显示了运行中的 Spring AOP 代理的高级视图。在图 5-1 中，Spring 有两个代理实现：JDK 动态代理和 CGLIB 代理。默认情况下，当被通知的目标对象实现一个接口时，Spring 将使用 JDK 动态代理来创建目标的代理实例。但是，当被通知目标对象没有实现接口(例如，它是一个具体类)时，将使用 CGLIB 来创建代理实例。一个主要原因是 JDK 动态代理仅支持接口代理。关于代理的内容，将在 5.6 节"了解代理"中详细讨论。

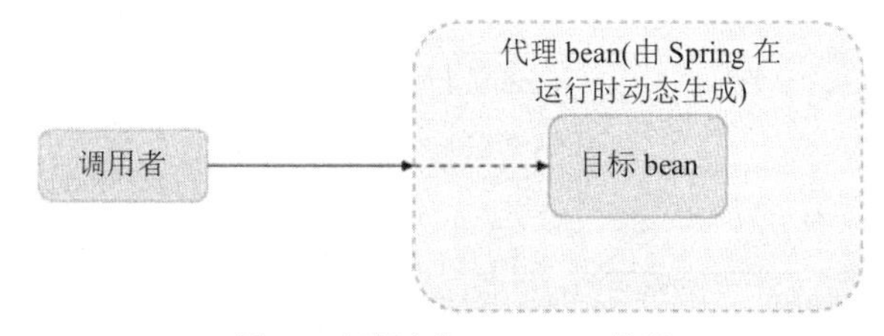

图 5-1　运行中的 Spring AOP 代理

5.4.1 Spring 中的连接点

Spring AOP 中最明显的简化是只支持一种连接点类型：方法调用。乍一看，如果你熟悉其他的 AOP 实现(比如支持更多连接点的 AspectJ)，可能会认为这是一个严重的限制，但实际上这使得 Spring 更易于访问。

方法调用连接点是迄今为止最有用的连接点，使用它可以实现 AOP 在日常编程中许多有用的任务。请记住，如果需要在除方法调用外的连接点通知一些代码，则可以一起使用 Spring 和 AspectJ。

5.4.2 Spring 中的切面

在 Spring AOP 中，切面由实现了 Advisor 接口的类的实例表示。Spring 提供了可以在应用程序中重复使用的、便捷的 Advisor 实现，从而无须创建自定义的 Advisor 实现。Advisor 有两个子接口：PointcutAdvisor 和 IntroductionAdvisor。

所有的 Advisor 实现都实现了 PointcutAdvisor 接口，这些实现使用切入点来控制应用于连接点的通知。在 Spring 中，引言被视为特殊类型的通知，通过使用 IntroductionAdvisor 接口，可以控制将引言引用于哪些类。

我们将在后续的 5.5 节“在 Spring 中使用顾问和切入点”中详细讨论 PointcutAdvisor 实现。

5.4.3 关于 ProxyFactory 类

ProxyFactory 类控制 Spring AOP 中的织入和代理创建过程。在创建代理之前，必须指定被通知对象或目标对象。正如你之前所看到的那样，可以使用 setTarget()方法执行此操作。在内部，ProxyFactory 将代理创建过程委托给 DefaultAopProxyFactory 的一个实例，该实例又转而委托给 Cglib2AopProxy 或 JdkDynamicAopProxy，具体是哪一个取决于应用程序的设置。本章后面将更详细地讨论代理创建。

ProxyFactory 类提供了你在前面的代码示例中看到的 addAdvice()方法，主要用于将通知应用于类中所有的方法调用，而不是有选择地应用。在内部，addAdvice()将传入的通知封装到 DefaultPointcutAdvisor(DefaultPointcutAdvisor 是 PointcutAdvisor 的标准实现)的一个实例中，并使用默认包含所有方法的切入点对其进行配置。当想要更多地控制所创建的 Advisor 或想要向代理添加引入时，可以自己创建 Advisor 并使用 ProxyFactory 的 addAdvisor()方法。

可以使用相同的 ProxyFactory 实例来创建多个代理，每个代理都有不同的切面。为了帮助实现该过程，ProxyFactory 提供了 removeAdvice()和 removeAdvisor()方法，这些方法允许从 ProxyFactory 中删除之前传入的任何通知或顾问。如果想要检查 ProxyFactory 是否附有特定的通知，可以调用 adviceIncluded()，传入要检查的通知对象。

5.4.4 在 Spring 中创建通知

Spring 支持六种通知，如表 5-1 所示。

表 5-1 Spring 中的通知类型

通知名称	接口	描述
前置通知	org.springframework.aop.MethodBeforeAdvice	通过使用前置通知，可以在连接点执行之前完成自定义处理。因为 Spring 中的连接点就是方法调用，所以通常允许方法执行之前执行预处理。虽然前置通知可以完全访问方法调用的目标以及传递给方法的参数，但却无法控制方法本身的执行。如果前置通知抛出异常，那么拦截器链(以及目标方法)的进一步执行将被中止，并且异常将传回拦截器链
后置返回通知	org.springframework.aop.AfterReturningAdvice	在连接点的方法调用完成执行并返回一个值后执行后置返回通知。后置返回通知可以访问方法调用的目标、传递给方法的参数以及返回值。由于方法在调用后置返回通知时已经执行，因此根本无法控制方法调用。如果目标方法抛出异常，则不会运行后置返回通知，并且异常将照常传回调用堆栈
后置通知	org.springframework.aop.AfterAdvice	仅当被通知方法正常完成时才执行后置通知。然而，无论被通知方法的结果如何，后置通知都会被执行。即使被通知方法失败并抛出异常，后置通知也会执行
环绕通知	org.aopalliance.intercept.MethodInterceptor	在 Spring 中，环绕通知使用方法拦截器的 AOP Alliance 标准进行建模。环绕通知允许在方法调用之前和之后执行，并且可以控制允许进行方法调用的点。如果需要，可以选择完全绕过方法，从而提供自己的逻辑实现

(续表)

通知名称	接口	描述
异常通知	org.springframework.aop.ThrowsAdvice	异常通知在方法调用返回后执行，但只有在该调用抛出异常时才执行。异常通知只能捕获特定的异常，如果使用异常通知，则可以访问抛出异常的方法、传递给调用的参数以及调用的目标
引入通知	org.springframework.aop.IntroductionInterceptor	Spring 将引入建模为特殊类型的拦截器。通过使用引入拦截器，可以指定由引入通知引入的方法的实现

5.4.5 通知的接口

从之前关于 ProxyFactory 类的讨论中可以看出，可以直接(通过使用 addAdvice()方法)或间接(通过使用带有 addAdvisor()方法的 Advisor 实现)将通知添加到代理中。通知和顾问之间的主要区别在于，顾问可以携带相关切入点的通知，从而更细致地控制在哪个连接点上拦截通知。关于通知，Spring 为 Advice 接口创建了一个定义明确的层次结构。该层次结构基于 AOP Alliance 接口，如图 5-2 所示。

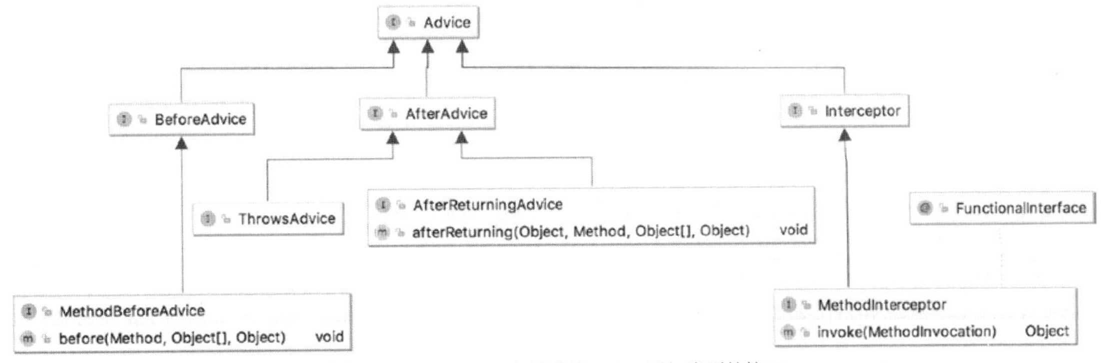

图 5-2 IntelliJ IDEA 中描述的 Spring 通知类型的接口

这种层次结构不仅具有良好的面向对象设计的优点，而且还能够用于处理通知类型，例如，通过使用 ProxyFactory 上的 addAdvice()方法，可以轻松添加新的通知类型，而无须修改 ProxyFactory 类。

5.4.6 创建前置通知

*前置通知(Before advice)*是 Spring 中最有用的通知类型之一。该通知可以修改传递给方法的参数，并可以通过抛出异常来阻止方法的执行。在本节中，将演示两个使用前置通知的简单示例：一个在方法执行前将包含方法名称的消息写入控制台输出，另一个则可用于限制对对象方法的访问。在下面的代码片段中，可以看到 SimpleBeforeAdvice 类的代码：

```
package com.apress.prospring5.ch5;

import java.lang.reflect.Method;

import org.springframework.aop.MethodBeforeAdvice;
import org.springframework.aop.framework.ProxyFactory;

public class SimpleBeforeAdvice implements MethodBeforeAdvice {
    public static void main(String... args) {
        Guitarist johnMayer = new Guitarist();

        ProxyFactory pf = new ProxyFactory();
        pf.addAdvice(new SimpleBeforeAdvice());
        pf.setTarget(johnMayer);

        Guitarist proxy = (Guitarist) pf.getProxy();

        proxy.sing();
```

```
    }
    @Override
    public void before(Method method, Object[] args, Object target)
            throws Throwable {
        System.out.println("Before '" + method.getName() + "', tune guitar.");
    }
}
```

Guitarist 类非常简单，只有一种方法 sing()，负责在控制台上打印出歌词。该类扩展了 Singer 接口，这个接口在本书后面还会用到。

```
package com.apress.prospring5.ch5;
import com.apress.prospring5.ch2.common.Singer;

public class Guitarist implements Singer {

    private String lyric="You're gonna live forever in me";
@Override
    public void sing(){
        System.out.println(lyric);
    }
}
```

基本上，该通知确保了 johnMayer bean 在唱歌之前调整吉他。在上述代码中，可以看到使用 SimpleBeforeAdvice 类的一个实例创建了 Guitarist 类的一个实例。MethodBeforeAdvice 接口由 SimpleBeforeAdvice 实现，它定义了一个方法 before()，AOP 框架将在调用连接点的方法之前调用该方法。请记住，此时使用 addAdvice()方法所提供的默认切入点，该切入点匹配类中的所有方法。before()方法被传入三个参数：要调用的方法、将传递给该方法的参数以及作为调用目标的 Object。SimpleBeforeAdvice 类使用 before()方法的 Method 参数将消息写入控制台输出，其中包含要调用方法的名称。运行示例，产生以下输出：

```
Before 'sing', tune guitar.
You're gonna live forever in me
```

如你所见，这里显示了对 sing()调用的输出，但在此之前，可以看到由 SimpleBeforeAdvice 生成的输出。

5.4.7 通过使用前置通知保护方法访问

本节中实现的前置通知可以在允许进行方法调用之前检查用户凭证。如果用户凭证无效，那么通知会抛出异常，从而阻止方法执行。本节中的示例很简单。它允许用户使用任何密码进行身份验证，并且只允许一个硬编码的用户访问受保护的方法。但是，该例说明了使用 AOP 来实现横切关注点(比如安全性)是非常容易的。

> 这只是一个示例，演示了前置通知的用法。为了确保 Spring bean 的方法安全执行，Spring Security 项目已经提供全面的支持；因此不需要自己实现这些功能。

以下代码片段显示了 SecureBean 类。这是要使用 AOP 进行安全保护的类。

```
package com.apress.prospring5.ch5;

public class SecureBean {
    public void writeSecureMessage() {
        System.out.println("Every time I learn something new, "
            + "it pushes some old stuff out of my brain");
    }
}
```

SecureBean 类被赋予了来自 Homer Simpson 的“智慧”，不让每个人都看到是非常明智的。因为该例要求用户进行身份验证，所以需要存储他们的详细信息。以下代码片段显示了可以用来存储用户凭证的 UserInfo 类：

```
package com.apress.prospring5.ch5;

public class UserInfo {
    private String userName;
    private String password;

    public UserInfo(String userName, String password) {
        this.userName = userName;
        this.password = password;
    }

    public String getPassword() {
        return password;
```

```
    }
    public String getUserName() {
        return userName;
    }
}
```

该类只保存关于用户的数据，以便可以用它们做一些有用的事情。以下代码片段显示了 SecurityManager 类，该类负责验证用户并存储其凭据供以后检索：

```
package com.apress.prospring5.ch5;

public class SecurityManager {
    private static ThreadLocal<UserInfo>
            threadLocal = new ThreadLocal<>();
    public void login(String userName, String password) {
        threadLocal.set(new UserInfo(userName, password));
    }

    public void logout() {
        threadLocal.set(null);
    }

    public UserInfo getLoggedOnUser() {
        return threadLocal.get();
    }
}
```

应用程序使用 SecurityManager 类对用户进行身份验证，稍后还会检索当前已通过身份验证的用户的详细信息。应用程序使用 login()方法对用户进行身份验证。这只是模拟实现。在实际的应用程序中，login()方法可能会根据数据库或 LDAP 目录检查所提供的凭据，但此时可以假设所有用户都被允许进行身份验证。login()方法为用户创建一个 UserInfo 对象，并使用 ThreadLocal 将其存储在当前线程中。logout()方法则将存储在 ThreadLocal 中的任何值设置为 null。最后，getLoggedOnUser()方法返回当前通过身份验证的用户的 UserInfo 对象。如果没有用户通过身份验证，则此方法返回 null。

为了检查用户是否通过身份验证，以及如果通过，是否允许用户访问 SecureBean 上的方法，需要创建一个通知，该通知在方法之前执行并检查 SecurityManager.getLoggedOnUser()针对一组允许用户的凭据所返回的 UserInfo 对象。通知 SecurityAdvice 的代码如下所示：

```
package com.apress.prospring5.ch5;

import java.lang.reflect.Method;

import org.springframework.aop.MethodBeforeAdvice;

public class SecurityAdvice implements MethodBeforeAdvice {
    private SecurityManager securityManager;

    public SecurityAdvice() {
        this.securityManager = new SecurityManager();
    }

    @Override
    public void before(Method method, Object[] args, Object target)
            throws Throwable {
        UserInfo user = securityManager.getLoggedOnUser();

        if (user == null) {
            System.out.println("No user authenticated");
            throw new SecurityException(
                "You must login before attempting to invoke the method: "
                + method.getName());
        } else if ("John".equals(user.getUserName())) {
            System.out.println("Logged in user is John - OKAY!");
        } else {
            System.out.println("Logged in user is " + user.getUserName()
                + " NOT GOOD :(");
            throw new SecurityException("User " + user.getUserName()
                + " is not allowed access to method " + method.getName());
        }
    }
}
```

SecurityAdvice 类首先在其构造函数中创建了一个 SecurityManager 实例，然后将此实例存储在一个字段中。你应该注意到，应用程序和 SecurityAdvice 不需要共享相同的 SecurityManager 实例，因为所有数据都通过使用 ThreadLocal 存储在当前线程中。在 before()方法中，执行一个简单的检查，查看经过身份验证的用户的用户名是否为 John。如果是 John，则允许用户访问；否则会抛出异常。另外请注意，还应该检查 UserInfo 对象是否为 null，null 表示当前用户未通过身份验证。

在以下代码片段中，可以看到使用 SecurityAdvice 类来保护 SecureBean 类的示例应用程序：

```
package com.apress.prospring5.ch5;

import org.springframework.aop.framework.ProxyFactory;

public class SecurityDemo {
    public static void main(String... args) {
        SecurityManager mgr = new SecurityManager();

        SecureBean bean = getSecureBean();

        mgr.login("John", "pwd");
        bean.writeSecureMessage();
        mgr.logout();

        try {
            mgr.login("invalid user", "pwd");
            bean.writeSecureMessage();
        } catch(SecurityException ex) {
            System.out.println("Exception Caught: " + ex.getMessage());
        } finally {
            mgr.logout();
        }

        try {
            bean.writeSecureMessage();
        } catch(SecurityException ex) {
            System.out.println("Exception Caught: " + ex.getMessage());
        }
    }

    private static SecureBean getSecureBean() {
        SecureBean target = new SecureBean();

        SecurityAdvice advice = new SecurityAdvice();

        ProxyFactory factory = new ProxyFactory();
        factory.setTarget(target);
        factory.addAdvice(advice);

        SecureBean proxy = (SecureBean)factory.getProxy();

        return proxy;
    }
}
```

在 getSecureBean()方法中，使用 SecurityAdvice 的一个实例创建被通知 SecureBean 类的代理。该代理被返回给调用者。当调用者调用此代理的任何方法时，调用首先被路由到 SecurityAdvice 的实例以进行安全检查。在 main()方法中，共测试了两种方案，首先使用两组用户凭证调用 SecureBean.writeSecureMessage()方法，然后是在没有用户凭证的情况下调用 SecureBean.writeSecureMessage()方法。由于 SecurityAdvice 只允许当前认证用户为 John 时执行方法调用，因此可以判断出上述代码中只有第一种方案会成功。运行该例，产生以下输出：

```
Logged in user is John - OKAY!
Every time I learn something new, it pushes some old stuff out of my brain
Logged in user is invalid user NOT GOOD :(
Exception Caught: User invalid user is not allowed access to method
          writeSecureMessage
No user authenticated
Exception Caught: You must login before attempting to invoke the method:
          writeSecureMessage
```

正如你所看到的，只有 SecureBean.writeSecureMessage()的第一次调用被允许执行。其余调用都因为 SecurityAdvice 抛出的 SecurityException 异常而被阻止。虽然该例很简单，但却很好地演示了前置通知的用处。安全性是前置通知的典型示例，同时当要求修改传递给方法的参数时，你会发现它也很有用。

5.4.8 创建后置返回通知

后置返回通知在连接点返回方法调用后执行。由于方法已经执行，因此不能更改传递给方法的参数。虽然可以读取这些参数，但不能更改执行路径，也不能阻止该方法执行。这些限制都在预料之中；但是，没有预料到的是，无法在后置返回通知中修改返回值。当使用后置返回通知时，只能添加处理。尽管后置返回通知不能修改方法调用的返回值，但却可以抛出可以发送到堆栈的异常(而不是返回值)。

在本节中，将会演示在应用程序中使用后置返回通知的两个示例。第一个示例简单地在调用方法后将消息写入控制台输出。第二个示例演示了如何使用后置返回通知将错误检查添加到方法中。假设有一个类 KeyGenerator，它为加密目的而生成密钥。许多加密算法都存在一个问题，即少数密钥被认为是弱密钥。弱密钥是指那些在不知道密钥的情况下可以轻松获得原始消息的密钥。对于 DES 算法，总共有 256 个可能的密钥。在这个密钥空间中，有 4 个密钥被认为是弱密钥，另外 12 个密钥被认为是半弱密钥。尽管随机生成这些密钥之一的机会很小(252 分之 1)，但对密钥进行测试非常简单，因此值得这样做。在本节的第二个示例中，创建了后置返回通知，用于检查由 KeyGenerator 生成的弱密钥，如果找到，则引发异常。

有关弱密钥和密码学的更多信息，建议访问 William Stallings 的网站 http://williamstallings.com/Cryptography/。

在下面的代码片段中，可以看到 SimpleAfterReturningAdvice 类，该类通过在方法返回后向控制台输出写入消息来演示如何使用后置返回通知。此时重复使用了前面介绍的 Guitarist 类。

```
package com.apress.prospring5.ch5;

import java.lang.reflect.Method;

import org.springframework.aop.AfterReturningAdvice;
import org.springframework.aop.framework.ProxyFactory;

public class SimpleAfterReturningAdvice implements
            AfterReturningAdvice {
    public static void main(String... args) {
        Guitarist target = new Guitarist();

        ProxyFactory pf = new ProxyFactory();

        pf.addAdvice(new SimpleAfterReturningAdvice());
        pf.setTarget(target);

        Guitarist proxy = (Guitarist) pf.getProxy();
        proxy.sing();
    }

    @Override
    public void afterReturning(Object returnValue, Method method,
          Object[] args, Object target) throws Throwable {
        System.out.println("After '" + method.getName()+ "' put down guitar.");
    }
}
```

请注意，AfterReturningAdvice 接口声明了一个方法 afterReturning()，为它传入方法调用的返回值、对所调用方法的引用、传递给方法的参数以及调用的目标。运行此例将生成以下输出：

```
You're gonna live forever in me
After 'sing' put down guitar.
```

该输出与前置通知示例的输出类似，不同之处在于，后置返回通知所写入的消息是在 writeMessage()方法所写入的消息之后出现的。后置返回通知的一个很好用法是在方法可能返回无效值时执行一些额外的错误检查。

在前面描述的场景中，密钥生成器有可能生成对于特定算法来说被认为较弱的密钥。在理想情况下，密钥生成器会检查这些弱密钥，但由于这些密钥的出现机会通常很小，因此许多生成器并不会进行检查。通过使用后置返回通知，可以通知方法生成密钥并执行这种额外检查。以下是一个非常原始的密钥生成器：

```
package com.apress.prospring5.ch5;

import java.util.Random;

public class KeyGenerator {
    protected static final long WEAK_KEY = 0xFFFFFFF0000000L;
```

```
    protected static final long STRONG_KEY = 0xACDF03F590AE56L;
    private Random rand = new Random();

    public long getKey() {
        int x = rand.nextInt(3);

        if (x == 1) {
            return WEAK_KEY;
        }

        return STRONG_KEY;
    }
}
```

该密钥生成器不应该被认为是安全的。该例被故意设计得非常简单，并且有三分之一的概率产生弱密钥。在下面的代码片段中，可以看到 WeakKeyCheckAdvice，它检查 getKey()方法的结果是否是弱密钥：

```
package com.apress.prospring5.ch5;

import java.lang.reflect.Method;
import org.springframework.aop.AfterReturningAdvice;

import static com.apress.prospring5.ch5.KeyGenerator.WEAK_KEY;

public class WeakKeyCheckAdvice implements AfterReturningAdvice {
    @Override
    public void afterReturning(Object returnValue, Method method,
            Object args,Object target) throws Throwable {

        if ((target instanceof KeyGenerator)
                && ("getKey".equals(method.getName()))) {
            long key = ((Long) returnValue).longValue();

            if (key == WEAK_KEY) {
                throw new SecurityException(
                    "Key Generator generated a weak key. Try again");
            }
        }
    }
}
```

在 afterReturning()方法中，首先检查在连接点上执行的方法是否是 getKey()方法。如果是，则检查结果值，看看它是否是弱密钥。如果发现 getKey()方法的结果是一个弱密钥，那么抛出 SecurityException 异常来通知该方法的调用代码。以下代码片段显示了一个简单的应用程序，演示了如何使用后置返回通知：

```
package com.apress.prospring5.ch5;

import org.springframework.aop.framework.ProxyFactory;

public class AfterAdviceDemo {
    private static KeyGenerator getKeyGenerator() {
        KeyGenerator target = new KeyGenerator();
        ProxyFactory factory = new ProxyFactory();
        factory.setTarget(target);
        factory.addAdvice(new WeakKeyCheckAdvice());

        return (KeyGenerator)factory.getProxy();
    }

    public static void main(String... args) {
        KeyGenerator keyGen = getKeyGenerator();

        for(int x = 0; x < 10; x++) {
            try {
                long key = keyGen.getKey();
                System.out.println("Key: " + key);
            } catch(SecurityException ex) {
                System.out.println("Weak Key Generated!");
            }
        }
    }
}
```

在创建了 KeyGenerator 目标的被通知代理后，AfterAdviceDemo 类会尝试生成十个密钥。如果在每次生成密钥的过程中都抛出 SecurityException 异常，则会向控制台写入消息，通知用户生成了弱密钥；否则，显示所生成的密钥。将示例在我们的机器上运行一次，生成以下输出：

```
Key: 48658904092028502
Weak Key Generated!
Key: 48658904092028502
Weak Key Generated!
Weak Key Generated!
Weak Key Generated!
Key: 48658904092028502
Key: 48658904092028502
Key: 48658904092028502
Key: 48658904092028502
```

如你所见，KeyGenerator 类有时会和预期的一样生成弱密钥，并且 WeakKeyCheckAdvice 可确保在遇到弱密钥时引发 SecurityException 异常。

5.4.9 创建环绕通知

环绕通知(*around advice*)功能类似于前置通知和后置通知功能的组合，但存在一个很大的区别：可以修改返回值。不仅如此，还可以阻止方法执行。这意味着通过使用环绕通知，基本上可以用新代码替换整个方法的实现。通过使用 MethodInterceptor 接口，可以将 Spring 中的环绕通知模型化为拦截器。环绕通知有许多用途，你会发现 Spring 的许多功能都是使用方法拦截器创建的，例如远程代理支持和事务管理功能。方法拦截还是分析应用程序执行的一种很好的机制，它构成了本节中示例的基础。

在此，我们并不会演示如何为方法拦截建立一个简单的示例；相反，将引用前面第一个使用了 Agent 类的示例，演示如何使用基本方法拦截器在方法调用的任何一侧写入消息。在前面的示例中，可以看到，MethodInterceptor 接口的 invoke()方法没有提供与 MethodBeforeAdvice 和 AfterReturningAdvice 相同的参数集合。该方法未被传入调用的目标、方法或所使用的参数。但是，通过使用传递给 invoke()的 MethodInvocation 对象可以访问此类数据。在下面的示例中你会看到相关的演示。

对于本例，我们想要通过某种方式来通知一个类，以便获取有关其方法的运行时性能的基本信息。具体而言，想知道该方法执行多长时间。为了达到这个目的，可以使用 Spring 中包含的 StopWatch 类，并且需要一个 MethodInterceptor，因为需要在方法调用之前启动 StopWatch 并在调用之后停止它。

下面的代码片段显示了将要通过 StopWatch 类和环绕通知进行配置的 WorkerBean 类：

```java
package com.apress.prospring5.ch5;

public class WorkerBean {
    public void doSomeWork(int noOfTimes) {
        for(int x = 0; x < noOfTimes; x++) {
            work();
        }
    }

    private void work() {
        System.out.print("");
    }
}
```

这是一个非常简单的类。DoSomeWork()方法接收一个参数 noOfTimes，并且完全按照此方法指定的次数调用 work()方法。work()方法只是对 System.out.print()进行了一次虚拟调用，它传入一个空的 String。这样可以防止编译器优化 work()方法，从而直接调用 work()。

在下面的代码片段中，可以看到使用 StopWatch 类来分析方法调用时间的 ProfilingInterceptor 类。此时，使用该拦截器来分析前面显示的 WorkerBean 类。

```java
package com.apress.prospring5.ch5;

import java.lang.reflect.Method;

import org.aopalliance.intercept.MethodInterceptor;
import org.aopalliance.intercept.MethodInvocation;
import org.springframework.util.StopWatch;

public class ProfilingInterceptor implements MethodInterceptor {
    @Override
    public Object invoke(MethodInvocation invocation) throws Throwable {
        StopWatch sw = new StopWatch();
```

```
        sw.start(invocation.getMethod().getName());

        Object returnValue = invocation.proceed();
        sw.stop();
        dumpInfo(invocation, sw.getTotalTimeMillis());
        return returnValue;
    }

    private void dumpInfo(MethodInvocation invocation, long ms) {
        Method m = invocation.getMethod();
        Object target = invocation.getThis();
        Object args = invocation.getArguments();

        System.out.println("Executed method: " + m.getName());
        System.out.println("On object of type: " +
            target.getClass().getName());

        System.out.println("With arguments:");
        for (int x = 0; x < args.length; x++) {
            System.out.print(" > " + argsx);
        }
        System.out.print("\n");

        System.out.println("Took: " + ms + " ms");
    }
}
```

在 invoke()方法(MethodInterceptor 接口中唯一的方法)中，首先创建一个 StopWatch 实例，然后立即运行它，允许方法调用继续调用 MethodInvocation.proceed()。一旦方法调用结束并且捕获返回值，就停止 StopWatch 并将所得的总毫秒数与 MethodInvocation 对象一起传递给 dumpInfo()方法。最后，返回 MethodInvocation.proceed()所返回的 Object，以便调用者获得正确的返回值。此时，不希望以任何方式破坏调用堆栈；我们只是作为方法调用的侦听者。但如果需要，可以完全更改调用堆栈，将方法调用重定向到另一个对象或远程服务，也可以简单地重新实现拦截器内的方法逻辑并返回不同的返回值。

dumpInfo()方法只是将有关方法调用的一些信息写入控制台输出，其中包括方法执行的时间。在 dumpInfo()的前三行中，可以看到如何使用 MethodInvocation 对象来确定被调用的方法、调用的原始目标以及使用的参数。

以下代码示例显示了 ProfilingDemo 类，它首先通过 ProfilingInterceptor 通知 WorkerBean 的一个实例，然后分析 doSomeWork()方法。

```
package com.apress.prospring5.ch5;

import org.springframework.aop.framework.ProxyFactory;

public class ProfilingDemo {
    public static void main(String... args) {
        WorkerBean bean = getWorkerBean();
        bean.doSomeWork(10000000);
    }

    private static WorkerBean getWorkerBean() {
        WorkerBean target = new WorkerBean();
        ProxyFactory factory = new ProxyFactory();
        factory.setTarget(target);
        factory.addAdvice(new ProfilingInterceptor());

        return (WorkerBean)factory.getProxy();
    }
}
```

运行该例会产生以下输出：

```
Executed method: doSomeWork
On object of type: com.apress.prospring5.ch5.WorkerBean
With arguments:
        > 10000000
Took: 1139 ms
```

从输出可以看到哪个方法被执行、目标类是什么、传入了哪些参数以及调用需要多长时间。

5.4.10 创建异常通知

*异常通知*类似于后置返回通知，因为它也在连接点之后执行，也是一个方法调用，但只有当方法抛出异常时异

常通知才会执行。此外，异常通知与后置返回通知还有一点相似，因为它几乎不能控制程序的执行。如果使用了异常通知，则不能选择忽略引发的异常而返回方法的值。可以对程序流进行的唯一修改是更改抛出的异常类型。这是一个相当强大的概念，可以使应用程序开发更简单。考虑一种情况，假设有一个 API 抛出了一组定义不明确的异常。通过使用异常通知，可以通知该 API 中的所有类，并对异常层次结构重新进行分类，使其更易于管理和描述。当然，也可以使用异常通知在整个应用程序中提供集中式错误日志记录，从而减少散布在应用程序中的错误日志代码数量。

正如从图 5-2 所示的图表中看到的那样，ThrowsAdvice 接口实现了异常通知。与你在前面看到的接口不同，ThrowsAdvice 没有定义任何方法；相反，它只是 Spring 使用的标记接口。其原因是 Spring 允许类型化的异常通知，从而能够准确地定义异常通知应该捕获哪些异常类型。Spring 通过使用反射来检测具有特定签名的方法，从而实现了上述功能。Spring 寻找两种不同的方法签名。最好的方法是用一个简单的示例来说明。下面的代码片段演示了一个简单的 bean，它包含两个方法，两者都会简单地抛出不同类型的异常：

```
package com.apress.prospring5.ch5;

public class ErrorBean {
    public void errorProneMethod() throws Exception {
        throw new Exception("Generic Exception");
    }

    public void otherErrorProneMethod() throws IllegalArgumentException {
        throw new IllegalArgumentException("IllegalArgument Exception");
    }
}
```

下面的代码显示了 SimpleThrowsAdvice 类，它演示了 Spring 在异常通知上查找的两个方法签名：

```
package com.apress.prospring5.ch5;

import java.lang.reflect.Method;

import org.springframework.aop.ThrowsAdvice;
import org.springframework.aop.framework.ProxyFactory;

public class SimpleThrowsAdvice implements ThrowsAdvice {
    public static void main(String... args) throws Exception {
        ErrorBean errorBean = new ErrorBean();

        ProxyFactory pf = new ProxyFactory();
        pf.setTarget(errorBean);
        pf.addAdvice(new SimpleThrowsAdvice());

        ErrorBean proxy = (ErrorBean) pf.getProxy();
        try {
            proxy.errorProneMethod();
        } catch (Exception ignored) {

        }

        try {
            proxy.otherErrorProneMethod();
        } catch (Exception ignored) {

        }
    }

    public void afterThrowing(Exception ex) throws Throwable {
        System.out.println("***");
        System.out.println("Generic Exception Capture");
        System.out.println("Caught: " + ex.getClass().getName());
        System.out.println("***\n");
    }

    public void afterThrowing(Method method, Object args, Object target,
       IllegalArgumentException ex) throws Throwable {
       System.out.println("***");
       System.out.println("IllegalArgumentException Capture");
       System.out.println("Caught: " + ex.getClass().getName());
       System.out.println("Method: " + method.getName());
           System.out.println("***\n");
    }
}
```

Spring 在异常通知中寻找的第一方法是一个或多个被称为 afterThrowing()的公共方法。方法的返回类型并不重要，但最好是 void，因为此方法不能返回任何有意义的值。SimpleThrowsAdvice 类中的第一个 afterThrowing()方法有一个 Exception 类型的参数。可以指定任何类型的 Exception 作为参数，当不关心抛出异常的方法或传入的参数时，该方法是比较理想的选择。请注意，此方法可以捕获 Exception 以及 Exception 的任何子类型，除非相关类型有自己的 afterThrowing()方法。

在第二个 afterThrowing()方法中，声明四个参数来捕获抛出异常的方法、传入方法的参数以及方法调用的目标。此方法中参数的顺序很重要，必须指定全部四个参数。请注意，第二个 afterThrowing()方法捕获 IllegalArgumentException 类型(或其子类型)的异常。运行此例将生成以下输出：

```
***
Generic Exception Capture
Caught: java.lang.Exception
***

***
IllegalArgumentException Capture
Caught: java.lang.IllegalArgumentException
Method: otherErrorProneMethod
***
```

正如你所看到的，当抛出常见的旧的 Exception 时，会调用第一个 afterThrowing()方法；但是当抛出 IllegalArgumentException 时，则会调用第二个 afterThrowing()方法。Spring 仅为每个 Exception 调用 afterThrowing()方法，正如你在类 SimpleThrowsAdvice 的示例中所看到的，Spring 将使用其签名包含 Exception 类型的最佳匹配方法。当后置异常通知有两个 afterThrowing()方法，并且两者都用相同的 Exception 类型声明，但是其中一个方法只有一个参数，而另一个方法有四个参数时，Spring 会调用带有四个参数的 afterThrowing()方法。

后置异常通知在各种情况下都很有用；它允许对整个 Exception 层次结构进行重新分类，并为应用程序构建集中式异常日志记录。我们发现，在调试实时应用程序时，后置异常通知特别有用，因为它允许添加额外的日志记录代码而无须修改应用程序代码。

5.4.11 选择通知类型

一般来说，选择哪种通知类型是根据应用程序的需求确定的，应该根据需要选择最具体的通知类型。也就是说，能够使用前置通知，就不要使用环绕通知。大多数情况下，虽然环绕通知可以完成其他三种通知类型可以完成的所有内容，但可能对于想要实现的内容来说，使用环绕通知有点过度了。如果使用最具体的通知类型，则可以使代码的意图更清晰，并且还可以减少错误发生的可能性。应该考虑使用计算方法调用的通知。当使用前置通知时，需要编写的代码就是计数器，但是当使用环绕通知时，需要记住调用方法并将值返回给调用者。这些小事情可以让虚假的错误蔓延到整个应用程序中。应该保持通知类型尽可能集中，从而减少错误的范围。

5.5 在 Spring 中使用顾问和切入点

到目前为止，你所看到的所有示例都使用了 ProxyFactory 类。该类提供了一种简单的方法来获取和配置自定义用户代码中的 AOP 代理实例。ProxyFactory.addAdvice()方法用于配置代理的通知。此方法在后台委托给 addAdvisor()，创建 DefaultPointcutAdvisor 的实例并使用指向所有方法的切入点对其进行配置。通过这种方式，通知被视为适用于目标上的所有方法。在某些情况下，例如当使用 AOP 进行记录时，可能需要这么做；但在更多情况下，可能需要限制通知适用的方法。

当然，可以简单地在通知中检查被通知的方法是否正确，但这种方法有几个缺点。首先，将可接受的方法列表硬编码到通知中会降低通知的可重用性。通过使用切入点，可以配置通知适用的方法，而无须将相关代码放入通知中；这明显提高了通知的可重用性。将方法列表硬编码到通知中还会对应用程序性能产生影响。为了检查通知中被通知的方法，每次调用目标上的任何方法时都需要执行检查。这明显降低了应用程序的性能。当使用切入点时，对每种方法执行一次检查，并将结果缓存起来供以后使用。不使用切入点来限制通知适用的方法所带来的另一个与性能相关的缺点是，Spring 可以在创建代理时对未通知方法进行优化，从而加快对未通知方法的调用。在本章后面讨论代理时，会更加详细地讨论这些优化。

强烈建议避免将硬编码的方法检查放入通知，而是尽可能使用切入点来控制通知对目标上方法的适用性。也就是说，在某些情况下，可以将检查代码硬编码到通知中。思考一下前面的后置返回通知示例，该通知旨在捕获由KeyGenerator类生成的弱密钥。这种类型的通知与所通知的类紧密耦合，明智的做法是在通知内部进行检查以确保它被应用于正确的类型。我们将通知与目标之间的这种耦合称为*目标关联性(target affinity)*。一般来说，当通知具有很少或没有目标关联性时，应该使用切入点。也就是说，通知可以适用于任何类型或范围广泛的类型。当通知具有较强的目标关联性时，应该在通知内检查通知是否被正确使用；这有助于在通知被误用时减少出现令人头痛的错误。此外，还建议避免不必要地通知方法。正如稍后将看到的，这样做会导致调用速度明显下降，从而可能对应用程序的整体性能产生巨大影响。

5.5.1 Pointcut 接口

Spring 中的切入点通过实现 Pointcut 接口来创建，如下所示：

```
package org.springframework.aop;

public interface Pointcut {
    ClassFilter getClassFilter ();
    MethodMatcher getMethodMatcher();
}
```

正如从这段代码中看到的那样，Pointcut接口定义了两个方法——getClassFilter()和getMethodMatcher()，它们分别返回ClassFilter和MethodMatcher的实例。显然，如果选择实现Pointcut接口，则需要实现这些方法。但值得庆幸的是，正如将在下一节中看到的那样，通常并不需要这么做，因为Spring提供了一些可供选择的Pointcut实现，它们覆盖了大部分(但不是全部)用例。

当确定 Pointcut 是否适用于特定的方法时，Spring 首先使用 Pointcut.getClassFilter()返回的 ClassFilter 实例检查Pointcut接口是否适用于该方法的类。下面显示了ClassFilter接口：

```
org.springframework.aop;

public interface ClassFilter {
    boolean matches(Class<?> clazz);
}
```

正如你所看到的，ClassFilter接口定义了单个方法matches()，并传入一个Class实例来表示需要检查的类。毫无疑问，如果切入点适用于类，matches()方法返回true，否则返回false。

MethodMatcher接口比ClassFilter接口更复杂，如下所示：

```
package org.springframework.aop;

public interface MethodMatcher {
    boolean matches(Method m, Class<?> targetClass);
    boolean isRuntime();
    boolean matches(Method m, Class<?> targetClass, Object[] args);
}
```

Spring支持两种类型的MethodMatcher——静态和动态MethodMatcher，具体是哪种类型由isRuntime()的返回值决定。在使用MethodMatcher之前，Spring调用isRuntime()来确定MethodMatcher是静态的还是动态的，返回值为false表示是静态的，返回值为true表示是动态的。

对于静态切入点，Spring会针对目标上的每个方法调用一次MethodMatcher的matches(Method，Class<T>)方法，并缓存返回值，以便这些方法在后续调用。这样一来，只会对每个方法进行一次方法适用性检查，方法的后续调用将不会再调用matches()。

即使使用动态切入点，在第一次调用方法来确定方法的整体适用性时，Spring也仍然通过使用matches(Method，Class<T>)执行静态检查。如果静态检查返回true，那么Spring将使用matches(Method，Class<T>，Object[])方法对每个方法的调用执行进一步检查。这样一来，动态MethodMatcher可以根据特定的方法调用(而不仅仅是方法本身)来确定切入点是否应该应用。例如，只有当参数是一个值大于100的Integer时才需要应用切入点。在这种情况下，可以编写matches(Method，Class<T>，Object[])方法来对每个调用的参数进行进一步检查。

显然，静态切入点比动态切入点执行得更好，因为它们避免了每次调用时所要进行的额外检查；而动态切入点在决定是否应用通知方面提供了更大的灵活性。一般来说，建议尽可能使用静态切入点。但是，如果所使用的通知增加了大量开销，那么通过使用动态切入点来避免任何不必要的通知调用可能是比较明智的做法。

一般来说，很少需要从头开始创建自己的 Pointcut 实现，因为 Spring 为静态和动态切入点都提供了抽象基类。接下来的几节将介绍这些基类以及其他 Pointcut 实现。

5.5.2 可用的切入点实现

从版本 4.0 开始，Spring 提供了八个 Pointcut 接口的实现：两个用作创建静态和动态切入点的便捷类的抽象类，以及六个具体类，其中每一个具体类完成以下操作。

- 一起构成多个切入点
- 处理控制流切入点
- 执行简单的基于名称的匹配
- 使用正则表达式定义切入点
- 使用 AspectJ 表达式定义切入点
- 定义在类或方法级别查找特定注解的切入点

表 5-2 总结了这八个 Pointcut 接口的实现。

表 5-2 八个 Pointcut 接口的实现

实现类	描述
org.springframework.aop.support.annotation.AnnotationMatchingPointcut	此实现在类或方法上查找特定的 Java 注解。该类需要 JDK 5 或更高版本
org.springframework.aop.aspectj.AspectJExpressionPointcut	此实现使用 AspectJ 织入器以 AspectJ 语法评估切入点表达式
org.springframework.aop.support.ComposablePointcut	ComposablePointcut 类使用诸如 union()和 intersection()等操作组合两个或更多个切入点
org.springframework.aop.support.ControlFlowPointcut	ControlFlowPointcut 是一种特殊的切入点，它们匹配另一个方法的控制流中的所有方法，即任何作为另一个方法的结果而直接或间接调用的方法
org.springframework.aop.support.DynamicMethodMatcherPointcut	此实现旨在作为构建动态切入点的基类
org.springframework.aop.support.JdkRegexpMethodPointcut	该实现允许使用 JDK 1.4 正则表达式支持定义切入点。该类需要 JDK 1.4 或更高版本
org.springframework.aop.support.NameMatchMethodPointcut	通过使用 NameMatchMethodPointcut，可以创建一个切入点，对方法名称列表执行简单匹配
org.springframework.aop.support.StaticMethodMatcherPointcut	StaticMethodMatcherPointcut 类用作构建静态切入点的基础

图 5-3 显示了 Pointcut 实现类的统一建模语言(UML)[1]图。

5.5.3 使用 DefaultPointcutAdvisor

在使用任何 Pointcut 实现之前，必须先创建 Advisor 接口的实例，或者更具体地说创建一个 PointcutAdvisor 接口。请记住，从前面的讨论中可以看出，Advisor 是 Spring 中某个切面的表示(具体内容请参见上一节中的“Spring 中的切面”部分)，它是通知和切入点的结合体，规定了应该通知哪些方法以及如何通知。Spring 提供了许多 PointcutAdvisor 的实现，但现在只重点关注一个，DefaultPointcutAdvisor。这是一个简单的 PointcutAdvisor，用于将单个 Pointcut 与单个 Advice 相关联。

1 UML 在开发过程中是非常重要的，因为它是简化应用程序逻辑并使其可视化的一种方式，可以在编写代码之前轻松检测问题。在新成员加入团队时，它也可以用作文档，使他们尽快投入开发中。可以在 www.uml.org/上找到更多信息。

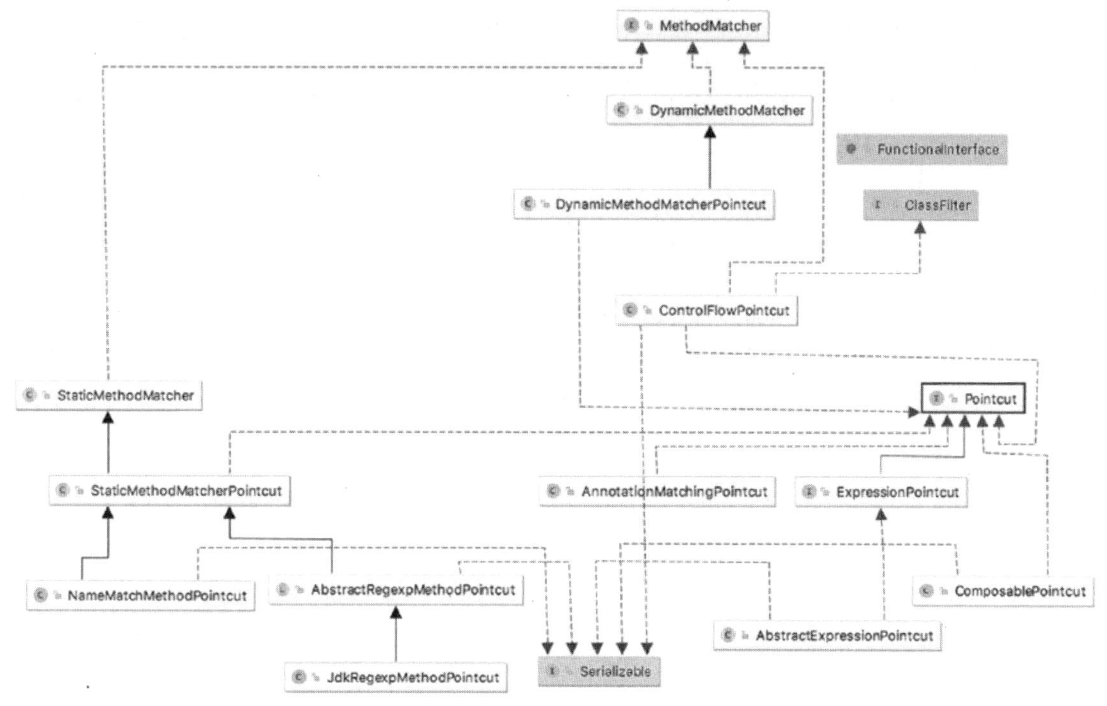

图 5-3　在 Intellij IDEA 中表示为 UML 图的切入点实现类

5.5.4　使用 StaticMethodMatcherPointcut 创建静态切入点

在本节中，将通过扩展抽象类 StaticMethodMatcherPointcut 来创建一个简单的静态切入点。由于 StaticMethodMatcherPointcut 类扩展了实现 MethodMatcher 接口的 StaticMethodMatcher 类(也是一个抽象类)，因此需要实现方法 matches(Method，Class<?>)。其他 Pointcut 实现自动处理。虽然 matches()是需要实现的唯一方法(当扩展 StaticMethodMatcherPointcut 类时)，但也可以重写 getClassFilter()方法，如本例中所示，以确保仅通知正确类型的方法。

对于本例，有两个类，分别是 GoodGuitarist 和 GreatGuitarist，它们都定义了相同的方法，并且都是 Singer 接口中方法的实现。

```
package com.apress.prospring5.ch5;
import com.apress.prospring5.ch2.common.Singer;

public class GoodGuitarist implements Singer {

    @Override public void sing() {
       System.out.println("Who says I can't be free \n" +
             "From all of the things that I used to be");
    }
}

public class GreatGuitarist implements Singer {

    @Override public void sing() {
       System.out.println("I shot the sheriff, \n" +
             "But I did not shoot the deputy");
    }
}
```

在该例中，希望能够通过使用相同的 DefaultPointcutAdvisor 创建这两个类的代理，但只将通知应用于 GoodGuitarist 类的 sing()方法。为此，创建了 SimpleStaticPointcut 类，如下所示：

```
package com.apress.prospring5.ch5;

import java.lang.reflect.Method;

import org.springframework.aop.ClassFilter;
```

```
import org.springframework.aop.support.StaticMethodMatcherPointcut;

public class SimpleStaticPointcut extends StaticMethodMatcherPointcut {
    @Override
    public boolean matches(Method method, Class<?> cls) {
        return ("sing".equals(method.getName()));
    }

    @Override
    public ClassFilter getClassFilter() {
        return cls -> (cls == GoodGuitarist.class);
    }
}
```

可以看到，此时实现了 StaticMethodMatcher 抽象类所需的 matches(Method，Class<?>)方法。如果方法的名称是 sing，则简单地返回 true；否则，返回 false。通过使用 lambda 表达式，在上面的代码示例中隐藏了在 getClassFilter()方法中实现 ClassFilter 的匿名类的创建。扩展的 lambda 表达式如下所示：

```
public ClassFilter getClassFilter() {
    return new ClassFilter() {
        public boolean matches(Class<?> cls) {
            return (cls == GoodGuitarist.class);
        }
    };
}
```

请注意，还重写了 getClassFilter()方法，并返回一个 ClassFilter 实例，该实例的 matches()方法仅针对 GoodGuitarist 类返回 true。有了这个静态切入点，可以说只有 GoodGuitarist 类的方法才会匹配，更具体地说，只有该类的 sing()方法才会匹配。

下面的代码片段显示了 SimpleAdvice 类，该类简单地在方法调用的任意一侧写出消息：

```
package com.apress.prospring5.ch5;

import org.aopalliance.intercept.MethodInterceptor;
import org.aopalliance.intercept.MethodInvocation;

public class SimpleAdvice implements MethodInterceptor {
    @Override
    public Object invoke(MethodInvocation invocation) throws Throwable {
        System.out.println(">> Invoking " + invocation.getMethod().getName());
        Object retVal = invocation.proceed();
        System.out.println(">> Done\n");
        return retVal;
    }
}
```

在下面的代码片段中，可以看到一个简单的驱动应用程序，该例使用 SimpleAdvice 和 SimpleStaticPointcut 类创建了 DefaultPointcutAdvisor 实例。另外，因为这两个类实现了相同的接口，所以可以基于接口而不是具体的类来创建代理。

```
package com.apress.prospring5.ch5;

import com.apress.prospring5.ch2.common.Singer;
import org.aopalliance.aop.Advice; import org.springframework.aop.Advisor;
import org.springframework.aop.Pointcut;
import org.springframework.aop.framework.ProxyFactory;
import org.springframework.aop.support.DefaultPointcutAdvisor;

public class StaticPointcutDemo {
    public static void main(String... args) {
        GoodGuitarist johnMayer = new GoodGuitarist();
        GreatGuitarist ericClapton = new GreatGuitarist();

        Singer proxyOne;
        Singer proxyTwo;
        Pointcut pc = new SimpleStaticPointcut();
        Advice advice = new SimpleAdvice();
        Advisor advisor = new DefaultPointcutAdvisor(pc, advice);

        ProxyFactory pf = new ProxyFactory();
        pf.addAdvisor(advisor);
        pf.setTarget(johnMayer);
        proxyOne = (Singer)pf.getProxy();

        pf = new ProxyFactory();
```

```
        pf.addAdvisor(advisor);
        pf.setTarget(ericClapton);
        proxyTwo = (Singer)pf.getProxy();

        proxyOne.sing();
        proxyTwo.sing();
    }
}
```

请注意，DefaultPointcutAdvisor 实例还被用来创建两个代理：一个用于 GoodGuitarist 的实例，另一个用于 EricClapton 的实例。最后，在两个代理上调用 sing()方法。运行此例将生成以下输出：

```
>> Invoking sing
Who says I can't be free
From all of the things that I used to be
>> Done

I shot the sheriff,
But I did not shoot the deputy
```

如你所见，SimpleAdvice 实际调用的唯一方法是 GoodGuitarist 类的 sing()方法，与预期完全相同。正如在讨论代理选项时你所看到的，限制通知适用的方法非常简单，并且对于应用程序获得最佳性能至关重要。

5.5.5 使用 DyanmicMethodMatcherPointcut 创建动态切入点

创建动态切入点与创建静态切入点没有多大区别，因此在本例中，将使用下面所示的类创建动态切入点：

```
package com.apress.prospring5.ch5;

public class SampleBean {
    public void foo(int x) {
        System.out.println("Invoked foo() with: " + x);
    }
    public void bar() {
        System.out.println("Invoked bar()");
    }
}
```

对于本例，只想通知 foo()方法，但与前面示例不同的是，只有在传入的 int 参数大于或小于 100 时才通知该方法。

与静态切入点一样，Spring 为创建动态切入点提供了一个方便的基类：DynamicMethodMatcherPointcut。DynamicMethodMatcherPointcut 类有一个必须实现的抽象方法 matches(Method,Class<?>,Object[])(通过所实现的 MethodMatcher 接口)，但是你会看到，通过实现 matches(Method,Class<?>)方法来控制静态检查行为是比较谨慎的做法。以下代码片段显示了 SimpleDynamicPointcut 类：

```
package com.apress.prospring5.ch5;

import java.lang.reflect.Method;

import org.springframework.aop.ClassFilter;
import org.springframework.aop.support.DynamicMethodMatcherPointcut;

public class SimpleDynamicPointcut
        extends DynamicMethodMatcherPointcut {
    @Override
    public boolean matches(Method method, Class<?> cls) {
        System.out.println("Static check for " + method.getName());
        return ("foo".equals(method.getName()));
    }

    @Override
    public boolean matches(Method method, Class<?> cls, Object args) {
        System.out.println("Dynamic check for " + method.getName());

        int x = ((Integer) args0).intValue();

        return (x != 100);
    }

    @Override
    public ClassFilter getClassFilter() {
        return cls -> (cls == SampleBean.class);
    }
}
```

从上面的代码示例中可以看出，这里使用与上一节相似的方式重写了 getClassFilter()方法。此时，无须在方法匹配方法中对类进行检查，这对于动态检查特别重要。虽然只需要实现动态检查，但该例仍然实现了静态检查。原因是 bar()方法永远不会被通知。通过使用静态检查来指明这一点，这样 Spring 就不必为此方法执行动态检查。这是因为当执行静态检查方法时，Spring 将首先检查 bar()方法，如果检查结果不匹配，Spring 将停止进一步的动态检查。此外，静态检查的结果将被缓存以获得更好的性能。但是如果省略静态检查，Spring 则会在每次调用 bar()方法时执行动态检查。推荐的做法是，在 getClassFilter()方法中执行类检查，在 matches(Method,Class<?>)方法中执行方法检查以及在 matches(Method,Class<?>,Object[])方法中执行参数检查。这样就会让切入点更容易理解和维护，并且性能也会更好。

在 matches(Method,Class<?>,Object[])方法中，如果传递给 foo()方法的 int 参数的值不等于 100，则返回 false；否则，返回 true。请注意，在动态检查中，我们知道正在处理的是 foo()方法，因为没有其他方法可以通过静态检查。在下面的代码片段中，可以看到这个切入点的实例：

```
package com.apress.prospring5.ch5;

import org.springframework.aop.Advisor;
import org.springframework.aop.framework.ProxyFactory;
import org.springframework.aop.support.DefaultPointcutAdvisor;

public class DynamicPointcutDemo {
    public static void main(String... args) {
        SampleBean target = new SampleBean();

        Advisor advisor = new DefaultPointcutAdvisor(
           new SimpleDynamicPointcut(), new SimpleAdvice());

        ProxyFactory pf = new ProxyFactory();
        pf.setTarget(target);
           pf.addAdvisor(advisor);
        SampleBean proxy = (SampleBean)pf.getProxy();

        proxy.foo(1);
        proxy.foo(10);
        proxy.foo(100);

        proxy.bar();
        proxy.bar();
        proxy.bar();
    }
}
```

请注意，此时使用了与静态切入点示例中相同的通知类。然而，在本例中，只应该通知前两次 foo()调用。动态检查将阻止第三次 foo()调用被通知，而静态检查则阻止通知 bar()方法。运行该例将产生以下输出：

```
Static check for bar
Static check for foo
Static check for toString
Static check for clone
Static check for foo
Dynamic check for foo
>> Invoking foo
Invoked foo() with: 1
>> Done
Dynamic check for foo
>> Invoking foo
Invoked foo() with: 10
>> Done

Dynamic check for foo
Invoked foo() with: 100
Static check for bar
Invoked bar()
Invoked bar()
Invoked bar()
```

正如所预料的，只有前两个 foo()方法的调用被通知。请注意，bar()调用没有进行动态检查，这主要是因为在 bar()方法上所做的静态检查。此时需要注意的一点是，foo()方法进行了两次静态检查：一次是在初始阶段，当所有方法被检查时；另一次是在第一次被调用时。

正如你所看到的，动态切入点能比静态切入点提供更大的灵活性，但是由于它们需要额外的运行时间开销，因此只有在绝对必要时才应该使用动态切入点。

5.5.6 使用简单名称匹配

通常，在创建切入点时，希望仅根据方法名称进行匹配，而忽略方法签名和返回类型。在这种情况下，应该避免创建 StaticMethodMatcherPointcut 的子类，并使用 NameMatchMethodPointcut(它是 StaticMethodMatcherPointcut 的子类)与方法名称列表进行匹配。当使用 NameMatchMethodPointcut 时，无须考虑方法签名，所以对于方法 sing()和 sing(guitar)来说，它们都会与名称 foo 相匹配。

在下面的代码片段中，可以看到 GrammyGuitarist 类，它是 Singer 的另一个实现，这位格莱美奖歌手可以用他的声音唱歌，同时还可以使用吉他，并且可以会谈和休息。

```
package com.apress.prospring5.ch5;

import com.apress.prospring5.ch2.common.Guitar;
import com.apress.prospring5.ch2.common.Singer;

public class GrammyGuitarist implements Singer {

    @Override public void sing() {
       System.out.println("sing: Gravity is working against me\n" +
    "And gravity wants to bring me down");
    }

    public void sing(Guitar guitar) {
       System.out.println("play: " + guitar.play());
    }

    public void rest(){
       System.out.println("zzz");
    }

    public void talk(){
       System.out.println("talk");
    }
}

//chapter02/hello-world/src/main/java/com/apress/prospring5/ch2/common/Guitar.java
package com.apress.prospring5.ch2.common;

public class Guitar {
    public String play(){
       return "G C G C Am D7";
    }
}
```

对于本例，希望通过使用 NameMatchMethodPointcut 来匹配 sing()、sing(guitar)和 rest()方法，从而转换为匹配名称 foo 和 bar，如以下代码片段所示：

```
package com.apress.prospring5.ch5;

import com.apress.prospring5.ch2.common.Guitar;
import org.springframework.aop.Advisor;
import org.springframework.aop.framework.ProxyFactory;
import org.springframework.aop.support.DefaultPointcutAdvisor;
import org.springframework.aop.support.NameMatchMethodPointcut;

public class NamePointcutDemo {

    public static void main(String... args) {
       GrammyGuitarist johnMayer = new GrammyGuitarist();

       NameMatchMethodPointcut pc = new NameMatchMethodPointcut();
       pc.addMethodName("sing");
       pc.addMethodName("rest");

       Advisor advisor = new DefaultPointcutAdvisor(pc, new SimpleAdvice());
       ProxyFactory pf = new ProxyFactory();
       pf.setTarget(johnMayer);
       pf.addAdvisor(advisor);

       GrammyGuitarist proxy = (GrammyGuitarist) pf.getProxy();
       proxy.sing();
```

```
            proxy.sing(new Guitar());
            proxy.rest();
            proxy.talk();
        }
    }
```

没必要为切入点创建一个类；可以简单地创建 NameMatchMethodPointcut 的一个实例，并且根据需要使用就可以了。请注意，此时使用 addMethodName()方法为切入点添加了两个方法名称：sing 和 rest。运行此例将生成以下输出：

```
>> Invoking sing
sing: Gravity is working against me
And gravity wants to bring me down
>> Done

>> Invoking sing
play: G C G C Am D7
>> Done

>> Invoking rest
zzz
>> Done

talk
```

正如预期的那样，由于存在切入点，sing()、sing(Guitar)和 rest()方法被通知，但 talk()方法却没有被通知。

5.5.7 用正则表达式创建切入点

在上一节中，讨论了如何针对预定义的方法列表执行简单匹配。但是如果事先不知道所有方法的名称，而是只知道名称所遵循的模式，又该怎么做呢？例如，如果想匹配名称以 get 开头的所有方法，该怎么办？在这种情况下，可以使用正则表达式切入点 JdkRegexpMethodPointcut 来匹配基于正则表达式的方法名称。下面的示例将演示另一个 Guitarist 类，它包含三个方法：

```
package com.apress.prospring5.ch5;

import com.apress.prospring5.ch2.common.Singer;

public class Guitarist implements Singer {

    @Override public void sing() {
        System.out.println("Just keep me where the light is");
    }

    public void sing2() {
        System.out.println("Just keep me where the light is");
    }

    public void rest() {
        System.out.println("zzz");
    }
}
```

通过使用基于正则表达式的切入点，可以匹配类中名称以 String 开头的所有方法，如下所示：

```
package com.apress.prospring5.ch5;

import org.springframework.aop.Advisor;
import org.springframework.aop.framework.ProxyFactory;
import org.springframework.aop.support.DefaultPointcutAdvisor;
import org.springframework.aop.support.JdkRegexpMethodPointcut;

public class RegexpPointcutDemo {
    public static void main(String... args) {
        Guitarist johnMayer = new Guitarist();

        JdkRegexpMethodPointcut pc = new JdkRegexpMethodPointcut();
        pc.setPattern(".*sing.*");
        Advisor advisor = new DefaultPointcutAdvisor(pc, new SimpleAdvice());

        ProxyFactory pf = new ProxyFactory();
        pf.setTarget(johnMayer);
        pf.addAdvisor(advisor);
        Guitarist proxy = (Guitarist) pf.getProxy();
```

```
        proxy.sing();
        proxy.sing2();
        proxy.rest();
    }
}
```

请注意，无须为切入点创建一个类；相反，只需要创建一个 JdkRegexpMethodPointcut 实例并指定要匹配的模式就可以了。值得注意的是模式。当进行方法名称匹配时，Spring 将匹配方法的完全限定名称，因此对于 sing1()，Spring 针对 com.apress.prospring5.ch5.Guitarist.sing1 进行匹配，这就是为什么在模式中以*开头的原因。这是一个非常重要的概念，因为它允许匹配给定包中的所有方法，而不必确切知道该包中有哪些类以及方法名称。运行该例产生以下输出：

```
>> Invoking sing
Just keep me where the light is
>> Done

>> Invoking sing2
Oh gravity, stay the hell away from me
>> Done

zzz
```

正如期望的那样，只有 sing()和 sing2()方法被通知，因为 rest()方法与正则表达式模式不匹配。

5.5.8 使用 AspectJ 切入点表达式创建切入点

除了 JDK 正则表达式，还可以使用 AspectJ 的切入点表达式语言进行切入点声明。在本章的后面你将会看到，当使用 aop 名称空间在 XML 配置中声明切入点时，Spring 默认使用 AspectJ 的切入点语言。而且当使用 Spring 的@AspectJ 注解类型的 AOP 支持时，也需要使用 AspectJ 的切入点语言。所以当想要使用表达式语言来声明切入点时，使用 AspectJ 切入点表达式是最好的方式。Spring 提供了 AspectJExpressionPointcut 类来通过 AspectJ 的表达式语言定义切入点。要在 Spring 中使用 AspectJ 切入点表达式，需要在项目的类路径中包含两个 AspectJ 库文件，分别是 aspectjrt.jar 和 aspectjweaver.jar。依赖项及其版本则在主 build.gradle 配置文件中配置(并配置为 chapter05 项目中所有模块的依赖项)。

```
ext {
    aspectjVersion = '1.9.0.BETA-5'
...
    misc = [
        ...
        Aspectjweaver     : "org.aspectj:aspectjweaver:$aspectjVersion",
        Aspectjrt         : "org.aspectj:aspectjrt:$aspectjVersion"
    ]
...
```

鉴于前面 Guitarist 类的实现，可以使用 AspectJ 表达式来实现用 JDK 正则表达式所实现的相同功能，代码如下所示：

```
package com.apress.prospring5.ch5;

import org.springframework.aop.Advisor;
import org.springframework.aop.aspectj.AspectJExpressionPointcut;
import org.springframework.aop.framework.ProxyFactory;
import org.springframework.aop.support.DefaultPointcutAdvisor;

public class AspectjexpPointcutDemo {
    public static void main(String... args) {
        Guitarist johnMayer = new Guitarist();

        AspectJExpressionPointcut pc = new AspectJExpressionPointcut();
        pc.setExpression("execution(* sing*(..))");
        Advisor advisor = new DefaultPointcutAdvisor(pc, new SimpleAdvice());

        ProxyFactory pf = new ProxyFactory();
        pf.setTarget(johnMayer);
        pf.addAdvisor(advisor);
        Guitarist proxy = (Guitarist) pf.getProxy();

        proxy.sing();
        proxy.sing2();
```

```
        proxy.rest();
    }
}
```

请注意，此时使用 AspectJExpressionPointcut 的 setExpression()方法来设置匹配标准。表达式 excution(* sing*(..))意味着通知应该适用于任何以sing 开头、具有任何参数并返回任何类型的方法的执行。运行该程序将获得与使用JDK正则表达式的上一个示例相同的结果。

5.5.9 创建注解匹配切入点

如果应用程序基于注解，那么可能希望使用自己指定的注解来定义切入点，也就是将通知逻辑应用于所有具有特定注解的方法或类型。Spring 提供了 AnnotationMatchingPointcut 类来定义使用注解的切入点。接下来，让我们重新使用前面的示例，看看在使用注解作为切入点时应该如何做。

首先定义一个名为 AdviceRequired 的注解，它是将被用于声明切入点的注解。以下代码片段显示了该注解类：

```
package com.apress.prospring5.ch5;

import java.lang.annotation.ElementType;
import java.lang.annotation.Retention;
import java.lang.annotation.RetentionPolicy;
import java.lang.annotation.Target;

@Retention(RetentionPolicy.RUNTIME)
@Target({ElementType.TYPE, ElementType.METHOD})
public @interface AdviceRequired {
}
```

在上面的代码示例中，可以看到通过使用@interface 作为类型将接口声明为注解，并且使用@Target 注解定义了该注解可以在类型或方法级别上应用。下面的代码片段显示了另一个 Guitarist 类实现，其中的一个方法使用了前面定义的注解：

```
package com.apress.prospring5.ch5;

import com.apress.prospring5.ch2.common.Guitar;
import com.apress.prospring5.ch2.common.Singer;

public class Guitarist implements Singer {

    @Override public void sing() {
       System.out.println("Dream of ways to throw it all away");
    }

    @AdviceRequired
    public void sing(Guitar guitar) {
       System.out.println("play: " + guitar.play());
    }

    public void rest(){
       System.out.println("zzz");
    }
}
```

以下代码片段显示了测试程序：

```
package com.apress.prospring5.ch5;

import com.apress.prospring5.ch2.common.Guitar;
import org.springframework.aop.Advisor;
import org.springframework.aop.framework.ProxyFactory;
import org.springframework.aop.support.DefaultPointcutAdvisor;
import org.springframework.aop.support.annotation.AnnotationMatchingPointcut;

public class AnnotationPointcutDemo {
    public static void main(String... args) {
       Guitarist johnMayer = new Guitarist();

       AnnotationMatchingPointcut pc = AnnotationMatchingPointcut
           .forMethodAnnotation(AdviceRequired.class);
       Advisor advisor = new DefaultPointcutAdvisor(pc, new SimpleAdvice());

       ProxyFactory pf = new ProxyFactory();
       pf.setTarget(johnMayer);
       pf.addAdvisor(advisor);
```

```
        Guitarist proxy = (Guitarist) pf.getProxy();

        proxy.sing(new Guitar());
        proxy.rest();
    }
}
```

在上面的代码片段中，通过调用静态方法 forMethodAnnotation()并传入注解类型来获取 AnnotationMatchingPointcut 的实例。这表明希望将通知应用于所有使用给定注解进行注解的方法。也可以通过调用 forClassAnnotation()方法来指定在类级别上应用的注解。以下显示了程序运行时的输出：

```
>> Invoking sing
play: G C G C Am D7
>> Done

zzz
```

正如你所看到的，因为注解了 sing()方法，所以只有该方法被通知。

5.5.10 便捷的 Advisor 实现

对于许多 Pointcut 实现来说，Spring 还提供了一个便捷的 Advisor 实现来充当切入点。例如，可以简单地使用 NameMatchMethodPointcutAdvisor，而不是像前面的示例那样配合使用 NameMatchMethodPointcut 与 DefaultPointcutAdvisor，如下面的代码片段所示：

```
package com.apress.prospring5.ch5;
...
import org.springframework.aop.support.NameMatchMethodPointcutAdvisor;
public class NamePointcutUsingAdvisor {
    public static void main(String... args) {
        GrammyGuitarist johnMayer = new GrammyGuitarist();

        NameMatchMethodPointcut pc = new NameMatchMethodPointcut();
        pc.addMethodName("sing");
        pc.addMethodName("rest");

        Advisor advisor =
           new NameMatchMethodPointcutAdvisor(new SimpleAdvice());
        ProxyFactory pf = new ProxyFactory();
        pf.setTarget(johnMayer);
        pf.addAdvisor(advisor);

        GrammyGuitarist proxy = (GrammyGuitarist) pf.getProxy();
        proxy.sing();
        proxy.sing(new Guitar());
        proxy.rest();
        proxy.talk();
    }
}
```

请注意，此时并没有创建 NameMatchMethodPointcut 实例，而是在 NameMatchMethodPointcutAdvisor 实例上配置切入点的详细信息。通过这种方式，NameMatchMethodPointcutAdvisor 充当了顾问和切入点。

通过浏览 org.springframework.aop.support 包的 Javadoc 文档，可以找到不同 Advisor 实现的完整详细信息。上述两种方法之间没有明显的性能差异，除第二个示例中的代码略少外，实际编码方法中的差异很小。但推荐使用第一种方法，因为代码中的意图更清晰。但最终使用哪种方法取决于个人偏好。

5.6 了解代理

到目前为止，你只是粗略了解了一下 ProxyFactory 所生成的代理。前面提到过，在 Spring 中有两种类型的代理：使用 JDK Proxy 类创建的 JDK 代理以及使用 CGLIB Enhancer 类创建的基于 CGLIB 的代理。你可能想知道这两种代理之间有什么区别，以及为什么 Spring 需要两种代理类型。在本节中，将详细研究代理之间的差异。

代理的核心目标是拦截方法调用，并在必要时执行适用于特定方法的通知链。通知的管理和调用基本上是独立于代理的，由 Spring AOP 框架管理。而代理主要负责拦截对所有方法的调用，并将它们根据需要传递给 AOP 框架，以便应用通知。

除上述核心功能外，代理还必须支持一组附加功能。可以通过 AopContext 类(这是一个抽象类)配置代理以公开自己，以便可以检索代理并从目标对象调用代理上的被通知方法。当通过 ProxyFactory.setExposeProxy()启用该功能时，代理负责确保代理类被适当地公开。另外，所有代理类默认实现 Advised 接口，从而允许在创建代理之后更改通知链。代理还必须确保任何返回代理类(即返回代理目标)的方法实际上返回的是代理而不是目标。

正如你所看到的，典型的代理需要执行很多工作，并且所有这些逻辑都在 JDK 和 CGLIB 代理中实现。

5.6.1 使用 JDK 动态代理

*JDK 代理*是 Spring 中最基本的代理类型。与 CGLIB 代理不同，JDK 代理只能生成接口的代理，而不能生成类的代理。这样一来，想要代理的任何对象都必须至少实现一个接口，并且生成的代理将是实现该接口的对象。图 5-4 显示了这种代理的抽象模式。

一般来说，为类使用接口是一种很好的设计，但并不总是可行的，尤其是当使用第三方或旧代码时。在这种情况下，必须使用 CGLIB 代理。当使用 JDK 代理时，所有方法调用都会被 JVM 拦截并路由到代理的 invoke()方法。然后由 invoke()方法确定是否通知有关方法(根据由切入点定义的规则)，如果确定想要通知，则通过使用反射调用通知链，然后调用方法本身。除此之外，invoke()方法还可以执行前一节中讨论的所有逻辑。

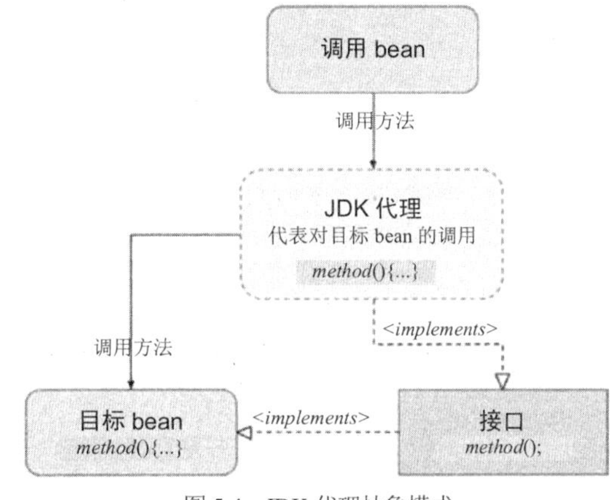

图 5-4 JDK 代理抽象模式

在调用 invoke()之前，JDK 代理无法区分被通知方法和未被通知方法。这意味着对于代理上的未被通知方法，invoke()方法仍然会被调用，所有检查仍然会执行，并且仍然可以通过使用反射进行调用。显然，每次调用方法时，都会导致运行时开销，即使代理不会执行额外的处理，而只是通过反射调用未被通知的方法。

通过使用 setInterfaces()(位于由 ProxyFactory 类间接扩展的 AdvisedSupport 类中)指定要代理的接口列表，从而指示 ProxyFactory 使用 JDK 代理。

5.6.2 使用 CGLIB 代理

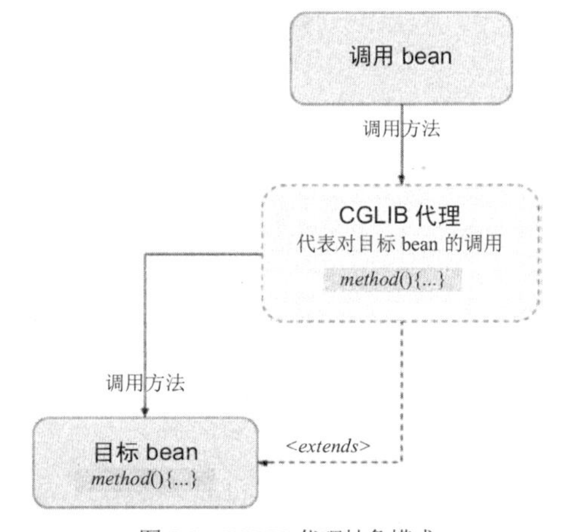

图 5-5 CGLIB 代理抽象模式

如果使用 JDK 代理，那么在每次调用 invoke()方法时，有关如何处理特定方法调用的决策都会在运行时做出。而当使用 CGLIB 时，CGLIB 会为每个代理动态生成新类的字节码，并尽可能重用已生成的类。在这种情况下，所生成的代理类型将是目标对象类的子类。图 5-5 显示了这种代理的抽象模式。

当首次创建 CGLIB 代理时，CGLIB 会询问 Spring 如何处理每个方法。这意味着每次调用 JDK 代理上的 invoke()时所执行的许多决策对于 CGLIB 代理来说只会执行一次。由于 CGLIB 生成实际的字节码，因此在处理方法的方式上有更多的灵活性。例如，CGLIB 代理可以生成适当的字节码来直接调用任何未被通知的方法，从而减少代理所带来的开销。另外，CGLIB 代理可以确定一个方法是否返回代理；如果不返回，则允许直接调用方法调用，从而进一步减少运行时间开销。

CGLIB 代理还以不同于 JDK 代理的方式处理固定通知链。固定通知链是在代理生成后不会更改的链。默认情况下，即使在创建代理后，也可以更改代理上的顾问和通知，虽然很少有必要这么做。CGLIB 代理以特定方式处理固定通知链，从而减少执行通知链的运行时间开销。

5.6.3 比较代理性能

到目前为止，只是以较宽松的方式讨论了代理类型之间实现的差异。在本节中，将运行一个简单的测试来比较一下 CGLIB 代理与 JDK 代理的性能。

首先，创建一个名为 DefaultSimpleBean 的类，将其用作代理的目标对象。下面显示了 SimpleBean 接口和 DefaultSimpleBean 类：

```
ppackage com.apress.prospring5.ch5;

public interface SimpleBean {
    void advised();
    void unadvised();
}

public class DefaultSimpleBean implements SimpleBean {
    private long dummy = 0;
    @Override
    public void advised() {
       dummy = System.currentTimeMillis();
    }

    @Override
    public void unadvised() {
       dummy = System.currentTimeMillis();
    }
}
```

在以下示例中，将显示 TestPointcut 类，该类用于对通知的方法进行静态检查：

```
package com.apress.prospring5.ch5;

import java.lang.reflect.Method;

import org.springframework.aop.support.StaticMethodMatcherPointcut;

public class TestPointcut extends StaticMethodMatcherPointcut {
    @Override
    public boolean matches(Method method, Class cls) {
       return ("advise".equals(method.getName()));
    }
}
```

下一个代码片段描述了 NoOpBeforeAdvice 类，它是一个简单的前置通知，没有任何操作：

```
package com.apress.prospring5.ch5;

import java.lang.reflect.Method;

import org.springframework.aop.MethodBeforeAdvice;

public class NoOpBeforeAdvice implements MethodBeforeAdvice {
    @Override
    public void before(Method method, Object args, Object target)
       throws Throwable {
       // no-op
    }
}
```

在下面的代码片段中，可以看到用于测试不同代理类型的代码：

```
package com.apress.prospring5.ch5;

import org.springframework.aop.Advisor;
import org.springframework.aop.framework.Advised;
import org.springframework.aop.framework.ProxyFactory;
import org.springframework.aop.support.DefaultPointcutAdvisor;

public class ProxyPerfTest {
    public static void main(String... args) {
       SimpleBean target = new DefaultSimpleBean();

       Advisor advisor = new DefaultPointcutAdvisor(new TestPointcut(),
          new NoOpBeforeAdvice());

       runCglibTests(advisor, target);
       runCglibFrozenTests(advisor, target);
```

```
        runJdkTests(advisor, target);
    }

    private static void runCglibTests(Advisor advisor, SimpleBean target) {
        ProxyFactory pf = new ProxyFactory();
        pf.setProxyTargetClass(true);
        pf.setTarget(target);
        pf.addAdvisor(advisor);

        SimpleBean proxy = (SimpleBean)pf.getProxy();
        System.out.println("Running CGLIB (Standard) Tests");
        test(proxy);
    }

    private static void runCglibFrozenTests(Advisor advisor, SimpleBean target) {
        ProxyFactory pf = new ProxyFactory();
        pf.setProxyTargetClass(true);
        pf.setTarget(target);
        pf.addAdvisor(advisor);
        pf.setFrozen(true);

        SimpleBean proxy = (SimpleBean) pf.getProxy();
        System.out.println("Running CGLIB (Frozen) Tests");
        test(proxy);
    }

    private static void runJdkTests(Advisor advisor, SimpleBean target) {
        ProxyFactory pf = new ProxyFactory();
        pf.setTarget(target);
        pf.addAdvisor(advisor);
        pf.setInterfaces(new Class{SimpleBean.class});

        SimpleBean proxy = (SimpleBean)pf.getProxy();
        System.out.println("Running JDK Tests");
        test(proxy);
    }

    private static void test(SimpleBean bean) {
        long before = 0;
        long after = 0;
        System.out.println("Testing Advised Method");
        before = System.currentTimeMillis();
        for(int x = 0; x < 500000; x++) {
            bean.advised();
        }
        after = System.currentTimeMillis();

        System.out.println("Took " + (after - before) + " ms");

        System.out.println("Testing Unadvised Method");
        before = System.currentTimeMillis();
        for(int x = 0; x < 500000; x++) {
            bean.unadvised();
        }
        after = System.currentTimeMillis();

        System.out.println("Took " + (after - before) + " ms");

        System.out.println("Testing equals() Method");
        before = System.currentTimeMillis();
        for(int x = 0; x < 500000; x++) {
            bean.equals(bean);
        }
        after = System.currentTimeMillis();

        System.out.println("Took " + (after - before) + " ms");

        System.out.println("Testing hashCode() Method");
        before = System.currentTimeMillis();
        for(int x = 0; x < 500000; x++) {
            bean.hashCode();
        }
        after = System.currentTimeMillis();

        System.out.println("Took " + (after - before) + " ms");

        Advised advised = (Advised)bean;
```

```
        System.out.println("Testing Advised.getProxyTargetClass() Method");
        before = System.currentTimeMillis();
        for(int x = 0; x < 500000; x++) {
            advised.getTargetClass();
        }
        after = System.currentTimeMillis();

        System.out.println("Took " + (after - before) + " ms");

        System.out.println(">>>\n");
    }
}
```

在这段代码中，可以看到正在测试三种代理：

- 一个标准的 CGLIB 代理。
- 具有冻结通知链(frozen advice chain)的 CGLIB 代理(即，当通过调用 ProxyConfig 类中的 setFrozen()方法来冻结代理时，CGLIB 将执行进一步的优化；但是，不允许更改通知)。
- 一个 JDK 代理。

对于每种代理类型，运行以下五个测试用例：

- **被通知方法(测试 1)**：这是被通知的方法。测试中所使用的通知类型不执行任何处理的前置通知，因此减少了通知对性能测试的影响。
- **未被通知方法(测试 2)**：这是代理上未被通知的方法。通常代理有很多未被通知的方法。该测试主要测试针对不同代理未被通知方法的执行情况。
- **equals()方法(测试 3)**：该测试主要测试调用 equals()方法的开销。当使用代理作为 HashMap 或类似集合中的键时，这一点尤其重要。
- **hashCode()方法(测试 4)**：与 equals()方法一样，当使用 HashMaps 或类似集合时，hashCode()方法很重要。
- **执行 Advised 接口上的方法(测试 5)**：如前所述，代理默认实现了 Advised 接口，允许在创建后修改代理并查询有关代理的信息。该测试主要测试使用不同的代理类型访问 Advised 接口上方法的速度。

表 5-3 显示了这些测试的结果。

表 5-3　代理性能测试结果(单位：毫秒)

	CGLIB(标准)	CGLIB(冻结)	JDK
被通知方法	245	135	224
未被通知方法	92	42	78
equals()	9	6	77
hashCode()	29	13	23
Advised.getProxyTargetClass()	9	6	15

如你所见，标准 CGLIB 和 JDK 动态代理在被通知方法和未被通知方法之间的性能差异不大。一如既往，这些数字将根据硬件和正在使用的 JDK 而有所不同。

但是，当使用具有冻结通知链的 CGLIB 代理时，性能上存在显著差异。对于 equals()和 hashCode()方法也是一样，在使用 CGLIB 代理时速度明显更快。而对于 Advised 接口上的方法，可以看到在 CGLIB 冻结代理上运行得更快。原因是被通知方法早就在 intercept()方法中进行了处理，从而避免了其他方法所需的大部分逻辑。

5.6.4 选择要使用的代理

决定使用哪个代理通常很容易。CGLIB 代理可以代理类和接口，而 JDK 代理只能代理接口。在性能方面，除非在冻结模式下使用 CGLIB，否则 JDK 和 CGLIB 标准模式之间没有显著差异(至少在运行被通知和未被通知方法时没有显著差异)。在这种情况下，通知链不能更改且 CGLIB 在冻结模式下会进行进一步优化。当需要代理类时，CGLIB 代理是默认选择，因为它是唯一能够生成类代理的代理。如果想要在代理接口时使用 CGLIB 代理，必须使用 setOptimize()方法将 ProxyFactory 中的 optimize 标志的值设置为 true。

5.7 切入点的高级使用

在本章前面，你了解了 Spring 所提供的六个基本 Pointcut 的实现；在大多数情况下，这些切入点可以满足应用程序的需求。但是，有时在定义切入点时可能需要更大的灵活性。Spring 提供了两个额外的 Pointcut 实现，分别是 ComposablePointcut 和 ControlFlowPointcut，它们提供了所需的灵活性。

5.7.1 使用控制流切入点

由 ControlFlowPointcut 类实现的 Spring 控制流切入点类似于许多其他 AOP 实现中可用的 cflow 构造，尽管功能上没有那么强大。本质上，Spring 中的控制流切入点适用于类中给定方法或所有方法下的所有方法调用。该过程很难形象化表示，所以最好的解释是使用一个示例。

下面的代码片段显示了一个 SimpleBeforeAdvice 类，它写出了一条用于描述它所通知方法的消息：

```
package com.apress.prospring5.ch5;

import java.lang.reflect.Method;

import org.springframework.aop.MethodBeforeAdvice;

public class SimpleBeforeAdvice implements MethodBeforeAdvice {
    @Override
    public void before(Method method, Object args, Object target)
          throws Throwable {
       System.out.println("Before method: " + method);
    }
}
```

该通知类允许我们查看 ControlFlowPointcut 应用于哪些方法。接下来，看看简单的 TestBean 类：

```
package com.apress.prospring5.ch5;

public class TestBean {
    public void foo() {
       System.out.println("foo()");
    }
}
```

可以看到，想要通知的是简单的 foo()方法。但是，此时有一个特殊的要求：只有在从另一个特定的方法调用 foo()方法时才会通知该方法。下面的代码片段显示了一个简单的用于该例的驱动程序：

```
package com.apress.prospring5.ch5;

import org.springframework.aop.Advisor;
import org.springframework.aop.Pointcut;
import org.springframework.aop.framework.ProxyFactory;
import org.springframework.aop.support.ControlFlowPointcut;
import org.springframework.aop.support.DefaultPointcutAdvisor;

public class ControlFlowDemo {
    public static void main(String... args) {
       ControlFlowDemo ex = new ControlFlowDemo();
       ex.run();
    }

    public void run() {
       TestBean target = new TestBean();

       Pointcut pc = new ControlFlowPointcut(ControlFlowDemo.class,
          "test");
       Advisor advisor = new DefaultPointcutAdvisor(pc,
          new SimpleBeforeAdvice());

       ProxyFactory pf = new ProxyFactory();
       pf.setTarget(target);
       pf.addAdvisor(advisor);

       TestBean proxy = (TestBean) pf.getProxy();
       System.out.println("\tTrying normal invoke");
       proxy.foo();
       System.out.println("\n\tTrying under ControlFlowDemo.test()");
       test(proxy);
```

```
    }
    private void test(TestBean bean) {
        bean.foo();
    }
}
```

在上面的代码片段中，首先将对通知代理与 ControlFlowPointcut 进行组装，然后调用 foo()方法两次，一次直接从 main()方法调用，另一次从 test()方法调用。此时，有趣的是以下所示的代码行：

```
Pointcut pc = new ControlFlowPointcut(ControlFlowDemo.class, "test");
```

在该行中，为 ControlFlowDemo 类的 test()方法创建了一个 ControlFlowPointcut 实例。实际上，这就是说："切入从 ControlFlowExample.test()方法调用的所有方法"。请注意，尽管说的是"切入所有方法"，但实际上意味着"切入代理对象上的所有方法，而该代理对象由与 ControlFlowPointcut 的实例相对应的 Advisor 通知"。运行前面的示例将在控制台中产生以下结果：

```
    Trying normal invoke
foo()

    Trying under ControlFlowDemo.test()
Before method: public void com.apress.prospring5.ch5.TestBean.foo()
foo()
```

正如你所看到的，当 sing()方法首次在 test()方法的控制流之外被调用时，它是不会被通知的。当它再次执行时，它位于 test()方法的控制流内，ControlFlowPointcut 指示其相关通知适用于该方法，因此会通知该方法。请注意，如果在 test()方法内调用另一个不在被通知代理中的方法，那么该方法不会被通知。

控制流切入点非常有用，允许仅在另一个上下文中执行某个对象时才有选择地通知它。但是，请注意，在其他切入点上使用控制流切入点会大大降低性能。

再来看一个示例。假设有一个事务处理系统，它包含一个 TransactionService 接口以及一个 AccountService 接口。此时，希望应用后置通知，以便当 TransactionService. ReverseTransaction()调用 AccountService.updateBalance()方法更新账户余额时，将电子邮件通知发送给客户。但在其他情况下则不会发送电子邮件。在这种情况下，控制流切入点是非常有用的。图 5-6 显示了该场景的 UML 序列图。

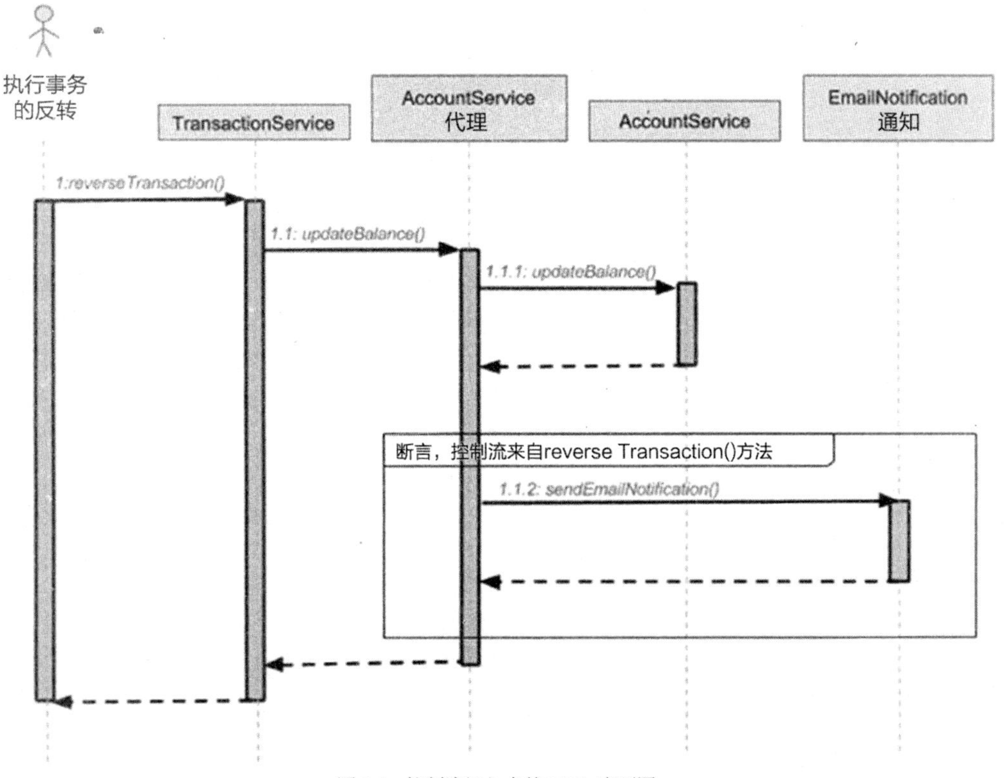

图 5-6　控制流切入点的 UML 序列图

5.7.2 使用组合切入点

在前面的切入点示例中，仅为每个 Advisor 使用了一个切入点。在大多数情况下，这通常是足够的，但在某些情况下，可能需要将两个或更多个切入点组合在一起以实现所需的目标。假设想要切入 bean 上的所有 getter 和 setter 方法。虽然 getter 方法有一个切入点，setter 方法也有一个切入点，但是没有同时适用于这两种方法的切入点。当然，可以用新的逻辑创建另一个切入点，但更好的方法是使用 ComposablePointcut 将两个切入点组合成一个切入点。

ComposablePointcut 支持两种方法：union()和 intersection()。默认情况下，ComposablePointcut 是通过一个匹配所有类的 ClassFilter 以及一个匹配所有方法的 MethodMatcher 来创建的，当然，也可以在构建过程中提供自己的初始 ClassFilter 和 MethodMatcher。union()和 intersection()方法都被重载以接收 ClassFilter 和 MethodMatcher 参数。

可以通过传入 ClassFilter、MethodMatcher 或 Pointcut 接口的实例来调用 ComposablePointcut.union()方法。联合操作的结果是，ComposablePointcut 将在其调用链中添加“或”条件，以便与连接点进行匹配。对于 ComposablePointcut.intersection()方法也是如此，但添加的是“和”条件，这意味着如果想要应用通知，ComposablePointcut 中的所有 ClassFilter、MethodMatcher 和 Pointcut 定义都应该匹配。可以将其想象为 SQL 查询中的 WHERE 子句，其中 union()方法类似于“or”运算符，而 intersection()方法则类似于“and”运算符。

与控制流切入点一样，上述过程很难直观显示，所以通过示例更容易理解。以下示例显示了前一个示例所使用的带有四个方法的 GrammyGuitarist 类：

```
package com.apress.prospring5.ch5;

import com.apress.prospring5.ch2.common.Guitar;
import com.apress.prospring5.ch2.common.Singer;

public class GrammyGuitarist implements Singer {

    @Override public void sing() {
       System.out.println("sing: Gravity is working against me\n" +
              "And gravity wants to bring me down");
    }

    public void sing(Guitar guitar) {
       System.out.println("play: " + guitar.play());
    }

    public void rest(){
       System.out.println("zzz");
    }

    public void talk(){
       System.out.println("talk");
    }
}
```

在该例中，将通过使用相同的 ComposablePointcut 实例来生成三个代理，但每次都会通过使用 union()或 intersection()方法来修改 ComposablePointcut。然后，调用 target bean 代理上的三个方法，并查看哪些方法被通知。具体代码示例如下所示：

```
package com.apress.prospring5.ch5;

import java.lang.reflect.Method;

import org.springframework.aop.Advisor;
import org.springframework.aop.ClassFilter;
import org.springframework.aop.framework.ProxyFactory;
import org.springframework.aop.support.ComposablePointcut;
import org.springframework.aop.support.DefaultPointcutAdvisor;
import org.springframework.aop.support.StaticMethodMatcher;

public class ComposablePointcutExample {
    public static void main(String... args) {
       GrammyGuitarist johnMayer = new GrammyGuitarist();

       ComposablePointcut pc = new ComposablePointcut(ClassFilter.TRUE,
          new SingMethodMatcher());
       System.out.println("Test 1 >> ");
       GrammyGuitarist proxy = getProxy(pc, johnMayer);
       testInvoke(proxy);
```

```
        System.out.println();

        System.out.println("Test 2 >> ");
        pc.union(new TalkMethodMatcher());
        proxy = getProxy(pc, johnMayer);
        testInvoke(proxy);
        System.out.println();

        System.out.println("Test 3 >> ");
        pc.intersection(new RestMethodMatcher());
        proxy = getProxy(pc, johnMayer);
        testInvoke(proxy);
    }

    private static GrammyGuitarist getProxy(ComposablePointcut pc,
            GrammyGuitarist target) {
        Advisor advisor = new DefaultPointcutAdvisor(pc,
            new SimpleBeforeAdvice());

        ProxyFactory pf = new ProxyFactory();
        pf.setTarget(target);
        pf.addAdvisor(advisor);
        return (GrammyGuitarist) pf.getProxy();
    }

    private static void testInvoke(GrammyGuitarist proxy) {
        proxy.sing();
        proxy.sing(new Guitar());
        proxy.talk();
        proxy.rest();
    }

    private static class SingMethodMatcher extends StaticMethodMatcher {
        @Override
        public boolean matches(Method method, Class<?> cls) {
            return (method.getName().startsWith("si"));
        }
    }

    private static class TalkMethodMatcher extends StaticMethodMatcher {
        @Override
        public boolean matches(Method method, Class<?> cls) {
            return "talk".equals(method.getName());
        }
    }

    private static class RestMethodMatcher extends StaticMethodMatcher {
    @Override
        public boolean matches(Method method, Class<?> cls) {
            return (method.getName().endsWith("st"));
        }
    }
}
```

在该例中，首先要注意的是三个私有 MethodMatcher 实现的集合。SingMethodMatcher 匹配所有以 get 开头的方法。这是用来组合 ComposablePointcut 的默认 MethodMatcher。因此，GrammyGuitarist 方法的第一轮调用只会导致 sing()方法被通知。

TalkMethodMatcher 匹配所有名为 talk 的方法，并且在第二轮调用中通过使用 union()与 ComposablePointcut 进行组合。此时，组合了两个 MethodMatchers，一个匹配所有以 si 开头的方法，另一个匹配所有名为 talk 的方法。现在预计第二轮中的所有调用都会被通知。TalkMethodMatcher 更加具体，只匹配 talk()方法。在第三轮调用中，此 MethodMatcher 与 ComposablePointcut 通过使用 intersection()相组合。

因为 RestMethodMatcher 是通过使用 intersection()组合而成的，所以预计在第三轮调用中没有任何方法被通知，因为没有与所组合的 MethodMatchers 相匹配的方法。

运行此例将生成以下输出：

```
Test 1 >>
Before method: public void
    com.apress.prospring5.ch5.GrammyGuitarist.sing()
sing: Gravity is working against me
And gravity wants to bring me down
Before method: public void com.apress.prospring5.ch5.
    GrammyGuitarist.sing(com.apress.prospring5.ch2.common.Guitar)
```

```
play: G C G C Am D7
talk
zzz
Test 2 >>
Before method: public void
    com.apress.prospring5.ch5.GrammyGuitarist.sing()
sing: Gravity is working against me
And gravity wants to bring me down
Before method: public void
    com.apress.prospring5.ch5.GrammyGuitarist.talk()
Before method: public void com.apress.prospring5.ch5.
    GrammyGuitarist.sing(com.apress.prospring5.ch2.common.Guitar)
play: G C G C Am D7
talk
zzz

Test 3 >>
sing: Gravity is working against me
And gravity wants to bring me down
talk
zzz
```

虽然此例仅在组合过程中演示了如何使用 MethodMatchers，但在构建切入点时使用 ClassFilter 也同样简单。事实上，在构建组合切入点时，可以使用 MethodMatchers 和 ClassFilters 的组合。

5.7.3 组合和切入点接口

在前面的小节中，介绍了如何使用多个 MethodMatchers 和 ClassFilters 创建组合切入点。此外，还可以通过使用其他实现了 Pointcut 接口的对象来创建组合切入点。

构建组合切入点的另一种方法是使用 org.springframework.aop.support.Pointcuts 类。该类提供了三个静态方法：intersection()和 union()方法都接收两个切点作为参数来构造一个组合切入点；同时，提供的 matches(Pointcut, Method, Class, Object [])方法用来快速检查切入点是否与所提供的方法、类和方法参数相匹配。

Pointcuts 类仅支持两个切入点上的操作。因此，如果想要将 MethodMatcher 和 ClassFilter 与 Pointcut 相结合，则需要使用 ComposablePointcut 类。但是，如果只需要组合两个切入点，那么 Pointcuts 类更加方便。

5.7.4 切入点小结

Spring 提供了一组强大的 Pointcut 实现，它们可以满足大部分(但不是全部)应用程序的需求。请记住，如果找不到合适的切入点实现来满足需求，则可以通过实现 Pointcut、MethodMatcher 和 ClassFilter 从头开始创建自己的实现。

可以使用两种模式来组合切入点和顾问。第一种模式(即到目前为止所使用的模式)涉及切入点实现与顾问的解耦。在前面所示的代码中，创建了 Pointcut 实现的实例，然后使用 DefaultPointcutAdvisor 实现将通知和 Pointcut 一起添加到代理中。

第二种模式是 Spring 文档中的许多示例所采用的方式，也就是将 Pointcut 封装到自己的 Advisor 实现中。这样一来，就有了一个实现 Pointcut 和 PointcutAdvisor 的类，而 PointcutAdvisor.getPointcut()方法只是简单地返回该类。这是 Spring 中的许多类所使用的一种方法，例如 StaticMethodMatcherPointcutAdvisor。你会发现第一种模式最灵活，允许在不同的 Advisor 实现中使用不同的切入点实现。然而，当需要在应用程序的不同部分或跨多个应用程序使用相同的 Pointcut 和 Advisor 组合时，第二种模式就非常有用了。

当每个 Advisor 必须有一个单独的 Pointcut 实例，需要让 Advisor 负责创建切入点时，第二种模式也是很有用的。如果回想一下前一章中关于代理性能的讨论，你会记得，未被通知的方法比被通知的方法执行得更好。出于这个原因，应该确保通过使用切入点，仅通知绝对必要的方法。这样就可以通过使用 AOP 来减少添加到应用程序的不必要开销。

5.8 引入入门

引入(Introduction)是 Spring 中可用的 AOP 功能集的重要组成部分。通过使用引入，可以动态地向现有对象引入新功能。在 Spring 中，可以将任何接口的实现引入现有对象。你可能很好奇为什么引入这么有用。当可以简单地在开发时添加一个功能时，为什么要在运行时动态添加该功能呢？该问题的答案很简单。当一个功能是横切的并且使用传统通知不易实现时，就需要动态添加该功能。

5.8.1 引入的基础知识

Spring 将*引入*视为一种特殊类型的通知，更具体地说，将其作为一种特殊类型的环绕通知。由于引入仅适用于类级别，因此不能在引入时使用切入点；在语义上讲，两者不匹配。引入将新的接口实现添加到类中，而切入点定义了通知适用于哪些方法。可以通过实现 IntroductionInterceptor 接口来创建引入，该接口扩展了 MethodInterceptor 和 DynamicIntroductionAdvice 接口。图 5-7 显示了此结构以及这两个接口的方法，如 IntelliJ IDEA UML 插件所示。如你所见，MethodInterceptor 接口定义了一个 invoke()方法。通过使用此方法，可以为所引入的接口提供实现，并根据需要对任何其他方法执行截取操作。在单个方法内部实现接口的所有方法可能会非常麻烦，并且很可能会产生大量的代码，必须通过了解这些代码才能决定调用哪个方法。但值得庆幸的是，Spring 提供了一个名为 DelegatingIntroductionInterceptor 的 IntroductionInterceptor 的默认实现，这使得创建引入更加简单。如果想要使用 DelegatingIntroductionInterceptor 创建引入，可以创建一个既继承了 DelegatingIntroductionInterceptor，又实现了想要引入的接口的类。然后，DelegatingIntroductionInterceptor 实现简单地将所有引入方法的调用委托给相应的方法。如果对上述介绍还不太明白，别担心；在下一节你将会看到一个具体的示例。

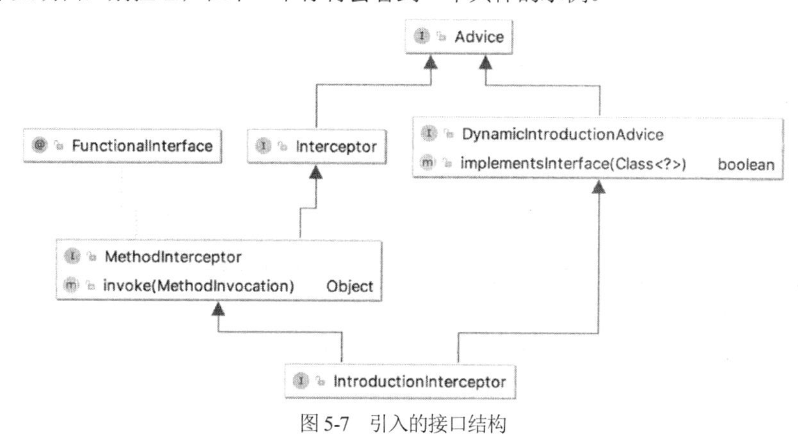

图 5-7 引入的接口结构

就像在使用切入点通知时需要使用 PointcutAdvisor 一样，需要使用 IntroductionAdvisor 向代理添加引入。IntroductionAdvisor 的默认实现是 DefaultIntroductionAdvisor，它可以满足大多数(但不是全部)的引入需求。你应该知道，不允许使用 ProxyFactory.addAdvice()来添加引用，否则会导致抛出 AopConfigException 异常。相反，应该使用 addAdvisor()方法并传入 IntroductionAdvisor 接口的实例。

当使用标准通知(即不是引入)时，可能会将相同的通知实例用于多个对象。Spring 文档将此称为*基于类型的生命周期(per-class life cycle)*，尽管可以为许多类使用单个通知实例。对于引入来说，引入通知构成了被通知对象的状态的一部分，因此，针对每个被通知对象都必须有一个不同的通知实例。这被称为*基于实例的生命周期(per-instance life cycle)*。因为必须确保每个被通知对象都有一个独立的引入实例，所以通常最好创建 DefaultIntroductionAdvisor 的一个子类，它负责创建引入通知。这样一来，只需要确保为每个对象创建一个新的顾问类实例就可以了，因为 DefaultIntroductionAdvisor 的子类会自动创建新的引入实例。例如，希望将前置通知应用于 Contact 类的所有实例上的 setFirstName()方法。图 5-8 显示了适用于 Contact 类型的所有对象的相同通知。接下来假设想要将一个引入混合到 Contact 类的所有实例中，并且引入将为每个 Contact 实例提供信息(例如，指示特定实例是否被修改的 isModified 属性)。

在这种情况下，将为每个 Contact 实例创建引入并绑定到该特定实例，如图 5-9 所示。图 5-9 涵盖了创建引入的基础知识。接下来将讨论如何使用引入来解决对象修改检测问题。

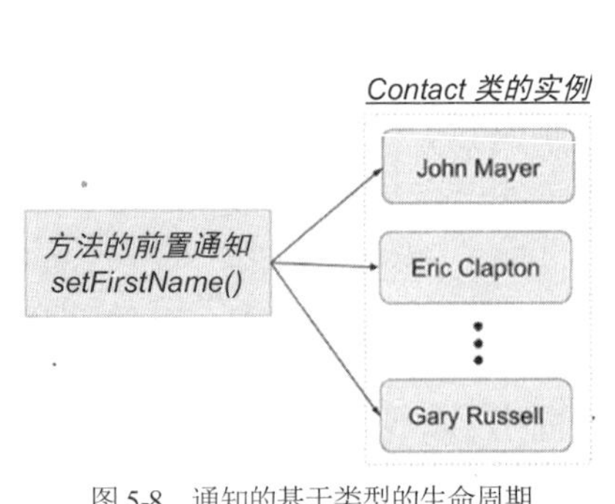

图 5-8 通知的基于类型的生命周期

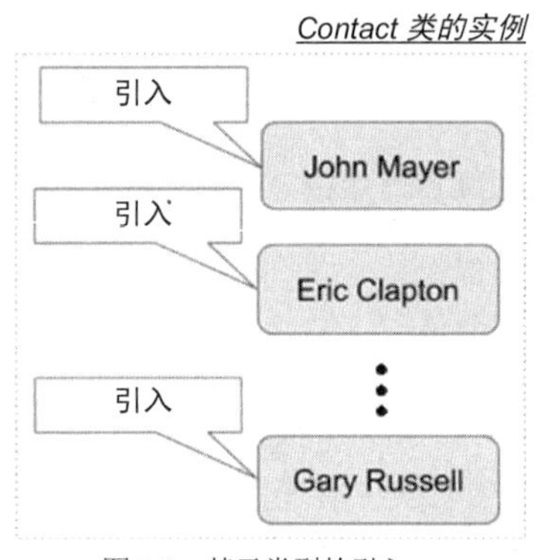

图 5-9 基于类型的引入

5.8.2 使用引入进行对象修改检测

由于许多原因，*对象修改检测*(*object modification detection*)是一种非常用的技术。通常，应用修改检测技术来防止在持久化对象数据时不必要的数据库访问。如果对象被传递给待修改但还没有修改的方法，那么向数据库发出一条更新语句就没有多大意义。此时，使用修改检测技术可以真正提高应用程序的吞吐量，特别是当数据库已经超负荷或位于远程网络中，从而使通信成为昂贵的操作时，更是如此。

但遗憾的是，这种功能很难手动实现，因为需要将修改检测功能添加到每个可以修改对象状态的方法中，以便检查对象状态是否真正被修改。当考虑进行 null 检查以及检查值是否确实发生变化时，每个方法大约需要查看八行代码。虽然可以将其重构为单一方法，但每次需要执行检查时仍要调用此方法。如果在一个使用了许多类的典型应用程序中进行修改检测，就会发生意想不到的灾难。

此时，引入显然可以提供帮助。我们不希望每个需要修改检测的类都继承一些基本实现，这样会失去唯一的继承机会，同时也不希望将检查代码添加到每个状态更改方法中。通过使用引入，可以为修改检测问题提供灵活的解决方案，而无须编写大量重复且易出错的代码。

在本例中，将使用引入构建一个完整的修改检测框架。修改检测逻辑由 IsModified 接口封装，其实现将被引入到适当的对象中，连同拦截逻辑一起自动执行修改检测。针对本例，将使用 JavaBeans 规范，因为我们认为修改是对 setter 方法的任何调用。当然，并不是将 setter 方法的所有调用都视为修改；将会检查传递给 setter 方法的值是否不同于当前存储在对象中的值。该解决方案的唯一缺陷是，如果对象上的任何一个值发生了变化，即使是将对象设置回原始状态，也会被视为修改。例如，有一个具有 firstName 属性的 Contact 对象。假设在处理过程中，将 firstName 属性从 Peter 更改为 John。此时，该对象被标记为已修改。但是，即使在稍后的处理中将值从 John 更改回原始值 Peter，该对象也仍将被标记为已修改。跟踪此类变化的一种方法是存储对象整个生命周期中的全部变化历史。然而，本例的实现是十分平常的，并且能够满足大多数的需求。实施更完整的解决方案会导致一个过于复杂的示例。

使用 IsModified 接口

修改检测解决方案的核心是 IsModified 接口，虚拟应用程序使用该接口做出关于对象持久化的智能决策。此时，并不打算介绍应用程序如何使用 IsModified；相反，将重点介绍引入的实现。以下代码片段显示了 IsModified 接口：

```
package com.apress.prospring5.ch5.introduction;

public interface IsModified {
    boolean isModified();
}
```

上述代码没有什么特别之处——只是一个方法 isModified()，用于指示对象是否已被修改。

创建 mixin

下一步是创建实现 IsModified 接口并将其引入到对象的代码，这被称为 mixin。正如前面提到的，相比于直接实现 IntroductionInterceptor 接口，通过继承 DelegatingIntroductionInterceptor 创建 mixin 要简单得多。mixin 类 IsModifiedMixin 是 DelegatingIntroductionInterceptor 的子类，它实现了 IsModified 接口，具体代码如下所示：

```
package com.apress.prospring5.ch5.introduction;
```

```
import java.lang.reflect.Method;
import java.util.HashMap;
import java.util.Map;
import org.aopalliance.intercept.MethodInvocation;
import org.springframework.aop.support.DelegatingIntroductionInterceptor;

public class IsModifiedMixin extends DelegatingIntroductionInterceptor
        implements IsModified {
    private boolean isModified = false;

    private Map<Method, Method> methodCache = new HashMap<>();

    @Override
    public boolean isModified() {
        return isModified;
    }

    @Override
    public Object invoke(MethodInvocation invocation) throws Throwable {
        if (!isModified) {
            if ((invocation.getMethod().getName().startsWith("set"))
                && (invocation.getArguments().length == 1)) {

                Method getter = getGetter(invocation.getMethod());

                if (getter != null) {
                    Object newVal = invocation.getArguments()0;
                    Object oldVal = getter.invoke(invocation.getThis(), null);
                    if((newVal == null) && (oldVal == null)) {
                        isModified = false;
                    } else if((newVal == null) && (oldVal != null)) {
                        isModified = true;
                    } else if((newVal != null) && (oldVal == null)) {
                        isModified = true;
                    } else {
                        isModified = !newVal.equals(oldVal);
                    }
                }
            }
        }

        return super.invoke(invocation);
    }

    private Method getGetter(Method setter) {
        Method getter = methodCache.get(setter);

        if (getter != null) {
            return getter;
        }

        String getterName = setter.getName().replaceFirst("set", "get");
        try {
            getter = setter.getDeclaringClass().getMethod(getterName, null);
            synchronized (methodCache) {
                methodCache.put(setter, getter);
            }
            return getter;
        } catch (NoSuchMethodException ex) {
            return null;
        }
    }
}
```

首先要注意的是 IsModified 接口的实现，它由私有修改字段和 isModified()方法组成。该例强调了为什么每个被通知对象必须有一个 mixin 实例，mixin 不仅向对象引入方法，还引入状态。如果在多个对象上共享此 mixin 的单个实例，那么也将共享状态，这意味着在第一次修改单个对象时，所有对象都显示为已修改。

实际上不必为 mixin 实现 invoke()方法，但在本例中，这样做可以在修改发生时进行自动检测。只有当对象仍未修改时，才开始执行修改检测；一旦知道对象已被修改，就不需要进行修改检测。下一步，查看该方法是否是 setter 方法，如果是，则检索相应的 getter 方法。请注意，此时对 getter/setter 方法进行缓存，以便将来进行更快速的检索。最后，将 getter 方法返回的值和传递给 setter 方法的值进行比较，以确定是否发生了修改。请注意，该例检查了 null 的不同可能组合，并适当地设置修改。请务必记住，当使用 DelegatingIntroductionInterceptor 时，必须在重写 invoke()

时调用 super.invoke()，因为是由 DelegatingIntroductionInterceptor 将调用分派到正确的位置(被通知对象或 mixin 本身)。

可以在 mixin 中实现尽可能多的接口，每个接口都会自动引入到被通知对象中。

创建顾问

下一步是创建一个 Advisor 来封装 mixin 类的创建。虽然这一步是可选的，但它有助于确保为每个被通知对象使用一个新的 mixin 实例。以下代码片段显示了 IsModifiedAdvisor 类：

```
package com.apress.prospring5.ch5.introduction;

import org.springframework.aop.support.DefaultIntroductionAdvisor;

public class IsModifiedAdvisor extends DefaultIntroductionAdvisor {
    public IsModifiedAdvisor() {
        super(new IsModifiedMixin());
    }
}
```

请注意，此时已经扩展了 DefaultIntroductionAdvisor 来创建 IsModifiedAdvisor。该顾问的实现很简单，非常容易理解。

把所有内容都放在一起

现在已经有了一个 mixin 类和一个 Advisor 类，接下来可以测试修改检测框架。此时要使用的类是前面所提到的 Contact 类，它是 common 包的一部分。由于可重用性方面的原因，这个类通常用作本书中项目的依赖项。该类的内容如下所示：

```
package com.apress.prospring5.ch2.common;

public class Contact {

    private String name;
    private String phoneNumber;
    private String email;

    public String getName() {
        return name;
    }

    public void setName(String name) {
        this.name = name;
    }

    // getters and setter for other fields
    ...
}
```

虽然该 bean 有一组属性，但只有 name 属性用于测试修改检测 mixin。以下代码片段首先显示了如何组装被通知代理，然后是测试修改检测代码：

```
package com.apress.prospring5.ch5;

import com.apress.prospring5.ch2.common.Contact;
import com.apress.prospring5.ch5.introduction.IsModified;
import com.apress.prospring5.ch5.introduction.IsModifiedAdvisor;
import org.springframework.aop.IntroductionAdvisor;
import org.springframework.aop.framework.ProxyFactory;

public class IntroductionDemo {
    public static void main(String... args) {
        Contact target = new Contact();
        target.setName("John Mayer");

        IntroductionAdvisor advisor = new IsModifiedAdvisor();

        ProxyFactory pf = new ProxyFactory();
        pf.setTarget(target);
        pf.addAdvisor(advisor);
        pf.setOptimize(true);
        Contact proxy = (Contact) pf.getProxy();
        IsModified proxyInterface = (IsModified)proxy;

        System.out.println("Is Contact?: " + (proxy instanceof Contact));
        System.out.println("Is IsModified?: " + (proxy instanceof IsModified));
```

```
        System.out.println("Has been modified?: " +
            proxyInterface.isModified());

        proxy.setName("John Mayer");

        System.out.println("Has been modified?: " +
            proxyInterface.isModified());

        proxy.setName("Eric Clapton");

        System.out.println("Has been modified?: " +
            proxyInterface.isModified());
    }
}
```

请注意，当创建代理时，将 optimize 标志设置为 true，从而强制使用 CGLIB 代理。原因在于，当使用 JDK 代理引入 mixin 时，所生成的代理并不是对象类(在本例中为 Contact)的实例；该代理仅实现 mixin 接口，而不是原始类。通过 CGLIB 代理，原始类由代理与 mixin 接口一起实现。

还需要注意，在代码中，首先进行测试以查看代理是否为 Contact 类的实例，然后查看代理是否是 IsModified 接口的实例。当使用 CGLIB 代理时，两个测试都会返回 true；而如果使用 JDK 代理，则只有 IsModified 测试返回 true。最后一步，先将 name 属性设置为其当前值，再设置一个新值来检验修改检测代码，每次都会检查 isModified 标志的值。运行此例将生成以下输出：

```
Is Contact?: true
Is IsModified?: true
Has been modified?: false
Has been modified?: false
Has been modified?: true
```

如你所料，两个 instanceof 测试都返回 true。请注意，在发生任何修改之前，第一次调用 isModified()将返回 false。在将 name 属性值设置为相同的值后，调用 isModified()也返回 false。但是，在将 name 属性值设置为新值后，isModified()方法返回 true，表示该对象实际上已被修改。

5.8.3 引入小结

引入是 Spring AOP 最强大的功能之一；它们不仅可以扩展现有方法的功能，还可以动态扩展接口和对象实现。使用引入是通过定义良好的接口来实现应用程序交互的横切逻辑的完美方式。一般来说，这种逻辑都以声明方式而非编程方式应用。通过使用本例中定义的 IsModifiedMixin 和下一节中讨论的框架服务，可以以声明的方式定义哪些对象能够进行修改检测，并且无须修改这些对象的实现。

显然，由于引入是通过代理工作的，它们会增加一定的开销。代理上的所有方法都被认为是被通知方法，因为切入点不能与引入结合使用。但是，如果可以通过使用诸如对象修改检测之类的引入来实现许多服务，那么这种性能开销对于实现服务所需代码的减少，以及完全集中服务逻辑所带来的稳定性和可维护性来说则是很小的代价。

5.9 AOP 的框架服务

到目前为止，不得不编写大量代码来通知对象并为它们生成代理。尽管这本身并不是一个多大的问题，但却意味着所有的通知配置都被硬编码到应用程序中，从而无法获取通过透明地为方法实现提供通知所带来的一些好处。但值得庆幸的是，Spring 提供了额外的框架服务，允许在应用程序配置中创建一个被通知代理，然后将此代理注入到目标 bean 中，就像任何其他依赖项一样。

对 AOP 配置使用声明式方法优于手动编程机制。当使用声明性机制时，不仅可以将通知的配置外部化，还可以减少编码错误。此外，可以利用 DI 和 AOP 的组合来启用 AOP，以便可以在完全透明的环境中使用它。

5.9.1 以声明的方式配置 AOP

在使用 Spring AOP 的声明式配置时，存在三个选项。

- 使用 ProxyFactoryBean：在 Spring AOP 中，当根据定义的 Spring bean 创建 AOP 代理时，ProxyFactoryBean 提供了一种声明方式来配置 Spring 的 ApplicationContext(以及底层的 BeanFactory)。

- **使用 Spring aop 名称空间**：在 Spring 2.0 中引入的 aop 名称空间提供了一种简化的方式(与 ProxyFactoryBean 相比)来定义 Spring 应用程序中的切面及其 DI 需求。但是，aop 名称空间在后台也使用了 ProxyFactoryBean。
- **使用@AspectJ 样式注解**：除基于 XML 的 aop 名称空间外，还可以在类中使用@AspectJ 样式注解来配置 Spring AOP。尽管使用的语法基于 AspectJ，并且在使用此选项时需要包含一些 AspectJ 库，但在引导 ApplicationContext 时，Spring 仍然使用代理机制(即为目标创建代理对象)。

5.9.2 使用 ProxyFactoryBean

ProxyFactoryBean 类是 FactoryBean 的一个实现，它允许指定一个 bean 作为目标，并且为该 bean 提供一组通知和顾问(这些通知和顾问最终被合并到一个 AOP 代理中)。ProxyFactoryBean 用于将拦截器逻辑应用于现有的目标 bean，做法是当调用该 bean 上的方法时，在方法调用之前和之后执行拦截器。因为可以同时使用顾问和通知，所以不仅可以以声明的方式配置通知，还可以配置切入点。

ProxyFactoryBean 与 ProxyFactory 共享一个公共接口(org.springframework.aop.framework.Advised 接口)，这两个类都间接实现了 org.springframework.aop.framework.AdvisedSupport 类，而 org.springframework.aop.framework.AdvisedSupport 类又实现了 Advised 接口，因此公开了许多相同的标志，比如 frozen、optimize 和 exposeProxy。这些标志的值直接被传递给底层的 ProxyFactory，从而允许以声明的方式配置工厂。

使用 ProxyFactoryBean

使用 ProxyFactoryBean 很简单。首先定义一个将成为目标 bean 的 bean，然后通过使用 ProxyFactoryBean 定义应用程序将访问的 bean，并使用目标 bean 作为代理目标。在可能的情况下，应该将目标 bean 定义为代理 bean 声明内的匿名 bean，从而可以防止应用程序意外访问未被通知的 bean。但是，在某些情况下(例如接下来将要演示的示例)，可能需要为同一个 bean 创建多个代理，此时应该使用普通的顶级 bean。

对于下面的示例，想象一下如下场景：有一位歌手和一位纪录片导演一起制作了一部巡回纪录片。这种情况下，Documentarist 依赖于 Singer 实现。此时的 Singer 实现将使用前面介绍的 GrammyGuitarist，具体内容如下所示：

```
package com.apress.prospring5.ch5;

import com.apress.prospring5.ch2.common.Guitar;
import com.apress.prospring5.ch2.common.Singer;

public class GrammyGuitarist implements Singer {

    @Override public void sing() {
        System.out.println("sing: Gravity is working against me\n" +
                "And gravity wants to bring me down");
    }

    public void sing(Guitar guitar) {
        System.out.println("play: " + guitar.play());
    }

    public void rest(){
        System.out.println("zzz");
    }

    public void talk(){
        System.out.println("talk");
    }
}
```

Documentarist 类主要用于告诉歌手在拍摄纪录片时应该做什么：

```
package com.apress.prospring5.ch5;

public class Documentarist {

    private GrammyGuitarist guitarist;

    public void execute() {
        guitarist.sing();
        guitarist.talk();
    }

    public void setDep(GrammyGuitarist guitarist) {
```

```
        this.guitarist = guitarist;
    }
}
```

在本例中，将为单个 GrammySinger 实例创建两个代理，这两个代理都有相同的基本通知，如下所示：

```
package com.apress.prospring5.ch5;

import org.aspectj.lang.JoinPoint;

public class AuditAdvice {
    public void simpleBeforeAdvice(JoinPoint joinPoint) {
        System.out.println("Executing: " +
              joinPoint.getSignature().getDeclaringTypeName() + " "
              + joinPoint.getSignature().getName());
    }
}
```

第一个代理将直接使用通知来通知目标；因此，所有的方法将被通知。而对于第二个代理，将配置 AspectJExpressionPointcut 和 DefaultPointcutAdvisor，以便只有 GrammySinger 类的 sing()方法被通知。为了对通知进行测试，将创建两个 Documentarist 类型的 bean 定义，每个定义都将使用不同的代理注入。然后，调用这些 bean 的 execute()方法，并观察当依赖项上的被通知方法被调用时会发生什么。图 5-10 显示了此例的配置(app-context-xml.xml)。之所以使用图像来描述这种配置，是因为它可能看起来有点混乱，通过图像可以更容易看到每个 bean 被注入的位置。在本例中，只是使用 Spring 的 DI 功能设置了前面在代码中设置的属性。唯一让人感兴趣的是使用匿名 bean 作为切入点，并且使用了 ProxyFactoryBean 类。我们更喜欢在不共享时使用匿名 bean 作为切入点，因为它能够让可直接访问的 bean 集合尽可能小，并且尽可能与应用程序相关。在使用 ProxyFactoryBean 时要认识到的重要一点是，ProxyFactoryBean 声明是被公开给应用程序的声明，也是在实现依赖项时所使用的声明。不建议使用底层的目标 bean 声明，所以只有当想绕过 AOP 框架时才应该使用该 bean，虽然一般来说应用程序并不应该知道 AOP 框架，并且也因此不应该绕过它。出于以上原因，应该尽可能使用匿名 bean 来避免应用程序被意外访问。

```
<?xml version="1.0" encoding="UTF-8"?>

<beans xmlns="http://www.springframework.org/schema/beans"
       xmlns:xsi="http://www.w3.org/2001/XMLSchema-instance"
       xmlns:p="http://www.springframework.org/schema/p"
       xmlns:util="http://www.springframework.org/schema/util"
       xsi:schemaLocation="http://www.springframework.org/schema/beans
       http://www.springframework.org/schema/beans/spring-beans.xsd
       http://www.springframework.org/schema/util
       http://www.springframework.org/schema/util/spring-util.xsd">

    <bean id="johnMayer"class="com.apress.prospring5.ch5.GrammyGuitarist"/>
    <bean id="advice" class="com.apress.prospring5.ch5.AuditAdvice"/>

    <bean id="documentaristOne" class="com.apress.prospring5.ch5.Documentarist"
          p:guitarist-ref="proxyOne"/>

    <bean id="proxyOne" class="org.springframework.aop.framework.ProxyFactoryBean"
          p:target-ref="johnMayer"
          p:interceptorNames-ref="interceptorAdviceNames"/>

    <util:list id="interceptorAdviceNames">
          <value>advice</value>
    </util:list>


    <bean id="documentaristTwo" class="com.apress.prospring5.ch5.Documentarist"
          p:guitarist-ref="proxyTwo"/>

    <bean id="proxyTwo" class="org.springframework.aop.framework.ProxyFactoryBean"
          p:target-ref="johnMayer"
          p:interceptorNames-ref="interceptorAdvisorNames"/>

    <util:list id="interceptorAdvisorNames">
          <value>advisor</value>
    </util:list>

    <bean id="advisor" class="org.springframework.aop.support.DefaultPointcutAdvisor"
          p:advice-ref="advice">
          <property name="pointcut">
                <bean class="org.springframework.aop.aspectj.AspectJExpressionPointcut"
                      p:expression="execution(* sing*(..))"/>
          </property>
    </bean>

</beans>
```

图 5-10 声明式 AOP 配置

以下代码片段显示了一个简单的类，它首先从 ApplicationContext 获取两个 Documentarist 实例，然后运行每个实例的 execute()方法：

```
package com.apress.prospring5.ch5;
import org.springframework.context.support.GenericXmlApplicationContext;
    public class ProxyFactoryBeanDemo {
        public static void main(String... args) {
            GenericXmlApplicationContext ctx =
        new GenericXmlApplicationContext();
        ctx.load("spring/app-context-xml.xml");
        ctx.refresh();

        Documentarist documentaristOne =
            ctx.getBean("documentaristOne", Documentarist.class);
        Documentarist documentaristTwo =
            ctx.getBean("documentaristTwo", Documentarist.class);

        System.out.println("Documentarist One >>");
        documentaristOne.execute();

        System.out.println("\nDocumentarist Two >> ");
        documentaristTwo.execute();
    }
}
```

运行此例将生成以下输出：

```
Documentarist One >>
Executing: public void com.apress.prospring5.ch5.GrammyGuitarist.sing()
sing: Gravity is working against me
And gravity wants to bring me down
Executing: public void com.apress.prospring5.ch5.GrammyGuitarist.talk()
talk

Documentarist Two >>
Executing: public void com.apress.prospring5.ch5.GrammyGuitarist.sing()
sing: Gravity is working against me
And gravity wants to bring me down
talk
```

正如预期的那样，第一个代理中的 sing()和 talk()方法都被通知了，因为它的配置中没有使用切入点。然而，对于第二个代理，由于在配置中使用了切入点，因此只有 sing()方法被通知。

使用 ProxyFactoryBean 进行引入

不仅可以使用 ProxyFactoryBean 类来通知对象，还可以将 mixin 引入对象中。在前面介绍引入时讲过，必须使用 IntroductionAdvisor 添加引入，而不能直接添加引入。在使用 ProxyFactoryBean 和引入时也适用同样的规则。当使用 ProxyFactoryBean 时，如果为 mixin 创建了自定义 Advisor，那么配置代理会变得更加容易。以下配置代码片段显示了本章前面针对 IsModifiedMixin 引入的示例配置(app-context-xml.xml)：

```
<beans ...>

    <bean id="guitarist"
        class="com.apress.prospring5.ch2.common.Contact"
        p:name="John Mayer"/>
    <bean id="advisor"
          class="com.apress.prospring5.ch5.introduction.IsModifiedAdvisor"/>

    <util:list id="interceptorAdvisorNames">
        <value>advisor</value>
    </util:list>

    <bean id="bean"
        class="org.springframework.aop.framework.ProxyFactoryBean"
        p:target-ref="guitarist"
        p:interceptorNames-ref="interceptorAdvisorNames"
        p:proxyTargetClass="true">
    </bean>
</beans>
```

从配置中可以看到，使用 IsModifiedAdvisor 类作为 ProxyFactoryBean 的顾问，因为不需要为同一个目标对象创建另一个代理，所以针对目标 bean 使用了匿名声明。下面的代码片段显示了对前面的引入示例所做的修改，该例从 ApplicationContext 获得代理：

```
package com.apress.prospring5.ch5;

import com.apress.prospring5.ch2.common.Contact;
import com.apress.prospring5.ch5.introduction.IsModified;
import org.springframework.context.support.GenericXmlApplicationContext;

public class IntroductionConfigDemo {
    public static void main(String... args) {
        GenericXmlApplicationContext ctx = new GenericXmlApplicationContext();
        ctx.load("classpath:spring/app-context-xml.xml");
        ctx.refresh();

        Contact bean = (Contact) ctx.getBean("bean");
        IsModified mod = (IsModified) bean;

        System.out.println("Is Contact?: " + (bean instanceof Contact));
        System.out.println("Is IsModified?: " + (bean instanceof IsModified));

        System.out.println("Has been modified?: " + mod.isModified());
        bean.setName("John Mayer");

        System.out.println("Has been modified?: " + mod.isModified());
        bean.setName("Eric Clapton");

        System.out.println("Has been modified?: " + mod.isModified());
    }
}
```

运行该例将产生与前面的引入例子完全相同的输出，但是这一次，代理从 ApplicationContext 获得，并且在应用程序代码中没有任何配置。

由于已经介绍了 Java 配置，因此之前描述的 XML 配置可以替换为配置类，如下所示：

```
package com.apress.prospring5.ch5.config;

import com.apress.prospring5.ch2.common.Contact;
import com.apress.prospring5.ch5.introduction.IsModifiedAdvisor;
import org.springframework.aop.Advisor;
import org.springframework.aop.framework.ProxyFactoryBean;
import org.springframework.context.annotation.Bean;
import org.springframework.context.annotation.Configuration;
@Configuration
public class AppConfig {

    @Bean
    public Contact guitarist() {
        Contact guitarist = new Contact();
        guitarist.setName("John Mayer");
        return guitarist;
    }

    @Bean
    public Advisor advisor() {
        return new IsModifiedAdvisor();
    }

    @Bean ProxyFactoryBean bean() {
        ProxyFactoryBean proxyFactoryBean = new ProxyFactoryBean();
        proxyFactoryBean.setTarget(guitarist());
        proxyFactoryBean.setProxyTargetClass(true);
        proxyFactoryBean.addAdvisor(advisor());
        return proxyFactoryBean;
    }
}
```

为了测试前面的类是否真正起作用，在类 IntroductionConfigDemo 的 main()方法中，用下面的代码替换初始化上下文的代码行：

```
GenericApplicationContext ctx =
    new AnnotationConfigApplicationContext(AppConfig.class);
```

配置类的区别在于，不需要通过名称引用 advisor bean，也可以将它添加到列表中，以便将其作为参数提供给 ProxyFactoryBean，因为可以直接调用 addAdvisor(..)，并且 advisor bean 可以作为参数提供。这显然简化了配置。

ProxyFactoryBean 小结

当使用 ProxyFactoryBean 时，可以配置 AOP 代理，它提供了编程方法的所有灵活性，同时无须将应用程序耦合到 AOP 配置。除非需要在运行时做出关于如何创建代理的决策，否则最好使用代理配置的声明式方法而不是编程式方法。接下来还要学习声明式 Spring AOP 的另外两个选项，这两个选项都是基于 Spring 2.0 或其更高版本(带有 JDK 5 或其更高版本)的应用程序的首选选项。

5.9.3 使用 aop 名称空间

aop 名称空间为声明式 Spring AOP 配置提供了非常简化的语法。为了演示该名称空间是如何工作的，接下来重复使用前面的 ProxyFactoryBean 示例，并稍微进行一点修改以演示其用法。仍然使用上一个示例中的 GrammyGuitarist 类，但 Documentarist 将被扩展为使用 Guitar 参数调用 sing()方法。

```
package com.apress.prospring5.ch5;

import com.apress.prospring5.ch2.common.Guitar;

public class NewDocumentarist extends Documentarist {

    @Override
    public void execute() {
       guitarist.sing();
       guitarist.sing(new Guitar());
       guitarist.talk();
    }
}
```

对通知类的更改如下所示：

```
package com.apress.prospring5.ch5;

import org.aspectj.lang.JoinPoint;

public class SimpleAdvice {

    public void simpleBeforeAdvice(JoinPoint joinPoint) {
       System.out.println("Executing: " +
          joinPoint.getSignature().getDeclaringTypeName() + " "
          + joinPoint.getSignature().getName());
    }
}
```

你将看到，通知类不再需要实现 MethodBeforeAdvice 接口。此外，前置通知接收连接点作为参数，而不是方法、对象和参数。实际上，对于通知类来说，该参数是可选的，所以可以使用不带参数的方法。但是，如果在通知中需要访问被通知连接点的相关信息(此时想要转储调用类型和方法名称的信息)，就需要定义用于接收的参数。当为该方法定义参数时，Spring 会自动将连接点传递给方法，以便进行处理。以下是来自 app-context-xml-01.xml 文件的 Spring XML 配置，其中包含了 aop 名称空间：

```
<?xml version="1.0" encoding="UTF-8"?>

<beans xmlns="http://www.springframework.org/schema/beans"
    xmlns:xsi="http://www.w3.org/2001/XMLSchema-instance"
     xmlns:p="http://www.springframework.org/schema/p"
    xmlns:aop="http://www.springframework.org/schema/aop"
    xsi:schemaLocation="http://www.springframework.org/schema/beans
       http://www.springframework.org/schema/beans/spring-beans.xsd
       http://www.springframework.org/schema/aop
       http://www.springframework.org/schema/aop/spring-aop.xsd">

    <aop:config>
       <aop:pointcut id="singExecution"
          expression="execution(
    * com.apress.prospring5.ch5..sing*(com.apress.prospring5.ch2.common.Guitar)
)"/>

       <aop:aspect ref="advice">
          <aop:before pointcut-ref="singExecution"
                method="simpleBeforeAdvice"/>
       </aop:aspect>
    </aop:config>

    <bean id="advice"
          class="com.apress.prospring5.ch5.SimpleAdvice"/>
    <bean id="johnMayer"
          class="com.apress.prospring5.ch5.GrammyGuitarist"/>
    <bean id="documentarist"
          class="com.apress.prospring5.ch5.NewDocumentarist"
          p:guitarist-ref="johnMayer"/>
</beans>
```

首先，需要在<beans>标记中声明 aop 名称空间。其次，所有的 Spring AOP 配置都放在<aop:config>标记下。在<aop:config>下，可以定义切入点、切面、顾问等，并像往常一样引用其他 Spring bean。

在上面所示的配置中，使用 ID singExecution 定义了一个切入点。如下表达式意味着通知所有前缀为 sing 的方法，并且这些类在包 com.apress.prospring5.ch5(包括所有子包)中定义：

```
"execution(*
com.apress.prospring5.ch5..sing*(com.apress.prospring5.ch2.common.Guitar)
)"
```

另外，sing()方法应该接收一个 Guitar 类型的参数。之后，通过使用<aop:aspect>标记声明切面，并且通知类通过 ID advice(即 SimpleAdvice 类)引用 Spring bean。pointcutref 属性通过 ID singExecution 引用所定义的切入点，前置通知(使用<aop:before>标记声明)为通知 bean 内的 simpleBeforeAdvice()方法。要测试前面的配置，可以使用下面的类：

```
package com.apress.prospring5.ch5;

import org.springframework.context.support.GenericXmlApplicationContext;

public class AopNamespaceDemo {
    public static void main(String... args) {
        GenericXmlApplicationContext ctx =
            new GenericXmlApplicationContext();
        ctx.load("classpath:spring/app-context-xml-01.xml");
        ctx.refresh();
        NewDocumentarist documentarist =
            ctx.getBean("documentarist", NewDocumentarist.class);
        documentarist.execute();
        ctx.close();
    }
}
```

在该例中，像往常一样，简单地初始化 ApplicationContext，然后检索 bean 并调用其 execute()方法。运行该程序将产生以下输出：

```
sing: Gravity is working against me
And gravity wants to bring me down
Executing: com.apress.prospring5.ch5.GrammyGuitarist sing
play: G C G C Am D7
talk
```

正如你所看到的，只有通过 Guitar 参数调用的 sing(..)方法才被通知。没有参数的 sing()方法和 talk()方法都没有被通知。这与预期的完全相同，并且与 ProxyFactoryBean 配置相比，配置更加简化。

接下来进一步修改前面的示例，使其更复杂一些。现在假设仅通知 ID 以 john 开头、类型为 Guitar 且 brand 属性被设置为 Gibson 的带有 Spring bean 的方法。

为此，首先必须更改 Guitar 类以添加 brand 属性。该属性是非强制性的，并使用默认值填充，以使前面的示例正常工作。

```
package com.apress.prospring5.ch2.common;

public class Guitar {
    private String brand =" Martin";

    public String play(){
        return "G C G C Am D7";
    }

    public String getBrand() {
        return brand;
    }

    public void setBrand(String brand) {
        this.brand = brand;
    }
}
```

然后，需要让 NewDocumentarist 用一种特殊的吉他品牌来调用 sing()方法。

```
package com.apress.prospring5.ch5;

import com.apress.prospring5.ch2.common.Guitar;

public class NewDocumentarist extends Documentarist {
```

```
    @Override
    public void execute() {
        guitarist.sing();
        Guitar guitar = new Guitar();
        guitar.setBrand("Gibson");
        guitarist.sing(guitar);
        guitarist.talk();
    }
}
```

现在需要一种新的、更复杂的通知类型。参数 guitar 被添加到前置通知的签名中。其次，在通知中，只有在参数的 brand 属性等于 Gibson 时才检查和执行逻辑。

```
package com.apress.prospring5.ch5;

import com.apress.prospring5.ch2.common.Guitar;
import org.aspectj.lang.JoinPoint;

public class ComplexAdvice {
    public void simpleBeforeAdvice(JoinPoint joinPoint, Guitar value) {
        if(value.getBrand().equals("Gibson")) {
            System.out.println("Executing: " +
                joinPoint.getSignature().getDeclaringTypeName() + " "
                + joinPoint.getSignature().getName());
        }
    }
}
```

此外，XML 配置也需要修改，因为需要使用新的通知类型并更新切入点表达式(可以在 app-context-xml-02.xml 中找到完整的配置，除下面显示的行外，其他所有内容与 app-context-xml-01.xml 的内容相同，因此在此不再描述)。

```
<beans ..>
...

    <bean id="advice"
        class="com.apress.prospring5.ch5.ComplexAdvice"/>

    <aop:config>
        <aop:pointcut id="singExecution"
            expression="execution(* sing*(com.apress.prospring5.ch2.common.Guitar))
                and args(value) and bean(john*)"/>
</beans>
```

需要将两条指令添加到切入点表达式中。首先，args(value)指令指示 Spring 将该参数与名称 value 一起传递给前置通知。其次，bean(john *)指令指示 Spring 仅通知 ID 以 john 作为前缀的 bean。这是一个十分强大的功能；如果有一个定义良好的 Spring bean 命名结构，就可以轻松地通知想要的对象。例如，可以通过使用 bean(*DAO*)将通知应用于所有 DAO bean，或者通过使用 bean(* Service *)将通知应用于所有服务层 bean，而不是使用完全限定的类名进行匹配。使用新的配置文件 app-context-xml02.xml 运行相同的测试程序会产生以下输出：

```
sing: Gravity is working against me
And gravity wants to bring me down
Executing: com.apress.prospring5.ch5.GrammyGuitarist sing
play: G C G C Am D7
talk
```

可以看到，只有带有 Guitar 参数且 brand 属性等于 Gibson 的 sing()方法才被通知。

接下来再看一个使用 aop 名称空间来处理环绕通知的示例。可以简单地向 ComplexAdvice 类添加一个新的方法，而不是创建另一个实现 MethodInterceptor 接口的类。以下代码示例显示了修改后的 ComplexAdvice 类中名为 simpleAroundAdvice()的新方法：

```
//ComplexAdvice.java
public Object simpleAroundAdvice(ProceedingJoinPoint pjp,
      Guitar value) throws Throwable {
    System.out.println("Before execution: " +
        pjp.getSignature().getDeclaringTypeName() + " "
            + pjp.getSignature().getName()
            + " argument: " + value.getBrand());

        Object retVal = pjp.proceed();

        System.out.println("After execution: " +
            pjp.getSignature().getDeclaringTypeName() + " "
                + pjp.getSignature().getName()
```

```
                + " argument: " + value.getBrand());

        return retVal;
    }
```

新添加的 simpleAroundAdvice()方法需要至少接收一个 ProceedingJoinPoint 类型的参数，以便可以继续调用目标对象。此外，还添加了 value 参数，以便在通知中显示该值。必须调整<aop:aspect>的 XML 配置以添加新的通知(可以在 app-context-xml-03.xml 中找到完整的配置，除下面显示的行外，其他所有内容与 app-context-xml-02.xml 中的内容完全相同，因此不再描述)。

```
<beans ..>
...
    <aop:config>
        <aop:pointcut id="singExecution"
            expression="execution(
                    * sing*(com.apress.prospring5.ch2.common.Guitar))
                  and args(value) and bean(john*)"
        />

        <aop:aspect ref="advice">
            <aop:before pointcut-ref="singExecution"
                method="simpleBeforeAdvice"/>
        <aop:around pointcut-ref="singExecution"
            method="simpleAroundAdvice"/>
        </aop:aspect>
    </aop:config>
</beans>
```

此时只是添加新的标记<aop:around>来声明环绕通知并引用相同的切入点。下面，再次修改 NewDocumentarist.execute()方法以包含带有默认 Guitar 的 sing()调用，从而获得想要分析的行为。

```
package com.apress.prospring5.ch5;

import com.apress.prospring5.ch2.common.Guitar;

public class NewDocumentarist extends Documentarist {

    @Override
    public void execute() {
        guitarist.sing();
        Guitar guitar = new Guitar();
        guitar.setBrand("Gibson");
        guitarist.sing(guitar);
        guitarist.sing(new Guitar());
        guitarist.talk();
    }
}
```

再次运行测试程序，将获得以下输出：

```
sing: Gravity is working against me
And gravity wants to bring me down

Executing: com.apress.prospring5.ch5.GrammyGuitarist sing
Before execution: com.apress.prospring5.ch5.GrammyGuitarist sing argument: Gibson
play: G C G C Am D7
After execution: com.apress.prospring5.ch5.GrammyGuitarist sing argument: Gibson

Before execution: com.apress.prospring5.ch5.GrammyGuitarist sing argument: Martin
play: G C G C Am D7
After execution: com.apress.prospring5.ch5.GrammyGuitarist sing argument: Martin
talk
```

这里有两个有趣的地方。首先，你会发现环绕通知适用于两个带有 Guitar 参数的 sing(..)方法的调用，因为并没有检查参数。其次，对于以"Gibson" Guitar 作为参数的 sing()方法，前置通知和环绕通知都被执行，默认情况下前置通知优先。

在使用 aop 名称空间或@AspectJ 样式注解时，可以使用两种类型的后置通知。一个是后置返回通知(使用<aop:after-returning>标记)，仅在目标方法正常完成时适用。另一个是后置通知(使用<aop:after>标记)，这发生在方法正常完成或者方法遇到错误并抛出异常时。如果无论目标方法的执行结果如何都需要执行通知，那么应该使用后置通知。

5.10 使用@AspectJ 样式注解

在 JDK 5 或更新版本中使用 Spring AOP 时，还可以使用@AspectJ 样式注解来声明通知。然而，如前所述，Spring 仍然使用自己的代理机制来通知目标方法，而不是使用 AspectJ 的织入机制。

在本节中，将通过使用@AspectJ 样式注解来了解如何实现与 aop 名称空间中相同的切面。AspectJ 是一种通用的面向方面的 Java 扩展，主要用来解决传统编程方法中无法很好解决的问题，换句话说就是横切关注点。对于本节中的示例，还将使用其他 Spring bean 的注解，并且使用 Java 配置类。

以下示例描述了使用注解声明 bean 的 GrammyGuitarist 类：

```
package com.apress.prospring5.ch5;

import com.apress.prospring5.ch2.common.Guitar;
import com.apress.prospring5.ch2.common.Singer;
import org.springframework.stereotype.Component;

@Component("johnMayer")
public class GrammyGuitarist implements Singer {

    @Override public void sing() {
       System.out.println("sing: Gravity is working against me\n" +
                 "And gravity wants to bring me down");
    }

    public void sing(Guitar guitar) {
       System.out.println("play: " + guitar.play());
    }

    public void rest(){
       System.out.println("zzz");
    }

    public void talk(){
       System.out.println("talk");
    }
}
```

NewDocumentarist 类也需要调整。

```
package com.apress.prospring5.ch5;

import com.apress.prospring5.ch2.common.Guitar;
import org.springframework.beans.factory.annotation.Autowired;
import org.springframework.beans.factory.annotation.Qualifier;
import org.springframework.stereotype.Component;

@Component("documentarist")
public class NewDocumentarist {
    protected GrammyGuitarist guitarist;
    public void execute() {
       guitarist.sing();
       Guitar guitar = new Guitar();
       guitar.setBrand("Gibson");
       guitarist.sing(guitar);
       guitarist.talk();
    }

    @Autowired
    @Qualifier("johnMayer")
    public void setGuitarist(GrammyGuitarist guitarist) {
       this.guitarist = guitarist;
    }
}
```

此时使用@Component 注解来注解这两个类，并为它们分配相应的名称。在 GrammyGuitarist 类中，属性 guitarist 的 setter 方法被@Autowired 注解为由 Spring 自动注入。

现在看一下使用了@AspectJ 样式注解的 AnnotationAdvice 类。此时一次性实现了切入点、前置通知和环绕通知。以下代码片段显示了 AnnotationAdvice 类：

```
package com.apress.prospring5.ch5;

import com.apress.prospring5.ch2.common.Guitar;
```

```
import org.aspectj.lang.JoinPoint;
import org.aspectj.lang.ProceedingJoinPoint;
import org.aspectj.lang.annotation.Around;
import org.aspectj.lang.annotation.Aspect;
import org.aspectj.lang.annotation.Before;
import org.aspectj.lang.annotation.Pointcut;
import org.springframework.stereotype.Component;

@Component
@Aspect
public class AnnotatedAdvice {
    @Pointcut("execution(*
          com.apress.prospring5.ch5..sing*(com.apress.prospring5.ch2.common.Guitar))
        && args(value)")
    public void singExecution(Guitar value) {
    }

    @Pointcut("bean(john*)")
    public void isJohn() {
    }

    @Before("singExecution(value) && isJohn()")
    public void simpleBeforeAdvice(JoinPoint joinPoint, Guitar value) {
        if(value.getBrand().equals("Gibson")) {
        System.out.println("Executing: " +
            joinPoint.getSignature().getDeclaringTypeName() + " "
            + joinPoint.getSignature().getName() + " argument: " + value.getBrand());
        }
    }

    @Around("singExecution(value) && isJohn()")
    public Object simpleAroundAdvice(ProceedingJoinPoint pjp,
        Guitar value) throws Throwable {
        System.out.println("Before execution: " +
            pjp.getSignature().getDeclaringTypeName() + " "
            + pjp.getSignature().getName()
            + " argument: " + value.getBrand());

        Object retVal = pjp.proceed();

        System.out.println("After execution: " +
            pjp.getSignature().getDeclaringTypeName() + " "
            + pjp.getSignature().getName()
            + " argument: " + value.getBrand());

        return retVal;
    }
}
```

你会注意到上述代码结构与在 aop 名称空间中使用的代码结构非常相似，只是使用了注解。但是，仍然有几点值得注意。

- 使用@Component 和@Aspect 来注解 AnnotatedAdvice 类。@Aspect 注解用于声明它是一个切面类。如果在 XML 配置中使用了<context:component-scan>标记，那么为了让 Spring 扫描组件，还需要使用@Component 来注解类。
- 切入点被定义为返回 void 的方法。在该类中，定义了两个切入点；两个都用@Pointcut 注解。此时故意将 aop 名称空间示例中的切入点表达式分成两部分。一部分(由 singExecution(Guitar value)方法表示)定义了执行包 com.apress.prospring4.ch5 下所有类中带有参数 guitar 的 sing*()方法的切入点。参数(value)也会被传递给通知。另一部分(由 isJohn()方法表示)定义了另一个切入点，执行名称以 john 为前缀的 Spring bean 中的所有方法。还要注意，需要使用&&来定义切入点表达式中的“和”条件，而对于 aop 名称空间，则需要使用 and 运算符。
- 使用@Before 注解前置通知方法，而环绕通知则使用@Around 注解。对于这两种通知类型，都传入使用类中定义的两个切入点的值。值 singExecution(value) && isJohn()意味着两个切入点的条件都应该匹配，以便应用通知，这与 ComposablePointcut 中的交集操作是相同的。
- 前置通知逻辑和环绕通知逻辑与 aop 名称空间示例中的相同。

在完成所有注解后，XML 配置变得非常简单。

```
<?xml version="1.0" encoding="UTF-8"?>

<beans xmlns="http://www.springframework.org/schema/beans"
```

```
    xmlns:xsi="http://www.w3.org/2001/XMLSchema-instance"
    xmlns:aop="http://www.springframework.org/schema/aop"
    xmlns:context="http://www.springframework.org/schema/context"
    xsi:schemaLocation="http://www.springframework.org/schema/aop
       http://www.springframework.org/schema/aop/spring-aop.xsd
       http://www.springframework.org/schema/beans
       http://www.springframework.org/schema/beans/spring-beans.xsd
       http://www.springframework.org/schema/context
       http://www.springframework.org/schema/context/spring-context.xsd">

    <aop:aspectj-autoproxy/>

    <context:component-scan
       base-package="com.apress.prospring5.ch5"/>
</beans>
```

此时只声明了两个标记。<aop:aspect-autoproxy>标记用于通知 Spring 扫描@AspectJ 样式注解，而<context:component-scan>标记仍然要求 Spring 扫描通知所在包中的 Spring bean。此外，还需要用@Component 注解通知类，以表明它是一个 Spring 组件。

以下代码片段描述了测试此配置的类：

```
package com.apress.prospring5.ch5;

import org.springframework.context.support.GenericXmlApplicationContext;

public class AspectJAnnotationDemo {
    public static void main(String... args) {
        GenericXmlApplicationContext ctx =
           new GenericXmlApplicationContext();
        ctx.load("classpath:spring/app-context-xml.xml");
        ctx.refresh();

        NewDocumentarist documentarist =
           ctx.getBean("documentarist", NewDocumentarist.class);
        documentarist.execute();
    }
}
```

如果运行该例，将会收获一些惊喜，因为在控制台中会出现以下内容：

```
Exception in thread "main"
org.springframework.beans.factory.UnsatisfiedDependencyException:
Error creating bean with name 'documentarist': Unsatisfied dependency
expressed through method 'setGuitarist' parameter 0; nested exception is
  org.springframework.beans.factory.BeanNotOfRequiredTypeException:
  Bean named 'johnMayer' is expected to be of type
     'com.apress.prospring5.ch5.GrammyGuitarist' but was actually of
     type 'com.sun.proxy.$Proxy18'
...
```

那么，究竟发生了什么事呢？GrammyGuitarist 实现了 Singer 接口，并且默认情况下会创建基于接口的 JDK 动态代理。但是 NewDocumentarist 严格要求依赖项的类型是 Grammy-Guitarist 或是该类型的扩展。因此，示例抛出了上面所示的异常。那么如何解决该问题呢？有两种方法：一种方法是修改 NewDocumentarist 以接收 Singer 依赖项，但该方法并不适合目前的示例，因为我们想要访问的是 GrammyGuitarist 类中的方法，而不是在 Singer 接口中定义的方法实现。另一种方法是要求 Spring 生成 CGLIB——一种基于类的代理。在 XML 中，可以通过修改<aop:aspectj-autoproxy/>标记的配置并将 proxy-target-class 属性值设置为 true 来完成此操作。

Java 配置类比这更简单：

```
@Configuration
@ComponentScan(basePackages = {"com.apress.prospring5.ch5"})
@EnableAspectJAutoProxy(proxyTargetClass = true)
public class AppConfig {
}
```

请注意@EnableAspectJAutoProxy注解。它相当于<aop:aspectj-autoproxy />，并且还有一个类似于proxy-target-class 属性的 proxyTargetClass 属性。该注解支持处理用 AspectJ 的@Aspect 注解的组件，并且也可用于用@Configuration 注解的类。

以下是测试程序。它被设计为一个JUnit测试用例，因此XML和Java配置示例可以放在同一个类中。IntelliJ IDEA 等智能编辑器可以单独执行每种测试方法。

```
package com.apress.prospring5.ch5;

import org.junit.Test;
import org.springframework.context.annotation.AnnotationConfigApplicationContext;
import org.springframework.context.support.GenericApplicationContext;
import org.springframework.context.support.GenericXmlApplicationContext;

public class AspectJAnnotationTest {

    @Test
    public void xmlTest() {
        GenericXmlApplicationContext ctx =
            new GenericXmlApplicationContext();
        ctx.load("classpath:spring/app-context-xml.xml");
        ctx.refresh();

        NewDocumentarist documentarist =
            ctx.getBean("documentarist", NewDocumentarist.class);
        documentarist.execute();

        ctx.close();
    }

    @Test
    public void configTest() {
        GenericApplicationContext ctx =
            new AnnotationConfigApplicationContext(AppConfig.class);

        NewDocumentarist documentarist =
            ctx.getBean("documentarist", NewDocumentarist.class);
        documentarist.execute();
        ctx.close();
    }
}
```

运行这些测试方法中的任何一个，如果通过，则产生以下输出：

```
sing: Gravity is working against me
And gravity wants to bring me down
Before execution: com.apress.prospring5.ch5.GrammyGuitarist sing argument: Gibson
Executing: com.apress.prospring5.ch5.GrammyGuitarist sing argument: Gibson
play: G C G C Am D7
After execution: com.apress.prospring5.ch5.GrammyGuitarist sing argument: Gibson
talk
```

Spring Boot 提供了一个特殊的 AOP 启动程序库，可以避免一些配置麻烦。可以与往常一样，在 pro-spring-15/build.properties 文件中配置该库，并可以作为依赖项添加(使用子项目配置文件 aspectj-boot / build.gradle 中给定的名称)。

```
//pro-spring-15/build.properties
ext {
    bootVersion = '2.0.0.BUILD-SNAPSHOT'

    ...
    boot = [
        springBootPlugin:
            "org.springframework.boot:spring-boot-gradle-plugin:$bootVersion",
    ...
    starterAop:
        "org.springframework.boot:spring-boot-starter-aop:$bootVersion"
   ]
}
//aspectj-boot/build.gradle
buildscript {
        ...
    dependencies {
        classpath boot.springBootPlugin
    }
}

apply plugin: 'org.springframework.boot'

dependencies {
    compile boot.starterAop
}
```

在图 5-11 中，可以看到作为 Spring Boot 项目的依赖项添加的一组库。将库作为依赖项添加到应用程序后，就不再需要@EnableAspectJAutoProxy(proxyTarget Class = true)注解，因为默认情况下已启用 AOP Spring 支持。不必在任何地方设置属性，因为 Spring Boot 会自动检测所需要的代理类型。针对前面的示例，可以删除 AppConfig 类并将其替换为典型的 Spring Boot 应用程序类。

```
package com.apress.prospring5.ch5;

import org.slf4j.Logger;
import org.slf4j.LoggerFactory;
import org.springframework.boot.SpringApplication;
import org.springframework.boot.autoconfigure.SpringBootApplication;
import org.springframework.context.ConfigurableApplicationContext;

@SpringBootApplication
public class Application {

    private static Logger logger = LoggerFactory.getLogger(Application.class);

    public static void main(String args) throws Exception {
       ConfigurableApplicationContext ctx =
          SpringApplication.run(Application.class, args);
       assert (ctx != null);

       NewDocumentarist documentarist =
          ctx.getBean("documentarist", NewDocumentarist.class);
       documentarist.execute();

       System.in.read();
       ctx.close();
    }
}
```

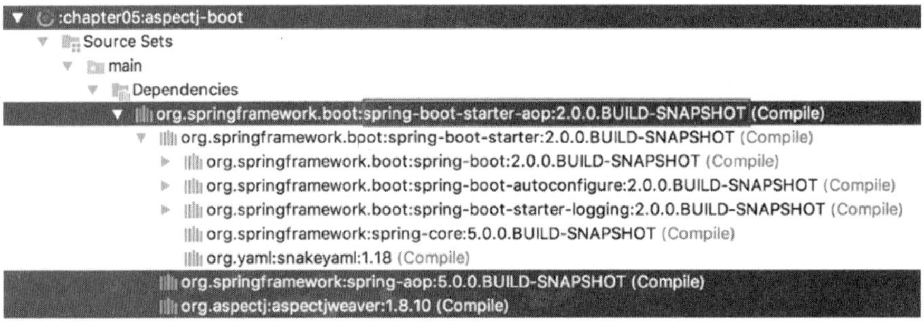

:chapter05:aspectj-boot
Source Sets
main
Dependencies
org.springframework.boot:spring-boot-starter-aop:2.0.0.BUILD-SNAPSHOT (Compile)
org.springframework.boot:spring-boot-starter:2.0.0.BUILD-SNAPSHOT (Compile)
org.springframework.boot:spring-boot:2.0.0.BUILD-SNAPSHOT (Compile)
org.springframework.boot:spring-boot-autoconfigure:2.0.0.BUILD-SNAPSHOT (Compile)
org.springframework.boot:spring-boot-starter-logging:2.0.0.BUILD-SNAPSHOT (Compile)
org.springframework:spring-core:5.0.0.BUILD-SNAPSHOT (Compile)
org.yaml:snakeyaml:1.18 (Compile)
org.springframework:spring-aop:5.0.0.BUILD-SNAPSHOT (Compile)
org.aspectj:aspectjweaver:1.8.10 (Compile)

图 5-11　IntelliJ IDEA 中描述的 Spring Boot AOP 启动程序可交换依赖项

声明式 Spring AOP 配置的注意事项

到目前为止，已经讨论了三种声明 Spring AOP 配置的方法，包括 ProxyFactoryBean、aop 名称空间和@AspectJ 样式注解。aop 名称空间比 ProxyFactoryBean 简单得多。所以，常见的问题是，是使用 aop 名称空间还是使用@AspectJ 样式注解？

如果 Spring 应用程序基于 XML 配置，那么使用 aop 名称空间是十分自然的选择，因为它能够让 AOP 和 DI 配置样式保持一致。另一方面，如果应用程序主要基于注解，那么请使用@AspectJ 样式注解。不管用哪种方法，应该根据应用程序的需求选择合适的配置方法，并尽可能保持一致。

而且，aop 名称空间和@AspectJ 样式注解还有其他一些不同之处。

- 切入点表达式语法有一些细微差别(例如，在前面的讨论中曾讲过，需要在 aop 名称空间中使用 and，但在@AspectJ 样式注解中使用&&)。
- aop 名称空间仅支持“单例”切面实例化模型。
- 在 aop 名称空间中，不能“组合”多个切入点表达式；而在使用@AspectJ 的示例中，可以在前置通知和环绕通知中组合两个切入点定义(即 singExecution(value) && isJohn())。当使用 aop 名称空间并且需要创建一个新的结合了匹配条件的切入点表达式时，需要使用 ComposablePointcut 类。

5.11 AspectJ 集成

AOP 为许多基于 OOP 的应用程序中出现的常见问题提供了强大的解决方案。在使用 Spring AOP 时，可以利用 AOP 功能的可选择子集，在大多数情况下，可以解决在应用程序中遇到的问题。但是，在某些情况下，可能需要使用 Spring AOP 范围之外的某些 AOP 功能。

从连接点的角度来看，Spring AOP 仅支持与执行公共非静态方法相匹配的切入点。但是，在某些情况下，可能需要向受保护/私有的方法应用通知，比如对象构建或字段访问期间等。

在这些情况下，需要使用更全面的功能集来查看 AOP 实现。此时，我们偏爱使用 AspectJ，因为可以使用 Spring 配置 AspectJ 切面，AspectJ 成为 Spring AOP 的完美补充。

5.11.1 关于 AspectJ

AspectJ 是全功能的 AOP 实现，它使用织入过程(编译时或加载时织入)将各个切面引入到代码中。在 AspectJ 中，切面和切入点都是使用 Java 类似语法构建的，从而降低了 Java 开发人员的学习难度。我们不会花太多时间学习 AspectJ 及其工作原理，因为这不在本书的讨论范围之内。相反，将介绍一些简单的 AspectJ 示例，并演示如何使用 Spring 进行配置。有关 AspectJ 的更多信息，可以阅读 Ramnivas Laddad 编写的 *AspectJ in Action: Enterprise AOP with Spring Applications*(Manning 出版社，2009 出版)。

我们并不打算介绍如何将 AspectJ 方面织入应用程序。有关详细信息，请参阅 AspectJ 文档或查看第 5 章的 aspectj-aspects 项目中构建的 Gradle。

5.11.2 使用单例切面

默认情况下，AspectJ 切面是单例，这意味着为每个类加载器获取单个实例。Spring 在使用 AspectJ 切面时所面临的问题是不能创建切面实例，因为该过程已经由 AspectJ 本身处理。但是，每个切面都公开了一个名为 org.aspectj.lang.Aspects.aspectOf()的方法，可用于访问切面实例。通过使用 aspectOf()方法以及 Spring 配置的一项特殊功能，可以让 Spring 帮助我们对切面进行配置。通过这种支持，可以充分利用 AspectJ 强大的 AOP 功能集，同时又不会失去 Spring 优秀的 DI 和配置功能。这也意味着应用程序不需要两种单独的配置方法；可以对 Spring 管理的所有 bean 和 AspectJ 切面使用相同的 Spring ApplicationContext 方法。

为了在 Spring 应用程序中支持切面，需要将 Gradle 插件添加到配置中。可以从以下网址找到如何在 Gradle 应用程序中使用 Gradle 插件的源代码和说明：https://github.com/eveoh/gradle-aspectj。下面的代码显示了 chapter05/aspectj-aspects/build.gradle 的内容：

```
buildscript {
    repositories {
       mavenLocal()
       mavenCentral()
       maven { url "http://repo.spring.io/release" }
       maven { url "http://repo.spring.io/milestone" }
       maven { url "http://repo.spring.io/snapshot" }
       maven { url "https://repo.spring.io/libs-snapshot" }
       maven { url "https://maven.eveoh.nl/content/repositories/releases" }
    }
    dependencies {
       classpath "nl.eveoh:gradle-aspectj:1.6"
    }
}

apply plugin: 'aspectj'

jar {
    manifest {
       attributes(
          'Main-Class': 'com.apress.prospring5.ch5.AspectJDemo',
          "Class-Path": configurations.compile.collect { it.getName() }.join(' '))
    }
}
```

在下面的代码片段中，可以看到一个基本类 MessageWriter，我们将使用 AspectJ 进行通知：

```
package com.apress.prospring5.ch5;

public class MessageWriter {
    public void writeMessage() {
        System.out.println("foobar!");
    }
    public void foo() {
        System.out.println("foo");
    }
}
```

在该例中，将使用 AspectJ 来通知 writeMessage()方法，并在方法调用之前和之后写出一条消息。这些消息将使用 Spring 进行配置。下面的代码示例显示了 MessageWrapper 切面(文件名为 MessageWrapper.aj，它是一个 AspectJ 文件，而不是标准的 Java 类)：

```
package com.apress.prospring5.ch5;

public aspect MessageWrapper {
    private String prefix;
    private String suffix;

    public void setPrefix(String prefix) {
        this.prefix = prefix;
    }

    public String getPrefix() {
        return this.prefix;
    }

    public void setSuffix(String suffix) {
        this.suffix = suffix;
    }

    public String getSuffix() {
        return this.suffix;
    }

    pointcut doWriting() :
        execution(*
    com.apress.prospring5.ch5.MessageWriter.writeMessage());
    before() : doWriting() {
        System.out.println(prefix);
    }

    after() : doWriting() {
        System.out.println(suffix);
    }
}
```

基本上，上述代码创建了一个名为 MessageWrapper 的切面，就像使用普通的 Java 类一样，为该切面赋予两个属性 suffix 和 prefix，我们将在通知 writeMessage()方法时使用它们。接下来，为单个连接点定义一个命名切入点 doWriting()，此时，执行 writeMessage()方法。虽然 AspectJ 有大量的连接点，但它们超出了该例的讨论范围。最后，定义两个通知：一个在 doWriting()切入点之前执行，另一个在该切入点之后执行。以下配置代码片段显示了如何在 Spring 中配置该切面(app-config-xml.xml)：

```
<?xml version="1.0" encoding="UTF-8"?>

<beans xmlns="http://www.springframework.org/schema/beans"
    xmlns:xsi="http://www.w3.org/2001/XMLSchema-instance"
    xmlns:p="http://www.springframework.org/schema/p"
    xsi:schemaLocation="http://www.springframework.org/schema/beans
      http://www.springframework.org/schema/beans/spring-beans.xsd">

    <bean id="aspect" class="com.apress.prospring5.ch5.MessageWrapper"
        factory-method="aspectOf" p:prefix="The Prefix" p:suffix="The Suffix"/>
</beans>
```

如你所见，切面 bean 的许多配置与标准 bean 的配置相似。唯一的区别是使用了<bean>标记的 factory-method 属性。factory-method 属性旨在允许遵循传统工厂模式的类无缝集成到 Spring 中。例如，如果有一个具有私有构造函数的类 Foo，并且还有一个静态工厂方法 getInstance()，那么使用 factory-method 属性可以允许此类的 bean 由 Spring 进行管理。每个 AspectJ 切面所公开的 aspectOf()方法都允许访问切面的实例，从而允许 Spring 设置切面的属性。下面是用于该例的一个简单的驱动程序：

```
package com.apress.prospring5.ch5;
import org.springframework.context.support.GenericXmlApplicationContext;

public class AspectJDemo {
    public static void main(String... args) {
        GenericXmlApplicationContext ctx =
            new GenericXmlApplicationContext();
        ctx.load("classpath:spring/app-context-xml.xml");
        ctx.refresh();

        MessageWriter writer = new MessageWriter();
        writer.writeMessage();
        writer.foo();
    }
}
```

请注意，首先加载 ApplicationContext，以便允许 Spring 配置该切面。然后，创建 MessageWriter 的一个实例，最后调用 writeMessage()和 foo()方法。该例的输出如下所示：

```
The Prefix
foobar!
The Suffix
foo
```

如你所见，MessageWrapper 切面中的通知已被应用于 writeMessage()方法，并且在写出消息时，通知将使用 ApplicationContext 配置中指定的前缀和后缀值。

5.12 小结

在本章中，介绍了大量的 AOP 核心概念，并探讨了这些概念如何转换为 Spring AOP 实现。讨论了 Spring AOP 中已经实现和尚未实现的功能，针对那些 Spring 尚未实现的功能，使用 AspectJ 作为 AOP 解决方案。另外，花了一些时间来详细解释 Spring 中可用的通知类型，以及四种通知类型的实例。研究了如何通过使用切入点来限制通知适用的方法，尤其是研究了 Spring 提供的六个基本的切入点实现。详细讨论了如何构建 AOP 代理、不同的选项以及它们的与众不同之处。比较了三种代理类型之间的性能，并重点强调在 JDK 代理与 CGLIB 代理之间进行选择的一些主要差异和限制。介绍了切入点的高级选项，以及如何使用引入来扩展由对象实现的接口集合。还介绍了 Spring 框架服务以声明式地配置 AOP，从而避免将 AOP 代理构建逻辑硬编码到代码中。最后讨论了如何将 Spring 和 AspectJ 集成在一起，以便在不失去 Spring 灵活性的情况下使用 AspectJ 的附加功能。当然本章也包含了很多关于 AOP 的内容！

在下一章中，将讨论另一个完全不同的主题——如何使用 Spring 的 JDBC 支持来从根本上简化基于 JDBC 的数据访问代码的创建。

第6章

Spring JDBC 支持

到目前为止，你已经看到了构建完全由 Spring 管理的应用程序是多么容易，并且应该对 bean 配置和面向方面编程(AOP)有了深刻的理解。然而，还缺少一部分内容：如何获取驱动应用程序的数据？

除了简单的一次性命令行实用程序之外，几乎每个应用程序都需要将数据保存到某种数据存储中。最常见和最方便的数据存储是关系数据库。

以下是 2017 年排名前七名的关系数据库：

- Oracle 数据库
- Microsoft SQL Server
- IBM DB2
- SAP Sybase ASE
- PostgreSQL
- MariaDB Enterprise
- MySQL

如果你所在的公司无法承受获取前四种数据库的许可证所需的昂贵费用，那么可能正在使用上述列表中的后三种数据库。最常用的开源关系数据库是 MySQL(http://mysql.com)和 PostgreSQL(postgresql.org)。MySQL 通常更广泛地用于 Web 应用程序开发，特别是在 Linux 平台上[1]。另一方面，PostgreSQL 对 Oracle 开发人员更友好，因为它的过程语言 PLpgSQL 非常接近 Oracle 的 PL/SQL 语言。

即使选择最快、最可靠的数据库，也可能会因为使用设计和实现欠佳的数据访问层而失去速度和灵活性。应用程序往往非常频繁地使用数据访问层；因此，数据访问代码中的任何不必要的瓶颈都会影响整个应用程序，无论应用程序设计得有多好。

在本章中，将演示如何使用 Spring 来简化使用 JDBC 的数据访问代码的实现。首先看一下在没有使用 Spring 的情况下所需编写的大量且重复的代码，然后将其与使用 Spring 数据访问类实现的类进行比较。结果是惊人的，因为 Spring 可以充分使用人工调整 SQL 查询的全部功能，同时使需要实现的支持代码量最小化。具体来说，主要讨论以下内容：

- **比较传统 JDBC 代码和 Spring JDBC 支持**：探讨 Spring 如何在保持相同功能的前提下简化旧式 JDBC 代码。你还将看到 Spring 如何访问低级 JDBC API 以及如何将低级 JDBC API 映射到便利类(如 JdbcTemplate)。
- **连接到数据库**：尽管没有深入讨论数据库连接管理的每个细节，但还是演示了简单的 Connection 和 DataSource 之间的根本区别。当然，还将讨论 Spring 如何管理数据源以及可以在应用程序中使用哪些数据源。
- **检索数据并将其映射到 Java 对象**：将演示如何检索数据以及如何有效地将所选数据映射到 Java 对象。此外，你还会了解到 Spring JDBC 是对象-关系映射(ORM)工具的可行替代方案。
- **插入、更新和删除数据**：最后，将讨论如何使用 Spring 执行这些类型的查询来实现插入、更新和删除操作。

6.1 介绍 Lambda 表达式

Java 8 支持 lambda 表达式以及许多其他功能。lambda 表达式可以很好地替代匿名内部类，并且是使用 Spring

1 WordPress 是一个使用广泛的博客平台，它使用 MySQL 或 MariaDB 来存储数据。

JDBC 支持的理想方法。使用 lambda 表达式需要 Java 8。本书是在 Java 8 的预发布版本和第一个通用版本发布期间编写的，所以可能并不是所有人都在使用 Java 8。鉴于此，章节代码示例和源代码下载在适用的地方显示这两种风格。大多数使用了模板或回调的 Spring API 都可以使用 lambda 表达式，而不仅限于 JDBC。本章并不会详细讨论 lambda 表达式本身，因为它们是 Java 语言功能，你应该熟悉 lambda 概念和语法。有关更多信息，请参阅 http://docs.oracle.com/javase/tutorial/java/javaOO/lambdaexpressions.html 上的 lambda 表达式教程。

6.2 示例代码的示例数据模型

在继续后面的讨论之前，先介绍一个简单的数据模型，该模型将用于本章以及接下来几章中的示例，后续几章主要讨论其他数据访问技术(我们将根据各章的主题需求相应地扩展该模型)。

该模型是一个包含两个表的简单音乐数据库。第一个表是 SINGER，里面存储歌手的信息；另一个表是 ALBUM，里面存储歌手发布的专辑。每位歌手可以有零张或多张专辑；换句话说，SINGER 和 ALBUM 表之间是一对多的关系。歌手的信息包括他们的姓名和出生日期。图 6-1 显示了该数据库的实体关系(ER)图。

图 6-1　用于示例代码的简单数据模型

如你所见，这两个表都有 ID 列，在插入过程中数据库会自动分配 ID 列。对于 ALBUM 表，存在与 SINGER 表的外键关系，SINGER 表通过 SINGER_ID 列与 SINGER 表的主键(即 ID 列)链接起来。

在本章的某些示例中，将使用开源数据库 MySQL 来演示与真实数据库的交互。这就需要有一个可用的 MySQL 实例。本章并不会介绍如何安装 MySQL。你可以使用所选择的其他数据库，但需要修改模式和函数定义。此外，我们还会介绍嵌入式数据库的使用，它不需要 MySQL 数据库。

如果想要使用 MySQL，可以在官方网站上找到关于安装和配置 MySQL 的非常好的教程。在下载并安装了 MySQL[1]之后，可以使用 root 账户来访问它。通常，当开发一个应用程序时，需要一个新的模式和用户。对于本章中的代码示例，模式被命名为 MUSICDB，访问它的用户为 prospring5。下面介绍用于创建它们的 SQL 代码，可以在 plain-jdbc 项目的 resources 目录下的 ddl.sql 文件中找到这些代码，里面包括 MySQL Community Server 版本 5.17.18 中的错误修正，这是本书编写时的最新版本。

```
CREATE USER 'prospring5'@'localhost' IDENTIFIED BY 'prospring5';

CREATE SCHEMA MUSICDB;
GRANT ALL PRIVILEGES ON MUSICDB . * TO 'prospring5'@'localhost';
FLUSH PRIVILEGES;

/*in case of java.sql.SQLException: The server timezone value 'UTC'
    is unrecognized or represents more than one timezone. */
SET GLOBAL time_zone = '+3:00';
```

以下代码片段描述了创建前面提到的两个表所需的 SQL 代码。这些代码位于 plain-jdbc 项目的 resources 目录下的 schema.sql 文件中。

```
CREATE TABLE SINGER (

     ID INT NOT NULL AUTO_INCREMENT
   , FIRST_NAME VARCHAR(60) NOT NULL
   , LAST_NAME VARCHAR(40) NOT NULL
   , BIRTH_DATE DATE
   , UNIQUE UQ_SINGER_1 (FIRST_NAME, LAST_NAME)
   , PRIMARY KEY (ID)
);

CREATE TABLE ALBUM (
```

1　从 https://dev.mysql.com/downloads/下载 MySQL Community Server。

```
    ID INT NOT NULL AUTO_INCREMENT
  , SINGER_ID INT NOT NULL
  , TITLE VARCHAR(100) NOT NULL
  , RELEASE_DATE DATE
  , UNIQUE UQ_SINGER_ALBUM_1 (SINGER_ID, TITLE)
  , PRIMARY KEY (ID)
  , CONSTRAINT FK_ALBUM FOREIGN KEY (SINGER_ID)
        REFERENCES SINGER (ID)
);
```

如果使用 IntelliJ IDEA 之类的智能编辑器，则可以使用数据库视图来检查模式和表。在图 6-2 中，可以看到 IntelliJ IDEA 中描述的 MUSICDB 模式的内容。

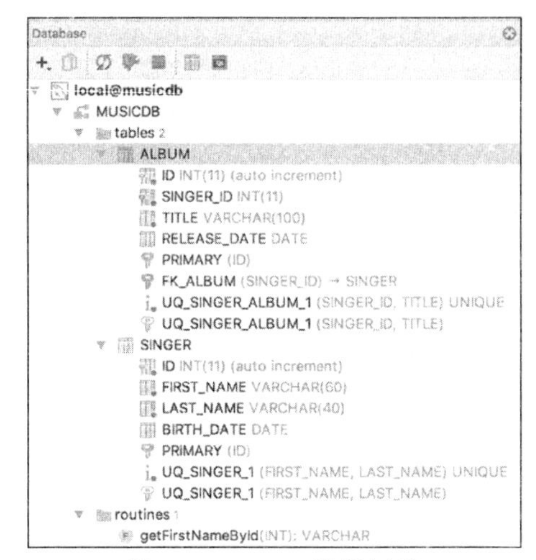

图 6-2　MUSICDB 模式的内容

因为需要数据来测试 JDBC 的使用情况，所以还提供了一个名为 test-data.sql 的文件，里面包含一组用来填充这两个表的 INSERT 语句。

```
insert into singer (first_name, last_name, birth_date)
    values ('John', 'Mayer', '1977-10-16');
insert into singer (first_name, last_name, birth_date)
    values ('Eric', 'Clapton', '1945-03-30');
insert into singer (first_name, last_name, birth_date)
    values ('John', 'Butler', '1975-04-01');

insert into album (singer_id, title, release_date)
    values (1, 'The Search For Everything', '2017-01-20');
insert into album (singer_id, title, release_date)
    values (1, 'Battle Studies', '2009-11-17');
insert into album (singer_id, title, release_date)
    values (2, ' From The Cradle ', '1994-09-13');
```

在本章后面的内容中，你将看到通过 JDBC 从数据库中检索数据并直接将结果集映射到 Java 对象(即 POJO)的示例。这些映射到表中记录的类也称为*实体*(*entity*)。对于 SINGER 表，将创建 Singer 类，该类将被实例化以创建映射到歌手记录的 Java 对象。

```
package com.apress.prospring5.ch6.entities;

import java.io.Serializable;
import java.sql.Date;
import java.util.List;

public class Singer implements Serializable {

    private Long id;
    private String firstName;
    private String lastName;
    private Date birthDate;
    private List<Album> albums;

    public void setId(Long id) {
       this.id = id;
    }

    public Long getId() {
       return this.id;
    }

    public void setFirstName(String firstName) {
       this.firstName = firstName;
    }

    public String getFirstName() {
       return this.firstName;
    }

    public void setLastName(String lastName) {
       this.lastName = lastName;
    }

    public String getLastName() {
       return this.lastName;
    }

    public boolean addAbum(Album album) {
```

```
        if (albums == null) {
            albums = new ArrayList<>();
            albums.add(album);
            return true;
        } else {
            if (albums.contains(album)) {
                                return false;
            }
        }
        albums.add(album);
        return true;
        }

        public void setAlbums(List<Album> albums) {
            this.albums = albums;
        }

        public List<Album> getAlbums() {
            return albums;
        }

        public void setBirthDate(Date birthDate) {
            this.birthDate = birthDate;
        }

        public Date getBirthDate() {
            return birthDate;
        }

        public String toString() {
                return "Singer - Id: " + id + ", First name: " + firstName
                    + ", Last name: " + lastName + ", Birthday: " + birthDate;
        }
}
```

以类似的方式，创建 Album 类。

```
package com.apress.prospring5.ch6.entities;

import java.io.Serializable;
import java.sql.Date;

public class Album implements Serializable {
        private Long id;
        private Long singerId;
        private String title;
        private Date releaseDate;

        public void setId(Long id) {
            this.id = id;
        }

        public Long getId() {
            return this.id;
        }

        public void setSingerId(Long singerId) {
            this.singerId = singerId;
        }

        public Long getSingerId() {
            return this.singerId;
        }

        public void setTitle(String title) {
            this.title = title;
        }

        public String getTitle() {
            return this.title;
        }

        public void setReleaseDate(Date releaseDate) {
            this.releaseDate = releaseDate;
        }

        public Date getReleaseDate() {
            return this.releaseDate;
```

```
    }

    @Override
    public String toString() {
        return "Album - Id: " + id + ", Singer id: " + singerId
            + ", Title: " + title + ", Release Date: " + releaseDate;
    }
}
```

让我们从一个简单的 SingerDao 接口开始，它封装了歌手信息的所有数据访问服务。代码如下：

```
package com.apress.prospring5.ch6.dao;

import com.apress.prospring5.ch6.entities.Singer;

import java.util.List;

public interface SingerDao {
    List<Singer> findAll();
    List<Singer> findByFirstName(String firstName);
    String findLastNameById(Long id);
    String findFirstNameById(Long id);
    void insert(Singer singer);
    void update(Singer singer);
    void delete(Long singerId);
    List<Singer> findAllWithDetail();
    void insertWithDetail(Singer singer);
}
```

在上面所示的接口中，分别定义了两个查找方法以及插入、更新和删除方法。它们对应于术语 CRUD(创建[Creat]、读取[Read]、更新[Updata]、删除[Delete])。

最后，为了便于测试，还需要修改一下 logback.xml 配置文件，将日志级别变为 DEBUG，以供所有类使用。在 DEBUG 级别，Spring JDBC 模块将所有被触发的 SQL 语句输出到数据库，这样就能知道到底发生了什么；这对于诊断 SQL 语法错误特别有用。以下配置示例描述了启用 DEBUG 级别的 logback.xml 文件的内容(位于 src/main/resources 下 Chapter 6 项目的源代码文件中)。

```
<?xml version="1.0" encoding="UTF-8"?>
<configuration>

    <contextListener class="ch.qos.logback.classic.jul.LevelChangePropagator">
        <resetJUL>true</resetJUL>
    </contextListener>

    <appender name="console" class="ch.qos.logback.core.ConsoleAppender">
        <encoder>

    <pattern>%d{HH:mm:ss.SSS} [%thread] %-5level %logger{5} - %msg%n</pattern>

        </encoder>
    </appender>

    <logger name="com.apress.prospring5.ch5" level="debug"/>

    <logger name="org.springframework" level="off"/>

    <root level="debug">
        <appender-ref ref="console" />
    </root>
</configuration>
```

6.3 研究 JDBC 基础结构

JDBC 为 Java 应用程序访问存储在数据库中的数据提供了一种标准方式。JDBC 基础结构的核心是针对每个数据库的驱动程序,即允许 Java 代码访问数据库的驱动程序。

一旦加载驱动程序，就会注册 java.sql.DriverManager 类。该类管理驱动程序列表并提供建立与数据库连接的静态方法。DriverManager 的 getConnection()方法返回驱动程序实现的 java.sql.Connection 接口。该接口允许针对数据库运行 SQL 语句。

JDBC 框架非常复杂且经过严格测试；然而，这种复杂性也带来一定的开发难度。第一层的复杂性在于确保代码管理与数据库的连接。连接是一种稀缺资源，建立起来非常昂贵。通常，数据库创建一个线程或为每个连接生成一个子进程。而且并发连接的数量通常是有限的，处于打开状态的连接过多会减慢数据库的速度。

下面将演示 Spring 如何帮助管理这种复杂性，但在继续之前，先演示一下如何在纯 JDBC 中选择、删除和更新数据。

本章介绍的所有项目都需要特殊的数据库库作为依赖项：mysql-connector、spring-jdbc、dbcp 等。可以查看每个项目的 build.gradle 配置文件，并从 pro-spring-15/build.gradle 了解所使用的版本和库。

接下来创建 SingerDao 接口的一个简单实现，用于通过纯 JDBC 与数据库进行交互。请记住，我们已经知道有关数据库连接的相关信息，下面在为每条语句创建连接时都会采取谨慎且昂贵(就性能而言)的方法。这极大降低了 Java 的性能，并为数据库增加了额外的压力，因为必须为每个查询建立连接。此外，如果保持连接处于打开状态，则可能导致数据库服务器停止。下面的代码片段显示了管理 JDBC 连接所需的代码，以 MySQL 为例：

```
package com.apress.prospring5.ch6.dao;
...
public class PlainSingerDao implements SingerDao {

    private static Logger logger =
       LoggerFactory.getLogger(PlainSingerDao.class);

    static {
       try {
          Class.forName("com.mysql.cj.jdbc.Driver");
       } catch (ClassNotFoundException ex) {
          logger.error("Prblem loadng DB dDiver!", ex);
       }
    }

    private Connection getConnection() throws SQLException {
       return DriverManager.getConnection(
          "jdbc:mysql://localhost:3306/musicdb?useSSL=true",
          "prospring5", "prospring5");
    }

    private void closeConnection(Connection connection) {
       if (connection == null) {
          return;
       }
       try {
          connection.close();
       } catch (SQLException ex) {
          logger.error("Problem closing connection to the database!",ex);
       }
    }
...
```

代码远不止这些，但上面显示的代码可以让我们了解管理 JDBC 连接所需的步骤。上述代码甚至不涉及连接池，这是一种更有效地管理数据库连接的常用技术。此时不讨论连接池(连接池在本章后面的 6.5 节“数据库连接和数据源”中讨论)；相反，下面的代码片段使用普通 JDBC 显示了 SingerDao 接口的 findAll()、insert()和 delete()方法的实现：

```
package com.apress.prospring5.ch6.dao;
...
public class PlainSingerDao implements SingerDao {
    @Override
    public List<Singer> findAll() {
       List<Singer> result = new ArrayList<>();
       Connection connection = null;
       try {
          connection = getConnection();
          PreparedStatement statement =
             connection.prepareStatement("select * from singer");
          ResultSet resultSet = statement.executeQuery();
          while (resultSet.next()) {
             Singer singer = new Singer();
             singer.setId(resultSet.getLong("id"));
             singer.setFirstName(resultSet.getString("first_name"));
             singer.setLastName(resultSet.getString("last_name"));
             singer.setBirthDate(resultSet.getDate("birth_date"));
             result.add(singer);
          }
         statement.close();
      } catch (SQLException ex) {
      logger.error("Problem when executing SELECT!",ex);
```

```
        } finally {
        closeConnection(connection);
        }
        return result;
    }

    @Override
    public void insert(Singer singer) {
        Connection connection = null;
            try {
                connection = getConnection();
                PreparedStatement statement = connection.prepareStatement(
                    "insert into Singer (first_name, last_name, birth_date)
                     values (?, ?, ?)"
                    , Statement.RETURN_GENERATED_KEYS);
                statement.setString(1, singer.getFirstName());
                statement.setString(2, singer.getLastName());
                statement.setDate(3, singer.getBirthDate());
                statement.execute();
                ResultSet generatedKeys = statement.getGeneratedKeys();
                if (generatedKeys.next()) {
                    singer.setId(generatedKeys.getLong(1));
                }
            statement.close();
            } catch (SQLException ex) {
            logger.error("Prblem executing INSERT", ex);
            } finally {
            closeConnection(connection);
        }
    }

    @Override
    public void delete(Long singerId) {
        Connection connection = null;
        try {
            connection = getConnection();
            PreparedStatement statement = connection.prepareStatement
                 ("delete from singer where id=?");
            statement.setLong(1, singerId);
            statement.execute();
            statement.close();
        } catch (SQLException ex) {
            logger.error("Prblem executing DELETE", ex);
        } finally {
            closeConnection(connection);
        }
    }
    ...
}
```

要测试 PlainSingerDao 类，可以使用以下类：

```
package com.apress.prospring5.ch6;

import com.apress.prospring5.ch6.dao.PlainSingerDao;
import com.apress.prospring5.ch6.dao.SingerDao;
import com.apress.prospring5.ch6.entities.Singer;
import org.slf4j.Logger;
import org.slf4j.LoggerFactory;

import java.sql.Date;
import java.util.GregorianCalendar;
import java.util.List;

public class PlainJdbcDemo {
    private static SingerDao singerDao = new PlainSingerDao();
    private static Logger logger = LoggerFactory.getLogger(PlainJdbcDemo.class);

    public static void main(String... args) {
        logger.info("Listing initial singer data:");

        listAllSingers();

        logger.info("-------------");
        logger.info("Insert a new singer");

        Singer singer = new Singer();
        singer.setFirstName("Ed");
        singer.setLastName("Sheeran");
```

```
        singer.setBirthDate(new Date
            ((new GregorianCalendar(1991, 2, 1991)).getTime().getTime()));
        singerDao.insert(singer);

        logger.info("Listing singer data after new singer created:");
        listAllSingers();

        logger.info("-------------");
        logger.info("Deleting the previous created singer");

        singerDao.delete(singer.getId());

        logger.info("Listing singer data after new singer deleted:");
        listAllSingers();
    }

    private static void listAllSingers() {
        List<Singer> singers = singerDao.findAll();

        for (Singer singer: singers) {
            logger.info(singer.toString());
        }
    }
}
```

可以注意到，现在使用一个记录器将信息打印到控制台。运行上面的程序会产生以下结果(假设有一个名为 MUSICDB 的本地安装的 MySQL 数据库，其用户名和密码设置为 prospring5 并且样本数据已加载)：

```
INFO    c.a.p.c.PlainJdbcDemo - Listing initial singer data:
INFO    c.a.p.c.PlainJdbcDemo - Singer - Id: 1, First name: John, Last name: Mayer,
                                 Birthday: 1977-10-15
INFO    c.a.p.c.PlainJdbcDemo - Singer - Id: 2, First name: Eric, Last name: Clapton,
                                 Birthday: 1945-03-29
INFO    c.a.p.c.PlainJdbcDemo - Singer - Id: 3, First name: John, Last name: Butler,
                                 Birthday: 1975-03-31

INFO    c.a.p.c.PlainJdbcDemo - -------------
INFO    c.a.p.c.PlainJdbcDemo - Insert a new singer
INFO    c.a.p.c.PlainJdbcDemo - Listing singer data after new singer created:
INFO    c.a.p.c.PlainJdbcDemo - Singer - Id: 1, First name: John, Last name: Mayer,
                                 Birthday: 1977-10-15
INFO    c.a.p.c.PlainJdbcDemo - Singer - Id: 2, First name: Eric, Last name: Clapton,
                                 Birthday: 1945-03-29
INFO    c.a.p.c.PlainJdbcDemo - Singer - Id: 3, First name: John, Last name: Butler,
                                 Birthday: 1975-03-31
INFO    c.a.p.c.PlainJdbcDemo - Singer - Id: 5, First name: Ed, Last name: Sheeran,
                                 Birthday: 1996-08-10

INFO    c.a.p.c.PlainJdbcDemo - -------------
INFO    c.a.p.c.PlainJdbcDemo - Deleting the previous created singer
INFO    c.a.p.c.PlainJdbcDemo - Listing singer data after new singer deleted:
INFO    c.a.p.c.PlainJdbcDemo - Singer - Id: 1, First name: John, Last name: Mayer,
                                 Birthday: 1977-10-15
INFO    c.a.p.c.PlainJdbcDemo - Singer - Id: 2, First name: Eric, Last name: Clapton,
                                 Birthday: 1945-03-29
INFO    c.a.p.c.PlainJdbcDemo - Singer - Id: 3, First name: John, Last name: Butler,
                                 Birthday: 1975-03-31
```

如输出所示，第一段代码显示了初始数据，第二段代码显示添加了新的记录，最后一段代码显示新创建的歌手(Ed Sheeran)被删除。

正如你在代码示例中看到的，需要将很多代码移动到辅助类中，或者更糟糕的是，每个 DAO 类中都会重复该代码。从应用程序开发人员的角度来看，这是 JDBC 的主要缺点；没有过多时间在每个 DAO 类中编写重复的代码。相反，我们更希望专注于编写真正需要 DAO 类的代码：选择、更新和删除数据。需要编写的辅助代码越多，需要处理的异常检查次数也就越多，并且可能会在代码中引入更多的错误。

这正是 DAO 框架和 Spring 的用武之地。DAO 框架清除了实际上没有执行任何定制逻辑的代码，并且程序员不需要再考虑所有需要执行的内务管理。另外，Spring 对 JDBC 的支持也使开发变得更加轻松。

6.4 Spring JDBC 基础结构

本章之前讨论的代码不是很复杂，但却很乏味，而且由于编写了这么多代码，因此编码出错的可能性也非常高。

接下来看一下 Spring 如何让事情变得更轻松、优雅。

概述以及要使用的包

Spring 中的 JDBC 支持被分为表 6-1 所示的五个包；每个包负责处理 JDBC 访问的不同方面。

表 6-1　Spring JDBC 包

包	描述
org.springframework.jdbc.core	该包包含 Spring 中 JDBC 类的基础，包括核心 JDBC 类 JdbcTemplate，它简化了使用 JDBC 编写数据库操作的过程。几个子包提供了 JDBC 数据访问支持，具有更多特定的用途(例如，支持命名参数的 JdbcTemplate 类)以及相关的支持类
org.springframework.jdbc.datasource	该包包含辅助类和 DataSource 实现，可用来在 JEE 容器外运行 JDBC 代码。几个子包提供了对嵌入式数据库、数据库初始化和各种数据源查找机制的支持
org.springframework.jdbc.object	该包包含有助于将数据库返回的数据转换为对象或对象列表的类。这些对象和列表是纯 Java 对象，因此与数据库断开连接
org.springframework.jdbc.support	该包中最重要的类是 SQLException 翻译支持。它允许 Spring 识别数据库所使用的错误代码并将它们映射到更高级别的异常
org.springframework.jdbc.config	该包包含支持 Spring 的 ApplicationContext 中 JDBC 配置的类。例如，它包含用于 jdbc 名称空间(例如，<jdbc:embeddeddatabase>标记)的处理程序类

下面通过查看最低级别的功能来开始讨论 Spring JDBC 支持。在运行 SQL 查询之前需要完成的第一件事是建立与数据库的连接。

6.5　数据库连接和数据源

可以提供一个实现了 javax.sql.DataSource 的 bean，从而使用 Spring 来帮助管理数据库连接。DataSource 和 Connection 之间的区别在于 DataSource 可以提供并管理 Connection。

DriverManagerDataSource(位于 org.springframework.jdbc.datasource 包中)是 DataSource 的最简单实现。通过查看类名，可以猜测它只是调用 DriverManager 来获得连接。由于 DriverManagerDataSource 不支持数据库连接池，因此该类不适用于除测试外的其他应用。DriverManagerDataSource 的配置非常简单，可以在下面的代码片段中看到；只需要提供驱动程序类名、连接 URL、用户名和密码(drivermanager-cfg-01.xml)即可。

```
<?xml version="1.0" encoding="UTF-8"?>
<beans xmlns="http://www.springframework.org/schema/beans"
       xmlns:xsi="http://www.w3.org/2001/XMLSchema-instance"
       xmlns:context="http://www.springframework.org/schema/context"
       xmlns:p="http://www.springframework.org/schema/p"
       xsi:schemaLocation="http://www.springframework.org/schema/beans
        http://www.springframework.org/schema/beans/spring-beans.xsd
        http://www.springframework.org/schema/context
        http://www.springframework.org/schema/context/spring-context.xsd">

    <bean id="dataSource"
       class="org.springframework.jdbc.datasource.DriverManagerDataSource"
       p:driverClassName="${jdbc.driverClassName}"
       p:url="${jdbc.url}" p:username="${jdbc.username}"
       p:password="${jdbc.password}"/>

       <context:property-placeholder location="classpath:db/jdbc.properties"/>
</beans>
```

你可能已经认出上述代码中的属性。它们通常表示传递给 JDBC 以获取 Connection 接口的值。一般来说，数据库连接信息存储在属性文件中，以便在不同的部署环境中进行维护和替换。以下代码片段显示了一个示例属性文件 jdbc.properties，Spring 的属性占位符将从中加载连接信息：

```
jdbc.driverClassName=com.mysql.cj.jdbc.Driver jdbc.url=jdbc:mysql://localhost:3306/
musicdb?useSSL=true
jdbc.username=prospring5
```

```
jdbc.password=prospring5
```

通过在 mix 中添加 util 名称空间，也可以像下面所示的那样编写相同的配置(drivermanager-cfg-02.xml)：

```
<?xml version="1.0" encoding="UTF-8"?>
<beans xmlns="http://www.springframework.org/schema/beans"
      xmlns:xsi="http://www.w3.org/2001/XMLSchema-instance"
      xmlns:util="http://www.springframework.org/schema/util"
      xmlns:p="http://www.springframework.org/schema/p"
      xsi:schemaLocation="http://www.springframework.org/schema/beans
       http://www.springframework.org/schema/beans/spring-beans.xsd
       http://www.springframework.org/schema/util
       http://www.springframework.org/schema/util/spring-util.xsd">

   <bean id="dataSource"
         class="org.springframework.jdbc.datasource.DriverManagerDataSource"
         p:driverClassName="#{jdbc.driverClassName}"
         p:url="#{jdbc.url}"
         p:username="#{jdbc.username}"
         p:password="#{jdbc.password}"/>

   <util:properties id="jdbc" location="classpath:db/jdbc2.properties"/>

</beans>
```

此时需要做一下修改。这些属性被加载到一个名为 jdbc 的 java.util.Properties bean 中，因此需要更改属性文件中属性的名称，以便能够使用 jdbc 前缀访问它们。下面所示的是 jdbc2.properties 文件的内容：

```
driverClassName=com.mysql.cj.jdbc.Driver
url=jdbc:mysql://localhost:3306/musicdb?useSSL=true
username=prospring5
password=prospring5
```

由于本书更侧重于使用 Java 配置类，因此还需要一个配置类：

```
package com.apress.prospring5.ch6.config;

import org.springframework.beans.factory.annotation.Value;
import org.springframework.context.annotation.Bean;
import org.springframework.context.annotation.Configuration;
import org.springframework.context.annotation.Lazy;
import org.springframework.context.annotation.PropertySource;
import org.springframework.context.support.PropertySourcesPlaceholderConfigurer;
import org.springframework.jdbc.datasource.SimpleDriverDataSource;
import org.springframework.jdbc.datasource.init.DatabasePopulatorUtils;

import javax.sql.DataSource;
import java.sql.Driver;

@Configuration
@PropertySource("classpath:db/jdbc2.properties")
public class DbConfig {

   @Value("${driverClassName}")
   private String driverClassName;
   @Value("${url}")
   private String url;
   @Value("${username}")
   private String username;
   @Value("${password}")
   private String password;

   @Bean
   public static PropertySourcesPlaceholderConfigurer
         propertySourcesPlaceholderConfigurer() {
      return new PropertySourcesPlaceholderConfigurer();
   }

   @Lazy
   @Bean
   public DataSource dataSource() {
      try {
         SimpleDriverDataSource dataSource =
            new SimpleDriverDataSource();
         Class<? extends Driver> driver =
             (Class<? extends Driver>) Class.forName(driverClassName);
         dataSource.setDriverClass(driver);
         dataSource.setUrl(url);
         dataSource.setUsername(username);
```

```
            dataSource.setPassword(password);
            return dataSource;
            } catch (Exception e) {
               return null;
            }
        }
    }
```

要测试这些类，可以使用以下测试类：

```
package com.apress.prospring5.ch6;
import com.apress.prospring5.ch6.config.DbConfig;
import org.junit.Test;
...
import static org.junit.Assert.assertNotNull;
import static org.junit.Assert.assertTrue;

public class DbConfigTest {

   private static Logger logger = LoggerFactory.getLogger(DbConfigTest.class);

   @Test
   public void testOne() throws SQLException {
      GenericXmlApplicationContext ctx = new GenericXmlApplicationContext();
      ctx.load("classpath:spring/drivermanager-cfg-01.xml");
      ctx.refresh();
      DataSource dataSource = ctx.getBean("dataSource", DataSource.class);
      assertNotNull(dataSource);
      testDataSource(dataSource);

      ctx.close();
   }

   @Test
   public void testTwo() throws SQLException {
      GenericApplicationContext ctx =
         new AnnotationConfigApplicationContext(DbConfig.class);
      DataSource dataSource = ctx.getBean("dataSource", DataSource.class);
      assertNotNull(dataSource);
      testDataSource(dataSource);

      ctx.close();
   }

   private void testDataSource(DataSource dataSource) throws SQLException{
      Connection connection = null;
      try {
         connection = dataSource.getConnection();
         PreparedStatement statement =
            connection.prepareStatement("SELECT 1");
         ResultSet resultSet = statement.executeQuery();
         while (resultSet.next()) {
            int mockVal = resultSet.getInt("1");
            assertTrue(mockVal== 1);
         }
         statement.close();
      } catch (Exception e) {
         logger.debug("Something unexpected happened.", e);
      } finally {
         if (connection != null) {
            connection.close();
         }
      }
   }
}
```

这里再次使用了一个测试类，因为重用一些代码更实用一些，并且还可以学习如何使用 JUnit 快速为所编写的任何代码编写测试代码。第一个方法 testOne()用于测试 XML 配置，第二个方法用于测试 DbConfig 配置类。在从任何配置获取 dataSource bean 后，使用模拟查询 SELECT 1 测试与 MySQL 数据库的连接。

在实际应用程序中，可以使用 Apache Commons BasicDataSource[1]或由 JEE 应用程序服务器(例如，JBoss、WebSphere、WebLogic 或 GlassFish)实现的 DataSource，从而进一步提高应用程序的性能。也可以在纯 JDBC 代码中使用 DataSource 并获得相同的池化优势；但是，在大多数情况下，会选择在中心位置配置 DataSource。另一方面，

1 官方网站为 http://commons.apache.org/dbcp。

Spring 允许声明一个 dataSource bean，并在 ApplicationContext 定义文件中设置连接属性。请参阅以下配置示例；文件名是 datasource-dbcp.xml：

```
<beans ...>

    <bean id="dataSource"
        class="org.apache.commons.dbcp2.BasicDataSource"
        destroy-method="close"
        p:driverClassName="#{jdbc.driverClassName}"
        p:url="#{jdbc.url}"
        p:username="#{jdbc.username}"
        p:password="#{jdbc.password}"/>

    <util:properties id="jdbc" location="classpath:db/jdbc2.properties"/>

</beans>
```

这个特定的 Spring 管理的 DataSource 在 org.apache.commons.dbcp.BasicDataSource 中实现。最重要的是 dataSource bean 实现了 javax.sql.DataSource，并且可以在数据访问类中使用它。

另一种配置 dataSource bean 的方法是使用 JNDI。如果正在开发的应用程序要在 JEE 容器中运行，那么可以利用容器所管理的连接池。要想使用基于 JNDI 的数据源，需要更改 dataSource bean 声明，如以下示例(datasource-jndi.xml)所示：

```
<beans ...>
    <bean id="dataSource"
        class="org.springframework.jndi.JndiObjectFactoryBean"

        p:jndiName="java:comp/env/jdbc/musicdb"/>
</beans>
```

在上面的示例中，使用了 Spring 的 JndiObjectFactoryBean，以便通过 JNDI 查找来获取数据源。从版本 2.5 开始，Spring 提供了 jee 名称空间，从而进一步简化了配置。下面的代码显示了使用 jee 名称空间的相同 JNDI 数据源配置(datasource-jee.xml)：

```
<?xml version="1.0" encoding="UTF-8"?>
<beans xmlns="http://www.springframework.org/schema/beans"
    xmlns:xsi="http://www.w3.org/2001/XMLSchema-instance"
    xmlns:jee="http://www.springframework.org/schema/jee"
    xsi:schemaLocation="http://www.springframework.org/schema/beans
        http://www.springframework.org/schema/beans/spring-beans.xsd
        http://www.springframework.org/schema/jee
        http://www.springframework.org/schema/jee/spring-jee.xsd">

    <jee:jndi-lookup jndi-name="java:comp/env/jdbc/prospring5ch6"/>
</beans>
```

上面所示的配置代码片段首先在<beans>标记中声明了 jee 名称空间，然后在<jee:jndi-lookup>标记中声明数据源。如果采用 JNDI 方法，则不要忘记在应用程序描述符文件中添加资源引用(resource-ref)。请参阅以下代码片段：

```
<root-node>
    <resource-ref>
        <res-ref-name>jdbc/musicdb</res-ref-name>
        <res-type>javax.sql.DataSource</res-type>
        <res-auth>Container</res-auth>
        </resource-ref>
    </root-node>
```

<root-node>是一个占位符值；需要根据模块的打包方式进行更改。例如，如果应用程序是 Web 模块，那么在 Web 部署描述符(WEB-INF/web.xml)中变为<web-app>。很可能还需要在特定于应用程序服务器的描述符文件中配置 resource-ref。但是请注意，resource-ref 元素配置 jdbc/musicdb 引用名称，并且 dataSource bean 的 jndiName 被设置为 java:comp/env/jdbc/musicdb。

正如你所看到的，Spring 允许以你喜欢的任何方式配置 DataSource，并且对应用程序的其他代码隐藏了数据源的实际实现或位置。换句话说，DAO 类不知道，也不需要知道 DataSource 指向的地方。

连接管理也被委托给 dataSource bean，后者完成管理工作或使用 JEE 容器来完成所有工作。

6.6 嵌入数据库支持

从版本 3.0 开始，Spring 还提供了嵌入式数据库支持，该支持会自动启动嵌入式数据库并将其作为应用程序的

DataSource 公开。以下配置代码片段显示了嵌入式数据库的配置(embedded-h2-cfg.xml)：

```
<?xml version="1.0" encoding="UTF-8"?>
<beans xmlns="http://www.springframework.org/schema/beans"
   xmlns:xsi="http://www.w3.org/2001/XMLSchema-instance"
   xmlns:jdbc="http://www.springframework.org/schema/jdbc"
   xmlns:p="http://www.springframework.org/schema/p"
   xmlns:c="http://www.springframework.org/schema/c"
   xsi:schemaLocation="http://www.springframework.org/schema/beans
      http://www.springframework.org/schema/beans/spring-beans.xsd
      http://www.springframework.org/schema/jdbc
      http://www.springframework.org/schema/jdbc/spring-jdbc.xsd">

   <jdbc:embedded-database id="dataSource" type="H2">
      <jdbc:script location="classpath:db/h2/schema.sql"/>
      <jdbc:script location="classpath:db/h2/test-data.sql"/>
   </jdbc:embedded-database>

</beans>
```

在上面的代码清单中，首先在<beans>标记中声明了 jdbc 名称空间。之后，使用<jdbc:embedded-database>来声明嵌入式数据库并将其 ID 设置为 dataSource。在该标记内，还指示 Spring 执行指定的脚本以创建数据库模式并相应地填充测试数据。请注意，脚本的顺序非常重要，包含数据定义语言(DDL)的文件应该始终第一个显示，然后是带有数据操作语言(DML)的文件。对于 type 属性，指定要使用的嵌入式数据库的类型。从版本 4.0 开始，Spring 支持 HSQL(默认)、H2 和 DERBY。

嵌入式数据库也可以使用 Java 配置类进行配置。要使用的类是 EmbeddedDatabaseBuilder，它使用数据库创建和加载数据脚本作为参数来创建实现了 DataSource 的 EmbeddedDatabase 实例。

```
package com.apress.prospring5.ch6.config;

import org.slf4j.Logger;
import org.slf4j.LoggerFactory;
import org.springframework.context.annotation.Bean;
import org.springframework.context.annotation.Configuration;
import org.springframework.jdbc.datasource.embedded.EmbeddedDatabaseBuilder;
import org.springframework.jdbc.datasource.embedded.EmbeddedDatabaseType;

import javax.sql.DataSource;

@Configuration
public class EmbeddedJdbcConfig {

   private static Logger logger =
      LoggerFactory.getLogger(EmbeddedJdbcConfig.class);

   @Bean
   public DataSource dataSource() {
      try {
         EmbeddedDatabaseBuilder dbBuilder =
            new EmbeddedDatabaseBuilder();
         return dbBuilder.setType(EmbeddedDatabaseType.H2)
      .addScripts("classpath:db/h2/schema.sql",
      "classpath:db/h2/test-data.sql").build();
      } catch (Exception e) {
         logger.error("Embedded DataSource bean cannot be created!", e);
         return null;
      }
   }
...
}
```

对于本地开发或单元测试来说，嵌入式数据库支持是非常有用的。在本章的其余部分，将使用嵌入式数据库来运行示例代码，因此计算机不需要安装数据库即可运行示例。

不仅可以通过 JDBC 名称空间来利用嵌入式数据库支持，还可以初始化在其他位置运行的数据库实例，例如 MySQL、Oracle 等。无须指定 type 和 embedded-database，只需要使用 initialize-database 即可，并且脚本将针对预期的 dataSource 执行，就像嵌入式数据库一样。

6.7 在 DAO 类中使用 DataSource

数据访问对象(DAO)模式用于将低级数据访问 API 或操作与高级业务服务相分离。数据访问对象模式需要以下组件：

- **DAO 接口**：该接口定义了在模型对象(或多个对象)上执行的标准操作。
- **DAO 实现**：该类提供了 DAO 接口的具体实现。通常使用 JDBC 连接或数据源来处理模型对象。
- **模型对象也称为数据对象或实体**：这是映射到数据表记录的简单 POJO。

接下来创建 SingerDao 接口，如下所示：

```
package com.apress.prospring5.ch6.dao;

public interface SingerDao {
    String findNameById(Long id);
}
```

对于简单的实现，首先将一个 dataSource 属性添加到 JdbcSingerDao 实现类。将 dataSource 属性添加到实现类而不是接口的原因应该非常明显：接口不需要知道数据将如何进行检索和更新。将 DataSource mutator 方法添加到接口中，理想情况下，这会迫使实现声明 getter 和 setter 存根(stub)。显然，这并不是很好的设计实践。看一下下面所示的简单 JdbcSingerDao 类：

```
import com.apress.prospring5.ch6.dao.SingerDao;
import com.apress.prospring5.ch6.entities.Singer;
import org.apache.commons.lang3.NotImplementedException;
import org.springframework.beans.factory.BeanCreationException;
import org.springframework.beans.factory.InitializingBean;
import org.springframework.jdbc.core.JdbcTemplate;

import javax.sql.DataSource;
import java.util.List;

public class JdbcSingerDao implements SingerDao, InitializingBean {

    private DataSource dataSource;

    public void setDataSource(DataSource dataSource) {
        this.dataSource = dataSource;
    }

    public void afterPropertiesSet() throws Exception {
        if (dataSource == null) {
            throw new BeanCreationException(
                "Must set dataSource on SingerDao");
        }
    }
...
}
```

现在可以指示 Spring 通过使用 JdbcSingerDao 实现来配置 singerDao bean 并设置 dataSource 属性，如以下 EmbeddedJdbcConfig 配置类中所示：

```
package com.apress.prospring5.ch6.config;

...
@Configuration
public class EmbeddedJdbcConfig {

    private static Logger logger =
        LoggerFactory.getLogger(EmbeddedJdbcConfig.class);

    @Bean
    public DataSource dataSource() {
        try {
            EmbeddedDatabaseBuilder dbBuilder =
                new EmbeddedDatabaseBuilder();
            return dbBuilder.setType(EmbeddedDatabaseType.H2)
        .addScripts("classpath:db/h2/schema.sql",
        "classpath:db/h2/test-data.sql").build();
        } catch (Exception e) {
            logger.error("Embedded DataSource bean cannot be created!", e);
            return null;
        }
}

    @Bean
    public SingerDao singerDao(){
        JdbcSingerDao dao = new JdbcSingerDao();
        dao.setDataSource(dataSource());
```

```
        return dao;
    }
}
```

Spring 支持相当多的嵌入式数据库，但它们必须作为依赖项添加到项目中。在 pro-spring-15\gradle.build 中可以看到一些被配置的特定数据库的库：

```
ext {
    derbyVersion = '10.13.1.1'
    dbcpVersion = '2.1'
    mysqlVersion = '6.0.6'
    h2Version = '1.4.194'
    ...

    db = [
        mysql: "mysql:mysql-connector-java:$mysqlVersion",
        derby: "org.apache.derby:derby:$derbyVersion",
        dbcp : "org.apache.commons:commons-dbcp2:$dbcpVersion",
        h2 : "com.h2database:h2:$h2Version"
    ]
}
```

现在，Spring 通过将 dataSource 属性设置为 dataSource bean 来实例化 JdbcSingerDao 类，从而创建 singerDao bean。确保设置 bean 的所有必需属性是一种非常好的做法。最简单的方法是实现 InitializingBean 接口并提供 afterPropertiesSet()方法的实现。这样就可以确保在 JdbcSingerDao 上设置了所有必需的属性。有关 bean 初始化的更多讨论，请参阅第 4 章。

你在前面所看到的代码使用 Spring 来管理数据源并引入 SingerDao 接口及其 JDBC 实现。此外，还在 Spring ApplicationContext 文件的 JdbcSingerDao 类中设置了 dataSource 属性。最后，通过将实际的 DAO 操作添加到接口和实现来扩展代码。

6.8 异常处理

由于 Spring 提倡使用运行时异常而不是检查型异常，因此需要一种机制将检查型 SQL 异常转换为运行时 Spring JDBC 异常。因为 Spring 的 SQL 异常是运行时异常，所以它们可能比检查型异常更精细。虽然根据定义，这并不是运行时异常的特征，但是在 throws 子句中声明一长串检查型异常是极不方便的；因此，检查型异常往往比等价的运行时异常粒度更粗。

Spring 提供了 SQLExceptionTranslator 接口的默认实现，该接口负责将通用 SQL 错误代码转换为 Spring JDBC 异常。大多数情况下，该实现足够了，但也可以扩展 Spring 的默认实现并将新的 SQLExceptionTranslator 实现设置为在 JdbcTemplate 中使用，如以下代码示例所示：

```
package com.apress.prospring5.ch6;

import java.sql.SQLException;

import org.springframework.dao.DataAccessException;
import org.springframework.dao.DeadlockLoserDataAccessException;
import org.springframework.jdbc.support.SQLErrorCodeSQLExceptionTranslator;

public class MySQLErrorCodesTranslator extends
                                        SQLErrorCodeSQLExceptionTranslator {
    @Override
    protected DataAccessException customTranslate(String task,
            String sql, SQLException sqlex) {
        if (sqlex.getErrorCode() == -12345) {
            return new DeadlockLoserDataAccessException(task, sqlex);
        }
        return null;
    }
}
```

同时，还需要将依赖项 spring-jdbc 添加到项目中。可以看到 spring-jdbc 被配置为在 pro-spring-15/gradle.build 中使用：

```
ext {
    springVersion = '5.0.0.M4'
    ...

    spring = [
```

```
        jdbc : "org.springframework:spring-jdbc:$springVersion",
        ...
    ]
}
```

该库将作为依赖项添加到用作本章示例的所有 JDBC 项目中。

```
//chapter06/spring-jdbc-embedded/build.gradle
dependencies {
    compile spring.jdbc
    compile db.h2, db.derby
}
```

要想使用自定义转换器，需要将其传递到 DAO 类的 JdbcTemplate 中。以下代码片段显示了增强型 JdbcSingerDao.setDataSource()方法的部分代码，从而说明其用法：

```
...
public class JdbcSingerDao implements SingerDao, InitializingBean {

    private DataSource dataSource;
    private JdbcTemplate jdbcTemplate;

    public void setDataSource(DataSource dataSource) {
        this.dataSource = dataSource;
        JdbcTemplate jdbcTemplate = new JdbcTemplate();
        jdbcTemplate.setDataSource(dataSource);
        MySQLErrorCodesTranslator errorTranslator =
                new MySQLErrorCodesTranslator();
        errorTranslator.setDataSource(dataSource);
        jdbcTemplate.setExceptionTranslator(errorTranslator);
        this.jdbcTemplate = jdbcTemplate;
    }
...
}
```

当使用自定义 SQL 异常转换器时，如果 Spring 在对数据库执行 SQL 语句时检测到 SQL 异常，就会调用它，并且在错误代码为-12345 时执行自定义异常转换。而对于其他错误，Spring 将回退到默认的异常转换机制。显然，也可以将 SQLExceptionTranslator 创建为 Springmanaged bean，并在 DAO 类中使用 JdbcTemplate bean。即便不了解 JdbcTemplate 类，也不要担心；接下来会更详细地讨论它。

6.9 JdbcTemplate 类

该类代表 Spring JDBC 支持的核心。它可以执行所有类型的 SQL 语句。用最简单的观点来看，可以对数据定义和数据操作语句进行分类。数据定义语句用于创建各种数据库对象(表、视图、存储过程等)，而数据操作语句则操作数据，可以分类为选择和更新语句。选择语句通常返回一组行，每行都有相同的一组列。更新语句修改数据库中的数据，但不返回任何结果。

JdbcTemplate 类允许向数据库发出任何类型的 SQL 语句并返回任何类型的结果。

在本节中，将通过 JdbcTemplate 类在 Spring 中完成 JDBC 编程的几个常见用例。

6.9.1 在 DAO 类中初始化 JdbcTemplate

在讨论如何使用 JdbcTemplate 之前，先看一下如何准备 JdbcTemplate，以便在 DAO 类中使用。方法很简单，大多数时候只需要通过传入数据源对象(应该由 Spring 注入到 DAO 类中)来构造类。以下代码片段显示了如何初始化 JdbcTemplate 对象：

```
public class JdbcSingerDao implements SingerDao, InitializingBean {

    private DataSource dataSource;
    private JdbcTemplate jdbcTemplate;

    public void setDataSource(DataSource dataSource) {
        this.dataSource = dataSource;
        JdbcTemplate jdbcTemplate = new JdbcTemplate();
        jdbcTemplate.setDataSource(dataSource);
    }
    ...
}
```

一般的做法是在 setDataSource 方法中初始化 JdbcTemplate，这样一旦 Spring 注入数据源，JdbcTemplate 也将被初始化并备用。

配置完毕后，JdbcTemplate 是线程安全的。这意味着也可以选择在 Spring 的配置中初始化一个 JdbcTemplate 实例，并将其注入到所有的 DAO bean 中。下面的代码描述了该配置：

```
package com.apress.prospring5.ch6.config;
...
@Configuration
public class EmbeddedJdbcConfig {

   private static Logger logger =
      LoggerFactory.getLogger(EmbeddedJdbcConfig.class);

   @Bean
      public DataSource dataSource() {
         try {
            EmbeddedDatabaseBuilder dbBuilder =
               new EmbeddedDatabaseBuilder();
            return dbBuilder.setType(EmbeddedDatabaseType.H2)
         .addScripts("classpath:db/h2/schema.sql",
         "classpath:db/h2/test-data.sql").build();
         } catch (Exception e) {
            logger.error("Embedded DataSource bean cannot be created!", e);
            return null;
         }
   }

   @Bean public JdbcTemplate jdbcTemplate(){
      JdbcTemplate jdbcTemplate = new JdbcTemplate();
      jdbcTemplate.setDataSource(dataSource());
      return jdbcTemplate;
   }

   @Bean
   public SingerDao singerDao() {
      JdbcSingerDao dao = new JdbcSingerDao();
      dao.setJdbcTemplate(jdbcTemplate());
      return dao;
   }
}
```

使用 JdbcTemplate 检索单值

首先，从一个返回单值的简单查询开始。比如，希望能够通过 ID 来检索歌手的姓名。通过使用 jdbcTemplate，可以轻松地检索值。以下代码片段显示了 JdbcSingerDao 类中 findNameById()方法的实现。而对于其他方法，则创建空的实现。

```
...
public class JdbcSingerDao implements SingerDao, InitializingBean {

   private JdbcTemplate jdbcTemplate;

   public void setJdbcTemplate(JdbcTemplate jdbcTemplate) {
      this.jdbcTemplate = jdbcTemplate;
   }

   @Override public String findNameById(Long id) {
      return jdbcTemplate.queryForObject(
         "select first_name || ' ' || last_name from singer where id = ?",
            new Object{id}, String.class);
   }

   @Override public void insert(Singer singer) {
      throw new NotImplementedException("insert");
   }

...
}
```

在上面的代码清单中，使用 JdbcTemplate 的 queryForObject()方法来检索名字的值。第一个参数是 SQL 字符串，第二个参数由若干个参数组成，这些参数将被传递给 SQL 以便以对象数组格式进行参数绑定。最后一个参数是要返回的类型，此时为 String。除 Object 外，还可以查询其他类型，例如 Long 和 Integer。接下来看看结果。以下代码片段显示了测试程序。同样，将使用 JUnit 测试类，因为这允许单独运行测试方法，并且在执行 glade build 时运行测试，也可确保构建保持稳定。

```
package com.apress.prospring5.ch6;
...
import static org.junit.Assert.assertNotNull;
import static org.junit.Assert.assertTrue;

public class JdbcCfgTest {

   @Test
   public void testH2DB() {
      GenericXmlApplicationContext ctx =
         new GenericXmlApplicationContext();
      ctx.load("classpath:spring/embedded-h2-cfg.xml");
      ctx.refresh();
      testDao(ctx.getBean(SingerDao.class));
      ctx.close();
   }

   private void testDao(SingerDao singerDao) {
      assertNotNull(singerDao);
      String singerName = singerDao.findNameById(1l);
      assertTrue("John Mayer".equals(singerName));
}
}
```

在执行测试方法testH2DB()时，希望singerDao.findNameById(1l)调用返回字符串John Mayer，并且使用assertTrue()方法测试此假设。如果初始化数据库时出现任何问题，此测试将失败。

6.9.2 通过 NamedParameterJdbcTemplate 使用命名参数

在前面的示例中，使用普通的占位符(? 字符)作为查询参数，并且需要将参数值作为 Object 数组传递。当使用普通占位符时，顺序非常重要，将参数放入数组的顺序应与查询中参数的顺序相同。

一些开发人员更喜欢使用命名参数来确保每个参数都完全按照预期进行绑定。在 Spring 中，名为 NamedParameterJdbcTemplate(位于 org.springframework.jdbc.core.namedparam 包中)的 JdbcTemplate 类的扩展提供了对命名参数的支持。

NamedParameterJdbcTemplate 的初始化与 JdbcTemplate 相同，所以只需要声明一个 NamedParameterJdbcTemplate 类型的 bean 并将其注入到 Dao 类中即可。在下面的代码中，可以看到新的和改进的 JdbcSingerDao:

```
package com.apress.prospring5.ch6;

public class JdbcSingerDao implements
      SingerDao, InitializingBean {

   private NamedParameterJdbcTemplate namedParameterJdbcTemplate;
   @Override
   public String findNameById(Long id) {
      String sql = "SELECT first_name ||' '|| last_name
         FROM singer WHERE id = :singerId";
      Map<String, Object> namedParameters = new HashMap<>();
      namedParameters.put("singerId", id);
      return namedParameterJdbcTemplate.queryForObject(sql,
         namedParameters, String.class);
   }

   public void setNamedParameterJdbcTemplate
       (NamedParameterJdbcTemplate namedParameterJdbcTemplate) {
         this.namedParameterJdbcTemplate = namedParameterJdbcTemplate;
   }

   @Override
   public void afterPropertiesSet() throws Exception {
      if (namedParameterJdbcTemplate == null) {
      throw new BeanCreationException
          ("Null NamedParameterJdbcTemplate on SingerDao");
      }
   }
   ...
}
```

可以看到，此时使用了命名参数:singerId(以冒号而不是占位符？为前缀)。以下代码片段用于测试新的JdbcSingerDao：

```
public class NamedJdbcCfgTest {

   @Test
    public void testCfg(){
       GenericApplicationContext ctx =
          new AnnotationConfigApplicationContext(NamedJdbcCfg.class);
       SingerDao singerDao = ctx.getBean(SingerDao.class);
       assertNotNull(singerDao);
       String singerName = singerDao.findNameById(1l);
       assertTrue("John Mayer".equals(singerName));

       ctx.close();
   }
}
```

当执行测试方法 testCfg()时，希望 singerDao.findNameById(1l)调用返回字符串 John Mayer，并使用 assertTrue()方法来测试该假设。一旦初始化数据库时出现任何问题，此测试将失败。

6.9.3 使用 RowMapper<T>检索域对象

在大多数情况下，希望查询一行或多行，然后将每行转换为相应的域对象或实体，而不是检索单个值。Spring 的 RowMapper<T>接口(位于 org.springframework.jdbc.core 包中)提供了一种简单的方法来完成从 JDBC 结果集到 POJO 的映射。接下来通过实现 SingerDao(使用了 RowMapper<T>接口)的 findAll()方法来看看该接口的实际应用。在下面的代码片段中，可以看到 findAll()方法的实现：

```
package com.apress.prospring5.ch6;
...
public class JdbcSingerDao implements
   SingerDao, InitializingBean {

   private NamedParameterJdbcTemplate namedParameterJdbcTemplate;

   public void setNamedParameterJdbcTemplate(
         NamedParameterJdbcTemplate namedParameterJdbcTemplate) {
      this.namedParameterJdbcTemplate = namedParameterJdbcTemplate;
   }

   @Override
   public List<Singer> findAll() {
      String sql = "select id, first_name, last_name, birth_date from singer";
      return namedParameterJdbcTemplate.query(sql, new SingerMapper());
   }

   private static final class SingerMapper
         implements RowMapper<Singer> {

      @Override
      public Singer mapRow(ResultSet rs, int rowNum)
            throws SQLException {
         Singer singer = new Singer();
         singer.setId(rs.getLong("id"));
         singer.setFirstName(rs.getString("first_name"));
         singer.setLastName(rs.getString("last_name"));
         singer.setBirthDate(rs.getDate("birth_date"));
         return singer;
      }
   }

   @Override
   public void afterPropertiesSet() throws Exception {
      if (namedParameterJdbcTemplate == null) {
         throw new BeanCreationException(
            "Null NamedParameterJdbcTemplate on SingerDao");
      }
   }
}
```

在上面的代码片段中，定义了一个名为 SingerMapper 的静态内部类，它实现了 RowMapper <Singer>接口。该类需要提供 mapRow()实现，以便将 ResultSet 的特定记录中的值转换为所需的域对象。之所以创建一个静态内部类，是为了在多个查找器方法中共享 RowMapper <Singer>。

可以使用 Java 8 lambda 表达式完全跳过 SingerMapper 类的显式实现；因此，findAll()方法按照如下方式重构：

```
public List<Singer> findAll() {
   String sql = "select id, first_name, last_name, birth_date from singer";
   return namedParameterJdbcTemplate.query(sql, (rs, rowNum) -> {
      Singer singer = new Singer();
      singer.setId(rs.getLong("id"));
      singer.setFirstName(rs.getString("first_name"));
      singer.setLastName(rs.getString("last_name"));
      singer.setBirthDate(rs.getDate("birth_date"));
      return singer;
   });
}
```

之后，findAll()方法只需要调用查询方法并传入查询字符串和行映射器即可。如果查询需要参数，那么 query()方法还提供了一个接收查询参数的重载方法。以下测试类包含 findAll()方法的测试方法：

```
public class RowMapperTest {

   @Test
   public void testRowMapper() {
      GenericApplicationContext ctx =
      new AnnotationConfigApplicationContext(NamedJdbcCfg.class);
      SingerDao singerDao = ctx.getBean(SingerDao.class);
      assertNotNull(singerDao);
      List<Singer> singers = singerDao.findAll();
      assertTrue(singers.size() == 3);

      singers.forEach(singer -> {
         System.out.println(singer);
         if (singer.getAlbums() != null) {
            for (Album album :
                  singer.getAlbums()) {
               System.out.println("---" + album);
            }
         }
      });

      ctx.close();
   }
}
```

如果运行 testRowMapper 方法，则测试必须通过，并且必须生成以下内容：

```
Singer - Id: 1, First name: John, Last name: Mayer, Birthday: 1977-10-16
Singer - Id: 2, First name: Eric, Last name: Clapton, Birthday: 1945-03-30
Singer - Id: 3, First name: John, Last name: Butler, Birthday: 1975-04-01
```

由于 RowMapper<Singer>实现并未在返回的 Singer 实例上实际设置专辑信息，因此未打印相关信息。

6.10 使用 ResultSetExtractor 检索嵌套域对象

下面看一个更复杂的例子，在该例中，需要用连接从父表(SINGER)和子表(ALBUM)中检索数据，然后将数据转换回嵌套对象(Singer 中的 Album)。

前面提到的 RowMapper<T>仅适用于将行映射到单个域对象；对于更复杂的对象结构，则需要使用 ResultSetExtractor 接口。为了演示它的用法，在 SingerDao 接口中添加一个方法 findAllWithAlbums()。该方法应该使用专辑信息填充歌手名单。

以下代码片段显示了向接口添加 findAllWithAlbums()方法以及使用 ResultSetExtractor 实现该方法的过程：

```
package com.apress.prospring5.ch6;
...
public class JdbcSingerDao implements SingerDao, InitializingBean {

   private NamedParameterJdbcTemplate namedParameterJdbcTemplate;

   public void setNamedParameterJdbcTemplate(
      NamedParameterJdbcTemplate namedParameterJdbcTemplate) {
      this.namedParameterJdbcTemplate = namedParameterJdbcTemplate;
   }

   @Override
```

```
    public List<Singer> findAllWithAlbums() {
        String sql = "select s.id, s.first_name, s.last_name, s.birth_date" +
                ", a.id as a.album_id, a.title, a.release_date from singer s " +
                "left join album a on s.id = a.singer_id";
        return namedParameterJdbcTemplate.query(sql, new SingerWithDetailExtractor());
    }

    private static final class SingerWithDetailExtractor implements
            ResultSetExtractor<List<Singer>> {

        @Override
        public List<Singer> extractData(ResultSet rs) throws SQLException,
                DataAccessException {
            Map<Long, Singer> map = new HashMap<>();
            Singer singer;
            while (rs.next()) {
                Long id = rs.getLong("id");
                singer = map.get(id);
                if (singer == null) {
                    singer = new Singer();
                    singer.setId(id);
                    singer.setFirstName(rs.getString("first_name"));
                    singer.setLastName(rs.getString("last_name"));
                    singer.setBirthDate(rs.getDate("birth_date"));
                    singer.setAlbums(new ArrayList<>());
                    map.put(id, singer);
                }
                Long albumId = rs.getLong("singer_tel_id");
                if (albumId > 0) {
                    Album album = new Album();
                    album.setId(albumId);
                    album.setSingerId(id);
                    album.setTitle(rs.getString("title"));
                    album.setReleaseDate(rs.getDate("release_date"));
                    singer.addAbum(album);
                }
            }
            return new ArrayList<>(map.values());
        }
    }

    @Override
    public void afterPropertiesSet() throws Exception {
        if (namedParameterJdbcTemplate == null) {
            throw new BeanCreationException(
                "Null NamedParameterJdbcTemplate on SingerDao");
        }
    }
}
```

上述代码看起来很像RowMapper示例，但此时声明了一个实现ResultSetExtractor的内部类。然后实现extractData()方法，将结果集转换为对应的 Singer 对象列表。而对于 findAllWithDetail()方法，查询使用左连接将两个表连接起来，以便没有专辑的歌手也会被检索。最终的查询结果是两张表的笛卡尔积。最后，使用 JdbcTemplate.query()方法，并传入查询字符串和结果集提取器。

当然，SingerWithDetailExtractor 内部类实际上并不是必需的，因为使用了 lambda 表达式。以下显示了使用 Java 8 lambda 表达式的 findAllWithAlbums()版本：

```
public List<Singer> findAllWithAlbums() {
    String sql = "select s.id, s.first_name, s.last_name, s.birth_date" +
        ", a.id as a.album_id, a.title, a.release_date from singer s " +
        "left join album a on s.id = a.singer_id";
    return namedParameterJdbcTemplate.query(sql, rs -> {
        Map<Long, Singer> map = new HashMap<>();
        Singer singer;
        while (rs.next()) {
            Long id = rs.getLong("id");
            singer = map.get(id);
            if (singer == null) {
                singer = new Singer();
                singer.setId(id);
                singer.setFirstName(rs.getString("first_name"));
                singer.setLastName(rs.getString("last_name"));
                singer.setBirthDate(rs.getDate("birth_date"));
                singer.setAlbums(new ArrayList<>());
                map.put(id, singer);
```

```
            }
            Long albumId = rs.getLong("singer_tel_id");
            if (albumId > 0) {
                Album album = new Album();
                album.setId(albumId);
                album.setSingerId(id);
                album.setTitle(rs.getString("title"));
                album.setReleaseDate(rs.getDate("release_date"));
                singer.addAbum(album);
            }
        }
        return new ArrayList<>(map.values());
    });
}
```

以下测试类包含针对 findAllWithAlbums()方法的测试方法：

```
public class ResultSetExtractorTest {

    @Test
    public void testResultSetExtractor() {
        GenericApplicationContext ctx =
            new AnnotationConfigApplicationContext(NamedJdbcCfg.class);
        SingerDao singerDao = ctx.getBean(SingerDao.class);
        assertNotNull(singerDao);
        List<Singer> singers = findAllWithAlbums();
        assertTrue(singers.size() == 3);

        singers.forEach(singer -> {
        System.out.println(singer);
            if (singer.getAlbums() != null) {
                for (Album album :
                        singer.getAlbums()) {
                    System.out.println("\t--> " + album);
                }
            }
        });

        ctx.close();
    }
}
```

如果运行 testResultSetExtractor()方法，则必须通过测试，并且必须生成以下内容：

```
Singer - Id: 1, First name: John, Last name: Mayer, Birthday: 1977-10-16
    --> Album - Id: 2, Singer id: 1, Title: Battle Studies,
            Release Date: 2009-11-17
    --> Album - Id: 1, Singer id: 1, Title: The Search For Everything,
            Release Date: 2017-01-20
Singer - Id: 2, First name: Eric, Last name: Clapton, Birthday: 1945-03-30
    --> Album - Id: 3, Singer id: 2, Title: From The Cradle,
            Release Date: 1994-09-13
Singer - Id: 3, First name: John, Last name: Butler, Birthday: 1975-04-01
```

可以看到，歌手和他们的专辑信息都已相应列出。这些数据来自数据填充脚本，可以在 resources/db/test-data.sql 中找到每个 JDBC 示例项目的脚本。到目前为止，已经介绍了如何使用 JdbcTemplate 执行一些常见的查询操作。JdbcTemplate(以及 NamedParameterJdbcTemplate 类)提供了一些支持数据更新操作(包括插入、更新、删除等)的重载 update()方法。然而，update()方法很容易理解，所以把它留作练习，供你研究。另一方面，在后面的章节中将使用 Spring 提供的 SqlUpdate 类来执行数据更新操作。

6.11 建模 JDBC 操作的 Spring 类

在前面的章节中，你看到了 JdbcTemplate 和相关的数据映射器实用程序类如何极大地简化使用 JDBC 开发数据访问逻辑的编程模型。Spring 建立在 JdbcTemplate 的基础之上，还提供了许多有用的类来模拟 JDBC 数据操作，从而让开发人员以更面向对象的方式将 ResultSet 中的查询和转换逻辑维护到域对象。具体来说，本节介绍以下类：

- MappingSqlQuery <T>：MappingSqlQuery <T>类允许将查询字符串和 mapRow()方法一起封装到一个类中。
- SqlUpdate：SqlUpdate 类能够封装任何 SQL 更新语句，同时还提供了许多有用的功能，以便绑定 SQL 参数，在插入新的记录后检索 RDBMS 生成的键等。

- BatchSqlUpdate：顾名思义，该类允许执行批量更新操作。例如，可以遍历 Java List 对象并让 BatchSqlUpdate 对记录进行排序，然后批量提交更新语句。可以随时设置批量大小并刷新操作。
- SqlFunction<T>：SqlFunction<T>类允许使用参数和返回类型调用数据库中的存储函数。此外，还可以使用另一个类 StoredProcedure 来帮助调用存储过程。
- 使用注解来设置 JDBC DAO。

首先看看如何使用注解来设置 DAO 实现类。以下示例代码实现了 SingerDao 接口方法(至此，得到完整的 SingerDao 实现)，显示了 SingerDao 接口类及其提供的数据访问服务的完整列表：

```
package com.apress.prospring5.ch6.dao;

import com.apress.prospring5.ch6.entities.Singer;
import java.util.List;

public interface SingerDao {

    List<Singer> findAll();
    List<Singer> findByFirstName(String firstName);
    String findNameById(Long id);
    String findLastNameById(Long id);
    String findFirstNameById(Long id);
    List<Singer> findAllWithAlbums();

    void insert(Singer singer);
    void update(Singer singer);
    void delete(Long singerId);
    void insertWithAlbum(Singer singer);
}
```

在本书的开头部分，介绍了构造型注解，并将@Repository 作为@Component 注解的特化版加以详细学习，该注解旨在用于处理数据库操作的 bean[1]。以下代码片段显示了数据源属性的初始声明以及如何使用 JSR-250 注解将其注入到@Repository 注解的 DAO 类中：

```
package com.apress.prospring5.ch6.dao;
...
@Repository("singerDao")
public class JdbcSingerDao implements SingerDao {

    private static final Log logger =
        LogFactory.getLog(JdbcSingerDao.class);
    private DataSource dataSource;

    @Resource(name = "dataSource")
    public void setDataSource(DataSource dataSource) {
        this.dataSource = dataSource;
    }

    public DataSource getDataSource() {
        return dataSource;
    }
    ...
}
```

在上面的代码清单中，使用@Repository 来声明名为 singerDao 的 Spring bean，并且由于该类包含数据访问代码，因此@Repository 还指示 Spring 在更适用于应用程序的 DataAccessException 层次结构中执行特定于数据库的 SQL 异常。

此外，为了在应用程序中记录消息，还使用 SL4J 日志组件声明了日志变量。为 dataSource 属性使用 JSR-250 的@Resource，从而让 Spring 使用 dataSource 的名称来注入数据源。

在下面的代码示例中，显示了当使用注解声明处理 bean 的 DAO 时所需的 Java 配置类：

```
package com.apress.prospring5.ch6.config;
import org.apache.commons.dbcp2.BasicDataSource;
...
@Configuration
@PropertySource("classpath:db/jdbc2.properties")
@ComponentScan(basePackages = "com.apress.prospring5.ch6")
public class AppConfig {
```

---

1 这表明被注解的类是一个存储库，最初由 *Domain-Driven Design*(由 Evans 于 2003 年出版)定义为“封装存储、检索和模拟对象集合的搜索行为的机制”。

```
    private static Logger logger =
        LoggerFactory.getLogger(AppConfig.class);

    @Value("${driverClassName}")
    private String driverClassName;
    @Value("${url}")
    private String url;
    @Value("${username}")
    private String username;
    @Value("${password}")
    private String password;

    @Bean
    public static PropertySourcesPlaceholderConfigurer
        propertySourcesPlaceholderConfigurer() {
        return new PropertySourcesPlaceholderConfigurer();
    }

    @Bean(destroyMethod = "close")
    public DataSource dataSource() {
        try {
            BasicDataSource dataSource = new BasicDataSource();
            dataSource.setDriverClassName(driverClassName);
            dataSource.setUrl(url);
            dataSource.setUsername(username);
            dataSource.setPassword(password);
            return dataSource;
        } catch (Exception e) {
            logger.error("DBCP DataSource bean cannot be created!", e);
            return null;
        }
    }
}
```

在该配置中，声明了一个 MySQL 数据库，该数据库正在使用一个可重用的 BasicDataSource 进行访问，并使用组件扫描来自动发现 Spring bean。本章的开头部分已经介绍了如何安装和设置 MySQL 数据库并创建 musicdb 模式。在建立了基础架构后，可以继续执行 JDBC 操作。

6.12 使用 MappingSqlQuery<T>查询数据

Spring 提供了 MappingSqlQuery<T>类来对查询操作进行建模。基本上，首先使用数据源和查询字符串来构造 MappingSqlQuery<T>类。然后，实现 mapRow()方法，从而将每条 ResultSet 记录映射到相应的域对象。

首先创建 SelectAllSingers 类(它表示用于选择所有歌手的查询操作)，该类扩展了 MappingSqlQuery<T>抽象类。SelectAllSingers 类如下所示：

```
package com.apress.prospring5.ch6;

import java.sql.ResultSet;
import java.sql.SQLException;

import javax.sql.DataSource;

import com.apress.prospring5.ch6.entities.Singer;
import org.springframework.jdbc.object.MappingSqlQuery;

public class SelectAllSingers extends MappingSqlQuery<Singer> {
    private static final String SQL_SELECT_ALL_SINGER =
        "select id, first_name, last_name, birth_date from singer";

    public SelectAllSingers(DataSource dataSource) {
        super(dataSource, SQL_SELECT_ALL_SINGER);
    }

    protected Singer mapRow(ResultSet rs, int rowNum)
        throws SQLException {
        Singer singer = new Singer();

        singer.setId(rs.getLong("id"));
        singer.setFirstName(rs.getString("first_name"));
        singer.setLastName(rs.getString("last_name"));
        singer.setBirthDate(rs.getDate("birth_date"));
```

```
        return singer;
    }
}
```

在 SelectAllSingers 类中，声明了用于选择所有歌手的 SQL。在类的构造函数中，通过调用 super()方法(使用了 DataSource 和 SQL 语句)来构造类。此外，还实现了 MappingSqlQuery<T>.mapRow()方法，从而提供结果集到 Singer 域对象的映射。

通过使用 SelectAllSingers 类，可以在 JdbcSingerDao 类中实现 findAll()方法。以下代码片段描述了 JdbcSingerDao 类的部分代码：

```
package com.apress.prospring5.ch6;
@Repository("singerDao")
public class JdbcSingerDao implements SingerDao {
    private DataSource dataSource;
    private SelectAllSingers selectAllSingers;
    @Resource(name = "dataSource")
     public void setDataSource(DataSource dataSource) {
        this.dataSource = dataSource;
        this.selectAllSingers = new SelectAllSingers(dataSource);
    }

    @Override
    public List<Singer> findAll() {
        return selectAllSingers.execute();
    }

    ...
}
```

在 setDataSource()方法中，首先注入了 DataSource，然后构造 SelectAllSingers 类的一个实例。在 findAll()方法中，只需要调用 execute()方法即可，该方法间接继承了 SqlQuery <T>抽象类。以上就是需要完成的所有工作。下面的代码片段显示了用于测试 findAll()方法的方法：

```
com.apress.prospring5.ch6;
public class AnnotationJdbcTest {

    private GenericApplicationContext ctx;
    private SingerDao singerDao;

    @Before
    public void setUp() {
        ctx = new AnnotationConfigApplicationContext(AppConfig.class);
        singerDao = ctx.getBean(SingerDao.class);
        assertNotNull(singerDao);
    }

    @Test
    public void testFindAll() {
        List<Singer> singers = singerDao.findAll();
        assertTrue(singers.size() == 3);
        singers.forEach(singer -> {
            System.out.println(singer);
            if (singer.getAlbums() != null) {
                for (Album album : singer.getAlbums()) {
                    System.out.println("\t--> " + album);
                }
            }
        });
        ctx.close();
    }

    @After
    public void tearDown() {
        ctx.close();
    }
}
```

运行测试方法，如果测试通过，则产生以下输出：

```
Singer - Id: 1, First name: John, Last name: Mayer, Birthday: 1977-10-15
Singer - Id: 2, First name: Eric, Last name: Clapton, Birthday: 1945-03-29
Singer - Id: 3, First name: John, Last name: Butler, Birthday: 1975-03-31
```

如果通过编辑 logback-test.xml 配置文件并添加以下元素为 org.springframework.jdbc 包启用 DEBUG 日志记录：

```
<logger name="org.springframework.jdbc" level="debug"/>
```

那么在控制台中，还会看到 Spring 提交的查询，如下所示：

```
DEBUG o.s.j.c.JdbcTemplate - Executing prepared SQL query
DEBUG o.s.j.c.JdbcTemplate - Executing prepared SQL statement
    [select id, first_name, last_name, birth_date from singer]
```

接下来继续实现 findByFirstName()方法，该方法使用了一个命名参数。与前面的示例一样，创建完成操作所需的 SelectSingerByFirstName 类，如下所示：

```
package com.apress.prospring5.ch6;
...
import org.springframework.jdbc.core.SqlParameter;

public class SelectSingerByFirstName extends MappingSqlQuery<Singer> {

   private static final String SQL_FIND_BY_FIRST_NAME =
      "select id, first_name, last_name, birth_date from
         singer where first_name = :first_name";

   public SelectSingerByFirstName(DataSource dataSource) {
      super(dataSource, SQL_FIND_BY_FIRST_NAME);
      super.declareParameter(new SqlParameter("first_name", Types.VARCHAR));
   }

   protected Singer mapRow(ResultSet rs, int rowNum) throws SQLException {
      Singer singer = new Singer();

      singer.setId(rs.getLong("id"));
      singer.setFirstName(rs.getString("first_name"));
      singer.setLastName(rs.getString("last_name"));
      singer.setBirthDate(rs.getDate("birth_date"));

      return singer;
   }
}
```

SelectSingerByFirstName 类与 SelectAllSingers 类相似。首先，SQL 语句不同，并带有一个名为 first_name 的命名参数。在构造函数方法中，调用 declareParameter()方法(它间接地继承自 org.springframework.jdbc.object.DrbmsOperation 抽象类)。下面在 JdbcSingerDao 类中实现 findByFirstName()方法。此时可以看到更新的代码：

```
package com.apress.prospring5.ch6.dao;
...
@Repository("singerDao")
public class JdbcSingerDao implements SingerDao {

   private static Logger logger = LoggerFactory.getLogger(JdbcSingerDao.class);
   private DataSource dataSource;
   private SelectSingerByFirstName selectSingerByFirstName;

   @Resource(name = "dataSource")
   public void setDataSource(DataSource dataSource) {

      this.dataSource = dataSource;
      this.selectSingerByFirstName =
         new SelectSingerByFirstName(dataSource);
   }

   @Override
   public List<Singer> findByFirstName(String firstName) {
      Map<String, Object> paramMap = new HashMap<>();
      paramMap.put("first_name", firstName);
      return selectSingerByFirstName.executeByNamedParam(paramMap);
   }
...
}
```

在注入数据源之后，构造 SelectSingerByFirstName 的一个实例。然后，在 findByFirstName()方法中，使用命名参数和值构造了 HashMap。最后，调用 executeByNamedParam()方法(它间接继承了 SqlQuery<T>抽象类)。通过执行下面所示的 testFindByFirstName()测试方法来测试该方法：

```
package com.apress.prospring5.ch6;
...
import org.junit.After;
import org.junit.Before;
import org.junit.Test;
```

```
import static org.junit.Assert.assertNotNull;
import static org.junit.Assert.assertTrue;

public class AnnotationJdbcTest {

    private GenericApplicationContext ctx;
    private SingerDao singerDao;

    @Before
    public void setUp() {
        ctx = new AnnotationConfigApplicationContext(AppConfig.class);
        singerDao = ctx.getBean(SingerDao.class);
        assertNotNull(singerDao);
    }

    @Test
    public void testFindByFirstName() {
        List<Singer> singers = singerDao.findByFirstName("John");
        assertTrue(singers.size() == 1);
        listSingers(singers);
        ctx.close();
    }

    private void listSingers(List<Singer> singers){
        singers.forEach(singer -> {
            System.out.println(singer);
            if (singer.getAlbums() != null) {
                for (Album album : singer.getAlbums()) {
                    System.out.println("\t--> " + album);
                }
            }
        });
    }

    @After
    public void tearDown() {
        ctx.close();
    }
}
```

运行测试方法，如果测试通过，则产生以下输出：

```
Singer - Id: 1, First name: John, Last name: Mayer, Birthday: 1977-10-15
```

此处值得注意的一点是，MappingSqlQuery<T>仅适用于将单个行映射到域对象。对于嵌套对象，则需要将JdbcTemplate与ResultSetExtractor一起使用，就像JdbcTemplate类中提供的示例方法findAllWithAlbums()一样。

使用 SqlUpdate 更新数据

为了更新数据，Spring 提供了 SqlUpdate 类。以下代码片段显示了用于更新操作的 UpdateSinger 类(扩展自SqlUpdate类)：

```
package com.apress.prospring5.ch6;

import java.sql.Types;

import javax.sql.DataSource;

import org.springframework.jdbc.core.SqlParameter;
import org.springframework.jdbc.object.SqlUpdate;

public class UpdateSinger extends SqlUpdate {
    private static final String SQL_UPDATE_SINGER =
        "update singer set first_name=:first_name, last_name=:last_name,
        birth_date=:birth_date where id=:id";

    public UpdateSinger(DataSource dataSource) {
        super(dataSource, SQL_UPDATE_SINGER);
        super.declareParameter(new SqlParameter("first_name", Types.VARCHAR));
        super.declareParameter(new SqlParameter("last_name", Types.VARCHAR));
        super.declareParameter(new SqlParameter("birth_date", Types.DATE));
        super.declareParameter(new SqlParameter("id", Types.INTEGER));
    }
}
```

你现在应该很熟悉上面的代码清单了。上述代码使用查询构造 SqlUpdate 类的一个实例，同时声明了命名参数。以下代码片段显示了 JdbcSingerDao 类中 update()方法的实现：

```
package com.apress.prospring5.ch6.dao;
...
@Repository("singerDao")
public class JdbcSingerDao implements SingerDao {

   private static Logger logger =
      LoggerFactory.getLogger(JdbcSingerDao.class);
   private DataSource dataSource;
   private UpdateSinger updateSinger;

   @Override
   public void update(Singer singer) {
      Map<String, Object> paramMap = new HashMap<String, Object>();
      paramMap.put("first_name", singer.getFirstName());
      paramMap.put("last_name", singer.getLastName());
      paramMap.put("birth_date", singer.getBirthDate());
      paramMap.put("id", singer.getId());
      updateSinger.updateByNamedParam(paramMap);
      logger.info("Existing singer updated with id: " + singer.getId());
   }

   @Resource(name = "dataSource")
   public void setDataSource(DataSource dataSource) {
      this.dataSource = dataSource;
      this.updateSinger = new UpdateSinger(dataSource);
   }
   ...
}
```

在注入数据源之后，构建 UpdateSinger 的一个实例。在 update()方法中，通过传入的 Singer 对象构造命名参数 HashMap，然后调用 updateByNamedParam()来更新联系人记录。为了测试该操作，在 AnnotationJdbcTest 中添加一个新的方法。

```
package com.apress.prospring5.ch6;
...
public class AnnotationJdbcTest {

   private GenericApplicationContext ctx;
   private SingerDao singerDao;

   @Before
   public void setUp() {
      ctx = new AnnotationConfigApplicationContext(AppConfig.class);
      singerDao = ctx.getBean(SingerDao.class);
      assertNotNull(singerDao);
   }

   @Test
   public void testSingerUpdate() {
      Singer singer = new Singer();
      singer.setId(1L);
      singer.setFirstName("John Clayton");
      singer.setLastName("Mayer");
      singer.setBirthDate(new Date(
            (new GregorianCalendar(1977, 9, 16)).getTime().getTime()));
      singerDao.update(singer);

      List<Singer> singers = singerDao.findAll();
      listSingers(singers);
   }

   private void listSingers(List<Singer> singers){
      singers.forEach(singer -> {
         System.out.println(singer);
         if (singer.getAlbums() != null) {
            for (Album album : singer.getAlbums()) {
               System.out.println("\t--> " + album);
            }
         }
      });
   }

   @After
```

```
    public void tearDown() {
        ctx.close();
    }
}
```

此时，只需要构造一个 Singer 对象，然后调用 update()方法。运行该程序，会从 listSingers()方法产生输出。运行测试方法，如果测试通过，则产生以下输出：

```
Singer - Id: 1, First name: John Clayton, Last name: Mayer, Birthday: 1977-10-16
Singer - Id: 2, First name: Eric, Last name: Clapton, Birthday: 1945-03-29
Singer - Id: 3, First name: Jimi, Last name: Hendrix, Birthday: 1942-11-26
```

6.13 插入数据并检索生成的键

为了插入数据，也可以使用 SqlUpdate 类。一个有趣的问题是：主键(通常是 ID 列)是如何生成的。该值仅在插入语句完成后才可用，这是因为 RDBMS 在插入时会为记录生成标识值。列 ID 使用属性 AUTO_INCREMENT 进行声明，并且是主键，同时在插入操作期间将由 RDBMS 分配 ID 值。如果使用的是 Oracle，则会首先从 Oracle 序列中获取唯一 ID，然后使用查询结果执行插入语句。

在旧版的 JDBC 中，这种方法有点棘手。例如，如果使用 MySQL，则需要执行 SQL select last_insert_id()；而对于 Microsoft SQL Server，则使用@@ IDENTITY 语句。

幸运的是，从 JDBC 版本 3.0 开始，添加了一项新功能，允许以统一的方式检索 RDBMS 生成的键。以下代码片段显示了 insert()方法的实现，该方法还检索新插入的联系人记录的生成键。该代码片段可以在大多数(但不是全部)数据库中工作；只需要确保所使用的是与 JDBC 3.0 或更新版本兼容的 JDBC 驱动程序即可。

首先为插入操作创建 InsertSinger 类，它扩展了 SqlUpdate 类。

```
package com.apress.prospring5.ch6;

import java.sql.Types;

import javax.sql.DataSource;

import org.springframework.jdbc.core.SqlParameter;
import org.springframework.jdbc.object.SqlUpdate;

public class InsertSinger extends SqlUpdate {
    private static final String SQL_INSERT_SINGER =
        "insert into singer (first_name, last_name, birth_date)
            values (:first_name, :last_name, :birth_date)";

    public InsertSinger(DataSource dataSource) {
        super(dataSource, SQL_INSERT_SINGER);
        super.declareParameter(new SqlParameter("first_name", Types.VARCHAR));
        super.declareParameter(new SqlParameter("last_name", Types.VARCHAR));
        super.declareParameter(new SqlParameter("birth_date", Types.DATE));
        super.setGeneratedKeysColumnNames(new String {"id"});
        super.setReturnGeneratedKeys(true);
    }
}
```

InsertSinger 类与 UpdateSinger 类几乎相同，但需要做两件额外的事情。在构造 InsertSinger 类时，调用方法 SqlUpdate.SetGeneratedKeysColumnNames()以声明 ID 列的名称。然后使用方法 SqlUpdate.setReturnGeneratedKeys()指示底层的 JDBC 驱动程序检索生成的键。在下面的代码中，可以看到 JdbcSingerDao 类中 insert()方法的实现：

```
package com.apress.prospring5.ch6.dao;
...
@Repository("singerDao")
public class JdbcSingerDao implements SingerDao {
    private static Logger logger =
        LoggerFactory.getLogger(JdbcSingerDao.class);
    private DataSource dataSource;
    private InsertSinger insertSinger;

    @Override
    public void insert(Singer singer) {
        Map<String, Object> paramMap = new HashMap<>();
        paramMap.put("first_name", singer.getFirstName());
        paramMap.put("last_name", singer.getLastName());
        paramMap.put("birth_date", singer.getBirthDate());
```

```
        KeyHolder keyHolder = new GeneratedKeyHolder();
        insertSinger.updateByNamedParam(paramMap, keyHolder);
        singer.setId(keyHolder.getKey().longValue());
        logger.info("New singer inserted with id: " + singer.getId());
    }

    @Resource(name = "dataSource")
    public void setDataSource(DataSource dataSource) {
        this.dataSource = dataSource;
        this.insertSinger = new InsertSinger(dataSource);
    }
...
}
```

在注入数据源之后，构建一个 InsertSinger 实例。在 insert()方法中，也使用了 SqlUpdate.updateByNamedParam()方法。另外，将 KeyHolder 的一个实例传递给方法，用来保存生成的 ID。在插入数据后，可以从 KeyHolder 中检索生成的键。接下来看看测试方法如何查找 insert()。

```
package com.apress.prospring5.ch6;
...
public class AnnotationJdbcTest {
    private GenericApplicationContext ctx;
    private SingerDao singerDao;

    @Before
    public void setUp() {
        ctx = new AnnotationConfigApplicationContext(AppConfig.class);
        singerDao = ctx.getBean(SingerDao.class);
        assertNotNull(singerDao);
    }

    @Test
    public void testSingerInsert(){
        Singer singer = new Singer();
        singer.setFirstName("Ed");
        singer.setLastName("Sheeran");
        singer.setBirthDate(new Date(
                (new GregorianCalendar(1991, 1, 17)).getTime().getTime()));
        singerDao.insert(singer);
        List<Singer> singers = singerDao.findAll();
        listSingers(singers);
    }

    private void listSingers(List<Singer> singers){
        singers.forEach(singer -> {
            System.out.println(singer);
        if (singer.getAlbums() != null) {
            for (Album album : singer.getAlbums()) {
                System.out.println("\t--> " + album);
                }
            }
        });
    }

    @After
    public void tearDown() {
        ctx.close();
    }
}
```

运行测试方法，如果测试通过，则产生以下输出：

```
Singer - Id: 1, First name: John Clayton, Last name: Mayer, Birthday: 1977-10-16
Singer - Id: 2, First name: Eric, Last name: Clapton, Birthday: 1945-03-29
Singer - Id: 3, First name: Jimi, Last name: Hendrix, Birthday: 1942-11-26
Singer - Id: 6, First name: Ed, Last name: Sheeran, Birthday: 1991-02-17
```

6.14 使用 BatchSqlUpdate 进行批处理操作

对于批处理操作，可以使用 BatchSqlUpdate 类。新的 insertWithAlbum()方法可以将歌手及其发行的专辑插入到数据库中。为了能够插入专辑记录，需要创建 InsertSingerAlbum 类，如下所示：

```
package com.apress.prospring5.ch6;

import java.sql.Types;
```

```
import javax.sql.DataSource;

import org.springframework.jdbc.core.SqlParameter;
import org.springframework.jdbc.object.BatchSqlUpdate;
public class InsertSingerAlbum extends BatchSqlUpdate {

   private static final String SQL_INSERT_SINGER_ALBUM =
      "insert into album (singer_id, title, release_date)
         values (:singer_id, :title, :release_date)";

   private static final int BATCH_SIZE = 10;
   public InsertSingerAlbum(DataSource dataSource) {
      super(dataSource, SQL_INSERT_SINGER_ALBUM);

      declareParameter(new SqlParameter("singer_id", Types.INTEGER));
      declareParameter(new SqlParameter("title", Types.VARCHAR));
      declareParameter(new SqlParameter("release_date", Types.DATE));

      setBatchSize(BATCH_SIZE);
   }
}
```

请注意，在构造函数中，调用 BatchSqlUpdate.setBatchSize()方法来设置 JDBC 插入操作的批处理大小。接下来看一下 JdbcSingerDao 类中 insertWithAlbum()方法的实现：

```
package com.apress.prospring5.ch6.dao;
...
@Repository("singerDao")
public class JdbcSingerDao implements SingerDao {

   private static Logger logger =
      LoggerFactory.getLogger(JdbcSingerDao.class);
   private DataSource dataSource;
   private InsertSingerAlbum insertSingerAlbum;

   @Override
   public void insertWithAlbum(Singer singer) {
      insertSingerAlbum = new InsertSingerAlbum(dataSource);
      Map<String, Object> paramMap = new HashMap<>();
      paramMap.put("first_name", singer.getFirstName());
      paramMap.put("last_name", singer.getLastName());
      paramMap.put("birth_date", singer.getBirthDate());
      KeyHolder keyHolder = new GeneratedKeyHolder();
      insertSinger.updateByNamedParam(paramMap, keyHolder);
      singer.setId(keyHolder.getKey().longValue());
      logger.info("New singer inserted with id: " + singer.getId());
      List<Album> albums = singer.getAlbums();
      if (albums != null) {
            for (Album album : albums) {
               paramMap = new HashMap<>();
               paramMap.put("singer_id", singer.getId());
               paramMap.put("title", album.getTitle());
               paramMap.put("release_date", album.getReleaseDate());
               insertSingerAlbum.updateByNamedParam(paramMap);
            }
         }
         insertSingerAlbum.flush();
      }
   @Resource(name = "dataSource")
   public void setDataSource(DataSource dataSource) {
      this.dataSource = dataSource;
   }

   @Override
   public List<Singer> findAllWithAlbums() {
      JdbcTemplate jdbcTemplate = new JdbcTemplate(getDataSource());
      String sql = "SELECT s.id, s.first_name, s.last_name, s.birth_date" +
         ", a.id AS album_id, a.title, a.release_date FROM singer s " +
         "LEFT JOIN album a ON s.id = a.singer_id";
      return jdbcTemplate.query(sql, new SingerWithAlbumExtractor());
   }

   private static final class SingerWithAlbumExtractor
         implements ResultSetExtractor<List<Singer>> {
      public List<Singer> extractData(ResultSet rs) throws
            SQLException, DataAccessException {
         Map<Long, Singer> map = new HashMap<>();
```

```
            Singer singer;
            while (rs.next()) {
                Long id = rs.getLong("id");
                singer = map.get(id);
                if (singer == null) {
                    singer = new Singer();
                    singer.setId(id);
                    singer.setFirstName(rs.getString("first_name"));
                    singer.setLastName(rs.getString("last_name"));
                    singer.setBirthDate(rs.getDate("birth_date"));
                    singer.setAlbums(new ArrayList<>());
                    map.put(id, singer);
                }
                Long albumId = rs.getLong("album_id");
                if (albumId > 0) {
                    Album album = new Album();
                    album.setId(albumId);
                    album.setSingerId(id);
                    album.setTitle(rs.getString("title"));
                    album.setReleaseDate(rs.getDate("release_date"));
                    singer.getAlbums().add(album);
                }
            }
            return new ArrayList<>(map.values());
        }
    }
...
}
```

每次调用 insertWithAlbum()方法时，都会创建一个新的 InsertSingerAlbum 实例，因为 BatchSqlUpdate 类不是线程安全的。然后就像使用 SqlUpdate 一样使用 BatchSqlUpdate，主要区别在于 BatchSqlUpdate 类对插入操作进行排序并批量提交给数据库。每当记录数量等于批处理大小时，Spring 将为挂起的记录对数据库执行批量插入操作。另一方面，完成批处理操作后，将调用 BatchSqlUpdate.flush()方法来指示 Spring 清除所有挂起的操作(即正在排队但尚未达到批处理大小的插入操作)。最后，遍历 Singer 对象中的 Album 对象列表并调用 BatchSqlUpdate.updateByNamedParam()方法。为了便于测试，还实现了 insertWithAlbum()方法。该方法的代码量非常大，可以通过使用 Java 8 lambda 表达式来减少代码。

```
public List<Singer> findAllWithAlbums() {
    JdbcTemplate jdbcTemplate = new JdbcTemplate(getDataSource());
    String sql = "SELECT s.id, s.first_name, s.last_name, s.birth_date" +
            ", a.id AS album_id, a.title, a.release_date FROM singer s " +
            "LEFT JOIN album a ON s.id = a.singer_id";
    return jdbcTemplate.query(sql, rs -> {
        Map<Long, Singer> map = new HashMap<>();
        Singer singer;
        while (rs.next()) {
            Long id = rs.getLong("id");
            singer = map.get(id);
            if (singer == null) {
                singer = new Singer();
                singer.setId(id);
                singer.setFirstName(rs.getString("first_name"));
                singer.setLastName(rs.getString("last_name"));
                singer.setBirthDate(rs.getDate("birth_date"));
                singer.setAlbums(new ArrayList<>());
                map.put(id, singer);
            }
            Long albumId = rs.getLong("album_id");
            if (albumId > 0) {
                Album album = new Album();
                album.setId(albumId);
                album.setSingerId(id);
                album.setTitle(rs.getString("title"));
                album.setReleaseDate(rs.getDate("release_date"));
                singer.getAlbums().add(album);
            }
        }
        return new ArrayList<>(map.values());
    });
}
```

接下来看看测试方法如何查找 insertWithAlbum()。

```
package com.apress.prospring5.ch6;
...
```

```
public class AnnotationJdbcTest {

    private GenericApplicationContext ctx;
    private SingerDao singerDao;

    @Before
    public void setUp() {
        ctx = new AnnotationConfigApplicationContext(AppConfig.class);
        singerDao = ctx.getBean(SingerDao.class);
        assertNotNull(singerDao);
    }
    @Test
        public void testSingerInsertWithAlbum(){
        Singer singer = new Singer();
        singer.setFirstName("BB");
        singer.setLastName("King");
        singer.setBirthDate(new Date(
                (new GregorianCalendar(1940, 8, 16)).getTime().getTime()));

        Album album = new Album();
        album.setTitle("My Kind of Blues");
        album.setReleaseDate(new Date(
                (new GregorianCalendar(1961, 7, 18)).getTime().getTime()));
        singer.addAbum(album);

        album = new Album();
        album.setTitle("A Heart Full of Blues");
        album.setReleaseDate(new Date(
                (new GregorianCalendar(1962, 3, 20)).getTime().getTime()));
        singer.addAbum(album);

        singerDao.insertWithAlbum(singer);

        List<Singer> singers = singerDao.findAllWithAlbums();
        listSingers(singers);
    }

    private void listSingers(List<Singer> singers){
        singers.forEach(singer -> {
            System.out.println(singer);
            if (singer.getAlbums() != null) {
                for (Album album : singer.getAlbums()) {
                    System.out.println("\t--> " + album);
                }
            }
        });
    }

    @After
    public void tearDown() {
        ctx.close();
    }
}
```

运行测试方法，如果测试通过，则产生以下输出：

```
Singer - Id: 1, First name: John, Last name: Mayer, Birthday: 1977-10-15
    --> Album - Id: 1, Singer id: 1, Title: The Search For Everything,
        Release Date: 2017-01-19
    --> Album - Id: 2, Singer id: 1, Title: Paradise Valley,
        Release Date: 2013-08-19
    --> Album - Id: 3, Singer id: 1, Title: Born and Raised,
        Release Date: 2012-05-22
    --> Album - Id: 4, Singer id: 1, Title: Battle Studies,
        Release Date: 2009-11-16
Singer - Id: 2, First name: Eric, Last name: Clapton, Birthday: 1945-03-29
    --> Album - Id: 5, Singer id: 2, Title: From The Cradle,
        Release Date: 1994-09-13
Singer - Id: 3, First name: Jimi, Last name: Hendrix, Birthday: 1942-11-26
Singer - Id: 4, First name: BB, Last name: King, Birthday: 1940-09-15
    --> Album - Id: 6, Singer id: 4, Title: My Kind of Blues,
        Release Date: 1961-08-17
    --> Album - Id: 7, Singer id: 4, Title: A Heart Full of Blues,
        Release Date: 1962-04-19
```

6.15 使用 SqlFunction 调用存储函数

Spring 还提供了一些类来简化执行使用了 JDBC 的存储过程/函数。在本节中，将演示如何通过使用 SqlFunction 类来执行一个简单的函数。同时，还演示了如何使用 MySQL 作为数据库，创建一个存储函数，并通过使用 SqlFunction<T>类来调用它。

假设有一个模式名称为 musicdb 的 MySQL 数据库，其用户名和密码均等于 prospring5(与 6.3 节“研究 JDBC 基础结构”中的示例相同)。创建一个名为 getFirstNameById()的存储函数，该函数接收联系人的 ID 并返回联系人的名字。以下代码显示了在 MySQL 中创建存储函数的脚本(stored-function.sql)。对 MySQL 数据库运行该脚本。

```
DELIMITER //
CREATE FUNCTION getFirstNameById(in_id INT)
   RETURNS VARCHAR(60)
   BEGIN
      RETURN (SELECT first_name FROM singer WHERE id = in_id);
   END //
DELIMITER ;
```

该存储函数只接收 ID 并返回带有该 ID 的歌手记录的名字。接下来，创建 StoredFunctionFirstNameById 类来表示该存储函数操作，该类扩展了 SqlFunction<T>类。可以在以下代码片段中看到该类的内容：

```
package com.apress.prospring5.ch6;

import java.sql.Types;

import javax.sql.DataSource;

import org.springframework.jdbc.core.SqlParameter;
import org.springframework.jdbc.object.SqlFunction;

public class StoredFunctionFirstNameById extends SqlFunction<String> {
   private static final String SQL = "select getfirstnamebyid(?)";

   public StoredFunctionFirstNameById (DataSource dataSource) {
      super(dataSource, SQL);
      declareParameter(new SqlParameter(Types.INTEGER));
      compile();
   }
}
```

该类扩展了 SqlFunction <T>并传入 String 类型，它表示函数的返回类型。然后声明 SQL 来调用 MySQL 中的存储函数。最后，在构造函数中声明参数，并编译操作。现在，该类已经做好了在实现类中使用的准备。以下代码片段显示了更新后的使用了存储函数的 JdbcSingerDao 类：

```
package com.apress.prospring5.ch6.dao;
...
@Repository("singerDao")
public class JdbcSingerDao implements SingerDao {

   private static Logger logger = LoggerFactory.getLogger(JdbcSingerDao.class);
   private DataSource dataSource;
   private StoredFunctionFirstNameById storedFunctionFirstNameById;

   @Override
   public String findFirstNameById(Long id) {
      List<String> result = storedFunctionFirstNameById.execute(id);
      return result.get(0);
   }
   ...
}
```

在注入数据源之后，构造 StoredFunctionFirstNameById 的一个实例。然后在 findFirstNameById()方法中调用它的 execute()方法，并传入联系人 ID。该方法将返回一个字符串列表，此时只需要第一个字符串即可，因为结果集中只应该有一条记录返回。测试该功能非常简单。

```
package com.apress.prospring5.ch6;
...
public class AnnotationJdbcTest {

   private GenericApplicationContext ctx;
   private SingerDao singerDao;
```

```
    @Before
    public void setUp() {
        ctx = new AnnotationConfigApplicationContext(AppConfig.class);
        singerDao = ctx.getBean(SingerDao.class);
        assertNotNull(singerDao);
    }

    @Test
    public void testFindFirstNameById(){
        String firstName = singerDao.findFirstNameById(2L);
        assertEquals("Eric", firstName);
        System.out.println("Retrieved value: " + firstName);
    }

    @After
    public void tearDown() {
        ctx.close();
    }
}
```

在上述程序中，将 ID 2 传入存储函数。此时将返回 Eric，如果针对 MySQL 数据库运行 test-data.sql，那么 Eric 即为这 ID 等于 2 的记录对应的名字。运行上述程序会产生以下输出：

```
o.s.j.c.JdbcTemplate - Executing prepared SQL query
o.s.j.c.JdbcTemplate - Executing prepared SQL statement
   [select getfirstnamebyid?]
o.s.j.d.DataSourceUtils - Fetching JDBC Connection from DataSource
o.s.j.d.DataSourceUtils - Returning JDBC Connection to DataSource
Retrieved value: Eric
```

可以看到：检索到了正确的名字。这里只是给出一个简单的示例来演示 Spring JDBC 模块的功能。Spring 还提供了其他类(例如，StoredProcedure)来调用复杂的存储过程，并返回复杂的数据类型。如果需要使用 JDBC 访问存储过程，建议参考 Spring 的参考手册。

6.16 Spring Data 项目：JDBC Extensions

近年来，随着许多专用数据库的兴起，数据库技术发展非常迅速，现在 RDBMS 已经不是应用程序后端数据库的唯一选择。为了适应这种数据库技术发展和开发人员社区的需求，Spring 创建了 Spring Data 项目(http://springsource.org/spring-data)。该项目的主要目标是在 Spring 的核心数据访问功能之上提供有用的扩展，以便与传统 RDBMS 之外的数据库进行交互。

Spring Data 项目带有各种扩展。本节介绍的是 JDBC Extensions(http://springsource.org/spring-data/jdbc-extensions)。顾名思义，该扩展提供了一些高级功能，以便使用 Spring 开发 JDBC 应用程序。下面列出了 JDBC Extensions 提供的主要功能：

- **QueryDSL 支持**：QueryDSL(http://querydsl.com)是一种适用于特定领域的语言，为开发类型安全查询提供了一个框架。Spring Data 的 JDBC Extensions 提供了 QueryDslJdbcTemplate，以便利用 QueryDSL 而不是 SQL 语句来开发 JDBC 应用程序。
- **对 Oracle 数据库的高级支持**：该扩展为 Oracle 数据库用户提供高级功能。在数据库连接方面，它支持 Oracle 特定的会话设置，以及在使用 Oracle RAC 时的快速连接故障切换(Fast Connection Failover)技术。此外，还提供了与 Oracle Advanced Queuing 集成的类。在数据类型方面，提供对 Oracle 的 XML 类型、STRUCT 和 ARRAY 等的本地支持。

如果正在使用 Spring 和 Oracle 数据库开发 JDBC 应用程序，那么 JDBC Extensions 非常值得一看。

6.17 使用 JDBC 的注意事项

借助于丰富的功能集，可以看到当使用 JDBC 与底层 RDBMS 进行交互时 Spring 会让开发变得更加轻松。但仍然需要开发相当多的代码，特别是在将结果集转换为相应的域对象时。

在 JDBC 基础之上开发了许多开源库，以帮助缩小关系数据结构和 Java 的 OO 模型之间的差距。例如，iBATIS 是一个流行的 DataMapper 框架(也基于 SQL 映射)。iBATIS 允许将带有存储过程或查询的对象映射到 XML 描述符文件。像 Spring 一样，iBATIS 提供了一种声明式的方式来查询对象映射，从而大大节省了维护 SQL 查询所花费的时间，这些 SQL 查询可能分散在各种 DAO 类中。

还有许多其他专注于对象模型而不是查询的 ORM 框架。比较流行的包括 Hibernate、EclipseLink(也称为 TopLink)和 OpenJPA。所有这些都符合 JCP 的 JPA 规范。

近年来，这些 ORM 工具和映射框架已经变得更加成熟，所以大多数开发人员都会选择使用其中的一种，而不是直接使用 JDBC。但是，如果需要对提交给数据库的查询进行绝对控制(例如，在 Oracle 中使用分层查询)，那么 Spring JDBC 实际上是一个可行的选择。而在使用 Spring 时，一个很大的优点是可以混合搭配不同的数据访问技术。例如，可以将 Hibernate 用作主 ORM，然后将 JDBC 用作一些复杂查询逻辑或批处理操作的补充；可以在单个业务操作中将它们混合搭配，并将它们封装在同一个数据库事务中。Spring 可以帮助你轻松处理这些情况。

6.18 Spring Boot JDBC

由于已经介绍了用于 Web 和简单控制台应用程序的 Spring Boot，因此在本书中介绍适用于 JDBC 的 Spring Boot 启动器库是合乎逻辑的。它可以帮助我们删除样板配置并直接跳转到实现。

当把 spring-boot-starter-jdbc 作为依赖项添加到项目中时，会将一组库添加到项目中，但数据库驱动程序并没有添加。采取哪种数据库驱动程序必须由开发者决定。本节将介绍的项目是 spring-boot-jdbc，它是 chapter06 项目的子模块。它的 Gradle 依赖项和版本在父 build.gradle 文件中指定：

```
ext {
   h2Version = '1.4.194'
   bootVersion = '2.0.0.BUILD-SNAPSHOT'

   boot = [
      springBootPlugin:
            "org.springframework.boot:spring-boot-gradle-plugin:$bootVersion",
      Starter :
            "org.springframework.boot:spring-boot-starter:$bootVersion",
      starterWeb :
            "org.springframework.boot:spring-boot-starter-web:$bootVersion",
      Actuator :
            "org.springframework.boot:spring-boot-starter-actuator:$bootVersion",
      starterTest :
            "org.springframework.boot:spring-boot-starter-test:$bootVersion",
      starterAop :
            "org.springframework.boot:spring-boot-starter-aop:$bootVersion",
      starterJdbc :
            "org.springframework.boot:spring-boot-starter-jdbc:$bootVersion"
   ]

   db = [
      h2 : "com.h2database:h2:$h2Version",
      ..
   ]
   ...
}
```

它们只能通过使用它们的属性名称，在 spring-boot-jdbc\build.gradle 配置文件中被声明为依赖项。

```
buildscript {
   repositories {
      mavenLocal() mavenCentral()
      maven { url "http://repo.spring.io/release" }
      maven { url "http://repo.spring.io/milestone" }
      maven { url "http://repo.spring.io/snapshot" }
      maven { url "https://repo.spring.io/libs-snapshot" }
   }
   dependencies {
      classpath boot.springBootPlugin
   }
}

apply plugin: 'org.springframework.boot'

dependencies {
   compile project(':chapter06:plain-jdbc')
   compile boot.starterJdbc, db.h2
}
```

IntelliJ IDEA Gradle 项目视图中的自动配置库如图 6-3 所示。spring-boot-starter-jdbc 库使用 tomcat-jdbc 来配置 DataSource bean。因此，如果没有显式配置 DataSource bean 并且类路径中有嵌入式数据库驱动程序，那么 Spring Boot 将使用内存数据库设置自动注册 DataSource bean。此外，Spring Boot 还会自动注册以下 bean：

- JdbcTemplate bean
- NamedParameterJdbcTemplate bean
- PlatformTransactionManager(DataSourceTransactionManager) bean

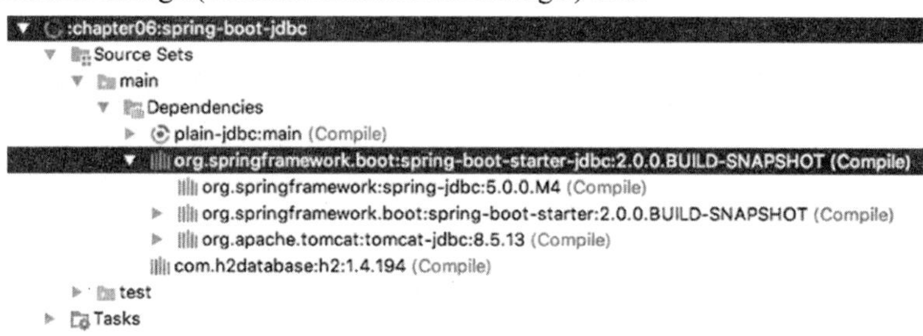

图 6-3　Spring Boot JDBC 启动器依赖项

以下是其他一些可能会减少环境设置工作量的有趣事情。

- Spring Boot 在 src/main/resources 下查找嵌入式数据库的初始化文件。它期望找到一个名为 schema.sql 的文件，其中包含 SQL DDL 语句(例如，CREATE TABLE 语句)，以及一个名为 data.sql 且包含 DML 语句(例如 INSERT 语句)的文件。可以使用此文件在启动时初始化数据库。
- 这些文件的位置和名称可以在 application.properties 文件(位于 src/main/resources 下)中配置。下面显示了一个允许 Spring Boot 应用程序使用 SQL 文件的示例配置文件：

```
spring.datasource.schema=db/schema.sql
spring.datasource.data=db/test-data.sql
```

- 默认情况下，Spring Boot 会在启动时初始化数据库，但也可以通过将 spring.datasource.initialize = false 属性添加到 application.properties 文件中来进行更改。

除了前面提到的内容之外，Spring Boot 还提供了一些域类或实体，以及一个 DAO bean。Singer bean 与本章中随处可见的那些 Singer bean 相同，其实现驻留在 chapter06/plain-jdbc 项目中，该项目可作为依赖项添加到任何地方。下面显示了要使用的 JdbcSingerDao 类：

```
package com.apress.prospring5.ch6;

import com.apress.prospring5.ch6.dao.SingerDao;
import com.apress.prospring5.ch6.entities.Singer;
import org.apache.commons.lang3.NotImplementedException;
import org.springframework.beans.factory.BeanCreationException; import org.springframework.
beans.factory.InitializingBean; import org.springframework.beans.factory.annotation.
Autowired; import org.springframework.jdbc.core.JdbcTemplate;
import org.springframework.stereotype.Component;

import java.util.List;

@Component
public class JdbcSingerDao implements SingerDao, InitializingBean {
   private JdbcTemplate jdbcTemplate;

   @Autowired
   public void setJdbcTemplate(JdbcTemplate jdbcTemplate) {
      this.jdbcTemplate = jdbcTemplate;
   }

   @Override public String findNameById(Long id) {
      return jdbcTemplate.queryForObject(
            "SELECT first_name || ' ' || last_name FROM singer WHERE id = ?",
            new Object{id}, String.class);
   }
   ...
}
```

以下代码示例显示了 Spring Boot 的入口点类，即使用@SpringBootApplication 注解的类。请注意，该类非常简单。

```
package com.apress.prospring5.ch6;
import com.apress.prospring5.ch6.dao.SingerDao;
import org.slf4j.Logger;
import org.slf4j.LoggerFactory;
import org.springframework.boot.SpringApplication;
import org.springframework.boot.autoconfigure.SpringBootApplication;
import org.springframework.context.ConfigurableApplicationContext;

@SpringBootApplication
public class Application {

   private static Logger logger =
      LoggerFactory.getLogger(Application.class);
   public static void main(String... args) throws Exception {
         ConfigurableApplicationContext ctx =
            SpringApplication.run(Application.class, args);
         assert (ctx != null);

         SingerDao singerDao = ctx.getBean(SingerDao.class);
         String singerName = singerDao.findNameById(1L);
         logger.info("Retrieved singer: " + singerName);

         System.in.read();
         ctx.close();
   }
}
```

6.19 小结

本章演示了如何使用 Spring 来简化 JDBC 编程。学习了如何连接到数据库并执行选择、更新、删除和插入操作，以及调用数据库存储函数。详细讨论了如何使用核心 Spring JDBC 类 JdbcTemplate。介绍了构建在 JdbcTemplate 之上的其他 Spring 类，它们可以帮助对各种 JDBC 操作进行建模。演示了如何在合适的情况下使用 Java 8 中新的 lambda 表达式。此外，还介绍了 Spring Boot JDBC，因为不管是有助于更多地关注应用程序的业务逻辑实现，还是减少配置，它都是一个很好的工具。在接下来的几章中，将讨论如何在开发数据访问逻辑时使用 Spring 以及流行的 ORM 技术。

第7章

在 Spring 中使用 Hibernate

在前一章中，介绍了如何在 Spring 应用程序中使用 JDBC。然而，尽管 Spring 在简化 JDBC 开发方面已经有了长足的进步，但仍然需要编写很多的代码。在本章中，将介绍一个名为 Hibernate 的对象关系映射(ORM)库。

如果你有使用 EJB 实体 bean(在 EJB 3.0 之前)开发数据访问应用程序的经验，那么可能会对这个痛苦的过程记忆犹新。在开发企业级 Java 应用程序时，烦琐的映射配置、事务划分以及每个 bean 中管理其生命周期的许多样板代码极大降低了开发效率。

就像 Spring 被开发出来以支持基于 POJO 的开发和声明式配置管理，从而摆脱 EJB 沉重而笨拙的设置一样，开发人员社区意识到通过使用更简单、轻量级和基于 POJO 的框架可以简化数据访问逻辑的开发。从那时起，出现了许多库；它们通常被称为 *ORM 库*。ORM 库的主要目标是缩小关系数据库管理系统(RDBMS)中的关系数据结构与 Java 中面向对象(OO)模型之间的差距，以便开发人员可以专注于使用对象模型进行编程，同时轻松执行与持久化相关的操作。

在开源社区提供的 ORM 库中，Hibernate 是最成功的。它的主要功能已经赢得主流 Java 开发人员社区的关注，比如基于 POJO 的方法、易于开发以及对复杂关系定义的支持等。

Hibernate 的普及也影响到了 JCP，后者开发了 Java 数据对象(JDO)规范，并作为 Java EE 中的标准 ORM 技术之一。从 EJB 3.0 开始，EJB 实体 bean 甚至被 Java Persistence API(JPA)取代。JPA 的许多概念都受到流行 ORM 库(比如 Hibernate、TopLink 和 JDO)的影响。Hibernate 和 JPA 之间的关系也非常接近。Hibernate 的创始人 Gavin King 代表 JBoss 成为定义 JPA 规范的 JCP 专家组成员之一。从 3.2 版开始，Hibernate 提供了 JPA 的实现。这意味着当使用 Hibernate 开发应用程序时，可以选择使用 Hibernate 自己的 API 或 JPA API 作为持久化服务提供程序。

在了解了 Hibernate 的简要历史之后，本章将介绍如何在开发数据访问逻辑时使用 Spring 与 Hibernate。Hibernate 是一个涉猎如此广泛的 ORM 库，仅仅在一章中就涵盖方方面面的内容是不可能的，并且已有大量的书籍详细讨论了 Hibernate。

本章将介绍 Spring 中 Hibernate 的基本概念和主要用例，特别讨论以下几个主题：

- **配置 Hibernate SessionFactory**：Hibernate 的核心概念围绕由 SessionFactory 管理的 Session 接口进行。我们将演示如何配置 Hibernate 的会话工厂，以便在 Spring 应用程序中使用。
- **使用 Hibernate 的 ORM 的主要概念**：除了介绍如何使用 Hibernate 将 POJO 映射到底层关系数据库结构的主要概念，还讨论一些常用的关系，包括一对多关系和多对多关系。
- **数据操作**：举例说明如何在 Spring 环境中使用 Hibernate 执行数据操作(查询、插入、更新、删除)。在使用 Hibernate 时，其 Session 接口是与之交互的主要接口。

> 当定义对象到关系的映射时，Hibernate 支持两种配置样式。一种是在 XML 文件中配置映射信息，另一种是在实体类中使用 Java 注解(在 ORM 或 JPA 世界中，被映射到底层关系数据库结构的 Java 类被称为实体类)。本章重点介绍使用注解方法进行对象关系映射。对于映射注解，使用 JPA 标准(例如，在 javax.persistence 包下)，因为它们可以与 Hibernate 自己的注解互换，从而便于将来迁移到 JPA 环境。

7.1 示例代码的示例数据模型

图 7-1 显示了本章中使用的数据模型。

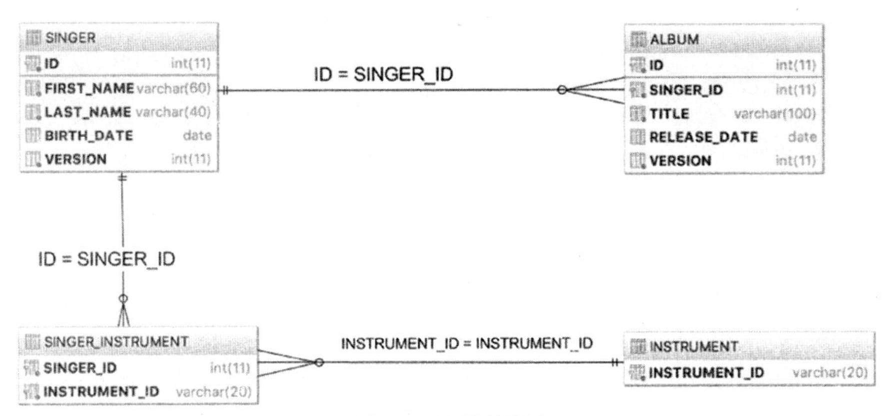

图7-1　示例数据模型

如该数据模型所示，主要添加了两个新表，即INSTRUMENT和SINGER_INSTRUMENT(连接表)。SINGER_INSTRUMENT模拟SINGER表和INSTRUMENT表之间的多对多关系。在SINGER和ALBUM表中添加了VERSION列用于乐观锁(optimistic locking)，相关内容将在后面详细讨论。在本章的示例中，将使用嵌入式H2数据库，因此不需要数据库名称。以下是创建本章示例所需表的脚本：

```
CREATE TABLE SINGER (
    ID INT NOT NULL AUTO_INCREMENT
    , FIRST_NAME VARCHAR(60) NOT NULL
    , LAST_NAME VARCHAR(40) NOT NULL
    , BIRTH_DATE DATE
    , VERSION INT NOT NULL DEFAULT 0
    , UNIQUE UQ_SINGER_1 (FIRST_NAME, LAST_NAME)
    , PRIMARY KEY (ID)
);

CREATE TABLE ALBUM (
    ID INT NOT NULL AUTO_INCREMENT
    , SINGER_ID INT NOT NULL
    , TITLE VARCHAR(100) NOT NULL
    , RELEASE_DATE DATE
    , VERSION INT NOT NULL DEFAULT 0
    , UNIQUE UQ_SINGER_ALBUM_1 (SINGER_ID, TITLE)
    , PRIMARY KEY (ID)
    , CONSTRAINT FK_ALBUM_SINGER FOREIGN KEY (SINGER_ID)
    REFERENCES SINGER (ID)
);

CREATE TABLE INSTRUMENT (
        INSTRUMENT_ID VARCHAR(20) NOT NULL
        , PRIMARY KEY (INSTRUMENT_ID)
    );

CREATE TABLE SINGER_INSTRUMENT (
    SINGER_ID INT NOT NULL
    , INSTRUMENT_ID VARCHAR(20) NOT NULL
    , PRIMARY KEY (SINGER_ID, INSTRUMENT_ID)
    , CONSTRAINT FK_SINGER_INSTRUMENT_1 FOREIGN KEY (SINGER_ID)
    REFERENCES SINGER (ID) ON DELETE CASCADE
    , CONSTRAINT FK_SINGER_INSTRUMENT_2 FOREIGN KEY (INSTRUMENT_ID)
    REFERENCES INSTRUMENT (INSTRUMENT_ID)
);
```

以下SQL是用于数据填充的脚本：

```
insert into singer (first_name, last_name, birth_date)
   values ('John', 'Mayer', '1977-10-16');
insert into singer (first_name, last_name, birth_date)
   values ('Eric', 'Clapton', '1945-03-30');
insert into singer (first_name, last_name, birth_date)
   values ('John', 'Butler', '1975-04-01');

insert into album (singer_id, title, release_date)
   values (1, 'The Search For Everything', '2017-01-20');
insert into album (singer_id, title, release_date)
   values (1, 'Battle Studies', '2009-11-17');
```

```
insert into album (singer_id, title, release_date)
   values (2, 'From The Cradle ', '1994-09-13');

insert into instrument (instrument_id) values ('Guitar');
insert into instrument (instrument_id) values ('Piano');
insert into instrument (instrument_id) values ('Voice');
insert into instrument (instrument_id) values ('Drums');
insert into instrument (instrument_id) values ('Synthesizer');
insert into singer_instrument(singer_id, instrument_id) values (1, 'Guitar');
insert into singer_instrument(singer_id, instrument_id) values (1, 'Piano');
insert into singer_instrument(singer_id, instrument_id) values (2, 'Guitar');
```

7.2 配置 Hibernate 的 SessionFactory

正如本章前面提到的，Hibernate 的核心概念基于 Session 接口，Session 接口是从 SessionFactory 获得的。Spring 提供了一些类来支持将 Hibernate 的会话工厂配置为具有所需属性的 Spring bean。要想使用 Hibernate，必须将 Hibernate 依赖项添加到项目中。下面显示了本章项目中使用的 Gradle 配置：

```
//pro-spring-15/build.gradle
ext {
   hibernateVersion = '5.2.10.Final'
   ...
   hibernate = [
      validator: "org.hibernate:hibernate-validator:5.1.3.Final",
      ehcache : "org.hibernate:hibernate-ehcache:$hibernateVersion",
       [ em] : "org.hibernate:hibernate-entitymanager:$hibernateVersion"
   ]
   ...
}
//chapter07.gradle
dependencies {
   //we specify these dependencies for all submodules,
      except the boot module, that defines its own
   if !project.name.contains"boot" {
      compile spring.contextSupport, spring.orm,
      misc.slf4jJcl, misc.logback, db.h2, misc.lang3, [hibernate.em]
   }
   testCompile testing.junit
}
```

在以下配置中，可以看到配置本章应用程序示例所需的 XML 元素：

```
<?xml version="1.0" encoding="UTF-8"?>
<beans xmlns="http://www.springframework.org/schema/beans"
   xmlns:xsi="http://www.w3.org/2001/XMLSchema-instance"
   xmlns:context="http://www.springframework.org/schema/context"
   xmlns:tx="http://www.springframework.org/schema/tx"
   xmlns:p="http://www.springframework.org/schema/p"
   xmlns:jdbc="http://www.springframework.org/schema/jdbc"
   xmlns:util="http://www.springframework.org/schema/util"
   xsi:schemaLocation="
      http://www.springframework.org/schema/jdbc
      http://www.springframework.org/schema/jdbc/spring-jdbc.xsd
      http://www.springframework.org/schema/beans
      http://www.springframework.org/schema/beans/spring-beans.xsd
      http://www.springframework.org/schema/tx
      http://www.springframework.org/schema/tx/spring-tx.xsd
      http://www.springframework.org/schema/util
      http://www.springframework.org/schema/util/spring-util.xsd
      http://www.springframework.org/schema/context
      http://www.springframework.org/schema/context/spring-context.xsd">

   <jdbc:embedded-database id="dataSource" type="H2">
      <jdbc:script location="classpath:sql/schema.sql"/>
      <jdbc:script location="classpath:sql/test-data.sql"/>
   </jdbc:embedded-database>

   <bean id="transactionManager"
      class="org.springframework.orm.hibernate5.HibernateTransactionManager"
      p:sessionFactory-ref="sessionFactory"/>

   <tx:annotation-driven/>

   <context:component-scan base-package=
         "com.apress.prospring5.ch7"/>
```

```
    <bean id="sessionFactory"
        class="org.springframework.orm.hibernate5.LocalSessionFactoryBean"
         p:dataSource-ref="dataSource"
         p:packagesToScan="com.apress.prospring5.ch7.entities"
         p:hibernateProperties-ref="hibernateProperties"/>

    <util:properties id="hibernateProperties">
        <prop key="hibernate.dialect">org.hibernate.dialect.H2Dialect</prop>
        <prop key="hibernate.max_fetch_depth">3</prop>
        <prop key="hibernate.jdbc.fetch_size">50</prop>
        <prop key="hibernate.jdbc.batch_size">10</prop>
        <prop key="hibernate.hbm2ddl.auto">create-drop</prop>
        <prop key="hibernate.format_sql">true</prop>
        <prop key="hibernate.use_sql_comments">true</prop>
    </util:properties>
</beans>
```

下面是等效的 Java 配置类，稍后介绍这两个配置的组件：

```
package com.apress.prospring5.ch7.config;

import com.apress.prospring5.ch6.CleanUp;
import org.hibernate.SessionFactory;
import org.slf4j.Logger;
import org.slf4j.LoggerFactory;
import org.springframework.context.annotation.Bean;
import org.springframework.context.annotation.Configuration;
import org.springframework.jdbc.core.JdbcTemplate;
import org.springframework.jdbc.datasource.embedded.EmbeddedDatabaseBuilder;
import org.springframework.jdbc.datasource.embedded.EmbeddedDatabaseType;
import org.springframework.orm.hibernate5.HibernateTransactionManager;
import org.springframework.orm.hibernate5.LocalSessionFactoryBean;
import org.springframework.transaction.PlatformTransactionManager;
import org.springframework.transaction.annotation.EnableTransactionManagement;
import javax.sql.DataSource;
import java.io.IOException;
import java.util.Properties;

@Configuration
@ComponentScan(basePackages =
        "com.apress.prospring5.ch7")
@EnableTransactionManagement
public class AppConfig {

    private static Logger logger =
        LoggerFactory.getLogger(AppConfig.class);

    @Bean
    public DataSource dataSource() {
        try {
        EmbeddedDatabaseBuilder dbBuilder =
            new EmbeddedDatabaseBuilder();
        return dbBuilder.setType(EmbeddedDatabaseType.H2)
              .addScripts("classpath:sql/schema.sql",
              "classpath:sql/test-data.sql").build();
        } catch (Exception e) {
        logger.error("Embedded DataSource bean cannot be created!", e);
        return null;
        }
    }

    private Properties hibernateProperties() {
        Properties hibernateProp = new Properties();
        hibernateProp.put("hibernate.dialect", "org.hibernate.dialect.H2Dialect");
        hibernateProp.put("hibernate.format_sql", true);
        hibernateProp.put("hibernate.use_sql_comments", true);
        hibernateProp.put("hibernate.show_sql", true);
        hibernateProp.put("hibernate.max_fetch_depth", 3);
        hibernateProp.put("hibernate.jdbc.batch_size", 10);
        hibernateProp.put("hibernate.jdbc.fetch_size", 50);
        return hibernateProp;
    }

    @Bean public SessionFactory sessionFactory()
        throws IOException {
        LocalSessionFactoryBean sessionFactoryBean = new LocalSessionFactoryBean();
        sessionFactoryBean.setDataSource(dataSource());
```

```
        sessionFactoryBean.setPackagesToScan("com.apress.prospring5.ch7.entities");
        sessionFactoryBean.setHibernateProperties(hibernateProperties());
        sessionFactoryBean.afterPropertiesSet();
        return sessionFactoryBean.getObject();
    }

    @Bean public PlatformTransactionManager transactionManager()
        throws IOException {
        return new HibernateTransactionManager(sessionFactory());
    }
}
```

在上面的配置中，声明了几个能够支持 Hibernate 会话工厂的 bean。主要配置如下所示。

- dataSource bean：所使用的数据库是一个 H2 嵌入式数据库，如第 6 章所述。
- transactionManager bean：Hibernate会话工厂需要事务管理器来进行事务性数据访问。Spring在包org.springframework.orm.hibernate5.HibernateTransactionManager中声明了一个针对Hibernate 5的事务管理器。该bean在声明时使用了ID transactionManager。默认情况下，当使用XML配置时，无论何时需要事务管理，Spring都将在其ApplicationContext中使用名称transactionManager查找bean。相比于按名称搜索，当按类型搜索bean时，Java配置稍微灵活一点。我们将在第9章中详细讨论事务。另外，声明标记<tx:annotation-driven>以支持使用注解声明事务划分要求。与Java配置等价的是@EnableTransactionManagement注解。
- **组件扫描**：你应该对此标记以及@ComponentScan 注解非常熟悉了。它们用来指示 Spring 扫描包 com.apress.prospring5.ch7 中的组件，以便检测用@Repository 注解的 bean。
- Hibernate sessionFactory bean：sessionFactory bean 是最重要的部分。在该 bean 中提供了几个属性。首先，需要将 dataSource bean 注入会话工厂。其次，指示 Hibernate 扫描包 com.apress.prospring5.ch.entities 中的域对象。最后，hibernateProperties 属性提供了 Hibernate 的配置细节。配置参数有很多，此时只定义几个应该为每个应用程序提供的重要属性。表 7-1 列出了 Hibernate 会话工厂的主要配置参数。

表 7-1　Hibernate 属性

属性	描述
hibernate.dialect	为 Hibernate 应该使用的查询指定数据库方言。Hibernate 支持许多数据库的 SQL 方言。这些方言都是 org.hibernate.dialect.Dialect 的子类，主要的方言包括 H2Dialect、Oracle10gDialect、PostgreSQLDialect、MySQLDialect、SQLServerDialect 等
hibernate.max_fetch_depth	当一个映射对象与其他映射对象相关联时，用来声明外部连接的“深度”。该设置可以防止 Hibernate 通过过多嵌套关联获取太多数据。常用的值是 3
hibernate.jdbc.fetch_size	指定来自底层 JDBC ResultSet 的记录数，Hibernate 每次只能从数据库中提取指定数量的记录。例如，查询已提交给数据库，并且 ResultSet 包含 500 条记录。如果获取大小为 50，那么 Hibernate 需要获取 10 次才能得到所有数据
ibernate.jdbc.batch_size	指示 Hibernate 应该将更新操作的数量组合到一个批处理中。这对于在 Hibernate 中执行批处理作业是很有用的。显然，当正在执行批处理作业以更新数十万条记录时，希望 Hibernate 能够批量分组查询，而不是逐个提交更新
hibernate.show_sql	指示 Hibernate 是否应将 SQL 查询输出到日志文件或控制台。应该在开发环境中启用此功能，这对测试和故障排除过程有很大帮助
hibernate.format_sql	指示日志或控制台中的 SQL 输出是否应格式化
hibernate..use_sql_comments	如果设置为 true，Hibernate 将在 SQL 内部生成注释以便于调试

有关 Hibernate 所支持的完整属性列表，请参阅 Hibernate 的 ORM 用户指南中的第 23 节，网址为 https://docs.jboss.org/hibernate/orm/5.2/userguide/html_single/Hibernate_User_Guide.html。

7.3　使用 Hibernate 注解的 ORM 映射

完成配置后，下一步就是对 Java POJO 实体类及其对底层关系数据结构的映射进行建模。

映射有两种方法。第一种方法是设计对象模型，然后根据对象模型生成数据库脚本。例如，对于会话工厂配置，可以传入 Hibernate 属性 hibernate.hbm2ddl.auto，从而让 Hibernate 自动将模式 DDL 导出到数据库。第二种方法是从数据模型开始，然后用所需的映射对 POJO 进行建模。我们往往更喜欢使用后一种方法，因为可以更好地控制数据模型，这对于优化数据访问的性能非常有用。而第一种方法将在本章后面介绍，从而演示用 Hibernate 配置 Spring 应用程序的不同方式。根据前面的数据模型，图 7-2 显示了具有类图的相应 OO 模型。

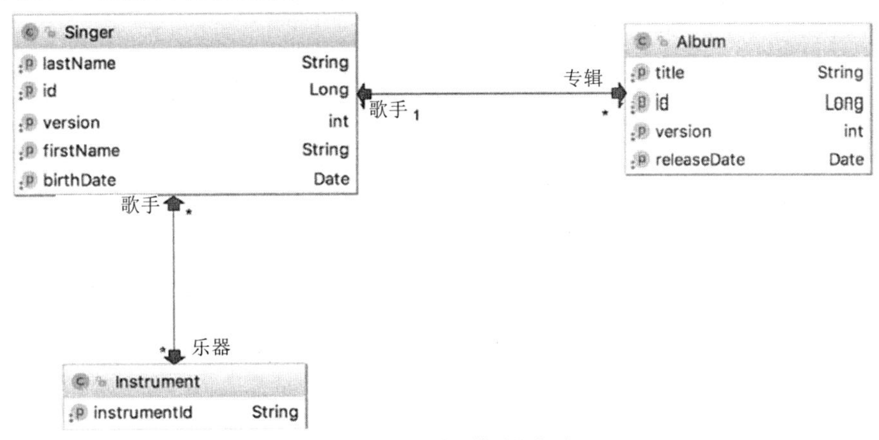

图 7-2　示例数据模型的类图

可以看到 Singer 和 Album 之间存在一对多关系，而 Single 和 Instrument 对象之间存在多对多关系。

7.3.1　简单的映射

首先从映射类的简单属性开始。以下代码片段显示了带有映射注解的 Singer 类：

```
package com.apress.prospring5.ch7.entities;

import javax.persistence.*;
import java.io.Serializable;
import java.util.Date;
import java.util.HashSet;
import java.util.Set;

import static javax.persistence.GenerationType.IDENTITY;

@Entity
@Table(name = "singer")
public class Singer implements Serializable {

    private Long id;
    private String firstName;
    private String lastName;
    private Date birthDate;
    private int version;

    public void setId(Long id) {
        this.id = id;
    }

@Id
@GeneratedValue(strategy = IDENTITY)
@Column(name = "ID")
public Long getId() {
    return this.id;
}

@Version
@Column(name = "VERSION")
public int getVersion() {
    return version;
}

@Column(name = "FIRST_NAME")
public String getFirstName() {
```

```
        return this.firstName;
    }

    @Column(name = "LAST_NAME")
    public String getLastName() {
        return this.lastName;
    }

    @Temporal(TemporalType.DATE)
    @Column(name = "BIRTH_DATE")
    public Date getBirthDate() {
        return birthDate;
    }

    public void setFirstName(String firstName) {
        this.firstName = firstName;
    }

    public void setLastName(String lastName) {
        this.lastName = lastName;
    }

    public void setBirthDate(Date birthDate) {
        this.birthDate = birthDate;
    }

    public void setVersion(int version) {
        this.version = version;
    }

    public String toString() {
        return "Singer - Id: " + id + ", First name: " + firstName
            + ", Last name: " + lastName + ", Birthday: " + birthDate;
    }
}
```

首先用@Entity 注解类型，这意味着这是一个映射的实体类。@Table 注解定义了数据库中该实体映射到的表名。对于每个映射属性，都使用@Column 对其进行注解，并提供列名。

> 如果类型和属性名称与表名和列名相同，则可以跳过表名和列名。

关于映射，需要强调以下几点：

- 对于 birthDate 属性，应该使用@Temporal 注解它，并使用 TemporalType.DATE 值作为参数，这意味着想要将数据类型从 Java 日期类型(java.util.Date)映射到 SQL 日期类型(java.sql.Date)。这样就可以像平常一样，在应用程序中使用 java.util.Date 来访问 Singer 对象中的 birthDate 属性。
- 对于id属性，使用@Id对它进行注解。这意味着它是对象的主键。在管理会话中的联系人实体实例时，Hibernate将使用它作为唯一标识符。另外，@GeneratedValue注解告诉Hibernate如何生成id值。IDENTITY策略意味着id值是在插入过程中由后端生成的。
- 对于 version 属性，使用@Version 对其进行注解。这指示 Hibernate 希望使用乐观锁机制，并使用 version 属性作为版本控制。每次 Hibernate 更新记录时，它都会将实体实例的版本与数据库中记录的版本进行比较。如果两个版本相同，则意味着之前没有人更新数据，此时 Hibernate 将更新数据并增大版本字段值。但是，如果版本不一样，则意味着有人已经更新过记录，Hibernate 会抛出 StaleObjectStateException 异常，该异常将被 Spring 转换为 HibernateOptimisticLockingFailureException 异常。本例中将使用整数来进行版本控制。除整数外，Hibernate 还支持使用时间戳。不过，建议使用整数进行版本控制，因为 Hibernate 每次更新后都会将版本号加 1。如果使用时间戳，那么 Hibernate 将在每次更新后更新最新的时间戳。时间戳不太安全，因为两个并发事务可能同时在加载和更新同一项目。

另一个映射对象是 Album，如下所示：

```
package com.apress.prospring5.ch7.entities;

import javax.persistence.*;
import java.io.Serializable;
import java.util.Date;

import static javax.persistence.GenerationType.IDENTITY;
```

```
@Entity
@Table(name = "album")
public class Album implements Serializable {

   private Long id;
   private String title;
   private Date releaseDate;
   private int version;
public void setId(Long id) {
   this.id = id;
}

@Id
@GeneratedValue(strategy = IDENTITY)
@Column(name = "ID")
public Long getId() {
   return this.id;
}

@Version
@Column(name = "VERSION")
public int getVersion() {
   return version;
}

@Column
public String getTitle() {
   return this.title;
}

@Temporal(TemporalType.DATE)
@Column(name = "RELEASE_DATE")
public Date getReleaseDate() {
   return this.releaseDate;
}

public void setTitle(String title) {
   this.title = title;
}

public void setReleaseDate(Date releaseDate) {
   this.releaseDate = releaseDate;
}

public void setVersion(int version) {
   this.version = version;
}
@Override

public String toString() {
   return "Album - Id: " + id + ", Title: " +
      title + ", Release Date: " + releaseDate;
   }
}
```

以下是本章示例中使用的第三个实体类:

```
package com.apress.prospring5.ch7.entities;
import javax.persistence.*;
import java.io.Serializable;
import java.util.HashSet;
import java.util.Set;

@Entity
@Table(name = "instrument")
public class Instrument implements Serializable {
   private String instrumentId;

   @Id
   @Column(name = "INSTRUMENT_ID")
   public String getInstrumentId() {
      return this.instrumentId;
   }

   public void setInstrumentId(String instrumentId) {
      this.instrumentId = instrumentId;
   }
```

```
    @Override
    public String toString() {
        return "Instrument :" + getInstrumentId();
    }
}
```

7.3.2 一对多映射

Hibernate 能够建立多种关联模型。最常见的关联是一对多和多对多。每位歌手都可以拥有零张或多张专辑，因此两者之间是一对多关联(用 ORM 术语来说，一对多关联用于对数据结构中的零对多关系和一对多关系进行建模)。以下代码片段描述了定义 Singer 和 Album 实体之间一对多关系所需的属性和方法：

```
package com.apress.prospring5.ch7.entities;

import javax.persistence.*;
import java.io.Serializable;
import java.util.Date;
import java.util.HashSet;
import java.util.Set;

import static javax.persistence.GenerationType.IDENTITY;

@Entity
@Table(name = "singer")
public class Singer implements Serializable {

    private Long id;
    private String firstName;
    private String lastName;
    private Date birthDate;
    private int version;
    private Set<Album> albums = new HashSet<>();
    ...
    @OneToMany(mappedBy = "singer", cascade=CascadeType.ALL,
        orphanRemoval=true)

    public Set<Album> getAlbums() {
        return albums;
    }

    public boolean addAbum(Album album) {
        album.setSinger(this);
        return getAlbums().add(album);
    }

    public void removeAlbum(Album album) {
        getAlbums().remove(album);
    }

    public void setAlbums(Set<Album> albums) {
        this.albums = albums;
    }
    ...
}
```

属性 contactTelDetails 的 getter 方法用@OneToMany 进行了注解，它用于指示与 Album 类的一对多关系。多个属性被传递给该注解。mappedBy 属性指明 Album 类中提供关联的属性(即，通过 FK_ALBUM_SINGER 表中的外键定义进行链接)。cascade 属性意味着更新操作应该“级联”到子级。orphanRemoval 属性意味着在专辑更新后，应该从数据库中删除不再存在于数据集中的条目。以下代码片段显示了 Album 类中用于关联映射的更新代码：

```
package com.apress.prospring5.ch7.entities;

import javax.persistence.*;
import java.io.Serializable;
import java.util.Date;

import static javax.persistence.GenerationType.IDENTITY;
@E
ntity
@Table(name = "album")
public class Album implements Serializable {

    private Long id;
    private String title;
```

```
    private Date releaseDate;
    private int version;

    private Singer singer;
    @ManyToOne
    @JoinColumn(name = "SINGER_ID")
    public Singer getSinger() {
        return this.singer;
    }

    public void setSinger(Singer singer) {
        this.singer = singer;
    }
    ...
}
```

此时，使用@ManyToOne注解singer属性的getter方法，表明与Singer关联的另一方。同时，还为潜在的外键列名指定@JoinColumn注解。最后，重写toString()方法，以便稍后将输出打印到控制台，从而进行测试。

7.3.3 多对多映射

每位歌手都可以演奏零种或多种乐器，而每种乐器也与零位或多位歌手相关联，这意味着两者之间是多对多映射。多对多映射需要一个连接表，即SINGER_INSTRUMENT。以下代码示例显示了需要添加到Singer类以实现此关联的代码：

```
package com.apress.prospring5.ch7.entities;

import javax.persistence.*;
import java.io.Serializable;
import java.util.Date;
import java.util.HashSet;
import java.util.Set;

import static javax.persistence.GenerationType.IDENTITY;

@Entity
@Table(name = "singer")
public class Singer implements Serializable {

    private Long id;
    private String firstName;
    private String lastName;
    private Date birthDate;
    private int version;

    private Set<Instrument> instruments = new HashSet<>();

    @ManyToMany
    @JoinTable(name = "singer_instrument",
        joinColumns = @JoinColumn(name = "SINGER_ID"),
        inverseJoinColumns = @JoinColumn(name = "INSTRUMENT_ID"))
    public Set<Instrument> getInstruments() {
        return instruments;
    }
    public void setInstruments(Set<Instrument> instruments) {
        this.instruments = instruments;
    }
    ...
}
```

Singer类中instruments属性的getter方法用@ManyToMany进行了注解。此外，还使用@JoinTable来指示Hibernate应该查找的底层连接表。名称是连接表的名称，joinColumns定义了作为SINGER表外键的列，而inverseJoinColumns定义了作为关联的另一方(即INSTRUMENT表)的外键列。以下是Instrument类中用来实现此关联的代码：

```
package com.apress.prospring5.ch7.entities;

import javax.persistence.*;
import java.io.Serializable;
import java.util.HashSet;
import java.util.Set;

@Entity
@Table(name = "instrument")
public class Instrument implements Serializable {
```

```
    private String instrumentId;
    private Set<Singer> singers = new HashSet<>();

    @Id
    @Column(name = "INSTRUMENT_ID")
    public String getInstrumentId() {
       return this.instrumentId;
    }

    @ManyToMany
    @JoinTable(name = "singer_instrument",
       joinColumns = @JoinColumn(name = "INSTRUMENT_ID"),
       inverseJoinColumns = @JoinColumn(name = "SINGER_ID"))
    public Set<Singer> getSingers() {
       return this.singers;
    }

    public void setSingers(Set<Singer> singers) {
       this.singers = singers;
    }

    public void setInstrumentId(String instrumentId) {
       this.instrumentId = instrumentId;
    }

    @Override
    public String toString() {
       return "Instrument :" + getInstrumentId();
    }
}
```

上述映射与 Singer 的映射大致相同，但 joinColumns 和 inverseJoinColumns 属性相反。

7.4 Hibernate 会话接口

在 Hibernate 中，当与数据库交互时，需要处理的主要接口是 Session 接口，它是从 SessionFactory 获得的。以下代码片段显示了本章示例中使用的 SingerDaoImpl 类，并将已配置的 Hibernate SessionFactory 注入类中：

```
package com.apress.prospring5.ch7.dao;

import com.apress.prospring5.ch7.entities.Singer;
import org.apache.commons.logging.Log;
import org.apache.commons.logging.LogFactory;
import org.hibernate.SessionFactory;
import org.springframework.stereotype.Repository;
import org.springframework.transaction.annotation.Transactional;

import javax.annotation.Resource;
import java.util.List;

@Transactional
@Repository("singerDao")
public class SingerDaoImpl implements SingerDao {

    private SessionFactory sessionFactory;

    public SessionFactory getSessionFactory() {
       return sessionFactory;
    }

    @Resource(name = "sessionFactory")
    public void setSessionFactory(SessionFactory sessionFactory) {
       this.sessionFactory = sessionFactory;
    }
...
}
```

像往常一样，使用@Repository 注解将 DAO 类声明为 Spring bean。@Transactional 注解定义了事务需求(相关内容将在第 9 章中进一步讨论)。通过使用@Resource 注解注入 sessionFactory 属性。

```
package com.apress.prospring5.ch7.dao;

import com.apress.prospring5.ch7.entities.Singer;

import java.util.List;
```

```
public interface SingerDao {
    List<Singer> findAll();
    List<Singer> findAllWithAlbum();
    Singer findById(Long id);
    Singer save(Singer contact);
    void delete(Singer contact);
}
```

SingerDao 接口比较简单；里面包含三个查找方法、一个保存方法和一个删除方法。save()方法将执行插入和更新操作。

7.4.1 使用 Hibernate 查询语言查询数据

Hibernate 与其他 ORM 工具(比如 JDO 和 JPA)一样都是围绕对象模型进行设计的。所以，在定义了映射之后，不再需要构建 SQL 与数据库进行交互。相反，对于 Hibernate 来说，可以使用 Hibernate 查询语言(HQL)来定义查询。当与数据库交互时，Hibernate 会代替我们将查询转换为 SQL 语句。

在编写 HQL 查询时，其语法与 SQL 很相似。但是，需要考虑的是对象方面而不是数据库方面。接下来将介绍几个示例。

7.4.2 使用延迟获取进行简单查询

接下来从实现 findAll()方法开始，它只是从数据库中检索所有联系人。以下代码示例显示了用于此功能的更新后的代码：

```
package com.apress.prospring5.ch7.dao;

import com.apress.prospring5.ch7.entities.Singer;
import org.apache.commons.logging.Log;
import org.apache.commons.logging.LogFactory;
import org.hibernate.SessionFactory;
import org.springframework.stereotype.Repository;
import org.springframework.transaction.annotation.Transactional;

import javax.annotation.Resource;
import java.util.List;

@Transactional
@Repository("singerDao")
public class SingerDaoImpl implements SingerDao {

    private static final Log logger = LogFactory.getLog(SingerDaoImpl.class);
    private SessionFactory sessionFactory;

    @Transactional(readOnly = true)
    public List<Singer> findAll() {
        return sessionFactory.getCurrentSession()
            .createQuery("from Singer s").list();
    }
    ...
}
```

SessionFactory.getCurrentSession()方法获取 Hibernate 的 Session 接口。然后，调用 Session.createQuery()方法，并传入 HQL 语句。from Singer 语句表明从数据库中检索所有联系人。该语句的另一种替换语法是 select s from Singer s。@Transactional(readOnly = true)注解表明希望将事务设置为只读。为只读方法设置该属性可以产生更好的性能。

以下代码片段显示了一个简单的 SingerDaoImpl 测试程序：

```
package com.apress.prospring5.ch7;

import com.apress.prospring5.ch7.config.AppConfig;
import com.apress.prospring5.ch7.dao.SingerDao;
import com.apress.prospring5.ch7.entities.Singer;
import org.slf4j.Logger;
import org.slf4j.LoggerFactory;
import org.springframework.context.annotation.AnnotationConfigApplicationContext;
import org.springframework.context.support.GenericApplicationContext;

import java.util.List;
```

```
public class SpringHibernateDemo {

    private static Logger logger =
        LoggerFactory.getLogger(SpringHibernateDemo.class);

    public static void main(String... args) {
        GenericApplicationContext ctx =
            new AnnotationConfigApplicationContext(AppConfig.class);

        SingerDao singerDao = ctx.getBean(SingerDao.class);
        singerDao.delete(singer);
        listSingers(singerDao.findAll());

        ctx.close();
    }

    private static void listSingers(List<Singer> singers) {
        logger.info(" ---- Listing singers:");
        for (Singer singer : singers) {
        logger.info(singer.toString());
        }
    }
}
```

运行上述类将产生以下输出：

```
---- Listing singers:
Singer - Id: 1, First name: John, Last name: Mayer, Birthday: 1977-10-16
Singer - Id: 3, First name: John, Last name: Butler, Birthday: 1975-04-01
Singer - Id: 2, First name: Eric, Last name: Clapton, Birthday: 1945-03-30
```

虽然获取了歌手记录，那么专辑和乐器呢？让我们修改测试类来打印详细信息。在下面的代码片段中，可以看到 listSingers()被 listSingersWithAlbum()替换：

```
package com.apress.prospring5.ch7;

import com.apress.prospring5.ch7.config.AppConfig;
import com.apress.prospring5.ch7.dao.SingerDao;
import com.apress.prospring5.ch7.entities.Album;
import com.apress.prospring5.ch7.entities.Instrument;
import com.apress.prospring5.ch7.entities.Singer;
import org.slf4j.Logger;
import org.slf4j.LoggerFactory;
import org.springframework.context.annotation.AnnotationConfigApplicationContext;
import org.springframework.context.support.GenericApplicationContext;

import java.util.List;

public class SpringHibernateDemo {

    private static Logger logger =
        LoggerFactory.getLogger(SpringHibernateDemo.class);

    public static void main(String... args) {
        GenericApplicationContext ctx =
            new AnnotationConfigApplicationContext(AppConfig.class);

        SingerDao singerDao = ctx.getBean(SingerDao.class);
        Singer singer = singerDao.findById(2l);
        singerDao.delete(singer);
        listSingersWithAlbum(singerDao.findAllWithAlbum());

        ctx.close();
    }

    private static void listSingersWithAlbum(List<Singer> singers) {
        logger.info(" ---- Listing singers with instruments:");
        for (Singer singer : singers) {
        logger.info(singer.toString());
        if (singer.getAlbums() != null) {
            for (Album album :
                singer.getAlbums()) {
                logger.info("\t" + album.toString());
            }
        }
        if (singer.getInstruments() != null) {
            for (Instrument instrument : singer.getInstruments()) {
```

```
                logger.info("\t" + instrument.getInstrumentId());
            }
        }
        }
    }
}
```

再次运行程序，你会看到下面所示的异常：

```
---- Listing singers with instruments:
Singer - Id: 1, First name: John, Last name: Mayer, Birthday: 1977-10-16
org.hibernate.LazyInitializationException: failed to lazily initialize a
   collection of role: com.apress.prospring5.ch7.entities.Singer.albums,
   could not initialize proxy - no Session
```

当尝试访问关联时，你会看到 Hibernate 抛出 LazyInitializationException 异常。

这是因为，默认情况下，Hibernate 会*延迟(lazily)*获取该关联，这意味着 Hibernate 不会连接关联表(即 ALBUM)以获取记录。这主要出于性能上的考虑；正如你所想象的那样，如果一个查询检索到数千条记录并且所有的关联都被检索到，那么大量的数据传输将会降低性能。

7.4.3 使用关联获取进行查询

要想让 Hibernate 从关联中获取数据，有两种选择。第一种选择是，可以使用获取模式 EAGER 来定义关联，例如@ManyToMany(fetch = FetchType.EAGER)。这告诉 Hibernate 在每个查询中获取关联记录。但是，如上所述，这将影响数据检索性能。

另一种选择是，强制 Hibernate 在需要时获取查询中的关联记录。如果使用 Criteria 查询，则可以调用 Criteria.setFetchMode()函数来指示 Hibernate 急切地获取关联。当使用 NamedQuery 时，可以使用 fetch 操作符来指示 Hibernate 急切地获取关联。

接下来看看 findAllWithAlbum()方法的实现，它将检索所有联系人信息以及他们的电话详细信息和业余爱好。该例将使用 NamedQuery 方式。NamedQuery 可以被外化为 XML 文件或者在实体类中使用注解来声明。下面显示了使用注解定义的命名查询修订的 Singer 域对象：

```
package com.apress.prospring5.ch7.entities;

import javax.persistence.Entity;
import javax.persistence.Table;
import javax.persistence.NamedQueries;
import javax.persistence.NamedQuery;
...

@Entity
@Table(name = "singer")
@NamedQueries({
   @NamedQuery(name="Singer.findAllWithAlbum",
      query="select distinct s from Singer s " +
         "left join fetch s.albums a " +
         "left join fetch s.instruments i")
})
public class Singer implements Serializable {
   ...
}
```

首先定义一个名为 Singer.findAllWithAlbum 的 NamedQuery 实例。然后在 HQL 中定义查询。请注意子句 left join fetch，它指示 Hibernate 获取关联。此外，还需要使用 select distinct；否则，Hibernate 将返回重复的对象(如果单个歌手有两张与之关联的专辑，则会返回两个 Singer 对象)。

以下是 findAllWithAlbum()方法的实现：

```
package com.apress.prospring5.ch7.dao;
...

@Transactional
@Repository("singerDao")
public class SingerDaoImpl implements SingerDao {
   @Transactional(readOnly = true)
   public List<Singer> findAllWithAlbum() {
      return sessionFactory.getCurrentSession().
         getNamedQuery("Singer.findAllWithAlbum").list();
      }
}
```

此时，使用了 Session.getNamedQuery()方法，并传入 NamedQuery 实例的名称。修改测试程序(SpringHibernateDemo)，以便调用 singerDao.findAllWithAlbum()，并产生以下输出：

```
---- Listing singers with instruments:
Singer - Id: 1, First name: John, Last name: Mayer, Birthday: 1977-10-16
   Album - Id: 2, Singer id: 1, Title: Battle Studies, Release Date: 2009-11-17
   Album - Id: 1, Singer id: 1, Title: The Search For Everything,
         Release Date: 2017-01-20
      Instrument: Guitar
   Instrument: Piano
Singer - Id: 3, First name: John, Last name: Butler, Birthday: 1975-04-01
Singer - Id: 2, First name: Eric, Last name: Clapton, Birthday: 1945-03-30
   Album - Id: 3, Singer id: 2, Title: From The Cradle , Release Date: 1994-09-13
   Instrument: Guitar
```

现在所有歌手的细节信息都被正确地检索出来了。接下来看一看带有参数的 NamedQuery 的另一个示例。这一次将实现 findById()方法，并获取关联。以下代码片段显示添加了新命名查询的 Singer 类：

```
package com.apress.prospring5.ch7.entities;

import javax.persistence.*;
import java.io.Serializable;
import java.util.Date;
import java.util.HashSet;
import java.util.Set;

import static javax.persistence.GenerationType.IDENTITY;

@Entity
@Table(name = "singer")
@NamedQueries({
   @NamedQuery(name="Singer.findById",
    query="select distinct s from Singer s " +
       "left join fetch s.albums a " +
       "left join fetch s.instruments i " +
       "where s.id = :id"),
   @NamedQuery(name="Singer.findAllWithAlbum",
    query="select distinct s from Singer s " +
       "left join fetch s.albums a " +
       "left join fetch s.instruments i")
})
public class Singer implements Serializable {
    ...
}
```

在名为 Singer.findById 的命名查询中，声明了一个命名参数：id。下面显示了 SingerDaoImpl 中 findById()方法的实现：

```
package com.apress.prospring5.ch7.dao;

import com.apress.prospring5.ch7.entities.Singer;
import org.apache.commons.logging.Log;
import org.apache.commons.logging.LogFactory;
import org.hibernate.SessionFactory;
import org.springframework.stereotype.Repository;
import org.springframework.transaction.annotation.Transactional;

import javax.annotation.Resource;
import java.util.List;

@Transactional
@Repository("singerDao")
public class SingerDaoImpl implements SingerDao {

    private static final Log logger = LogFactory.getLog(SingerDaoImpl.class);
    private SessionFactory sessionFactory;

    @Transactional(readOnly = true)
    public Singer findById(Long id) {
        return (Singer) sessionFactory.getCurrentSession().
            getNamedQuery("Singer.findById").
            setParameter("id", id).uniqueResult();
    }
    ...
}
```

在上面的代码中，使用了相同的 Session.getNamedQuery()方法。同时调用了 setParameter()方法，并传入带有值的命名参数。对于多个参数，可以使用 Query 接口的 setParameterList()或 setParameters()方法。

还有一些更高级的查询方法，比如本地查询和标准查询，我们在下一章讨论 JPA 时讨论这些方法。为了测试上述方法，必须相应地修改 SpringHibernateDemo 类。

```
package com.apress.prospring5.ch7;
...

public class SpringHibernateDemo {

    private static Logger logger =
        LoggerFactory.getLogger(SpringHibernateDemo.class);

    public static void main(String... args) {
        GenericApplicationContext ctx =
            new AnnotationConfigApplicationContext(AppConfig.class);
            SingerDao singerDao = ctx.getBean(SingerDao.class);
            Singer singer = singerDao.findById(2l);
            logger.info(singer.toString());

        ctx.close();
    }
}
```

运行该程序会产生以下输出：

```
Singer - Id: 1, First name: John, Last name: Mayer, Birthday: 1977-10-16
```

7.5 插入数据

使用 Hibernate 插入数据很简单。另外一件奇特的事情是检索数据库生成的主键。在前面关于 JDBC 的章节中，首先需要明确声明想要检索生成的键，然后传入 KeyHolder 实例，最后在执行 insert 语句之后从中获取键值。有了 Hibernate，所有这些操作就都不需要了。在插入操作之后，Hibernate 将检索生成的键并填充域对象。以下代码片段显示了 save()方法的实现：

```
package com.apress.prospring5.ch7.dao;

import com.apress.prospring5.ch7.entities.Singer;
import org.apache.commons.logging.Log;
import org.apache.commons.logging.LogFactory;
import org.hibernate.SessionFactory;
import org.springframework.stereotype.Repository;
import org.springframework.transaction.annotation.Transactional;

import javax.annotation.Resource;
import java.util.List;

@Transactional
@Repository("singerDao")
public class SingerDaoImpl implements SingerDao {
    private static final Log logger = LogFactory.getLog(SingerDaoImpl.class);
    private SessionFactory sessionFactory;

    public Singer save(Singer singer) {
        sessionFactory.getCurrentSession().saveOrUpdate(singer);
        logger.info("Singer saved with id: " + singer.getId());
        return singer;
    }
    ...
}
```

只需要调用 Session.saveOrUpdate()方法，它可以用于插入和更新操作。另外，我们还记录了保存对象后将由 Hibernate 填充的 Singer 对象的 ID。以下代码片段显示了在 SINGER 表中插入新的歌手记录的代码，其中 ALBUM 表中有两条子记录，并且测试了插入是否成功。另外，因为正在修改表的内容，所以 JUnit 类更适合单独测试每个操作。

```
package com.apress.prospring5.ch7;

import com.apress.prospring5.ch7.config.AppConfig;
import com.apress.prospring5.ch7.dao.SingerDao;
```

```
import com.apress.prospring5.ch7.entities.Album;
import com.apress.prospring5.ch7.entities.Instrument;
import com.apress.prospring5.ch7.entities.Singer;
import org.junit.Before;
import org.junit.Test;
import org.slf4j.Logger;
import org.slf4j.LoggerFactory;
import org.springframework.context.annotation.AnnotationConfigApplicationContext;
import org.springframework.context.support.GenericApplicationContext;

import java.util.Date;
import java.util.GregorianCalendar;
import java.util.List;

import static org.junit.Assert.assertEquals;
import static org.junit.Assert.assertNotNull;

public class SingerDaoTest {
   private static Logger logger =
      LoggerFactory.getLogger(SingerDaoTest.class);

   private GenericApplicationContext ctx;
   private SingerDao singerDao;

   @Before
   public void setUp(){
      ctx = new AnnotationConfigApplicationContext(AppConfig.class);
      singerDao = ctx.getBean(SingerDao.class);
      assertNotNull(singerDao);
   }
@Test
public void testInsert(){
   Singer singer = new Singer();
   singer.setFirstName("BB");
   singer.setLastName("King");
   singer.setBirthDate(new Date(
       (new GregorianCalendar(1940, 8, 16)).getTime().getTime()));

   Album album = new Album();
   album.setTitle("My Kind of Blues");
   album.setReleaseDate(new java.sql.Date(
       (new GregorianCalendar(1961, 7, 18)).getTime().getTime()));
   singer.addAbum(album);

   album = new Album();
   album.setTitle("A Heart Full of Blues");
   album.setReleaseDate(new java.sql.Date(
       (new GregorianCalendar(1962, 3, 20)).getTime().getTime()));
   singer.addAbum(album);

   singerDao.save(singer);
   assertNotNull(singer.getId());

   List<Singer> singers = singerDao.findAllWithAlbum();
   assertEquals(4, singers.size());
   listSingersWithAlbum(singers);
}

@Test
public void testFindAll(){
   List<Singer> singers = singerDao.findAll();
   assertEquals(3, singers.size());
   listSingers(singers);
}

@Test
public void testFindAllWithAlbum(){
   List<Singer> singers = singerDao.findAllWithAlbum();
   assertEquals(3, singers.size());
   listSingersWithAlbum(singers);
}

@Test
public void testFindByID(){
   Singer singer = singerDao.findById(1L);
   assertNotNull(singer);
   logger.info(singer.toString());
}
```

```
private static void listSingers(List<Singer> singers) {
    logger.info(" ---- Listing singers:");
    for (Singer singer : singers) {
    logger.info(singer.toString());
    }
}
    private static void listSingersWithAlbum(List<Singer> singers) {
        logger.info(" ---- Listing singers with instruments:");
        for (Singer singer : singers) {
        logger.info(singer.toString());
        if (singer.getAlbums() != null) {
            for (Album album :
                singer.getAlbums()) {
                logger.info("\t" + album.toString());
            }
        }
        if (singer.getInstruments() != null) {
            for (Instrument instrument : singer.getInstruments()) {
                logger.info("\tInstrument: " + instrument.getInstrumentId());
            }
        }
        }
    }
    @After
    public void tearDown(){
        ctx.close();
    }
}
```

如前面的代码所示，在 testInsert()方法中，添加了两张专辑，并保存了对象。之后，通过调用 listSingersWithAlbum 再次列出所有歌手的信息。运行 testInsert()方法会产生以下输出：

```
...
INFO o.h.d.Dialect - HHH000400:
   Using dialect: org.hibernate.dialect.H2Dialect
INFO o.h.h.i.QueryTranslatorFactoryInitiator - HHH000397:
   Using ASTQueryTranslatorFactory
Hibernate:
   /* insert com.apress.prospring5.ch7.entities.Singer
      */ insert
      into
         singer
          (ID, BIRTH_DATE, FIRST_NAME, LAST_NAME, VERSION)
      values
          (null, ?, ?, ?, ?)
Hibernate:
   /* insert com.apress.prospring5.ch7.entities.Album
      */ insert
      into
         album
          (ID, RELEASE_DATE, SINGER_ID, title, VERSION)
      values
          (null, ?, ?, ?, ?)
Hibernate:
   /* insert com.apress.prospring5.ch7.entities.Album
      */ insert
      into
         album
       (ID, RELEASE_DATE, SINGER_ID, title, VERSION)
      values
       (null, ?, ?, ?, ?)
INFO c.a.p.c.d.SingerDaoImpl - Singer saved with id: 4
...
INFO - ---- Listing singers with instruments:
INFO - Singer - Id: 4, First name: BB, Last name: King,
  Birthday: 1940-09-16
INFO -  Album - Id: 5, Singer id: 4, Title: A Heart Full of Blues,
   Release Date: 1962-04-20
INFO -  Album - Id: 4, Singer id: 4, Title: My Kind of Blues,
   Release Date: 1961-08-18
INFO - Singer - Id: 1, First name: John, Last name: Mayer,
Birthday: 1977-10-16
INFO -  Album - Id: 2, Singer id: 1, Title: Battle Studies,
   Release Date: 2009-11-17
INFO -  Album - Id: 1, Singer id: 1, Title: The Search For Everything,
Release Date: 2017-01-20
INFO -  Instrument: Piano
```

```
INFO -  Instrument: Guitar
INFO - Singer - Id: 3, First name: John, Last name: Butler,
   Birthday: 1975-04-01
INFO - Singer - Id: 2, First name: Eric, Last name: Clapton,
   Birthday: 1945-03-30
INFO -  Album - Id: 3, Singer id: 2, Title: From The Cradle ,
   Release Date: 1994-09-13
INFO -  Instrument: Guitar
```

日志记录配置已被修改，以便打印更详细的 Hibernate 信息。从 INFO 日志记录中，可以看到正确填充了新保存的联系人的 ID。Hibernate 还会显示所有正在执行的 SQL 语句，以便知道幕后发生了什么。

7.6 更新数据

更新数据与插入数据一样简单。假设歌手的 ID 为 1，想要更新其名字并删除一张专辑。为了测试更新操作，以下代码片段显示了 testUpdate()方法：

```
package com.apress.prospring5.ch7;
...
public class SingerDaoTest {

   private GenericApplicationContext ctx;
   private SingerDao singerDao;
   ...
   @Test
   public void testUpdate(){
      Singer singer = singerDao.findById(1L);
      //making sure such singer exists
      assertNotNull(singer);

      //making sure we got expected singer
      assertEquals("Mayer", singer.getLastName());

      //retrieve the album
      Album album = singer.getAlbums().stream().filter(
         a -> a.getTitle().equals("Battle Studies")).findFirst().get();

      singer.setFirstName("John Clayton");
      singer.removeAlbum(album);
      singerDao.save(singer);

      // test the update
      listSingersWithAlbum(singerDao.findAllWithAlbum());
   }
   ...
}
```

如前面的代码示例所示，首先检索 ID 为 1 的记录，并更改名字。然后循环遍历 Album 对象，检索标题为 *Battle Studies* 的专辑，并将其从歌手的 albums 属性中删除。最后，再次调用 singerDao.save()方法。运行该程序，将看到以下输出：

```
INFO o.h.h.i.QueryTranslatorFactoryInitiator - HHH000397:
   Using ASTQueryTranslatorFactory
...
INFO - Singer saved with id: 1
Hibernate:
   /* update
      com.apress.prospring5.ch7.entities.Album */ update
         album
      set
         RELEASE_DATE=?,
         SINGER_ID=?,
         title=?,
         VERSION=?
      where
         ID=?
         and VERSION=?
Hibernate:
   /* delete com.apress.prospring5.ch7.entities.Album */ delete
      from
         album
      where
         ID=?
```

```
        and VERSION=?
INFO ----- Listing singers with instruments:
INFO - Singer - Id: 1, First name: John Clayton, Last name: Mayer,
   Birthday: 1977-10-16
INFO - Album - Id: 1, Singer id: 1, Title: The Search For Everything,
   Release Date: 2017-01-20
INFO - Instrument: Guitar
INFO - Instrument: Piano
INFO - Singer - Id: 2, First name: Eric, Last name: Clapton,
   Birthday: 1945-03-30
INFO - Album - Id: 3, Singer id: 2, Title: From The Cradle ,
   Release Date: 1994-09-13
INFO - Instrument: Guitar
INFO - Singer - Id: 3, First name: John, Last name: Butler,
   Birthday: 1975-04-01
```

此时你会看到名字被更新，*Battle Studies* 专辑被删除。之所以可以删除专辑信息，是因为向一对多关联中传入了 orphanRemoval = true 属性，该属性会指示 Hibernate 删除那些数据库中虽然存在，但保存后却不在对象中的所有孤立记录。

7.7 删除数据

删除数据也很简单。只需要调用 Session.delete()方法并传入 Contact 对象即可。以下代码片段显示了用于删除数据的代码：

```
package com.apress.prospring5.ch7.dao;
...
@Transactional
@Repository("singerDao")
public class SingerDaoImpl implements SingerDao {

   private static final Log logger =
      LogFactory.getLog(SingerDaoImpl.class);
   private SessionFactory sessionFactory;

   public void delete(Singer singer) {
      sessionFactory.getCurrentSession().delete(singer);
      logger.info("Singer deleted with id: " + singer.getId());
   }
   ...
}
```

当在映射中定义了 cascade = CascadeType.ALL 时，删除操作将删除歌手记录及其所有相关信息，包括专辑和乐器信息。以下代码片段显示了用于测试删除方法 testDelete()的代码：

```
package com.apress.prospring5.ch7;
...
public class SingerDaoTest {
   private static Logger logger =
      LoggerFactory.getLogger(SingerDaoTest.class);
   private GenericApplicationContext ctx;
   private SingerDao singerDao;

   @Test
   public void testDelete(){
      Singer singer = singerDao.findById(2l);
      //making sure such singer exists
      assertNotNull(singer);
      singerDao.delete(singer);

      listSingersWithAlbum(singerDao.findAllWithAlbum());
   }
}
```

上述代码首先检索 ID 为 2 的歌手，然后调用删除方法以删除该歌手的信息。运行程序，将会产生如下输出：

```
INFO o.h.h.i.QueryTranslatorFactoryInitiator - HHH000397:
   Using ASTQueryTranslatorFactory
...
INFO c.a.p.c.d.SingerDaoImpl - Singer deleted with id: 2
Hibernate:
   /* delete collection com.apress.prospring5.ch7.entities.
      Singer.instruments */ delete
      from
```

```
        singer_instrument
    where
        SINGER_ID=?
Hibernate:
    /* delete com.apress.prospring5.ch7.entities.Album */ delete
        from
            album
        where
            ID=?
            and VERSION=?
Hibernate:
    /* delete com.apress.prospring5.ch7.entities.Singer */ delete
        from
            singer
        where
            ID=?
            and VERSION=?
INFO - ---- Listing singers with instruments:
INFO - Singer - Id: 1, First name: John, Last name: Mayer,
   Birthday: 1977-10-16
INFO - Album - Id: 1, Singer id: 1, Title: The Search For Everything,
   Release Date: 2017-01-20
INFO - Album - Id: 2, Singer id: 1, Title: Battle Studies,
   Release Date: 2009-11-17
INFO - Instrument: Piano
INFO - Instrument: Guitar
INFO - Singer - Id: 3, First name: John, Last name: Butler,
   Birthday: 1975-04-01
```

可以看到 ID 为 2 的歌手以及 ALBUM 和 SINGER_INSTRUMENT 表中的子记录都被删除了。

7.8 配置 Hibernate 以便从实体生成表

在使用了 Hibernate 的启动应用程序中，常见的行为是首先编写实体类，然后根据其内容生成数据库表。这通常是使用 Hibernate 属性 hibernate.hbm2ddl.auto 来完成的。当应用程序第一次启动时，该属性的值被设置为 create；这样就会让 Hibernate 根据使用 JPA 和 Hibernate 注解定义的关系扫描实体并生成表和键(主键、外键和唯一键)。

如果实体配置正确并且得到的数据库对象与预期完全一致，就应该将该属性的值更改为 update，从而告诉 Hibernate 使用实体上执行的任何更改来更新现有数据库，并保留原始数据库和插入到其中的任何数据。

在生产应用程序中，比较可行的办法是编写在伪数据库上运行的单元和集成测试，在执行完所有测试用例之后，就丢弃该测试数据库。通常，该测试数据库是内存数据库，可以告诉 Hibernate 先创建数据库，并在执行完测试后丢弃它，方法是将 hibernate.hbm2ddl.auto 的值设置为 create-drop。

可以在 Hibernate 官方文档中找到 hibernate.hbm2ddl.auto 属性值的完整列表[1]。

以下代码片段显示了 Java 配置类 AdvancedConfig。如你所见，引入了 hibernate.hbm2ddl.auto，并且使用的数据源是 DBCP 池化数据源(pooled data source)。

```
package com.apress.prospring5.ch7.config;

import com.apress.prospring5.ch6.CleanUp;
import org.apache.commons.dbcp2.BasicDataSource;
import org.hibernate.SessionFactory;
import org.slf4j.Logger;
import org.slf4j.LoggerFactory;
import org.springframework.beans.factory.annotation.Value;
import org.springframework.context.annotation.Bean;
import org.springframework.context.annotation.ComponentScan;
import org.springframework.context.annotation.Configuration;
import org.springframework.context.annotation.PropertySource;
import org.springframework.context.support.PropertySourcesPlaceholderConfigurer;
import org.springframework.jdbc.core.JdbcTemplate;
import org.springframework.orm.hibernate5.HibernateTransactionManager;
import org.springframework.orm.hibernate5.LocalSessionFactoryBuilder;
import org.springframework.transaction.PlatformTransactionManager;
import org.springframework.transaction.annotation.EnableTransactionManagement;

import javax.sql.DataSource;
import java.io.IOException;
```

1 参见 https://docs.jboss.org/hibernate/orm/5.0/manual/en-US/html/ch03.html 页面上的表 5-7。

```
import java.util.Properties;

@Configuration
@ComponentScan(basePackages = "com.apress.prospring5.ch7")
@EnableTransactionManagement
@PropertySource("classpath:db/jdbc.properties")
public class AdvancedConfig {
private static Logger logger =
    LoggerFactory.getLogger(AdvancedConfig.class);

@Value("${driverClassName}")
private String driverClassName;
@Value("${url}")
private String url;
@Value("${username}")
private String username;
@Value("${password}")
private String password;

@Bean
public static PropertySourcesPlaceholderConfigurer
        propertySourcesPlaceholderConfigurer() {
    return new PropertySourcesPlaceholderConfigurer();
}

@Bean(destroyMethod = "close")
public DataSource dataSource() {
    try {
        BasicDataSource dataSource = new BasicDataSource();
        dataSource.setDriverClassName(driverClassName);
        dataSource.setUrl(url);
        dataSource.setUsername(username);
        dataSource.setPassword(password);
        return dataSource;
    } catch (Exception e) {
        logger.error("DBCP DataSource bean cannot be created!", e);
        return null;
    }
}

private Properties hibernateProperties() {
    Properties hibernateProp = new Properties();
    hibernateProp.put("hibernate.dialect", "org.hibernate.dialect.H2Dialect");
    hibernateProp.put("hibernate.hbm2ddl.auto", "create-drop");
    hibernateProp.put("hibernate.format_sql", true);
    hibernateProp.put("hibernate.use_sql_comments", true);
    hibernateProp.put("hibernate.show_sql", true);
    hibernateProp.put("hibernate.max_fetch_depth", 3);
    hibernateProp.put("hibernate.jdbc.batch_size", 10);
    hibernateProp.put("hibernate.jdbc.fetch_size", 50);
    return hibernateProp;
}

@Bean
public SessionFactory sessionFactory() {
    return new LocalSessionFactoryBuilder(dataSource())
            .scanPackages("com.apress.prospring5.ch7.entities")
            .addProperties(hibernateProperties())
            .buildSessionFactory();
}

    @Bean public PlatformTransactionManager transactionManager()
            throws IOException {
        return new HibernateTransactionManager(sessionFactory());
    }
}
```

jdbc.properties 文件包含访问内存数据库所需的属性。

```
driverClassName=org.h2.Driver
url=jdbc:h2:musicdb
username=prospring5
password=prospring5
```

但在这种情况下，如何填充初始数据呢？可以使用 DatabasePopulator 实例，它是一个类似于 DbUnit 的库[1]，或

1 可以在 http://dbunit.sourceforge.net/上找到 DbUnit 官方站点。

者使用类似于 DbIntializer bean 的自定义填充 bean，如下所示：

```
package com.apress.prospring5.ch7.config;

import com.apress.prospring5.ch7.dao.InstrumentDao;
import com.apress.prospring5.ch7.dao.SingerDao;
import com.apress.prospring5.ch7.entities.Album;
import com.apress.prospring5.ch7.entities.Instrument;
import com.apress.prospring5.ch7.entities.Singer;
import org.slf4j.Logger;
import org.slf4j.LoggerFactory;
import org.springframework.beans.factory.annotation.Autowired;
import org.springframework.stereotype.Service;

import javax.annotation.PostConstruct;
import java.util.Date;
import java.util.GregorianCalendar;

@Service
public class DBInitializer {
   private Logger logger =
      LoggerFactory.getLogger(DBInitializer.class);

   @Autowired SingerDao singerDao;
   @Autowired InstrumentDao instrumentDao;

   @PostConstruct
   public void initDB(){
      logger.info("Starting database initialization...");

      Instrument guitar = new Instrument();
      guitar.setInstrumentId("Guitar");
      instrumentDao.save(guitar);
      ...
      Singer singer = new Singer();
      singer.setFirstName("John");
      singer.setLastName("Mayer");
      singer.setBirthDate(new Date(
              (new GregorianCalendar(1977, 9, 16)).getTime().getTime()));
      singer.addInstrument(guitar);
      singer.addInstrument(piano);

      Album album1 = new Album();
      album1.setTitle("The Search For Everything");
      album1.setReleaseDate(new java.sql.Date(
              (new GregorianCalendar(2017, 0, 20)).getTime().getTime()));
      singer.addAbum(album1);

      Album album2 = new Album();
      album2.setTitle("Battle Studies");
      album2.setReleaseDate(new java.sql.Date(
              (new GregorianCalendar(2009, 10, 17)).getTime().getTime()));
      singer.addAbum(album2);

      singerDao.save(singer);
      ...
      logger.info("Database initialization finished.");
   }
}
```

DbIntializer 只是一个简单的 bean，其中存储库被注入为依赖项，并且具有由注解@PostConstruct 定义的初始化方法，在该方法中创建对象并将其保存到数据库中。该 bean 使用@Service 注解进行标注，从而将其标记为提供初始化数据库内容服务的 bean。该 bean 将在创建 ApplicationContext 时创建，并且执行初始化方法，从而确保在使用上下文之前填充数据库。

使用 AdvancedConfig 配置类，以前运行的相同测试集都可以顺利通过。

7.9 注解方法或字段

在前面的示例中，实体在其 getter 方法上使用了 JPA 注解。但 JPA 注解也可以直接在字段上使用，主要有以下几个优点：

- 实体配置更清晰，主要位于字段部分，而不是散布在整个类中。显然，所编写的代码应该遵循整洁代码建议，将所有字段声明放在类的同一连续部分。
- 注解实体字段并不强制提供 setter/getter 方法。这对于@Version 注解字段来说非常有用，该字段不应该手动修改；只有 Hibernate 才能访问它。
- 注解字段允许在 setter 方法中执行额外的处理(例如，在从数据库中加载数据后对其进行加密/计算)。而属性访问的问题在于，在加载对象时也会调用 setter 方法。

互联网上有很多关于哪种方法更好的讨论。从性能角度来看，两者没有任何区别。最终使用哪种方法取决于开发者，因为在一些情况下，对注解访问器更有意义。但请记住，实际上数据库中保存了对象的状态，而对象的状态是由其字段的值定义的，而不是由访问器返回的值。这也意味着可以精确地按照数据库的方式重新创建对象。所以，从某种意义上说，在 getter 方法上设置注解可能会破坏封装性。

接下来，你会看到 Singer 实体类被重写为带注解的字段，并扩展了抽象类 AbstractEntity，该抽象类包含应用程序中对所有 Hibernate 实体类通用的两个字段：

```
// AbstractEntity.java
package com.apress.prospring5.ch7.entities;

import javax.persistence.*;
import java.io.Serializable;

@MappedSuperclass
public abstract class AbstractEntity implements Serializable {

   @Id
   @GeneratedValue(strategy = GenerationType.AUTO)
   @Column(updatable = false)
   protected Long id;

   @Version
   @Column(name = "VERSION")
   private int version;

   public Long getId() {
      return id;
   }

   public void setId(Long id) {
      this.id = id;
   }
}

//Singer.java
@Entity
@Table(name = "singer")
@NamedQueries({
      @NamedQuery(name=Singer.FIND_SINGER_BY_ID,
         query="select distinct s from Singer s " +
            "left join fetch s.albums a " +
            "left join fetch s.instruments i " +
            "where s.id = :id"),
      @NamedQuery(name=Singer.FIND_ALL_WITH_ALBUM,
         query="select distinct s from Singer s " +
            "left join fetch s.albums a " +
            "left join fetch s.instruments i")
})
public class Singer extends AbstractEntity {

   public static final String FIND_SINGER_BY_ID = "Singer.findById";
   public static final String FIND_ALL_WITH_ALBUM = "Singer.findAllWithAlbum";

   @Column(name = "FIRST_NAME")
   private String firstName;

   @Column(name = "LAST_NAME")
   private String lastName;

   @Temporal(TemporalType.DATE)
   @Column(name = "BIRTH_DATE")
   private Date birthDate;

   @OneToMany(mappedBy = "singer", cascade=CascadeType.ALL,
```

```
        orphanRemoval=true)
    private Set<Album> albums = new HashSet<>();

    @ManyToMany

    @JoinTable(name = "singer_instrument",
        joinColumns = @JoinColumn(name = "SINGER_ID"),
        inverseJoinColumns = @JoinColumn(name = "INSTRUMENT_ID"))
    private Set<Instrument> instruments = new HashSet<>();
    ...
}
```

7.10 使用 Hibernate 时的注意事项

如本章的示例所示，一旦正确定义所有的对象-关系映射、关联和查询，Hibernate 就可以提供一个环境，以便开发人员可以更专注于使用对象模型进行编程，而不是为每个操作编写 SQL 语句。在过去的几年中，Hibernate 一直在迅速发展，被 Java 开发人员广泛应用，作为开源社区和企业的数据访问层库。

但是，你需要牢记一些要点。首先，由于无法控制所生成的 SQL，因此在定义映射时应特别小心，尤其是关联以及获取策略。其次，应该观察由 Hibernate 生成的 SQL 语句，以确保所有的执行都符合预期的要求。

理解 Hibernate 如何管理会话的内部机制也很重要，特别是在批处理作业操作中。Hibernate 将在会话中托管被管理对象，并定期刷新和清除它们。设计不佳的数据访问逻辑可能会导致 Hibernate 频繁刷新会话并极大地影响性能。如果想要对查询进行绝对控制，可以使用本地查询，相关内容将在下一章中讨论。

最后，设置(批处理大小、获取大小等)在调整 Hibernate 性能方面起着重要作用。应该在会话工厂中定义它们，并在加载测试应用程序时调整它们以确定最佳值。

不管怎样，对于那些正在寻找一种 OO 方式来实现数据访问逻辑的 Java 开发人员来说，Hibernate 以及将在下一章中讨论的 JPA 支持都是自然而然的选择。

7.11 小结

在本章中，首先讨论了 Hibernate 的基本概念以及如何在 Spring 应用程序中对其进行配置。然后介绍了定义 ORM 映射的常用技巧，介绍了关联以及如何使用 HibernateTemplate 类来执行各种数据库操作。关于 Hibernate，本章只介绍了它的一小部分功能和特性。对于那些有兴趣在 Spring 中使用 Hibernate 的人，强烈建议学习 Hibernate 的标准文档。此外，许多书籍也都详细讨论了 Hibernate。推荐阅读由 Joseph Ottinger、Jeff Linwood 和 Dave Minter 编写的 *Beginning Hibernate: For Hibernate 5*(Apress 出版社，2016 出版)[1]以及由 Mike Keith 和 Merrick Schincariol 编写的 *Pro JPA 2*(Apress 出版社，2013 出版)[2]。在下一章中，将学习 JPA 以及如何在使用 Spring 时使用它。Hibernate 对 JPA 提供了极好的支持，在下一章将继续使用 Hibernate 作为示例的持久化提供程序。对于查询和更新操作，JPA 的行为与 Hibernate 相似。在下一章中，还将讨论一些高级主题，包括本地查询和标准查询以及如何使用 Hibernate 及其 JPA 支持。

---

1 从 Apress 官方网站下载电子书，网址为 http://apress.com/us/book/9781484223185。
2 从 Apress 官方网站下载电子书，网址为 http://apress.com/us/book/9781430249269。

第 8 章

在 Spring 中使用 JPA 2 进行数据访问

在前一章中，讨论了如何在使用 ORM 方法实现数据访问逻辑时使用 Hibernate 和 Spring，演示了如何在 Spring 的配置中配置 Hibernate 的 SessionFactory，以及如何使用 Session 接口进行各种数据访问操作。但是，这只是使用 Hibernate 的一种方式。在 Spring 应用程序中使用 Hibernate 的另一种方式是使用 Hibernate 作为标准 Java 持久化 API(JPA)的持久化提供程序。

Hibernate 的 POJO 映射及其强大的查询语言(HQL)取得了巨大的成功，也影响了 Java 世界中数据访问技术标准的发展。在 Hibernate 之后，JCP 开发了 Java 数据对象(JDO)标准，然后开发了 JPA。

在编写本书时，JPA 已经发展到 2.1 版本，并提供了标准化的概念，如 PersistenceContext、EntityManager 和 Java 持久化查询语言(JPQL)。这些标准化为开发人员在 JPA 持久化提供程序(比如 Hibernate、EclipseLink、Oracle TopLink 和 Apache OpenJPA)之间切换提供了一种方法。因此，大多数新的 JEE 应用程序都采用 JPA 作为数据访问层。

Spring 还对 JPA 提供了极好的支持。例如，提供了许多 EntityManagerFactoryBean 实现来启动 JPA 实体管理器，并支持前面提到的所有 JPA 提供程序。Spring Data 项目还提供了一个名为 Spring Data JPA 的子项目，它为在 Spring 应用程序中使用 JPA 提供了高级支持。Spring Data JPA 项目的主要功能包括存储库的概念和规范以及对查询域特定语言(QueryDSL)的支持。

本章将介绍如何在 Spring 中使用 JPA 2.1，并将 Hibernate 用作基础的持久化提供程序。首先学习如何使用 JPA 的 EntityManager 接口和 JPQL 来实现各种数据库操作。然后介绍 Spring Data JPA 如何进一步帮助简化 JPA 开发。最后讨论与 ORM 相关的高级主题，包括本地查询和标准查询。

具体来说，主要讨论以下主题：

- **Java 持久化 API(JPA)的核心概念**：介绍 JPA 的一些主要概念。
- **配置 JPA 实体管理器**：讨论 Spring 所支持的 EntityManagerFactory 的类型以及如何在 Spring 的 XML 配置中配置最常用的一种类型，即 LocalContainerEntityManagerFactoryBean。
- **数据操作**：演示如何在 JPA 中实现基本的数据库操作，这与使用 Hibernate 时的概念非常相似。
- **高级查询操作**：讨论如何在 JPA 中使用本地查询以及强类型标准 API 来实现更灵活的查询操作。
- **介绍 Spring 数据 Java 持久化 API(JPA)**：将讨论 Spring Data JPA 项目，并演示它如何帮助简化数据访问逻辑的开发。
- **跟踪实体更改和审计**：在数据库更新操作中，跟踪实体创建或上次更新的日期以及由谁进行更改是一项常见要求。此外，对于诸如客户之类的关键信息，通常需要使用历史表来存储实体的每个版本。本章将讨论 Spring Data JPA 和 Hibernate Envers(Hibernate Entity Versioning System)如何帮助简化这种逻辑的开发。

和 Hibernate 一样，JPA 支持在 XML 或 Java 注解中定义映射。本章重点介绍映射的注解类型，因为它的用法往往比 XML 样式更受欢迎。

8.1 JPA 2.1 介绍

与其他的 Java 规范请求(JSR)一样，JPA 2.1 规范(JSR-338)的目标是在 JSE 和 JEE 环境中对 ORM 编程模型进行标准化。它定义了 JPA 持久化提供程序应该实现的一组通用概念、注解、接口和其他服务。当按照 JPA 标准进行编程时，开发人员可以随意切换底层提供程序，就像切换到另一台符合 JEE 标准的应用程序服务器一样。

在 JPA 中，核心概念是 EntityManager 接口，它是来自 EntityManagerFactory 类型的工厂。EntityManager 的主要工作是维护一个持久化上下文，在该上下文中存储由其管理的所有实体实例。EntityManager 的配置被定义为一个持久化单元，并且在应用程序中可以有多个持久化单元。如果使用的是 Hibernate，那么可以像使用 Session 接口一样使用持久化上下文。同样，EntityManagerFactory 等同于 SessionFactory。在 Hibernate 中，托管实体存储在会话中，可以通过 Hibernate 的 SessionFactory 或 Session 接口直接与会话进行交互。但是，在 JPA 中，不能直接与持久化上下文进行交互。相反，需要依靠 EntityManager 来完成相关工作。

JPQL 与 HQL 类似，所以如果你之前使用过 HQL，那么应该很容易接受 JPQL。但是，在 JPA 2 中，引入了强类型的 Criteria API，它依赖映射实体的元数据来构造查询。鉴于此，任何错误将在编译时而不是运行时被发现。

关于 JPA 2 的详细讨论，推荐阅读由 Mike Keith 和 Merrick Schincariol 编写的 *Pro JPA 2*(Apress 出版社，2013 年出版)[1]。在本节中，将讨论 JPA 的基本概念、本章所使用的示例数据模型以及如何配置 Spring 的 ApplicationContext 来支持 JPA。

8.1.1 示例代码的示例数据模型

在本章中，将使用与第 7 章中相同的数据模型。但是，当讨论如何实现审计功能时，将添加几列数据和历史表以供演示。首先从上一章中使用的相同的数据库创建脚本开始。如果跳过了第 7 章，那么请查看 7.1 节“示例代码的示例数据模型”中介绍的数据模型，这将帮助你了解本章中的示例代码。

8.1.2 配置 JPA 的 EntityManagerFactory

正如本章前面提到的，为了在 Spring 中使用 JPA，需要配置 EntityManagerFactory，就像在 Hibernate 中使用的 SessionFactory 一样。Spring 支持三种类型的 EntityManagerFactory 配置。

第一种类型使用 LocalEntityManagerFactoryBean 类。这是最简单的一种，只需要持久化单元名称即可。但是，由于不支持 DataSource 的注入，因此无法参与全局事务，也就是说这种类型仅适用于简单的开发目的。

第二种类型用于 JEE 兼容的容器，其中应用程序服务器根据部署描述符中的信息启动 JPA 持久化单元。这样就允许 Spring 通过 JNDI 查找来查找实体管理器。以下配置代码片段描述了通过 JNDI 查找实体管理器所需的元素：

```
<beans ...>
    <jee:jndi-lookup id="prospring5Emf"
        jndi-name="persistence/prospring5PersistenceUnit"/>
</beans>
```

在 JPA 规范中，应该在配置文件 META-INF/persistence.xml 中定义持久化单元。但是，从 Spring 3.1 开始，增加了一项新功能，从而消除了这种要求；我们将在本章稍后介绍如何使用该功能。

第三种类型(也是本章最常用的)类型使用 LocalContainerEntityManagerFactoryBean 类，它支持 DataSource 的注入并可以参与本地和全局事务。以下配置代码片段显示了相应的 XML 配置文件(app-context-annotation.xml)：

```
<?xml version="1.0" encoding="UTF-8"?>
<beans xmlns="http://www.springframework.org/schema/beans"
    xmlns:xsi="http://www.w3.org/2001/XMLSchema-instance"
    xmlns:context="http://www.springframework.org/schema/context"
    xmlns:jdbc="http://www.springframework.org/schema/jdbc"
    xmlns:tx="http://www.springframework.org/schema/tx"
    xsi:schemaLocation="http://www.springframework.org/schema/jdbc
        http://www.springframework.org/schema/jdbc/spring-jdbc.xsd
        http://www.springframework.org/schema/beans
        http://www.springframework.org/schema/beans/spring-beans.xsd
        http://www.springframework.org/schema/tx
        http://www.springframework.org/schema/tx/spring-tx.xsd
        http://www.springframework.org/schema/context
        http://www.springframework.org/schema/context/spring-context.xsd">

    <jdbc:embedded-database id="dataSource" type="H2">
        <jdbc:script location="classpath:sql/schema.sql"/>
        <jdbc:script location="classpath:sql/test-data.sql"/>
</jdbc:embedded-database>

<bean id="transactionManager" class=
    "org.springframework.orm.jpa.JpaTransactionManager">
```

1 可从 www.apress.com/us/book/9781430249269 在线获取。

```
    <property name="entityManagerFactory" ref="emf"/>
</bean>
    <tx:annotation-driven transaction-manager="transactionManager" />

    <bean id="emf" class=
        "org.springframework.orm.jpa.LocalContainerEntityManagerFactoryBean">
        <property name="dataSource" ref="dataSource" />
        <property name="jpaVendorAdapter">
            <bean class=
            "org.springframework.orm.jpa.vendor.HibernateJpaVendorAdapter" />
        </property>
        <property name="packagesToScan" value="com.apress.prospring5.ch8.entities"/>
        <property name="jpaProperties">
            <props>
        <prop key="hibernate.dialect">
    org.hibernate.dialect.H2Dialect
        </prop>
    <prop key="hibernate.max_fetch_depth">3</prop>
    <prop key="hibernate.jdbc.fetch_size">50</prop>
    <prop key="hibernate.jdbc.batch_size">10</prop>
    <prop key="hibernate.show_sql">true</prop>
            </props>
        </property>
    </bean>

    <context:component-scan base-package="com.apress.prospring5.ch8" />
</beans>
```

你可能希望使用 Java 配置类进行等效配置，如下所示：

```
package com.apress.prospring5.ch8.config;

import org.slf4j.Logger;
import org.slf4j.LoggerFactory;
import org.springframework.context.annotation.Bean;
import org.springframework.context.annotation.ComponentScan;
import org.springframework.context.annotation.Configuration;
import org.springframework.jdbc.datasource.embedded.EmbeddedDatabaseBuilder;
import org.springframework.jdbc.datasource.embedded.EmbeddedDatabaseType;
import org.springframework.orm.jpa.JpaTransactionManager;
import org.springframework.orm.jpa.JpaVendorAdapter;
import org.springframework.orm.jpa.LocalContainerEntityManagerFactoryBean;
import org.springframework.orm.jpa.vendor.HibernateJpaVendorAdapter;
import org.springframework.transaction.PlatformTransactionManager;
import org.springframework.transaction.annotation.EnableTransactionManagement;

import javax.persistence.EntityManagerFactory;
import javax.sql.DataSource;
import java.util.Properties;

@Configuration
@EnableTransactionManagement
@ComponentScan(basePackages = {"com.apress.prospring5.ch8.service"})
public class JpaConfig {

    private static Logger logger = LoggerFactory.getLogger(JpaConfig.class);

    @Bean
    public DataSource dataSource() {
        try {
            EmbeddedDatabaseBuilder dbBuilder =
        new EmbeddedDatabaseBuilder();
            return dbBuilder.setType(EmbeddedDatabaseType.H2)
        .addScripts("classpath:db/schema.sql", "classpath:db/test-data.sql").build();
            } catch (Exception e) {
                logger.error("Embedded DataSource bean cannot be created!", e);
                return null;
        }
    }

    @Bean
    public PlatformTransactionManager transactionManager() {
        return new JpaTransactionManager(entityManagerFactory());
    }

    @Bean
    public JpaVendorAdapter jpaVendorAdapter() {
        return new HibernateJpaVendorAdapter();
```

```
    }

    @Bean
    public Properties hibernateProperties() {
        Properties hibernateProp = new Properties();
        hibernateProp.put("hibernate.dialect", "org.hibernate.dialect.H2Dialect");
        hibernateProp.put("hibernate.format_sql", true);
        hibernateProp.put("hibernate.use_sql_comments", true);
        hibernateProp.put("hibernate.show_sql", true);
        hibernateProp.put("hibernate.max_fetch_depth", 3);
        hibernateProp.put("hibernate.jdbc.batch_size", 10);
        hibernateProp.put("hibernate.jdbc.fetch_size", 50);
        return hibernateProp;
    }

    @Bean
    public EntityManagerFactory entityManagerFactory() {
        LocalContainerEntityManagerFactoryBean factoryBean =
            new LocalContainerEntityManagerFactoryBean();
        factoryBean.setPackagesToScan("com.apress.prospring5.ch8.entities");
        factoryBean.setDataSource(dataSource());
        factoryBean.setJpaVendorAdapter(new HibernateJpaVendorAdapter());
        factoryBean.setJpaProperties(hibernateProperties());
        factoryBean.setJpaVendorAdapter(jpaVendorAdapter());
        factoryBean.afterPropertiesSet();
        return factoryBean.getNativeEntityManagerFactory();
    }
}
```

在上面的配置中，为了能够使用 Hibernate 作为持久化提供程序，声明了几个 bean 来支持 LocalContainerEntityManagerFactoryBean 的配置。主要配置如下。

- **dataSource bean**：使用 H2 声明一个嵌入式数据库作为数据源。因为是嵌入式数据库，所以数据库名称不是必需的。
- **transactionManager bean**：EntityManagerFactory需要一个事务管理器来访问事务数据。Spring为JPA提供了一个事务管理器(org.springframework.orm.jpa.JpaTransactionManager)。所声明的bean的ID为transactionManager。我们将在第9章详细讨论事务。此外，还声明了标记<tx:annotation-driven>，以便支持使用注解声明事务划界的需求。它的等价注解是@EnableTransactionManagement，该注解必须放在用@Configuration注解的类上。
- **组件扫描**：对该标记你应该非常熟悉了。它指示 Spring 扫描包 com.apress.prospring5.ch8 中的组件。
- **JPA EntityManagerFactory bean**：emf bean 是最重要的部分。首先，声明该 bean 使用 LocalContainerEntityManagerFactoryBean。在该 bean 中，提供了几个属性。首先，正如你可能预料的那样，需要注入 dataSource bean。其次，使用类 HibernateJpaVendorAdapter 来配置属性 jpaVendorAdapter，因为所使用的是 Hibernate。然后，指示实体工厂扫描包 com.apress.prospring5.ch8(由<property name = "packagesToScan">标记指定)中带有 ORM 注解的域对象。请注意，此功能自 Spring 3.1 起才可用，并且在域类扫描的支持下，可以跳过 META-INF/persistence.xml 文件中持久化单元的定义。最后，jpaProperties 属性提供了持久化提供程序 Hibernate 的配置细节。相关的配置选项与第 7 章中使用的配置选项相同，因此不再做详细介绍。

8.1.3 使用 JPA 注解进行 ORM 映射

Hibernate 在许多方面都影响了 JPA 的设计。对于映射注解，它们非常接近，以至于在第 7 章中用于将域对象映射到数据库的注解与在 JPA 中使用的注解是相同的。如果仔细查看一下第 7 章中域类的源代码，就会看到所有映射注解都位于 javax.persistence 包中，这意味着这些注解已经与 JPA 兼容。

一旦正确配置了 EntityManagerFactory，将其注入到自己的类中就很简单了。以下代码片段显示了 SingerServiceImpl 类的代码，我们将用它作为使用 JPA 执行数据库操作的示例：

```
package com.apress.prospring5.ch8.service;

import com.apress.prospring5.ch8.entities.Singer;
import org.slf4j.Logger;
import org.slf4j.LoggerFactory;
import org.springframework.stereotype.Service;
import org.springframework.stereotype.Repository;
import org.springframework.transaction.annotation.Transactional;
```

```
import org.apache.commons.lang3.NotImplementedException;
import java.util.List;

import javax.persistence.PersistenceContext;
import javax.persistence.EntityManager;
import javax.persistence.TypedQuery;

import org.apache.commons.logging.Log;
import org.apache.commons.logging.LogFactory;

@Service("jpaSingerService")
@Repository
@Transactional
public class SingerServiceImpl implements SingerService {
   final static String ALL_SINGER_NATIVE_QUERY =
      "select id, first_name, last_name, birth_date, version from singer";

   private static Logger logger =
      LoggerFactory.getLogger(SingerServiceImpl.class);

   @PersistenceContext
   private EntityManager em;

   @Transactional(readOnly=true)
   @Override
   public List<Singer> findAll() {
      throw new NotImplementedException("findAll");
   }

   @Transactional(readOnly=true)
   @Override
   public List<Singer> findAllWithAlbum() {
      throw new NotImplementedException("findAllWithAlbum");
   }

   @Transactional(readOnly=true)
   @Override
   public Singer findById(Long id) {
      throw new NotImplementedException("findById");
   }

   @Override
   public Singer save(Singer singer) {
      throw new NotImplementedException("save");
   }

   @Override
   public void delete(Singer singer) {
      throw new NotImplementedException("delete");
   }

   @Transactional(readOnly=true)
   @Override
   public List<Singer> findAllByNativeQuery() {
      throw new NotImplementedException("findAllByNativeQuery");
   }
}
```

多个注解被应用于该类。@Service 注解用于将类标识为 Spring 组件，该组件为其他层提供业务服务，Spring bean 的名称为 jpaSingerService。@Repository 注解表明该类包含数据访问逻辑，并指示 Spring 将特定于供应商的异常转换为 Spring 的 DataAccessException 层次结构。正如你所熟悉的那样，@Transactional 注解用于定义事务需求。

为了注入到 EntityManager，需要使用@PersistenceContext 注解，这是标准的用于实体管理器注入的 JPA 注解。此时，你可能会问为什么使用名称@PersistenceContext 来注入实体管理器，但如果思考一下持久化上下文本身是由 EntityManager 管理的，注解命名就非常有意义了。如果应用程序中有多个持久化单元，那么可以将 unitName 属性添加到注解中以指定要注入到哪个持久化单元。通常，持久化单元表示独立的后端 DataSource。

8.2 使用 JPA 执行数据库操作

本节介绍如何在 JPA 中执行数据库操作。以下代码片段显示了 SingerService 接口，它表明将提供的歌手信息服务：

```
package com.apress.prospring5.ch8.service;
```

```
import com.apress.prospring5.ch8.entities.Singer;

import java.util.List;

public interface SingerService {
   List<Singer> findAll();
   List<Singer> findAllWithAlbum();
   Singer findById(Long id);
   Singer save(Singer singer);
   void delete(Singer singer);
   List<Singer> findAllByNativeQuery();
}
```

该接口非常简单；里面三个查找方法，一个保存方法和一个删除方法。其中保存方法save()用于插入和更新操作。

8.2.1 使用 Java 持久化查询语言来查询数据

JPQL 和 HQL 的语法相似，事实上，可以重用第 7 章中所有的 HQL 查询来实现 SingerService 接口中的三个查找方法。为了使用 JPA 和 Hibernate，需要将以下依赖项添加到项目中：

```
//pro-spring-15/build.gradle
ext {
   hibernateVersion = '5.2.10.Final'
   hibernateJpaVersion = '1.0.0.Final'
   ..
   hibernate = [
      em :
         "org.hibernate:hibernate-entitymanager:$hibernateVersion",
      jpaApi :
         "org.hibernate.javax.persistence:hibernate-jpa-2.1-api:$hibernateJpaVersion"
   ]
}

//chapter08.gradle
dependencies {
      //we specify these dependencies for all submodules,
         //except the boot module, that defines its own
      if !project.name.contains"boot" {
         compile spring.contextSupport, spring.orm, spring.context,
      misc.slf4jJcl, misc.logback, db.h2, misc.lang3,
   hibernate.em, hibernate.jpaApi
      }
      testCompile testing.junit
}
```

以下代码片段是第 7 章中 Singer 域对象模型类的代码：

```
//Singer.java
package com.apress.prospring5.ch8.entities;

import static javax.persistence.GenerationType.IDENTITY;

import java.io.Serializable;
import java.util.Date;
import java.util.HashSet;
import java.util.Set;
import javax.persistence.Entity;
import javax.persistence.Table;
import javax.persistence.Id;
import javax.persistence.GeneratedValue;
import javax.persistence.Column;
import javax.persistence.Version;
import javax.persistence.Temporal;
import javax.persistence.TemporalType;
import javax.persistence.OneToMany;
import javax.persistence.ManyToMany;
import javax.persistence.JoinTable;
import javax.persistence.JoinColumn;
import javax.persistence.CascadeType;
import javax.persistence.NamedQueries;
import javax.persistence.NamedQuery;
import javax.persistence.SqlResultSetMapping;
import javax.persistence.EntityResult;

@Entity
```

```
@Table(name = "singer")
@NamedQueries({
      @NamedQuery(name=Singer.FIND_ALL, query="select s from Singer s"),
      @NamedQuery(name=Singer.FIND_SINGER_BY_ID,
   query="select distinct s from Singer s " +
   "left join fetch s.albums a " +
   "left join fetch s.instruments i " +
   "where s.id = :id"),
      @NamedQuery(name=Singer.FIND_ALL_WITH_ALBUM,
   query="select distinct s from Singer s " +
   "left join fetch s.albums a " +
   "left join fetch s.instruments i")
})
@SqlResultSetMapping(
   name="singerResult",
   entities=@EntityResult(entityClass=Singer.class)
)
public class Singer implements Serializable {

   public static final String FIND_ALL = "Singer.findAll";
   public static final String FIND_SINGER_BY_ID = "Singer.findById";
   public static final String FIND_ALL_WITH_ALBUM = "Singer.findAllWithAlbum";

   @Id
   @GeneratedValue(strategy = IDENTITY)
   @Column(name = "ID")
   private Long id;

   @Version
   @Column(name = "VERSION")
   private int version;

   @Column(name = "FIRST_NAME")
   private String firstName;

   @Column(name = "LAST_NAME")
   private String lastName;

   @Temporal(TemporalType.DATE)
   @Column(name = "BIRTH_DATE")
   private Date birthDate;

   @OneToMany(mappedBy = "singer", cascade=CascadeType.ALL,
      orphanRemoval=true)
   private Set<Album> albums = new HashSet<>();

   @ManyToMany
   @JoinTable(name = "singer_instrument",
      joinColumns = @JoinColumn(name = "SINGER_ID"),
      inverseJoinColumns = @JoinColumn(name = "INSTRUMENT_ID"))
   private Set<Instrument> instruments = new HashSet<>();
   //setters and getters

   @Override
   public String toString() {
      return "Singer - Id: " + id + ", First name: " + firstName
         + ", Last name: " + lastName + ", Birthday: " + birthDate;
   }
}
// Album.java
package com.apress.prospring5.ch8.entities;

import static javax.persistence.GenerationType.IDENTITY;

import java.io.Serializable;
import java.text.SimpleDateFormat;
import java.util.Date;
import javax.persistence.*;

@Entity
@Table(name = "album")
public class Album implements Serializable {
   @Id
   @GeneratedValue(strategy = IDENTITY)
   @Column(name = "ID")
```

```
    private Long id;

    @Version
    @Column(name = "VERSION")
    private int version;

    @Column
    private String title;

    @Temporal(TemporalType.DATE)
    @Column(name = "RELEASE_DATE")

    private Date releaseDate;

    @ManyToOne
    @JoinColumn(name = "SINGER_ID")
    private Singer singer;

    public Album() {
    //needed byJPA
    }

    public Album(String title, Date releaseDate) {
        this.title = title;
        this.releaseDate = releaseDate;
    }
    //setters and getters
}
//Instrument.java
package com.apress.prospring5.ch8.entities;

import java.io.Serializable;
import javax.persistence.Entity;
import javax.persistence.Table;
import javax.persistence.Column;
import javax.persistence.Id;
import javax.persistence.ManyToMany;
import javax.persistence.JoinTable;
import javax.persistence.JoinColumn;
import java.util.Set;
import java.util.HashSet;

@Entity
@Table(name = "instrument")
public class Instrument implements Serializable {
    @Id
    @Column(name = "INSTRUMENT_ID")
    private String instrumentId;

    @ManyToMany
    @JoinTable(name = "singer_instrument",
        joinColumns = @JoinColumn(name = "INSTRUMENT_ID"),
        inverseJoinColumns = @JoinColumn(name = "SINGER_ID"))
    private Set<Singer> singers = new HashSet<>();

    //setters and getters
}
```

如果分析一下使用@NamedQuery 定义的查询，你会发现 HQL 和 JPQL 似乎没有区别。首先从 findAll()方法开始，它只是从数据库中检索所有歌手。

```
package com.apress.prospring5.ch8.service;
...
@Service("jpaSingerService")
@Repository
@Transactional
public class SingerServiceImpl implements SingerService {
    final static String ALL_SINGER_NATIVE_QUERY =
        "select id, first_name, last_name, birth_date, version from singer";

    private static Logger logger =
        LoggerFactory.getLogger(SingerServiceImpl.class);

    @PersistenceContext
    private EntityManager em;
```

```
    @Transactional(readOnly=true)
    @Override
    public List<Singer> findAll() {
        return em.createNamedQuery(Singer.FIND_ALL, Singer.class)
            .getResultList();
    }
    ...
}
```

如上述代码清单所示，我们使用了 EntityManager.createNamedQuery()方法，并传入查询的名称和预期的返回类型。此时，EntityManager 将返回一个 TypedQuery <X>接口。然后调用 TypedQuery.getResultList()方法来检索歌手。为了测试该方法的实现，将使用一个测试类，在该类中，包含针对每个将要实现的 JPA 方法的测试方法。

```
package com.apress.prospring5.ch8;

import com.apress.prospring5.ch8.config.JpaConfig;
import com.apress.prospring5.ch8.entities.Singer;
import com.apress.prospring5.ch8.service.SingerService;
import org.junit.After;
import org.junit.Before;
import org.junit.Test;
import org.slf4j.Logger;
import org.slf4j.LoggerFactory;
import org.springframework.context.annotation.AnnotationConfigApplicationContext;
import org.springframework.context.support.GenericApplicationContext;

import java.util.List;

import static org.junit.Assert.assertEquals;
import static org.junit.Assert.assertNotNull;

public class SingerJPATest {
    private static Logger logger = LoggerFactory.getLogger(SingerJPATest.class);

    private GenericApplicationContext ctx;
    private SingerService singerService;

    @Before
    public void setUp(){
        ctx = new AnnotationConfigApplicationContext(JpaConfig.class);
        singerService = ctx.getBean(SingerService.class);
        assertNotNull(singerService);
    }

    @Test
    public void testFindAll(){
        List<Singer> singers = singerService.findAll();
        assertEquals(3, singers.size());
        listSingers(singers);
    }

    private static void listSingers(List<Singer> singers) {
        logger.info(" ---- Listing singers:");
        for (Singer singer : singers) {
            logger.info(singer.toString());
        }
    }

    @After
    public void tearDown(){
        ctx.close();
    }
}
```

如果 assertEquals 没有抛出异常(即测试失败)，运行 testFindAll()测试方法将会生成以下输出：

```
---- Listing singers:
Singer - Id: 1, First name: John, Last name: Mayer, Birthday: 1977-10-16
Singer - Id: 2, First name: Eric, Last name: Clapton, Birthday: 1945-03-30
Singer - Id: 3, First name: John, Last name: Butler, Birthday: 1975-04-01
```

对于关联，JPA 规范指出，默认情况下，持久化提供程序必须立即获取关联。但是，对于 Hibernate 的 JPA 实现，默认的获取策略仍然是延迟获取。所以，当使用 Hibernate 的 JPA 实现时，不需要显式地将关联定义为延迟获取。Hibernate 的默认获取策略与 JPA 规范不同。

接下来实现 findAllWithAlbum()方法，该方法将获取所有相关的专辑和乐器信息，实现如下所示：

```
package com.apress.prospring5.ch8.service;
...
@Service("jpaSingerService")
@Repository
@Transactional
public class SingerServiceImpl implements SingerService {
   final static String ALL_SINGER_NATIVE_QUERY =
      "select id, first_name, last_name, birth_date, version from singer";

   private static Logger logger =
      LoggerFactory.getLogger(SingerServiceImpl.class);

   @PersistenceContext
   private EntityManager em;

   @Transactional(readOnly=true)
   @Override
   public List<Singer> findAllWithAlbum() {
      List<Singer> singers = em.createNamedQuery
          (Singer.FIND_ALL_WITH_ALBUM, Singer.class).getResultList();
      return singers;
   }
   ...
}
```

findAllWithAlbum()方法与 findAll()方法相同，但前者使用了一个不同的命名查询，并启用了 left join fetch。下面所示的方法用于测试 findAllWithAlbum()并打印相关条目：

```
package com.apress.prospring5.ch8;
...
public class SingerJPATest {
   private static Logger logger =
      LoggerFactory.getLogger(SingerJPATest.class);

   private GenericApplicationContext ctx;
   private SingerService singerService;

   @Before
   public void setUp(){
      ctx = new AnnotationConfigApplicationContext(JpaConfig.class);
      singerService = ctx.getBean(SingerService.class);
      assertNotNull(singerService);
   }

   @Test
   public void testFindAllWithAlbum(){
      List<Singer> singers = singerService.findAllWithAlbum();
      assertEquals(3, singers.size());
      listSingersWithAlbum(singers);
   }

   private static void listSingersWithAlbum(List<Singer> singers) {
      logger.info(" ---- Listing singers with instruments:");
      for (Singer singer : singers) {
         logger.info(singer.toString());
         if (singer.getAlbums() != null) {
      for (Album album :
   singer.getAlbums()) {
      logger.info("\t" + album.toString());
   }
      }
      if (singer.getInstruments() != null) {
   for (Instrument instrument : singer.getInstruments()) {
      logger.info("\tInstrument: " + instrument.getInstrumentId());
   }
         }
      }
   }
   @After
   public void tearDown(){
      ctx.close();
   }
}
```

如果 assertEquals 没有抛出异常(即测试失败)，那么运行 testFindAllWithAlbum()测试方法将生成以下输出：

```
INFO o.h.h.i.QueryTranslatorFactoryInitiator - HHH000397:
   Using ASTQueryTranslatorFactory
Hibernate:
   /* Singer.findAllWithAlbum */ select
      distinct singer0_.ID as ID1_2_0_,
      albums1_.ID as ID1_0_1_,
      instrument3_.INSTRUMENT_ID as INSTRUME1_1_2_,
      singer0_.BIRTH_DATE as BIRTH_DA2_2_0_,
      singer0_.FIRST_NAME as FIRST_NA3_2_0_,
      singer0_.LAST_NAME as LAST_NAM4_2_0_,
      singer0_.VERSION as VERSION5_2_0_,
      albums1_.RELEASE_DATE as RELEASE_2_0_1_,
      albums1_.SINGER_ID as SINGER_I5_0_1_,
      albums1_.title as title3_0_1_,
      albums1_.VERSION as VERSION4_0_1_,
      albums1_.SINGER_ID as SINGER_I5_0_0__,
      albums1_.ID as ID1_0_0_ ,
      instrument2_.SINGER_ID as SINGER_I1_3_1__,
      instrument2_.INSTRUMENT_ID as INSTRUME2_3_1__
   from
      singer singer0_
   left outer join
      album albums1_
         on singer0_.ID=albums1_.SINGER_ID
   left outer join
      singer_instrument instrument2_
         on singer0_.ID=instrument2_.SINGER_ID
   left outer join
      instrument instrument3_
         on instrument2_.INSTRUMENT_ID=instrument3_.INSTRUMENT_ID
INFO ----- Listing singers with instruments:
INFO - Singer - Id: 1, First name: John, Last name: Mayer, Birthday: 1977-10-16
INFO - Album - id: 2, Singer id: 1, Title: Battle Studies,
   Release Date: 2009-11-17
INFO - Album - id: 1, Singer id: 1, Title: The Search For Everything,
   Release Date: 2017-01-20
INFO - Instrument: Guitar
INFO - Instrument: Piano
INFO - Singer - Id: 3, First name: John, Last name: Butler, Birthday: 1975-04-01
INFO - Singer - Id: 2, First name: Eric, Last name: Clapton, Birthday: 1945-03-30
INFO - Album - id: 3, Singer id: 2, Title: From The Cradle ,
   Release Date: 1994-09-13
INFO - Instrument: Guitar
```

如果为 Hibernate 启用日志记录功能，那么还可以看到用来从数据库中提取所有数据的本地查询。

现在来看看 findById()方法，该方法演示了如何在 JPA 中使用具有命名参数的命名查询。关联也会被提取。以下代码片段显示了实现过程：

```
package com.apress.prospring5.ch8.service;
...
@Service("jpaSingerService")
@Repository
@Transactional
public class SingerServiceImpl implements SingerService {
   final static String ALL_SINGER_NATIVE_QUERY =
      "select id, first_name, last_name, birth_date, version from singer";

   private static Logger logger =
      LoggerFactory.getLogger(SingerServiceImpl.class);

   @PersistenceContext
   private EntityManager em;

   @Transactional(readOnly=true)
   @Override
   public Singer findById(Long id) {
      TypedQuery<Singer> query = em.createNamedQuery
          (Singer.FIND_SINGER_BY_ID, Singer.class);

      query.setParameter("id", id);
      return query.getSingleResult();

   }
   ...
}
```

调用 EntityManager.createNamedQuery(java.lang.String name,java.lang.Class <T>resultClass)以获取 TypedQuery <T>

接口的实例，该实例确保查询结果必须是 Singer 类型。然后使用 TypedQuery<T>.setParameter()方法设置查询中命名参数的值并调用 getSingleResult()方法，最终结果应该只包含具有指定 ID 的单个 Singer 对象。我们把对该方法的测试作为一项练习留给读者来完成。

```
package com.apress.prospring5.ch8.service;
...
@Service("jpaSingerService")
@Repository
@Transactional
public class SingerServiceImpl implements SingerService {
    final static String ALL_SINGER_NATIVE_QUERY =
        "select id, first_name, last_name, birth_date, version from singer";
    private static Logger logger =
        LoggerFactory.getLogger(SingerServiceImpl.class);

    @PersistenceContext
    private EntityManager em;

    @Transactional(readOnly=true)
    @Override
    public Singer findById(Long id) {
        TypedQuery<Singer> query = em.createNamedQuery
        (Singer.FIND_SINGER_BY_ID, Singer.class);
        query.setParameter("id", id);
        return query.getSingleResult();
    }   ...
}
```

8.2.2 查询非类型化结果

在大多数情况下，都希望向数据库提交查询并随意操作结果，而不是将它们存储在映射的实体类中。一个典型的示例是基于网络的报告，它可以列出跨多个表格的一定数量的列。例如，假设有一个用来显示歌手信息和最近发布的专辑名称的网页。摘要信息将包含歌手的完整姓名及其最近发布的专辑目的。没有专辑的歌手将不会被列出。在这种情况下，可以通过一个查询并手动操作 ResultSet 对象来实现此目的。

创建一个名为 SingerSummaryUntypeImpl 的新类并命名方法 displayAllSingerSummary()。以下代码片段显示了该方法的典型实现：

```
package com.apress.prospring5.ch8.service;

import org.springframework.stereotype.Repository;
import org.springframework.stereotype.Service;
import org.springframework.transaction.annotation.Transactional;

import javax.persistence.EntityManager;
import javax.persistence.PersistenceContext;
import java.util.Iterator;
import java.util.List;

@Service("singerSummaryUntype")
@Repository
@Transactional
public class SingerSummaryUntypeImpl {

    @PersistenceContext
    private EntityManager em;

    @Transactional(readOnly = true)
    public void displayAllSingerSummary() {
        List result = em.createQuery(
        "select s.firstName, s.lastName, a.title from Singer s "
        + "left join s.albums a "
        + "where a.releaseDate=(select max(a2.releaseDate) "
        + "from Album a2 where a2.singer.id = s.id)")
        .getResultList();
        int count = 0;
        for (Iterator i = result.iterator(); i.hasNext(); ) {
    Object[] values = (Object[]) i.next();
    System.out.println(++count + ": " + values[0] + ", "
    + values[1] + ", " + values[2]);
        }
    }
}
```

如前面的代码示例所示，首先使用 EntityManager.createQuery()方法创建查询，并传入 JPQL 语句，然后获取结果列表。

当在 JPQL 中明确指定要选择的列时，JPA 将返回一个迭代器，该迭代器中的每一项都是一个对象数组。遍历该迭代器，并显示对象数组中每个元素的值。每个对象数组都对应 ResultSet 对象中的一条记录。以下代码片段显示了测试程序：

```
package com.apress.prospring5.ch8;

import com.apress.prospring5.ch8.config.JpaConfig;
import com.apress.prospring5.ch8.service.SingerSummaryUntypeImpl;
import org.junit.After;
import org.junit.Before;
import org.junit.Test;
import org.slf4j.Logger;
import org.slf4j.LoggerFactory;
import org.springframework.context.annotation.AnnotationConfigApplicationContext;
import org.springframework.context.support.GenericApplicationContext;

import java.util.List;

import static org.junit.Assert.assertEquals;
import static org.junit.Assert.assertNotNull;

public class SingerSummaryJPATest {

    private static Logger logger =
        LoggerFactory.getLogger(SingerSummaryJPATest.class);
    private GenericApplicationContext ctx;
    private SingerSummaryUntypeImpl singerSummaryUntype;

    @Before
    public void setUp() {
        ctx = new AnnotationConfigApplicationContext(JpaConfig.class);
        singerSummaryUntype = ctx.getBean(SingerSummaryUntypeImpl.class);
        assertNotNull(singerSummaryUntype);
    }
    @Test
    public void testFindAllUntype() {
        singerSummaryUntype.displayAllSingerSummary();
    }

    @After
    public void tearDown() {
        ctx.close();
    }
}
```

运行测试程序会产生以下输出：

```
1: John, Mayer, The Search For Everything
2: Eric, Clapton, From The Cradle
```

在 JPA 中，有一个更优雅的解决方案，而不仅仅是简单操作查询返回的对象数组，相关内容将在下一节讨论。

8.3 使用构造函数表达式查询自定义结果类型

在 JPA 中，当查询像上一节那样的自定义结果时，可以指示 JPA 直接从每条记录构建一个 POJO。此 POJO 又称为*视图*，因为它包含来自多个表的数据。对于上一节中的示例，可以创建一个名为 SingerSummary 的 POJO，它存储了歌手摘要的查询结果。以下代码片段显示了该类：

```
package com.apress.prospring5.ch8.view;

import java.io.Serializable;

public class SingerSummary implements Serializable {
    private String firstName;
    private String lastName;
    private String latestAlbum;

    public SingerSummary(String firstName, String lastName,
    String latestAlbum) {
```

```
        this.firstName = firstName;
        this.lastName = lastName;
        this.latestAlbum = latestAlbum;
    }

    public String getFirstName() {
        return firstName;
    }

    public String getLastName() {
        return lastName;
    }

    public String getLatestAlbum() {
        return latestAlbum;
    }

    public String toString() {
        return "First name: " + firstName + ", Last Name: " + lastName
    + ", Most Recent Album: " + latestAlbum;
    }
}
```

上面所示的 SingerSummary 类具有每个歌手摘要的属性，并带有一个接收所有属性的构造函数。定义了 SingerSummary 类之后，可以修改 findAll()方法并在查询中使用构造函数表达式来指示 JPA 提供程序将 ResultSet 映射到 SingerSummary 类。首先为 SingerSummary 服务创建一个接口。以下代码片段显示了该接口：

```
package com.apress.prospring5.ch8.service;

import com.apress.prospring5.ch8.view.SingerSummary;

import java.util.List;

public interface SingerSummaryService {
    List<SingerSummary> findAll();
}
```

现在，可以查看 SingerSummaryImpl.findAll()方法的实现，其中针对 ResultSet 映射使用了构造函数表达式：

```
package com.apress.prospring5.ch8.service;

import com.apress.prospring5.ch8.view.SingerSummary;
import org.springframework.stereotype.Repository;
import org.springframework.stereotype.Service;
import org.springframework.transaction.annotation.Transactional;

import javax.persistence.EntityManager;
import javax.persistence.PersistenceContext;
import java.util.List;

@Service("singerSummaryService")
@Repository
@Transactional
public class SingerSummaryServiceImpl implements SingerSummaryService {

    @PersistenceContext
    private EntityManager em;

    @Transactional(readOnly = true)
    @Override
    public List<SingerSummary> findAll() {
        List<SingerSummary> result = em.createQuery(
        "select new com.apress.prospring5.ch8.view.SingerSummary("
        + "s.firstName, s.lastName, a.title) from Singer s "
        + "left join s.albums a "
        + "where a.releaseDate=(select max(a2.releaseDate):
        + "from Album a2 where a2.singer.id = s.id)",
        SingerSummary.class).getResultList();
        return result;
    }
}
```

在 JPQL 语句中，指定了 new 关键字以及 POJO 类的完全限定名称，POJO 类将存储结果并传入所选择的属性作为每个 SingerSummary 类的构造函数参数。最后，将 SingerSummary 类传递给 createQuery()方法以指示结果类型。以下代码片段显示了测试程序：

```
package com.apress.prospring5.ch8;
```

```
import com.apress.prospring5.ch8.config.JpaConfig;
import com.apress.prospring5.ch8.service.SingerSummaryService;
import com.apress.prospring5.ch8.view.SingerSummary;
import org.junit.After;
import org.junit.Before;
import org.junit.Test;
import org.slf4j.Logger;
import org.slf4j.LoggerFactory;
import org.springframework.context.annotation.AnnotationConfigApplicationContext;
import org.springframework.context.support.GenericApplicationContext;

import java.util.List;

import static org.junit.Assert.assertEquals;
import static org.junit.Assert.assertNotNull;

public class SingerSummaryJPATest {

    private static Logger logger =
        LoggerFactory.getLogger(SingerSummaryJPATest.class);
    private GenericApplicationContext ctx;
    private SingerSummaryService singerSummaryService;

    @Before
    public void setUp() {
        ctx = new AnnotationConfigApplicationContext(JpaConfig.class);
        singerSummaryService = ctx.getBean(SingerSummaryService.class);
        assertNotNull(singerSummaryService);
    }

@Test
public void testFindAll() {
    List<SingerSummary> singers = singerSummaryService.findAll();
    listSingerSummary(singers);
    assertEquals(2, singers.size());
}
    private static void listSingerSummary(List<SingerSummary> singers) {
        logger.info(" ---- Listing singers summary:");
        for (SingerSummary singer : singers) {
    logger.info(singer.toString());
        }
    }

    @After
    public void tearDown() {
        ctx.close();
    }
}
```

再次执行 testFindAll()方法，此时将会为列表中的每个 SingerSummary 对象生成输出，如下所示(其他输出被省略)：

```
INFO  ---- Listing singers summary:
INFO  - First name: John, Last Name: Mayer, Most Recent Album: The Search For Everything
INFO  - First name: Eric, Last Name: Clapton, Most Recent Album: From The Cradle
```

如你所见，如果想要将自定义查询的结果映射为 POJO 以进行进一步处理，那么构造函数表达式是非常有用的。

8.3.1 插入数据

使用 JPA 插入数据很简单。和 Hibernate 一样，JPA 也支持检索数据库生成的主键。以下代码片段显示了 save()方法：

```
package com.apress.prospring5.ch8.service;
...
@Service("jpaSingerService")
@Repository
@Transactional
public class SingerServiceImpl implements SingerService {
    final static String ALL_SINGER_NATIVE_QUERY =
        "select id, first_name, last_name, birth_date, version from singer";

    private static Logger logger =
        LoggerFactory.getLogger(SingerServiceImpl.class);
```

```
    @PersistenceContext
    private EntityManager em;

    @Override
    public Singer save(Singer singer) {
        if (singer.getId() == null) {
    logger.info("Inserting new singer");
    em.persist(singer);
        } else {
    em.merge(singer);
    logger.info("Updating existing singer");
        }
        logger.info("Singer saved with id: " + singer.getId());

        return singer;
    }
    ...
}
```

如上所示，save()方法首先通过检查 id 值来检查对象是否是新的实体实例。如果 id 为 null(即尚未分配)，那么该对象就是一个新的实体实例，此时调用 EntityManager.persist()方法。当调用 persist()方法时，EntityManager 将持久化该实体实例并使其成为当前持久化上下文中的托管实例。如果 id 值存在，则表示正在执行更新操作，将调用 EntityManager.merge()方法。当调用 merge()方法时，EntityManager 将实体的状态合并到当前的持久化上下文中。

以下代码片段显示了插入新歌手记录的代码。这一切都是在测试方法中完成的，因为我们想要测试数据插入操作是否成功。

```
package com.apress.prospring5.ch8;
...
public class SingerJPATest {
    private static Logger logger =
        LoggerFactory.getLogger(SingerJPATest.class);

    private GenericApplicationContext ctx;
    private SingerService singerService;

    @Before
    public void setUp(){
        ctx = new AnnotationConfigApplicationContext(JpaConfig.class);
        singerService = ctx.getBean(SingerService.class);
        assertNotNull(singerService);
    }

    @Test
    public void testInsert(){
        Singer singer = new Singer();
        singer.setFirstName("BB");
        singer.setLastName("King");
        singer.setBirthDate(new Date(
         (new GregorianCalendar(1940, 8, 16)).getTime().getTime()));

        Album album = new Album();
        album.setTitle("My Kind of Blues");
        album.setReleaseDate(new java.sql.Date(
        (new GregorianCalendar(1961, 7, 18)).getTime().getTime()));
        singer.addAbum(album);

        album = new Album();
        album.setTitle("A Heart Full of Blues");
        album.setReleaseDate(new java.sql.Date(
        (new GregorianCalendar(1962, 3, 20)).getTime().getTime()));
        singer.addAbum(album);
        singerService.save(singer);
        assertNotNull(singer.getId());

        List<Singer> singers = singerService.findAllWithAlbum();
        assertEquals(4, singers.size());
        listSingersWithAlbum(singers);
    }
    ...

    @After
    public void tearDown(){
        ctx.close();
    }
}
```

如上所示，创建了一位新歌手，添加了两张专辑并保存对象。然后再次列出所有的歌手，在验证表中拥有正确的记录数之后，运行该程序会产生以下输出：

```
INFO - ---- Listing singers with instruments:
INFO - Singer - Id: 4, First name: BB, Last name: King, Birthday: 1940-09-16
INFO -    Album - id: 5, Singer id: 4, Title: A Heart Full of Blues,
   Release Date: 1962-04-20
INFO -    Album - id: 4, Singer id: 4, Title: My Kind of Blues,
   Release Date: 1961-08-18
INFO - Singer - Id: 1, First name: John, Last name: Mayer, Birthday: 1977-10-16
INFO -    Album - id: 1, Singer id: 1, Title: The Search For Everything,
   Release Date: 2017-01-20
INFO -    Album - id: 2, Singer id: 1, Title: Battle Studies,
   Release Date: 2009-11-17
INFO -    Instrument: Piano
INFO -    Instrument: Guitar
INFO - Singer - Id: 3, First name: John, Last name: Butler, Birthday: 1975-04-01
INFO - Singer - Id: 2, First name: Eric, Last name: Clapton, Birthday: 1945-03-30
INFO -    Album - id: 3, Singer id: 2, Title: From The Cradle,
   Release Date: 1994-09-13
INFO -    Instrument: Guitar
```

通过 INFO 日志记录，可以看到新保存歌手的 ID 已正确填充。此外，Hibernate 还会显示所有触发数据库的 SQL 语句。

8.3.2 更新数据

更新数据与插入数据一样简单。接下来看一个示例。假设想要更新 ID 为 1 的歌手的名字并删除一张专辑。为了测试更新操作，以下代码片段显示了 testUpdate()方法：

```
package com.apress.prospring5.ch8;
...
public class SingerJPATest {
   private static Logger logger =
      LoggerFactory.getLogger(SingerJPATest.class);
   private GenericApplicationContext ctx;
   private SingerService singerService;

   @Before
   public void setUp(){
      ctx = new AnnotationConfigApplicationContext(JpaConfig.class);
      singerService = ctx.getBean(SingerService.class);
      assertNotNull(singerService);
   }

   @Test
   public void testUpdate(){
      Singer singer = singerService.findById(1L);
      //making sure such singer exists assertNotNull(singer);
      //making sure we got expected record assertEquals("Mayer", singer.getLastName());
      //retrieve the album
      Album album = singer.getAlbums().stream()
         .filter(a -> a.getTitle().equals("Battle Studies")).findFirst().get();

      singer.setFirstName("John Clayton");
      singer.removeAlbum(album);
      singerService.save(singer);

      listSingersWithAlbum(singerService.findAllWithAlbum());
   }
   ...

   @After
   public void tearDown(){
      ctx.close();
   }
}
```

首先检索 ID 为 1 的记录，并更改名字。然后，循环遍历专辑对象并检索名为 *Battle Studies* 的专辑，将其从歌手的 albums 属性中移除。最后，再次调用 SingerService.save()方法。当运行程序时，你会看到下面的输出(其他输出被省略)：

```
---- Listing singers with instruments:
```

```
Singer - Id: 1, First name: John Clayton, Last name: Mayer, Birthday: 1977-10-16
   Album - id: 1, Singer id: 1, Title: The Search For Everything,
      Release Date: 2017-01-20
   Instrument: Piano
   Instrument: Guitar
Singer - Id: 2, First name: Eric, Last name: Clapton, Birthday: 1945-03-30
   Album - id: 3, Singer id: 2, Title: From The Cradle ,
      Release Date: 1994-09-13
   Instrument: Guitar
Singer - Id: 3, First name: John, Last name: Butler, Birthday: 1975-04-01
```

此时可以看到名字已更新，专辑也被删除。之所以可以删除专辑信息，是因为在一对多关联中定义的 orphanRemoval = true 属性，该属性指示 JPA 提供程序(Hibernate)删除虽然在数据库中存在但在对象中不再存在的所有孤立记录。

```
@OneToMany(mappedBy = "singer", cascade=CascadeType.ALL, orphanRemoval=true)
```

8.3.3 删除数据

删除数据同样很简单。只需要调用 EntityManager.remove()方法并传入歌手对象即可。以下代码片段显示了删除歌手的更新后的代码：

```
package com.apress.prospring5.ch8.service;
...
@Service("jpaSingerService")
@Repository
@Transactional
public class SingerServiceImpl implements SingerService {
   final static String ALL_SINGER_NATIVE_QUERY =
      "select id, first_name, last_name, birth_date, version from singer";

   private static Logger logger =
      LoggerFactory.getLogger(SingerServiceImpl.class);

   @PersistenceContext
   private EntityManager em;

      @Override
   public void delete(Singer singer) {
      Singer mergedSinger = em.merge(singer);
      em.remove(mergedSinger);
      logger.info("Singer with id: " + singer.getId() + " deleted successfully");
   }
    ...
}
```

首先调用 EntityManager.merge()方法，将实体的状态合并到当前的持久化上下文中。merge()方法返回被托管的实体实例。然后调用 EntityManager.remove()方法，并传入被托管的歌手实体实例。当在映射中定义了 cascade = CascadeType.ALL 时，删除操作将删除歌手记录及其所有相关信息，包括专辑和乐器。要测试删除操作，可以使用 testDelete()方法，如以下代码片段所示：

```
package com.apress.prospring5.ch8;
...
public class SingerJPATest {
   private static Logger logger =
      LoggerFactory.getLogger(SingerJPATest.class);

   private GenericApplicationContext ctx;
   private SingerService singerService;

   @Before
   public void setUp(){
      ctx = new AnnotationConfigApplicationContext(JpaConfig.class);
      singerService = ctx.getBean(SingerService.class);
      assertNotNull(singerService);
   }

      @Test
   public void testDelete(){
      Singer singer = singerService.findById(2l);
      //making sure such singer exists
      assertNotNull(singer);
      singerService.delete(singer);
```

```
        listSingersWithAlbum(singerService.findAllWithAlbum());
    }
    ...

    @After
    public void tearDown(){
        ctx.close();
    }
}
```

上面的代码首先检索了 ID 为 2 的歌手，然后调用 delete()方法以删除歌手信息。运行该程序会产生以下输出：

```
  ---- Listing singers with instruments:
Singer - Id: 1, First name: John, Last name: Mayer, Birthday: 1977-10-16
    Album - id: 1, Singer id: 1, Title: The Search For Everything,
        Release Date: 2017-01-20
    Album - id: 2, Singer id: 1, Title: Battle Studies,
        Release Date: 2009-11-17
    Instrument: Piano
    Instrument: Guitar
Singer - Id: 3, First name: John, Last name: Butler, Birthday: 1975-04-01
```

此时可以看到，ID 为 1 的歌手已被删除。

8.4 使用本地查询

在讨论了使用 JPA 执行简单的数据库操作之后，现在让我们继续讨论一些更高级的主题。有时候你可能想要对提交给数据库的查询进行绝对控制，比如在 Oracle 数据库中使用分级查询。这种特定于数据库的查询被称为*本地查询(native query)*。

JPA 支持执行本地查询；EntityManager 按照原样将查询提交给数据库，而不执行任何映射或转换。使用 JPA 本地查询的一个主要好处是能将 ResultSet 映射回 ORM 映射的实体类。以下两节讨论如何使用本地查询来检索所有歌手并将 ResultSet 直接映射回 Singer 对象。

8.5 使用简单的本地查询

为了演示如何使用本地查询，实现一个从数据库中检索所有歌手的新方法。以下代码片段显示了必须添加到 SingerServiceImpl 中的新方法：

```
package com.apress.prospring5.ch8.service;
...
@Service("jpaSingerService")
@Repository
@Transactional
public class SingerServiceImpl implements SingerService {
    final static String ALL_SINGER_NATIVE_QUERY =
        "select id, first_name, last_name, birth_date, version from singer";

    private static Logger logger =
        LoggerFactory.getLogger(SingerServiceImpl.class);
     @Transactional(readOnly=true)
    @Override
    public List<Singer> findAllByNativeQuery() {
        return em.createNativeQuery(ALL_SINGER_NATIVE_QUERY,
            Singer.class).getResultList();
    }
    ...
}
```

可以看到，本地查询只是一条简单的 SQL 语句，用于从 SINGER 表中检索所有列。为了创建并执行查询，首先调用 EntityManager.createNativeQuery()方法，并传入查询字符串以及结果类型。结果类型应该是映射的实体类(此时是 Singer 类)。createNativeQuery()方法返回一个 Query 接口，该接口提供了 getResultList()操作来获取结果列表。JPA 提供程序将执行查询并根据实体类中定义的 JPA 映射将 ResultSet 对象转换为实体实例。执行前面的方法将产生与 findAll()方法相同的结果。

8.6 使用 SQL ResultSet 映射进行本地查询

除了映射的域对象之外，还可以传入一个表示 SQL ResultSet 映射名称的字符串。SQL ResultSet 映射是在实体类级别通过使用@SqlResultSetMapping 注解定义的。一个 SQL ResultSet 映射可以有一个或多个实体和列映射。

```
package com.apress.prospring5.ch8.entities;

import javax.persistence.Entity;
import javax.persistence.Table;
import javax.persistence.SqlResultSetMapping;
import javax.persistence.EntityResult;
...
@Entity
@Table(name = "singer")
@SqlResultSetMapping(
   name="singerResult",
   entities=@EntityResult(entityClass=Singer.class)
)
public class Singer implements Serializable {
  ...
}
```

名为 singerResult 的 SQL ResultSet 映射是为实体类定义的，同时使用了 Singer 类中的 entityClass 属性。JPA 支持多个实体的更复杂的映射，并支持向下映射到列级别的映射。

在定义 SQL ResultSet 映射之后，可以使用 ResultSet 映射的名称调用 findAllByNativeQuery()方法。以下代码片段显示了更新后的 findAllByNativeQuery()方法：

```
package com.apress.prospring5.ch8.service;
...
@Service("jpaSingerService")
@Repository
@Transactional
public class SingerServiceImpl implements SingerService {
   final static String ALL_SINGER_NATIVE_QUERY =
      "select id, first_name, last_name, birth_date, version from singer";

   private static Logger logger =
      LoggerFactory.getLogger(SingerServiceImpl.class);

    @Transactional(readOnly=true)
   @Override
   public List<Singer> findAllByNativeQuery() {
      return em.createNativeQuery(ALL_SINGER_NATIVE_QUERY,
         "singerResult").getResultList();
   }
   ...
}
```

如你所见，JPA 为执行本地查询提供了强大支持，同时还提供了灵活的 SQL ResultSet 映射工具。

使用 JPA 2 Criteria API 进行条件查询

大多数应用程序会为用户提供前端来搜索信息。有时可能会存在以下情况：应用程序显示了大量可搜索的字段，而用户只在其中的一些字段中输入信息并进行搜索。要准备大量查询是很困难的，用户可以选择输入不同的参数组合。在这种情况下，Criteria API 查询就派上用场了。

在 JPA 2 中，引入的一个主要新功能是强类型的 Criteria API 查询。在新的 Criteria API 中，传递到查询中的条件是基于映射的实体类的元模型。因此，所指定的每个条件都是强类型的，并且在编译而不是运行时时会发现错误。

在 JPA Criteria API 中，实体类的元模型由具有下划线(_)后缀的实体类名称表示。例如，Singer 实体类的元模型类是 Singer_。以下代码片段显示了 Singer_元模型类：

```
package com.apress.prospring5.ch8;

import java.util.Date;
import javax.annotation.Generated;
import javax.persistence.metamodel.SetAttribute;
import javax.persistence.metamodel.SingularAttribute;
import javax.persistence.metamodel.StaticMetamodel;
```

```
@Generated(value = "org.hibernate.jpamodelgen.JPAMetaModelEntityProcessor")
@StaticMetamodel(Singer.class)
public abstract class Singer_ {

    public static volatile SingularAttribute<Singer, String> firstName;
    public static volatile SingularAttribute<Singer, String> lastName;
    public static volatile SetAttribute<Singer, Album> albums;
    public static volatile SetAttribute<Singer, Instrument> instruments;
    public static volatile SingularAttribute<Singer, Long> id;
    public static volatile SingularAttribute<Singer, Integer> version;
    public static volatile SingularAttribute<Singer, Date> birthDate;
}
```

元模型类用@StaticMetamodel 注解，属性是映射的实体类。类中是每个属性及其相关类型的声明。

编写和维护这些元模型类是非常单调乏味的。但是，可以使用相关工具根据实体类中的 JPA 映射自动生成这些元模型类。Hibernate 提供的工具被称为 Hibernate 元模型生成器(www.hibernate.org/subprojects/jpamodelgen.html)。

如何生成元模型类取决于正在使用什么工具来开发和构建项目。建议阅读文档的“用法”部分(http://docs.jboss.org/hibernate/jpamodelgen/1.3/reference/en-US/html_single/#chapter-usage)以了解具体的细节。书中的示例代码将使用 Gradle 生成元模型类。生成元模型类所需的依赖项是 hibernate-jpamodelgen 库。该依赖项在 pro-spring-15/build.gradle 文件中用对应版本进行了配置。

```
ext {
    ...

    //persistency libraries
    hibernateVersion = '5.2.10.Final'
    hibernateJpaVersion = '1.0.0.Final'

    hibernate = [
    ...
    jpaModelGen: "org.hibernate:hibernate-jpamodelgen:$hibernateVersion",
    jpaApi     : "org.hibernate.javax.persistence:hibernate-jpa-2.1-api:
       $hibernateJpaVersion",
    querydslapt: "com.mysema.querydsl:querydsl-apt:2.7.1"
    ]
    ...
```

上面所示的是用来生成元模型类的主要库。在编译模块之前，generateQueryDSL Gradle 任务在 chapter08/jpa-criteria/build.gradle 中用它来生成元模型类。chapter08/jpa-criteria/build.gradle 配置如下所示：

```
sourceSets {
    generated
}

sourceSets.generated.java.srcDirs = ['src/main/generated']

configurations {
    querydslapt
}

dependencies {
    compile hibernate.querydslapt, hibernate.jpaModelGen
}

task generateQueryDSL(type: JavaCompile, group: 'build',
    description: 'Generates the QueryDSL query types') {
    source = sourceSets.main.java
    classpath = configurations.compile + configurations.querydslapt

    options.compilerArgs = [
    "-proc:only",
    "-processor", "org.hibernate.jpamodelgen.JPAMetaModelEntityProcessor"
    ]
    destinationDir = sourceSets.generated.java.srcDirs.iterator.next
}
compileJava.dependsOn generateQueryDSL
```

随着类生成策略的建立，让我们定义一个查询，它接收要搜索歌手的名字和姓氏。以下代码片段显示了 SingerService 接口中新方法 findByCriteriaQuery()的定义：

```
package com.apress.prospring5.ch8;

import java.util.List;
```

```
public interface SingerService {
    List<Singer> findAll();
    List<Singer> findAllWithAlbum();
    Singer findById(Long id);
    Singer save(Singer singer);
    void delete(Singer singer);
    List<Singer> findAllByNativeQuery();
    List<Singer> findByCriteriaQuery(String firstName, String lastName);
}
```

下一个代码片段显示了使用 JPA 2 Criteria API 查询的 findByCriteriaQuery()方法的实现:

```
package com.apress.prospring5.ch8;

import org.springframework.stereotype.Service;
import org.springframework.stereotype.Repository;
import org.springframework.transaction.annotation.Transactional;

import java.util.List;

import javax.persistence.PersistenceContext;
import javax.persistence.EntityManager;
import javax.persistence.TypedQuery;
import javax.persistence.criteria.CriteriaBuilder;
import javax.persistence.criteria.CriteriaQuery;
import javax.persistence.criteria.Root;
import javax.persistence.criteria.JoinType;
import javax.persistence.criteria.Predicate;

import org.apache.commons.logging.Log;
import org.apache.commons.logging.LogFactory;

@Service("jpaSingerService")
@Repository
@Transactional
public class SingerServiceImpl implements SingerService {
    final static String ALL_SINGER_NATIVE_QUERY =
        "select id, first_name, last_name, birth_date, version from singer";

    private Log log =
        LogFactory.getLog(SingerServiceImpl.class);

    @PersistenceContext
    private EntityManager em;
    ...

    @Transactional(readOnly=true)
    @Override
    public List<Singer> findByCriteriaQuery(String firstName, String lastName) {
        log.info("Finding singer for firstName: " + firstName
            + " and lastName: " + lastName);
        CriteriaBuilder cb = em.getCriteriaBuilder();
        CriteriaQuery<Singer> criteriaQuery = cb.createQuery(Singer.class);
        Root<Singer> singerRoot = criteriaQuery.from(Singer.class);
        singerRoot.fetch(Singer_.albums, JoinType.LEFT);
        singerRoot.fetch(Singer_.instruments, JoinType.LEFT);

        criteriaQuery.select(singerRoot).distinct(true);
        Predicate criteria = cb.conjunction();

        if (firstName != null) {
    Predicate p = cb.equal(singerRoot.get(Singer_.firstName),
        firstName);
    criteria = cb.and(criteria, p);
        }

        if (lastName != null) {
    Predicate p = cb.equal(singerRoot.get(Singer_.lastName),
        lastName);
    criteria = cb.and(criteria, p);
        }

        criteriaQuery.where(criteria);
        return em.createQuery(criteriaQuery).getResultList();
    }
}
```

下面，让我们来分解一下 Criteria API 的用法：

- 调用 EntityManager.getCriteriaBuilder()来检索 CriteriaBuilder 的一个实例。
- 使用 CriteriaBuilder.createQuery()创建类型化查询，并传入 Singer 作为结果类型。
- 调用 CriteriaQuery.from()方法，并传入实体类。返回结果是与指定实体相对应的查询根对象(Root <Singer>接口)。查询根对象构成查询中路径表达式的基础。
- 调用两次 Root.fetch()方法，立即提取与专辑和乐器相关的关联。JoinType.LEFT 参数指定了一个外部连接。使用 JoinType.LEFT 作为第二个参数调用 Root.fetch()方法等效于在 JPQL 中指定左连接读取连接(Left join fetch join)操作。
- 调用 CriteriaQuery.select()方法，并传入根查询对象作为结果类型。使用 distinct(true)方法意味着应该消除重复的记录。
- Predicate 实例是通过调用 CriteriaBuilder.conjunction()方法获得的，这意味着会产生一个或多个限制条件的组合。Predicate 实例可以是简单谓词或复合谓词，谓词用于表示由表达式定义的选择条件。
- 检查名字和姓氏参数。对于每个非空参数，都会使用 CriteriaBuilder()方法(即 CriteriaBuilder.and()方法)构造一个新的 Predicate 实例。方法 equal()用来指定相等限制，在该方法中调用 Root.get()并传入限制适用的实体类元模型的相应属性。然后将构造的谓词通过调用 CriteriaBuilder.and()方法与现有谓词(由变量条件存储)“合并”。
- Predicate 实例是使用所有条件和限制构造的，并通过调用 CriteriaQuery.where()方法作为 where 子句传递给查询。
- 最后，CriteriaQuery 被传递给 EntityManager。然后，EntityManager 根据传入的 CriteriaQuery 值构造查询，执行查询并返回结果。

要测试条件查询操作，以下代码片段显示了更新后的 SingerJPATest 类：

```
package com.apress.prospring5.ch8;

import com.apress.prospring5.ch8.config.JpaConfig;
import com.apress.prospring5.ch8.Album;
import com.apress.prospring5.ch8.Instrument;
import com.apress.prospring5.ch8.Singer;
import com.apress.prospring5.ch8.SingerService;
import org.junit.After;
import org.junit.Before;
import org.junit.Test;
import org.slf4j.Logger;
import org.slf4j.LoggerFactory;
import org.springframework.context.annotation.AnnotationConfigApplicationContext;
import org.springframework.context.support.GenericApplicationContext;

import java.util.List;

import static org.junit.Assert.assertEquals;
import static org.junit.Assert.assertNotNull;

public class SingerJPATest {
    private static Logger logger =
        LoggerFactory.getLogger(SingerJPATest.class);

    private GenericApplicationContext ctx;
    private SingerService singerService;

    @Before
    public void setUp(){
        ctx = new AnnotationConfigApplicationContext(JpaConfig.class);
        singerService = ctx.getBean("jpaSingerService", SingerService.class);
        assertNotNull(singerService);
    }
    @Test
    public void tesFindByCriteriaQuery(){
        List<Singer> singers = singerService.findByCriteriaQuery("John", "Mayer");
        assertEquals(1, singers.size());
        listSingersWithAlbum(singers);
    }
    private static void listSingersWithAlbum(List<Singer> singers) {
        logger.info(" ---- Listing singers with instruments:");
        singers.forEach(s -> {
```

```
            logger.info(s.toString());
            if (s.getAlbums() != null) {
                s.getAlbums().forEach(a -> logger.info("\t" + a.toString()));
            }
            if (s.getInstruments() != null) {
                s.getInstruments().forEach(i -> logger.info
                    ("\tInstrument: " + i.getInstrumentId()));
            }
        });
    }

    @After
    public void tearDown(){
        ctx.close();
    }
}
```

运行该程序会产生以下输出(其他输出被省略，但生成的查询被保留):

```
INFO  o.h.h.i.QueryTranslatorFactoryInitiator -
      HHH000397: Using ASTQueryTranslatorFactory
INFO  c.a.p.c.SingerServiceImpl -
      Finding singer for firstName: John and lastName: Mayer
Hibernate:
   select
    distinct singer0_.ID as ID1_2_0_,
    albums1_.ID as ID1_0_1_,
    instrument3_.INSTRUMENT_ID as INSTRUME1_1_2_,
    singer0_.BIRTH_DATE as BIRTH_DA2_2_0_,
    singer0_.FIRST_NAME as FIRST_NA3_2_0_,
    singer0_.LAST_NAME as LAST_NAM4_2_0_,
    singer0_.VERSION as VERSION5_2_0_,
    albums1_.RELEASE_DATE as RELEASE_2_0_1_,
    albums1_.SINGER_ID as SINGER_I5_0_1_,
    albums1_.title as title3_0_1_,
    albums1_.VERSION as VERSION4_0_1_,
    albums1_.SINGER_ID as SINGER_I5_0_0__,
    albums1_.ID as ID1_0_0__,
    instrument2_.SINGER_ID as SINGER_I1_3_1__,
    instrument2_.INSTRUMENT_ID as INSTRUME2_3_1__
        from
    singer singer0_
        left outer join
    album albums1_
        on singer0_.ID=albums1_.SINGER_ID
        left outer join
    singer_instrument instrument2_

        on singer0_.ID=instrument2_.SINGER_ID
        left outer join
    instrument instrument3_

        on instrument2_.INSTRUMENT_ID=instrument3_.INSTRUMENT_ID
        where
    1=1
    and singer0_.FIRST_NAME=?
    and singer0_.LAST_NAME=?
INFO  c.a.p.c.SingerJPATest - ---- Listing singers with instruments:
INFO  c.a.p.c.SingerJPATest - Singer - Id: 1, First name: John, Last name: Mayer,
    Birthday: 1977-10-16
INFO  c.a.p.c.SingerJPATest -  Album - id: 2, Singer id: 1,
    Title: Battle Studies, Release Date: 2009-11-17
INFO  c.a.p.c.SingerJPATest -  Album - id: 1, Singer id: 1,
    Title: The Search For Everything, Release Date: 2017-01-20
INFO  c.a.p.c.SingerJPATest -  Instrument: Guitar
INFO  c.a.p.c.SingerJPATest -  Instrument: Piano
```

可以尝试其他组合或将空值传递给任意参数，并观察输出。

8.7 Spring Data JPA 介绍

Spring Data JPA 项目是大型项目 Spring Data 下的一个子项目。该项目的主要目标是提供附加功能来简化使用 JPA 开发应用程序。

Spring Data JPA 提供了几个主要功能。在本节中，主要讨论两个功能。第一个功能是 Repository 抽象，第二个功能是用于跟踪实体类的基本审计信息的实体监听器。

8.7.1 添加 Spring Data JPA 库依赖项

要使用 Spring Data JPA，需要将依赖项添加到项目中。下面显示了使用 Spring Data JPA 所需的 Gradle 配置：

```
//pro-spring-15/build.gradle
ext {
   //spring libs
   springVersion = '5.0.0.M5'
   bootVersion = '2.0.0.BUILD-SNAPSHOT'
   springDataVersion = '2.0.0.M2'
   ...

   spring = [
      data : "org.springframework.data:spring-data-jpa:$springDataVersion",
      ...
   ]
   ...
}

//chapter08/spring-data-jpa/build.gradle
dependencies {
   compile spring.aop, spring.data, misc.guava
}
```

8.7.2 使用 Spring Data JPA Repository 抽象进行数据库操作

Spring Data 及其所有子项目的一个主要概念是 Repository 抽象，它属于 Spring Data Commons 项目(https://github.com/spring-projects/spring-datacommons)。在编写本书时，它的版本是 2.0.0 M2。在 Spring Data JPA 中，Repository 抽象封装了底层的 JPA EntityManager，并为基于 JPA 的数据访问提供了一个更简单的接口。Spring Data 的中心接口是 org.springframework.data.repository.Repository<T,ID extends Serializable>，它是属于 Spring Data Commons 发行版的标记接口。Spring Data 提供了 Repository 接口的各种扩展；其中之一是 org.springframework.data.repository.CrudRepository 接口(它也属于 Spring Data Commons 项目)，我们将在本节中讨论该接口。

CrudRepository 接口提供了许多常用的方法。以下代码片段显示了来自 Spring Data Commons 项目源代码的接口声明：

```
package org.springframework.data.repository;

import java.io.Serializable;

@NoRepositoryBean
public interface CrudRepository<T, ID extends Serializable>
   extends Repository<T, ID> {
   long count();
   void delete(ID id);
   void delete(Iterable<? extends T> entities);
   void delete(T entity);
   void deleteAll();
   boolean exists(ID id);
   Iterable<T> findAll();
   T findOne(ID id);
   Iterable<T> save(Iterable<? extends T> entities);
   T save(T entity);
}
```

虽然方法命名是不言自明的，但最好通过一个简单的示例来演示一下 Repository 抽象是如何工作的。稍微修改一下 SingerService 接口，使其仅包含三个查找方法。以下代码片段显示了修订后的 SingerService 接口：

```
package com.apress.prospring5.ch8;

import com.apress.prospring5.ch8.entities.Singer;

import java.util.List;

public interface SingerService {
   List<Singer> findAll();
   List<Singer> findByFirstName(String firstName);
   List<Singer> findByFirstNameAndLastName(String firstName, String lastName);
}
```

下一步是准备 SingerRepository 接口，该接口扩展了 CrudRepository 接口。以下代码片段显示了 SingerRepository 接口：

```
package com.apress.prospring5.ch8;
import java.util.List;

import com.apress.prospring5.ch8.entities.Singer;
import org.springframework.data.repository.CrudRepository;

public interface SingerRepository extends CrudRepository<Singer, Long> {
    List<Singer> findByFirstName(String firstName);
    List<Singer> findByFirstNameAndLastName(String firstName, String lastName);
}
```

只需要在该接口中声明两个方法，因为 findAll()方法已经由 CrudRepository.findAll()方法提供。如上面的代码清单所示，SingerRepository 接口扩展了 CrudRepository 接口，并传入实体类(Singer)和 ID 类型(Long)。Spring Data 的 Repository 抽象的一个奇妙之处在于，当使用 findByFirstName()和 findByFirstNameAndLastName()的通用命名约定时，不需要为 Spring Data JPA 提供命名查询。相反，Spring Data JPA 将根据方法名称进行“推断”并构建查询。例如，对于 findByFirstName()方法，Spring Data JPA 将自动准备查询 select s from Singer s where s.firstName = :firstName 并根据方法参数设置命名参数 firstName。

要想使用 Repository 抽象，必须在 Spring 的配置中定义它。以下代码片段显示了配置文件(app-context-annotation.xml)：

```
<?xml version="1.0" encoding="UTF-8"?>

<beans xmlns="http://www.springframework.org/schema/beans"
    xmlns:xsi="http://www.w3.org/2001/XMLSchema-instance"
    xmlns:context="http://www.springframework.org/schema/context"
    xmlns:jdbc="http://www.springframework.org/schema/jdbc"
    xmlns:jpa="http://www.springframework.org/schema/data/jpa"
    xmlns:tx="http://www.springframework.org/schema/tx"
    xsi:schemaLocation="http://www.springframework.org/schema/beans
        http://www.springframework.org/schema/beans/spring-beans.xsd
        http://www.springframework.org/schema/context
        http://www.springframework.org/schema/context/spring-context.xsd
        http://www.springframework.org/schema/jdbc
        http://www.springframework.org/schema/jdbc/spring-jdbc.xsd
        http://www.springframework.org/schema/data/jpa
        http://www.springframework.org/schema/data/jpa/spring-jpa.xsd
        http://www.springframework.org/schema/tx
        http://www.springframework.org/schema/tx/spring-tx.xsd">

    <jdbc:embedded-database id="dataSource" type="H2">
        <jdbc:script location="classpath:db/schema.sql"/>
        <jdbc:script location="classpath:db/test-data.sql"/>
    </jdbc:embedded-database>

    <bean id="transactionManager"
        class="org.springframework.orm.jpa.JpaTransactionManager">
        <property name="entityManagerFactory" ref="emf"/>
    </bean>
    <tx:annotation-driven transaction-manager="transactionManager" />

    <bean id="emf"
      class="org.springframework.orm.jpa.LocalContainerEntityManagerFactoryBean">
        <property name="dataSource" ref="dataSource" />
        <property name="jpaVendorAdapter">
        <bean class="org.springframework.orm.jpa.vendor.HibernateJpaVendorAdapter" />
        </property>
        <property name="packagesToScan"
            value="com.apress.prospring5.ch8.entities"/>
        <property name="jpaProperties">
            <props>
                <prop key="hibernate.dialect">
                    org.hibernate.dialect.H2Dialect
                </prop>
                <prop key="hibernate.max_fetch_depth">3</prop>
                <prop key="hibernate.jdbc.fetch_size">50</prop>
                <prop key="hibernate.jdbc.batch_size">10</prop>
                <prop key="hibernate.show_sql">true</prop>
            </props>
        </property>
    </bean>
```

```
    <context:annotation-config/>

    <context:component-scan base-package="com.apress.prospring5.ch8" >
        <context:exclude-filter type="annotation"
            expression="org.springframework.context.annotation.Configuration" />
    </context:component-scan>

    <jpa:repositories base-package="com.apress.prospring5.ch8"
                      entity-manager-factory-ref="emf"
                      transaction-manager-ref="transactionManager"/>
</beans>
```

首先，需要在配置文件中添加jpa 名称空间。然后，使用<jpa:repositories>标记配置 Spring Data JPA 的 Repository 抽象。指示 Spring 对包 com.apress.prospring5.ch8 进行扫描以获取存储库接口，并传入 EntityManagerFactory 和事务管理器。

此时你可能没有注意到，在<context:component-scan>定义中有一个<context:exclude-filter>，它指定了用@Configuration 注解的类。引入该元素是为了避免扫描那些可用来替换先前描述的 XML 配置的 Java 配置类。该类如下所示：

```
package com.apress.prospring5.ch8.config;

import org.slf4j.Logger;
import org.slf4j.LoggerFactory;
import org.springframework.context.annotation.Bean;
import org.springframework.context.annotation.ComponentScan;
import org.springframework.context.annotation.Configuration;
import org.springframework.data.jpa.repository.config.EnableJpaRepositories;
import org.springframework.jdbc.datasource.embedded.EmbeddedDatabaseBuilder;
import org.springframework.jdbc.datasource.embedded.EmbeddedDatabaseType;
import org.springframework.orm.jpa.JpaTransactionManager;
import org.springframework.orm.jpa.JpaVendorAdapter;
import org.springframework.orm.jpa.LocalContainerEntityManagerFactoryBean;
import org.springframework.orm.jpa.vendor.HibernateJpaVendorAdapter;
import org.springframework.transaction.PlatformTransactionManager;
import org.springframework.transaction.annotation.EnableTransactionManagement;

import javax.persistence.EntityManagerFactory;
import javax.sql.DataSource;
import java.util.Properties;

@Configuration
@EnableTransactionManagement
@ComponentScan(basePackages = {"com.apress.prospring5.ch8"})
@EnableJpaRepositories(basePackages = {"com.apress.prospring5.ch8"})
public class DataJpaConfig {

    private static Logger logger = LoggerFactory.getLogger(DataJpaConfig.class);

    @Bean
    public DataSource dataSource() {
        try {
            EmbeddedDatabaseBuilder dbBuilder = new EmbeddedDatabaseBuilder();
            return dbBuilder.setType(EmbeddedDatabaseType.H2)
                    .addScripts("classpath:db/schema.sql",
                    "classpath:db/test-data.sql").build();
            } catch (Exception e) {
                logger.error("Embedded DataSource bean cannot be created!", e);
                return null;
            }
    }

    @Bean
    public PlatformTransactionManager transactionManager() {
        return new JpaTransactionManager(entityManagerFactory());
    }

    @Bean
    public JpaVendorAdapter jpaVendorAdapter() {
        return new HibernateJpaVendorAdapter();
    }
    @Bean
        public Properties hibernateProperties() {
        Properties hibernateProp = new Properties();
        hibernateProp.put("hibernate.dialect", "org.hibernate.dialect.H2Dialect");
```

```
        hibernateProp.put("hibernate.format_sql", true);
        hibernateProp.put("hibernate.use_sql_comments", true);
        hibernateProp.put("hibernate.show_sql", true);
        hibernateProp.put("hibernate.max_fetch_depth", 3);
        hibernateProp.put("hibernate.jdbc.batch_size", 10);
        hibernateProp.put("hibernate.jdbc.fetch_size", 50);
        return hibernateProp;
    }

    @Bean
    public EntityManagerFactory entityManagerFactory() {
        LocalContainerEntityManagerFactoryBean factoryBean =
            new LocalContainerEntityManagerFactoryBean();
        factoryBean.setPackagesToScan("com.apress.prospring5.ch8.entities");
        factoryBean.setDataSource(dataSource());
        factoryBean.setJpaVendorAdapter(new HibernateJpaVendorAdapter());
        factoryBean.setJpaProperties(hibernateProperties());
        factoryBean.setJpaVendorAdapter(jpaVendorAdapter());
        factoryBean.afterPropertiesSet();
        return factoryBean.getNativeEntityManagerFactory();
    }
}
```

此时用来支持 Spring Data JPA 存储库的唯一配置元素是@EnableJpaRepositories 注解。通过使用 basePackages 属性，可以扫描自定义 Repository 扩展所在的包并创建存储库 bean。其余的依赖项(emf 和 transactionManager bean)由 Spring 容器自动注入。

以下代码片段显示了 SingerService 接口的三个查找方法的实现：

```
package com.apress.prospring5.ch8;
import com.apress.prospring5.ch8.entities.Singer;
import org.springframework.stereotype.Service;
import org.springframework.transaction.annotation.Transactional;

import org.springframework.beans.factory.annotation.Autowired;

import java.util.List;
import com.google.common.collect.Lists;

@Service("springJpaSingerService")
@Transactional
public class SingerServiceImpl implements SingerService {
    @Autowired
    private SingerRepository singerRepository;

    @Transactional(readOnly=true)
    public List<Singer> findAll() {
        return Lists.newArrayList(singerRepository.findAll());
    }

    @Transactional(readOnly=true)
    public List<Singer> findByFirstName(String firstName) {
        return singerRepository.findByFirstName(firstName);
    }

    @Transactional(readOnly=true)
    public List<Singer> findByFirstNameAndLastName(
        String firstName, String lastName) {
        return singerRepository.findByFirstNameAndLastName(
            firstName, lastName);
    }
}
```

可以看到，只需要将根据 SingerRepository 接口生成的 singerRepository bean 注入到服务类中，Spring Data JPA 就会为我们完成所有低级别的工作，而无须使用 EntityManager。在下面的代码片段中，可以看到一个测试类，对该测试类的内容你应该比较熟悉了：

```
package com.apress.prospring5.ch8;

import com.apress.prospring5.ch8.config.DataJpaConfig;
import com.apress.prospring5.ch8.entities.Singer;
import org.junit.After;
import org.junit.Before;
import org.junit.Test;
import org.slf4j.Logger;
import org.slf4j.LoggerFactory;
```

```
import org.springframework.context.annotation.AnnotationConfigApplicationContext;
import org.springframework.context.support.GenericApplicationContext;

import java.util.List;

import static org.junit.Assert.assertEquals;
import static org.junit.Assert.assertNotNull;
import static org.junit.Assert.assertTrue;

public class SingerDataJPATest {

    private static Logger logger =
        LoggerFactory.getLogger(SingerDataJPATest.class);

    private GenericApplicationContext ctx;
    private SingerService singerService;

    @Before
    public void setUp(){
        ctx = new AnnotationConfigApplicationContext(DataJpaConfig.class);
        singerService = ctx.getBean(SingerService.class);
        assertNotNull(singerService);
    }

    @Test
    public void testFindAll(){
        List<Singer> singers = singerService.findAll();
        assertTrue(singers.size() > 0);
        listSingers(singers);
    }

    @Test
    public void testFindByFirstName(){
        List<Singer> singers = singerService.findByFirstName("John");
        assertTrue(singers.size() > 0);
        assertEquals(2, singers.size());
        listSingers(singers);
    }
    @Test
    public void testFindByFirstNameAndLastName(){
        List<Singer> singers =
            singerService.findByFirstNameAndLastName("John", "Mayer");
        assertTrue(singers.size() > 0);
        assertEquals(1, singers.size());
        listSingers(singers);
    }

    private static void listSingers(List<Singer> singers) {
        logger.info(" ---- Listing singers:");
        for (Singer singer : singers) {
            logger.info(singer.toString());
        }
    }

    @After
    public void tearDown() {
        ctx.close();
    }
}
```

运行该测试类，所有测试都应该通过，并将期望的数据打印到控制台中。

8.8 使用 JpaRepository

除 CrudRepository 外，还有一个更高级的 Spring 接口，可以让创建自定义存储库变得更容易；它被称为 JpaRepository 接口，该接口提供批处理、分页和排序操作。图 8-1 显示了 JpaRepository 和 CrudRepository 接口之间的关系。根据应用程序的复杂性，可以选择使用 CrudRepository 或 JpaRepository。从图 8-1 可以看出，JpaRepository 扩展了 CrudRepository；因此，该接口提供 CrudRepository 接口所具有的所有功能。

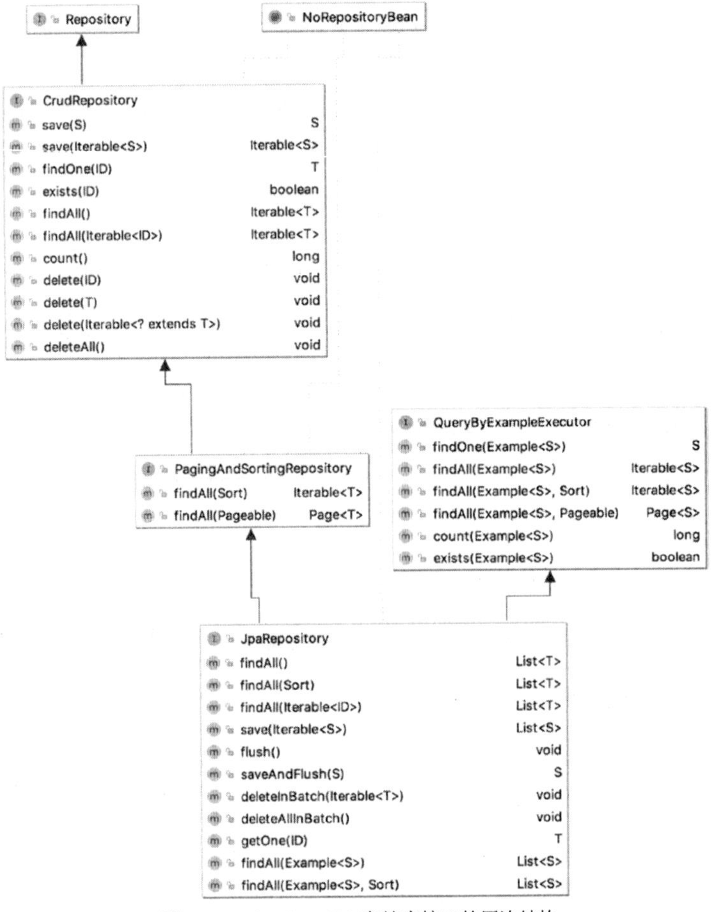

图 8-1　Spring Data JPA 存储库接口的层次结构

8.9　带有自定义查询的 Spring Data JPA

在复杂的应用程序中，可能需要使用一些 Spring 无法“推断”的自定义查询。在这种情况下，必须使用@Query 注解显式定义查询。接下来使用该注解来搜索所有名称中包含 *The* 的音乐专辑。以下代码片段描述了 AlbumRepository 接口：

```
package com.apress.prospring5.ch8.repos;

import com.apress.prospring5.ch8.entities.Album;
import com.apress.prospring5.ch8.entities.Singer;
import org.springframework.data.jpa.repository.JpaRepository;
import org.springframework.data.jpa.repository.Query;
import org.springframework.data.repository.query.Param;

import java.util.List;

public interface AlbumRepository extends JpaRepository<Album, Long> {
    List<Album> findBySinger(Singer singer);

    @Query("select a from Album a where a.title like %:title%")
    List<Album> findByTitle(@Param("title") String t);
}
```

上述查询中有一个名为 title 的命名参数。当命名参数的名称与使用@Query 注解的方法中的参数名称相同时，就不需要@Param 注解了。但是如果方法参数具有不同的名称，则需要使用@Param 注解来告诉 Spring，该参数的值将被注入到查询的指定参数中。

AlbumServiceImpl 非常简单，只使用 albumRepository bean 来调用它的方法。

```
//AlbumService.java
package com.apress.prospring5.ch8.services;

import com.apress.prospring5.ch8.entities.Album;
import com.apress.prospring5.ch8.entities.Singer;
import java.util.List;

public interface AlbumService {
    List<Album> findBySinger(Singer singer);
    List<Album> findByTitle(String title);
}

//AlbumServiceImpl.java
package com.apress.prospring5.ch8.services;

import com.apress.prospring5.ch8.entities.Album;
import com.apress.prospring5.ch8.entities.Singer;
import com.apress.prospring5.ch8.repos.AlbumRepository;
import org.springframework.beans.factory.annotation.Autowired;
import org.springframework.stereotype.Service;
import org.springframework.transaction.annotation.Transactional;
import java.util.List;
@Service("springJpaAlbumService")
@Transactional
public class AlbumServiceImpl implements AlbumService {
    @Autowired
    private AlbumRepository albumRepository;

    @Transactional(readOnly=true)
    @Override public List<Album> findBySinger(Singer singer) {
        return albumRepository.findBySinger(singer);
    }

    @Override public List<Album> findByTitle(String title) {
        return albumRepository.findByTitle(title);
    }
}
```

要测试 findByTitle(..)方法，可以使用以下测试类：

```
package com.apress.prospring5.ch8;

import com.apress.prospring5.ch8.config.DataJpaConfig;
import com.apress.prospring5.ch8.entities.Album;
import com.apress.prospring5.ch8.services.AlbumService;
import org.junit.After;
import org.junit.Before;
import org.junit.Test;
import org.slf4j.Logger;
import org.slf4j.LoggerFactory;
import org.springframework.context.annotation.AnnotationConfigApplicationContext;
import org.springframework.context.support.GenericApplicationContext;

import java.util.List;

import static org.junit.Assert.assertEquals;
import static org.junit.Assert.assertNotNull;
import static org.junit.Assert.assertTrue;

public class SingerDataJPATest {

    private static Logger logger =
        LoggerFactory.getLogger(SingerDataJPATest.class);

    private GenericApplicationContext ctx;
    private AlbumService albumService;

    @Before
    public void setUp(){
        ctx = new AnnotationConfigApplicationContext(DataJpaConfig.class);
        albumService = ctx.getBean(AlbumService.class);
        assertNotNull(albumService);
    }

    @Test
    public void testFindByTitle(){
        List<Album> albums = albumService.findByTitle("The");
        assertTrue(albums.size() > 0);
```

```
        assertEquals(2, albums.size());
        albums.forEach(a -> logger.info(a.toString() + ", Singer: "
                + a.getSinger().getFirstName() + " "
            + a.getSinger().getLastName()));
    }

    @After
    public void tearDown() {
        ctx.close();
    }
}
```

如果运行上述测试类，testFindByTitle 将通过测试，并将两张专辑的详细信息打印到控制台中。

```
INFO c.a.p.c.SingerDataJPATest - Album - id: 1, Singer id: 1,
   Title: The Search For Everything, Release Date: 2017-01-20, Singer: John Mayer
INFO c.a.p.c.SingerDataJPATest - Album - id: 3, Singer id: 2,
   Title: From The Cradle, Release Date: 1994-09-13, Singer: Eric Clapton
```

跟踪实体类的变化

在大多数应用程序中，需要跟踪用户维护的业务数据的基本审计活动。审计信息通常包括创建数据的用户、创建日期、上次修改数据的日期以及用户。

Spring Data JPA 项目以 JPA 实体监听器的形式提供了上述功能，可帮助自动跟踪审计信息。在 Spring 4 之后，要想使用该功能，实体类需要实现 Auditable<U, ID extends Serializable, T extends TemporalAccessor> extends Persistable<ID>接口(属于 Spring Data Commons)或扩展任何实现该接口的类。以下代码片段显示了从 Spring Data 的参考文档中提取的 Auditable 接口：

```
package org.springframework.data.domain;

import java.io.Serializable;
import java.time.temporal.TemporalAccessor;
import java.util.Optional;

public interface Auditable<U, ID extends Serializable,
    T extends TemporalAccessor> extends Persistable<ID> {
    Optional<U> getCreatedBy();

    void setCreatedBy(U createdBy);

    Optional<T> getCreatedDate();

    void setCreatedDate(T creationDate);

    Optional<U> getLastModifiedBy();

    void setLastModifiedBy(U lastModifiedBy);

    Optional<T> getLastModifiedDate();

    void setLastModifiedDate(T lastModifiedDate);
}
```

为了说明该接口是如何工作的，接下来在数据库模式中创建一个名为 SINGER_AUDIT 的新表，该表基于 SINGER 表，并添加了四个与审计相关的列。以下代码片段显示了表创建脚本(schema.sql)：

```
CREATE TABLE SINGER_AUDIT (
    ID INT NOT NULL AUTO_INCREMENT
  , FIRST_NAME VARCHAR(60) NOT NULL
  , LAST_NAME VARCHAR(40) NOT NULL
  , BIRTH_DATE DATE
  , VERSION INT NOT NULL DEFAULT 0
  , CREATED_BY VARCHAR(20)
  , CREATED_DATE TIMESTAMP
  , LAST_MODIFIED_BY VARCHAR(20)
  , LAST_MODIFIED_DATE TIMESTAMP
  , UNIQUE UQ_SINGER_AUDIT_1 (FIRST_NAME, LAST_NAME)
  , PRIMARY KEY (ID)
);
```

从前面所示的 Auditable 接口的定义可以看出，日期类型列仅限于扩展 java.time.temporal.TemporalAccessor 的类型。

从 Spring 5 开始，实现 Auditable<U,ID extends Serializable>不再是必需的，因为所有内容都可以由注解替代。四个带下划线的列表示与审计相关。请注意@CreatedBy、@CreatedDate、@LastModifiedBy 和@LastModifiedDate 注解。使用了这些注解，对日期列的类型限制就不再适用了。下一步是创建名为 SingerAudit 的实体类。以下代码片段显示了 SingerAudit 类：

```
package com.apress.prospring5.ch8.entities;

import org.springframework.data.annotation.CreatedBy;
import org.springframework.data.annotation.CreatedDate;
import org.springframework.data.annotation.LastModifiedBy;
import org.springframework.data.annotation.LastModifiedDate;
import org.springframework.data.jpa.domain.support.AuditingEntityListener;

import javax.persistence.*;
import java.io.Serializable;
import java.util.Date;
import java.util.Optional;

import static javax.persistence.GenerationType.IDENTITY;

@Entity
@EntityListeners(AuditingEntityListener.class)
@Table(name = "singer_audit")
public class SingerAudit implements Serializable {

    @Id
    @GeneratedValue(strategy = IDENTITY)
    @Column(name = "ID")
    private Long id;

    @Version
    @Column(name = "VERSION")
    private int version;

    @Column(name = "FIRST_NAME")
    private String firstName;

    @Column(name = "LAST_NAME")
    private String lastName;

    @Temporal(TemporalType.DATE)
    @Column(name = "BIRTH_DATE")
    private Date birthDate;

    @CreatedDate
    @Column(name = "CREATED_DATE")
    @Temporal(TemporalType.TIMESTAMP)
    private Date createdDate;

    @CreatedBy
    @Column(name = "CREATED_BY")
    private String createdBy;

    @LastModifiedBy
    @Column(name = "LAST_MODIFIED_BY")
    private String lastModifiedBy;

    @LastModifiedDate
    @Column(name = "LAST_MODIFIED_DATE")
    @Temporal(TemporalType.TIMESTAMP)
    private Date lastModifiedDate;

    public Long getId() {
        return this.id;
    }

    public String getFirstName() {
        return this.firstName;
    }

    public void setFirstName(String firstName) {
        this.firstName = firstName;
    }

    public String getLastName() {
        return this.lastName;
```

```
    }

    public void setLastName(String lastName) {
        this.lastName = lastName;
    }

    public Date getBirthDate() {
        return this.birthDate;
    }

    public void setBirthDate(Date birthDate) {
        this.birthDate = birthDate;
    }

    public Optional<String> getCreatedBy() {
        return Optional.of(createdBy);
    }

    public void setCreatedBy(String createdBy) {
        this.createdBy = createdBy;
    }

    public Optional<Date> getCreatedDate() {
        return Optional.of(createdDate);
    }

    public void setCreatedDate(Date createdDate) {
        this.createdDate = createdDate;
    }

    public Optional<String> getLastModifiedBy() {
        return Optional.of(lastModifiedBy);
    }

    public void setLastModifiedBy(String lastModifiedBy) {
        this.lastModifiedBy = lastModifiedBy;
    }

    public Optional<Date> getLastModifiedDate() {
        return Optional.of(lastModifiedDate);
    }

    public void setLastModifiedDate(Date lastModifiedDate) {
        this.lastModifiedDate = lastModifiedDate;
    }

    public String toString() {
        return "Singer - Id: " + id + ", First name: " + firstName
            + ", Last name: " + lastName + ", Birthday: " + birthDate
            + ", Created by: " + createdBy + ", Create date: " + createdDate
            + ", Modified by: " + lastModifiedBy + ", Modified date: "
            + lastModifiedDate;
    }
}
```

@Column 注解用于审计字段，以映射到表中的实际列。@EntityListeners(AuditingEntityListener.class)注解将AuditingEntityListener 注册为仅用于持久化上下文中的实体。在需要多个实体类的更复杂的示例中，可以将审计功能分离到用@MappedSuperclass 注解的抽象类中，并且该类也用@EntityListeners(AuditingEntityListener.class)进行了注解。如果 SingerAudit 是此类层次结构的一部分，那么它必须扩展以下类：

```
package com.apress.prospring5.ch8.entities;

import org.springframework.data.annotation.CreatedBy;
import org.springframework.data.annotation.CreatedDate;
import org.springframework.data.annotation.LastModifiedBy;
import org.springframework.data.annotation.LastModifiedDate;
import org.springframework.data.jpa.domain.support.AuditingEntityListener;

import javax.persistence.*;
import java.io.Serializable;
import java.util.Date;
import java.util.Optional;

@MappedSuperclass
@EntityListeners(AuditingEntityListener.class)
public abstract class AuditableEntity<U> implements Serializable {
```

```
    @CreatedDate
    @Column(name = "CREATED_DATE")
    @Temporal(TemporalType.TIMESTAMP)
    protected Date createdDate;

    @CreatedBy
    @Column(name = "CREATED_BY")
    protected String createdBy;

    @LastModifiedBy
    @Column(name = "LAST_MODIFIED_BY")
    protected String lastModifiedBy;

    @LastModifiedDate
    @Column(name = "LAST_MODIFIED_DATE")
    @Temporal(TemporalType.TIMESTAMP)
    protected Date lastModifiedDate;

    public Optional<String> getCreatedBy() {
        return Optional.of(createdBy);
    }

    public void setCreatedBy(String createdBy) {
        this.createdBy = createdBy;
    }

    public Optional<Date> getCreatedDate() {
        return Optional.of(createdDate);
    }

    public void setCreatedDate(Date createdDate) {
        this.createdDate = createdDate;
    }

    public Optional<String> getLastModifiedBy() {
        return Optional.of(lastModifiedBy);
    }

    public void setLastModifiedBy(String lastModifiedBy) {
        this.lastModifiedBy = lastModifiedBy;
    }

    public Optional<Date> getLastModifiedDate() {
        return Optional.of(lastModifiedDate);
    }

    public void setLastModifiedDate(Date lastModifiedDate) {
        this.lastModifiedDate = lastModifiedDate;
    }
}
```

这样就大大减少了 SingerAudit 类的代码量，如下所示：

```
package com.apress.prospring5.ch8.entities;

import javax.persistence.*;
import java.util.Date;

import static javax.persistence.GenerationType.IDENTITY;

@Entity
@Table(name = "singer_audit")
public class SingerAudit extends AuditableEntity<SingerAudit> {

    @Id
    @GeneratedValue(strategy = IDENTITY)
    @Column(name = "ID")
    private Long id;

    @Version
    @Column(name = "VERSION")
    private int version;

    @Column(name = "FIRST_NAME")
    private String firstName;
    @Column(name = "LAST_NAME")
    private String lastName;
```

```
    @Temporal(TemporalType.DATE)
    @Column(name = "BIRTH_DATE")
    private Date birthDate;

    public Long getId() {
        return this.id;
    }

    public String getFirstName() {
        return this.firstName;
    }

    public void setFirstName(String firstName) {
        this.firstName = firstName;
    }

    public String getLastName() {
        return this.lastName;
    }

    public void setLastName(String lastName) {
        this.lastName = lastName;
    }

    public Date getBirthDate() {
        return this.birthDate;
    }

    public void setBirthDate(Date birthDate) {
        this.birthDate = birthDate;
    }

    public String toString() {
        return "Singer - Id: " + id + ", First name: " + firstName
            + ", Last name: " + lastName + ", Birthday: " + birthDate
            + ", Created by: " + createdBy + ", Create date: " + createdDate
            + ", Modified by: " + lastModifiedBy + ", Modified date: "
            + lastModifiedDate;
    }
}
```

下面的代码片段显示了 SingerAuditService 接口，此时只定义了几个用来演示审计功能的方法：

```
package com.apress.prospring5.ch8.services;

import com.apress.prospring5.ch8.entities.SingerAudit;

import java.util.List;

public interface SingerAuditService {
    List<SingerAudit> findAll();
    SingerAudit findById(Long id);
    SingerAudit save(SingerAudit singer);
}
```

SingerAuditRepository 接口只是扩展了 CrudRepository，后者已经实现了将用于 SingerAuditService 的所有方法。findById()方法由 CrudRepository.findOne()方法实现。以下代码片段显示了服务实现类 SingerAuditServiceImpl：

```
package com.apress.prospring5.ch8.services;

import com.apress.prospring5.ch8.entities.SingerAudit;
import com.apress.prospring5.ch8.repos.SingerAuditRepository;
import org.springframework.stereotype.Service;
import org.springframework.transaction.annotation.Transactional;
import org.springframework.beans.factory.annotation.Autowired;

import java.util.List;
import com.google.common.collect.Lists;

@Service("singerAuditService")
@Transactional
public class SingerAuditServiceImpl implements SingerAuditService {

    @Autowired
    private SingerAuditRepository singerAuditRepository;

    @Transactional(readOnly=true)
    public List<SingerAudit> findAll() {
```

```
        return Lists.newArrayList(singerAuditRepository.findAll());
    }

    public SingerAudit findById(Long id) {
        return singerAuditRepository.findOne(id).get();
    }

    public SingerAudit save(SingerAudit singer) {
        return singerAuditRepository.save(singer);
    }
}
```

此外，还需要完成一些配置工作。通过使用 XML 配置，可以将提供审计服务的 AuditingEntityListener<T> JPA 实体监听器声明到一个名为/src/main/resources/META-INF/orm.xml(必须使用该文件名，因为这是 JPA 规范中的规定)的文件中，而该文件位于项目根文件夹下。声明监听器，如下所示：

```
<?xml version="1.0" encoding="UTF-8" ?>

<entity-mappings xmlns="http://java.sun.com/xml/ns/persistence/orm"
    xmlns:xsi="http://www.w3.org/2001/XMLSchema-instance"
    xsi:schemaLocation="http://java.sun.com/xml/ns/persistence/orm
        http://java.sun.com/xml/ns/persistence/orm_2_0.xsd"
        version="2.0">
    <description>JPA</description>
    <persistence-unit-metadata>
        <persistence-unit-defaults>
            <entity-listeners>
                <entity-listener class="org.springframework.data.jpa.domain.support.
                AuditingEntityListener" />
        </entity-listeners>
        </persistence-unit-defaults>
    </persistence-unit-metadata>
</entity-mappings>
```

通过使用注解配置，该文件被替换为@EntityListeners(AuditingEntityListener.class)注解。JPA 提供程序将在持久化操作(保存和更新事件)期间发现用于审计字段处理的监听器。Spring 配置的其余部分与前面的内容几乎完全相同，但有如下例外：添加了启用审计功能的新的配置注解。

```
package com.apress.prospring5.ch8.com.apress.prospring5.ch8.config;

import org.slf4j.Logger;
import org.slf4j.LoggerFactory;
import org.springframework.context.annotation.Bean;
import org.springframework.context.annotation.ComponentScan;
import org.springframework.context.annotation.Configuration;
import org.springframework.data.jpa.repository.config.EnableJpaAuditing;
import org.springframework.data.jpa.repository.config.EnableJpaRepositories;
import org.springframework.jdbc.datasource.embedded.EmbeddedDatabaseBuilder;
import org.springframework.jdbc.datasource.embedded.EmbeddedDatabaseType;
import org.springframework.orm.jpa.JpaTransactionManager;
import org.springframework.orm.jpa.JpaVendorAdapter;
import org.springframework.orm.jpa.LocalContainerEntityManagerFactoryBean;
import org.springframework.orm.jpa.vendor.HibernateJpaVendorAdapter;
import org.springframework.transaction.PlatformTransactionManager;
import org.springframework.transaction.annotation.EnableTransactionManagement;

import javax.persistence.EntityManagerFactory;
import javax.sql.DataSource;
import java.util.Properties;

@Configuration
@EnableTransactionManagement
@ComponentScan(basePackages = {"com.apress.prospring5.ch8"})
@EnableJpaRepositories(basePackages = {"com.apress.prospring5.ch8.repos"})
@EnableJpaAuditing(auditorAwareRef = "auditorAwareBean")
public class AuditConfig {
    // same content as the configuration in the previous section
    ...
}
```

@EnableJpaAuditing(auditorAwareRef ="auditorAwareBean")注解等价于通过使用 XML 配置启用 JPA 审计功能的<jpa:auditing auditor-aware-ref="auditorAwareBean"/>元素。auditorAwareBean bean 提供了用户信息。以下代码片段显示了 AuditorAwareBean 类：

```
package com.apress.prospring5.ch8;
```

```
import org.springframework.data.domain.AuditorAware;
import org.springframework.stereotype.Component;

import java.util.Optional;

@Component
public class AuditorAwareBean implements AuditorAware<String> {
    public Optional<String> getCurrentAuditor() {
        return Optional.of("prospring5");
    }
}
```

AuditorAwareBean 实现了 AuditorAware<T>接口，并传入 String 类型。在实际情况中，传入的应该是用户信息的实例，例如，表示正在执行数据更新操作的登录用户的 User 类。此时使用 String 只是为了简单起见。在 AuditorAwareBean 类中，实现了 getCurrentAuditor()方法，并将值硬编码为 prospring5。而在实际情况中，应该从底层的安全基础设施中获取用户。例如，在 Spring Security 中，可以从 SecurityContextHolder 类检索用户信息。

现在所有的实现工作都已完成，下面的代码片段显示了 SpringAuditJPADemo 测试程序：

```
package com.apress.prospring5.ch8;

import java.util.GregorianCalendar;
import java.util.List;
import java.util.Date;

import com.apress.prospring5.ch8.config.AuditConfig;
import com.apress.prospring5.ch8.entities.SingerAudit;
import com.apress.prospring5.ch8.services.SingerAuditService;
import org.springframework.context.annotation.AnnotationConfigApplicationContext;
import org.springframework.context.support.GenericApplicationContext;

public class SpringAuditJPADemo {
    public static void main(String... args) {
        GenericApplicationContext ctx =
            new AnnotationConfigApplicationContext(AuditConfig.class);

        SingerAuditService singerAuditService = ctx.getBean(SingerAuditService.class);

        List<SingerAudit> singers = singerAuditService.findAll();
        listSingers(singers);
        System.out.println("Add new singer");
        SingerAudit singer = new SingerAudit();
        singer.setFirstName("BB");
        singer.setLastName("King");
        singer.setBirthDate(new Date(
                (new GregorianCalendar(1940, 8, 16)).getTime().getTime()));
        singerAuditService.save(singer);
        singers = singerAuditService.findAll();
        listSingers(singers);

        singer = singerAuditService.findById(4l);
        System.out.println("");
        System.out.println("Singer with id 4:" + singer);
        System.out.println("");

        System.out.println("Update singer");
        singer.setFirstName("John Clayton");
        singerAuditService.save(singer);
        singers = singerAuditService.findAll();
        listSingers(singers);

        ctx.close();
    }

    private static void listSingers(List<SingerAudit> singerAudits) {
        System.out.println("");
        System.out.println("Listing singers without details:");
        for (SingerAudit audit: singerAudits) {
            System.out.println(audit);
            System.out.println();
        }
    }
}
```

在 main()方法中，列出了插入新歌手以及更新信息之后的歌手审计信息。运行该程序会产生以下输出：

```
Add new singer
Listing singers without details:
// other singers ...
Singer - Id: 4, First name: BB, Last name: King, Birthday: 1940-09-16,
    Created by: prospring5, Create date: 2017-05-07 14:19:02.96,
    Modified by: prospring5, Modified date: 2017-05-07 14:19:02.96

Update singer
Listing singers without details:
// other singers ...
Singer - Id: 4, First name: Riley B., Last name: King, Birthday: 1940-09-16,
  Created by: prospring5, Create date: 2017-05-07 14:33:15.645,
  Modified by: prospring5, Modified date: 2017-05-07 14:33:15.663
```

在上面的输出中，可以看到在创建新歌手之后，创建日期和最后修改日期是相同的。但是，在更新之后，最后修改日期会更新。审计功能是 Spring Data JPA 提供的另一个便利功能，因此不需要亲自实现逻辑。

8.10 通过使用 Hibernate Envers 保存实体版本

在企业级应用程序中，对于关键业务数据，始终需要保留每个实体的版本。例如，在客户关系管理(CRM)系统中，当每次插入、更新或删除客户记录时，应将先前版本保存在历史记录或审计表中，以满足公司的审计或其他合规要求。

要做到这一点，有两种常见的选择。第一种选择是创建数据库触发器，在执行任何更新操作之前将更新前的记录克隆到历史记录表中。第二种选择是在数据访问层(例如，通过使用 AOP)开发相关的逻辑。但是，这两种选择都有各自的缺点。触发器方法与数据库平台绑定，而手动实现逻辑相当不灵活且容易出错。

Hibernate Envers(“entity versioning system”(实体版本控制系统)的缩写)是一个 Hibernate 模块，专门用于实现实体版本的自动化。在本节中，将讨论如何使用 Envers 实现 SingerAudit 实体的版本控制。

Hibernate Envers 并不是 JPA 的功能。之所以在本节介绍它，是因为我们认为在讨论了 Spring Data JPA 中可用的一些基本审计功能之后，介绍 Hibernate Envers 是比较合适的。

Envers 支持两种审计策略，如表 8-1 所示。

表 8-1 Envers 审计策略

审计策略	描述
默认策略	Envers 维护一个用于修改记录的列。每次插入或更新记录时，会使用从数据库序列或表中检索到的版本号将新记录插入到历史记录表中
有效性审计策略	该策略将存储每条历史记录的开始和最终版本。每次插入或更新记录时，都会使用开始版本号将新记录插入到历史记录表中。同时，以前的记录用最终版本号进行更新。也可以配置 Envers，记录将最终版本更新到以前的历史记录中的时间戳

在本节中，将演示如何应用有效性审计策略。尽管这会触发更多的数据库更新，但检索历史记录的速度会更快。由于最终版本时间戳也被写入了历史记录，因此在查询数据时可以更容易地在特定时间点识别记录的快照。

8.10.1 为实体版本控制添加表

为了支持实体版本控制，需要添加一些表。首先，对于实体(本例中为 SingerAudit 实体类)将进行版本控制的每个表，都需要创建相应的历史记录表；而对于 SINGER_AUDIT 表中记录的版本控制，则创建一个名为 SINGER_AUDIT_H 的历史记录表。以下代码片段显示了表创建脚本(schema.sql)：

```
CREATE TABLE SINGER_AUDIT_H (
    ID INT NOT NULL AUTO_INCREMENT
  , FIRST_NAME VARCHAR(60) NOT NULL
  , LAST_NAME VARCHAR(40) NOT NULL
  , BIRTH_DATE DATE
  , VERSION INT NOT NULL DEFAULT 0
  , CREATED_BY VARCHAR(20)
  , CREATED_DATE TIMESTAMP
```

```
    , LAST_MODIFIED_BY VARCHAR(20)
    , LAST_MODIFIED_DATE TIMESTAMP
    , AUDIT_REVISION INT NOT NUL
    , ACTION_TYPE INT
    , AUDIT_REVISION_END INT
    , AUDIT_REVISION_END_TS TIMESTAMP
    , UNIQUE UQ_SINGER_AUDIT_H_1 (FIRST_NAME, LAST_NAME)
    , PRIMARY KEY (ID, AUDIT_REVISION)
);
```

为了支持有效性审计策略，需要为每个历史记录表添加四列，如上面脚本中含有下划线的代码所示。表 8-2 显示了这些列及其用途。此外，Hibernate Envers 还需要另一个表来跟踪版本号和创建每个版本的时间戳。表名应该是 REVINFO。

表 8-2　Envers 审计策略

审计策略	数据	描述
AUDIT_REVISION	INT	历史记录的开始版本
ACTION_TYPE	INT	操作类型，包含以下可能值：0 表示添加，1 表示修改，2 表示删除
AUDIT_REVISION_END	INT	历史记录的最终版本
AUDIT_REVISION_END_TS	TIMESTAMP	最终版本被更新的时间戳

以下代码片段显示了表创建脚本(schema.sql)：

```
CREATE TABLE REVINFO (
   REVTSTMP BIGINT NOT NULL
   , REV INT NOT NULL AUTO_INCREMENT
   , PRIMARY KEY (REVTSTMP, REV)
);
```

REV 列用于存储每个版本号，当创建新的历史记录时它将自动递增。REVTSTMP 列存储创建版本时的时间戳(以数字格式显示)。

8.10.2　为实体版本控制配置 EntityManagerFactory

Hibernate Envers 以 EJB 监听器的形式实现。可以在 LocalContainerEntityManagerFactory bean 中配置这些监听器。下面显示了 Java 配置类。显示 XML 配置没有意义，因为本节的唯一区别是添加了很多额外的 Hibernate 特定属性。

```
package com.apress.prospring5.ch8.config;

import org.hibernate.envers.boot.internal.EnversServiceImpl;
import org.hibernate.envers.event.spi.EnversPostUpdateEventListenerImpl;
import org.hibernate.event.spi.PostUpdateEventListener;
import org.slf4j.Logger;
import org.slf4j.LoggerFactory;
import org.springframework.context.annotation.Bean;
import org.springframework.context.annotation.ComponentScan;
import org.springframework.context.annotation.Configuration;
import org.springframework.data.jpa.domain.support.AuditingEntityListener;
import org.springframework.data.jpa.repository.config.EnableJpaAuditing;
import org.springframework.data.jpa.repository.config.EnableJpaRepositories;
import org.springframework.jdbc.datasource.embedded.EmbeddedDatabaseBuilder;
import org.springframework.jdbc.datasource.embedded.EmbeddedDatabaseType;
import org.springframework.orm.jpa.JpaTransactionManager;
import org.springframework.orm.jpa.JpaVendorAdapter;
import org.springframework.orm.jpa.LocalContainerEntityManagerFactoryBean;
import org.springframework.orm.jpa.vendor.HibernateJpaVendorAdapter;
import org.springframework.transaction.PlatformTransactionManager;
import org.springframework.transaction.annotation.EnableTransactionManagement;

import javax.persistence.EntityManagerFactory;
import javax.sql.DataSource;
import java.util.Properties;

@Configuration
@EnableTransactionManagement
@ComponentScan(basePackages = {"com.apress.prospring5.ch8"})
@EnableJpaRepositories(basePackages = {"com.apress.prospring5.ch8.repos"})
@EnableJpaAuditing(auditorAwareRef = "auditorAwareBean")
public class EnversConfig {
```

```
    private static Logger logger = LoggerFactory.getLogger(EnversConfig.class);

    @Bean
    public DataSource dataSource() {
        try {
            EmbeddedDatabaseBuilder dbBuilder = new EmbeddedDatabaseBuilder();
            return dbBuilder.setType(EmbeddedDatabaseType.H2)
            .addScripts("classpath:db/schema.sql", "classpath:db/test-data.sql").build();
        } catch (Exception e) {
            logger.error("Embedded DataSource bean cannot be created!", e);
            return null;
        }
    }

    @Bean
    public PlatformTransactionManager transactionManager() {
        return new JpaTransactionManager(entityManagerFactory());
    }

    @Bean
    public JpaVendorAdapter jpaVendorAdapter() {
        return new HibernateJpaVendorAdapter();
    }

    @Bean
    public Properties hibernateProperties() {
        Properties hibernateProp = new Properties();
        hibernateProp.put("hibernate.dialect", "org.hibernate.dialect.H2Dialect");
        hibernateProp.put("hibernate.format_sql", true);
        hibernateProp.put("hibernate.show_sql", true);
        hibernateProp.put("hibernate.max_fetch_depth", 3);
        hibernateProp.put("hibernate.jdbc.batch_size", 10);
        hibernateProp.put("hibernate.jdbc.fetch_size", 50);
        //Properties for Hibernate Envers
        hibernateProp.put("org.hibernate.envers.audit_table_suffix", "_H");
        hibernateProp.put("org.hibernate.envers.revision_field_name",
            "AUDIT_REVISION");
        hibernateProp.put("org.hibernate.envers.revision_type_field_name",
            "ACTION_TYPE");
        hibernateProp.put("org.hibernate.envers.audit_strategy",

            "org.hibernate.envers.strategy.ValidityAuditStrategy");
        hibernateProp.put(

            "org.hibernate.envers.audit_strategy_validity_end_rev_field_name",
            "AUDIT_REVISION_END");
        hibernateProp.put(

            "org.hibernate.envers.audit_strategy_validity_store_revend_timestamp",
            "True");
        hibernateProp.put(

        "org.hibernate.envers.audit_strategy_validity_revend_timestamp_field_name",
                "AUDIT_REVISION_END_TS");
        return hibernateProp;
    }

    @Bean
    public EntityManagerFactory entityManagerFactory() {
        LocalContainerEntityManagerFactoryBean factoryBean =
            new LocalContainerEntityManagerFactoryBean();
        factoryBean.setPackagesToScan("com.apress.prospring5.ch8.entities");
        factoryBean.setDataSource(dataSource());
        factoryBean.setJpaVendorAdapter(new HibernateJpaVendorAdapter());
        factoryBean.setJpaProperties(hibernateProperties());
        factoryBean.setJpaVendorAdapter(jpaVendorAdapter());
        factoryBean.afterPropertiesSet();
        return factoryBean.getObject();
    }
}
```

Envers 审计事件监听器 org.hibernate.envers.event.AuditEventListener 被附加到各种持久化事件中。监听器将拦截 post-insert、post-update 或 post-delete 事件，并将实体类的更新前快照克隆到历史记录表中。此外，监听器还被附加到那些关联的更新事件(pre-collection-update、pre-collection-remove 和 pre-collection-recreate)，以便处理与实体类关联的更新操作。Envers 能够保持关联(例如一对多关联或多对多关联)中实体的历史。这里还为 Hibernate Envers 定义了一些属性，在表 8-3 中对这些属性进行了总结(为了简单起见，省略了属性的前缀 org.hibernate.envers)[1]。

表 8-3 Hibernate Envers 属性表

属性	描述
audit_table_suffix	版本化实体的表名后缀。例如，对于映射到 SINGER_AUDIT 表的实体类 SingerAudit，由于为该属性定义了值_H，因此 Envers 将在表 SINGER_AUDIT_H 中保留该实体类的历史记录
revision_field_name	历史记录表的列，用于存储每条历史记录的版本号
revision_type_field_name	历史记录表的列，用于存储更新操作类型
audit_strategy	用于实体版本控制的审计策略
audit_strategy_validity_end_rev_field_name	历史记录表的列，用于存储每条历史记录的最终版本号。仅在使用有效性审计策略时才需要该属性
audit_strategy_validity_store_revend_timestamp	是否在更新每条历史记录的最终版本号时存储时间戳。仅在使用有效性审计策略时才需要该属性
audit_strategy_validity_revend_timestamp_field_name	历史记录表的列，用于存储更新每条历史记录的最终版本号时的时间戳。仅在使用有效性审计策略并且 audit_strategy_validity_store_revend_timestamp 属性设置为 true 时才需要该属性

8.10.3 启用实体版本控制和历史检索

想要启用实体版本控制，只需要使用@Audited 注解实体类即可。该注解可以在类级别使用，然后对所有字段的更改进行审计。如果想要从审计中忽略某些字段，则可以在这些字段上使用@NotAudited。下面所示的 SingerAudit 实体类应用了该注解：

```
package com.apress.prospring5.ch8.entities;

import org.hibernate.envers.Audited;
import org.springframework.data.jpa.domain.support.AuditingEntityListener;
...

@Entity
@Table(name = "singer_audit")
@Audited
@EntityListeners(AuditingEntityListener.class)
public class SingerAudit implements Serializable {
    ...
}
```

实体类用@Audited 注解，Envers 监听器将检查并执行更新实体的版本控制。默认情况下，Envers 会尝试保留关联的历史记录；如果想避免这种情况，应该使用前面提到的@NotAudited 注解。

为了检索历史记录，Envers 提供了 org.hibernate.envers.AuditReader 接口，它可以从 AuditReaderFactory 类中获取。接下来在 SingerAuditService 接口中添加一个名为 findAuditByRevision()的新方法，以便通过版本号检索 SingerAudit 历史记录。以下代码片段显示了 SingerAuditService 接口：

```
package com.apress.prospring5.ch8.services;

import com.apress.prospring5.ch8.entities.SingerAudit;

import java.util.List;

public interface SingerAuditService {
    List<SingerAudit> findAll();
    SingerAudit findById(Long id);
    SingerAudit save(SingerAudit singerAudit);
    SingerAudit findAuditByRevision(Long id, int revision);
}
```

要检索历史记录，一种选择是传入实体的 ID 和版本号。以下代码片段显示了用于提取修订版本的方法的实现：

```
package com.apress.prospring5.ch8.services;

import com.apress.prospring5.ch8.entities.SingerAudit;
```

1 可以在 Hibernate 官方文档中找到完整的 Hibernate 属性列表，网址为 https://docs.jboss.org/hibernate/orm/5.2/userguide/html_single/Hibernate_User_Guide.html#envers-configuration。

```
import com.apress.prospring5.ch8.repos.SingerAuditRepository;
import com.google.common.collect.Lists;
import org.hibernate.envers.AuditReader;
import org.hibernate.envers.AuditReaderFactory;
import org.springframework.beans.factory.annotation.Autowired;
import org.springframework.stereotype.Service;
import org.springframework.transaction.annotation.Transactional;

import javax.persistence.EntityManager;
import javax.persistence.PersistenceContext;
import java.util.List;

@Service("singerAuditService")
@Transactional
public class SingerAuditServiceImpl implements SingerAuditService {

   @Autowired
   private SingerAuditRepository singerAuditRepository;

   @PersistenceContext
   private EntityManager entityManager;

   @Transactional(readOnly = true)
   public List<SingerAudit> findAll() {
      return Lists.newArrayList(singerAuditRepository.findAll());
   }

   public SingerAudit findById(Long id) {
      return singerAuditRepository.findOne(id).get();
   }

   public SingerAudit save(SingerAudit singer) {
      return singerAuditRepository.save(singer);
   }

   @Transactional(readOnly = true)
   @Override
   public SingerAudit findAuditByRevision(Long id, int revision) {
      AuditReader auditReader = AuditReaderFactory.get(entityManager);
      return auditReader.find(SingerAudit.class, id, revision);
   }
}
```

首先将 EntityManager 注入到传递给 AuditReaderFactory 的类中，以检索 AuditReader 的实例。然后可以调用 AuditReader.find()方法来检索特定版本的 SingerAudit 实体的实例。

8.10.4 测试实体版本控制

接下来看看实体版本控制的工作原理。以下代码片段显示了测试代码；启动 ApplicationContext 和 listSingers()方法的代码与 SpringJpaDemo 类中的代码相同。

```
package com.apress.prospring5.ch8;

import java.util.GregorianCalendar;
import java.util.List;
import java.util.Date;

import com.apress.prospring5.ch8.config.EnversConfig;
import com.apress.prospring5.ch8.entities.SingerAudit;
import com.apress.prospring5.ch8.services.SingerAuditService;
import org.springframework.context.annotation.AnnotationConfigApplicationContext;
import org.springframework.context.support.GenericApplicationContext;
import org.springframework.context.support.GenericXmlApplicationContext;

public class SpringEnversJPADemo {
   public static void main(String... args) {
      GenericApplicationContext ctx =
         new AnnotationConfigApplicati onContext(EnversConfig.class);
      SingerAuditService singerAuditService =
         ctx.getBean(SingerAuditService.class);

      System.out.println("Add new singer");
      SingerAudit singer = new SingerAudit();
      singer.setFirstName("BB");
      singer.setLastName("King");
```

```
        singer.setBirthDate(new Date(
                (new GregorianCalendar(1940, 8, 16)).getTime().getTime()));
        singerAuditService.save(singer);
        listSingers(singerAuditService.findAll());

        System.out.println("Update singer");
        singer.setFirstName("Riley B.");
        singerAuditService.save(singer);
        listSingers(singerAuditService.findAll());
        SingerAudit oldSinger = singerAuditService.findAuditByRevision(4l, 1);
        System.out.println("");
        System.out.println("Old Singer with id 1 and rev 1:" + oldSinger);
        System.out.println("");
        oldSinger = singerAuditService.findAuditByRevision(4l, 2);
        System.out.println("");
        System.out.println("Old Singer with id 1 and rev 2:" + oldSinger);
        System.out.println("");

        ctx.close();
    }

    private static void listSingers(List<SingerAudit> singers) {
        System.out.println("");
        System.out.println("Listing singers:");
        for (SingerAudit singer: singers) {
            System.out.println(singer);
            System.out.println();
        }
    }
}
```

首先创建一位新歌手，然后更新。接着，分别检索版本 1 和版本 2 的 SingerAudit 实体。运行代码将生成以下输出：

```
Listing singers:
...//
Singer - Id: 4, First name: BB, Last name: King, Birthday: 1940-09-16,
    Created by: prospring5, Create date: 2017-05-11 23:50:14.778,
    Modified by: prospring5, Modified date: 2017-05-11 23:50:14.778

Old Singer with id 1 and rev 1:Singer - Id: 4,
    First name: BB, Last name: King, Birthday: 1940-09-16,
    Created by: prospring5, Create date: 2017-05-11 23:50:14.778,
    Modified by: prospring5, Modified date: 2017-05-11 23:50:14.778

Old Singer with id 1 and rev 2:Singer - Id: 4,
    First name: Riley B., Last name: King, Birthday: 1940-09-16,
    Created by: prospring5, Create date: 2017-05-11 23:50:14.778,
    Modified by: prospring5, Modified date: 2017-05-11 23:50:15.0
```

从上面的输出可以看到，在更新操作之后，SingerAudit 的名字更改为 Riley B。但是，在查看历史记录时，版本 1 中的名字是 BB；而在版本 2 中，名字变成了 Riley B。你还注意到，版本 2 的最后修改日期正确反映了更新的日期和时间。

8.11 Spring Boot JPA

到现在为止，已经配置了一切，包括实体、数据库、存储库和服务。此时，你应该可以想到，应该有一个 Spring Boot 启动器来帮助更快地开发项目并最大限度地减少配置工作。Spring Boot JPA 启动器依赖 Spring Boot JPA，因此它附带了预配置的嵌入式数据库；你需要做的是将依赖项放在类路径上。此外，它还带有 Hibernate 来抽象持久化层。Spring Repository 接口会被自动检测到。所以，开发人员所需要做的就是提供实体、存储库扩展以及将两者结合在一起使用的 Application 类。最后，还可以开发一个类来填充数据库。所有这些工作都将在本节中完成并解释。

首先，需要将 Spring Boot JPA 启动器作为依赖项添加。具体操作与之前的所有其他库一样。在根文件夹的 build.gradle 文件中添加版本、组 ID 和工件 ID，并在 chapter08/boot-jpa/build.gradle 文件中定义依赖项。下面的代码显示了这两个配置片段：

```
//pro-spring-15/build.gradle
ext {
    //spring libs
    springVersion = '5.0.0.M5'
    bootVersion = '2.0.0.BUILD-SNAPSHOT'
    springDataVersion = '2.0.0.M2'
```

```
    boot = [
           ...
           starterJdbc :
              "org.springframework.boot:spring-boot-starter-jdbc:$bootVersion",
           starterJpa :
              "org.springframework.boot:spring-boot-starter-data-jpa:$bootVersion"
    ]

    testing = [
           junit: "junit:junit:$junitVersion"
    ]

    db = [
           ...
           h2 : "com.h2database:h2:$h2Version"
    ]
}

//chapter08/boot-jpa/build.gradle
buildscript {
    repositories {
       mavenLocal
       mavenCentral
       maven { url "http://repo.spring.io/release" }
       maven { url "http://repo.spring.io/milestone" }
       maven { url "http://repo.spring.io/snapshot" }
       maven { url "https://repo.spring.io/libs-snapshot" }
    }
    dependencies {
       classpath boot.springBootPlugin
    }
}
    apply plugin: 'org.springframework.boot'

    dependencies {
       compile boot.starterJpa, db.h2
    }
}
```

在添加了这些依赖项并刷新项目后，在 IntelliJ IDEA Gradle 项目视图中应显示整个 spring-bootstarter-data-jpa 依赖关系树，如图 8-2 所示。

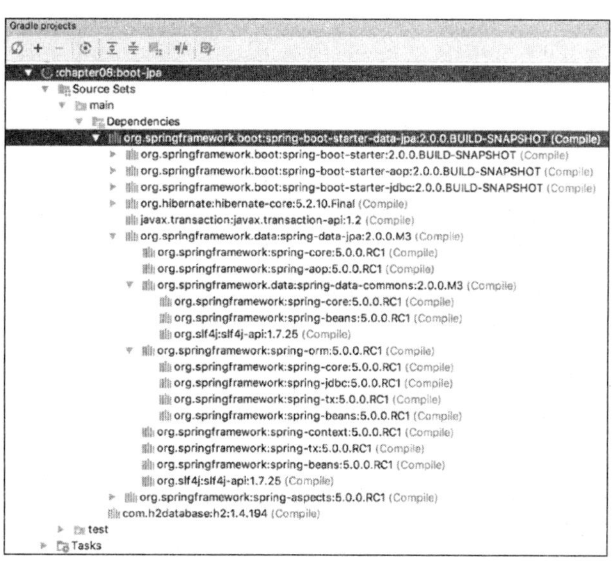

图 8-2 Spring Boot 启动器的依赖关系树

这些实体与前面使用的实体(Singer、Album 和 Instrument)相似，因此不需要再对它们进行描述。初始化这些实体的 bean 是 DBInitializer 类型，它是一个服务类，它使用 Spring Boot 提供的 SingerRepository 和 InstrumentRepository bean 将一组对象保存到数据库中。其内容如下所示：

```
package com.apress.prospring5.ch8.config;
```

```
import com.apress.prospring5.ch8.InstrumentRepository;
import com.apress.prospring5.ch8.SingerRepository;
import com.apress.prospring5.ch8.entities.Album;
import com.apress.prospring5.ch8.entities.Instrument;
import com.apress.prospring5.ch8.entities.Singer;
import org.slf4j.Logger;
import org.slf4j.LoggerFactory;
import org.springframework.beans.factory.annotation.Autowired;
import org.springframework.stereotype.Service;

import javax.annotation.PostConstruct;
import java.util.Date;
import java.util.GregorianCalendar;

@Service
public class DBInitializer {
   private Logger logger = LoggerFactory.getLogger(DBInitializer.class);

   @Autowired
   SingerRepository singerRepository;

   @Autowired
   InstrumentRepository instrumentRepository;

   @PostConstruct
   public void initDB(){
      logger.info("Starting database initialization...");

      Instrument guitar = new Instrument();
      guitar.setInstrumentId("Guitar");
      instrumentRepository.save(guitar);

      Instrument piano = new Instrument();
      piano.setInstrumentId("Piano");
      instrumentRepository.save(piano);

      Instrument voice = new Instrument();
      voice.setInstrumentId("Voice");
      instrumentRepository.save(voice);

      Singer singer = new Singer();
      singer.setFirstName("John");
      singer.setLastName("Mayer");
      singer.setBirthDate(new Date(
              (new GregorianCalendar(1977, 9, 16)).getTime().getTime()));
      singer.addInstrument(guitar);
      singer.addInstrument(piano);

      Album album1 = new Album();
      album1.setTitle("The Search For Everything");
      album1.setReleaseDate(new java.sql.Date(
              (new GregorianCalendar(2017, 0, 20)).getTime().getTime()));
      singer.addAbum(album1);

      Album album2 = new Album();
      album2.setTitle("Battle Studies");
      album2.setReleaseDate(new java.sql.Date(
              (new GregorianCalendar(2009, 10, 17)).getTime().getTime()));
      singer.addAbum(album2);

      singerRepository.save(singer);

      singer = new Singer();
      singer.setFirstName("Eric");
      singer.setLastName("Clapton");
      singer.setBirthDate(new Date(
              (new GregorianCalendar(1945, 2, 30)).getTime().getTime()));
      singer.addInstrument(guitar);

      Album album = new Album();
      album.setTitle("From The Cradle");
      album.setReleaseDate(new java.sql.Date(
              (new GregorianCalendar(1994, 8, 13)).getTime().getTime()));
      singer.addAbum(album);

      singerRepository.save(singer);

      singer = new Singer();
```

```
        singer.setFirstName("John");
        singer.setLastName("Butler");
              singer.setBirthDate(new Date(
         (new GregorianCalendar(1975, 3, 1)).getTime().getTime()));

        singer.addInstrument(guitar);
        singerRepository.save(singer);
     logger.info("Database initialization finished.");
  }
}
```

本例中的 SingerRepository 和 InstrumentRepository 接口非常简单，如下所示：

```
//InstrumentRepository.java
package com.apress.prospring5.ch8;

import com.apress.prospring5.ch8.entities.Instrument;
import org.springframework.data.repository.CrudRepository;

public interface InstrumentRepository
   extends CrudRepository<Instrument, Long> {

}

//SingerRepository.java
package com.apress.prospring5.ch8;

import com.apress.prospring5.ch8.entities.Singer;
import org.springframework.data.repository.CrudRepository;

import java.util.List;

public interface SingerRepository
   extends CrudRepository<Singer, Long> {

   List<Singer> findByFirstName(String firstName);
   List<Singer> findByFirstNameAndLastName(String firstName, String lastName);
}
```

SingerRepository 将被直接注入到 Spring Boot 注解的 Application 类中，用于从数据库中检索所有歌手记录及其子记录并将它们记录在控制台中。Application 类如下所示：

```
package com.apress.prospring5.ch8;

import com.apress.prospring5.ch8.entities.Singer;
import org.slf4j.Logger;
import org.slf4j.LoggerFactory;
import org.springframework.beans.factory.annotation.Autowired;
import org.springframework.boot.CommandLineRunner;
import org.springframework.boot.SpringApplication;
import org.springframework.boot.autoconfigure.SpringBootApplication;
import org.springframework.context.ConfigurableApplicationContext;
import org.springframework.transaction.annotation.Transactional;

import java.util.List;

@SpringBootApplication(scanBasePackages = "com.apress.prospring5.ch8.config")
public class Application implements CommandLineRunner {

   private static Logger logger = LoggerFactory.getLogger(Application.class);
   @Autowired
   SingerRepository singerRepository;

   public static void main(String... args) throws Exception {
      ConfigurableApplicationContext ctx =
         SpringApplication.run(Application.class, args);
      System.in.read();
      ctx.close();
   }

   @Transactional(readOnly = true)
   @Override public void run(String... args) throws Exception {
      List<Singer> singers = singerRepository.findByFirstName("John");
      listSingersWithAlbum(singers);
   }

   private static void listSingersWithAlbum(List<Singer> singers) {
      logger.info(" ---- Listing singers with instruments:");
```

```
        singers.forEach(singer -> {
            logger.info(singer.toString());
            if (singer.getAlbums() != null) {
                singer.getAlbums().forEach(
                    album -> logger.info("\t" + album.toString()));
            }
            if (singer.getInstruments() != null) {
                singer.getInstruments().forEach(
                    instrument -> logger.info("\t" + instrument.getInstrumentId()));
            }
        });
    }
}
```

Application 类引入了一些新的东西，它实现了 CommandLineRunner 接口。该接口用来告诉 Spring Boot，当该 bean 被包含在 Spring 应用程序中时，应该执行 run()方法。

scanBasePackages="com.apress.prospring5.ch8.config"属性用来启动对参数指定的包进行组件扫描，从而创建包中包含的 bean 并添加到应用程序上下文中。当在不同于 Application 类的包中定义 bean 时，就需要这么做。

应该注意的另一件事是，不需要其他配置类；不再需要任何 SQL 脚本来初始化数据库，也不需要 Application 类上的任何其他注解。显然，如果想专注于应用程序的逻辑开发，Spring Boot 及其启动器依赖项是非常有用的。

如果运行上面的类，会得到预期的结果。(请注意，应用程序在退出之前等待按下某个键。)

```
INFO c.a.p.c.c.DBInitializer - Starting database initialization...
INFO c.a.p.c.c.DBInitializer - Database initialization finished.
INFO c.a.p.c.Application - ---- Listing singers with instruments:
INFO c.a.p.c.Application - Singer - Id: 1, First name: John, Last name: Mayer,
   Birthday: 1977-10-16
INFO c.a.p.c.Application - Album - id: 1, Singer id: 1, Title: Battle Studies,
   Release Date: 2009-11-17
INFO c.a.p.c.Application - Album - id: 2, Singer id: 1,
   Title: The Search For Everything, Release Date: 2017-01-20
INFO c.a.p.c.Application - Piano
INFO c.a.p.c.Application - Guitar
INFO c.a.p.c.Application - Singer - Id: 3, First name: John, Last name: Butler,
   Birthday: 1975-04-01
INFO c.a.p.c.Application - Guitar
INFO c.a.p.c.Application - Started Application in 3.464 seconds (JVM running for 4.0)
```

8.12 使用 JPA 时的注意事项

虽然前面介绍的内容比较多，但也只是讨论了 JPA 的一小部分。例如，使用 JPA 调用数据库存储过程就没有介绍。JPA 是完整且功能强大的 ORM 数据访问标准，在第三方库(比如 Spring Data JPA 和 Hibernate Envers)的帮助下，可以相对轻松地实现各种横切关注点。

JPA 是大多数主要开源社区以及商业供应商(JBoss、GlassFish、WebSphere、WebLogic 等)都支持的 JEE 标准。因此，采用 JPA 作为数据访问标准是一种非常有吸引力的选择。如果需要绝对控制查询，可以使用 JPA 的本地查询支持，而不是直接使用 JDBC。

总之，如果想要使用 Spring 开发 JEE 应用程序，推荐使用 JPA 来实现数据访问层。如果需要，仍然可以与 JDBC 混合使用，以满足某些特殊的数据访问需求。请始终记住，Spring 允许轻松地混合和匹配数据访问技术，并以透明方式处理事务管理。如果想简化开发，Spring Boot 可以通过预先配置的 bean 和自定义配置来实现。

8.13 小结

在本章中，介绍了 JPA 的基本概念，以及如何在 Spring 中通过使用 Hibernate 作为持久化服务提供程序来配置 JPA 的 EntityManagerFactory。然后讨论了如何使用 JPA 来执行基本的数据库操作。高级主题包括本地查询和强类型的 JPA Criteria API。

此外，本章还演示了 Spring Data JPA 的 Repository 抽象如何帮助简化 JPA 应用程序的开发，以及如何使用其实体监听器来跟踪实体类的基本审计信息。对于实体类的完整版本控制，可以使用 Hibernate Envers 来满足要求。

本章最后介绍了 JPA 应用程序的 Spring Boot，因为它简化了大量配置，并将重点放在开发所需的功能上。

在下一章中，将讨论 Spring 中的事务管理。

第9章

事务管理

事务是构建可靠企业级应用程序的最关键部分之一。最常见的事务类型是数据库操作。在典型的数据库更新操作中，首先数据库事务开始，然后数据被更新，最后提交或回滚事务(根据数据库操作的结果而定)。但是，在很多情况下，由于应用程序需求以及应用程序需要与之交互的后端资源(如 RDBMS、面向消息的中间件、ERP 系统等)的不同，事务管理可能更加复杂。

在 Java 应用程序开发的早期阶段(在 JDBC 创建之后，且在 JEE 标准或像 Spring 这样的应用程序框架出现之前)，开发人员通过程序控制和管理应用程序代码中的事务。当 JEE 特别是 EJB 标准出现时，开发人员能够使用容器管理事务(CMT)以声明方式管理事务。但是 EJB 部署描述符中复杂的事务声明很难维护，并且为事务处理引入了不必要的复杂性。一些开发人员喜欢对事务拥有更多的控制权，并选择 bean 管理事务(BMT)以编程方式管理事务。但是，使用 Java Transaction API(JTA)进行编程的复杂性阻碍了开发人员的工作效率。

正如第 5 章所讨论的，事务管理是横切关注点，不应该在业务逻辑中编码。实现事务管理的最合适方法是允许开发人员以声明方式定义事务需求，并让诸如 Spring、JEE 或 AOP 之类的框架代替我们织入事务处理逻辑。在本章中，将讨论 Spring 如何帮助简化事务处理逻辑的实现。Spring 支持声明式和编程式事务管理。

Spring 声明性事务提供极好的支持，这意味着不需要将事务管理代码和业务逻辑混淆起来。所要做的就是声明必须参与事务的相关方法(在类或层中)以及事务配置的细节，Spring 将负责处理事务管理。更具体地说，本章涵盖以下内容。

- **Spring 事务抽象层**：讨论 Spring 事务抽象类的基本组件，并解释如何使用这些类来控制事务的属性。
- **声明式事务管理**：演示如何使用 Spring 和简单的 Java 对象来实现声明式事务管理，提供使用 XML 配置文件以及 Java 注解的声明式事务管理示例。
- **编程式事务管理**：尽管编程式事务管理并不经常使用，但本章还是介绍了如何使用 Spring 提供的 TransactionTemplate 类，它可以让开发人员完全控制事务管理代码。
- **使用 JTA 实现全局事务**：对于需要跨越多个后端资源的全局事务，演示如何使用 JTA 在 Spring 中配置和实现全局事务。

9.1 研究 Spring 事务抽象层

在开发应用程序时，无论是否选择使用 Spring，在使用事务时，都必须选择是使用全局事务还是本地事务。本地事务特定于单个事务资源(例如，JDBC 连接)，而全局事务由容器管理并且可以跨越多个事务资源。

事务类型

本地事务易于管理，如果应用程序中的所有操作都需要与一个事务资源(如 JDBC 连接)进行交互，那么使用本地事务就足够了。但是，如果没有使用像 Spring 这样的应用程序框架，则需要编写大量的事务管理代码，并且如果将来事务的范围需要扩展到跨多个事务资源，就必须删除本地事务管理代码并重新编写代码，以便使用全局事务。

在 Java 世界中，全局事务是通过 JTA 实施的。在这种情况下，与 JTA 兼容的事务管理器通过各自的资源管理器连接到多个事务资源，而这些资源管理器能够通过 XA 协议(一种定义了分布式事务的开放标准)与事务管理器进行通信，并使用 2 Phase Commit(2PC)机制来确保所有后端数据源被更新或完全回滚。如果任意后端资源失败，则整个事务将回滚，因此对其他资源的更新也会回滚。

图 9-1 显示了使用 JTA 实现的全局事务的高级视图。如图 9-1 所示，主要有四部分参与了全局事务(通常也称为*分布式事务*)。第一部分是后端资源，比如 RDBMS、消息中间件、企业资源计划(ERP)系统等。

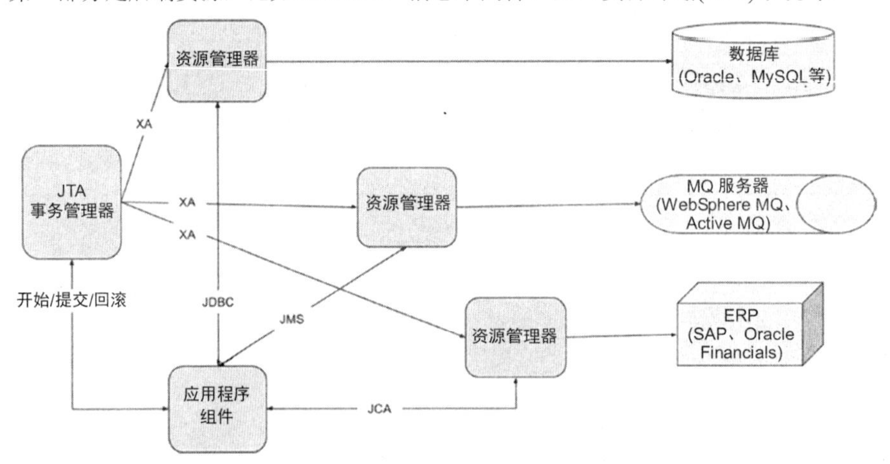

图 9-1　使用 JTA 实现的全局事务概述

第二部分是资源管理器，通常由后端资源供应商提供，并负责与后端资源进行交互。例如，当连接到 MySQL 数据库时，需要与 MySQL 的 Java 连接器提供的 MysqlXADataSource 类进行交互。其他后端资源(例如，MQ、ERP 等)也都提供了自己的资源管理器。

第三部分是 JTA 事务管理器，负责管理、协调和同步参与事务的所有资源管理器的事务状态。我们使用了 XA 协议，这是一种被广泛用于分布式事务处理的开放标准。JTA 事务管理器还支持 2PC，以便所有更改一并提交，并且当任何资源更新失败时，回滚整个事务，从而导致任何资源都没有被更新。整个机制由 Java Transaction Service(JTS) 规范指定。

最后一部分是应用程序。可以是应用程序本身，也可以是运行应用程序的底层容器或 Spring 框架，用来管理事务(开始、提交、回滚事务等)。同时，应用程序通过由 JEE 定义的各种标准与底层后端资源进行交互。如图 9-1 所示，应用程序通过 JDBC 连接到 RDBMS，通过 JMS 连接到 MQ，通过 Java EE Connector Architecture(JCA)连接到 ERP 系统。

所有兼容 JEE 的应用程序服务器(例如 JBoss、WebSphere、WebLogic 和 GlassFish)都支持 JTA，通过 JNDI 查找提供事务。对于独立应用程序或 Web 容器(例如 Tomcat 和 Jetty)，还存在开源和商业解决方案，这些环境(例如，Atomikos、Java Open Transaction Manager[JOTM]和 Bitronix)都提供对 JTA/XA 的支持。

9.2 PlatformTransactionManager 的实现

在 Spring 中，PlatformTransactionManager 接口使用 TransactionDefinition 和 TransactionStatus 接口来创建和管理事务。想要实现这些接口，必须详细了解事务管理器。

图 9-2 显示了 Spring 中 PlatformTransactionManager 的实现。

Spring 为 PlatformTransactionManager 接口提供了一组丰富的实现。CciLocalTransactionManager 类支持 JEE、JCA 和通用客户端接口(CCI)。DataSourceTransactionManager 类用于通用 JDBC 连接。对于 ORM 方面，有许多实现，包括 JPA(JpaTransactionManager 类)[1]和 Hibernate 5(HibernateTransactionManager)[2]。对于 JMS，该实现通过 JmsTransactionManager 类支持 JMS 2.0[3]。而对于 JTA，通用实现类是 JtaTransactionManager。Spring 还提供了几个特定于应用程序服务器的 JTA 事务管理器类。这些类为 WebSphere(WebSphereUowTransactionManager 类)、WebLogic(WebLogicJtaTransactionManager 类)和 Oracle OC4J(OC4JJtaTransactionManager 类)提供本地支持。

1　Spring 5 中删除了对 JDO 的支持；因此，类图中缺少 JdoTransactionManager。
2　Spring 5 仅使用 Hibernate 5；Hibernate 3 和 Hibernate 4 的实现已被删除。
3　Spring 5 中删除了对 JMS 1.1 的支持。

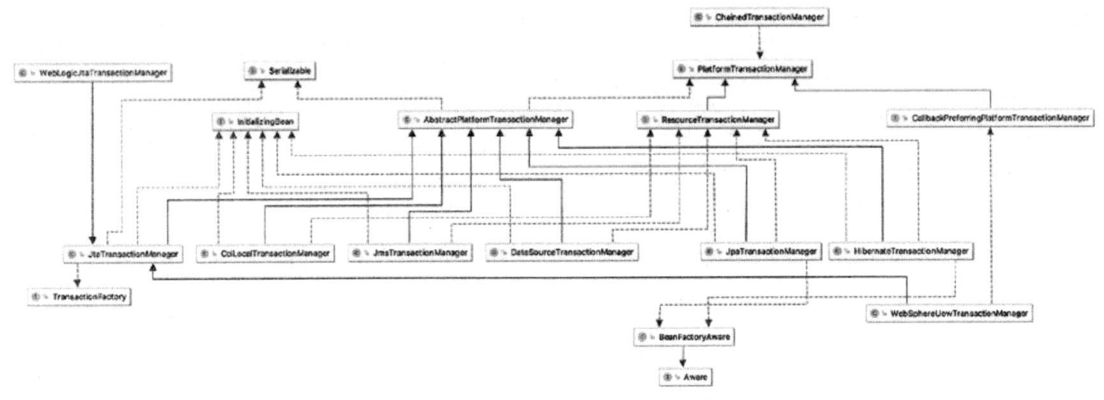

图 9-2 PlatformTransactionManager 的实现

9.3 分析事务属性

在本节中，将讨论 Spring 所支持的事务属性，重点关注与 RDBMS(作为后端资源)的交互。

事务具有四个众所周知的 ACID 属性(原子性、一致性、隔离性和持久性)，如何维护事务的这些属性取决于事务资源。对事务的原子性、一致性和持久性是无法控制的。但是，可以控制事务的传播和超时，以及配置事务是否应为只读并指定隔离级别。

Spring 将所有这些设置封装在 TransactionDefinition 接口中。该接口被用于 Spring 的事务支持的核心接口，即 PlatformTransactionManager 接口，该接口的实现在特定平台(如 JDBC 或 JTA)上执行事务管理。其核心方法 PlatformTransactionManager.getTransaction()将 TransactionDefinition 接口作为参数，并返回一个 TransactionStatus 接口。TransactionStatus 接口用于控制事务执行，更具体地说，是设置事务结果并检查事务是否完成或是否为新事务。

9.3.1 TransactionDefinition 接口

如前所述，TransactionDefinition 接口控制事务的属性。接下来详细地看一下如下所示的 TransactionDefinition 接口，并介绍它的方法：

```
package org.springframework.transaction;

import java.sql.Connection;

public interface TransactionDefinition {

    // Variable declaration statements omitted

    ...

    int getPropagationBehavior();

    int getIsolationLevel();

    int getTimeout();

    boolean isReadOnly();

    String getName();

}
```

该接口简单且显而易见的方法是 getTimeout()(返回事务必须完成的时间(以秒为单位))和 isReadOnly()(指示事务是否是只读的)。事务管理器可以使用此值来优化执行并进行检查以确保事务仅执行读取操作。getName()方法返回事务的名称。

另外两个方法需要详细讨论，即 getPropagationBehavior()和 getIsolationLevel()。首先从 getIsolationLevel()开始，它控制着其他事务能够看到的数据更改。表 9-1 列出了可以使用的事务隔离级别，并解释了其他事务可以访问当前事务中所做的哪些更改。可以使用 TransactionDefinition 接口中定义的静态值来表示隔离级别。

表 9-1 事务的隔离级别

隔离级别	描述
ISOLATION_DEFAULT	底层数据存储的默认隔离级别
ISOLATION_READ_UNCOMMITTED	最低的隔离级别；它几乎不是事务，因为它允许一个事务查看由其他未提交事务修改的数据
ISOLATION_READ_COMMITTED	大多数数据库的默认级别，它确保一个事务不能读取其他事务未提交的数据。但是，一旦数据可以被一个事务读取，那么数据就可以由其他事务更新
ISOLATION_REPEATABLE_READ	比 ISOLATION_READ_COMMITTED 更严格；它可以确保一旦选择了数据，就可以再次选择相同的数据集。即使其他事务插入新数据，也仍然可以选择新插入的数据
ISOLATION_SERIALIZABLE	最严格且最可靠的隔离级别，所有的事务都被视为一个接一个地执行

选择适当的隔离级别对于保持数据的一致性是非常重要的，此外选择的隔离级别也会对性能产生很大的影响。对于最高的隔离级别 ISOLATION_SERIALIZABLE，维护成本是特别昂贵的。

getPropagationBehavior()方法指定事务调用时所发生的事情，具体取决于是否存在活动事务。表 9-2 介绍了此方法的值。可以使用 TransactionDefinition 接口中定义的静态值表示传播类型。

表 9-2 事务的传播类型

传播类型	描述
PROPAGATION_REQUIRED	支持一个已经存在的事务。如果没有事务，则开始一个新的事务
PROPAGATION_SUPPORTS	支持一个已经存在的事务。如果没有事务，则以非事务方式执行
PROPAGATION_MANDATORY	支持一个已经存在的事务。如果没有活动事务，则抛出异常
PROPAGATION_REQUIRES_NEW	始终开始新的事务。如果活动事务已经存在，将其暂停
PROPAGATION_NOT_SUPPORTED	不支持活动事务的执行。始终以非事务方式执行并暂停任何现有事务
PROPAGATION_NEVER	即使存在活动事务，也始终以非事务方式执行。如果存在活动事务，则抛出异常
PROPAGATION_NESTED	如果存在活动事务，则在嵌套事务中运行。如果没有活动事务，则与 PROPAGATION_REQUIRED 相同

9.3.2 TransactionStatus 接口

下面所示的 TransactionStatus 接口允许事务管理器控制事务的执行。这些代码可以检查事务是新事务还是只读事务，并且可以启动回滚。

```
package org.springframework.transaction;

public interface TransactionStatus extends SavepointManager {

    boolean isNewTransaction();

    boolean hasSavepoint();

    void setRollbackOnly();

    boolean isRollbackOnly();

    void flush();

    boolean isCompleted();

}
```

TransactionStatus 接口中的方法非常容易理解；最值得注意的是 setRollbackOnly()方法，它会导致回滚并结束活动事务。

hasSavePoint()方法指示事务内部是否携带保存点(也就是说，事务是基于保存点而创建的嵌套事务)。如果可用(例如，与 Hibernate 一起使用时)，flush()方法会将底层会话存储到数据存储区。isCompleted()方法指示事务是否结束(即提交或回滚)。

9.4 示例代码的示例数据模型和基础结构

本节概述事务管理示例中所使用的示例数据模型和基础结构。使用 JPA 和 Hibernate 作为实现数据访问逻辑的持久化层。另外，使用 Spring Data JPA 及其存储库抽象简化基本数据库操作的开发。

9.4.1 创建一个带有依赖项的简单 Spring JPA 项目

首先从创建项目开始。因为正在使用 JPA，所以还需要为本章中的示例添加项目所需的依赖项。

```
//pro-spring-15/build.gradle
ext {
    //spring libs
    springVersion = '5.0.0.RC1'
    bootVersion = '2.0.0.BUILD-SNAPSHOT'
    springDataVersion = '2.0.0.M3'

    //logging libs
    slf4jVersion = '1.7.25'
    logbackVersion = '1.2.3'
    guavaVersion = '21.0'
    junitVersion = '4.12'

    aspectjVersion = '1.9.0.BETA-5'

    //database library
    h2Version = '1.4.194'

    //persistency libraries
    hibernateVersion = '5.2.10.Final'
    hibernateJpaVersion = '1.0.0.Final'
    atomikosVersion = '4.0.0M4'

    spring = [
       context          : "org.springframework:spring-context:$springVersion",
       aop              : "org.springframework:spring-aop:$springVersion",
       aspects          : "org.springframework:spring-aspects:$springVersion",
       tx               : "org.springframework:spring-tx:$springVersion",
       jdbc             : "org.springframework:spring-jdbc:$springVersion",
       contextSupport   : "org.springframework:spring-context-support:$springVersion",
       orm              : "org.springframework:spring-orm:$springVersion",
       data             : "org.springframework.data:spring-data-jpa:$springDataVersion",
       test             : "org.springframework:spring-test:$springVersion"
   ]

    hibernate = [
      ...
       em               : "org.hibernate:hibernate-entitymanager:$hibernateVersion",
       tx               : "com.atomikos:transactions-hibernate4:$atomikosVersion"
   ]

    boot = [
       ...
       springBootPlugin:
          "org.springframework.boot:spring-boot-gradle-plugin:$bootVersion",
          starterJpa :
             "org.springframework.boot:spring-boot-starter-data-jpa:$bootVersion"
   ]

    testing = [
       junit: "junit:junit:$junitVersion"
   ]

    misc = [
       ...
       slf4jJcl        : "org.slf4j:jcl-over-slf4j:$slf4jVersion",
          logback      : "ch.qos.logback:logback-classic:$logbackVersion",
       aspectjweaver   : "org.aspectj:aspectjweaver:$aspectjVersion",
          lang3        : "org.apache.commons:commons-lang3:3.5",
      guava            : "com.google.guava:guava:$guavaVersion"
   ]
```

```
    db = [
        ...
        h2 : "com.h2database:h2:$h2Version"
    ]
}

//chapter09/build.gradle
dependencies {
    //we specify these dependencies for all submodules, except
    // the boot module, that defines its own
    if !project.name.contains"boot" {
        //we exclude transitive dependencies, because spring-data
        //will take care of these
           compile spring.contextSupport {
           exclude module: 'spring-context'
           exclude module: 'spring-beans'
           exclude module: 'spring-core'
        }
        //we exclude the 'hibernate' transitive dependency
        //to have control over the version used
        compile hibernate.tx {
        exclude group: 'org.hibernate', module: 'hibernate'
    }
        compile spring.orm, spring.context, misc.slf4jJcl,
           misc.logback, db.h2, misc.lang3,
        hibernate.em
    }
    testCompile testing.junit
}
```

为了在修改事务属性时观察示例代码的详细行为，还需要在 logback.xml 文件中启用 DEBUG 级日志记录。以下代码片段显示了 logback.xml 文件：

```
<?xml version="1.0" encoding="UTF-8"?>
<configuration>

    <contextListener class="ch.qos.logback.classic.jul.LevelChangePropagator">
       <resetJUL>true</resetJUL>
    </contextListener>

    <appender name="console" class="ch.qos.logback.core.ConsoleAppender">
       <encoder>
          <pattern>%d{HH:mm:ss.SSS} %thread %-5level %logger{5} - %msg%n</pattern>
       </encoder>
    </appender>

    <logger name="com.apress.prospring5.ch8" level="debug"/>

    <logger name="org.springframework.transaction" level="info"/>

    <logger name="org.hibernate.SQL" level="debug"/>

    <root level="info">
       <appender-ref ref="console" />
    </root>
</configuration>
```

9.4.2 示例数据模型和通用类

为了简单起见，只使用两个表，即其他章中使用的 SINGER 和 ALBUM 表。无需 SQL 脚本来创建表，因为可以使用 Hibernate 属性 hibernate.hbm2ddl.auto 并将其设置为 create-drop，也就是说，每次进行测试时始终都有一个干净的数据库。这些表将根据 Singer 和 Album 实体生成。下面的代码描述了每个带注解的字段：

```
//Singer.java
package com.apress.prospring5.ch9.entities;
...

@Entity
@Table(name = "singer")
@NamedQueries({
        @NamedQuery(name=Singer.FIND_ALL, query="select s from Singer s"),
        @NamedQuery(name=Singer.COUNT_ALL, query="select count(s) from Singer s")
})
public class Singer implements Serializable {
```

```
    public static final String FIND_ALL = "Singer.findAll";
    public static final String COUNT_ALL = "Singer.countAll";

    @Id
    @GeneratedValue(strategy = IDENTITY)
    @Column(name = "ID")
    private Long id;

    @Version
    @Column(name = "VERSION")
    private int version;

    @Column(name = "FIRST_NAME")
    private String firstName;

    @Column(name = "LAST_NAME")
    private String lastName;

    @Temporal(TemporalType.DATE)
    @Column(name = "BIRTH_DATE")
    private Date birthDate;

    @OneToMany(mappedBy = "singer", cascade=CascadeType.ALL, orphanRemoval=true)
    private Set<Album> albums = new HashSet<>();

   ...
}
/Album.java
package com.apress.prospring5.ch9.entities;
...
@Entity
@Table(name = "album")
public class Album implements Serializable {
    @Id
    @GeneratedValue(strategy = IDENTITY)
    @Column(name = "ID")
    private Long id;

    @Version
    @Column(name = "VERSION")
    private int version;

    @Column
    private String title;
    @Temporal(TemporalType.DATE)
    @Column(name = "RELEASE_DATE")

    private Date releaseDate;

    @ManyToOne
    @JoinColumn(name = "SINGER_ID")
    private Singer singer;
    ...
}
```

这两个类将被隔离在名为base-dao的项目中，该项目将成为所有事务项目的依赖项。除了实体，存储库接口也在这个项目中定义。这些接口将在稍后介绍。此外，还需要定义DataSource bean的配置类。下面的代码显示了该配置类，出于实用和演示目的，数据库凭据、驱动程序和URL都被直接使用，而不是从外部文件读取。(但在实际应用中，希望你永远不要这样做。)

```
package com.apress.prospring5.ch9.config;
...

@Configuration
@EnableJpaRepositories(basePackages = {"com.apress.prospring5.ch9.repos"})
public class DataJpaConfig {

    private static Logger logger =
        LoggerFactory.getLogger(DataJpaConfig.class);

    @SuppressWarnings("unchecked")
    @Bean
    public DataSource dataSource() {
        try {
            SimpleDriverDataSource dataSource = new SimpleDriverDataSource();
            Class<? extends Driver> driver =
```

```
            (Class<? extends Driver>) Class.forName("org.h2.Driver");
         dataSource.setDriverClass(driver);
         dataSource.setUrl("jdbc:h2:musicdb");
         dataSource.setUsername("prospring5");
         dataSource.setPassword("prospring5");
         return dataSource;
      } catch (Exception e) {
         logger.error("Populator DataSource bean cannot be created!", e);
         return null;
      }
   }

   @Bean
   public Properties hibernateProperties() {
      Properties hibernateProp = new Properties();
      hibernateProp.put("hibernate.dialect", "org.hibernate.dialect.H2Dialect");
      hibernateProp.put("hibernate.hbm2ddl.auto", "create-drop");
      //hibernateProp.put("hibernate.format_sql", true);
      hibernateProp.put("hibernate.show_sql", true);
      hibernateProp.put("hibernate.max_fetch_depth", 3);
      hibernateProp.put("hibernate.jdbc.batch_size", 10);
      hibernateProp.put("hibernate.jdbc.fetch_size", 50);
      return hibernateProp;
   }

   @Bean
   public JpaVendorAdapter jpaVendorAdapter() {
      return new HibernateJpaVendorAdapter();
   }

   @Bean
   public EntityManagerFactory entityManagerFactory() {
      LocalContainerEntityManagerFactoryBean factoryBean =
         new LocalContainerEntityManagerFactoryBean();
      factoryBean.setPackagesToScan("com.apress.prospring5.ch9.entities");
      factoryBean.setDataSource(dataSource());
      factoryBean.setJpaVendorAdapter(new HibernateJpaVendorAdapter());
      factoryBean.setJpaProperties(hibernateProperties());
      factoryBean.setJpaVendorAdapter(jpaVendorAdapter());
      factoryBean.afterPropertiesSet();
      return factoryBean.getNativeEntityManagerFactory();
   }
}
```

此时，定义了一个嵌入式 H2 数据库。凭证直接在代码中设置，而 DataSource 实现是 SimpleDriverDataSource，它被设计为仅适用于简单型、测试型或演示型应用程序。

目前，使用注解是在 Spring 中定义事务需求的最常见方法。主要优点是，事务需求可以与详细事务属性(超时、隔离级别、传播行为等)一起在代码本身内定义，从而使应用程序更容易跟踪和维护。当然，也可以使用注解和 Java 配置类来完成配置。如果想要使用 XML 配置为 Spring 中的事务管理启用注解支持，需要在 XML 配置文件中添加 <tx:annotation-driven>标记。在以下配置代码片段中，可以看到事务配置代码以及事务匹配的名称空间。如果感兴趣，可以在项目中找到完整的配置。

```
<?xml version="1.0" encoding="UTF-8"?>
<beans xmlns="http://www.springframework.org/schema/beans"
    xmlns:xsi="http://www.w3.org/2001/XMLSchema-instance"
    ...
    xmlns:tx="http://www.springframework.org/schema/tx"
    xsi:schemaLocation="http://www.springframework.org/schema/beans
       http://www.springframework.org/schema/beans/spring-beans.xsd
       http://www.springframework.org/schema/tx
       http://www.springframework.org/schema/tx/spring-tx.xsd ...">

    <bean id="transactionManager"
       class="org.springframework.orm.jpa.JpaTransactionManager">
       <property name="entityManagerFactory" ref="emf"/>
    </bean>
    <tx:annotation-driven />

    <bean id="emf"
       class="org.springframework.orm.jpa.LocalContainerEntityManagerFactoryBean">
       ...
    </bean>

    <context:component-scan
```

```
        base-package="com.apress.prospring5.ch9" />

    <jpa:repositories base-package="com.apress.prospring5.ch9.repos"
                entity-manager-factory-ref="emf"
                transaction-manager-ref="transactionManager"/>
</beans>
```

因为使用的是JPA，所以定义了JpaTransactionManager bean。<tx:annotation-driven>标记指定使用注解进行事务管理。这个简单的定义指示Spring寻找类型为PlatformTransactionManager的名为transactionManager的bean。如果将事务bean命名为不同的名称，比如customTransactionManager，就必须使用transaction-manager属性声明元素定义，该属性必须接收事务管理bean的名称作为值。

```
<tx:annotation-driven transaction-manager="customTransactionManager"/>
```

然后定义EntityManagerFactory bean，紧接着定义<context:component-scan>标记以扫描服务层类。最后，使用<jpa:repositories>标记启用Spring Data JPA的存储库抽象。该元素在DataJpaConfiguration类中被替换为@EnableJpaRepositories注解。

在专业环境中，将持久化配置(DAO)与事务配置(服务)分开是常见的做法。这也就是之前介绍的XML内容在Java配置中被分成两个配置类的原因。前面介绍的DataJpaConfig仅包含数据访问bean，而下面将要介绍的ServicesConfig仅包含事务管理相关的bean：

```java
package com.apress.prospring5.ch9.config;

import org.springframework.beans.factory.annotation.Autowired;
import org.springframework.context.annotation.Bean;
import org.springframework.context.annotation.ComponentScan;
import org.springframework.context.annotation.Configuration;
import org.springframework.orm.jpa.JpaTransactionManager;
import org.springframework.transaction.PlatformTransactionManager;
import org.springframework.transaction.annotation.EnableTransactionManagement;

import javax.persistence.EntityManagerFactory;

@Configuration
@EnableTransactionManagement
@ComponentScan(basePackages = "com.apress.prospring5.ch9")
public class ServicesConfig {
    @Autowired EntityManagerFactory entityManagerFactory;

    @Bean
    public PlatformTransactionManager transactionManager() {
       return new JpaTransactionManager(entityManagerFactory);
    }
}
```

对于SingerService接口的实现，首先创建一个包含SingerService接口中所有方法的空的实现类。然后实现SingerService.findAll()方法。下面的代码片段显示了使用findAll()方法实现的SingerServiceImpl类：

```java
package com.apress.prospring5.ch9.services;

import com.apress.prospring5.ch9.entities.Singer;
import com.apress.prospring5.ch9.repos.SingerRepository;
import com.google.common.collect.Lists;
import org.springframework.beans.factory.annotation.Autowired;
import org.springframework.stereotype.Service;
import org.springframework.transaction.annotation.Propagation;
import org.springframework.transaction.annotation.Transactional;

import java.util.List;

@Service("singerService")
@Transactional
public class SingerServiceImpl implements SingerService {

    private SingerRepository singerRepository;

    @Autowired
    public void setSingerRepository(SingerRepository singerRepository) {
       this.singerRepository = singerRepository;
    }

    @Override
    @Transactional(readOnly = true)
```

```
    public List<Singer> findAll() {
        return Lists.newArrayList(singerRepository.findAll());
    }
}
```

当使用基于注解的事务管理时，需要处理的唯一注解是@Transactional。在前面的代码片段中，@Transactional 注解被应用在类级别，这意味着默认情况下，Spring 将确保事务在类中的每个方法执行之前存在。@Transactional 注解支持许多属性，可以修改这些属性来覆盖默认行为。表 9-3 显示了可用的属性以及可能值和默认值。

表 9-3 @Transactional 注解的属性

属性名称	默认值	可能值
propagation	Propagation.REQUIRED	Propagation.REQUIRED Propagation.SUPPORTS Propagation.MANDATORY Propagation.REQUIRES_NEW Propagation.NOT_SUPPORTED Propagation.NEVER Propagation.NESTED
isolation	Isolation.DEFAULT(底层资源的默认隔离级别)	Isolation.DEFAULT Isolation.READ_UNCOMMITTED Isolation.READ_COMMITTED Isolation.REPEATABLE_READ Isolation.SERIALIZABLE
timeout	TransactionDefinition.TIMEOUT_DEFAULT(基础资源的默认事务超时(以秒为单位))	大于零的整数值；指示超时的秒数
readOnly	false	{true, false}
rollbackFor	事务将被回滚的异常类	N/A
rollbackForClassName	事务将被回滚的异常类名称	N/A
noRollbackFor	事务不会回滚的异常类	N/A
noRollbackForClassName	事务不会回滚的异常类名称	N/A
value	"" (指定事务的限定符值)	N/A

因此，根据表 9-3，不带任何属性的@Transactional 注解意味着事务传播是必需的，隔离是默认的，超时是默认的，模式是可读写的。对于前面介绍的 findAll()方法，应该使用@Transactional(readOnly = true)注解该方法。这样就会覆盖在类级别应用的默认注解，而其他属性保持不变，但事务被设置为只读。以下代码片段显示了 findAll()方法的测试程序：

```
package com.apress.prospring5.ch9;

import java.util.List;
import com.apress.prospring5.ch9.config.DataJpaConfig;
import com.apress.prospring5.ch9.config.ServicesConfig;
import com.apress.prospring5.ch9.entities.Singer;
import com.apress.prospring5.ch9.services.SingerService;
import org.springframework.context.annotation.AnnotationConfigApplicationContext;
import org.springframework.context.support.GenericApplicationContext;
public class TxAnnotationDemo {
    public static void main(String... args) {
        GenericApplicationContext ctx =
        new AnnotationConfigApplicationContext(ServicesConfig.class,
              DataJpaConfig.class);

        SingerService singerService = ctx.getBean(SingerService.class);

        List<Singer> singers = singerService.findAll();
        singers.forEach(s -> System.out.println(s));
        ctx.close();
    }
}
```

启用适当的日志记录后，换句话说，设置<logger name ="org.springframework.orm.jpa" level ="debug"/>后，就能够在日志消息中看到与事务处理相关的消息。运行该程序会产生以下简要输出(有关完整详细信息，请参阅控制台中的调试日志记录)：

```
DEBUG o.s.o.j.JpaTransactionManager - Creating new transaction with name
    [com.apress.prospring5.ch9.services.SingerServiceImpl.findAll]:
       PROPAGATION_REQUIRED,ISOLATION_DEFAULT,readOnly; ''
DEBUG o.s.o.j.JpaTransactionManager - Participating in existing transaction
Hibernate: select singer0_.ID as ID1_1_, singer0_.BIRTH_DATE as BIRTH_DA2_1_,
    singer0_.FIRST_NAME as FIRST_NA3_1_, singer0_.LAST_NAME as LAST_NAM4_1_,
    singer0_.VERSION as VERSION5_1_ from singer singer0_
DEBUG o.s.o.j.JpaTransactionManager - Closing JPA EntityManager
  [...] after transaction

DEBUG o.s.o.j.JpaTransactionManager - Initiating transaction commit

Singer - Id: 1, First name: John, Last name: Mayer, Birthday: 1977-10-16
Singer - Id: 2, First name: Eric, Last name: Clapton, Birthday: 1945-03-30
Singer - Id: 3, First name: John, Last name: Butler, Birthday: 1975-04-01
```

如上面的输出所示，为了简洁起见，删除了不相关的输出语句。首先，在运行 findAll()方法之前，Spring 的 JpaTransactionManager 使用默认属性创建了一个新的事务(事务名称与方法的完全限定类名相同)，但事务被设置为只读(用方法级别的@Transactional 注解定义)。然后，提交查询，在完成且没有任何错误的情况下，事务被提交。JpaTransactionManager 负责处理事务的创建和提交操作。

接下来继续执行更新操作。需要在 SingerServiceImpl 类中实现 findById()和 save()方法。以下代码片段显示了实现：

```
package com.apress.prospring5.ch9.services;
...

@Service("singerService")
@Transactional
public class SingerServiceImpl implements SingerService {
    private SingerRepository singerRepository;

    @Autowired
    public void setSingerRepository(SingerRepository singerRepository) {
       this.singerRepository = singerRepository;
    }

    @Override
    @Transactional(readOnly = true)
    public List<Singer> findAll() {
       return Lists.newArrayList(singerRepository.findAll());
    }

    @Override
    @Transactional(readOnly = true)
    public Singer findById(Long id) {
       return singerRepository.findById(id).get();
    }

    @Override
    public Singer save(Singer singer) {
       return singerRepository.save(singer);
    }
}
```

findById()方法也使用@Transactional(readOnly = true)进行了注解。通常，readOnly = true 属性应该被应用于所有查找方法。主要原因是，大多数持久化提供程序会对只读事务执行一定程度的优化。例如，Hibernate 不会维护从数据库检索且只读的托管实例的快照。

对于 save()方法，只需要调用 CrudRepository.save()方法并且不提供任何注解。这意味着将使用类级别的注解，这是一个读写事务。修改 TxAnnotationDemo 类以测试 save()方法，如下面的代码所示：

```
package com.apress.prospring5.ch9;
...
public class TxAnnotationDemo {
    public static void main(String... args) {
       GenericApplicationContext ctx =
       new AnnotationConfigApplicationContext(ServicesConfig.class,
              DataJpaConfig.class);
```

```
        SingerService singerService = ctx.getBean(SingerService.class);

        List<Singer> singers = singerService.findAll();
        singers.forEach(s -> System.out.println(s));

        Singer singer = singerService.findById(1L);
        singer.setFirstName("John Clayton");
        singer.setLastName("Mayer");
        singerService.save(singer);
        System.out.println("Singer saved successfully: " + singer);

        ctx.close();
    }
}
```

检索 ID 为 1 的 Singer 对象，然后更新名字并保存到数据库中。运行代码将生成以下相关输出：

```
Singer saved successfully: Singer - Id: 1, First name: John Clayton,
    Last name: Mayer, Birthday: 1977-10-16
```

save()方法获取从类级别@Transactional 注解继承的默认属性。在完成更新操作后，Spring 的 JpaTransactionManager 触发事务提交，从而导致 Hibernate 刷新持久化上下文并将底层 JDBC 连接提交给数据库。最后，来看看 countAll()方法，我们主要介绍该方法的两种事务配置。尽管 CrudRepository.count()方法也可以实现相同的目的，但我们不会使用该方法。相反，为了演示需要，将实现另一个方法，主要是因为在 Spring Data 中，由 CrudRepository 接口定义的方法已经用适当的事务属性进行了标记。

以下代码片段显示了在 SingerRepository 接口中定义的新方法 countAllSingers()：

```
package com.apress.prospring5.ch9.repos;

import com.apress.prospring5.ch9.entities.Singer;
import org.springframework.data.jpa.repository.JpaRepository;
import org.springframework.data.jpa.repository.Query;
import org.springframework.data.repository.CrudRepository;

public interface SingerRepository extends
    CrudRepository<Singer, Long> {
        @Query("select count(s) from Singer s")
        Long countAllSingers();
}
```

对于新的 countAllSingers()方法，将应用@Query 注解，其值应该为计算联系人数量的 JPQL 语句。以下代码片段显示了 SingerServiceImpl 类中 countAll()方法的实现：

```
package com.apress.prospring5.ch9.services;
...

@Service("singerService")
@Transactional
public class SingerServiceImpl implements SingerService {

    private SingerRepository singerRepository;
    @Autowired
    public void setSingerRepository(SingerRepository singerRepository) {
        this.singerRepository = singerRepository;
    }

    @Override
    @Transactional(readOnly=true)
    public long countAll() {
        return singerRepository.countAllSingers();
    }
}
```

该注解与其他查找方法相同。要测试此方法，需要在 TxAnnotationDemo 类的 main()方法中添加 System.out.println("Singer count:"+ contactService.countAll());语句，并观察控制台。如果看到类似 Singer count:3 的消息，则表示该方法被正确执行。

在该输出中，可以看到 countAll()方法的事务是按照预期的方式创建的，是只读的。但是对于 countAll()方法，我们并不希望它被列入事务中。不需要由底层 JPA EntityManager 管理结果。相反，只是想要获取联系人数量，然后就将该方法忘掉。在这种情况下，可以将事务传播行为重写为 Propagation.NEVER。以下代码片段显示了修改后的 countAll()方法：

```
package com.apress.prospring5.ch9.services;
...

@Service("singerService")
@Transactional
public class SingerServiceImpl implements SingerService {
    ...
    @Override
    @Transactional(propagation = Propagation.NEVER)
    public long countAll() {
        return singerRepository.countAllSingers();
    }
}
```

再次运行测试代码，从调试输出中可以看到，不再为countAll()方法创建事务。

本节介绍了在日常处理事务时所涉及的一些主要配置。对于一些特殊情况，可能需要定义超时、隔离级别、针对特定异常的回滚(或不回滚)等。

Spring 的 JpaTransactionManager 不支持自定义隔离级别。相反，它总是使用底层数据存储的默认隔离级别。如果使用 Hibernate 作为 JPA 服务提供程序，那么对此可以使用如下解决方法：扩展 HibernateJpaDialect 类以支持自定义隔离级别。

9.4.3 使用 AOP 配置进行事务管理

声明式事务管理的另一种常见方法是使用 Spring 的 AOP 支持。在 Spring 2 之前，需要使用 TransactionProxyFactoryBean 类来定义 Spring bean 的事务需求。然而，从版本 2 开始，Spring 引入了 aop 名称空间并使用通用的 AOP 配置技术来定义事务需求，从而提供了一种更为简单的方法。当然，在引入注解之后，这种配置事务管理的方式也被弃用了。但是在某些情况下，这种方法也是很有用的，比如，需要封装不属于自己项目的事务代码，并且无法对代码进行编辑以添加@Transaction 注解。

在以下配置代码片段中，使用 XML 对上一节中的示例进行了配置，并使用了 aop 名称空间：

```
<?xml version="1.0" encoding="UTF-8"?>
<beans xmlns="http://www.springframework.org/schema/beans"
    xmlns:xsi="http://www.w3.org/2001/XMLSchema-instance"
    xmlns:context="http://www.springframework.org/schema/context"
    xmlns:tx="http://www.springframework.org/schema/tx"
    xmlns:aop="http://www.springframework.org/schema/aop"
    xsi:schemaLocation="
        http://www.springframework.org/schema/beans
        http://www.springframework.org/schema/beans/spring-beans.xsd
        http://www.springframework.org/schema/tx
        http://www.springframework.org/schema/tx/spring-tx.xsd
        http://www.springframework.org/schema/context
        http://www.springframework.org/schema/context/spring-context.xsd
        http://www.springframework.org/schema/aop
        http://www.springframework.org/schema/aop/spring-aop.xsd">

    <bean name="dataJpaConfig"
      class="com.apress.prospring5.ch9.config.DataJpaConfig" />

   <aop:config>
      <aop:pointcut id="serviceOperation" expression=
            "execution(* com.apress.prospring5.ch9.*ServiceImpl.*(..))"/>
   <aop:advisor pointcut-ref="serviceOperation" advice-ref="txAdvice"/>
</aop:config>

<tx:advice id="txAdvice">
   <tx:attributes>
      <tx:method name="find*" read-only="true"/>
      <tx:method name="count*" propagation="NEVER"/>
      <tx:method name="*"/>
   </tx:attributes>
</tx:advice>

<bean id="transactionManager" class="org.springframework.orm.jpa.JpaTransactionManager
   <property name="entityManagerFactory" ref="entityManagerFactory"/>
</bean>

<context:component-scan
```

```
        base-package="com.apress.prospring5.ch9.services" />
</beans>
```

该配置与本节开头介绍的 XML 非常相似。基本上，<tx:annotation-driven>标记被删除，并且<context:component-scan>标记被修改为用于声明式事务管理的包名。最重要的标记是<aop:config>和<tx:advice>。

在<aop:config>标记中，为服务层中的所有操作(即 com.apress.prospring5.ch9.services 包中的所有实现类)定义了一个切入点。该通知引用了 ID 为 txAdvice 的 bean(由<tx:advice>标记定义)。在<tx:advice>标记中，为想要参与事务的各种方法配置了事务属性。如上述标记所示，将所有查找方法(具有 find 前缀的方法)指定为只读的，并且将计数方法(具有前缀 count 的方法)指定为不会参与事务。对于其余的方法，则应用默认的事务行为。该配置与注解示例中的配置相同。

因为事务管理是通过 aop 显式完成的，所以在 SingerServiceImpl 类及其方法中不再需要@Transactional 注解。

要测试前面的配置，可以使用以下类：

```
package com.apress.prospring5.ch9;

import java.util.List;

import com.apress.prospring5.ch9.entities.Singer;
import com.apress.prospring5.ch9.services.SingerService;
import org.springframework.context.support.GenericXmlApplicationContext;

public class TxDeclarativeDemo {
    public static void main(String... args) {
        GenericXmlApplicationContext ctx = new GenericXmlApplicationContext();
        ctx.load("classpath:spring/tx-declarative-app-context.xml");
        ctx.refresh();

        SingerService singerService = ctx.getBean(SingerService.class);

    // Testing findAll()
        List<Singer> singers = singerService.findAll();
        singers.forEach(s -> System.out.println(s));

    // Testing save()
        Singer singer = singerService.findById(1L);
        singer.setFirstName("John Clayton");
        singerService.save(singer);
        System.out.println("Singer saved successfully: " + singer);

        // Testing countAll()
        System.out.println("Singer count: " + singerService.countAll());

        ctx.close();
    }
}
```

测试程序并观察 Spring 和 Hibernate 执行的与事务相关的操作的输出，基本上与注解示例的输出相同。

9.5 使用编程式事务

第三个选项是以编程方式控制事务行为。在这种情况下，有两种选择。第一种选择是将一个 PlatformTransactionManager 实例注入到 bean 中，并直接与事务管理器进行交互。另一种选择是使用 Spring 提供的 TransactionTemplate 类，它可以大大简化工作。在本节中，将演示如何使用 TransactionTemplate 类。为了简单起见，主要专注于实现 SingerServiceImpl.countAll()方法。以下代码片段描述了为使用编程式事务而进行修改后的 ServiceConfig 类：

```
package com.apress.prospring5.ch9.config;

import org.springframework.beans.factory.annotation.Autowired;
import org.springframework.context.annotation.Bean;
import org.springframework.context.annotation.ComponentScan;
import org.springframework.context.annotation.Configuration;
import org.springframework.orm.jpa.JpaTransactionManager;
import org.springframework.transaction.PlatformTransactionManager;
import org.springframework.transaction.TransactionDefinition;
import org.springframework.transaction.support.TransactionTemplate;

import javax.persistence.EntityManagerFactory;
```

```
@Configuration
@ComponentScan(basePackages = "com.apress.prospring5.ch9")
public class ServicesConfig {

    @Autowired EntityManagerFactory entityManagerFactory;

    @Bean
    public TransactionTemplate transactionTemplate() {
        TransactionTemplate tt = new TransactionTemplate();
        tt.setPropagationBehavior(TransactionDefinition.PROPAGATION_NEVER);
        tt.setTimeout(30);
        tt.setTransactionManager(transactionManager());
        return tt;
    }

    @Bean
    public PlatformTransactionManager transactionManager() {
        return new JpaTransactionManager(entityManagerFactory);
    }
}
```

此处 AOP 事务通知被删除。另外，还使用 org.springframework.transaction.support.TransactionTemplate 类定义了一个 transactionTemplate bean 以及一些事务属性。这里删除了@EnableTransactionManagement，因为现在事务管理没有显式执行。看一下下面所示的 countAll()方法的实现：

```
package com.apress.prospring5.ch9.services;

import com.apress.prospring5.ch9.entities.Singer;
import com.apress.prospring5.ch9.repos.SingerRepository;
import com.google.common.collect.Lists;
import org.springframework.beans.factory.annotation.Autowired;
import org.springframework.stereotype.Repository;
import org.springframework.stereotype.Service;
import org.springframework.transaction.support.TransactionTemplate;

import javax.persistence.EntityManager;
import javax.persistence.PersistenceContext;
import java.util.List;

@Service("singerService")
@Repository
public class SingerServiceImpl implements SingerService {
    @Autowired
    private SingerRepository singerRepository;

    @Autowired
    private TransactionTemplate transactionTemplate;

    @PersistenceContext
    private EntityManager em;

    @Override
    public long countAll() {
        return transactionTemplate.execute(
        transactionStatus -> em.createNamedQuery(Singer.COUNT_ALL,
      Long.class).getSingleResult());
    }
}
```

TransactionTemplate 类是从 Spring 注入的。然后在 countAll()方法中调用 TransactionTemplate.execute()方法，并传入一个实现了 TransactionCallback<T>接口的内部类的声明。最后用所需的逻辑重写 doInTransaction()方法。该逻辑将在由 transactionTemplate bean 定义的属性内运行。之所以没有在上述代码中清楚地看到该逻辑，是因为使用了 Java 8 lambda 表达式。以下代码是前面方法的扩展版本(因为是在引入 lambda 表达式之前编写的)：

```
public long countAll() {
        return transactionTemplate.execute(new TransactionCallback<Long>() {
            public Long doInTransaction(TransactionStatus transactionStatus) {
                return em.createNamedQuery(Singer.COUNT_ALL,
                        Long.class).getSingleResult();
            }
        });
    }
```

以下代码片段显示了测试程序：

```
package com.apress.prospring5.ch9;

import com.apress.prospring5.ch9.config.DataJpaConfig;
import com.apress.prospring5.ch9.config.ServicesConfig;
import com.apress.prospring5.ch9.services.SingerService;
import org.springframework.context.annotation.AnnotationConfigApplicationContext;
import org.springframework.context.support.GenericApplicationContext;

public class TxProgrammaticDemo {

    public static void main(String... args) {
        GenericApplicationContext ctx =
                new AnnotationConfigApplicationContext(ServicesConfig.class,
                DataJpaConfig.class);
        SingerService singerService = ctx.getBean(SingerService.class);
        System.out.println("Singer count: " + singerService.countAll());

        ctx.close();
    }
}
```

运行程序并观察结果。可以尝试调整事务属性并观察 countAll()方法的事务处理中会发生什么事情。

关于事务管理的几点考虑

在讨论实现事务管理的各种方法之后，应该使用哪一种方法呢？在一般情况下建议使用声明式方法，并且应尽可能避免在代码中实施事务管理。大多数情况下，一旦发现需要在应用程序中编写事务控制逻辑，就说明应用程序设计存在问题，此时，应该考虑将逻辑重构为可管理部分，并且在这些部分上以声明的方式定义事务需求。

对于声明式方法，使用 XML 和使用注解都有各自的优缺点。一些开发人员不愿意在代码中声明事务需求，而另一些开发人员则倾向于使用注解，以便进行简单维护，因为可以在代码中看到所有事务需求声明。再次强调一下，应该根据应用程序需求来做决定，并且一旦团队或公司对使用的方法进行了标准化，就请与配置风格保持一致。

9.6 使用 Spring 实现全局事务

许多企业级 Java 应用程序需要访问多个后端资源。例如，从外部业务合作伙伴收到的一条客户信息可能需要更新多个系统(CRM、ERP 等)的数据库。有些甚至可能需要生成一条消息，并通过 JMS 发送给 MQ 服务器，以便公司内对客户信息感兴趣的所有其他应用程序使用。跨越多个后端资源的事务被称为*全局(或分布式)事务*。

全局事务的一个主要特征是保证了原子性，这意味着所涉及的资源要么全部更新，要么都不更新。此外还包括应该由事务管理器处理的复杂协调和同步逻辑。在 Java 世界中，JTA 是实现全局事务的事实标准。

和本地事务一样，Spring 支持 JTA 事务，并从业务代码中隐藏相关逻辑。在本节中，将演示如何通过使用 JTA 和 Spring 来实现全局事务。

9.6.1 实现 JTA 示例的基础结构

本节将使用与前面示例相同的表。然而，嵌入式 H2 数据库并不完全支持 XA 标注(至少在编写本书时不完全支持)，所以在本例中，将使用 MySQL 作为后端数据库。

此外，我们还想演示如何在独立应用程序或 Web 容器环境中使用 JTA 实现全局事务。因此，在示例中，使用了 Atomikos(www.atomikos.com/Main/TransactionsEssentials)，它是一个被广泛使用的开源 JTA 事务管理器，主要用于非 JEE 环境。

为了说明全局事务是如何工作的，至少还需要两个后端资源。为了简单起见，将使用一个 MySQL 数据库，但使用两个 JPA 实体管理器来模拟用例。这样做的效果是相同的，因为可以使用多个 JPA 持久化单元来区分后端数据库。

在 MySQL 数据库中，创建了两个模式和相应的用户，如下面的 DDL 脚本所示：

```
CREATE USER 'prospring5_a'@'localhost' IDENTIFIED BY 'prospring5_a';
CREATE SCHEMA MUSICDB_A;
GRANT ALL PRIVILEGES ON MUSICDB_A . * TO 'prospring5_a'@'localhost';
   PRIVILEGES;
```

```
CREATE USER 'prospring5_b'@'localhost' IDENTIFIED BY 'prospring5_b';
CREATE SCHEMA MUSICDB_B;
GRANT ALL PRIVILEGES ON MUSICDB_B . * TO 'prospring5_b'@'localhost';
   PRIVILEGES;
```

设置完毕后，可以继续进行 Spring 配置和实现。

9.6.2 使用 JTA 实现全局事务

首先看看 Spring 的配置。以下代码片段描述了 XAJpaConfig 配置类，该类声明了访问两个数据库所需的 bean：

```
package com.apress.prospring5.ch9.config;

import com.atomikos.jdbc.AtomikosDataSourceBean;
import org.slf4j.Logger;
import org.slf4j.LoggerFactory;
import org.springframework.context.annotation.Bean;
import org.springframework.context.annotation.Configuration;
import org.springframework.data.jpa.repository.config.EnableJpaRepositories;
import org.springframework.jdbc.datasource.SimpleDriverDataSource;
import org.springframework.orm.jpa.JpaVendorAdapter;
import org.springframework.orm.jpa.LocalContainerEntityManagerFactoryBean;
import org.springframework.orm.jpa.vendor.HibernateJpaVendorAdapter;

import javax.persistence.EntityManagerFactory;
import javax.sql.DataSource;
import java.sql.Driver;
import java.util.Properties;

@Configuration
@EnableJpaRepositories
public class XAJpaConfig {

   private static Logger logger = LoggerFactory.getLogger(XAJpaConfig.class);
   @SuppressWarnings("unchecked")
   @Bean(initMethod = "init", destroyMethod = "close")
   public DataSource dataSourceA() {
      try {
         AtomikosDataSourceBean dataSource = new AtomikosDataSourceBean();
         dataSource.setUniqueResourceName("XADBMSA");
         dataSource.setXaDataSourceClassName(
                  "com.mysql.cj.jdbc.MysqlXADataSource");
         dataSource.setXaProperties(xaAProperties());
         dataSource.setPoolSize(1);
         return dataSource;
      } catch (Exception e) {
         logger.error("Populator DataSource bean cannot be created!", e);
         return null;
      }
   }

   @Bean
   public Properties xaAProperties() {
      Properties xaProp = new Properties();
      xaProp.put("databaseName", "musicdb_a");
      xaProp.put("user", "prospring5_a");
      xaProp.put("password", "prospring5_a");
      return xaProp;
   }

   @SuppressWarnings("unchecked")
   @Bean(initMethod = "init", destroyMethod = "close")
   public DataSource dataSourceB() {
      try {
         AtomikosDataSourceBean dataSource = new AtomikosDataSourceBean();
         dataSource.setUniqueResourceName("XADBMSB");
         dataSource.setXaDataSourceClassName(
               "com.mysql.cj.jdbc.MysqlXADataSource");
         dataSource.setXaProperties(xaBProperties());
         dataSource.setPoolSize(1);
         return dataSource;
      } catch (Exception e) {
         logger.error("Populator DataSource bean cannot be created!", e);
         return null;
      }
   }
```

```
    @Bean
    public Properties xaBProperties() {
        Properties xaProp = new Properties();
        xaProp.put("databaseName", "musicdb_b");
        xaProp.put("user", "prospring5_b");
        xaProp.put("password", "prospring5_b");
        return xaProp;
    }

    @Bean
    public Properties hibernateProperties() {
        Properties hibernateProp = new Properties();
        hibernateProp.put("hibernate.transaction.factory_class",
                "org.hibernate.transaction.JTATransactionFactory");
        hibernateProp.put("hibernate.transaction.jta.platform",
              "com.atomikos.icatch.jta.hibernate4.AtomikosPlatform");
        // required by Hibernate 5
        hibernateProp.put("hibernate.transaction.coordinator_class", "jta");
        hibernateProp.put("hibernate.dialect",
            "org.hibernate.dialect.MySQL5Dialect");
        // this will work only if users/schemas are created first,
        // use ddl.sql script for this
        hibernateProp.put("hibernate.hbm2ddl.auto", "create-drop");
        hibernateProp.put("hibernate.show_sql", true);
        hibernateProp.put("hibernate.max_fetch_depth", 3);
        hibernateProp.put("hibernate.jdbc.batch_size", 10);
        hibernateProp.put("hibernate.jdbc.fetch_size", 50);
        return hibernateProp;
    }

    @Bean
    public EntityManagerFactory emfA() {
        LocalContainerEntityManagerFactoryBean factoryBean =
            new LocalContainerEntityManagerFactoryBean();
        factoryBean.setPackagesToScan("com.apress.prospring5.ch9.entities");
        factoryBean.setDataSource(dataSourceA());
        factoryBean.setJpaVendorAdapter(new HibernateJpaVendorAdapter());
        factoryBean.setJpaProperties(hibernateProperties());
        factoryBean.setPersistenceUnitName("emfA");
        factoryBean.afterPropertiesSet();
        return factoryBean.getObject();
    }

    @Bean
    public EntityManagerFactory emfB() {
        LocalContainerEntityManagerFactoryBean factoryBean =
            new LocalContainerEntityManagerFactoryBean();
        factoryBean.setPackagesToScan("com.apress.prospring5.ch9.entities");
        factoryBean.setDataSource(dataSourceB());
        factoryBean.setJpaVendorAdapter(new HibernateJpaVendorAdapter());
        factoryBean.setJpaProperties(hibernateProperties());
        factoryBean.setPersistenceUnitName("emfB");
        factoryBean.afterPropertiesSet();
        return factoryBean.getObject();
    }
}
```

虽然该配置很长，但并不是太复杂。首先，定义两个 DataSource bean 来指示两个数据库资源。bean 名称分别是 dataSourceA 和 dataSourceB，它们分别连接到模式 musicdb_a 和 musicdb_a。这两个 DataSource bean 都使用了类 com.atomikos.jdbc.AtomikosDataSourceBean，该类支持符合 XA 标准的 DataSource，而在这两个 bean 的定义中，MySQL 的 XA DataSource 实现类被定义为 com.mysql.cj.jdbc.MysqlXADataSource，它是 MySQL 的资源管理器。然后，提供数据库连接信息。请注意，poolSize 属性定义了 Atomikos 需要维护的连接池内的连接数。该属性不是强制性的。但是，如果未提供该属性，那么 Atomikos 将使用默认值 1。

紧接着，定义两个 EntityManagerFactory bean，分别命名为 emfA 和 emfB。常见的 JPA 属性被封装在 hibernateProperties bean 中。这两个 bean 之间的唯一区别是，它们被注入不同的数据源(即 dataSourceA 被注入到 emfA 中，dataSourceB 被注入到 emfB 中)。因此，emfA 将通过 dataSourceA bean 连接到 MySQL 的 prospring5_a 模式，而 emfB 则通过 dataSourceB bean 连接到 MySQL 的 prospring5_b 模式。查看一下 emfBase bean 中的属性 hibernate.transaction.factory_class 和 hibernate.transaction.jta.platform。这两个属性非常重要，因为 Hibernate 使用它们来查找底层的 UserTransaction 和 TransactionManager bean，从而参与正在管理全局事务的持久化上下文。同样重要

的还有hibernate.transaction.coordinator_class，因为它是让Hibernate 4的Atomikos类使用Hibernate 5所必需的[1]。

以下代码片段描述了ServicesConfig，它声明了用于实现全局事务管理的bean：

```
package com.apress.prospring5.ch9.config;

import com.atomikos.icatch.config.UserTransactionService;
import com.atomikos.icatch.config.UserTransactionServiceImp;
import com.atomikos.icatch.jta.UserTransactionImp;
import com.atomikos.icatch.jta.UserTransactionManager;
import org.slf4j.Logger;
import org.slf4j.LoggerFactory;
import org.springframework.context.annotation.Bean;
import org.springframework.context.annotation.ComponentScan;
import org.springframework.context.annotation.Configuration;
import org.springframework.context.annotation.DependsOn;
import org.springframework.transaction.PlatformTransactionManager;
import org.springframework.transaction.annotation.EnableTransactionManagement;
import org.springframework.transaction.jta.JtaTransactionManager;
import javax.transaction.SystemException;
import javax.transaction.UserTransaction;
import java.util.Properties;

@Configuration
@EnableTransactionManagement
@ComponentScan(basePackages = "com.apress.prospring5.ch9.services")
public class ServicesConfig {

    private Logger logger = LoggerFactory.getLogger(ServicesConfig.class);
    @Bean(initMethod = "init", destroyMethod = "shutdownForce")
    public UserTransactionService userTransactionService(){
        Properties atProps = new Properties();
        atProps.put("com.atomikos.icatch.service",
            "com.atomikos.icatch.standalone.UserTransactionServiceFactory");
        return new UserTransactionServiceImp(atProps);
    }

    @Bean (initMethod = "init", destroyMethod = "close")
    @DependsOn("userTransactionService")
    public UserTransactionManager atomikosTransactionManager(){
        UserTransactionManager utm = new UserTransactionManager();
        utm.setStartupTransactionService(false);
        utm.setForceShutdown(true);
        return utm;
    }

    @Bean
    @DependsOn("userTransactionService")
    public UserTransaction userTransaction(){
        UserTransactionImp ut = new UserTransactionImp();
        try {
            ut.setTransactionTimeout(300);
        } catch (SystemException se) {
            logger.error("Configuration exception.", se);
            return null;
        }
        return ut;
    }

    @Bean
    public PlatformTransactionManager transactionManager(){
        JtaTransactionManager ptm = new JtaTransactionManager();
        ptm.setTransactionManager(atomikosTransactionManager());
        ptm.setUserTransaction(userTransaction());
        return ptm;
    }
}
```

对于Atomikos部分，定义了两个bean：atomikosTransactionManager和atomikosUserTransaction bean。实现类由Atomikos提供，分别实现了标准的Spring org.springframework.transaction.PlatformTransactionManager和javax.transaction.UserTransaction接口。这些bean提供JTA所需的事务协调和同步服务，并通过支持2PC的XA协议与资源管理器进行通信。然后，定义Spring的transactionManager bean(使用org.springframework.transaction.jta.JtaTransactionManager作

1 该配置得益于官方Atomikos文档(https://www.atomikos.com/Documentation/SpringIntegration)以及Stack Overflow社区(https://stackoverflow.com/questions/33127854/hibernate-5-with-spring-jta)提供的帮助。

为实现类)，并注入由 Atomikos 提供的两个事务 bean，从而指示 Spring 使用 Atomikos JTA 进行事务管理。另外，请注意 UserTransactionService bean，它用于配置 Atomikos 事务服务以管理挂起的事务[1]。

以下代码片段显示了 JTA 的 SingerServiceImpl 类。请注意，为了简单起见，只实现了 save()方法。

```
package com.apress.prospring5.ch9.services;

import com.apress.prospring5.ch9.entities.Singer;
import org.apache.commons.lang3.NotImplementedException;
import org.springframework.orm.jpa.JpaSystemException;
import org.springframework.stereotype.Repository;
import org.springframework.stereotype.Service;
import org.springframework.transaction.annotation.Transactional;

import javax.persistence.EntityManager;
import javax.persistence.PersistenceContext;
import javax.persistence.PersistenceException;
import java.util.List;

@Service("singerService")
@Repository
@Transactional
public class SingerServiceImpl implements SingerService {

   @PersistenceContext(unitName = "emfA")
   private EntityManager emA;
   @PersistenceContext(unitName = "emfB")
   private EntityManager emB;

   @Override
   @Transactional(readOnly = true)
   public List<Singer> findAll() {
      throw new NotImplementedException("findAll");
   }

   @Override
   @Transactional(readOnly = true)
   public Singer findById(Long id) {
      throw new NotImplementedException("findById");
   }

   @Override
   public Singer save(Singer singer) {
      Singer singerB = new Singer();
      singerB.setFirstName(singer.getFirstName());
      singerB.setLastName(singer.getLastName());
      if (singer.getId() == null) {
         emA.persist(singer);
         emB.persist(singerB);
         //throw new JpaSystemException(new PersistenceException());
      } else {
         emA.merge(singer);
         emB.merge(singer);
      }
      return singer;
   }

   @Override
   public long countAll() {
      return 0;
   }
}
```

所定义的两个实体管理器被注入到 SingerServiceImpl 类中。在 save()方法中，分别将联系人对象保存到两个模式。现在暂时忽略 throw exception 语句，稍后将用它来验证在保存到模式 prospring5_b 失败时事务是否已回滚。以下代码片段显示了测试程序：

```
package com.apress.prospring5.ch9;

import com.apress.prospring5.ch9.config.ServicesConfig;
import com.apress.prospring5.ch9.config.XAJpaConfig;
import com.apress.prospring5.ch9.entities.Singer;
import com.apress.prospring5.ch9.services.SingerService;
```

1 这是位于 https://www.atomikos.com/Documentation/SpringIntegration#The_Advanced_Case_40As_of_3.3_41 的 Atomikos 文档中作为示例给出的 XML 配置的注解配置调整版本。

```
import org.slf4j.Logger;
import org.slf4j.LoggerFactory;
import org.springframework.context.annotation.AnnotationConfigApplicationContext;
import org.springframework.context.support.GenericApplicationContext;

import java.util.Date;
import java.util.GregorianCalendar;

public class TxJtaDemo {
   private static Logger logger = LoggerFactory.getLogger(TxJtaDemo.class);
   public static void main(String... args) {
      GenericApplicationContext ctx =
         new AnnotationConfigApplicationContext(ServicesConfig.class,
            XAJpaConfig.class);
      SingerService singerService = ctx.getBean(SingerService.class);
      Singer singer = new Singer();
      singer.setFirstName("John");
      singer.setLastName("Mayer");
      singer.setBirthDate(new Date(
            (new GregorianCalendar(1977, 9, 16)).getTime().getTime()));
      singerService.save(singer);
      if (singer.getId() != null) {
         logger.info("--> Singer saved successfully");
      } else {
         logger.info("--> Singer was not saved, check the configuration!!");
      }
      ctx.close();
   }
}
```

该程序创建了一个新的联系人对象并调用 SingerService.save()方法。该实现将尝试将相同的对象保存到两个数据库中。如果一切顺利，运行该程序会产生以下输出(其他输出被省略)：

```
--> Singer saved successfully
```

Atomikos 创建了一个复合事务，并与 XA DataSource(在本例中为 MySQL)进行通信，同时执行同步、提交事务等操作。在数据库中，可以看到新的联系人被分别存储到数据库的两个模式中。但是，如果想检查代码是否完成了保存操作，可以实现 findAll()方法，如下所示：

```
package com.apress.prospring5.ch9.services;
...
@Service("singerService")
@Repository
@Transactional
public class SingerServiceImpl implements SingerService {

   private static final String FIND_ALL= "select s from Singer s";

   @PersistenceContext(unitName = "emfA")
   private EntityManager emA;
   @PersistenceContext(unitName = "emfB")
   private EntityManager emB;

   @Override
   @Transactional(readOnly = true)
   public List<Singer> findAll()
   {
      List<Singer> singersFromA = findAllInA();
      List<Singer> singersFromB = findAllInB();
      if (singersFromA.size()!= singersFromB.size()){
         throw new AsyncXAResourcesException("
            XA resources do not contain the same expected data.");
      }
      Singer sA = singersFromA.get(0);
      Singer sB = singersFromB.get(0);
      if (!sA.getFirstName().equals(sB.getFirstName())) {
         throw new AsyncXAResourcesException("
            XA resources do not contain the same expected data.");
      }
      List<Singer> singersFromBoth = new ArrayList<>();
      singersFromBoth.add(sA);
      singersFromBoth.add(sB);
      return singersFromBoth;
   }

   private List<Singer> findAllInA(){
      return emA.createQuery(FIND_ALL).getResultList();
```

```
    }

    private List<Singer> findAllInB(){
        return emB.createQuery(FIND_ALL).getResultList();
    }
    ...
}
```

因此，可以按照如下所示修改测试歌手是否被保存到两个数据库中的代码：

```
package com.apress.prospring5.ch9;
...
public class TxJtaDemo {
    private static Logger logger = LoggerFactory.getLogger(TxJtaDemo.class);

    public static void main(String... args) {
        GenericApplicationContext ctx =
        new AnnotationConfigApplicationContext(ServicesConfig.class,
                XAJpaConfig.class);
        SingerService singerService = ctx.getBean(SingerService.class);
        Singer singer = new Singer();
        singer.setFirstName("John");
        singer.setLastName("Mayer");
        singer.setBirthDate(new Date(
                (new GregorianCalendar(1977, 9, 16)).getTime().getTime()));
        singerService.save(singer);
        if (singer.getId() != null) {
            logger.info("--> Singer saved successfully");
        } else {
            logger.error("--> Singer was not saved, check the configuration!!");
        }
        // check saving in both databases
        List<Singer> singers = singerService.findAll();
        if (singers.size()!= 2) {
            logger.error("--> Something went wrong.");
        } else {
            logger.info("--> Singers form both DBs: " + singers);
        }

        ctx.close();
    }
}
```

接下来看看回滚是如何工作的。如下面的代码片段所示，此时没有调用 emB.persist()，而是抛出一个异常来模拟保存过程出现错误，因而导致数据不能保存在第二个数据库中。

```
package com.apress.prospring5.ch9.services;
...
@Service("singerService")
@Repository
@Transactional
public class SingerServiceImpl implements SingerService {

    private static final String FIND_ALL= "select s from Singer s";

    @PersistenceContext(unitName = "emfA")
    private EntityManager emA;
    @PersistenceContext(unitName = "emfB")
    private EntityManager emB;
    ...
    @Override
    public Singer save(Singer singer) {
        Singer singerB = new Singer();
        singerB.setFirstName(singer.getFirstName());
        singerB.setLastName(singer.getLastName());
        if (singer.getId() == null) {
            emA.persist(singer);
            if(true) {
                throw new JpaSystemException(new PersistenceException(
                    "Simulation of something going wrong."));
            }
            emB.persist(singerB);
        } else {
            emA.merge(singer);
            emB.merge(singer);
        }
        return singer;
    }
```

```
    @Override
    public long countAll() {
        return 0;
    }
}
```

再次运行该程序会产生以下结果：

```
...
INFO o.h.h.i.QueryTranslatorFactoryInitiator - HHH000397:
   Using ASTQueryTranslatorFactory
INFO o.s.o.j.LocalContainerEntityManagerFactoryBean - Initialized JPA
   EntityManagerFactory for persistence unit 'emfA'
INFO o.s.o.j.LocalContainerEntityManagerFactoryBean - Initialized JPA
   EntityManagerFactory for persistence unit 'emfB'
INFO o.s.t.j.JtaTransactionManager - Using JTA UserTransaction:
   com.atomikos.icatch.jta.UserTransactionImp@6da9dc6
INFO o.s.t.j.JtaTransactionManager - Using JTA TransactionManager:
   com.atomikos.icatch.jta.UserTransactionManager@2216effc
DEBUG o.s.t.j.JtaTransactionManager - Creating new transaction with name
[com.apress.prospring5.ch9.services.SingerServiceImpl.save]:
   PROPAGATION_REQUIRED,ISOLATION_DEFAULT; ''
DEBUG o.s.o.j.EntityManagerFactoryUtils - Opening JPA EntityManager
DEBUG o.s.o.j.EntityManagerFactoryUtils - Registering transaction synchronization
   for JPA EntityManager
Hibernate: insert into singer (BIRTH_DATE, FIRST_NAME, LAST_NAME, VERSION)
values (?, ?, ?, ?)
DEBUG o.s.o.j.EntityManagerFactoryUtils - Closing JPA EntityManager
DEBUG o.s.t.j.JtaTransactionManager - Initiating transaction rollback
WARN c.a.j.AbstractConnectionProxy - Forcing close of pending statement:
   com.mysql.cj.jdbc.PreparedStatementWrapper@3f685162
Exception in thread "main" org.springframework.orm.jpa.JpaSystemException:
   Simulation of something going wrong.;
...
Caused by: javax.persistence.PersistenceException:
   Simulation of something going wrong.
```

如上面的输出所示，第一位歌手被保存(注意 insert 语句)。但是，当保存到第二个 DataSource 时，由于抛出了异常，Atomikos 将回滚整个事务。可以查看模式 musicdb_a，你会看到第二个新歌手未被保存。

9.6.3 Spring Boot JTA

JTA 启动器的 Spring Boot 带有一组预配置的 bean，旨在帮助开发人员专注于代码的业务功能而不是环境设置。再次强调一下，无论组件是什么，都是所有 Spring Boot 启动器库要做的事情，所以前面的句子看起来似乎有点多余。JTA 的 Spring Boot 包含一个库以便使用 Atomikos，它提供相应的库并为你配置 Atomikos 组件。将前面的示例迁移到 Spring Boot 意味着将 DataSource 和事务管理器配置导入到 Spring Boot 应用程序中。但是本节的目的是演示 Spring Boot 如何通过提供的预配置 bean 来帮助加速涉及全局事务管理的应用程序的开发，所以必须编写另一个示例。假设将一条消息传递给一个消息队列，表示创建一个新的 Singer 实例。显然，如果将 Singer 记录保存到数据库的操作失败，就需要回滚事务并阻止发送消息。为此，需要执行以下操作：

● 为使用 JTA 和 JMS 而配置 Spring Boot Gradle 项目。配置如下所示：

```
//build.gradle
ext {
    ...
    bootVersion = '2.0.0.M1'
    atomikosVersion = '4.0.4'

    boot = [
            ...
     starterJpa :
        "org.springframework.boot:spring-boot-starter-data-jpa:$bootVersion",
     starterJta :
        "org.springframework.boot:spring-boot-starter-jta-atomikos:$bootVersion",
     starterJms :
        "org.springframework.boot:spring-boot-starter-artemis:$bootVersion"
    ]

    misc = [
        ...
        artemis : "org.apache.activemq:artemis-jms-server:2.1.0"
    ]

    db = [
```

```
        ...
        h2 : "com.h2database:h2:$h2Version"
    ]
}
//chapter09/boot-jta/build.gradle
buildscript {
    repositories {
        ...
    }

    dependencies {
        classpath boot.springBootPlugin
    }
}

apply plugin: 'org.springframework.boot'

dependencies {
    compile boot.starterJpa, boot.starterJta, boot.starterJms, db.h2
    compilemisc.artemis {
        exclude group: 'org.apache.geronimo.specs',
            module: 'geronimo-jms_2.0_spec'
    }
}
```

在图 9-3 中，可以看到之前声明的 Spring Boot 启动器库作为项目的依赖项，并被添加到项目中。

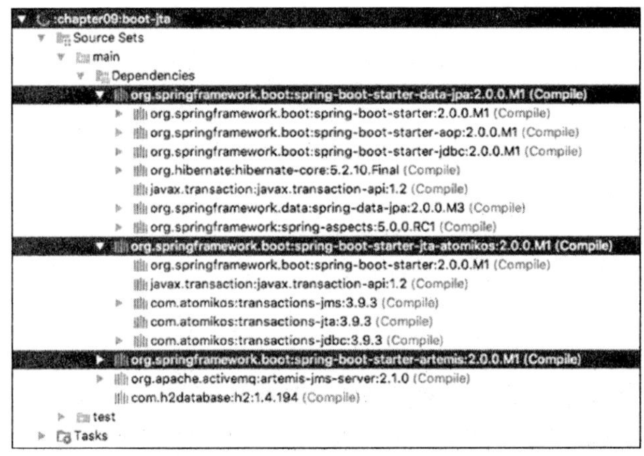

图 9-3　Spring Boot 启动器库及其依赖项

- 定义 Singer 实体类和处理它的存储库。Singer 实体类的结构与本节前面提到的结构相同，但没有任何关联实体。并且 SingerRepository 为空，因为在本例中只有 CrudRepository 提供的方法才能使用：save()和 count()。
- 定义一个保存 Singer 记录并发送确认信息的服务类。

```
package com.apress.prospring5.ch9.services;

import com.apress.prospring5.ch9.entities.Singer;
import com.apress.prospring5.ch9.ex.AsyncXAResourcesException;
import com.apress.prospring5.ch9.repos.SingerRepository;
import org.springframework.beans.factory.annotation.Autowired;
import org.springframework.jms.core.JmsTemplate;
import org.springframework.stereotype.Repository;
import org.springframework.stereotype.Service;
import org.springframework.transaction.annotation.Transactional;

import javax.persistence.EntityManager;
import javax.persistence.PersistenceContext;

@Service("singerService")
@Transactional
public class SingerServiceImpl implements SingerService {

    private SingerRepository singerRepository;
    private JmsTemplate jmsTemplate;
    public SingerServiceImpl(SingerRepository singerRepository,
        JmsTemplate jmsTemplate) {
        this.singerRepository = singerRepository;
```

```
        this.jmsTemplate = jmsTemplate;
    }

    @Override
    public Singer save(Singer singer) {
        jmsTemplate.convertAndSend("singers", "Just saved:" + singer);
        if(singer == null) {
            throw new AsyncXAResourcesException(
                "Simulation of something going wrong.");
        }
        singerRepository.save(singer);
        return singer;
    }

    @Override public long count() {
        return singerRepository.count();
    }
}
```

注入存储库 bean 或 JmsTemplate 都不需要@Autowired 注解。如果它们是唯一声明的 bean，那么 Spring Boot 像施魔法一样注入所需的 bean。

- 还需要一个 bean 来监听传递给 singers 队列的消息并将其打印出来。

```
package com.apress.prospring5.ch9;

import org.slf4j.Logger;
import org.slf4j.LoggerFactory;
import org.springframework.jms.annotation.JmsListener;
import org.springframework.stereotype.Component;

@Component
public class Messages {
    private static Logger logger = LoggerFactory.getLogger(Messages.class);

    @JmsListener(destination="singers")
    public void onMessage(String content){
        logger.info("--> Received content: " + content);
    }
}
```

- 配置 Artemis JMS 服务器以创建名为 singers 的嵌入式队列。该过程是通过将 application.properties 文件中的 spring.artemis.embedded.queues 属性设置为值 singers 来完成的，该文件可用于配置 Spring Boot 应用程序。

```
spring.artemis.embedded.queues=singers
spring.jta.log-dir=out
```

上面的配置片段描述了 application.properties 文件的内容。除 spring.artemis.embedded.queues 属性外，还使用 spring.jta.log-dir 来设置 Atomikos 应该写入 JTA 日志的位置，此时，设置为 out 目录。

- 下面所示的应用程序类封装了相关内容并进行测试：

```
package com.apress.prospring5.ch9;

import com.apress.prospring5.ch9.entities.Singer;
import com.apress.prospring5.ch9.services.SingerService;
import com.atomikos.jdbc.AtomikosDataSourceBean;
import org.h2.jdbcx.JdbcDataSource;
import org.slf4j.Logger;
import org.slf4j.LoggerFactory;
import org.springframework.beans.factory.annotation.Autowired;
import org.springframework.boot.CommandLineRunner;
import org.springframework.boot.SpringApplication;
import org.springframework.boot.autoconfigure.SpringBootApplication;
import org.springframework.context.ConfigurableApplicationContext;
import org.springframework.context.annotation.Bean;
import org.springframework.orm.jpa.LocalContainerEntityManagerFactoryBean;
import org.springframework.orm.jpa.vendor.HibernateJpaVendorAdapter;

import javax.persistence.EntityManagerFactory;
import javax.sql.DataSource;
import java.util.Date;
import java.util.GregorianCalendar;
import java.util.Properties;

import static org.hibernate.cfg.AvailableSettings.*;
import static org.hibernate.cfg.AvailableSettings.STATEMENT_FETCH_SIZE;
```

```
@SpringBootApplication(scanBasePackages = "com.apress.prospring5.ch9.services")
public class Application implements CommandLineRunner {

   private static Logger logger = LoggerFactory.getLogger(Application.class);

   public static void main(String... args) throws Exception {
      ConfigurableApplicationContext ctx =
         SpringApplication.run(Application.class, args);
      System.in.read();
      ctx.close();
   }

   @Autowired SingerService singerService;

   @Override public void run(String... args) throws Exception {
      Singer singer = new Singer();
      singer.setFirstName("John");
      singer.setLastName("Mayer");
      singer.setBirthDate(new Date(
            (new GregorianCalendar(1977, 9, 16)).getTime().getTime()));
      singerService.save(singer);

      long count = singerService.count();
      if (count == 1) {
         logger.info("--> Singer saved successfully");
      } else {
         logger.error("--> Singer was not saved, check the configuration!!");
      }

      try {
         singerService.save(null);
      } catch (Exception ex) {
         logger.error(ex.getMessage() + "Final count:" + singerService.count());
      }
   }
}
```

如果运行应用程序，将会看到如下所示的输出：

```
...
INFO c.a.j.AtomikosConnectionFactoryBean - AtomikosConnectionFactoryBean
   'jmsConnectionFactory': init...
INFO o.s.t.j.JtaTransactionManager - Using JTA UserTransaction:
   com.atomikos.icatch.jta.UserTransactionManager@408a247c
INFO c.a.j.AtomikosJmsXaSessionProxy - atomikos xa session proxy for resource
   jmsConnectionFactory: calling createQueue on JMS driver session...
INFO c.a.j.AtomikosJmsXaSessionProxy - atomikos xa session proxy for resource
jmsConnectionFactory: calling getTransacted on JMS driver session...
DEBUG o.s.t.j.JtaTransactionManager - Participating in existing transaction
DEBUG o.s.t.j.JtaTransactionManager - Initiating transaction commit
INFO c.a.d.x.XAResourceTransaction - XAResource.start ...
INFO c.a.d.x.XAResourceTransaction - XAResource.end ...
DEBUG o.s.t.j.JtaTransactionManager - Initiating transaction commit
DEBUG o.s.t.j.JtaTransactionManager - Creating new transaction with name
    [com.apress.prospring5.ch9.services.SingerServiceImpl.save]:
   PROPAGATION_REQUIRED,ISOLATION_DEFAULT; ''
INFO c.a.i.i.BaseTransactionManager - createCompositeTransaction ( 10000 ):
   created new ROOT transaction with id 127.0.0.1.tm0000200001
DEBUG o.s.t.j.JtaTransactionManager - Participating in existing transaction
INFO c.a.p.c.Application - --> Singer saved successfully
...//etc
```

从日志中可以清楚地看到针对每个操作所创建并重用的全局事务。如果按下任意键，再按 Enter 键退出应用程序，请耐心等待，因为应用程序需要一段时间才能正常关闭。

下面是关于使用 Spring Boot 创建 JTA 应用程序的一些结论：虽然看似很容易，但当使用多个数据源时，不可避免地要使用配置环境。另外，如果 JTA 提供程序是由 JEE 服务器提供的，那么事情就会变得非常复杂。但对于用于教育目的和测试的示例应用程序而言，它还是非常实用的。

9.6.4 使用 JTA 事务管理器的注意事项

是否将 JTA 用于全局事务管理目前还处在热议之中。例如，Spring 开发团队通常不推荐将 JTA 用于全局事务。

作为一般原则，当应用程序被部署到成熟的 JEE 应用程序服务器时，没有必要不使用 JTA，因为所有流行的 JEE 应用程序服务器的供应商都已经为其平台优化了 JTA 实现。当然，该功能是需要付费的。

对于独立或Web容器部署，则应该根据应用程序需求做出决定。最好尽早进行负载测试，以验证JTA是否会对性能产生影响。

好消息是，Spring可以在大多数主要Web和JEE容器中无缝地使用本地和全局事务，因此，当需要从一种事务管理策略切换到另一种时，通常不需要修改代码。如果决定在应用程序中使用 JTA，请确保使用 Spring 的JtaTransactionManager。

9.7 小结

事务管理是保证几乎任何类型应用程序中数据完整性的关键部分。在本章中，讨论了如何使用Spring来管理事务，同时对源代码几乎没有任何影响。你还学习了如何使用本地和全局事务。

本章提供了各种事务的实现示例，包括使用XML配置和注解的声明式方法以及编程式方法。

JEE应用程序服务器内部/外部都支持本地事务，并且只需要做简单配置即可在Spring中启用本地事务支持。但是，设置全局事务环境则需要完成更多工作，并且很大程度上取决于应用程序需要与哪个JTA提供程序以及相应的后端资源进行交互。

第 10 章

使用类型转换和格式化进行验证

在企业级应用程序中，验证至关重要。验证的目的是检查正在处理的数据是否满足所有预定义的业务需求，并确保数据在应用程序其他层中的完整性和有用性。

在应用程序开发中，数据验证通常与转换和格式化一起被提及。原因在于，数据源的格式很可能与应用程序中所使用的格式不同。例如，在 Web 应用程序中，用户在 Web 浏览器前端输入数据。当用户保存这些数据时，会将数据发送到服务器(在完成本地验证之后)。而在服务器端，执行数据绑定过程，在该过程中，首先从 HTTP 请求中提取数据，然后转换并根据每个属性定义的格式化规则(例如，日期格式为 yyyy-MM-dd)绑定到相应的域对象(例如，用户在 HTML 表单中输入歌手信息，然后绑定到服务器中的 Singer 对象)。数据绑定完成后，将验证规则应用于域对象以检查是否存在违反约束的行为。如果一切正常，则保存数据，并向用户显示验证成功的消息。否则，将填充验证错误的消息并显示给用户。

在本章的第一部分，你将学习 Spring 如何为类型转换、字段格式化以及验证提供复杂的支持。具体而言，本章涵盖以下主题：

- **Spring 类型转换系统和 Formatter 服务提供程序接口(SPI)**：将学习通用类型转换系统和 Formatter SPI，介绍如何使用新服务来替代以前的 PropertyEditor 支持，以及它们如何在任何 Java 类型之间进行转换。
- **Spring 中的验证**：讨论 Spring 如何支持域对象验证。首先，简要介绍 Spring 自己的 Validator 接口。然后，重点关注 JSR-349(bean 验证)支持。

10.1 依赖项

和前面几章一样，本章介绍的示例也需要一些依赖项，这些依赖项在以下配置代码段中描述。你可能注意到其中一个依赖项是 joda-time。如果运行的是 Java 8，那么 Spring 5 还支持 JSR-310，也就是 javax.time API。

```
//pro-spring-15/build.gradle
ext {
   //spring libs
   springVersion = '5.0.0.RC1'

   jodaVersion = '2.9.9'
   javaxValidationVersion = '2.0.0.Beta2'    //1.1.0.Final
   javaElVersion = '3.0.1-b04'               // 3.0.0
   glasshfishELVersion = '2.2.1-b05'         // 2.2

   hibernateValidatorVersion = '6.0.0.Beta2' //5.4.1.Final

   spring = [...]

   hibernate = [
      validator :
         "org.hibernate:hibernate-validator:$hibernateValidatorVersion",
      ...
   ]

   misc = [
      validation  :
         "javax.validation:validation-api:$javaxValidationVersion",
      joda        : "joda-time:joda-time:$jodaVersion",
      ...
   ]
```

```
    ...
}
//chapter10/build.gradle
dependencies {
      compile spring.contextSupport, misc.slf4jJcl, misc.logback,
      db.h2, misc.lang3, hibernate.em, hibernate.validator,
      misc.joda, misc.validation

      testCompile testing.junit
}
```

10.2 Spring 类型转换系统

在 Spring 3 中，引入了一种新的类型转换系统，它提供了一种在 Spring 应用程序中的任何 Java 类型之间进行转换的方法。下面将介绍这一新服务如何实现 PropertyEditor 支持所提供的相同功能，以及它如何支持任何 Java 类型之间的转换。此外，还将演示如何使用 Converter SPI 实现自定义类型转换器。

10.3 使用 PropertyEditors 从字符串进行转换

第 3 章介绍了 Spring 如何通过支持 PropertyEditor 来实现从属性文件中的字符串到 POJO 属性的转换。接下来快速回顾一下前面的内容，然后介绍 Spring 的 Converter SPI(Spring 3.0 以后可用)如何提供更强大的替换方法。

参见下面 Singer 类的简化版本：

```
package com.apress.prospring5.ch10;

import java.net.URL;
import java.text.SimpleDateFormat;

import org.joda.time.DateTime;

public class Singer {
   private String firstName;
   private String lastName;
   private DateTime birthDate;
   private URL personalSite;

   //getters and setters
   ...

   public String toString() {
      SimpleDateFormat sdf = new SimpleDateFormat("yyyy-MM-dd");
      return String.format("{First name: %s, Last name: %s,
      Birthday: %s, Site: %s}",
         firstName, lastName, sdf.format(birthDate.toDate()), personalSite);
   }
}
```

对于 birthDate 属性，我们使用了 JodaTime 的 DateTime 类。此外，如果适用，还可以使用一个 URL 类型的字段来表示歌手的个人网站。现在假设想在 Spring 的 ApplicationContext 中构造 Singer 对象，相关的值存储在 Spring 的配置文件或属性文件中。以下配置片段显示了 Spring XML 配置文件(prop-editor-app-context.xml)：

```
<?xml version="1.0" encoding="UTF-8"?>
<beans xmlns="http://www.springframework.org/schema/beans"
      xmlns:xsi="http://www.w3.org/2001/XMLSchema-instance"
      xmlns:context="http://www.springframework.org/schema/context"
      xmlns:util="http://www.springframework.org/schema/util"
      xmlns:p="http://www.springframework.org/schema/p"
      xsi:schemaLocation="http://www.springframework.org/schema/beans
         http://www.springframework.org/schema/beans/spring-beans.xsd
         http://www.springframework.org/schema/context
         http://www.springframework.org/schema/context/spring-context.xsd
         http://www.springframework.org/schema/util
         http://www.springframework.org/schema/util/spring-util.xsd">
   <context:annotation-config/>

   <context:property-placeholder location="classpath:application.properties"/>

   <bean id="customEditorConfigurer"
      class="org.springframework.beans.factory.config.CustomEditorConfigurer"
```

```
        p:propertyEditorRegistrars-ref="propertyEditorRegistrarsList"/>

    <util:list id="propertyEditorRegistrarsList">
        <bean class="com.apress.prospring5.ch10.DateTimeEditorRegistrar">
      <constructor-arg value="${date.format.pattern}"/>
        </bean>
    </util:list>

    <bean id="eric" class="com.apress.prospring5.ch10.Singer"
        p:firstName="Eric"
        p:lastName="Clapton"
        p:birthDate="1945-03-30"
        p:personalSite="http://www.ericclapton.com"/>

    <bean id="countrySinger" class="com.apress.prospring5.ch10.Singer"
        p:firstName="${countrySinger.firstName}"
        p:lastName="${countrySinger.lastName}"
        p:birthDate="${countrySinger.birthDate}"
        p:personalSite="${countrySinger.personalSite}"/>
</beans>
```

此时，构建了 Singer 类的两个不同的 bean。eric bean 是用配置文件中提供的值构造的；而对于 countrySinger bean，属性被外化为一个属性文件。此外，还定义了一个自定义编辑器，用于将 String 转换为 JodaTime 的 DateTime 类型，并且日期-时间格式模式也在属性文件中进行了外部化。以下代码片段显示了属性文件(application.properties)：

```
date.format.pattern=yyyy-MM-dd

countrySinger.firstName=John
countrySinger.lastName=Mayer
countrySinger.birthDate=1977-10-16
countrySinger.personalSite=http://johnmayer.com/
```

以下代码片段显示了将字符串值转换为 DateTime 类型 JodaTime 的自定义编辑器：

```
package com.apress.prospring5.ch10;

import org.joda.time.DateTime;
import org.joda.time.format.DateTimeFormat;
import org.joda.time.format.DateTimeFormatter;
import org.springframework.beans.PropertyEditorRegistrar;
import org.springframework.beans.PropertyEditorRegistry;
import java.beans.PropertyEditorSupport;

public class DateTimeEditorRegistrar implements PropertyEditorRegistrar {
    private DateTimeFormatter dateTimeFormatter;

    public DateTimeEditorRegistrar(String dateFormatPattern) {
        dateTimeFormatter = DateTimeFormat.forPattern(dateFormatPattern);
    }

    @Override
    public void registerCustomEditors(PropertyEditorRegistry registry) {
        registry.registerCustomEditor(DateTime.class,
            new DateTimeEditor(dateTimeFormatter));
    }
    private static class DateTimeEditor extends PropertyEditorSupport {
        private DateTimeFormatter dateTimeFormatter;

        public DateTimeEditor(DateTimeFormatter dateTimeFormatter) {
    this.dateTimeFormatter = dateTimeFormatter;
        }

        @Override
            public void setAsText(String text) throws IllegalArgumentException {
      setValue(DateTime.parse(text, dateTimeFormatter));
        }
    }
}
```

DateTimeEditorRegistrar 通过实现 PropertyEditorRegister 接口注册了自定义的 PropertyEditor。然后创建一个名为 DateTimeEditor 的内部类来处理 String 到 DateTime 类型的转换。在该例中使用了一个内部类，因为它只能通过 PropertyEditorRegistrar 实现来访问。现在进行测试。下面的代码片段显示了测试程序：

```
package com.apress.prospring5.ch10;

import org.slf4j.Logger;
import org.slf4j.LoggerFactory;
```

```
import org.springframework.context.support.GenericXmlApplicationContext;

public class PropEditorDemo {

   private static Logger logger =
      LoggerFactory.getLogger(PropEditorDemo.class);

   public static void main(String... args) {
      GenericXmlApplicationContext ctx =
     new GenericXmlApplicationContext();

      ctx.load("classpath:spring/prop-editor-app-context.xml");
      ctx.refresh();
      Singer eric = ctx.getBean("eric", Singer.class);
      logger.info("Eric info: " + eric);
      Singer countrySinger = ctx.getBean("countrySinger", Singer.class);
      logger.info("John info: " + countrySinger);

      ctx.close();
   }
}
```

正如你所看到的，这两个 Singer bean 在 ApplicationContext 中被检索并打印出来。运行该程序会产生以下输出：

```
[main] INFO c.a.p.c.PropEditorDemo - Eric info: {First name: Eric,
   Last name: Clapton, Birthday: 1945-03-30, Site: http://www.ericclapton.com}
[main] INFO c.a.p.c.PropEditorDemo - John info: {First name: John,
   Last name: Mayer, Birthday: 1977-10-16, Site: http://johnmayer.com/}
```

如输出所示，属性被转换并应用于 Singer bean。这里之所以使用 XML 而不是 Java 配置类，是因为被注入的值被声明为文本值，并且 Spring 在后台透明地进行转换。

10.4 Spring 类型转换介绍

在 Spring 3.0 中，引入了一个通用类型转换系统，该系统位于 org.springframework.core.convert 包中。除了提供 PropertyEditor 所支持的替代方法之外，还可以配置类型转换系统，从而在任何 Java 类型和 POJO 之间进行转换(而 PropertyEditor 专注于将属性文件中的 String 表示转换为 Java 类型)。

10.4.1 实现自定义转换器

为了查看实际运行中的类型转换系统，继续使用前面的示例并使用相同的 Singer 类。假设这次要使用类型转换系统将 String 格式的日期转换为 Singer 类的 birthDate 属性(该属性所属的类型是 JodaTime 的 DateTime)。为了支持转换，需要通过实现 org.springframework.core.convert.converter.Converter <S，T>接口来创建一个自定义转换器，而不是创建一个定制的 PropcrtyEditor。以下代码片段显示了该自定义转换器：

```
package com.apress.prospring5.ch10;

import org.joda.time.DateTime;
import org.joda.time.format.DateTimeFormat;
import org.joda.time.format.DateTimeFormatter;
import org.springframework.core.convert.converter.Converter;

import javax.annotation.PostConstruct;

public class StringToDateTimeConverter implements Converter<String, DateTime> {
   private static final String DEFAULT_DATE_PATTERN = "yyyy-MM-dd";
   private DateTimeFormatter dateFormat;
   private String datePattern = DEFAULT_DATE_PATTERN;

   public String getDatePattern() {
      return datePattern;
   }

   public void setDatePattern(String datePattern) {
      this.datePattern = datePattern;
   }

   @PostConstruct
   public void init() {
      dateFormat = DateTimeFormat.forPattern(datePattern);
```

```
    }

    @Override
    public DateTime convert(String dateString) {
        return dateFormat.parseDateTime(dateString);
    }
}
```

此时实现了接口 Converter <String, DateTime>，这意味着该自定义转换器负责将 String(源类型 S)转换为 DateTime 类型(目标类型 T)。对日期-时间模式的注入是可选的，可以通过调用 setter 方法 setDatePattern()来完成。如果未注入，则使用默认模式 yyyy-MM-dd。然后，在初始化方法(用@PostConstruct 注解的 init()方法)中，构造了 JodaTime 的 DateTimeFormat 类的一个实例，它将根据指定的模式执行转换。最后，实现 convert()方法来提供转换逻辑。

10.4.2 配置 ConversionService

如果想要使用转换服务而不是 PropertyEditor，则需要在 Spring 的 ApplicationContext 中配置 org.springframework.core.convert.ConversionService 接口的一个实例。以下代码片段显示了 Java 配置类：

```
package com.apress.prospring5.ch10.config;

import com.apress.prospring5.ch10.Singer;
import com.apress.prospring5.ch10.StringToDateTimeConverter;
import org.joda.time.DateTime;
import org.springframework.beans.factory.annotation.Value;
import org.springframework.context.annotation.Bean;
import org.springframework.context.annotation.ComponentScan;
import org.springframework.context.annotation.Configuration;
import org.springframework.context.annotation.PropertySource;
import org.springframework.context.support.ConversionServiceFactoryBean;
import org.springframework.context.support.PropertySourcesPlaceholderConfigurer;
import org.springframework.core.convert.converter.Converter;

import java.net.URL;
import java.util.HashSet;
import java.util.Set;
@PropertySource("classpath:application.properties")
@Configuration
@ComponentScan(basePackages = "com.apress.prospring5.ch10")
public class AppConfig {

    @Value("${date.format.pattern}")
    private String dateFormatPattern;

    @Bean
    public static PropertySourcesPlaceholderConfigurer
        propertySourcesPlaceholderConfigurer() {
         return new PropertySourcesPlaceholderConfigurer();
    }

    @Bean
    public Singer john(@Value("${countrySinger.firstName}") String firstName,
        @Value("${countrySinger.lastName}") String lastName,
        @Value("${countrySinger.personalSite}") URL personalSite,
        @Value("${countrySinger.birthDate}") DateTime birthDate)
            throws Exception {
          Singer singer = new Singer();
          singer.setFirstName(firstName);
          singer.setLastName(lastName);
          singer.setPersonalSite(personalSite);
          singer.setBirthDate(birthDate);
          return singer;
    }

    @Bean
    public ConversionServiceFactoryBean conversionService() {
        ConversionServiceFactoryBean conversionServiceFactoryBean =
      new ConversionServiceFactoryBean();
        Set<Converter> convs = new HashSet<>();
        convs.add(converter());
        conversionServiceFactoryBean.setConverters(convs);
        conversionServiceFactoryBean.afterPropertiesSet();
        return conversionServiceFactoryBean;
    }
```

```
    @Bean
    StringToDateTimeConverter converter(){
        StringToDateTimeConverter conv = new StringToDateTimeConverter();
        conv.setDatePattern(dateFormatPattern);
        conv.init();
        return conv;
    }
}
```

这些值都是从一个属性文件中读取的，内容与上一节中介绍的文件相同，并使用@Value 注解注入到所创建的 bean 中。

通过使用类 ConversionServiceFactoryBean 声明一个 conversionService bean，从而指示 Spring 使用类型转换系统。如果没有定义转换服务 bean，Spring 将使用基于 PropertyEditor 的系统。

默认情况下，类型转换服务支持常用类型之间的转换，包括字符串、数字、枚举、集合、映射等。另外，还支持在基于 PropertyEditor 的系统中将 String 转换为 Java 类型。

对于 conversionService bean，将自定义转换器配置为从 String 转换为 DateTime。测试程序如下所示：

```
package com.apress.prospring5.ch10;

import com.apress.prospring5.ch10.config.AppConfig;
import org.slf4j.Logger;
import org.slf4j.LoggerFactory;
import org.springframework.context.annotation.AnnotationConfigApplicationContext;
import org.springframework.context.support.GenericApplicationContext;

public class ConvServDemo {
    private static Logger logger = LoggerFactory.getLogger(ConvServDemo.class);

    public static void main(String... args) {
        GenericApplicationContext ctx =
            new AnnotationConfigApplicationContext(AppConfig.class);

        Singer john = ctx.getBean("john", Singer.class);
        logger.info("Singer info: " + john);
        ctx.close();
    }
}
```

运行测试程序会产生以下输出：

```
15:41:09.960 main INFO c.a.p.c.ConvServDemo - Singer info: {First name: John,
    Last name: Mayer, Birthday: 1977-10-16, Site: http://johnmayer.com/}
```

如你所见，john bean 的属性转换结果与使用 PropertyEditor 时的结果是相同的。

10.4.3 任意类型之间的转换

类型转换系统的真正优势是能够在任意类型之间进行转换。接下来看一下实际应用，假设有另一个类 AnotherSinger，它与 Singer 类相同。代码如下所示：

```
package com.apress.prospring5.ch10;

import java.net.URL;
import java.text.SimpleDateFormat;

import org.joda.time.DateTime;

public class AnotherSinger {
    private String firstName;
    private String lastName;
    private DateTime birthDate;
    private URL personalSite;
    //seters and getters
    ...

    public String toString() {
        SimpleDateFormat sdf = new SimpleDateFormat("yyyy-MM-dd");
        return String.format("{First name: %s, Last name: %s,
            Birthday: %s, Site: %s}", firstName, lastName,
            sdf.format(birthDate.toDate()), personalSite);
    }
}
```

我们希望能够将 Singer 类的任何实例转换为 AnotherSinger 类。在转换时，Singer 类的 firstName 和 lastName 值将分别成为 AnotherSinger 类的 lastName 和 firstName 值。下面实现另一个自定义转换器来执行该转换。以下代码片段显示了该自定义转换器：

```
package com.apress.prospring5.ch10;

import org.springframework.core.convert.converter.Converter;

public class SingerToAnotherSingerConverter
      implements Converter<Singer, AnotherSinger> {

   @Override
   public AnotherSinger convert(Singer singer) {
      AnotherSinger anotherSinger = new AnotherSinger();
      anotherSinger.setFirstName(singer.getLastName());
      anotherSinger.setLastName(singer.getFirstName());
      anotherSinger.setBirthDate(singer.getBirthDate());
      anotherSinger.setPersonalSite(singer.getPersonalSite());

      return anotherSinger;
   }
}
```

该类非常简单；只需要交换 Singer 和 AnotherSinger 类之间的 firstName 和 lastName 属性值即可。要将该自定义转换器注册到 ApplicationContext 中，请使用下面所示的代码片段替换 AppConfig 类中 conversionService bean 的定义：

```
package com.apress.prospring5.ch10.config;

import com.apress.prospring5.ch10.Singer;
import com.apress.prospring5.ch10.SingerToAnotherSingerConverter;
import com.apress.prospring5.ch10.StringToDateTimeConverter;
import org.springframework.context.annotation.Bean;
import org.springframework.context.annotation.ComponentScan;
import org.springframework.context.annotation.Configuration;
import org.springframework.context.support.ConversionServiceFactoryBean;
import org.springframework.core.convert.converter.Converter;

import java.net.URL;
import java.util.HashSet;
import java.util.Set;

@Configuration
@ComponentScan(basePackages = "com.apress.prospring5.ch10")
public class AppConfig {

   @Bean
   public Singer john() throws Exception {
      Singer singer = new Singer();
      singer.setFirstName("John");
      singer.setLastName("Mayer");
      singer.setPersonalSite(new URL("http://johnmayer.com/"));
      singer.setBirthDate(converter().convert("1977-10-16"));
      return singer;
   }

   @Bean
   public ConversionServiceFactoryBean conversionService() {
      ConversionServiceFactoryBean conversionServiceFactoryBean =
     new ConversionServiceFactoryBean();
      Set<Converter> convs = new HashSet<>();
      convs.add(converter());
      convs.add(singerConverter());
      conversionServiceFactoryBean.setConverters(convs);
      conversionServiceFactoryBean.afterPropertiesSet();
      return conversionServiceFactoryBean;
   }

   @Bean
   StringToDateTimeConverter converter() {
      return new StringToDateTimeConverter();
   }

   @Bean
   SingerToAnotherSingerConverter singerConverter() {
      return new SingerToAnotherSingerConverter();
   }
}
```

转换器属性中 bean 的顺序并不重要。为了测试转换，使用以下测试程序，也就是如下所示的 MultipleConvServDemo 类：

```
package com.apress.prospring5.ch10;

import org.slf4j.Logger;
import org.slf4j.LoggerFactory;
import org.springframework.context.annotation.AnnotationConfigApplicationContext;
import org.springframework.context.support.GenericApplicationContext;
import org.springframework.core.convert.ConversionService;

import com.apress.prospring5.ch10.config.AppConfig;
import java.util.ArrayList;
import java.util.HashSet;
import java.util.List;
import java.util.Set;

public class MultipleConvServDemo {
   private static Logger logger =
      LoggerFactory.getLogger(MultipleConvServDemo.class);

   public static void main(String... args) {
      GenericApplicationContext ctx =
         new AnnotationConfigApplicationContext(AppConfig.class);

      Singer john = ctx.getBean("john", Singer.class);

      logger.info("Singer info: " + john);

      ConversionService conversionService =
         ctx.getBean(ConversionService.class);

      AnotherSinger anotherSinger =
     conversionService.convert(john, AnotherSinger.class);
      logger.info("Another singer info: " + anotherSinger);

      String[] stringArray = conversionService.convert("a,b,c",
     String[].class);
      logger.info("String array: " + stringArray[0]
         + stringArray[1] + stringArray[2]);
      List<String> listString = new ArrayList<>();
      listString.add("a");
      listString.add("b");
      listString.add("c");

      Set<String> setString =
     conversionService.convert(listString, HashSet.class);

      for (String string: setString)
     System.out.println("Set: " + string);
   }
}
```

ConversionService 接口的句柄从 ApplicationContext 获得。因为已经用自定义转换器在 ApplicationContext 中注册了 ConversionService，所以可以使用它转换 Singer 对象，并且还可以在转换服务所支持的其他类型之间进行转换。如上述代码清单所示，出于演示目的，还添加了从 String(由逗号字符分隔)到 Array 的转换示例以及从 List 到 Set 的转换示例。运行该程序会产生以下输出：

```
[main] INFO c.a.p.c.MultipleConvServDemo - Singer info:
   {First name: John, Last name: Mayer, Birthday: 1977-10-16,
   Site: http://johnmayer.com/}
[main] INFO c.a.p.c.MultipleConvServDemo - Another singer info:
   {First name: Mayer, Last name: John, Birthday: 1977-10-16,
   Site: http://johnmayer.com/}
[main] INFO c.a.p.c.MultipleConvServDemo - String array: abc
Set: a
Set: b
Set: c
```

在输出中，可以看到已经正确完成了 Singer 和 AnotherSinger 之间的转换，以及 String 到 Array 和 List 到 Set 的转换。通过使用 Spring 的类型转换服务，可以轻松地创建自定义转换器，并在应用程序的任何层执行转换。一种可能的场景是有两个系统需要更新相同的歌手信息，但是数据库结构不同(例如，系统 A 中的姓氏对应着系统 B 中的名字，等等)。可以在保存之前使用类型转换系统来转换对象。

从 Spring 3.0 开始，Spring MVC 大量使用转换服务(以及下一节讨论的 Formatter SPI)。在 Web 应用程序上下文配置中，如果声明了标记<mvc:annotation-driven />或者在 Java 配置类中使用了 Spring 3.1 引入的@EnableWebMvc，那么将自动注册所有默认转换器(例如，StringToArrayConverter、StringToBooleanConverter 和 StringToLocaleConverter，它们都位于 org.springframework.core.convert.support 包中)和格式化器(例如 CurrencyFormatter、DateFormatter 和 NumberFormatter，它们都位于 org.springframework.format 包的各种子包中)。更多内容请参见第 16 章，第 16 章将讨论如何在 Spring 中开发 Web 应用程序。

10.5 Spring 中的字段格式化

除类型转换系统外，Spring 为开发人员带来的另一个重要功能是 Formatter SPI。正如预料的那样，Formatter SPI 可以帮助配置字段格式化。

在 Formatter SPI 中，实现格式化器的主要接口是 org.springframework.format.Formatter<T>。Spring 提供了一些常用类型的实现，包括 CurrencyFormatter、DateFormatter、NumberFormatter 和 PercentFormatter。

10.5.1 实现自定义格式化器

实现自定义格式化器也很容易。使用相同的 Singer 类并实现一个自定义格式化器，以便将 BirthDate 属性的 DateTime 类型转换为 String。

不过，此时会采取不同的方法：扩展 Spring 的 org.springframework.format.support.FormattingConversionServiceFactoryBean 类并提供自定义格式化器。FormattingConversionServiceFactoryBean 是一个工厂类，可以方便地访问底层的 FormattingConversionService 类(该类支持类型转换系统)，以及根据每个字段类型所定义的格式化规则完成字段格式化。

以下代码片段显示了一个自定义类，它扩展了 FormattingConversionServiceFactoryBean 类，并且定义了一个用于格式化 JodaTime 的 DateTime 类型的自定义格式器。

```
package com.apress.prospring5.ch10;

import org.joda.time.DateTime;
import org.joda.time.format.DateTimeFormat;
import org.joda.time.format.DateTimeFormatter;
import org.slf4j.Logger;
import org.slf4j.LoggerFactory;
import org.springframework.beans.factory.annotation.Autowired;
import org.springframework.format.Formatter;
import org.springframework.format.support.FormattingConversionServiceFactoryBean;
import org.springframework.stereotype.Component;

import javax.annotation.PostConstruct;
import java.text.ParseException;
import java.util.HashSet;
import java.util.Locale;
import java.util.Set;

@Component("conversionService")
public class ApplicationConversionServiceFactoryBean extends
        FormattingConversionServiceFactoryBean {
    private static Logger logger =
      LoggerFactory.getLogger(ApplicationConversionServiceFactoryBean.class);

    private static final String DEFAULT_DATE_PATTERN = "yyyy-MM-dd";
    private DateTimeFormatter dateFormat;
    private String datePattern = DEFAULT_DATE_PATTERN;
    private Set<Formatter<?>> formatters = new HashSet<>();
    public String getDatePattern() {
        return datePattern;
    }

    @Autowired(required = false)
    public void setDatePattern(String datePattern) {
        this.datePattern = datePattern;
    }

    @PostConstruct
```

```
    public void init() {
        dateFormat = DateTimeFormat.forPattern(datePattern);
        formatters.add(getDateTimeFormatter());
        setFormatters(formatters);
    }

    public Formatter<DateTime> getDateTimeFormatter() {
        return new Formatter<DateTime>() {
     @Override
     public DateTime parse(String dateTimeString, Locale locale)
         throws ParseException {
         logger.info("Parsing date string: " + dateTimeString);
         return dateFormat.parseDateTime(dateTimeString);
     }

     @Override
     public String print(DateTime dateTime, Locale locale) {
         logger.info("Formatting datetime: " + dateTime);
         return dateFormat.print(dateTime);
     }
         };
    }
}
```

在上面的代码清单中，自定义格式化器带有下划线。它实现了 Formatter <DateTime>接口以及该接口所定义的两个方法。parse()方法将 String 格式解析为 DateTime 类型(为了提供本地化支持，还传入了语言环境)，而 logger.info()方法将 DateTime 实例格式化为 String 类型。也可以将日期模式注入到 bean 中(或者默认使用 yyyy-MM-dd 格式)。另外，在 init()方法中，通过调用 setFormatters()方法来注册自定义格式器。可以根据需要添加尽可能多的格式化器。

10.5.2 配置 ConversionServiceFactoryBean

声明一个 FormattingConversionServiceFactoryBean 类型的 bean 可以大大减少 AppConfig 配置类的代码。

```
package com.apress.prospring5.ch10.config;

import com.apress.prospring5.ch10.ApplicationConversionServiceFactoryBean;
import com.apress.prospring5.ch10.Singer;
import org.springframework.beans.factory.annotation.Autowired;
import org.springframework.context.annotation.Bean;
import org.springframework.context.annotation.ComponentScan;
import org.springframework.context.annotation.Configuration;

import java.net.URL;
import java.util.Locale;

@Configuration
@ComponentScan(basePackages = "com.apress.prospring5.ch10")
public class AppConfig {

    @Autowired
    ApplicationConversionServiceFactoryBean conversionService;

    @Bean
    public Singer john() throws Exception {
        Singer singer = new Singer();
        singer.setFirstName("John");
        singer.setLastName("Mayer");
        singer.setPersonalSite(new URL("http://johnmayer.com/"));
        singer.setBirthDate(conversionService.
            getDateTimeFormatter().parse("1977-10-16", Locale.ENGLISH));
        return singer;
    }
}
```

测试程序如下所示：

```
package com.apress.prospring5.ch10;

import com.apress.prospring5.ch10.config.AppConfig;
import org.slf4j.Logger;
import org.slf4j.LoggerFactory;
import org.springframework.context.annotation.AnnotationConfigApplicationContext;
import org.springframework.context.support.GenericApplicationContext;
import org.springframework.core.convert.ConversionService;
```

```
public class ConvFormatServDemo {

   private static Logger logger =
      LoggerFactory.getLogger(ConvFormatServDemo.class);

   public static void main(String... args) {
      GenericApplicationContext ctx =
         new AnnotationConfigApplicationContext(AppConfig.class);

      Singer john = ctx.getBean("john", Singer.class);
      logger.info("Singer info: " + john);
      ConversionService conversionService =
         ctx.getBean("conversionService", ConversionService.class);
      logger.info("Birthdate of singer is : " +
         conversionService.convert(john.getBirthDate(), String.class));
      ctx.close();
   }
}
```

运行程序将产生如下输出：

```
Parsing date string: 1977-10-16
[main] INFO c.a.p.c.ConvFormatServDemo - Singer info: {
   First name: John, Last name: Mayer, Birthday: 1977-10-16,
   Site: http://johnmayer.com/}
Formatting datetime: 1977-10-16T00:00:00.000+03:00
[main] INFO c.a.p.c.ConvFormatServDemo -
   Birthdate of singer is : 1977-10-16
```

在输出中，可以看到 Spring 使用自定义格式化器的 parse()方法将属性从 String 转换为 birthDate 属性的 DateTime 类型。当调用 ConversionService.convert()方法并传入 birthDate 属性时，Spring 将调用 logger 的 info()方法来格式化输出。

10.6 Spring 中的验证

验证在任何应用程序中都十分关键。应用于域对象的验证规则可确保所有业务数据具有良好的结构并满足所有业务定义。在理想的情况下，所有验证规则都保存在某个集中的位置，同一组规则应用于同一类型的数据，而不管数据来自哪个源(例如，来自 Web 应用程序的用户输入，来自通过 Web 服务的远程应用程序，来自 JMS 消息或文件)。

在谈论验证时，转换和格式化也很重要，因为在验证一段数据之前，应根据每种类型定义的格式化规则将数据转换为所需的 POJO。例如，用户通过浏览器中的 Web 应用程序输入一些歌手信息，然后将这些数据提交给服务器。在服务器端，如果 Web 应用程序是在 Spring MVC 中开发的，那么 Spring 将从 HTTP 请求中提取数据，并根据格式规则将 String 转换为所需的类型(例如，将表示日期的 String 转换为 Date 字段，格式规则为 yyyy-MM-dd)。该过程被称为*数据绑定(data binding)*。当完成数据绑定并构建完域对象时，才会对该对象进行验证，同时返回任何错误并显示给用户。如果验证成功，该对象将被被持久保存到数据库。

Spring 支持两种主要类型的验证。第一种验证类型是由 Spring 提供的，可以通过实现 org.springframework.validation.Validator 接口来创建自定义验证器。另一种类型是通过 Spring 对 JSR-349(Bean Validation)的支持实现的。接下来将详细讨论这两种验证类型。

10.6.1 使用 Spring Validator 接口

可以通过创建一个实现了 Spring Validator 接口的类来开发一些验证逻辑。接下来看一下如何使用该接口。对于前面使用的 Singer 类，假设名字不能为空。为了根据此规则验证 Singer 对象，可以创建一个自定义验证器。以下代码片段显示了该自定义验证器类：

```
package com.apress.prospring5.ch10;

import org.springframework.stereotype.Component;
import org.springframework.validation.Errors;
import org.springframework.validation.ValidationUtils;
import org.springframework.validation.Validator;

@Component("singerValidator")
public class SingerValidator implements Validator {
   @Override
```

```
    public boolean supports(Class<?> clazz) {
        return Singer.class.equals(clazz);
    }

    @Override
    public void validate(Object obj, Errors e) {
        ValidationUtils.rejectIfEmpty(e, "firstName",
            "firstName.empty");
    }
}
```

该自定义验证器类实现了 Validator 接口及其两个方法。supports()方法指示验证器是否支持验证传入的类类型。validate()方法对传入的对象进行验证。验证结果将存储在 org.springframework.validation.Errors 接口的实例中。在 validate()方法中，仅对 firstName 属性进行检查，并使用方便的 ValidationUtils.rejectIfEmpty()方法确保歌手的名字不为空。最后一个参数是错误代码，可用于查找资源包中的验证消息以显示本地化的错误消息。

以下代码片段描述了配置类：

```
package com.apress.prospring5.ch10.config;

import org.springframework.context.annotation.ComponentScan;
import org.springframework.context.annotation.Configuration;
@Configuration
@ComponentScan(basePackages = "com.apress.prospring5.ch10")
public class AppConfig {
}
```

以下代码片段描述了测试程序：

```
package com.apress.prospring5.ch10;

import com.apress.prospring5.ch10.config.AppConfig;
import org.slf4j.Logger;
import org.slf4j.LoggerFactory;
import org.springframework.context.annotation.AnnotationConfigApplicationContext;
import org.springframework.context.support.GenericApplicationContext;
import org.springframework.validation.BeanPropertyBindingResult;
import org.springframework.validation.ObjectError;
import org.springframework.validation.ValidationUtils;
import org.springframework.validation.Validator;

import java.util.List;

public class SpringValidatorDemo {

    private static Logger logger =
        LoggerFactory.getLogger(SpringValidatorDemo.class);

    public static void main(String... args) {
        GenericApplicationContext ctx =
      new AnnotationConfigApplicationContext(AppConfig.class);

        Singer singer = new Singer();
        singer.setFirstName(null);
        singer.setLastName("Mayer");

        Validator singerValidator = ctx.getBean("singerValidator",
            Validator.class);
        BeanPropertyBindingResult result =
            new BeanPropertyBindingResult(singer, "John");

        ValidationUtils.invokeValidator(singerValidator, singer, result);

        List<ObjectError> errors = result.getAllErrors();
        logger.info("No of validation errors: " + errors.size());
        errors.forEach(e -> logger.info(e.getCode()));

        ctx.close();
    }
}
```

上述代码首先创建一个名字为 null 的 Singer 对象。然后，从 ApplicationContext 中检索验证器。为了保存验证结果，构造 BeanPropertyBindingResult 类的一个实例。为了进行验证，需要调用 ValidationUtils.invokeValidator()方法。最后检查验证错误。运行该程序会产生以下输出：

```
[main] INFO   c.a.p.c.SpringValidatorDemo - No of validation errors: 1
[main] INFO   c.a.p.c.SpringValidatorDemo - firstName.empty
```

此时，验证会产生错误，并正确显示错误代码。

10.6.2 使用 JSR-349 Bean Validation

从 Spring 4 开始，已经实现对 JSR-349(Bean Validation)的全面支持。Bean Validation API 在包 javax.validation.constraints 中以 Java 注解(例如@NotNull)的形式定义了一组可应用于域对象的约束。另外，可以使用注解开发和应用自定义验证器(例如，类级验证器)。

通过使用 Bean Validation API，可以避免耦合到特定的验证服务提供程序。可以使用标准注解和 API 来实现域对象的验证逻辑，而不必知道底层的验证服务提供程序。例如，Hibernate Validator V5(http://hibernate.org/subprojects/validator)是 JSR-349 参考实现。

Spring 对 Bean Validate API 提供了无缝支持，主要功能包括支持用于定义验证约束的 JSR-349 标准注解、自定义验证器以及在 Spring 的 ApplicationContext 中配置 JSR-349 验证。在下面的章节中将一一介绍这些功能。当类路径上使用 Hibernate Validator 版本 4 和 Validation API 的 1.0 版本时，Spring 仍然无缝地对 JSR-303 提供支持。

定义对象属性上的验证约束

首先，将验证约束应用于域对象属性。以下代码片段显示了更高级的 Singer 类，其中验证约束被应用于 firstName 和 genre 属性：

```
package com.apress.prospring5.ch10.obj;

import javax.validation.constraints.NotNull;
import javax.validation.constraints.Size;

public class Singer {

   @NotNull
   @Size(min=2, max=60)
   private String firstName;

   private String lastName;

   @NotNull
   private Genre genre;

   private Gender gender;

   //setters and getters
   ...
}
```

此处在所应用的验证约束条件下显示了下划线。对于 firstName 属性，应用了两个约束。第一个约束由@NotNull 注解控制，它表示该值不应该为空。此外，@Size 注解控制 firstName 属性的长度。@NotNull 注解还被应用于 genre 属性。

以下代码示例分别显示了 Genre 和 Gender 枚举类：

```
//Genre.java
package com.apress.prospring5.ch10.obj;
public enum Genre {
   POP("P"),
    JAZZ("J"),
   BLUES("B"),
   COUNTRY("C");
   private String code;

   private Genre(String code) {
      this.code = code;
   }

   public String toString() {
      return this.code;
   }
}

//Genfer.java
package com.apress.prospring5.ch10.obj;
```

```
public enum Gender {
   MALE("M"), FEMALE("F");
   private String code;

   Gender(String code) {
      this.code = code;
   }

   @Override
   public String toString() {
      return this.code;
   }
}
```

Genre 表示歌手所属的音乐类型，而 Gender 与音乐事业并不相关，因此可能为空。

10.6.3 在 Spring 中配置 Bean Validation 支持

为了在 Spring 的 ApplicationContext 中配置对 Bean Validation API 的支持，可以在 Spring 的配置中定义一个类型为 org.springframework.validation.beanvalidation.LocalValidatorFactoryBean 的 bean。以下代码片段描述了配置类：

```
package com.apress.prospring5.ch10.config;

import org.springframework.context.annotation.Bean;
import org.springframework.context.annotation.ComponentScan;
import org.springframework.context.annotation.Configuration;
import org.springframework.validation.beanvalidation.LocalValidatorFactoryBean;

@Configuration
@ComponentScan(basePackages = "com.apress.prospring5.ch10")
public class AppConfig {

   @Bean LocalValidatorFactoryBean validator() {
      return new LocalValidatorFactoryBean();
      }
}
```

声明一个类型为 LocalValidatorFactoryBean 的 bean 是必需的。默认情况下，Spring 会在类路径中搜索 Hibernate Validator 库，验证它是否存在。现在，我们创建一个为 Singer 类提供验证服务的服务类。验证器类如下所示：

```
package com.apress.prospring5.ch10;

import com.apress.prospring5.ch10.obj.Singer;
import org.springframework.beans.factory.annotation.Autowired;
import org.springframework.stereotype.Service;

import javax.validation.ConstraintViolation;
import javax.validation.Validator;
import java.util.Set;

@Service("singerValidationService")
public class SingerValidationService {

   @Autowired
   private Validator validator;

   public Set<ConstraintViolation<Singer>>
      validateSinger(Singer singer) {
      return validator.validate(singer);
   }
}
```

注入一个 javax.validation.Validator 实例(请注意与 Spring 提供的 Validator 接口不同，此时是 org.springframework.validation.Validator)。一旦定义了 LocalValidatorFactoryBean，就可以在应用程序中的任何位置创建 Validator 接口的句柄。要在 POJO 上执行验证，需要调用 Validator.validate()方法。验证结果将以 ConstraintViolation <T>接口的 List 形式返回。

测试程序如下所示：

```
package com.apress.prospring5.ch10;

import com.apress.prospring5.ch10.config.AppConfig;
import com.apress.prospring5.ch10.obj.Singer;
import org.slf4j.Logger;
```

```
import org.slf4j.LoggerFactory;
import org.springframework.context.annotation.AnnotationConfigApplicationContext;
import org.springframework.context.support.GenericApplicationContext;

import javax.validation.ConstraintViolation;
import java.util.Set;

public class Jsr349Demo {
   private static Logger logger =
      LoggerFactory.getLogger(Jsr349Demo.class);

   public static void main(String... args) {

      GenericApplicationContext ctx =
         new AnnotationConfigApplicationContext(AppConfig.class);

      SingerValidationService singerBeanValidationService =
         ctx.getBean( SingerValidationService.class);

      Singer singer = new Singer();
      singer.setFirstName("J");
      singer.setLastName("Mayer");
      singer.setGenre(null);
      singer.setGender(null);

      validateSinger(singer, singerBeanValidationService);

      ctx.close();
   }

   private static void validateSinger(Singer singer,
         SingerValidationService singerValidationService) {
      Set<ConstraintViolation<Singer>> violations =
         singerValidationService.validateSinger(singer);
      listViolations(violations);
   }

   private static void listViolations(
       Set<ConstraintViolation<Singer>> violations) {
      logger.info("No. of violations: " + violations.size());
      for (ConstraintViolation<Singer> violation : violations) {
         logger.info("Validation error for property: " +
         violation.getPropertyPath()
         + " with value: " + violation.getInvalidValue()
         + " with error message: " + violation.getMessage());
      }
   }
}
```

如上述代码清单所示，Singr 对象由违反了约束条件的 firstName 和 genra 构建。在 validateSinger()方法中，调用 SingerValidationService.validateSinger()方法，该方法又调用 JSR-349(Bean Validation)。运行该程序会产生以下输出：

```
[main] INFO o.h.v.i.u.Version - HV000001:
   Hibernate Validator 6.0.0.Beta2
[main] INFO c.a.p.c.Jsr349Demo - No. of violations: 2
[main] INFO c.a.p.c.Jsr349Demo - Validation error for property:
   firstName with value: J with error message: size must be between 2 and 60
[main] INFO c.a.p.c.Jsr349Demo - Validation error for property:
   genre with value: null with error message: may not be null
```

正如你所看到的，这里违反了两个约束条件，并显示了相关消息。在输出中，还可以看到 Hibernate Validator 已经构建了基于注解的默认验证错误消息。当然，也可以提供自己的验证错误消息，相关内容将在下一节中演示。

10.6.4 创建自定义验证器

除了进行属性级验证之外，还可以应用类级验证。例如，对于 Singer 类来说，如果是乡村歌手，那么可能需要确保 lastName 和 gender 属性不是 null(该需求并不是必需的，只是出于说明目的)。在这种情况下，可以开发一个自定义验证器来执行检查。在 Bean Validation API 中，开发一个自定义验证器分为两步。首先为验证器创建一个 Annotation 类型，如下面的代码片段所示。第二步是开发实现验证逻辑的类。

```
package com.apress.prospring5.ch10;

import java.lang.annotation.Documented;
import java.lang.annotation.ElementType;
```

```
import java.lang.annotation.Retention;
import java.lang.annotation.RetentionPolicy;
import java.lang.annotation.Target;

import javax.validation.Constraint;
import javax.validation.Payload;

@Retention(RetentionPolicy.RUNTIME)
@Target(ElementType.TYPE)
@Constraint(validatedBy=CountrySingerValidator.class)
@Documented
public @interface CheckCountrySinger {
    String message() default "Country Singer should
        have gender and last name defined";
    Class<?> groups() default {};
    Class<? extends Payload> payload() default {};
}
```

@Target(ElementType.TYPE)注解表明注解只应用于类级别。@Constraint 注解表明它是一个验证器，validatedBy 属性指定提供验证逻辑的类。在代码中，以方法的形式定义三个属性，如下所示：

- message 属性定义违反约束条件时返回的消息(或错误代码)。也可以在注解中提供默认消息。
- groups 属性指定适用的验证组。可以将验证器分配给不同的组，并对特定组执行验证。
- payload 属性指定其他有效载荷对象(即实现了 javax.validation.Payload 接口的类)。它允许将附加信息附加到约束上(例如，有效载荷对象可以指明违反约束的严重性)。

以下代码片段显示了提供验证逻辑的 CountrySingerValidator 类：

```
package com.apress.prospring5.ch10;

import com.apress.prospring5.ch10.obj.Singer;

import javax.validation.ConstraintValidator;
import javax.validation.ConstraintValidatorContext;

public class CountrySingerValidator implements
      ConstraintValidator<CheckCountrySinger, Singer> {

    @Override
    public void initialize(CheckCountrySinger constraintAnnotation) {
    }

    @Override
    public boolean isValid(Singer singer,
      ConstraintValidatorContext context) {
        boolean result = true;
        if (singer.getGenre() != null && (singer.isCountrySinger() &&
          (singer.getLastName() == null || singer.getGender() == null))) {
      result = false;
        }
        return result;
    }
}
```

CountrySingerValidator 类实现了 ConstraintValidator<CheckCountrySinger，Singer>接口，这意味着验证器检查 Singer 类的 CheckCountrySinger 注解。isValid()方法被实现，底层的验证服务提供程序(例如 Hibernate Validator)将验证中的实例传递给该方法。而在该方法中，验证如果歌手是乡村音乐歌手，那么 lastName 和 gender 属性不应为 null。结果是一个表示验证结果的布尔值。

要启用验证，请将@CheckCountrySinger 注解应用于 Singer 类，如下所示：

```
package com.apress.prospring5.ch10.obj;

import com.apress.prospring5.ch10.CheckCountrySinger;

import javax.validation.constraints.NotNull;
import javax.validation.constraints.Size;

@CheckCountrySinger
public class Singer {

    @NotNull
    @Size(min = 2, max = 60)
    private String firstName;
```

```
    private String lastName;

    @NotNull
    private Genre genre;

    private Gender gender;

    //getters and setter
    ...

    public boolean isCountrySinger() {
    return genre == Genre.COUNTRY;
    }
}
```

要测试自定义验证，请在测试类 Jsr349CustomDemo 中创建以下 Singer 实例：

```
public class Jsr349CustomDemo {
    ...

    Singer singer = new Singer();
    singer.setFirstName("John");
    singer.setLastName("Mayer");
    singer.setGenre(Genre.COUNTRY);
    singer.setGender(null);

    validateSinger(singer, singerValidationService);

    ...
}
```

运行该程序会产生以下输出(其他输出被省略)：

```
[main] INFO o.h.v.i.u.Version - HV000001: Hibernate Validator 6.0.0.Beta2
[main] INFO c.a.p.c.Jsr349CustomDemo - No. of violations: 1
[main] INFO c.a.p.c.Jsr349CustomDemo - Validation error for property:
    with value: com.apress.prospring5.ch10.obj.Singer@3116c353
    with error message: Country Singer should have gender and last name defined
```

在输出中，可以看到被检查的值(即 Singer 实例)违反了乡村歌手的验证规则，因为 gender 属性为 null。还要注意，在输出中，属性路径是空的，因为它是一个类级验证错误。

10.7 使用 AssertTrue 进行自定义验证

除实现自定义验证器外，在 Bean Validation API 中应用自定义验证的另一种方法是使用@AssertTrue 注解。接下来看一看它是如何工作的。此时对于 Singer 类，应该删除@CheckCountrySinger 注解，并修改 isCountrySinger()方法，如下所示：

```
public class Singer {

    @NotNull
    @Size(min = 2, max = 60)
    private String firstName;

    private String lastName;

    @NotNull
    private Genre genre;

    private Gender gender;
    ...

    @AssertTrue(message="ERROR! Individual customer should have
        gender and last name defined")
    public boolean isCountrySinger() {
        boolean result = true;

    if (genre!= null &&
        (genre.equals(Genre.COUNTRY) &&
            (gender == null || lastName == null))) {
        result = false;
    }

        return result;
    }
}
```

正如你所推断的那样，@CheckCountrySinger 注解和 CountrySingerValidator 类不再是必需的。

*is*CountrySinger()方法被添加到 Singer 类中，并使用@AssertTrue(位于包 javax.validation.constraints 中)进行了注解。当调用验证时，提供程序将调用检查并确保结果为 true。此外，JSR-349 还提供了@AssertFalse 注解来检查某些应该是 false 的条件。现在运行测试程序 Jsr349AssertTrueDemo，将获得与自定义验证器相同的输出。

10.8 自定义验证的注意事项

对于 JSR-349 中的自定义验证，应该使用哪种方法：自定义验证器还是@AssertTrue 注释？通常，@AssertTrue 方法实现起来更简单，可以在域对象的代码中看到验证规则。但是，对于具有更复杂逻辑的验证器(例如，假设需要注入一个服务类，访问数据库并检查有效值)，实现自定义验证器是不错的方法，因为你可能并不想将服务层对象添加到域对象中。而且，自定义验证器可以在相似的域对象中重用。

10.9 决定使用哪种验证 API

在讨论 Spring 的 Validator 接口以及 Bean Validation API 之后，应该在应用程序中使用哪一个？JSR-349 绝对是最好的选择。主要原因如下：

- JSR-349 是 JEE 标准，并得到许多前端/后端框架(例如，Spring、JPA 2、Spring MVC 和 GWT)的广泛支持。
- JSR-349 提供的标准验证 API 隐藏了底层提供程序，因此不受限于特定的提供程序。
- Spring 从版本 4 开始就与 JSR-349 紧密集成。例如，在 Spring MVC Web 控制器中，可以使用@Valid 注解(位于 javax.validation 包中)在方法中注解参数，Spring 将在数据绑定过程中自动调用 JSR-349 验证。此外，在 Spring MVC Web 应用程序上下文配置中，可以使用一个名为<mvc:annotation-driven />的简单标记将 Spring 配置为自动启用 Spring 类型转换系统和字段格式化，以及支持 JSR-349(Bean Validation)。
- 如果使用的是 JPA 2，那么提供程序会在持久化之前自动对实体执行 JSR-349 验证，从而提供了另一层保护。

有关将 Hibernate Validator 作为实现提供者使用 JSR-349(Bean Validation)的详细信息，请参阅 Hibernate Validator 的文档页面：http://docs.jboss.org/hibernate/validator/5.1/reference/en-US/HTML。

10.10 小结

在本章中，介绍了 Spring 类型转换系统以及字段 Formatter SPI。除了 PropertyEditors 支持，我们还学习了如何使用新的转换系统实现任意类型的转换。

本章还讨论了 Spring 中的验证支持、Spring 的 Validator 接口以及 Spring 中推荐的 JSR-349(Bean Validation)支持。

第 11 章

任务调度

任务调度是企业级应用程序中一项常见的功能。任务调度主要由三部分组成：任务(即需要在特定时间运行或定期运行的业务逻辑块)、触发器(指定任务应该执行的条件)以及调度程序(根据来自触发器的信息执行任务)。具体而言，本章涵盖以下主题。

- **Spring 中的任务调度**：将讨论 Spring 如何支持任务调度，重点介绍 Spring 3 中引入的 TaskScheduler 抽象。此外，还会介绍调度场景，如固定间隔调度和 cron 表达式。
- **异步任务执行**：演示如何在 Spring 中使用@Async 注解来异步执行任务。
- **Spring 中的任务执行**：简要讨论 Spring 的 TaskExecutor 接口以及如何执行任务。

11.1 任务调度示例的依赖项

可以在下面的 Gradle 配置代码片段中看到本章所需的依赖项：

```
//pro-spring-15/build.gradle
ext {
   springDataVersion = '2.0.0.M3'

   //logging libs
   slf4jVersion = '1.7.25'
   logbackVersion = '1.2.3'

   guavaVersion = '21.0'
   jodaVersion = '2.9.9'
   utVersion = '6.0.1.GA'

   junitVersion = '4.12'

   spring = [
      data :
         "org.springframework.data:spring-data-jpa:$springDataVersion",
      ...
   ]
   testing = [
      junit: "junit:junit:$junitVersion"
   ]

   misc = [
      slf4jJcl     : "org.slf4j:jcl-over-slf4j:$slf4jVersion",
      logback      : "ch.qos.logback:logback-classic:$logbackVersion",
      guava        : "com.google.guava:guava:$guavaVersion",
      joda         : "joda-time:joda-time:$jodaVersion",
      usertype     : "org.jadira.usertype:usertype.core:$utVersion",
      ...
   ]
   ...
}
...
//chapter11/build.gradle
dependencies {

   compile (spring.contextSupport) {
      exclude module: 'spring-context'
      exclude module: 'spring-beans'
      exclude module: 'spring-core'
   }
   compile misc.slf4jJcl, misc.logback, misc.lang3, spring.data,
```

```
        misc.guava, misc.joda, misc.usertype, db.h2

    testCompile testing.junit
}
```

11.2 Spring 中的任务调度

企业级应用程序通常需要调度任务。在许多应用程序中，需要定期运行各种任务(例如向客户发送电子邮件通知、运行日常作业、完成数据整理以及批量更新数据)，要么以固定的时间间隔执行(例如，每小时)、要么按照特定的时间表执行(例如，从周一至周五的晚上 8 点)。如前所述，任务调度由三部分组成：调度定义(触发器)、任务执行(调度程序)和任务本身。

在 Spring 应用程序中可以使用多种方法触发任务的执行。一种方法是通过已存在于应用程序部署环境中的调度系统从外部触发作业。例如，许多企业都使用一些商业系统(例如 Control-M 或 CA AutoSys)来调度任务。如果应用程序在 Linux/UNIX 平台上运行，则可以使用 crontab 调度程序。作业触发可以通过向 Spring 应用程序发送 RESTful-WS 请求并让 Spring 的 MVC 控制器触发任务来完成。

另一种方法是在 Spring 中使用任务调度支持。Spring 在任务调度方面提供了三个选项。

- **支持 JDK 定时器**：Spring 支持用于任务调度的 JDK 的 Timer 对象。
- **与 Quartz 集成**：Quartz Scheduler[1]是一个流行的开源调度库。
- **Spring 自己的 Spring TaskScheduler 抽象**：Spring 3 引入了 TaskScheduler 抽象，它提供了一种简单的方法来调度任务并支持大多数典型的需求。

本节将重点介绍如何使用 Spring 的 TaskScheduler 抽象来进行任务调度。

11.2.1 Spring TaskScheduler 抽象介绍

Spring 的 TaskScheduler 抽象主要有三个参与者。

- **Trigger 接口**：org.springframework.scheduling.Trigger 接口为定义触发机制提供了支持。Spring 提供了两个 Trigger 实现。CronTrigger 类支持基于 cron 表达式的触发，而 PeriodicTrigger 类支持基于初始延迟和固定间隔的触发。
- **任务**：任务是需要调度的业务逻辑。在 Spring 中，可以将任务指定为任何 Spring bean 中的方法。
- **TaskScheduler 接口**：org.springframework.scheduling.TaskScheduler 接口为任务调度提供支持。Spring 提供了 TaskScheduler 接口的三个实现类。TimerManagerTaskScheduler 类(在 org.springframework.scheduling.commonj 包中)封装了 CommonJ 的 commonj.timers.TimerManager 接口，该接口常用于商业 JEE 应用服务器(例如 WebSphere 和 WebLogic)。ConcurrentTaskScheduler 和 ThreadPoolTaskScheduler 类(都在 org.springframework.scheduling.concurrent 包中)封装了 java.util.concurrent.ScheduledThreadPoolExecutor 类，这两个类都支持从共享线程池执行任务。

图 11-1 显示了 Trigger 接口、TaskScheduler 接口和实现了 java.lang.Runnable 接口的任务实现之间的关系。如果想要通过使用 Spring 的 TaskScheduler 抽象来调度任务，有两种选择。一种是在 Spring 的 XML 配置中使用 task-namespace，另一种是使用注解。接下来分别介绍。

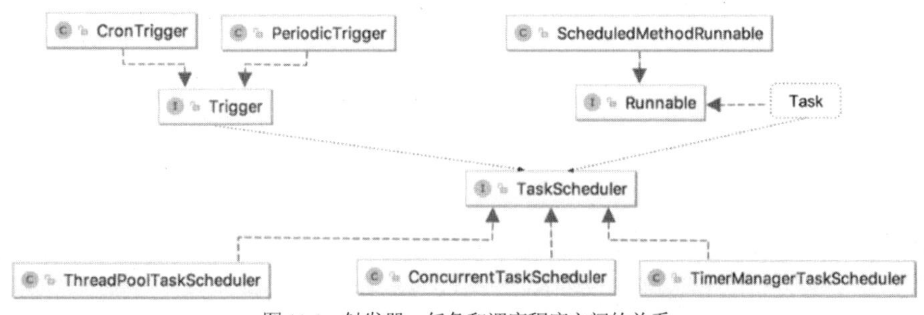

图 11-1　触发器、任务和调度程序之间的关系

1　官方网站为 www.quartz-scheduler.org。

11.2.2 研究示例任务

为了演示 Spring 中的任务调度，首先实现一个简单的任务，即维护汽车信息数据库的应用程序。以下代码片段显示了 Car 类，该类被实现为 JPA 实体类：

```
package com.apress.prospring5.ch11;

import static javax.persistence.GenerationType.IDENTITY;

import javax.persistence.Column;

import javax.persistence.Entity;
import javax.persistence.GeneratedValue;
import javax.persistence.Id;
import javax.persistence.Table;
import javax.persistence.Version;

import org.hibernate.annotations.Type;
import org.joda.time.DateTime;

@Entity
@Table(name="car")
public class Car {
    @Id
    @GeneratedValue(strategy = IDENTITY)
    @Column(name = "ID")
    private Long id;

    @Column(name="LICENSE_PLATE")
    private String licensePlate;

    @Column(name="MANUFACTURER")
    private String manufacturer;

    @Column(name="MANUFACTURE_DATE")
    @Type(type="org.jadira.usertype.dateandtime.joda.PersistentDateTime")
    private DateTime manufactureDate;
    @Column(name="AGE")
    private int age;

    @Version
    private int version;

    //getters and setters
    ...

     @Override
    public String toString() {
        SimpleDateFormat sdf = new SimpleDateFormat("yyyy-MM-dd");
        return String.format("{License: %s, Manufacturer: %s,
            Manufacture Date: %s, Age: %d}",
            licensePlate, manufacturer, sdf.format(manufactureDate.toDate()), age);
    }
}
```

把该实体类用作 Hibernate 生成的 CAR 表的模型。在本章中，数据访问和服务层的配置全部合二为一，由以下代码片段描述的 DataServiceConfig 类提供：

```
package com.apress.prospring5.ch11.config;

import org.slf4j.Logger;
import org.slf4j.LoggerFactory;
import org.springframework.context.annotation.Bean;

import org.springframework.context.annotation.ComponentScan;
import org.springframework.context.annotation.Configuration;
import org.springframework.data.jpa.repository.config.EnableJpaRepositories;
import org.springframework.jdbc.datasource.embedded.EmbeddedDatabaseBuilder;
import org.springframework.jdbc.datasource.embedded.EmbeddedDatabaseType;
import org.springframework.orm.jpa.JpaTransactionManager;
import org.springframework.orm.jpa.JpaVendorAdapter;
import org.springframework.orm.jpa.LocalContainerEntityManagerFactoryBean;
import org.springframework.orm.jpa.vendor.HibernateJpaVendorAdapter;
import org.springframework.transaction.PlatformTransactionManager;
```

```
import javax.persistence.EntityManagerFactory;
import javax.sql.DataSource;
import java.util.Properties;

@Configuration
@EnableJpaRepositories(basePackages = {"com.apress.prospring5.ch11.repos"})
@ComponentScan(basePackages = {"com.apress.prospring5.ch11"} )
public class DataServiceConfig {

    private static Logger logger =
        LoggerFactory.getLogger(DataServiceConfig.class);

    @Bean
    public DataSource dataSource() {
        try {
            EmbeddedDatabaseBuilder dbBuilder = new EmbeddedDatabaseBuilder();
            return dbBuilder.setType(EmbeddedDatabaseType.H2).build();
        } catch (Exception e) {
            logger.error("Embedded DataSource bean cannot be created!", e);
            return null;
        }
    }

    @Bean
    public Properties hibernateProperties() {
        Properties hibernateProp = new Properties();
        hibernateProp.put("hibernate.dialect", "org.hibernate.dialect.H2Dialect");
        hibernateProp.put("hibernate.hbm2ddl.auto", "create-drop");
        //hibernateProp.put("hibernate.format_sql", true);
        hibernateProp.put("hibernate.show_sql", true);
        hibernateProp.put("hibernate.max_fetch_depth", 3);
        hibernateProp.put("hibernate.jdbc.batch_size", 10);
        hibernateProp.put("hibernate.jdbc.fetch_size", 50);
        return hibernateProp;
    }

    @Bean
    public PlatformTransactionManager transactionManager() {
        return new JpaTransactionManager(entityManagerFactory());
    }

    @Bean
    public JpaVendorAdapter jpaVendorAdapter() {
        return new HibernateJpaVendorAdapter();
    }

    @Bean
    public EntityManagerFactory entityManagerFactory() {
        LocalContainerEntityManagerFactoryBean factoryBean =
            new LocalContainerEntityManagerFactoryBean();
        factoryBean.setPackagesToScan("com.apress.prospring5.ch11.entities");
        factoryBean.setDataSource(dataSource());
        factoryBean.setJpaVendorAdapter(new HibernateJpaVendorAdapter());
        factoryBean.setJpaProperties(hibernateProperties());
        factoryBean.setJpaVendorAdapter(jpaVendorAdapter());
        factoryBean.afterPropertiesSet();
        return factoryBean.getNativeEntityManagerFactory();
    }
}
```

DBInitializer 类负责填充 CAR 表。

```
package com.apress.prospring5.ch11.config;

import com.apress.prospring5.ch11.entities.Car;
import com.apress.prospring5.ch11.repos.CarRepository;
import org.joda.time.DateTime;
import org.joda.time.format.DateTimeFormat;
import org.joda.time.format.DateTimeFormatter;
import org.slf4j.Logger;
import org.slf4j.LoggerFactory;
import org.springframework.beans.factory.annotation.Autowired;
import org.springframework.stereotype.Service;
import javax.annotation.PostConstruct;

@Service
public class DBInitializer {
```

```
    private Logger logger = LoggerFactory.getLogger(DBInitializer.class);
    @Autowired CarRepository carRepository;

    @PostConstruct
    public void initDB() {
        logger.info("Starting database initialization...");
        DateTimeFormatter formatter = DateTimeFormat.forPattern("yyyy-MM-dd");

        Car car = new Car();
        car.setLicensePlate("GRAVITY-0405");
        car.setManufacturer("Ford");
        car.setManufactureDate(DateTime.parse("2006-09-12", formatter));
        carRepository.save(car);

        car = new Car();
        car.setLicensePlate("CLARITY-0432");
        car.setManufacturer("Toyota");
        car.setManufactureDate(DateTime.parse("2003-09-09", formatter));
        carRepository.save(car);

        car = new Car();
        car.setLicensePlate("ROSIE-0402");
        car.setManufacturer("Toyota");
        car.setManufactureDate(DateTime.parse("2017-04-16", formatter));
        carRepository.save(car);

        logger.info("Database initialization finished.");
    }
}
```

接下来为 Car 实体定义一个 DAO 层。我们将使用 Spring Data 的 JPA 及其存储库抽象支持。此时可以看到 CarRepository 接口，它只是 CrudRepository 的简单扩展，因为我们对任何特殊的 DAO 操作都不感兴趣。

```
package com.apress.prospring5.ch11.repos;

import com.apress.prospring5.ch11.entities.Car;
import org.springframework.data.repository.CrudRepository;

public interface CarRepository extends CrudRepository<Car, Long> {
}
```

服务层由 CarService 接口及其实现 CarServiceImpl 表示。

```
package com.apress.prospring5.ch11.services;
//CarService.jar
import com.apress.prospring5.ch11.entities.Car;

import java.util.List;

public interface CarService {
    List<Car> findAll();
    Car save(Car car);
    void updateCarAgeJob();
    boolean isDone();
}

//CarServiceImpl.jar
...
@Service("carService")
@Repository
@Transactional
public class CarServiceImpl implements CarService {
    public boolean done;

    final Logger logger = LoggerFactory.getLogger(CarServiceImpl.class);

    @Autowired
    CarRepository carRepository;

    @Override
    @Transactional(readOnly=true)
    public List<Car> findAll() {
        return Lists.newArrayList(carRepository.findAll());
    }

    @Override
    public Car save(Car car) {
        return carRepository.save(car);
```

```
    }

    @Override
    public void updateCarAgeJob() {
       List<Car> cars = findAll();

       DateTime currentDate = DateTime.now();
       logger.info("Car age update job started");

      cars.forEach(car -> {
           int age = Years.yearsBetween(car.getManufactureDate(),
              currentDate).getYears();

           car.setAge(age);
           save(car);
           logger.info("Car age update --> " + car);
       });
       logger.info("Car age update job completed successfully");
       done = true;
    }

    @Override
    public boolean isDone() {
       return done;
    }
}
```

这里总共提供了四个方法，如下所示：

- 第一个方法检索所有汽车的信息：List <Car> findAll()。
- 第二个方法保存更新的汽车对象：Car save(Car car)。
- 第三个方法是需要定期运行，以根据汽车制造日期和当前日期更新汽车使用年限的作业：void updateCarAgeJob()。
- 第四个方法，用于了解作业何时结束，以便应用程序可以正常关闭：boolean isDone()。

就像 Spring 中对其他名称空间的支持一样，task-namespace 通过使用 Spring 的 TaskScheduler 抽象为调度任务提供简化的配置。以下 XML 配置代码片段显示了 task-namespace-app-context.xml 文件的内容，其中包含调度任务的 Spring 应用程序的配置。使用 task-namespace 进行任务调度是非常简单的。

```
<?xml version="1.0" encoding="UTF-8"?>
<beans xmlns="http://www.springframework.org/schema/beans"
   xmlns:task="http://www.springframework.org/schema/task"
   xmlns:xsi="http://www.w3.org/2001/XMLSchema-instance"
   xsi:schemaLocation="http://www.springframework.org/schema/beans
      http://www.springframework.org/schema/beans/spring-beans.xsd
      http://www.springframework.org/schema/task
      http://www.springframework.org/schema/task/spring-task.xsd">

   <task:scheduler id="carScheduler" pool-size="10"/>

   <task:scheduled-tasks scheduler="carScheduler">
      <task:scheduled ref="carService"
         method="updateCarAgeJob" fixed-delay="10000"/>
   </task:scheduled-tasks>
</beans>
```

当遇到<task:scheduler>标记时，Spring 将实例化 ThreadPoolTaskScheduler 类的一个实例，并将属性 pool-size 指定为调度程序可以使用的线程池的大小。在<task:scheduled-tasks>标记内，可以调度一个或多个任务。在<task:scheduled>标记中，任务可以引用 Spring bean(本例中为 carService bean)中的特定方法(在本例中为 updateCarAgeJob()方法)。fixed-delay 属性指示 Spring 实例化 PeriodicTrigger 作为 TaskScheduler 的 Trigger 实现。

可以将任务调度配置与数据访问配置结合在一起，方法是声明一个新的配置类并导入使用@Import(针对配置类)和@ImportResource(针对 XML 配置)的两个配置。

```
package com.apress.prospring5.ch11.config;

import org.springframework.context.annotation.Configuration;
import org.springframework.context.annotation.Import;
import org.springframework.context.annotation.ImportResource;

@Configuration
@Import({ DataServiceConfig.class })
```

```
@ImportResource("classpath:spring/task-namespace-app-context.xml")
public class AppConfig {}
```

配置类(AppConfig)用于创建 Spring ApplicationContext 以测试 Spring 调度功能：

```
package com.apress.prospring5.ch11;

import com.apress.prospring5.ch11.config.AppConfig;
import com.apress.prospring5.ch11.services.CarService;
import com.apress.prospring5.ch11.services.CarServiceImpl;
import org.slf4j.Logger;
import org.slf4j.LoggerFactory;
import org.springframework.context.annotation.AnnotationConfigApplicationContext;
import org.springframework.context.support.GenericApplicationContext;

public class ScheduleTaskDemo {

   final static Logger logger = LoggerFactory.getLogger(CarServiceImpl.class);

   public static void main(String... args) throws Exception{
      GenericApplicationContext ctx =
         new AnnotationConfigApplicationContext(AppConfig.class);
      CarService carService = ctx.getBean("carService", CarService.class);

      while (!carService.isDone()) {
         logger.info("Waiting for scheduled job to end ...");
         Thread.sleep(250);
      }
      ctx.close();
   }
}
```

运行该程序会生成以下批量作业输出：

```
[main] INFO c.a.p.c.s.CarServiceImpl - Waiting for scheduled job to end ...
[carScheduler-1] INFO c.a.p.c.s.CarServiceImpl - Car age update job started
[carScheduler-1] INFO c.a.p.c.s.CarServiceImpl - Car age update --> {License:
   GRAVITY-0405, Manufacturer: Ford, Manufacture Date: 2006-09-12, Age: 10}
[carScheduler-1] INFO c.a.p.c.s.CarServiceImpl - Car age update --> {License:
   CLARITY-0432, Manufacturer: Toyota, Manufacture Date: 2003-09-09, Age: 13}
[carScheduler-1] INFO c.a.p.c.s.CarServiceImpl - Car age update --> {License:
   ROSIE-0402, Manufacturer: Toyota, Manufacture Date: 2017-04-16, Age: 0}
[carScheduler-1] INFO c.a.p.c.s.CarServiceImpl -
   Car age update job completed successfully
```

在上面的示例中，应用程序仅在调度任务运行一次后停止。正如所声明的那样，如果希望任务每 10 秒运行一次，方法是设置 fixed-delay ="10000"属性；也可以让应用程序一直运行，从而重复运行任务，直到用户按下一个键为止。ModifyScheduleTaskDemo 如下所示：

```
package com.apress.prospring5.ch11;

import com.apress.prospring5.ch11.config.AppConfig;
import org.springframework.context.annotation.AnnotationConfigApplicationContext;
import org.springframework.context.support.GenericApplicationContext;

public class ScheduleTaskDemo {

   public static void main(String... args) throws Exception {
      GenericApplicationContext ctx =
         new AnnotationConfigApplicationContext(AppConfig.class);

      System.in.read();
      ctx.close();
   }
}
```

通过输出可以看到汽车对象的 age 属性已经更新。除了固定的时间间隔之外，更灵活的调度机制是使用 cron 表达式。在 XML 配置文件中，可以将下面一行：

```
<task:scheduled ref="carService" method="updateCarAgeJob" fixed-delay="10000"/>
```

更改为：

```
<task:scheduled ref="carService" method="updateCarAgeJob" cron="0 * * * * *"/>
```

更改后，再次运行 ScheduleTaskDemo 类，并让应用程序运行超过一分钟，你会看到每分钟都会运行作业。

11.2.3 使用注解进行任务调度

使用 Spring 的 TaskScheduler 抽象来调度任务的另一种选择是使用注解。为此，Spring 提供了@Scheduled 注解。要为任务调度启用注解支持，需要在 Spring 的 XML 配置中提供<task:annotation-driven>标记。或者，如果使用配置类，则必须使用@EnableScheduling 进行注解。接下来讨论一下这种方法，并完全删除 XML。

```
package com.apress.prospring5.ch11.config;
import org.springframework.context.annotation.Configuration;
import org.springframework.context.annotation.Import;
import org.springframework.scheduling.annotation.EnableScheduling;

@Configuration
@Import({DataServiceConfig.class})
@EnableScheduling
public class AppConfig {
}
```

没错，以上就是所需的全部代码。甚至不再需要自己声明调度程序，因为 Spring 会帮我们代为处理。@Configuration 类中使用的@EnableScheduling 注解可以检测容器或其方法中 Spring 管理的任何 bean 上的@Scheduled 注解。有趣的是，使用@Scheduled 注解的方法甚至可以直接在@Configuration 类中声明。该注解告诉 Spring 查找关联的调度程序定义：要么是上下文中唯一的 TaskScheduler bean，要么是名为 taskScheduler 的 TaskScheduler bean 或 ScheduledExecutorService bean。如果没有找到，则在注册服务中创建和使用本地单线程的默认调度程序。

如果想在 Spring bean 中调度特定的方法，必须使用@Scheduled 对方法进行注解并传入调度需求。在下面的代码片段中，对 CarServiceImpl 类进行了扩展，并用于声明一个带有调度方法的新 bean，该方法重写了父类中的 updateCarAgeJob()方法以使用@Scheduled 注解：

```
package com.apress.prospring5.ch11.services;

import com.apress.prospring5.ch11.entities.Car;
import org.joda.time.DateTime;
import org.joda.time.Years;
import org.slf4j.Logger;
import org.slf4j.LoggerFactory;
import org.springframework.scheduling.annotation.Scheduled;
import org.springframework.stereotype.Repository;
import org.springframework.stereotype.Service;
import org.springframework.transaction.annotation.Transactional;

import java.util.List;

@Service("scheduledCarService")
@Repository
@Transactional

public class ScheduledCarServiceImpl extends CarServiceImpl{

    @Override
    @Scheduled(fixedDelay=10000)
    public void updateCarAgeJob() {
        List<Car> cars = findAll();

        DateTime currentDate = DateTime.now();
        logger.info("Car age update job started");

        cars.forEach(car -> {
            int age = Years.yearsBetween(
                car.getManufactureDate(), currentDate).getYears();
            car.setAge(age);
            save(car);
            logger.info("Car age update --> " + car);
        });

        logger.info("Car age update job completed successfully");
    }
}
```

测试程序如下所示：

```
package com.apress.prospring5.ch11;
```

```
import com.apress.prospring5.ch11.config.AppConfig;
import org.springframework.context.annotation.AnnotationConfigApplicationContext;
import org.springframework.context.support.GenericApplicationContext;

public class ScheduleTaskAnnotationDemo {

   public static void main(String... args) throws Exception {
      GenericApplicationContext ctx =
         new AnnotationConfigApplicationContext(AppConfig.class);

      System.in.read();

      ctx.close();
   }
}
```

运行程序所产生的结果与使用 task-namespace 几乎相同。可以通过更改@Scheduled 注解中的属性(即 fixedDelay、fixedRate、cron)来尝试不同的触发机制。可以自己随意测试一下。

```
[main] DEBUG o.s.s.a.ScheduledAnnotationBeanPostProcessor - Could not find default
   TaskScheduler bean
org.springframework.beans.factory.NoSuchBeanDefinitionException:
   No qualifying bean of type 'org.springframework.scheduling.TaskScheduler' available
... // more stacktrace here
[main] DEBUG o.s.s.a.ScheduledAnnotationBeanPostProcessor - Could not find default
   ScheduledExecutorService bean
org.springframework.beans.factory.NoSuchBeanDefinitionException:
   No qualifying bean of type 'java.util.concurrent.ScheduledExecutorService' available
... // more stacktrace here
[pool-1-thread-1] INFO c.a.p.c.s.CarServiceImpl - Car age update job started
[pool-1-thread-1] INFO c.a.p.c.s.CarServiceImpl - Car age update --> {License:
   GRAVITY-0405, Manufacturer: Ford, Manufacture Date: 2006-09-12, Age: 10}
[pool-1-thread-1] INFO c.a.p.c.s.CarServiceImpl - Car age update --> {License:
   CLARITY-0432, Manufacturer: Toyota, Manufacture Date: 2003-09-09, Age: 13}
[pool-1-thread-1] INFO c.a.p.c.s.CarServiceImpl - Car age update --> {License:
   ROSIE-0402, Manufacturer: Toyota, Manufacture Date: 2017-04-16, Age: 0}
[pool-1-thread-1] INFO c.a.p.c.s.CarServiceImpl - Car age update job
   completed successfully
```

此外，如果需要，还可以定义自己的 TaskScheduler bean。下面的示例声明了一个 ThreadPoolTaskScheduler bean，它等同于在上一节的 XML 配置中声明的 bean：

```
package com.apress.prospring5.ch11.config;

import org.springframework.context.annotation.Bean;
import org.springframework.context.annotation.Configuration;
import org.springframework.context.annotation.Import;
import org.springframework.scheduling.TaskScheduler;
import org.springframework.scheduling.annotation.EnableScheduling;
import org.springframework.scheduling.concurrent.DefaultManagedTaskScheduler;
import org.springframework.scheduling.concurrent.ThreadPoolTaskScheduler;

@Configuration
@Import({DataServiceConfig.class})
@EnableScheduling
public class AppConfig {

   @Bean TaskScheduler carScheduler() {
      ThreadPoolTaskScheduler carScheduler =
         new ThreadPoolTaskScheduler();
      carScheduler.setPoolSize(10);
      return carScheduler;
   }
}
```

如果运行测试示例，你会看到异常不再显示在日志中，并且执行该方法的调度程序的名称也变了，因为 TaskScheduler bean 被命名为 carScheduler。

```
[carScheduler-1] INFO c.a.p.c.s.CarServiceImpl - Car age update job started
[carScheduler-1] INFO c.a.p.c.s.CarServiceImpl - Car age update --> {License:
   GRAVITY-0405, Manufacturer: Ford, Manufacture Date: 2006-09-12, Age: 10}
[carScheduler-1] INFO c.a.p.c.s.CarServiceImpl - Car age update --> {License:
   CLARITY-0432, Manufacturer: Toyota, Manufacture Date: 2003-09-09, Age: 13}
[carScheduler-1] INFO c.a.p.c.s.CarServiceImpl - Car age update --> {License:
   ROSIE-0402, Manufacturer: Toyota, Manufacture Date: 2017-04-16, Age: 0}
[carScheduler-1] INFO c.a.p.c.s.CarServiceImpl - Car age update job
   completed successfully
```

11.2.4 Spring 中异步任务的执行

自 3.0 版以来，Spring 还支持使用注解来异步执行任务。要做到这一点，只需要用@Async 注解方法即可。

```
package com.apress.prospring5.ch11;

import java.util.concurrent.Future;

import org.slf4j.Logger;
import org.slf4j.LoggerFactory;
import org.springframework.scheduling.annotation.Async;
import org.springframework.scheduling.annotation.AsyncResult;
import org.springframework.stereotype.Service;

@Service("asyncService")
public class AsyncServiceImpl implements AsyncService {
    final Logger logger = LoggerFactory.getLogger(AsyncServiceImpl.class);

    @Async
    @Override
    public void asyncTask() {
        logger.info("Start execution of async. task");

        try {
            Thread.sleep(10000);
        } catch (Exception ex) {
            logger.error("Task Interruption", ex);
        }

        logger.info("Complete execution of async. task");
    }

    @Async
    @Override
    public Future<String> asyncWithReturn(String name) {
        logger.info("Start execution of async. task with return for "+ name);

        try {
            Thread.sleep(5000);
        } catch (Exception ex) {
            logger.error("Task Interruption", ex);
        }

        logger.info("Complete execution of async. task with return for " + name);

        return new AsyncResult<>("Hello: " + name);
    }
}
```

AsyncService 定义了两个方法。asyncTask()方法用于将信息记录到记录器。asyncWithReturn()方法接收一个 String 参数并返回 java.util.concurrent.Future <V>接口的一个实例。在完成 asyncWithReturn()后，结果将存储在 org.springframework.scheduling.annotation.AsyncResult <V>类的一个实例中，该类实现了 Future <V>接口，稍后调用者可以用它来检索执行结果。通过启用 Spring 的异步方法执行功能，可以拾取@Async 注解，而要想启用该功能，需要使用@EnableAsync 注解 Java 配置类。[1]

```
package com.apress.prospring5.ch11.config;

import org.springframework.context.annotation.ComponentScan;
import org.springframework.context.annotation.Configuration;
import org.springframework.context.annotation.Import;
import org.springframework.context.annotation.ImportResource;
import org.springframework.scheduling.annotation.EnableAsync;
import org.springframework.scheduling.annotation.EnableScheduling;

@Configuration
@EnableAsync
@ComponentScan(basePackages = {"com.apress.prospring5.ch11"} )
public class AppConfig {
}
```

1 在 XML 中，只需要声明<task:annotation-driven />即可。

测试程序如下所示：

```
package com.apress.prospring5.ch11;

import java.util.concurrent.Future;

import com.apress.prospring5.ch11.config.AppConfig;
import org.slf4j.Logger;
import org.slf4j.LoggerFactory;
import org.springframework.context.annotation.AnnotationConfigApplicationContext;
import org.springframework.context.support.GenericApplicationContext;

public class AsyncTaskDemo {
   private static Logger logger =
      LoggerFactory.getLogger(AsyncTaskDemo.class);

   public static void main(String... args) throws Exception{
      GenericApplicationContext ctx =
            new AnnotationConfigApplicationContext(AppConfig.class);

      AsyncService asyncService = ctx.getBean("asyncService",
         AsyncService.class);

      for (int i = 0; i < 5; i++) {
         asyncService.asyncTask();
      }
      Future<String> result1 = asyncService.asyncWithReturn("John Mayer");
      Future<String> result2 = asyncService.asyncWithReturn("Eric Clapton");
      Future<String> result3 = asyncService.asyncWithReturn("BB King");
      Thread.sleep(6000);

      logger.info("Result1: " + result1.get());
      logger.info("Result2: " + result2.get());
      logger.info("Result3: " + result3.get());

      System.in.read();
      ctx.close();
   }
}
```

首先调用 asyncTask()方法五次，然后用不同的参数调用 asyncWithReturn()方法三次，最后在休眠六秒后检索结果。运行该程序会产生以下输出：

```
...
17:55:31.851 [SimpleAsyncTaskExecutor-1] INFO c.a.p.c.AsyncServiceImpl -
   Start execution of async. task
17:55:31.851 [SimpleAsyncTaskExecutor-2] INFO c.a.p.c.AsyncServiceImpl -
   Start execution of async. task
17:55:31.851 [SimpleAsyncTaskExecutor-3] INFO c.a.p.c.AsyncServiceImpl -
   Start execution of async. task
17:55:31.851 [SimpleAsyncTaskExecutor-4] INFO c.a.p.c.AsyncServiceImpl -
   Start execution of async. task
17:55:31.852 [SimpleAsyncTaskExecutor-5] INFO c.a.p.c.AsyncServiceImpl -
   Start execution of async. task
17:55:31.852 [SimpleAsyncTaskExecutor-6] INFO c.a.p.c.AsyncServiceImpl -
   Start execution of async. task with return for John Mayer
17:55:31.852 [SimpleAsyncTaskExecutor-7] INFO c.a.p.c.AsyncServiceImpl -
   Start execution of async. task with return for Eric Clapton
17:55:31.852 [SimpleAsyncTaskExecutor-8] INFO c.a.p.c.AsyncServiceImpl -
   Start execution of async. task with return for BB King
17:55:36.856 [SimpleAsyncTaskExecutor-8] INFO c.a.p.c.AsyncServiceImpl -
   Complete execution of async. task with return for BB King
17:55:36.856 [SimpleAsyncTaskExecutor-6] INFO c.a.p.c.AsyncServiceImpl -
   Complete execution of async. task with return for John Mayer
17:55:36.856 [SimpleAsyncTaskExecutor-7] INFO c.a.p.c.AsyncServiceImpl -
   Complete execution of async. task with return for Eric Clapton
17:55:37.852 [main] INFO c.a.p.c.AsyncTaskDemo - Result1: Hello: John Mayer
17:55:37.853 [main] INFO c.a.p.c.AsyncTaskDemo - Result2: Hello: Eric Clapton
17:55:37.853 [main] INFO c.a.p.c.AsyncTaskDemo - Result3: Hello: BB King
17:55:41.852 [SimpleAsyncTaskExecutor-1] INFO c.a.p.c.AsyncServiceImpl -
   Complete execution of async. task
17:55:41.852 [SimpleAsyncTaskExecutor-4] INFO c.a.p.c.AsyncServiceImpl -
   Complete execution of async. task
17:55:41.852 [SimpleAsyncTaskExecutor-3] INFO c.a.p.c.AsyncServiceImpl -
   Complete execution of async. task
17:55:41.852 [SimpleAsyncTaskExecutor-5] INFO c.a.p.c.AsyncServiceImpl -
   Complete execution of async. task
17:55:41.852 [SimpleAsyncTaskExecutor-2] INFO c.a.p.c.AsyncServiceImpl -
   Complete execution of async. task
```

从输出可以看到，所有调用都是在同一时间开始的。带有返回值的三次调用首先完成并显示在控制台输出上。然后，对 asyncTask()方法的五次调用也完成了。

11.3 Spring 中任务的执行

Spring 自 2.0 版以来，就提供了一个通过 TaskExecutor 接口执行任务的抽象。TaskExecutor 所完成的操作就像它的名字一样：执行由 Java Runnable 实现表示的任务。此外，Spring 还提供了许多适合不同需求的 TaskExecutor 实现。可以在 http://docs.spring.io/spring/docs/current/javadoc-api/org/springframework/core/task/TaskExecutor.html 上找到完整的 TaskExecutor 实现列表。

下面列出了一些常用的 TaskExecutor 实现。

- SimpleAsyncTaskExecutor：在每次调用时创建新线程，不重用现有的线程。
- SyncTaskExecutor：不会异步执行，调用发生在调用线程中。
- SimpleThreadPoolTaskExecutor：Quartz 的 SimpleThreadPool 的子类，当需要 Quartz 和非 Quartz 组件之间共享线程池时使用。
- ThreadPoolTaskExecutor：TaskExecutor 的一种实现，提供了通过 bean 属性配置 ThreadPoolExecutor 并将其作为 Spring TaskExecutor 公开的功能。

虽然每种 TaskExecutor 实现都有自己的目的，但调用约定是相同的。唯一的变化是在配置中，需要定义所使用的 TaskExecutor 实现及其属性(如果有的话)。接下来看一个打印大量消息的简单示例。此时使用的 TaskExecutor 实现是 SimpleAsyncTaskExecutor。首先创建一个包含任务执行逻辑的 bean 类，如下所示：

```
package com.apress.prospring5.ch11;

import org.slf4j.Logger;
import org.slf4j.LoggerFactory;
import org.springframework.beans.factory.annotation.Autowired;
import org.springframework.core.task.TaskExecutor;
import org.springframework.stereotype.Component;

@Component
public class TaskToExecute {
    private final Logger logger =
        LoggerFactory.getLogger(TaskToExecute.class);

    @Autowired
    private TaskExecutor taskExecutor;
    public void executeTask() {
        for(int i=0; i < 10; ++ i) {
            taskExecutor.execute(() ->
                logger.info("Hello from thread: " +
                Thread.currentThread().getName()));
        }
    }
}
```

该类只是一个普通的 bean，需要将 TaskExecutor 作为依赖项注入并定义方法 executeTask()。executeTask()方法通过创建一个新的 Runnable 实例来调用所提供的 TaskExecutor 的 execute()方法，该实例包含针对此任务执行的逻辑。由于使用 lambda 表达式来创建 Runnable 实例，因此该过程看起来可能并不太明显。配置非常简单，与上一节中描述的配置相似。此时唯一需要考虑的是为 TaxExecutor bean 提供一个声明，该声明需要注入到 TaskToExecute bean 中。

```
package com.apress.prospring5.ch11.config;

import org.springframework.context.annotation.Bean;
import org.springframework.context.annotation.ComponentScan;
import org.springframework.context.annotation.Configuration;
import org.springframework.core.task.SimpleAsyncTaskExecutor;
import org.springframework.core.task.TaskExecutor;
import org.springframework.scheduling.TaskScheduler;
import org.springframework.scheduling.annotation.EnableAsync;
import org.springframework.scheduling.concurrent.ThreadPoolTaskScheduler;

@Configuration
@EnableAsync
@ComponentScan(basePackages = {"com.apress.prospring5.ch11"} )
public class AppConfig {
```

```
    @Bean TaskExecutor taskExecutor() {
        return new SimpleAsyncTaskExecutor();
    }
}
```

在前面的配置中声明了一个SimpleAsyncTaskExecutor类型的名为taskExecutor的简单bean。该bean由Spring IoC容器注入到TaskToExecute bean中。要测试执行情况，可以使用以下程序：

```
package com.apress.prospring5.ch11;

import com.apress.prospring5.ch11.config.AppConfig;
import org.springframework.context.annotation.AnnotationConfigApplicationContext;
import org.springframework.context.support.GenericApplicationContext;
public class TaskExecutorDemo {

    public static void main(String... args) throws Exception {
        GenericApplicationContext ctx =
            new AnnotationConfigApplicationContext(AppConfig.class);

        TaskToExecute taskToExecute = ctx.getBean(TaskToExecute.class);
        taskToExecute.executeTask();

        System.in.read();
        ctx.close();
    }
}
```

运行该例时，应该打印与以下内容类似的输出：

```
Hello from thread: SimpleAsyncTaskExecutor-1
Hello from thread: SimpleAsyncTaskExecutor-5
Hello from thread: SimpleAsyncTaskExecutor-3
Hello from thread: SimpleAsyncTaskExecutor-10
Hello from thread: SimpleAsyncTaskExecutor-8
Hello from thread: SimpleAsyncTaskExecutor-6
Hello from thread: SimpleAsyncTaskExecutor-2
Hello from thread: SimpleAsyncTaskExecutor-4
Hello from thread: SimpleAsyncTaskExecutor-9
Hello from thread: SimpleAsyncTaskExecutor-7
```

从输出可以看到，每个任务(正在打印的消息)都会在执行时显示，打印出消息、线程名称(默认为类名SimpleAsyncTaskExecutor)以及线程号。

11.4 小结

在本章中，介绍了 Spring 对任务调度的支持，重点介绍了 Spring 内置的 TaskScheduler 抽象，并演示了如何利用一个批处理数据更新作业来完成任务调度需求，还讨论了 Spring 如何支持用于异步执行任务的注解，简要介绍了 Spring 的 TaskExecutor 和通用实现。

不需要Spring Boot部分，因为任务注解的调度和异步执行是spring-context库的一部分，并且它们会与Spring Boot配置一起使用。另外，通过使用 Spring，可以非常容易地配置调度和异步任务；Spring Boot 在这方面没有太多可以改进的地方[1]。

1 但是，如果好奇并想将提供的项目转换为Spring Boot，那么可以访问以下网址，找到一个小教程：https://spring.io/guides/gs/scheduling-tasks/。

第12章

使用 Spring 远程处理

企业级应用程序通常需要与其他应用程序进行通信。以一家销售产品的公司为例；当客户下订单时，订单处理系统处理该订单并生成一个事务。在订单处理期间，需要对库存系统进行查询以检查产品是否有库存。当订单确认后，会向履行系统(fulfillment system)发送通知，以便向客户交付产品。最后，信息被发送到会计系统，生成发票并处理付款。

大多数情况下，这个业务流程并不是由一个应用程序来完成的，而是由许多应用程序一起工作来完成的。一些应用程序可能是内部开发的，而其他应用程序可能是从外部供应商处购买的。此外，应用程序可以运行在不同地点的不同计算机上，并且可以使用不同的技术和编程语言(例如Java、.NET或C++)来实现。在构建和实现应用程序时，在应用程序之间握手以构建高效的业务流程始终是一项关键任务。因此，应用程序需要通过各种协议和技术进行远程支持才能很好地适应企业环境。

在 Java 世界中，自 Java 首次创建以来，就一直支持远程处理。在早期(Java 1.x)，大多数远程需求是通过使用传统的 TCP 套接字或 Java 远程方法调用(Java Remote Method Invocation，RMI)来实现的。在 J2EE 出现后，EJB 和 JMS 成为应用程序服务器通信的常用选择。XML 和互联网的快速发展促成了在 HTTP 上使用 XML 实现远程支持，包括针对基于 XML 的 RPC(JAX-RPC)的 Java API、针对 XML Web Services(JAX-WS)的 Java API 以及基于 HTTP 的技术(例如 Hessian 和 Burlap)。此外，Spring 还提供了自己的基于 HTTP 的远程支持，称为 Spring HTTP 调用器。近年来，为了应对互联网的爆炸性增长以及对更具响应性的 Web 应用程序的需求(例如，通过 Ajax)，应用程序的更轻量级和高效的远程支持已成为企业成功的关键。因此，针对 RESTful Web Services(JAX-RS)的 Java API 被推出并迅速流行开来。其他协议，比如 Comet 和 HTML5 WebSocket，也吸引了很多开发人员。毋庸置疑，远程技术在不断地快速发展。

如前所述，在远程处理方面，Spring 提供了自己的支持(通过 Spring HTTP 调用器)，并且支持前面所提到的很多技术(例如 RMI、EJB、JMS、Hessian、Burlap[1]、JAX-RPC，JAX-WS 和 JAX-RS)。本章不可能涵盖所有这些内容。所以，我们将重点放在那些最常用的方面。具体而言，本章涵盖以下主题：

- **Spring HTTP 调用器**：如果两个需要通信的应用程序都是基于 Spring 的，那么 Spring HTTP 调用器提供了一种简单且有效的方法来调用其他应用程序所公开的服务。本章将演示如何使用 Spring HTTP 调用器在其服务层中公开服务，以及如何调用远程应用程序所提供的服务。
- **在 Spring 中使用 JMS**：Java 消息服务(JMS)提供了另一种在应用程序之间交换消息的异步且松散耦合的方法。本章将演示 Spring 如何通过 JMS 简化应用程序开发。
- **在 Spring 中使用 RESTful Web 服务**：RESTful Web 服务是专门围绕 HTTP 设计的，是为应用程序提供远程支持以及支持使用 Ajax 的高度交互式 Web 应用程序的最常用前端技术。本章将演示 Spring MVC 如何为使用 JAX-RS 公开服务提供全面支持以及如何使用 RestTemplate 类调用服务，还将讨论如何确保服务免遭未经授权的访问。
- **在 Spring 中使用 AMQP**：Spring 高级消息队列协议(AMQP)项目提供了围绕 AMQP 的、典型的、类似 Spring 的抽象以及 RabbitMQ 实现。该项目提供了一组丰富的功能，但在本章中，将通过 RPC 项目支持重点关注其远程处理功能。

1 当 Spring 4 发布时，有人提到 Burlap 不再处于积极的发展阶段，未来将完全放弃支持。

12.1 使用示例的数据模型

在本章的示例中，将使用一个简单的数据模型，其中只包含用于存储信息的 SINGER 表。该表由 Hibernate 根据下面显示的 Singer 类生成。该类及其属性用标准的 JPA 注解进行修饰。

```
package com.apress.prospring5.ch12.entities;

import javax.persistence.*;
import java.io.Serializable;
import java.util.Date;

import static javax.persistence.GenerationType.IDENTITY;

@Entity
@Table(name = "singer")
public class Singer implements Serializable {
   @Id
   @GeneratedValue(strategy = IDENTITY)
   @Column(name = "ID")
   private Long id;

   @Version
   @Column(name = "VERSION")
   private int version;

   @Column(name = "FIRST_NAME")
   private String firstName;

   @Column(name = "LAST_NAME")
   private String lastName;

   @Temporal(TemporalType.DATE)
   @Column(name = "BIRTH_DATE")
   private Date birthDate;

   //setters and getters
   ...
}
```

为了填充该表，使用一个初始化 bean。该类如下所示：

```
package com.apress.prospring5.ch12.services;

import com.apress.prospring5.ch12.entities.Singer;
import com.apress.prospring5.ch12.repos.SingerRepository;
import org.slf4j.Logger;
import org.slf4j.LoggerFactory;
import org.springframework.beans.factory.annotation.Autowired;
import org.springframework.stereotype.Service;

import javax.annotation.PostConstruct;
import java.util.Date;
import java.util.GregorianCalendar;

@Service
public class DBInitializer {

   private Logger logger = LoggerFactory.getLogger(DBInitializer.class);
   @Autowired
   SingerRepository singerRepository;

   @PostConstruct
   public void initDB() {
      logger.info("Starting database initialization...");
      Singer singer = new Singer();
      singer.setFirstName("John");
      singer.setLastName("Mayer");
      singer.setBirthDate(new Date(
            (new GregorianCalendar(1977, 9, 16)).getTime().getTime()));
      singerRepository.save(singer);

      singer = new Singer();
      singer.setFirstName("Eric");
      singer.setLastName("Clapton");
      singer.setBirthDate(new Date(
            (new GregorianCalendar(1945, 2, 30)).getTime().getTime()));
```

```
        singerRepository.save(singer);
        singer = new Singer();
        singer.setFirstName("John");
        singer.setLastName("Butler");
        singer.setBirthDate(new Date(
                (new GregorianCalendar(1975, 3, 1)).getTime().getTime()));

        singerRepository.save(singer);
        logger.info("Database initialization finished.");
    }
}
```

12.2 为 JPA 后端添加必需的依赖项

需要将所需的依赖项添加到项目中。以下配置代码片段显示了使用 JPA 2 和 Hibernate 作为持久化提供程序来实现服务层所需的依赖项。另外，还将使用 Spring Data JPA。

```
//pro-spring-15/build.gradle
ext {
   //spring libs
   springVersion = '5.0.0.RC1'
   springDataVersion = '2.0.0.M3'

   //logging libs
   slf4jVersion = '1.7.25'
   logbackVersion = '1.2.3'

   junitVersion = '4.12'

   //database library
   h2Version = '1.4.194'

   //persistency libraries
   hibernateVersion = '5.2.10.Final'
   hibernateJpaVersion = '1.0.0.Final'
   atomikosVersion = '4.0.0M4'

   spring = [
      context        : "org.springframework:spring-context:$springVersion",
      aop            : "org.springframework:spring-aop:$springVersion",
      aspects        : "org.springframework:spring-aspects:$springVersion",
      tx             : "org.springframework:spring-tx:$springVersion",
      jdbc           : "org.springframework:spring-jdbc:$springVersion",
      contextSupport : "org.springframework:spring-context-support:$springVersion",
      orm            : "org.springframework:spring-orm:$springVersion",
      data           : "org.springframework.data:spring-data-jpa:$springDataVersion",
      test           :"org.springframework:spring-test:$springVersion"
   ]
   hibernate = [
      ...
      em             : "org.hibernate:hibernate-entitymanager:$hibernateVersion",
      tx             : "com.atomikos:transactions-hibernate4:$atomikosVersion"
   ]

   testing = [
      junit: "junit:junit:$junitVersion"
   ]

   misc = [
      ...
      slf4jJcl       : "org.slf4j:jcl-over-slf4j:$slf4jVersion",
      logback        : "ch.qos.logback:logback-classic:$logbackVersion",
      lang3          : "org.apache.commons:commons-lang3:3.5",
      guava          : "com.google.guava:guava:$guavaVersion"
   ]

   db = [
      ...
      h2 : "com.h2database:h2:$h2Version"
   ]
}

//chapter12/spring-invoker/build.gradle
dependencies {
   //we specify these dependencies for all submodules, except
```

```
    // the boot module, that defines its own
    if (!project.name.contains("boot")) {
       //we exclude transitive dependencies, because spring-data
       //will take care of these
          compile (spring.contextSupport) {
          exclude module: 'spring-context'
          exclude module: 'spring-beans'
          exclude module: 'spring-core'
       }
       //we exclude the 'hibernate' transitive dependency
       //to have control over the version used
       compile (hibernate.tx) {
       exclude group: 'org.hibernate', module: 'hibernate'
    }
       compile misc.slf4jJcl, misc.logback, misc.lang3,
          hibernate.em, misc.guava
    }
    testCompile testing.junit
}
```

12.3 实现和配置 SingerService

了解了相关的依赖项后，接下来演示如何为本章中的示例实现和配置服务层。在下面的章节中，将首先讨论使用 JPA 2、Spring Data JPA 和 Hibernate 作为持久化服务提供程序的 SingerService 的实现。然后，介绍如何在 Spring 项目中配置服务层。

12.3.1 实现 SingerService

在该例中，演示如何向远程客户端公开歌手信息的各种操作服务；SingerService 接口如下所示。

```
package com.apress.prospring5.ch12.services;

import com.apress.prospring5.ch12.entities.Singer;

import java.util.List;

public interface SingerService {

   List<Singer> findAll();
   List<Singer> findByFirstName(String firstName);
   Singer findById(Long id);
   Singer save(Singer singer);
   void delete(Singer singer);
}
```

SingerService 接口中的方法应该是不言自明的。因为需要使用 Spring Data JPA 的存储库支持，所以实现了 SingerRepository 接口，如下所示：

```
package com.apress.prospring5.ch12.repos;

import com.apress.prospring5.ch12.entities.Singer;
import org.springframework.data.repository.CrudRepository;

import java.util.List;

public interface SingerRepository extends CrudRepository<Singer, Long> {
   List<Singer> findByFirstName(String firstName);
}
```

由于扩展了 CrudRepository<T,ID extends Serializable>接口，因此对于 SingerService 接口中的方法，只需要显式声明 findByFirstName()方法。

下面的代码片段显示了 SingerService 接口的实现类：

```
package com.apress.prospring5.ch12.services;

import com.apress.prospring5.ch12.entities.Singer;
import com.apress.prospring5.ch12.repos.SingerRepository;
import com.google.common.collect.Lists;
import org.springframework.beans.factory.annotation.Autowired;
import org.springframework.stereotype.Repository;
import org.springframework.stereotype.Service;
import org.springframework.transaction.annotation.Transactional;
```

```
import java.util.List;

@Service("singerService")
@Transactional
public class SingerServiceImpl implements SingerService {

    @Autowired
    private SingerRepository singerRepository;

    @Override
    @Transactional(readOnly = true)
    public List<Singer> findAll() {
        return Lists.newArrayList(singerRepository.findAll());
    }

    @Override
    @Transactional(readOnly = true)
    public List<Singer> findByFirstName(String firstName) {
        return singerRepository.findByFirstName(firstName);
    }

    @Override
    @Transactional(readOnly = true)
    public Singer findById(Long id) {
        return singerRepository.findById(id).get();
    }

    @Override
    public Singer save(Singer singer) {
        return singerRepository.save(singer);
    }

    @Override
    public void delete(Singer singer) {
        singerRepository.delete(singer);
    }
}
```

实现已基本完成，下一步是在 Web 项目中的 Spring 的 ApplicationContext 中配置服务，该内容将在下一节讨论。

12.3.2 配置 SingerService

对于数据访问和事务，可以使用一个简单的 Java 配置类，如前所述，代码如下所示：

```
package com.apress.prospring5.ch12.config;

import org.slf4j.Logger;
import org.slf4j.LoggerFactory;
import org.springframework.context.annotation.Bean;
import org.springframework.context.annotation.ComponentScan;
import org.springframework.context.annotation.Configuration;
import org.springframework.data.jpa.repository.config.EnableJpaRepositories;
import org.springframework.jdbc.datasource.embedded.EmbeddedDatabaseBuilder;
import org.springframework.jdbc.datasource.embedded.EmbeddedDatabaseType;
import org.springframework.orm.jpa.JpaTransactionManager;
import org.springframework.orm.jpa.JpaVendorAdapter;
import org.springframework.orm.jpa.LocalContainerEntityManagerFactoryBean;
import org.springframework.orm.jpa.vendor.HibernateJpaVendorAdapter;
import org.springframework.transaction.PlatformTransactionManager;

import javax.persistence.EntityManagerFactory;
import javax.sql.DataSource;
import java.util.Properties;

@Configuration
@EnableJpaRepositories(basePackages = {"com.apress.prospring5.ch12.repos"})
@ComponentScan(basePackages = {"com.apress.prospring5.ch12"} )
public class DataServiceConfig {

    private static Logger logger =
        LoggerFactory.getLogger(DataServiceConfig.class);

    @Bean
    public DataSource dataSource() {
        try {
            EmbeddedDatabaseBuilder dbBuilder =
```

```
            new EmbeddedDatabaseBuilder();
         return dbBuilder.setType(EmbeddedDatabaseType.H2).build();
      } catch (Exception e) {
         logger.error("Embedded DataSource bean cannot be created!", e);
         return null;
      }
   }

   @Bean
   public Properties hibernateProperties() {
      Properties hibernateProp = new Properties();
      hibernateProp.put("hibernate.dialect", "org.hibernate.dialect.H2Dialect");
      hibernateProp.put("hibernate.hbm2ddl.auto", "create-drop");
      hibernateProp.put("hibernate.show_sql", true);
      hibernateProp.put("hibernate.max_fetch_depth", 3);
      hibernateProp.put("hibernate.jdbc.batch_size", 10);
      hibernateProp.put("hibernate.jdbc.fetch_size", 50);
      return hibernateProp;
   }

   @Bean
   public PlatformTransactionManager transactionManager() {
      return new JpaTransactionManager(entityManagerFactory());
   }

   @Bean
   public JpaVendorAdapter jpaVendorAdapter() {
      return new HibernateJpaVendorAdapter();
   }

   @Bean
   public EntityManagerFactory entityManagerFactory() {
      LocalContainerEntityManagerFactoryBean factoryBean =
         new LocalContainerEntityManagerFactoryBean();
      factoryBean.setPackagesToScan("com.apress.prospring5.ch12.entities");
      factoryBean.setDataSource(dataSource());
      factoryBean.setJpaVendorAdapter(new HibernateJpaVendorAdapter());
      factoryBean.setJpaProperties(hibernateProperties());
      factoryBean.setJpaVendorAdapter(jpaVendorAdapter());
      factoryBean.afterPropertiesSet();
      return factoryBean.getNativeEntityManagerFactory();
   }
}
```

由于通过 Spring MVC 公开 HTTP 调用器，因此需要配置 Web 应用程序。为了在不使用任何 XML 的情况下配置 Spring Web MVC 应用程序，需要两个配置类，如下所示：

- 一个配置类实现 WebMvcConfigurer 接口。该接口在 Spring 3.1 中引入，并且定义了一些回调方法来为通过使用@EnableWebMvc 启用的 Spring MVC 定制基于 Java 的配置。因为此时只需要公开一个 HTTP 服务(不需要 Web 接口)，所以一个空的实现就足够了。使用@EnableWebMvc 注解的此接口的实现将用 mvc 名称空间替换 Spring XML 配置。更复杂的示例配置将在第 16 章中详细介绍。

```
package com.apress.prospring5.ch12.config;

import org.springframework.context.annotation.Configuration;
import org.springframework.web.servlet.config.annotation.EnableWebMvc;
import org.springframework.web.servlet.config.annotation.WebMvcConfigurer;

@Configuration
@EnableWebMvc
public class WebConfig implements WebMvcConfigurer {

}
```

- 另一个配置类实现了 org.springframework.web.WebApplicationInitializer 或者扩展了一个开箱即用的 Spring 实现。该接口需要在 Spring 3.0+环境中实现，以编程方式配置 ServletContext。这样就无须提供 web.xml 文件来配置 Web 应用程序。该类导入数据访问和事务的配置，并基于此创建根应用程序上下文。Web 应用程序上下文是使用 WebConfig 类以及为 HTTP 调用器服务定义配置的配置类一起创建的。也可以将这些类合并在一起，但在使用 Spring 时，最佳实践是将自定义服务和基础结构 bean 保留在不同的类中。

```
package com.apress.prospring5.ch12.config;

import org.springframework.web.servlet.support.
   AbstractAnnotationConfigDispatcherServletInitializer;
```

```
public class WebInitializer
    extends AbstractAnnotationConfigDispatcherServletInitializer {

    @Override
    protected Class<?>[] getRootConfigClasses() {
        return new Class<?>[]{
            DataServiceConfig.class
        };
    }

    @Override
    protected Class<?>[] getServletConfigClasses() {
        return new Class<?>[]{
            HttpInvokerConfig.class, WebConfig.class
        };
    }

    @Override
    protected String[] getServletMappings() {
        return new String[]{"/invoker/*"};
    }
}
```

由于这是一个 Spring MVC Web 应用程序，因此需要创建一个 WAR 文件并将其部署到 servlet 容器。可以使用多种方法实现这一点，例如，诸如 Tomcat 之类的独立容器、由 IDE 启动的 Tomcat 实例或者使用 Maven 等构建工具运行的嵌入式 Tomcat 实例。具体使用哪种方法取决于你的需求，但对于本地开发环境，建议使用从构建工具或直接从 IDE 启动的嵌入式 Tomcat 实例。在本书的代码中，使用了 Tomcat Server 9.x 版本，并且在 IntelliJ IDEA 中设置了一个启动器来启动 Web 应用程序。有关更多详细信息，请参阅本书的源代码。此时，应该通过所选择的方法来构建和部署 Web 应用程序。如果尝试在浏览器中加载 http://localhost:8080/URL，将会看到以下消息：

```
Spring Remoting: Simplifying Development of Distributed Applications

RMI services over HTTP should be correctly exposed when this page is visible.
```

这意味着 Web 应用程序已正确部署，现在可通过 URL http://localhost:8080/invoker/httpInvoker/singerService 访问 singerService bean。

12.3.3 公开服务

如果要与之通信的应用程序也是 Spring 驱动的，那么使用 Spring HTTP 调用器是不错的选择。它提供了一种非常简单的方法，可以将 Spring WebApplicationContext 中的服务公开给远程客户端，远程客户端也可以使用 Spring HTTP 调用器来调用服务。以下详细介绍公开和访问服务的过程。HttpInvokerConfig 类包含用于公开 HTTP 调用器服务的单个 bean。

```
package com.apress.prospring5.ch12.config;

import com.apress.prospring5.ch12.services.SingerService;
import org.springframework.beans.factory.annotation.Autowired;
import org.springframework.context.annotation.Bean;
import org.springframework.context.annotation.Configuration;
import org.springframework.remoting.httpinvoker.HttpInvokerServiceExporter;

@Configuration
public class HttpInvokerConfig {

    @Autowired
    SingerService singerService;

    @Bean(name = "/httpInvoker/singerService")
    public HttpInvokerServiceExporter httpInvokerServiceExporter() {
        HttpInvokerServiceExporter invokerService =
            new HttpInvokerServiceExporter();
        invokerService.setService(singerService);
        invokerService.setServiceInterface(SingerService.class);
        return invokerService;
    }
}
```

httpInvokerServiceExporter bean 是使用 HttpInvokerServiceExporter 类定义的，后者通过 HTTP 调用器将任何 Spring bean 作为服务导出。在 httpInvokerServiceExporter bean 中，定义了两个属性。第一个属性是 service，表示提供服务的 bean，针对该属性注入 singerService bean。第二个属性是要公开的接口类型，即 com.apress.prospring5.ch12.serviced.SingerService 接口。

现在，服务层已经完成并做好被远程客户端公开和使用的准备。

12.3.4 调用服务

通过 Spring HTTP 调用器调用服务是非常简单的。首先配置一个 Spring ApplicationContext，如下面的配置类中所示：

```
package com.apress.prospring5.ch12.config;

import com.apress.prospring5.ch12.services.SingerService;
import org.springframework.context.annotation.Bean;
import org.springframework.context.annotation.Configuration;
import org.springframework.remoting.httpinvoker.HttpInvokerProxyFactoryBean;

@Configuration
public class RmiClientConfig {

   @Bean
   public SingerService singerService() {
      HttpInvokerProxyFactoryBean factoryBean =
         new HttpInvokerProxyFactoryBean();
      factoryBean.setServiceInterface(SingerService.class);
      factoryBean.setServiceUrl(
         "http://localhost:8080/invoker/httpInvoker/singerService");
      factoryBean.afterPropertiesSet();
      return (SingerService) factoryBean.getObject();
   }
}
```

参见之前所示的客户端，声明一个 HttpInvokerProxyFactoryBean 类型的 bean，并设置两个属性。第一个属性 serviceUrl 指定远程服务的位置，即 http://localhost:8080/invoker/httpInvoker/singerService。第二个属性是服务的接口(即 SingerService)。如果正在为客户端开发另一个项目，那么需要在客户端应用程序的类路径中包含 SingerService 接口和 Singer 实体类。

以下代码片段显示了用于调用远程服务的测试类。此时使用一个测试类(使用了 RmiClientConfig 类)来创建测试上下文。SpringRunner 类是 SpringJUnit4ClassRunner 的别名，SpringJUnit4ClassRunner 是在 Spring 上下文中运行 JUnit 测试所必需的。更多内容将在第 13 章中详细介绍。

```
package com.apress.prospring5.ch12;

import com.apress.prospring5.ch12.config.RmiClientConfig;
import com.apress.prospring5.ch12.entities.Singer;
import com.apress.prospring5.ch12.services.DBInitializer;
import com.apress.prospring5.ch12.services.SingerService;
import org.junit.Test;
import org.junit.runner.RunWith;
import org.slf4j.Logger;
import org.slf4j.LoggerFactory;
import org.springframework.beans.factory.annotation.Autowired;
import org.springframework.test.context.ContextConfiguration;
import org.springframework.test.context.junit4.SpringRunner;
import java.util.List;

import static org.junit.Assert.assertEquals;

@ContextConfiguration(classes = RmiClientConfig.class)
@RunWith(SpringRunner.class)
public class RmiTests {
   private Logger logger = LoggerFactory.getLogger(RmiTests.class);

   @Autowired
   private SingerService singerService;

   @Test
   public void testRmiAll() {
      List<Singer> singers = singerService.findAll();
      assertEquals(3, singers.size());
      listSingers(singers);
```

```
    }

    @Test
    public void testRmiJohn() {
        List<Singer> singers = singerService.findByFirstName("John");
        assertEquals(2, singers.size());
        listSingers(singers);
    }

    private void listSingers(List<Singer> singers){
        singers.forEach(s -> logger.info(s.toString()));
    }
}
```

测试类应该在部署完 Web 应用程序之后执行。测试应该传递并列出由 singerService bean 返回的 Singer 实例。预期的输出如下所示：

```
//testRmiAll
INFO c.a.p.c.RmiTests - Singer - Id: 1, First name: John, Last name: Mayer,
    Birthday: 1977-10-16
INFO c.a.p.c.RmiTests - Singer - Id: 2, First name: Eric, Last name: Clapton,
    Birthday: 1945-03-30
INFO c.a.p.c.RmiTests - Singer - Id: 3, First name: John, Last name: Butler,
    Birthday: 1975-04-01
//testRmiJohn - all singers with firstName='John'
INFO c.a.p.c.RmiTests - Singer - Id: 1, First name: John, Last name: Mayer,
    Birthday: 1977-10-16
INFO c.a.p.c.RmiTests - Singer - Id: 3, First name: John, Last name: Butler,
    Birthday: 1975-04-01
```

12.4 在 Spring 中使用 JMS

使用面向消息的中间件(通常称为 MQ 服务器)是另一种支持应用程序间通信的流行方法。消息队列(MQ)服务器的主要优点在于为应用程序集成提供了异步和松耦合的方式。在 Java 世界中，JMS 是连接到 MQ 服务器以发送或接收消息的标准。MQ 服务器维护应用程序可以连接并发送和接收消息的队列及主题列表。以下是队列和主题之间差异的简要说明：

- **队列**：队列用于支持点对点消息交换模型。当一名生产者向队列发送消息时，MQ 服务器会将消息保留在队列中，并在消费者下次连接时将消息传递给一名消费者(且只能传递给一名消费者)。
- **主题**：主题用于支持发布-订阅模型。任何数量的客户端都可以订阅主题中的消息。当某条消息到达主题时，MQ 服务器会将其发送给订阅了该消息的所有客户端。当多个应用程序对同一条消息(例如，新闻提要)感兴趣时，此模型是非常有用的。

在 JMS 中，生产者连接到 MQ 服务器并将消息发送到队列或主题，而消费者也连接到 MQ 服务器并监听队列或主题中感兴趣的消息。在 JMS 1.1 中，API 是统一的，因此生产者和消费者无须处理不同的 API 来与队列和主题进行交互。在本节中，将重点介绍用于队列的点对点模式，这是企业内更常用的模式。

从 Spring Framework 4.0 开始，已经实现了对 JMS 2.0 的支持。只需要在类路径中添加 JMS 2.0 JAR 就可以使用 JMS 2.0 功能，同时保留 JMS 1.x 的向后兼容性。但从 Spring Framework 5.0 开始，就不再支持 JMS 1.1；因此，本书中的示例仅与 JMS 2.x 相关。

在编写本书时，ActiveMQ 还不支持 JMS 2.0；因此，将使用 HornetQ(包含从 2.4.0.Final 开始的 JMS 2.0 支持)作为示例中的消息代理，并使用独立服务器。下载和安装 HornetQ 的相关内容已经超出了本书的范围；请参阅 http://docs.jboss.org/hornetq/2.4.0.Final/docs/quickstart-guide/html/index.html[1]中的文档。

我们需要几个新的依赖项，并且所需的 Gradle 配置如下所示：

```
//pro-spring-15/build.gradle
ext {
    jmsVersion = '2.0'
    hornetqVersion = '2.4.0.Final'

    spring = [
    ...
```

1 如果更喜欢 Apache 产品，那么 Apache ActiveMQ Artemis 是具有非阻塞架构的 JMS 2 实现，它提供了十分出色的性能。可以在 https://activemq.apache.org/artemis/上找到更多的细节。此外，代码示例中还提供了一个使用 Artemis ActiveMQ 的项目。

```
    jms        : "org.springframework:spring-jms:$springVersion"
]
    misc = [
        ...
        Hornetq  : "org.hornetq:hornetq-jms-client:$hornetqVersion"
    ]
...
}
//chapter12/jms-hornetq/build.gradle
dependencies {
    compile spring.jms, misc.jms
}
```

在安装完服务器之后，需要在 HornetQ JMS 配置文件中创建一个队列。该文件位于解压 HornetQ 的目录下，位置是 config/stand-alone/non-clustered hornetq-jms.xml，需要添加如下所示的队列定义：

```
<configuration xmlns="urn:hornetq"
        xmlns:xsi="http://www.w3.org/2001/XMLSchema-instance"
        xsi:schemaLocation="urn:hornetq /schema/hornetq-jms.xsd">
        ...

    <queue name="prospring5">
        <entry name="/queue/prospring5"/>
    </queue>
</configuration>
```

现在，通过运行 run.sh 脚本(取决于使用的操作系统)启动 HornetQ 服务器，并确保服务器启动时没有任何错误。只需要确保在日志中看到如下所示的类似内容且没有任何异常消息即可，尤其是要确保出现如下加了下划线的内容：

```
...
00:36:21,171 INFO [org.hornetq.core.server] HQ221035: Live Server Obtained live lock
00:36:21,841 INFO [org.hornetq.core.server] HQ221003:
    trying to deploy queue jms.queue.DLQ
00:36:21,852 INFO [org.hornetq.core.server] HQ221003:
// the queue configured in the previous configuration sample
    trying to deploy queue jms.queue.prospring5
00:36:21,853 INFO [org.hornetq.core.server] HQ221003:
    trying to deploy queue jms.queue.ExpiryQueue
00:36:21,993 INFO [org.hornetq.core.server] HQ221020:
    Started Netty Acceptor version 4.0.13.Final localhost:5455
00:36:21,996 INFO [org.hornetq.core.server] HQ221020:
    Started Netty Acceptor version 4.0.13.Final localhost:5445
00:36:21,997 INFO [org.hornetq.core.server] HQ221007: Server is now live
```

现在必须提供 Spring 配置才能连接到此服务器并访问前面配置的 prospring5 队列。通常，应该有两个配置类，一个用于消息发送器，另一个用于消息监听器，但因为使用配置类的Spring JMS配置非常实用，并且不需要很多bean，所以可以将所有配置放在一个类中，如下所示：

```
package com.apress.prospring5.ch12.config;
import org.hornetq.api.core.TransportConfiguration;
import org.hornetq.core.remoting.impl.netty.NettyConnectorFactory;
import org.hornetq.core.remoting.impl.netty.TransportConstants;
import org.hornetq.jms.client.HornetQJMSConnectionFactory;
import org.hornetq.jms.client.HornetQQueue;
import org.springframework.context.annotation.Bean;
import org.springframework.context.annotation.ComponentScan;
import org.springframework.context.annotation.Configuration;
import org.springframework.jms.annotation.EnableJms;
import org.springframework.jms.config.DefaultJmsListenerContainerFactory;
import org.springframework.jms.config.JmsListenerContainerFactory;
import org.springframework.jms.core.JmsTemplate;
import org.springframework.jms.listener.DefaultMessageListenerContainer;

import javax.jms.ConnectionFactory;
import java.util.HashMap;
import java.util.Map;
@Configuration
@EnableJms
@ComponentScan("com.apress.prospring5.ch12")
public class AppConfig {

    @Bean HornetQQueue prospring5() {
        return new HornetQQueue("prospring5");
    }

    @Bean ConnectionFactory connectionFactory() {
    Map<String, Object> connDetails = new HashMap<>();
```

```
        connDetails.put(TransportConstants.HOST_PROP_NAME, "127.0.0.1");
        connDetails.put(TransportConstants.PORT_PROP_NAME, "5445");
        TransportConfiguration transportConfiguration = new TransportConfiguration(
            NettyConnectorFactory.class.getName(), connDetails);
        return new HornetQJMSConnectionFactory(false, transportConfiguration);
    }

    @Bean
    public JmsListenerContainerFactory<DefaultMessageListenerContainer>
            jmsListenerContainerFactory() {
        DefaultJmsListenerContainerFactory factory =
            new DefaultJmsListenerContainerFactory();
        factory.setConnectionFactory(connectionFactory());
        factory.setConcurrency("3-5");
        return factory;
    }

    @Bean JmsTemplate jmsTemplate() {
        JmsTemplate jmsTemplate = new JmsTemplate(connectionFactory());
        jmsTemplate.setDefaultDestination(prospring5());
        return jmsTemplate;
    }
}
```

javax.jms.ConnectionFactory 接口实现由 HornetQ Java 库(HornetQJMSConnectionFactory 类)提供,用于创建与 JMS 提供程序的连接。然后，声明一个 JmsListenerContainerFactory 类型的 bean，它将创建使用普通 JMS 客户端 API 接收 JMS 消息的消息监听器容器。jmsTemplate bean 用于将 JMS 消息发送到 prospring5 队列。

为了接收 JMS 消息，必须声明消息监听器组件，并提供目标(即 prospring5 队列)和 JMS 容器工厂 jmsListenerContainerFactory。

12.4.1 在 Spring 中实现 JMS 监听器

在 Spring 4.1 之前，为了开发消息监听器，需要创建一个实现了 javax.jms.MessageListener 接口并实现 onMessage()方法的类。Spring 4.1 中添加了@JmsListener 注解。该注解用于 bean 方法，从而将它们标记为处于指定目的地(队列或主题)的 JMS 消息监听器的目标。以下代码片段描述了 SimpleMessageListener 类和 bean 声明：

```
package com.apress.prospring5.ch12;

import org.slf4j.Logger;
import org.slf4j.LoggerFactory;
import org.springframework.jms.annotation.JmsListener;
import org.springframework.stereotype.Component;

import javax.jms.JMSException;
import javax.jms.Message;
import javax.jms.TextMessage;

@Component("messageListener")
public class SimpleMessageListener{
    private static final Logger logger =
        LoggerFactory.getLogger(SimpleMessageListener.class);

    @JmsListener(destination = "prospring5", containerFactory =
        "jmsListenerContainerFactory")
    public void onMessage(Message message) {
        TextMessage textMessage = (TextMessage) message;

        try {
            logger.info(">>> Received: " + textMessage.getText());
        } catch (JMSException ex) {
            logger.error("JMS error", ex);
        }
    }
}
```

此处将方法命名为 onMessage，以使其目的更加明显。在 onMessage()方法(使用@JmsListener 进行注解)中，传入了 javax.jms.Message 接口的一个实例。在该方法中，消息被转换为 javax.jms.TextMessage 接口的实例，并且使用 TextMessage.getText()方法检索文本中的消息正文。有关可能的消息格式列表，请参阅当前的 JEE 联机文档。

可通过在配置类上使用@EnableJms 或等效的 XML 元素声明(<jms:annotation-driven />)来处理@JmsListener 注解。

现在看一下如何将消息发送到 propring5 队列。

12.4.2 在 Spring 中发送 JMS 消息

接下来看看如何在 Spring 中使用 JMS 发送消息。为此，将使用类型为 org.springframework.jms.core.JmsTemplate 的便捷 bean jmsTemplate。首先，开发 MessageSender 接口及其实现类 SimpleMessageSender。以下代码片段分别显示了接口和类：

```
//MessageSender.java
package com.apress.prospring5.ch12;

public interface MessageSender {
    void sendMessage(String message);
}

//SimpleMessageSender.java
package com.apress.prospring5.ch12;

import javax.jms.JMSException;
import javax.jms.Message;
import javax.jms.Session;
import javax.jms.TextMessage;

import org.slf4j.Logger;
import org.slf4j.LoggerFactory;
import org.springframework.beans.factory.annotation.Autowired;
import org.springframework.jms.core.JmsTemplate;
import org.springframework.jms.core.MessageCreator;
import org.springframework.stereotype.Component;

@Component("messageSender")
public class SimpleMessageSender implements MessageSender {
    private static final Logger logger =
            LoggerFactory.getLogger(SimpleMessageSender.class);
    @Autowired
    private JmsTemplate jmsTemplate;
    @Override

    public void sendMessage(final String message) {
        jmsTemplate.setDeliveryDelay(5000L);
        this.jmsTemplate.send(new MessageCreator() {
            @Override
            public Message createMessage(Session session)
                    throws JMSException {
                TextMessage jmsMessage = session.createTextMessage(message);
                logger.info(">>> Sending: " + jmsMessage.getText());
                return jmsMessage;
            }
        });
    }
}
```

正如你所看到的，注入了 JmsTemplate 的一个实例。在 sendMessage()方法中，调用 JmsTemplate.send()方法，并构建 org.springframework.jms.core.MessageCreator 接口的实例。而在 MessageCreator 实例中，实现了 createMessage()方法以创建一个新的将被发送到 HornetQ 的新 TextMessage 实例。

消息监听器和消息发送器的 bean 声明都是使用组件扫描获取的。

现在将发送和接收行为绑定在一起，从而看看 JMS 是如何工作的。以下代码片段显示了发送消息和接收消息的主要测试程序：

```
package com.apress.prospring5.ch12;

import com.apress.prospring5.ch12.config.AppConfig;
import org.springframework.context.annotation.AnnotationConfigApplicationContext;
import org.springframework.context.support.GenericApplicationContext;
import org.springframework.context.support.GenericXmlApplicationContext;

import java.util.Arrays;

public class JmsHornetQSample {
    public static void main(String... args) throws Exception{
        GenericApplicationContext ctx =
            new AnnotationConfigApplicationContext(AppConfig.class);
```

```
        MessageSender messageSender =
           ctx.getBean("messageSender", MessageSender.class);

        for(int i=0; i < 10; ++i) {
           messageSender.sendMessage("Test message: " + i);
        }

        System.in.read();
        ctx.close();
    }
}
```

该程序很简单。运行程序，将消息发送到队列。SimpleMessageListener 类接收这些消息，并且可以在控制台中看到以下输出：

```
INFO c.a.p.c.SimpleMessageSender - >>> Sending: Test message: 0
INFO c.a.p.c.SimpleMessageSender - >>> Sending: Test message: 1
INFO c.a.p.c.SimpleMessageSender - >>> Sending: Test message: 2
INFO c.a.p.c.SimpleMessageSender - >>> Sending: Test message: 3
INFO c.a.p.c.SimpleMessageSender - >>> Sending: Test message: 4
INFO c.a.p.c.SimpleMessageSender - >>> Sending: Test message: 5
INFO c.a.p.c.SimpleMessageSender - >>> Sending: Test message: 6
INFO c.a.p.c.SimpleMessageSender - >>> Sending: Test message: 7
INFO c.a.p.c.SimpleMessageSender - >>> Sending: Test message: 8
INFO c.a.p.c.SimpleMessageSender - >>> Sending: Test message: 9
INFO c.a.p.c.SimpleMessageListener - >>> Received: Test message: 0
INFO c.a.p.c.SimpleMessageListener - >>> Received: Test message: 1
INFO c.a.p.c.SimpleMessageListener - >>> Received: Test message: 2
INFO c.a.p.c.SimpleMessageListener - >>> Received: Test message: 3
INFO c.a.p.c.SimpleMessageListener - >>> Received: Test message: 4
INFO c.a.p.c.SimpleMessageListener - >>> Received: Test message: 5
INFO c.a.p.c.SimpleMessageListener - >>> Received: Test message: 6
INFO c.a.p.c.SimpleMessageListener - >>> Received: Test message: 7
INFO c.a.p.c.SimpleMessageListener - >>> Received: Test message: 8
INFO c.a.p.c.SimpleMessageListener - >>> Received: Test message: 9
```

12.5 Spring Boot Artemis 启动器

在第 9 章曾经讲过，使用 Spring Boot 可以开发更加实用的 JMS 应用程序，并且给出了涉及数据库和队列的分布式事务示例。

当 Spring Boot 检测到 Artemis 在类路径中可用时，可以自动配置 javax.jms.ConnectionFactory bean。嵌入式 JMS 代理是自动启动和配置的。Artemis 可以在多种模式中使用，并且可以使用 application.properties 文件中设置的特殊 Artemis 属性进行配置。

Artemis 能够以 native 模式使用，可以连接到由 Netty 协议提供的代理。application.properties 文件可以如下所示：

```
spring.artemis.mode=native
spring.artemis.host=0.0.0.0
spring.artemis.port=61617
spring.artemis.user=prospring5
spring.artemis.password=prospring5
```

使用 Spring Boot 和 Artemis 创建 JMS 应用程序的最简单方法是使用嵌入式服务器，所需要的只是用于保存消息的队列名称。因此，application.properties 文件如下所示：

```
spring.artemis.mode=embedded
spring.artemis.embedded.queues=prospring5
```

本节的源代码将使用上述方法，因为需要最少的自定义配置。此时，所需要的只是 application.properties 配置文件和 Application 类，如下所示：

```
package com.apress.prospring5.ch12;

import org.slf4j.Logger;
import org.slf4j.LoggerFactory;
import org.springframework.boot.SpringApplication;
import org.springframework.boot.autoconfigure.SpringBootApplication;
import org.springframework.boot.autoconfigure.
   jms.DefaultJmsListenerContainerFactoryConfigurer;
import org.springframework.context.ConfigurableApplicationContext;
import org.springframework.context.annotation.Bean;
import org.springframework.jms.annotation.JmsListener;
```

```
import org.springframework.jms.config.DefaultJmsListenerContainerFactory;
import org.springframework.jms.config.JmsListenerContainerFactory;
import org.springframework.jms.core.JmsTemplate;

import javax.jms.ConnectionFactory;
import javax.jms.JMSException;
import javax.jms.Message;
import javax.jms.TextMessage;

@SpringBootApplication
public class Application {

   private static Logger logger =
      LoggerFactory.getLogger(Application.class);
   @Bean
   public JmsListenerContainerFactory<?>
      connectionFactory(ConnectionFactory connectionFactory,
         DefaultJmsListenerContainerFactoryConfigurer configurer) {
      DefaultJmsListenerContainerFactory factory =
         new DefaultJmsListenerContainerFactory();
      configurer.configure(factory, connectionFactory);
      return factory;
   }

   public static void main(String... args) throws Exception {
      ConfigurableApplicationContext ctx =
         SpringApplication.run(Application.class, args);
      JmsTemplate jmsTemplate = ctx.getBean(JmsTemplate.class);
      jmsTemplate.setDeliveryDelay(5000L);
      for (int i = 0; i < 10; ++i) {
         logger.info(">>> Sending: Test message: " + i);
         jmsTemplate.convertAndSend("prospring5", "Test message: " + i);
      }
      System.in.read();
      ctx.close();
   }

   @JmsListener(destination = "prospring5", containerFactory = "connectionFactory")
   public void onMessage(Message message) {
      TextMessage textMessage = (TextMessage) message;

      try {
         logger.info(">>> Received: " + textMessage.getText());
      } catch (JMSException ex) {
         logger.error("JMS error", ex);
      }
   }
}
```

当然，为了让示例正常运行，还必须将 Spring Boot JMS 启动器库用作依赖项，并且 Artemis 服务器必须位于类路径中。Gradle 配置如下所示：

```
//pro-spring-15/build.gradle
ext {
   bootVersion = '2.0.0.M1'
   artemisVersion = '2.1.0'

   boot = [
         ...
         starterJms :
            "org.springframework.boot:spring-boot-starter-artemis:$bootVersion"
   ]

   testing = [
         junit: "junit:junit:$junitVersion"
   ]

   misc = [
         ...
         artemisServer :
            "org.apache.activemq:artemis-jms-server:$artemisVersion"
   ]
...
}
//boot-jms/build.gradle
buildscript {
   repositories {
      ...
```

```
    }

    dependencies {
        classpath boot.springBootPlugin
    }
}
apply plugin: 'org.springframework.boot'

dependencies {
    compile boot.starterJms, misc.artemisServer
}
```

将 spring-boot-starter-artemis 声明为依赖项，从而无须使用@EnableJms 来处理用@JmsListener 注解的方法。jmsTemplate bean 由 Spring Boot 创建，默认配置由 application.properties 文件中设置的属性提供，它不仅可以发送消息，还可以接收消息(使用 receive()方法)，但这是同步完成的，意味着 jmsTemplate 会被阻塞。这就是使用显式配置的 JmsListenerContainerFactory bean 来创建 DefaultMessageListenerContainer 的原因——能够以最高的连接效率异步地使用消息。

如果运行 Application 类，将在控制台中得到一些输出，这与 HornetQ 输出非常相似。

12.6 在 Spring 中使用 RESTful-WS

如今，RESTful-WS 可能是使用最广泛的远程访问技术。从借助 HTTP 的远程服务调用到支持 Ajax 样式的交互式 Web 前端，RESTful-WS 正在被广泛采用。RESTful Web 服务之所以受欢迎，主要有以下几点原因。

- **易于理解**：RESTful Web 服务是围绕 HTTP 设计的。URL 和 HTTP 方法一起指定了请求的意图。例如，HTTP 方法为 GET 的 URL http://somedomain.com/restful/customer/1 意味着客户端想要检索客户信息，其中客户 ID 等于 1。
- **轻量级**：与基于 SOAP 的 Web 服务相比，RESTful 更轻量级，它包含大量的元数据来描述客户端希望调用的服务。对于 RESTful 请求和响应，作为 HTTP 请求和响应与其他任何 Web 应用程序一样。
- **防火墙友好**：由于 RESTful Web 服务被设计为可通过 HTTP(或 HTTPS)访问，因此应用程序变得更加防火墙友好并且易于被远程客户端访问。

在本节中，将通过 Spring MVC 模块介绍 RESTful-WS 的基本概念以及 Spring 对 RESTful-WS 的支持。

12.6.1 RESTful Web 服务介绍

RESTful-WS 中的 REST 是 REpresentational State Transfer 的缩写，是一种架构样式。REST 定义了一组架构约束，它们共同描述了一个访问资源的统一接口。这个统一接口通过不同的表达形式来识别和操作资源。为了识别资源，应该通过统一资源标识符(URI)访问一条信息。例如，URL http://somedomain.com/api/singer/1 是一个表示资源的 URI，即标识符为 1 的歌手信息。如果标识符为 1 的歌手不存在，客户端将收到 404 HTTP 错误，就像网站上的“找不到页面”错误一样。另一个示例 URL http://somedomain.com/api/singers 是一个表示歌手信息列表资源的 URI。这些可识别的资源可以通过各种表示形式进行管理，如表 12-1 所示。

表 12-1 操作资源的表示形式

HTTP 方法	描述
GET	检索资源的表示形式
HEAD	与 GET 相同，没有响应主体。通常用于获取标题
POST	创建新的资源
PUT	更新资源
DELETE	删除资源
OPTIONS	检索允许的 HTTP 方法

有关 RESTful Web 服务的详细介绍，推荐阅读由 Christian Gross 编写的 *Ajax and REST Recipes:A Problem-Solution Approach* 一书(Apress 出版社于 2006 年出版)。

12.6.2 为示例添加必需的依赖项

要构建 Spring REST 应用程序，需要添加一些新的依赖项。因为是从服务器发送对象到客户端，所以需要一个库来对它们进行序列化和反序列化。此外，还可以在同一个应用程序中使用多种类型的序列化，本例中将会使用 XML 和 JSON，因此需要这两种序列化的库。表 12-2 列出了依赖项及其用途。

表 12-2 RESTful Web 服务的依赖项

GroupId:Moduled	版本	用途
org.springframework:spring-oxm	5.0.0.RC1	Spring 的对象到 XML 映射模块
org.codehaus.jackson:jacksonDatabind	2.9.0.pr3	Jackson JSON 处理器支持 JSON 格式的数据
org.codehaus.castor:castor-xml	1.4.1	Castor XML 库将用于 XML 数据的编组和解组
org.apache.httpcomponents:httpclient	4.5.3	Apache HTTP Components 项目。HTTP 客户端库将用于 RESTful-WS 调用

12.6.3 设计 Singer RESTful Web 服务

在开发 RESTful-WS 应用程序时，第一步是设计服务结构，其中包括支持哪些 HTTP 方法以及针对不同操作的目标 URL。对于 Singer RESTful Web 服务，希望支持查询、创建、更新和删除操作；而对于查询，希望支持检索所有歌手或通过 ID 检索单名歌手。

这些服务将以 Spring MVC 控制器的形式实现，名为 SingerController 类，位于 com.apress.prospring5.ch12 包中。表 12-3 显示了 URL 模式、HTTP 方法、描述以及相应的控制器方法。对于这些 URL，全部都与 http://localhost:8080 相关。就数据格式而言，支持 XML 和 JSON。相应的格式将根据客户端的 HTTP 请求标头的可接受媒体类型提供。

表 12-3 XMLHttpRequest 方法和属性表

URL	HTTP 方法	描述	控制器方法
/singer/listdata	GET	检索所有歌手信息	listData
/singer/id	GET	根据 ID 检索单名歌手信息	findBySingerId(...)
/singer	POST	创建一名新歌手	create(...)
/singer/id	PUT	根据 ID 更新歌手信息	update(...)
/singer/id	DELETE	根据 ID 删除一名歌手信息	delete(...)

12.6.4 使用 Spring MVC 展示 REST 样式的 Web 服务

在本节中，将演示如何使用 Spring MVC 将歌手服务公开为 RESTful Web 服务，如前一节中所述。该例基于 Spring HTTP 调用器示例中所使用的一些 SingerService 类。

你应该对 Singer 类已经非常熟悉了，所以不再显示相关代码。但是为了序列化和反序列化歌手列表，需要将该类封装到一个容器中[1]，这里是如下所示的 Singler 类。它有一个属性，该属性是一个 Singer 对象列表。目的是支持将歌手列表(由 SingerController 类中的 listData()方法返回)转换为 XML 或 JSON 格式。

```
package com.apress.prospring5.ch12;

import com.apress.prospring5.ch12.entities.Singer;
import java.io.Serializable;
import java.util.List;

public class Singers implements Serializable {
    private List<Singer> singers;

    public Singers() {
    }
```

1 这是 XML 序列化所必需的，但 JSON 可以在没有容器类的情况下工作。

```
    public Singers(List<Singer> singers) {
        this.singers = singers;
    }

    public List<Singer> getSingers() {
        return singers;
    }

    public void setSingers(List<Singer> singers) {
        this.singers = singers;
    }
}
```

12.7 配置 Castor XML

为了支持将返回的歌手信息转换为XML格式，需要使用Castor XML库(http://castor.codehaus.org)。Castor支持POJO和XML之间转换的多种模式，在本例中，将使用XML文件来定义映射。以下XML片段显示了映射文件(oxm-mapping.xml)：

```
<mapping>
    <class name="com.apress.prospring5.ch12.Singers">
        <field name="singers"
            type="com.apress.prospring5.ch12.entities.Singer"
                collection="arraylist">
            <bind-xml name="singer"/>
        </field>
    </class>

    <class name="com.apress.prospring5.ch12.entities.Singer"
        identity="id">
        <map-to xml="singer" />

        <field name="id" type="long">
            <bind-xml name="id" node="element"/>
        </field>
        <field name="firstName" type="string">
            <bind-xml name="firstName" node="element" />
        </field>
        <field name="lastName" type="string">
            <bind-xml name="lastName" node="element" />
        </field>
        <field name="birthDate" type="string" handler="dateHandler">
            <bind-xml name="birthDate" node="element" />
        </field>
        <field name="version" type="integer">
            <bind-xml name="version" node="element" />
        </field>
    </class>

    <field-handler name="dateHandler"
        class="com.apress.prospring5.ch12.DateTimeFieldHandler">
        <param name="date-format" value="yyyy-MM-dd"/>
    </field-handler>
</mapping>
```

此处定义了两个映射。第一个<class>标记映射了Singers类，在该类中使用<bind-xml name ="singer"/>标记映射其歌手属性(Singer对象的列表)。然后映射Singer对象(使用第二个<class>标记内的<map-to xml ="singer"/>标记)。另外，为了支持从java.util.Date类型开始的转换(用于Singer的birthDate属性)，还实现了一个自定义的Castor字段处理程序。以下代码片段显示了该字段处理程序：

```
package com.apress.prospring5.ch12;

import org.exolab.castor.mapping.GeneralizedFieldHandler;
import org.exolab.castor.mapping.ValidityException;
import org.slf4j.Logger;
import org.slf4j.LoggerFactory;

import java.text.ParseException;
import java.text.SimpleDateFormat;
import java.util.Date;
import java.util.Properties;

public class DateTimeFieldHandler extends GeneralizedFieldHandler {
```

```
    private static Logger logger =
        LoggerFactory.getLogger(DateTimeFieldHandler.class);

    private static String dateFormatPattern;

    @Override
    public void setConfiguration(Properties config) throws ValidityException {
        dateFormatPattern = config.getProperty("date-format");
    }

    @Override
    public Object convertUponGet(Object value) {
        Date dateTime = (Date) value;
        return format(dateTime);
    }

    @Override
    public Object convertUponSet(Object value) {
        String dateTimeString = (String) value;
        return parse(dateTimeString);
    }

    @Override
    public Class<Date> getFieldType() {
        return Date.class;
    }

    protected static String format(final Date dateTime) {
        String dateTimeString = "";
        if (dateTime != null) {
            SimpleDateFormat sdf =
                new SimpleDateFormat(dateFormatPattern);
            dateTimeString = sdf.format(dateTime);
        }
        return dateTimeString;
    }

    protected static Date parse(final String dateTimeString) {
        Date dateTime = new Date();
        if (dateTimeString != null) {
            SimpleDateFormat sdf =
                new SimpleDateFormat(dateFormatPattern);
            try {
                dateTime = sdf.parse(dateTimeString);
            } catch (ParseException e) {
                logger.error("Not a valida date:" + dateTimeString, e);
            }
        }
        return dateTime;
    }
}
```

此处扩展了 Castor 的 org.exolab.castor.mapping.GeneralizedFieldHandler 类并实现了 convertUponGet()、convertUponSet()和 getFieldType()方法。在这些方法中，实现了执行 Date 和 String 类型之间转换的逻辑，以供 Castor 使用。

另外，还定义了一个用于 Castor 的属性文件。以下代码片段显示了该文件的内容(castor.properties):

```
org.exolab.castor.indent=true
```

该属性指示 Castor 生成带有缩进的 XML，以便在测试时更容易阅读。

12.7.1 实现 SingerController

下一步是实现控制器类 SingerController。下面的代码片段显示了该类的内容，它实现了表 12-3 中的所有方法:

```
package com.apress.prospring5.ch12;

import com.apress.prospring5.ch12.entities.Singer;
import com.apress.prospring5.ch12.services.SingerService;
import org.slf4j.Logger;
import org.slf4j.LoggerFactory;
import org.springframework.beans.factory.annotation.Autowired;
import org.springframework.http.HttpStatus;
import org.springframework.stereotype.Controller;
```

```
import org.springframework.web.bind.annotation.*;

@Controller
@RequestMapping(value="/singer")
public class SingerController {
    final Logger logger =
        LoggerFactory.getLogger(SingerController.class);

    @Autowired
    private SingerService singerService;
    @ResponseStatus(HttpStatus.OK)
    @RequestMapping(value = "/listdata", method = RequestMethod.GET)
    @ResponseBody
    public Singers listData() {
        return new Singers(singerService.findAll());
    }

    @RequestMapping(value="/{id}", method=RequestMethod.GET)
    @ResponseBody
    public Singer findSingerById(@PathVariable Long id) {
        return singerService.findById(id);
    }

    @RequestMapping(value="/", method=RequestMethod.POST)
    @ResponseBody
    public Singer create(@RequestBody Singer singer) {
        logger.info("Creating singer: " + singer);
        singerService.save(singer);
        logger.info("Singer created successfully with info: " + singer);
        return singer;
    }

    @RequestMapping(value="/{id}", method=RequestMethod.PUT)
    @ResponseBody
    public void update(@RequestBody Singer singer,
                       @PathVariable Long id) {
        logger.info("Updating singer: " + singer);
        singerService.save(singer);
        logger.info("Singer updated successfully with info: " + singer);
    }

    @RequestMapping(value="/{id}", method=RequestMethod.DELETE)
    @ResponseBody
    public void delete(@PathVariable Long id) {
        logger.info("Deleting singer with id: " + id);
        Singer singer = singerService.findById(id);
        singerService.delete(singer);
        logger.info("Singer deleted successfully");
    }
}
```

上述类的要点如下：

- 使用@Controller 来注解类，这表明它是一个 Spring MVC 控制器。
- 类级注解@RequestMapping(value ="/singer")表明此控制器将被映射到主 Web 上下文中的所有 URL。在本示例中，http://localhost:8080/singer 下的所有 URL 都将由此控制器处理。
- 自动将本章前面实现的服务层中的 SingerService 装入控制器。
- 每个方法的@RequestMapping 注解指示 URL 模式及其将要被映射到的相应 HTTP 方法。例如，listData()方法将使用 HTTP GET 方法映射到 http://localhost:8080/singer/listdata URL；而对于 update()方法，则使用 HTTP PUT 方法映射到 URL http://localhost:8080/singer/protect/T1/textbraceleftid/protect/T1/textbraceright。
- @ResponseBody 注解适用于所有方法，指示来自方法的所有返回值应直接写入 HTTP 响应流，并不是与视图匹配。
- 对于接收路径变量的方法(例如，findSingerById()方法)来说，路径变量用@PathVariable 进行注解，指示 Spring MVC 将 URL 内的路径变量(例如，http://localhost:8080/singer/1)绑定到 findSingerById()方法的 id 参数。请注意，对于 id 参数，类型是 Long，而 Spring 的类型转换系统会自动处理从 String 到 Long 的转换。
- 对于 create()和 update()方法，使用@RequestBody 对 Singer 参数进行注解，指示 Spring 自动将 HTTP 请求体内的内容绑定到 Singer 域对象。该转换将通过 HttpMessageConverter<Object>接口(在 org.springframework.http.converter 包中)的声明实例来完成以获得支持格式，本章后面将对此进行讨论。

从 Spring 4.0 开始，可以引入专用于 REST 的控制器注解@RestController。这是一个非常便利的注解，它本身用@Controller 和@ResponseBody 进行了注解。当在控制器类上使用时，所有使用@RequestMapping 注解的方法都会使用@ResponseBody 进行自动注解。本章后面会介绍使用该注解编写的 SingerController 版本。

12.7.2 配置 Spring Web 应用程序

由于需要 Spring Web 应用程序来解析客户端发送的 REST 请求，因此需要对其进行配置。本章前面介绍了一个简单的 Web 应用程序配置。现在必须使用 XML 和 JSON 的 HTTP 消息转换器 bean 来丰富该配置。

Spring Web 应用程序遵循 Front Controller 设计模式[1]，其中所有请求都由单个控制器接收，然后分派给相应的处理程序(控制器类)。中央调度程序是 org.springframework.web.servlet.DispatcherServlet 的一个实例，由 AbstractAnnotationConfigDispatcherServletInitializer 类进行注册，需要对该类进行扩展以替换 web.xml 配置。在示例中，由 WebInitializer 类完成相关操作，如下所示：

```
package com.apress.prospring5.ch12.init;

import com.apress.prospring5.ch12.config.DataServiceConfig;
import org.springframework.web.servlet.support.
   AbstractAnnotationConfigDispatcherServletInitializer;
public class WebInitializer extends
   AbstractAnnotationConfigDispatcherServletInitializer {

   @Override
   protected Class<?>[] getRootConfigClasses() {
      return new Class<?>[]{
         DataServiceConfig.class
      };
   }

   @Override
   protected Class<?>[] getServletConfigClasses() {
      return new Class<?>[]{
         WebConfig.class
   };
   }

   @Override
   protected String[] getServletMappings() {
      return new String[]{"/"};
   }
}
```

在 Spring MVC 中，每个 DispatchServlet 都将拥有自己的 WebApplicationContext(但是，DataServiceConfig.class 中定义的所有服务层 bean(称为根 WebApplicationContext)也可用作每个 servlet 的 WebApplicationContext)。

getServletMappings()方法指示 Web 容器(例如 Tomcat)将模式/下的所有 URL(例如，http://localhost:8080/singer)都交由 RESTful servlet 进行处理。当然，也可以在这里添加一个上下文，例如/ch12，但是对于本节中的示例，希望 URL 尽可能简短和清晰。

下面显示了带有 HTTP 消息转换器的 Spring MVC 配置类(WebConfig 类)：

```
package com.apress.prospring5.ch12.init;

import com.fasterxml.jackson.annotation.JsonInclude;
import com.fasterxml.jackson.databind.ObjectMapper;
import com.fasterxml.jackson.databind.SerializationFeature;
import org.springframework.beans.factory.annotation.Autowired;
import org.springframework.context.ApplicationContext;
import org.springframework.context.annotation.Bean;
import org.springframework.context.annotation.ComponentScan;
import org.springframework.context.annotation.Configuration;
import org.springframework.core.io.ClassPathResource;
import org.springframework.http.MediaType;
import org.springframework.http.converter.HttpMessageConverter;
import org.springframework.http.converter.json.MappingJackson2HttpMessageConverter;
import org.springframework.http.converter.xml.MarshallingHttpMessageConverter;
import org.springframework.oxm.castor.CastorMarshaller;
```

1 可以从以下网址找到对该设计模式的很好的解释：www.oracle.com/technetwork/java/frontcontroller-135648.html。

```
import org.springframework.web.servlet.config.annotation.DefaultServletHandlerConfigurer;
import org.springframework.web.servlet.config.annotation.EnableWebMvc;
import org.springframework.web.servlet.config.annotation.WebMvcConfigurer;
import org.springframework.web.servlet.config.annotation.WebMvcConfigurerAdapter;

import java.text.DateFormat;
import java.text.SimpleDateFormat;
import java.util.ArrayList;
import java.util.List;

@Configuration
@EnableWebMvc
@ComponentScan(basePackages = {"com.apress.prospring5.ch12"})
public class WebConfig extends WebMvcConfigurer {

    @Autowired ApplicationContext ctx;

    @Bean
    public MappingJackson2HttpMessageConverter

            mappingJackson2HttpMessageConverter() {
        MappingJackson2HttpMessageConverter
            mappingJackson2HttpMessageConverter =
              new MappingJackson2HttpMessageConverter();
        mappingJackson2HttpMessageConverter.setObjectMapper(objectMapper());
        return mappingJackson2HttpMessageConverter;
    }

    @Override
    public void configureDefaultServletHandling(
        DefaultServletHandlerConfigurer configurer) {
       configurer.enable();
    }

    @Bean
    public ObjectMapper objectMapper() {
        ObjectMapper objMapper = new ObjectMapper();
        objMapper.enable(SerializationFeature.INDENT_OUTPUT);
        objMapper.setSerializationInclusion(JsonInclude.Include.NON_NULL);
        DateFormat df = new SimpleDateFormat("yyyy-MM-dd");
        objMapper.setDateFormat(df);
        return objMapper;
    }

    @Override
    public void configureMessageConverters(List<HttpMessageConverter<?>> converters) {
        converters.add(mappingJackson2HttpMessageConverter());
        converters.add(singerMessageConverter());
    }

    @Bean MarshallingHttpMessageConverter singerMessageConverter() {
        MarshallingHttpMessageConverter mc = new MarshallingHttpMessageConverter();
        mc.setMarshaller(castorMarshaller());
        mc.setUnmarshaller(castorMarshaller());
        List<MediaType> mediaTypes = new ArrayList<>();
        MediaType mt = new MediaType("application", "xml");
        mediaTypes.add(mt);
        mc.setSupportedMediaTypes(mediaTypes);
        return mc;
    }

    @Bean CastorMarshaller castorMarshaller() {
        CastorMarshaller castorMarshaller = new CastorMarshaller();
        castorMarshaller.setMappingLocation(
            ctx.getResource( "classpath:spring/oxm-mapping.xml"));
        return castorMarshaller;
    }
}
```

该类的要点如下：

- @EnableWebMvc 注解[1]启用了对 Spring MVC 的注解(即@Controller 注解)支持，并注册了 Spring 的类型转换和格式化系统。另外，根据此注解的定义，还启用了 JSR-303 验证支持。

1 这相当于<mvc：annotation-driven>标记/。

- configureMessageConverters(...)方法[1]声明了用作支持格式的媒体转换的 HttpMessageConverter 实例。因为支持 JSON 和 XML 作为数据格式，所以声明了两个转换器。一个是 MappingJackson2HttpMessageConverter，这是 Spring 对 Jackson JSON 库的支持[2]。另一个是 MarshallingHttpMessageConverter，它是由 spring-oxm 模块为 XML 编组/解组而提供的。在 MarshallingHttpMessageConverter 内部，需要定义所使用的编组器和解组器，它们由 Castor 提供。
- 对于 castorMarshaller bean，使用了 Spring 提供的 org.springframework.oxm.castor.CastorMarshaller 类，它与 Castor 集成，同时还提供 Castor 处理所需的映射位置。
- @ComponentScan 注解[3]指示 Spring 扫描控制器类的指定包。

现在，服务器端服务已完成。接下来应该构建包含 Web 应用程序的 WAR 文件，或者如果使用的是 IDE(比如，IntelliJ IDEA 或 STS)，那么请启动 Tomcat 实例。

12.7.3 使用 curl 测试 RESTful-WS

现在，快速测试一下所实现的 RESTful Web 服务。一种简单的方法是使用 curl[4]，它是一种用 URL 语法传输数据的命令行工具。要使用该工具，只需要从网站上下载并将其解压到计算机上即可。

例如，要测试对所有歌手的检索，请在 Windows 或 UNIX/Linux 的终端中打开命令提示符，将 WAR 部署到服务器上，然后输入以下命令：

```
$ curl -v -H "Accept: application/json" http://localhost:8080/singer/listdata
* Trying ::1...
* Connected to localhost (::1) port 8080 (#0)
> GET /singer/listdata HTTP/1.1
> Host: localhost:8080
> User-Agent: curl/7.43.0
> Accept: application/json
>
< HTTP/1.1 200
< Content-Type: application/json;charset=UTF-8
< Transfer-Encoding: chunked
< Date: Sat, 17 Jun 2017 17:16:43 GMT
<
{
   "singers" : [ {
      "id" : 1,
      "version" : 0,
      "firstName" : "John",
      "lastName" : "Mayer",
      "birthDate" : "1977-10-16"
   }, {
      "id" : 2,
      "version" : 0,
      "firstName" : "Eric",
      "lastName" : "Clapton",
      "birthDate" : "1945-03-30"
   }, {
      "id" : 3,
      "version" : 0,
      "firstName" : "John",
      "lastName" : "Butler",
      "birthDate" : "1975-04-01"
   } ]
* Connection #0 to host localhost left intact
```

该命令向服务器的 RESTful Web 服务发送 HTTP 请求；此时，它调用 SingerController 中的 listData()方法来检索并返回所有歌手信息。另外，-H 选项声明一个 HTTP 标头属性，这意味着客户端想要以 JSON 格式接收数据。运行该命令，将针对返回的最初填充的歌手信息生成 JSON 格式的输出。接下来看看 XML 格式。命令和结果如下所示：

```
$ curl -v -H "Accept: application/xml" http://localhost:8080/singer/listdata
* Trying ::1...
* Connected to localhost (::1) port 8080 (#0)
> GET /singer/listdata HTTP/1.1
```

1 这相当于 Spring 3.1 中引入的<mvc:message-converters>标记。
2 Jackson JSON 库的官方网站为 http://jackson.codehaus.org。
3 这相当于<context:component-scan>标记。
4 请参见 http://curl.haxx.se。

```
> Host: localhost:8080
> User-Agent: curl/7.43.0
> Accept: application/xml
>
< HTTP/1.1 200
< Content-Type: application/xml
< Transfer-Encoding: chunked
< Date: Sat, 17 Jun 2017 17:18:22 GMT
<
<?xml version="1.0" encoding="UTF-8"?>
<singers>
   <singer>
      <id>1</id>
      <firstName>John</firstName>
      <lastName>Mayer</lastName>
      <birthDate>1977-10-16</birthDate>
      <version>0</version>
   </singer>
   <singer>
      <id>2</id>
      <firstName>Eric</firstName>
      <lastName>Clapton</lastName>
      <birthDate>1945-03-30</birthDate>
      <version>0</version>
   </singer>
   <singer>
      <id>3</id>
      <firstName>John</firstName>
      <lastName>Butler</lastName>
      <birthDate>1975-04-01</birthDate>
      <version>0</version>
   </singer>
</singers>
* Connection #0 to host localhost left intact
```

如你所见，两个示例之间存在一个不同点。所接收的媒体信息已从 JSON 更改为 XML。运行该命令会生成 XML 输出。这是因为在 RESTful servlet 的 WebApplicationContext 中定义了 HttpMessageConverter bean，而 Spring MVC 将根据客户端 HTTP 标头信息的接收媒体信息调用相应的消息转换器，并相应地写入 HTTP 响应。

12.7.4 使用 RestTemplate 访问 RESTful-WS

对于基于 Spring 的应用程序，可以使用 RestTemplate 类访问 RESTful Web 服务。在本节中，将演示如何使用该类访问服务器上的歌手服务。首先看看 Spring 的 RestTemplate 的基本 ApplicationContext 配置，如以下代码片段所示：

```
package com.apress.prosring5.ch12;

import com.apress.prospring5.ch12.CustomCredentialsProvider;
import org.apache.http.auth.Credentials;
import org.apache.http.auth.UsernamePasswordCredentials;
import org.apache.http.client.CredentialsProvider;
import org.apache.http.client.HttpClient;
import org.apache.http.impl.client.HttpClientBuilder;
import org.springframework.beans.factory.annotation.Autowired;
import org.springframework.context.ApplicationContext;
import org.springframework.context.annotation.Bean;
import org.springframework.context.annotation.Configuration;
import org.springframework.core.io.ClassPathResource;
import org.springframework.http.MediaType;
import org.springframework.http.client.HttpComponentsClientHttpRequestFactory;
import org.springframework.http.converter.HttpMessageConverter;
import org.springframework.http.converter.xml.MarshallingHttpMessageConverter;
import org.springframework.oxm.castor.CastorMarshaller;
import org.springframework.web.client.RestTemplate;

import java.util.ArrayList;
import java.util.List;

@Configuration
public class RestClientConfig {

   @Autowired ApplicationContext ctx;

   @Bean
   public HttpComponentsClientHttpRequestFactory httpRequestFactory() {
      HttpComponentsClientHttpRequestFactory httpRequestFactory =
```

```
            new HttpComponentsClientHttpRequestFactory();
        HttpClient httpClient = HttpClientBuilder.create().build();
        httpRequestFactory.setHttpClient(httpClient);
        return httpRequestFactory;
    }

    @Bean
    public RestTemplate restTemplate() {
        RestTemplate restTemplate = new RestTemplate(httpRequestFactory());
        List<HttpMessageConverter<?>> mcvs = new ArrayList<>();
        mcvs.add(singerMessageConverter());
        restTemplate.setMessageConverters(mcvs);
        return restTemplate;
    }

    @Bean MarshallingHttpMessageConverter singerMessageConverter() {
        MarshallingHttpMessageConverter mc =
            new MarshallingHttpMessageConverter();
        mc.setMarshaller(castorMarshaller());
        mc.setUnmarshaller(castorMarshaller());
        List<MediaType> mediaTypes = new ArrayList<>();
        MediaType mt = new MediaType("application", "xml");
        mediaTypes.add(mt);
        mc.setSupportedMediaTypes(mediaTypes);
        return mc;
    }

    @Bean CastorMarshaller castorMarshaller() {
        CastorMarshaller castorMarshaller = new CastorMarshaller();
        castorMarshaller.setMappingLocation(
            ctx.getResource( "classpath:spring/oxm-mapping.xml"));
        return castorMarshaller;
    }
}
```

此处通过使用 RestTemplate 类声明了 restTemplate bean，而该类使用 Castor 将 MarshallingHttpMessageConverter 实例注入属性 messageConverters，这与在服务器端是相同的。映射文件在服务器和客户端之间共享。另外，对于 restTemplate bean，在匿名类 MarshallingHttpMessageConverter 中，SupportedMediaTypes 属性被注入一个 MediaType 实例，表明唯一支持的媒体是 XML。因此，客户端始终期望以 XML 作为返回数据格式，并且 Castor 将帮助完成 POJO 和 XML 之间的转换。

如果想要测试 Web 应用程序所支持的所有 REST URL，那么 JUnit 类更适合，在由 RestClientConfig 定义的 Spring 应用程序上下文中执行。代码如下所示，在 IntelliJ IDEA 或 STS 等智能编辑器中，可以单独执行每个方法：

```
package com.apress.prosring5.ch12.test;

import com.apress.prospring5.ch12.Singers;
import com.apress.prospring5.ch12.entities.Singer;
import com.apress.prosring5.ch12.RestClientConfig;
import org.junit.Before;
import org.junit.Test;
import org.junit.runner.RunWith;
import org.slf4j.Logger;
import org.slf4j.LoggerFactory;
import org.springframework.beans.factory.annotation.Autowired;
import org.springframework.test.context.ContextConfiguration;
import org.springframework.test.context.junit4.SpringJUnit4ClassRunner;
import org.springframework.test.context.junit4.SpringRunner;
import org.springframework.web.client.RestTemplate;

import java.util.Date;
import java.util.GregorianCalendar;
import static org.junit.Assert.assertFalse;
import static org.junit.Assert.assertNotNull;
import static org.junit.Assert.assertTrue;

@RunWith(SpringJUnit4ClassRunner.class)
@ContextConfiguration(classes = {RestClientConfig.class})
public class RestClientTest {

    final Logger logger = LoggerFactory.getLogger(RestClientTest.class);
    private static final String URL_GET_ALL_SINGERS =
        "http://localhost:8080/singer/listdata";
    private static final String URL_GET_SINGER_BY_ID =
        "http://localhost:8080/singer/{id}";
```

```
private static final String URL_CREATE_SINGER =
    "http://localhost:8080/singer/";
private static final String URL_UPDATE_SINGER =
    "http://localhost:8080/singer/{id}";
private static final String URL_DELETE_SINGER =
    "http://localhost:8080/singer/{id}";

@Autowired RestTemplate restTemplate;

@Before
public void setUp() {
    assertNotNull(restTemplate);
}

@Test
public void testFindAll() {
    logger.info("--> Testing retrieve all singers");
    Singers singers = restTemplate.getForObject(URL_GET_ALL_SINGERS,
        Singers.class);
    assertTrue(singers.getSingers().size() == 3);
listSingers(singers);
}

@Test
public void testFindbyId() {
    logger.info("--> Testing retrieve a singer by id : 1");
    Singer singer = restTemplate.getForObject(URL_GET_SINGER_BY_ID,
        Singer.class, 1);
    assertNotNull(singer);
    logger.info(singer.toString());
}

@Test
public void testUpdate() {
    logger.info("--> Testing update singer by id : 1");
    Singer singer = restTemplate.getForObject(URL_UPDATE_SINGER,
        Singer.class, 1);
    singer.setFirstName("John Clayton");
    restTemplate.put(URL_UPDATE_SINGER, singer, 1);
    logger.info("Singer update successfully: " + singer);
}

@Test
public void testDelete() {
    logger.info("--> Testing delete singer by id : 3");
    restTemplate.delete(URL_DELETE_SINGER, 3);
    Singers singers = restTemplate.getForObject(URL_GET_ALL_SINGERS,
        Singers.class);
    Boolean found = false;
    for(Singer s: singers.getSingers()) {
        if(s.getId() == 3) {
            found = true;
        }
    };
    assertFalse(found);
    listSingers(singers);
}

@Test
public void testCreate() {
    logger.info("--> Testing create singer");
    Singer singerNew = new Singer();
    singerNew.setFirstName("BB");
    singerNew.setLastName("King");
    singerNew.setBirthDate(new Date(
            (new GregorianCalendar(1940, 8, 16)).getTime().getTime()));
    singerNew = restTemplate.postForObject(URL_CREATE_SINGER,
        singerNew, Singer.class);
    logger.info("Singer created successfully: " + singerNew);

    logger.info("Singer created successfully: " + singerNew);

    Singers singers = restTemplate.getForObject(URL_GET_ALL_SINGERS,
        Singers.class);
    listSingers(singers);
}

private void listSingers(Singers singers) {
```

```
        singers.getSingers().forEach(s -> logger.info(s.toString()));
    }
}
```

上述代码首先声明了访问各种操作的URL，这些URL将在稍后的示例中使用。然后注入RestTemplate的实例，并在testFindAll()方法中使用RestTemplate。最后调用getForObject()方法(对应于HTTP GET方法)，并传入URL和期望的返回类型(即包含歌手完整列表的Singers类)。

确保应用程序服务器正在运行。运行testFindAll()测试方法，测试应该通过并产生以下输出：

```
INFO c.a.p.c.t.RestClientTest - --> Testing retrieve all singers
INFO c.a.p.c.t.RestClientTest - Singer - Id: 1, First name: John, Last name: Mayer,
   Birthday: Sun Oct 16 00:00:00 EET 1977
INFO c.a.p.c.t.RestClientTest - Singer - Id: 2, First name: Eric, Last name: Clapton,
   Birthday: Fri Mar 30 00:00:00 EET 1945
INFO c.a.p.c.t.RestClientTest - Singer - Id: 3, First name: John, Last name: Butler,
   Birthday: Tue Apr 01 00:00:00 EET 1975
```

如你所见，在RestTemplate中注册的MarshallingHttpMessageConverter bean会自动将消息转换为POJO。接下来，通过ID检索歌手。在此方法中，使用了RestTemplate.getForObject()方法的一个变体，并传入想要检索的歌手的ID作为URL中的路径变量(即URL_GET_CONTACT_BY_ID中的{id}路径变量)。如果URL有多个路径变量，可以使用Map <String，Object>实例或通过方法的可变参数支持来传入路径变量。如果使用可变参数，则需要遵循URL中声明的路径变量的顺序。运行测试方法testFindbyId()。测试应该通过，并且可以看到以下输出：

```
INFO c.a.p.c.t.RestClientTest - --> Testing retrieve a singer by id : 1
INFO c.a.p.c.t.RestClientTest - Singer - Id: 1, First name: John, Last name: Mayer,
   Birthday: Sun Oct 16 00:00:00 EET 1977
```

正如你所看到的，正确的歌手信息被检索到。接下来是数据操作。首先检索想要更新的歌手。在更新歌手对象之后，使用与HTTP PUT方法对应的RestTemplate.put()方法，并传入更新URL、更新的歌手对象以及要更新歌手的ID。运行testUpdate()会产生以下输出(其他输出已省略)：

```
INFO c.a.p.c.t.RestClientTest - --> Testing update singer by id : 1
INFO c.a.p.c.t.RestClientTest - Singer update successfully: Singer - Id: 1,
   First name: John Clayton,
   Last name: Mayer, Birthday: Sun Oct 16 00:00:00 EET 1977
```

紧接着是删除操作。首先调用RestTemplate.delete()方法，该方法对应于HTTP DELETE方法，并传入URL和ID。然后，检索所有歌手信息并再次显示，从而验证删除操作。运行testDelete()测试方法会产生以下输出(其他输出已省略)：

```
INFO c.a.p.c.t.RestClientTest - --> Testing delete singer by id : 3
INFO c.a.p.c.t.RestClientTest - Singer - Id: 1,
   First name: John Clayton,
   Last name: Mayer, Birthday: Sun Oct 16 00:00:00 EET 1977
INFO c.a.p.c.t.RestClientTest - Singer - Id: 2, First name: Eric,
   Last name: Clapton, Birthday: Fri Mar 30 00:00:00 EET 1945
```

正如你所看到的，ID为3的歌手将被删除。最后是插入操作。首先构造Singer对象的一个新实例。然后调用RestTemplate.postForObject()方法，该方法对应于HTTP POST方法，并传入URL、要创建的Singer对象以及类类型。再次运行该程序会产生以下输出：

```
INFO c.a.p.c.t.RestClientTest - --> Testing create singer
INFO c.a.p.c.t.RestClientTest - Singer created successfully: Singer - Id: 4,
   First name: BB, Last name: King, Birthday: Mon Sep 16 00:00:00 EET 1940
  //listing all singers
INFO c.a.p.c.t.RestClientTest - Singer - Id: 1,
  First name: John Clayton, Last name: Mayer,
  Birthday: Sun Oct 16 00:00:00 EET 1977
INFO c.a.p.c.t.RestClientTest - Singer - Id: 2, First name: Eric,
  Last name: Clapton, Birthday: Fri Mar 30 00:00:00 EET 1945
INFO c.a.p.c.t.RestClientTest - Singer - Id: 4, First name: BB,
  Last name: King, Birthday: Mon Sep 16 00:00:00 EET 1940
```

12.7.5 使用 Spring Security 来保护 RESTful-WS

任何远程服务都需要相关的安全措施来限制未经授权的用户访问服务并检索商业信息或完成其他操作。RESTful-WS也不例外。在本节中，将演示如何使用Spring Security项目来保护服务器上的RESTful-WS。在本例中，将使用Spring Security 5.0.0.M2(编写本书时的最新稳定版本)，它为RESTful-WS提供一些有用的支持。

使用 Spring Security 来保护 RESTful-WS 共分为三步。首先，需要在 Web 应用程序部署描述符(web.xml)中添加一个名为 springSecurityFilterChain 的安全过滤器，但由于此时没有使用 XML 来配置应用程序，因此将过滤器替换为扩展了 AbstractSecurityWebApplicationInitializer 的类。此类注册了 DelegatingFilterProxy，以便在其他注册的过滤器之前使用 springSecurityFilterChain。下面所示的类是空的，因为还没有对它进行任何定制：

```
package com.apress.prospring5.ch12.init;

import org.springframework.security.web.
   context.AbstractSecurityWebApplicationInitializer;

public class SecurityWebApplicationInitializer
   extends AbstractSecurityWebApplicationInitializer {
}
```

现在需要为安全性添加一个 Spring 配置类，用于声明谁可以访问应用程序以及允许执行的操作。但在本例中，事情原本没有那么复杂：但因为教学目的而使用了基于内存的身份验证，所以添加一个名为 prospring5 的用户，密码为 prospring5，角色为 REMOTE。

```
package com.apress.prospring5.ch12.init;

import org.slf4j.Logger;
import org.slf4j.LoggerFactory;
import org.springframework.context.annotation.Configuration;
import org.springframework.security.config.annotation.authentication.
   builders.AuthenticationManagerBuilder;
import org.springframework.security.config.annotation.web.builders.HttpSecurity;
import org.springframework.security.config.annotation.web.configuration.
   EnableWebSecurity;
import org.springframework.security.config.annotation.web.configuration.
   WebSecurityConfigurerAdapter;

@Configuration
@EnableWebSecurity
public class SecurityConfig extends WebSecurityConfigurerAdapter {

   private static Logger logger = LoggerFactory.getLogger(SecurityConfig.class);

   @Autowired
   protected void configureGlobal(AuthenticationManagerBuilder auth)
      throws Exception {
     try {
        auth.inMemoryAuthentication()
          .withUser("prospring5")
          .password("prospring5")
          .roles("REMOTE");
      } catch (Exception e) {
          logger.error("Could not configure authentication!", e);
      }
   }

   @Override
   protected void configure(HttpSecurity http) throws Exception {
      http
          .sessionManagement()
       .sessionCreationPolicy(SessionCreationPolicy.STATELESS)
          .and()
          .authorizeRequests()
          .antMatchers("/**").permitAll()
          .antMatchers("/rest/**").hasRole("REMOTE").anyRequest().authenticated()
          .and()
          .formLogin()
          .and()
          .httpBasic()
       .and()
          .csrf().disable();
   }
}
```

该类使用了@EnableWebSecurity 进行注解，从而在 Spring Web 应用程序中启用安全行为。在 configure(...)方法中，声明 URL/rest/**下的资源应该受到保护。sessionCreationPolicy()方法用来允许配置是否在身份验证时创建 HTTP 会话。由于使用的 RESTful-WS 是无状态的，因此将值设置为 SessionCreationPolicy.STATELESS，指示 Spring Security 不要为所有 RESTful 请求创建 HTTP 会话，从而有助于提高 RESTful 服务的性能。

接下来，在 antMatchers("/rest/**")中，设置只有分配了 REMOTE 角色的用户才能访问 RESTful 服务。httpBasic()方法指定 RESTful 服务仅支持 HTTP 基本身份验证。

configureGlobal(AuthenticationManagerBuilder auth)方法定义了身份验证信息。此处定义了一个简单的身份验证提供程序，其中使用硬编码的用户和密码(都设置为 remote)，并分配了 REMOTE 角色。而在企业环境中，很可能通过数据库或 LDAP 查找完成身份验证。

.formLogin()方法用来告诉 Spring 生成基本登录表单，用于测试应用程序是否被正确保护。登录表单可通过 http://localhost:8080/login 访问。

过滤器 springSecurityFilterChain 能够让 Spring Security 拦截 HTTP 请求并进行身份验证和授权检查。因为只想保护 RESTful-WS，所以过滤器仅适用于 URL 模式/rest/*(请参阅 antMatchers(...)方法)。我们希望保护所有 REST URL，但允许用户查看应用程序的主页面(在浏览器中访问 http://localhost:8080 /时显示的简单 HTML 文件)，所以除了将 SecurityConfig 添加到根上下文应用程序中之外，还要添加 rest 应用程序上下文。

```
package com.apress.prospring5.ch12.init;

import com.apress.prospring5.ch12.config.DataServiceConfig;
import org.springframework.web.servlet.support.
   AbstractAnnotationConfigDispatcherServletInitializer;

public class WebInitializer extends
   AbstractAnnotationConfigDispatcherServletInitializer {

   @Override
   protected Class<?>[] getRootConfigClasses() {
      return new Class<?>[]{
         DataServiceConfig.class, SecurityConfig.class
      };
   }

   @Override
   protected Class<?>[] getServletConfigClasses() {
      return new Class<?>[]{
         WebConfig.class
      };
   }

   @Override
   protected String[] getServletMappings() {
      return new String[]{"/rest/**"};
   }
}
```

现在安全设置已经完成。如果重新部署项目并在 RestClientTest 下运行任何测试方法，则会有以下输出(其他输出已省略)：

```
Exception in thread "main" org.springframework.web.cient.HttpClientErrorException:
   401 Unauthorized
```

此时获得的 HTTP 状态码为 401，这意味着无权访问该服务。现在，配置客户端的 RestTemplate 以向服务器提供凭证信息。

```
package com.apress.prosring5.ch12;

import org.apache.http.HttpHost;
import org.apache.http.auth.AuthScope;
import org.apache.http.auth.Credentials;
import org.apache.http.auth.UsernamePasswordCredentials;
import org.apache.http.client.CredentialsProvider;
import org.apache.http.impl.client.BasicCredentialsProvider;
import org.apache.http.impl.client.CloseableHttpClient;
import org.apache.http.impl.client.HttpClientBuilder;
import org.apache.http.impl.client.HttpClients;
import org.springframework.beans.factory.annotation.Autowired;
import org.springframework.context.ApplicationContext;
import org.springframework.context.annotation.Bean;
import org.springframework.context.annotation.Configuration;
import org.springframework.http.MediaType;
import org.springframework.http.client.HttpComponentsClientHttpRequestFactory;
import org.springframework.http.converter.HttpMessageConverter;
import org.springframework.http.converter.xml.MarshallingHttpMessageConverter;
import org.springframework.oxm.castor.CastorMarshaller;
```

```
import org.springframework.web.client.RestTemplate;

import java.util.ArrayList;
import java.util.List;

@Configuration
public class RestClientConfig {

    @Autowired ApplicationContext ctx;

    @Bean Credentials credentials(){
        return new UsernamePasswordCredentials("prospring5", "prospring5");
    }

    @Bean
    CredentialsProvider provider() {
        BasicCredentialsProvider provider =
            new BasicCredentialsProvider();
        provider.setCredentials( AuthScope.ANY, credentials());
        return provider;
    }

    @Bean
    public HttpComponentsClientHttpRequestFactory httpRequestFactory() {
        CloseableHttpClient client = HttpClients.custom()
            .setDefaultCredentialsProvider(provider()).build();
        return new HttpComponentsClientHttpRequestFactory(client);
    }

    @Bean
    public RestTemplate restTemplate() {
        RestTemplate restTemplate = new RestTemplate();
        restTemplate.setRequestFactory(httpRequestFactory());
        List<HttpMessageConverter<?>> mcvs = new ArrayList<>();
        mcvs.add(singerMessageConverter());
        restTemplate.setMessageConverters(mcvs);
        return restTemplate;
    }

    @Bean MarshallingHttpMessageConverter singerMessageConverter() {
        MarshallingHttpMessageConverter mc = n
            ew MarshallingHttpMessageConverter();
        mc.setMarshaller(castorMarshaller());
        mc.setUnmarshaller(castorMarshaller());
        List<MediaType> mediaTypes = new ArrayList<>();
        MediaType mt = new MediaType("application", "xml");
        mediaTypes.add(mt);
        mc.setSupportedMediaTypes(mediaTypes);
        return mc;
    }

    @Bean CastorMarshaller castorMarshaller() {
        CastorMarshaller castorMarshaller = new CastorMarshaller();
        castorMarshaller.setMappingLocation(ctx.getResource(
            "classpath:spring/oxm-mapping.xml"));
        return castorMarshaller;
    }
}
```

在 restTemplate bean 中，注入一个引用 httpRequestFactory bean 的构造函数参数。对于 httpRequestFactory bean，则使用了 HttpComponentsClientHttpRequestFactory 类，这是 Spring 对 Apache HttpComponents HttpClient 库提供的支持，需要使用该库构造一个用来存储客户端证书的 CloseableHttpClient 实例。为了支持凭证的注入，可以创建一个类型为 UsernamePasswordCredentials 的简单 bean。UsernamePasswordCredentials 类由用户名 prospring5 和密码 prospring5 构造而成。通过使用构造并注入到 RestTemplate 中的 httpRequestFactory，所有使用该模板触发的 RESTful 请求都将携带所提供的凭证。现在可以在 RestClientTest 类中再次运行测试方法，你会看到服务像往常一样被调用。

12.8 使用 Spring Boot 开发 RESTful-WS

前面曾讲过，由于 Spring Boot 可以让开发变得更容易，因此单独用一节内容来介绍 Spring Boot 如何使 Spring RESTful 服务的开发变得更容易。Singer 实体、存储库和服务类与前面的相同，无须改变任何东西。为了让开发变得简单并尽可能多地使用默认的 Spring Boot 默认配置，XML 序列化也将被删除。JSON 序列化被默认支持。由于该应用程序是一个 Web 应用程序，因此配置与前面介绍的 Spring Boot Web 应用程序的配置相同，在此不再重复。Spring Boot 应用程序的 Application 类和入口点如下所示：

```
package com.apress.prospring5.ch12;

import org.slf4j.Logger;
import org.slf4j.LoggerFactory;
import org.springframework.boot.SpringApplication;
import org.springframework.boot.autoconfigure.SpringBootApplication;
import org.springframework.context.ConfigurableApplicationContext;

import java.io.IOException;

@SpringBootApplication(scanBasePackages = "com.apress.prospring5.ch12")
public class Application {
   private static Logger logger = LoggerFactory.getLogger(Application.class);

   public static void main(String args) throws IOException {
      ConfigurableApplicationContext ctx =
         SpringApplication.run(Application.class, args);
      assert (ctx != null);
      logger.info("Application Started ...");

      System.in.read();
      ctx.close();
   }
}
```

如之前所述，新改进的 SingerController 使用@RestController 和 Spring 4.3 中引入的特定于 HTTP 方法的映射注解进行了重写：

```
package com.apress.prospring5.ch12.controller;

import com.apress.prospring5.ch12.entities.Singer;
import com.apress.prospring5.ch12.services.SingerService;
import org.slf4j.Logger;
import org.slf4j.LoggerFactory;
import org.springframework.beans.factory.annotation.Autowired;
import org.springframework.http.HttpStatus;
import org.springframework.web.bind.annotation.*;

import java.util.List;
@RestController
@RequestMapping(value = "/singer")
public class SingerController {

   final Logger logger =
      LoggerFactory.getLogger(SingerController.class);

   @Autowired
   private SingerService singerService;

   @ResponseStatus(HttpStatus.OK)
   @GetMapping(value = "/listdata")
   public List<Singer> listData() {
      return singerService.findAll();
   }

   @ResponseStatus(HttpStatus.OK)
   @GetMapping(value = "/{id}")
   public Singer findSingerById(@PathVariable Long id) {
      return singerService.findById(id);
   }

   @ResponseStatus(HttpStatus.CREATED)
   @PostMapping(value="/")
   public Singer create(@RequestBody Singer singer) {
      logger.info("Creating singer: " + singer);
      singerService.save(singer);
      logger.info("Singer created successfully with info: " + singer);
      return singer;
   }

   @ResponseStatus(HttpStatus.OK)
```

```
    @PutMapping(value="/{id}")
    public void update(@RequestBody Singer singer,
          @PathVariable Long id) {
       logger.info("Updating singer: " + singer);
       singerService.save(singer);
       logger.info("Singer updated successfully with info: " + singer);
    }

    @ResponseStatus(HttpStatus.NO_CONTENT)
    @DeleteMapping(value="/{id}")
    public void delete(@PathVariable Long id) {
       logger.info("Deleting singer with id: " + id);
       Singer singer = singerService.findById(id);
       singerService.delete(singer);
       logger.info("Singer deleted successfully");
    }
}
```

Spring4.3 引入了一些与基本 HTTP 方法相匹配的@RequestMapping 注解的自定义形式。表 12-4 列出了新注解和旧式@RequestMapping 注解之间的等价关系。

表 12-4 Spring 4.3 引入的用于将 HTTP 方法请求映射到特定处理程序方法的注释

注解	旧式等价注解
@GetMapping	@RequestMapping(method = RequestMethod.GET)
@PostMapping	@RequestMapping(method = RequestMethod.POST)
@PutMapping	@RequestMapping(method = RequestMethod.PUT)
@DeleteMapping	@RequestMapping(method = RequestMethod.DELETE)

另外，因为使用了支持列表和数组的 JSON，所以不再需要类 Singers。

测试应用程序很简单，因为 RestTemplate 不需要任何配置。所需要做的只是通过调用默认构造函数来创建 RestTemplate 实例。测试方法与以前的相同。

```
package com.apress.prosring5.ch12.test;

import com.apress.prospring5.ch12.entities.Singer;
import org.junit.Before;
import org.junit.Test;
import org.slf4j.Logger;
import org.slf4j.LoggerFactory;
import org.springframework.web.client.RestTemplate;

import java.util.Arrays;
import java.util.Date;
import java.util.GregorianCalendar;

import static org.junit.Assert.*;

public class RestClientTest {

   final Logger logger =
      LoggerFactory.getLogger(RestClientTest.class);

   private static final String URL_GET_ALL_SINGERS =
      "http://localhost:8080/singer/listdata";
   ...
   RestTemplate restTemplate;

   @Before
   public void setUp() {
      restTemplate = new RestTemplate();
   }

   @Test
   public void testFindAll() {
      logger.info("--> Testing retrieve all singers");
      Singer singers = restTemplate.getForObject(
         URL_GET_ALL_SINGERS, Singer.class);
      assertTrue(singers.length == 3);
```

```
            listSingers(singers);
        }
        ...
    }
```

只需要运行 Application 类，然后逐个执行测试方法。

如果想确保应用程序真正工作，并且使用 JSON 格式序列化 Singer 实例，可以使用 curl 来测试这个 Spring Boot 应用程序。

```
curl -v -H "Accept: application/json" http://localhost:8080/singer/listdata
*   Trying ::1...
* Connected to localhost (::1) port 8080 (#0)
> GET /singer/listdata HTTP/1.1
> Host: localhost:8080
> User-Agent: curl/7.43.0
> Accept: application/json
>
< HTTP/1.1 200
< Content-Type: application/json;charset=UTF-8
< Transfer-Encoding: chunked
< Date: Sun, 18 Jun 2017 11:14:17 GMT
<
* Connection #0 to host localhost left intact
[{"id":1,"version":1,"firstName":"John Clayton","lastName":"Mayer",
"birthDate":245797200000},{"id":2,"version":0,"firstName":"Eric",
"lastName":"Clapton","birthDate":-781326000000},{"id":4, "version":0,"firstName":"BB","last
Name":"King",
"birthDate":-924404400000}]
```

如果对上面的输出感到困惑，那么请记住，在没有声明显式 JSON 消息转换器的情况下，Date 字段将显示为数字，并且响应也不会被格式化。

通过使用 Spring Boot，还可以非常容易地获取资源，这是第 16 章将要详细介绍的主题。

12.9 在 Spring 中使用 AMQP

远程处理也可以通过使用以高级消息队列协议(AMQP)作为传输协议的远程过程调用(RPC)通信来完成。AMQP 是实现面向消息中间件(MOM)的开放标准协议。

JMS 应用程序适用于任何操作系统环境，但仅支持 Java 平台。所以，所有通信应用程序都必须用 Java 开发。AMQP 标准可用于开发可轻松通信的多种语言的应用程序。

与使用 JMS 相似，AMQP 也使用消息代理来交换消息。在本例中，将使用 RabbitMQ[1]作为 AMQP 服务器。Spring 本身并没有在核心框架中提供远程处理功能。相反，而是由一个名为 Spring AMQP[2]的姊妹项目来处理，我们将其用作底层通信 API。Spring AMQP 项目提供了关于 AMQP 的基本抽象以及与 RabbitMQ 进行通信的实现。在本章中，并不会涵盖 AMQP 或 Spring AMQP 的所有功能，而只是介绍通过 RPC 通信提供的远程处理功能。

Spring AMQP 项目由两部分组成：spring-amqp 是基本抽象，springrabbit 是 RabbitMQ 实现。在编写本书时，Spring AMQP 的稳定版本是 2.0.0.M4。

首先，需要从 www.rabbitmq.com/download.html 获取 RabbitMQ 并启动服务器。RabbitMQ 可以满足我们的需求，并且不需要更改配置。一旦 RabbitMQ 运行，需要做的下一件事就是创建一个服务接口。在本例中，将创建一个简单的天气服务，返回根据所提供的状态码做出的天气预测。首先开始创建 WeatherService 接口，如下所示：

```
package com.apress.prospring5.ch12;

public interface WeatherService {
    String getForecast(String stateCode);
}
```

接下来，创建一个 WeatherService 实现，它将简单地回复所提供状态的天气预报，或者在没有任何可用天气预报的情况下返回不可用消息，如下所示：

```
package com.apress.prospring5.ch12;
import org.springframework.stereotype.Component;
```

1 参见 www.rabbitmq.org。

2 参见 http://projects.spring.io/spring-amqp。

```
@Component
public class WeatherServiceImpl implements WeatherService {
    @Override
    public String getForecast(String stateCode) {
        if ("FL".equals(stateCode)) {
            return "Hot";
        } else if ("MA".equals(stateCode)) {
            return "Cold";
        }

        return "Not available at this time";
    }
}
```

在确定所使用的天气服务代码之后，构建配置文件(amqp-rpc-app-context.xml)，该文件将配置 AMQP 连接并公开 WeatherService，如下所示：

```
<?xml version="1.0" encoding="UTF-8"?>
<beans xmlns="http://www.springframework.org/schema/beans"
       xmlns:xsi="http://www.w3.org/2001/XMLSchema-instance"
       xmlns:rabbit="http://www.springframework.org/schema/rabbit"
       xsi:schemaLocation="http://www.springframework.org/schema/beans
           http://www.springframework.org/schema/beans/spring-beans.xsd
           http://www.springframework.org/schema/rabbit
           http://www.springframework.org/schema/rabbit/spring-rabbit.xsd">
    <rabbit:connection-factory id="connectionFactory" host="localhost" />

    <rabbit:template id="amqpTemplate" connection-factory="connectionFactory"
                     reply-timeout="2000" routing-key="forecasts"
                     exchange="weather" />

    <rabbit:admin connection-factory="connectionFactory" />

    <rabbit:queue name="forecasts" />

    <rabbit:direct-exchange name="weather">
        <rabbit:bindings>
            <rabbit:binding queue="forecasts" key="forecasts" />
        </rabbit:bindings>
    </rabbit:direct-exchange>

    <bean id="weatherServiceProxy"
          class="org.springframework.amqp.remoting.client.AmqpProxyFactoryBean">
        <property name="amqpTemplate" ref="amqpTemplate" />
        <property name="serviceInterface"
              value="com.apress.prospring5.ch12.WeatherService" />
    </bean>

    <rabbit:listener-container connection-factory="connectionFactory">
        <rabbit:listener ref="weatherServiceExporter" queue-names="forecasts" />
    </rabbit:listener-container>
    <bean id="weatherServiceExporter"
          class="org.springframework.amqp.remoting.service.AmqpInvokerServiceExporter">
        <property name="amqpTemplate" ref="amqpTemplate" />
        <property name="serviceInterface"
          value="com.apress.prospring5.ch12.WeatherService" />
        <property name="service">
          <bean class="com.apress.prospring5.ch12.WeatherServiceImpl"/>
        </property>
    </bean>
</beans>
```

配置 RabbitMQ 连接以及交换和队列信息。然后，使用 AmqpProxyFactoryBean 类创建一个 bean，客户端将使用该类作为代理来发出 RPC 请求。对于响应，则使用 AmqpInvokerServiceExporter 类，它被连接到一个监听器容器。监听器容器负责监听 AMQP 消息并将它们交给天气服务。如你所见，连接、队列、监听器容器等方面的配置与 JMS 相似。虽然配置相似，但 JMS 和 AMQP 是两种完全不同的传输协议，建议访问 AMQP 网站[1]以获取协议的完整详细信息。

完成配置之后，创建一个示例类来执行 RPC 调用。

```
package com.apress.prospring5.ch12;

import org.slf4j.Logger;
```

1 请访问 AMQP 网站 www.amqp.org。

```
import org.slf4j.LoggerFactory;
import org.springframework.context.support.GenericXmlApplicationContext;

public class AmqpRpcDemo {
   private static Logger logger = LoggerFactory.getLogger(AmqpRpcDemo.class);
   public static void main(String... args) {
      GenericXmlApplicationContext ctx = new GenericXmlApplicationContext();
      ctx.load("classpath:spring/amqp-rpc-app-context.xml");
      ctx.refresh();

      WeatherService weatherService = ctx.getBean(WeatherService.class);
      logger.info("Forecast for FL: " + weatherService.getForecast("FL"));
      logger.info("Forecast for MA: " + weatherService.getForecast("MA"));
      logger.info("Forecast for CA: " + weatherService.getForecast("CA"));

      ctx.close();
   }
}
```

现在运行示例，应该得到以下输出：

```
INFO c.a.p.c.AmqpRpcDemo - Forecast for FL: Hot
INFO c.a.p.c.AmqpRpcDemo - Forecast for MA: Cold
INFO c.a.p.c.AmqpRpcDemo - Forecast for CA: Not available at this time
```

当然，XML 配置可以很容易地转换成 Java 配置类。但需要对其他涉及的类进行一些更改。WeatherServiceImpl 不再需要实现任何接口，因为它只会声明一个监听器方法，该方法将监听写入 forecasts 队列的消息。

```
package com.apress.prospring5.ch12;

import org.slf4j.Logger;
import org.slf4j.LoggerFactory;
import org.springframework.amqp.rabbit.annotation.RabbitListener;
import org.springframework.stereotype.Service;

@Service
public class WeatherServiceImpl {

   private static Logger logger =
      LoggerFactory.getLogger(WeatherServiceImpl.class);

   @RabbitListener(containerFactory="rabbitListenerContainerFactory",
      queues="forecasts")
   public void getForecast(String stateCode) {
      if ("FL".equals(stateCode)) {
         logger.info("Hot");
      } else if ("MA".equals(stateCode)) {
         logger.info("Cold");
      } else {
    logger.info("Not available at this time");
      }
   }
}
```

rabbitListenerContainerFactory bean 的类型为 RabbitListenerContainerFactory，用于创建常规的 SimpleMessageListenerContainer。接下来看看完整的Java配置。

```
package com.apress.prospring5.ch12.config;

import com.apress.prospring5.ch12.WeatherService;
import com.apress.prospring5.ch12.WeatherServiceImpl;
import org.springframework.amqp.core.*;
import org.springframework.amqp.rabbit.annotation.EnableRabbit;
import org.springframework.amqp.rabbit.config.SimpleRabbitListenerContainerFactory;
import org.springframework.amqp.rabbit.connection.CachingConnectionFactory;
import org.springframework.amqp.rabbit.core.RabbitAdmin;
import org.springframework.amqp.rabbit.core.RabbitTemplate;
import org.springframework.amqp.rabbit.listener.SimpleMessageListenerContainer;
import org.springframework.amqp.remoting.client.AmqpProxyFactoryBean;
import org.springframework.amqp.remoting.service.AmqpInvokerServiceExporter;
import org.springframework.beans.factory.annotation.Autowired;
import org.springframework.context.annotation.Bean;
import org.springframework.context.annotation.ComponentScan;
import org.springframework.context.annotation.Configuration;

@Configuration
@ComponentScan("com.apress.prospring5.ch12")
@EnableRabbit
```

```
public class RabbitMQConfig {

    final static String queueName = "forecasts";
    final static String exchangeName = "weather";

    @Bean CachingConnectionFactory connectionFactory() {
        return new CachingConnectionFactory("127.0.0.1");
    }

    @Bean RabbitTemplate amqpTemplate() {
        RabbitTemplate rabbitTemplate = new RabbitTemplate();
        rabbitTemplate.setConnectionFactory(connectionFactory());
        rabbitTemplate.setReplyTimeout(2000); rabbitTemplate.setRoutingKey(queueName);
        rabbitTemplate.setExchange(exchangeName);
        return rabbitTemplate;
    }

    @Bean Queue forecasts() {
        return new Queue(queueName, true);
    }

    @Bean Binding dataBinding(DirectExchange directExchange, Queue queue) {
        return BindingBuilder.bind(queue).to(directExchange).with(queueName);
    }

    @Bean RabbitAdmin admin() {
        RabbitAdmin admin = new RabbitAdmin(connectionFactory());
        admin.declareQueue(forecasts());
        admin.declareBinding(dataBinding(weather(), forecasts()));
        return admin;
    }

    @Bean DirectExchange weather() {
        return new DirectExchange(exchangeName, true, false);
    }

    @Bean
    public SimpleRabbitListenerContainerFactory
            rabbitListenerContainerFactory() {
        SimpleRabbitListenerContainerFactory factory =
                new SimpleRabbitListenerContainerFactory();
        factory.setConnectionFactory(connectionFactory());
        factory.setMaxConcurrentConsumers(5);
        return factory;
    }
}
```

上述配置中的所有 bean 都可以与它们各自的 XML 对应项相匹配。新元素是@EnableRabbit 注解。当在使用@Configuration 注解的类上使用该元素时，将启用由 RabbitListenerContainerFactory bean 在幕后创建的端点注解的 Rabbit 监听器。

要测试新的天气服务，还必须修改测试程序，并且使用 amqpTemplate 将消息发送到 foreecasts 队列，其中 WeatherServiceImpl.getForecast(...)将读取消息并打印天气预报。

```
package com.apress.prospring5.ch12;

import com.apress.prospring5.ch12.config.RabbitMQConfig;
import org.springframework.amqp.rabbit.core.RabbitTemplate;
import org.springframework.context.annotation.AnnotationConfigApplicationContext;
import org.springframework.context.support.GenericApplicationContext;

public class AmqpRpcDemo {

    public static void main(String... args) throws Exception {
        GenericApplicationContext ctx =
            new AnnotationConfigApplicationContext(RabbitMQConfig.class);
        RabbitTemplate rabbitTemplate = ctx.getBean(RabbitTemplate.class);
        rabbitTemplate.convertAndSend("FL");
        rabbitTemplate.convertAndSend("MA");
        rabbitTemplate.convertAndSend("CA");

        System.in.read();
        ctx.close();
    }
}
```

如果运行程序并启动 RabbitMQ 服务器，将看到以下输出：

```
[SimpleAsyncTaskExecutor-1] INFO c.a.p.c.WeatherServiceImpl - Hot
[SimpleAsyncTaskExecutor-1] INFO c.a.p.c.WeatherServiceImpl - Cold
[SimpleAsyncTaskExecutor-1] INFO c.a.p.c.WeatherServiceImpl - Not available at this time
```

在 Spring Boot 中使用 AMQP

Spring Boot 还可以帮助开发 AMQP 应用程序，它的启动器是 spring-boot-starter-amqp。配置简化了很多，不再需要定义 RabbitTemplate、RabbitAdmin 和 SimpleRabbitListenerContainerFactory bean，因为这些 bean 是由 Spring Boot 自动配置和创建的。WeatherServiceImpl 的实现没有多大变化，但由于 SimpleRabbitListenerContainerFactory bean 是由 Spring Boot 处理的，因此不需要将其添加为@RabbitListener 注解的值。

```
package com.apress.prospring5.ch12;

import org.slf4j.Logger;
import org.slf4j.LoggerFactory;
import org.springframework.amqp.rabbit.annotation.RabbitListener;
import org.springframework.stereotype.Service;

@Service
public class WeatherServiceImpl {

   private static Logger logger =
      LoggerFactory.getLogger(WeatherServiceImpl.class);

   @RabbitListener(queues="forecasts")
   public void getForecast(String stateCode) {
      if ("FL".equals(stateCode)) {
         logger.info("Hot");
      } else if ("MA".equals(stateCode)) {
         logger.info("Cold");
      } else {
         logger.info("Not available at this time");
      }
   }
}
```

使用@SpringBootApplication 注解的 Application 类也被用作配置类和运行器类。

```
package com.apress.prospring5.ch12;

import org.springframework.amqp.core.Binding;
import org.springframework.amqp.core.BindingBuilder;
import org.springframework.amqp.core.DirectExchange;
import org.springframework.amqp.core.Queue;
import org.springframework.amqp.rabbit.connection.CachingConnectionFactory;
import org.springframework.amqp.rabbit.core.RabbitTemplate;
import org.springframework.amqp.rabbit.listener.SimpleMessageListenerContainer;
import org.springframework.boot.SpringApplication;
import org.springframework.boot.autoconfigure.SpringBootApplication;
import org.springframework.context.ConfigurableApplicationContext;
import org.springframework.context.annotation.Bean;

@SpringBootApplication
public class Application {
   final static String queueName = "forecasts";
   final static String exchangeName = "weather";

   @Bean Queue forecasts() {
      return new Queue(queueName, true);
   }

   @Bean DirectExchange weather() {
      return new DirectExchange(exchangeName, true, false);
   }

   @Bean Binding dataBinding(DirectExchange directExchange, Queue queue) {
      return BindingBuilder.bind(queue).to(directExchange).with(queueName);
   }

   @Bean CachingConnectionFactory connectionFactory() {
      return new CachingConnectionFactory("127.0.0.1");
   }

   @Bean
   SimpleMessageListenerContainer messageListenerContainer() {
```

```
        SimpleMessageListenerContainer container =
            new SimpleMessageListenerContainer();
        container.setConnectionFactory(connectionFactory());
        container.setQueueNames(queueName);
        return container;
    }

    public static void main(String... args) throws java.lang.Exception {
        ConfigurableApplicationContext ctx =
            SpringApplication.run(Application.class, args);
        RabbitTemplate rabbitTemplate = ctx.getBean(RabbitTemplate.class);
        rabbitTemplate.convertAndSend(Application.queueName, "FL");
        rabbitTemplate.convertAndSend(Application.queueName, "MA");
        rabbitTemplate.convertAndSend(Application.queueName, "CA");

        System.in.read();
        ctx.close();
    }
}
```

正如你所看到的，不再需要@EnableRabbit 注解，尽管配置并没有减少太多，但这样做仍然是不错的改进。如果运行该类，则会得到与前面示例类似的输出结果。

```
DEBUG c.a.p.c.Application - Running with Spring Boot v2.0.0.M1, Spring v5.0.0.RC1
INFO c.a.p.c.Application - No active profile set, falling back to default profiles: default
INFO c.a.p.c.Application - Started Application in 2.211 seconds JVM running for 2.801
[SimpleAsyncTaskExecutor-1] INFO c.a.p.c.WeatherServiceImpl - Cold
[SimpleAsyncTaskExecutor-1] INFO c.a.p.c.WeatherServiceImpl - Hot
[SimpleAsyncTaskExecutor-1] INFO c.a.p.c.WeatherServiceImpl - Not available at this time
```

12.10 小结

在本章中，主要介绍了基于 Spring 的应用程序中最常用的远程处理技术。

如果两个应用程序都是用 Spring 构建的，那么使用 Spring HTTP 调用器是可行的选择。如果需要异步模式或松耦合的集成模式，那么 JMS 是一种常用的方法。本章讨论了如何在 Spring 中使用 RESTful-WS 来公开服务或使用 RestTemplate 类访问服务，最后讨论了如何使用 Spring AMQP 通过 RabbitMQ 进行 RPC 样式的远程处理。

对于每种技术(远程处理、REST、JMS 等)，Spring Boot 都是适用的，因为它就是我们一直在寻找的东西。

在下一章中，将讨论如何使用 Spring 来测试应用程序；详细阐述一些测试技术，以便让开发更加轻松。

第 13 章

Spring 测试

在为企业开发应用程序时，测试是确保所完成的应用程序按预期执行并满足各种需求(架构、安全性、用户需求等)的重要方法。每次进行更改时，都应确保所引入的更改不会影响现有的逻辑。维护持续构建和测试环境对于确保高质量应用程序至关重要。如果可以对所有代码进行高覆盖率的可重复测试，就可以高度自信地部署新的应用程序和更改应用程序。在企业开发环境中，可以针对企业级应用程序中的各个层进行多种测试，而每种测试都有自己的特点和要求。在本章中，将讨论各种应用程序层测试所涉及的基本概念，尤其是在测试 Spring 应用程序时。此外，还介绍 Spring 实现各个层的测试用例的方法。具体而言，本章涵盖以下主题：

- **企业测试框架**：简要描述企业测试框架，讨论各种测试及其目的，重点介绍针对各种应用程序层的单元测试。
- **逻辑单元测试**：最好的单元测试是仅对类中方法的逻辑进行测试，而所有其他依赖项都被正确的行为所“模拟”。在本章中，将讨论 Spring MVC 控制器类的逻辑单元测试的实现，并借助 Java 模拟库来模拟类的依赖项。
- **集成单元测试**：在企业测试框架中，集成测试指的是针对特定的业务逻辑测试不同应用程序层中一组类的交互。通常，在集成测试环境中，服务层应使用持久化层和后端数据库进行测试。然而，随着应用程序架构的发展和轻量级内存数据库的成熟，现在通常的做法是将服务层以及持久化层和后端数据库作为整体进行“单元测试”。例如，在本章中，将使用 JPA 2，同时使用 Hibernate 和 Spring Data JPA 作为持久化提供程序，将 H2 作为数据库。在该架构中，当测试服务层时，“模拟”Hibernate 和 Spring Data JPA 并不重要。因此，在本章中，将讨论服务层以及持久层和 H2 内存数据库的测试。这种测试通常被称为*集成单元测试*，它位于单元测试和全面集成测试之间。
- **前端单元测试**：即使测试了应用程序的每一层，在部署应用程序之后，也仍然需要确保整个应用程序按预期工作。更具体地说，对于 Web 应用程序，在部署到持续构建环境时，应该进行前端测试以确保用户界面正常工作。例如，对于歌手应用程序，应该确保正常功能的每个步骤都正常工作，并且还应该测试异常情况(例如，当信息未通过验证阶段时应用程序如何工作)。在本章中，将简要讨论前端测试框架。

13.1 测试类别介绍

企业测试框架是指整个应用程序生命周期中的测试活动。在不同的阶段，根据所定义的业务和技术要求，执行不同的测试活动来验证应用程序的功能是否按预期工作。

在每个阶段，执行不同的测试情况。其中一些是自动执行的，而另一些是手动执行的。在每种情况下，结果都由相应的人员(例如业务分析员、应用程序用户等)验证。表 13-1 描述了每种类型测试的特征和目标，以及用于实现测试的常用工具和库。

表 13-1 不同的测试类别在现实中的应用

测试类别	描述	常用工具
逻辑单元测试	逻辑单元测试需要一个对象并自行测试，而不必担心它在周围系统中所扮演的角色	单元测试：JUnit、TestNG 模拟对象：Mockito、EasyMock

(续表)

测试类别	描述	常用工具
集成单元测试	集成单元测试专注于在"接近真实"的环境中测试组件之间的交互。这些测试将执行与容器(嵌入式数据库、Web 容器等)的交互	嵌入式数据库：H2 数据库测试：DbUnit 内存 Web 容器：Jetty
前端单元测试	前端单元测试侧重于测试用户界面。目标是确保每个用户界面对用户的操作做出反应并按预期产生输出给用户	Selenium
持续构建和代码质量测试	应用程序代码库应该定期构建，以确保代码质量符合标准(例如，在适当的地方写上注释，没有空的异常捕获块等)。此外，测试的覆盖范围应尽可能广以确保开发的所有代码行都经过测试	代码质量：PMD、Check-style、FindBugs、Sonar 测试覆盖范围：Cobertura、EclEmma 构建工具：Gradle、Maven 持续构建：Hudson、Jenkins
系统集成测试	系统集成测试验证新系统中所有程序之间以及新系统与所有外部接口之间通信的准确性。系统集成测试还必须证明新系统可以在操作环境中根据功能规范和功能有效地执行，同时不会对其他系统产生不利影响	IBM Rational Functional Tester、HP Unified Functional Testing
系统质量测试	系统质量测试旨在确保所开发的应用程序满足那些非功能性需求。在大多数情况下，主要是测试应用程序的性能，以确保满足系统的并发用户和工作负载的目标需求。其他的非功能性需求包括安全性、高可用性功能等	Apache JMeter、HP LoadRunner
用户验收测试	用户验收测试模拟新系统的实际工作条件，包括用户手册和程序。大量的用户参与这个测试阶段，从而为用户提供操作新系统的宝贵经验。此外，还有利于程序员或设计人员了解新程序的用户体验。这种联合参与确保用户和操作人员都赞成对系统所做的修改	IBM Rational TextManager、HP Quality Center

在本章中，将重点介绍三种单元测试(逻辑单元测试、集成单元测试和前端单元测试)的实现，并演示 Spring TestContext 框架和其他支持工具及库如何帮助开发这些测试用例。

本章不会介绍 Spring 框架在测试领域提供的全部细节和类列表，而是介绍 Spring TestContext 框架中最常用的模式以及支持接口和类，演示如何实现测试用例。

13.2 使用 Spring 测试注解

在进行逻辑和集成测试之前，需要注意的是，除了标准注解(例如@Autowired 和@Resource)之外，Spring 还提供了特定于测试的注解。这些注释可用于逻辑和单元测试，并提供了各种功能，例如简化的上下文文件加载、配置文件、测试执行时间等。表 13-2 概述了注解及其用法。

表 13-2 企业测试框架的描述

注解	描述
@ContextConfiguration	类级注解，用于确定如何为集成测试加载和配置 ApplicationContext
@WebAppConfiguration	类级注解，用于指示加载的 ApplicationContext 应该是 WebApplicationContext
@ContextHierarchy	类级注解，用于指示哪个 bean 配置文件应该处于活动状态
@DirtiesContext	类级和方法级注释，用于指示上下文在执行测试期间以某种方式被修改或损坏，并且应该关闭和重新构建以供后续测试
@TestExecutionListeners	类级注解，用于配置应该使用 TestContextManager 注册的 TestExecutionListeners
@TransactionConfiguration	类级注解，用于指示事务配置，例如回滚设置和事务管理器(假设期望的事务管理器没有名为 transactionManager 的 bean)

(续表)

注解	描述
@Rollback	类级解和方法级注解，用于指示是否应该针对所注解的方法回滚事务。此外，它还是用于测试类的默认设置的类级注释
@BeforeTransaction	方法级注解，指示在为使用@Transactional 注解标记的测试方法启动事务之前，应该调用@BeforeTransaction 所注解的方法
@AfterTransaction	方法级注解，指示在为使用@Transactional 注解标记的测试方法结束事务之后，应该调用@AfterTransaction 所注解的方法
@IfProfileValue	类和方法级注解，用于指示应该为一组特定的环境条件启用测试方法
@ProfileValueSourceConfiguration	类级注解，用于指定@IfProfileValue 所使用的 ProfileValueSource。如果该注解未在测试中声明，则将 SystemProfileValueSource 用作默认值
@Timed	方法级注解，用于指示测试必须在指定的时间段内完成。
@Repeat	方法级注解，用于指示注解的测试方法应该重复指定的次数

13.3 实施逻辑单元测试

如前所述，逻辑单元测试是最好的测试级别，目的是验证单个类的行为，所有类的依赖项都会被预期的行为“模拟”。在本节中，通过实现 SingerController 类的测试用例来演示逻辑单元测试，其中服务层由预期的行为模拟。为了帮助模拟服务层的行为，还将展示如何使用 Mockito(http://site.mockito.org/)，这是一个流行的模拟框架。

Spring 框架为 spring-test 模块中的集成测试提供了一流支持。要为本节创建的集成测试提供测试上下文，需要使用 spring-test.jar 库。该库包含一些有价值的类，用于与 Spring 容器进行集成测试。

13.3.1 添加所需的依赖项

首先，需要将依赖项添加到项目中，如以下配置示例所示。此外，还将创建前面章节中创建的类和接口，比如 Singer、SingerService 等。

```
\\pro-spring-15\build.gradle
ext {
    //spring libs
    springVersion = '5.0.0.RC1'
    bootVersion = '2.0.0.M1'

    //testing libs
    mockitoVersion = '2.0.2-beta'
    junitVersion = '4.12'
    hamcrestVersion = '1.3'
    dbunitVersion = '2.5.3'
    poiVersion = '3.16'
    junit5Version = '5.0.0-M4'

    spring = [
        test            : "org.springframework:spring-test:$springVersion",
        ...
    ]

    boot = [
        starterTest :
            "org.springframework.boot:spring-boot-starter-test:$bootVersion",
        ...
    ]

    testing = [
        junit           : "junit:junit:$junitVersion",
        junit5          : "org.junit.jupiter:junit-jupiter-engine:$junit5Version",
        junitJupiter    : "org.junit.jupiter:junit-jupiter-api:$junit5Version",
        mockito         : "org.mockito:mockito-all:$mockitoVersion",
        easymock        : "org.easymock:easymock:3.4",
        jmock           : "org.jmock:jmock:2.8.2",
```

```
        hamcrestCore    : "org.hamcrest:hamcrest-core:$hamcrestVersion",
        hamcrestLib     : "org.hamcrest:hamcrest-library:$hamcrestVersion",
        dbunit           : "org.dbunit:dbunit:$dbunitVersion"
    ]
    misc = [
    ...
    poi : "org.apache.poi:poi:$poiVersion"
    ]
    ...
}
```

13.3.2 单元测试 Spring MVC 控制器

在表示层中，控制器类提供用户界面和服务层之间的集成。

控制器类中的方法将映射到 HTTP 请求。在这些方法中，首先对请求进行处理，然后绑定到模型对象，并与服务层(通过 Spring 的 DI 注入到控制器类中)交互以处理数据。完成后，控制器类将根据结果更新模型和视图状态(例如，用户消息、REST 服务的对象等)，并返回逻辑视图(或返回带有视图的模型)，以供 Spring MVC 解析要显示给用户的视图。

对于单元测试控制器类，主要目标是确保控制器方法正确更新模型和其他视图状态并返回正确的视图。由于我们只想测试控制器类的行为，因此需要用正确的行为来“模拟”服务层。

对于 SingerController 类，需要为 listData()和 create(Singer)方法开发测试用例。在下面的章节中，将讨论这些步骤。

测试 listData()方法

接下来为 singerController.listData()方法创建第一个测试用例。在该测试用例中，要确保当调用方法时，在从服务层检索到歌手列表之后，正确地将信息保存到模型中，并返回正确的对象。以下代码片段显示了该测试用例：

```
package com.apress.prospring5.ch13;

import static org.junit.Assert.assertEquals;
import static org.mockito.Mockito.mock;
import static org.mockito.Mockito.when;

import java.util.ArrayList;
import java.util.List;

import com.apress.prospring5.ch13.entities.Singer;
import com.apress.prospring5.ch13.entities.Singers;
import org.junit.Before;
import org.junit.Test;

import org.mockito.invocation.InvocationOnMock;
import org.mockito.stubbing.Answer;

import org.springframework.test.util.ReflectionTestUtils;
import org.springframework.ui.ExtendedModelMap;

public class SingerControllerTest {
    private final List<Singer> singers = new ArrayList<>();

    @Before
    public void initSingers() {
        Singer singer = new Singer();
        singer.setId(1l);
        singer.setFirstName("John");
        singer.setLastName("Mayer");
        singers.add(singer);
    }

    @Test
    public void testList() throws Exception {
        SingerService singerService = mock(SingerService.class);
        when(singerService.findAll()).thenReturn(singers);

        SingerController singerController = new SingerController();

        ReflectionTestUtils.setField(singerController,
            "singerService", singerService);
```

```
        ExtendedModelMap uiModel = new ExtendedModelMap();
        uiModel.addAttribute("singers", singerController.listData());

        Singers modelSingers = (Singers) uiModel.get("singers");

        assertEquals(1, modelSingers.getSingers().size());
    }
}
```

首先，该测试用例调用 initSingers()方法，该方法与@Before 注解一起应用，指示 JUnit 应该在运行每个测试用例之前运行该方法(如果想要在整个测试类之前运行某些逻辑，可以使用@BeforeClass 注解)。在该方法中，使用硬编码信息对歌手列表进行了初始化。

其次，将 testList()方法与@Test 注解一起应用，从而向 JUnit 表明这是 JUnit 应该运行的一个测试用例。在该测试用例中，通过使用 Mockito 的 Mockito.mock()方法(请注意 import static 语句)来模拟 SingerService 类型的私有变量 singerService。Mockito 还提供了 when()方法来模拟 SingerService.findAll()方法，该方法将由 SingerController 类使用。

然后，创建 SingerController 类的一个实例，并通过使用 Spring 提供的 ReflectionTestUtils 类的 setField()方法，使用模拟实例来设置 singerService 变量(该变量通常情况下由 Spring 注入)。ReflectionTestUtils 提供了一组基于反射的实用方法，可用于单元测试和集成测试场景中。另外，此处还构造了 ExtendedModelMap 类(它实现了 org.springframework.ui.Model 接口)的一个实例。

最后，调用 SingerController.listData()方法。在调用时，通过调用由 JUnit 提供的各种断言方法来验证结果，以确保歌手信息列表在视图使用的模型中得到正确保存。

现在可以运行该测试用例，它应该可以成功运行。可以通过自己的编译系统或 IDE 进行验证。现在可以继续使用 create()方法。

测试 create()方法

以下代码片段显示了测试 create()方法的代码：

```
package com.apress.prospring5.ch13;

import static org.junit.Assert.assertEquals;
import static org.mockito.Mockito.mock;
import static org.mockito.Mockito.when;
import java.util.ArrayList;
import java.util.List;

import com.apress.prospring5.ch13.entities.Singer;
import com.apress.prospring5.ch13.entities.Singers;
import org.junit.Before;
import org.junit.Test;

import org.mockito.invocation.InvocationOnMock;
import org.mockito.stubbing.Answer;

import org.springframework.test.util.ReflectionTestUtils;
import org.springframework.ui.ExtendedModelMap;

public class SingerControllerTest {
    private final List<Singer> singers = new ArrayList<>();

    @Test
    public void testCreate() {
        final Singer newSinger = new Singer();
        newSinger.setId(999l);
        newSinger.setFirstName("BB");
        newSinger.setLastName("King");

        SingerService singerService = mock(SingerService.class);
        when(singerService.save(newSinger)).thenAnswer(new Answer<Singer>() {
            public Singer answer(InvocationOnMock invocation) throws Throwable {
                singers.add(newSinger);
                return newSinger;
            }
        });

        SingerController singerController = new SingerController();
        ReflectionTestUtils.setField(singerController, "singerService",
        singerService);
```

```
        Singer singer = singerController.create(newSinger);
        assertEquals(Long.valueOf(999l), singer.getId());
        assertEquals("BB", singer.getFirstName());
        assertEquals("King", singer.getLastName());

        assertEquals(2, singers.size());
    }
}
```

首先，对 SingerService.save()方法进行模拟，以模拟在歌手列表中添加新的 Singer 对象。请注意 org.mockito.stubbing.Answer<T>接口的使用，该接口使用预期的逻辑模拟方法并返回一个值。

然后，调用 SingerController.create()方法，并调用断言操作来验证结果。再次运行，并注意测试用例结果。对于 create()方法，应该创建更多的测试用例来测试各种场景。例如，需要测试在保存操作期间何时遇到数据访问错误。

到目前为止，介绍的内容都可以通过 JMock(www.jmock.org/)完成，使用该库的 SingerControllerTest 类的一个版本是本节代码示例的一部分。在此不会详细介绍这些内容，因为模拟依赖项才是关注的焦点，而不是所使用的库[1]。

13.4 实现集成测试

在本节中，将实现服务层的集成测试。在 Singer 应用程序中，核心服务是 SingerServiceImpl 类，它是 SingerService 接口的 JPA 实现。

在对服务层进行单元测试时，将使用 H2 内存数据库来托管数据模型和测试数据，同时使用 JPA 提供程序(Hibernate 和 Spring Data JPA 的存储库抽象)。目标是确保 SingerServiceImpl 类正确执行业务功能。

在下面的章节中，将演示如何测试 SingerServiceImpl 类中的一些查找方法和保存操作。

13.4.1 添加所需的依赖项

为了使用数据库实现测试用例，还需要一个库，它可以在执行测试用例之前在数据库中填充所需的测试数据，并且可以轻松地执行必要的数据库操作。此外，为了更容易地准备测试数据，将支持以 Microsoft Excel 格式准备的测试数据。

为了实现上述目的，需要额外的库。在数据库方面，DbUnit(http://dbunit.sourceforge.net)是一个通用库，可以帮助实现与数据库相关的测试。此外，Apache POI(http://poi.apache.org)项目的库将用于解析在 Microsoft Excel 中准备的测试数据。

13.4.2 配置用于服务层测试的配置文件

Spring 3.1 中引入的 bean 定义配置文件功能对于通过测试组件的适当配置来实现测试用例是非常有用的。为了便于测试服务层，还将针对 ApplicationContext 配置使用配置文件功能。对于 Singer 应用程序，需要两个配置文件，如下所示。

- **开发配置文件(dev)**：包含开发环境配置的配置文件。例如，在开发系统中，后端 H2 数据库将同时执行数据库创建和初始数据填充脚本。
- **测试配置文件(test)**：包含测试环境配置的配置文件。例如，在测试环境中，后端 H2 数据库将仅执行数据库创建脚本，而数据则由测试用例填充。

接下来为歌手应用程序配置文件环境。对于歌手应用程序，在 XML 配置文件 datasource-tx-jpa.xml 中定义后端配置(即数据源、JPA、事务等)。此时只想在该文件中为 dev 配置文件配置数据源。为此，需要使用配置文件来封装数据源 bean。以下配置代码片段显示了所需要做的更改：

```
<?xml version="1.0" encoding="UTF-8"?>
<beans ...>

    <bean id="transactionManager"
        class="org.springframework.orm.jpa.JpaTransactionManager">
            <property name="entityManagerFactory" ref="emf"/>

    </bean>
```

1 另一种选择是 EasyMock，参见 http://easymock.org/。

```
    <tx:annotation-driven transaction-manager="transactionManager" />

    <bean id="emf"
       class="org.springframework.orm.jpa.LocalContainerEntityManagerFactoryBean">
       <property name="dataSource" ref="dataSource" />
       <property name="jpaVendorAdapter">
          <bean class=
             "org.springframework.orm.jpa.vendor.HibernateJpaVendorAdapter" />
       </property>
       <property name="packagesToScan" value="com.apress.prospring5.ch13"/>
       <property name="jpaProperties">
          <props>
             <prop key="hibernate.dialect">org.hibernate.dialect.H2Dialect</prop>
             <prop key="hibernate.max_fetch_depth">3</prop>
             <prop key="hibernate.jdbc.fetch_size">50</prop>
             <prop key="hibernate.jdbc.batch_size">10</prop>
             <prop key="hibernate.show_sql">true</prop>
          </props>
       </property>
    </bean>

    <context:annotation-config/>

    <jpa:repositories base-package="com.apress.prospring5.ch13"
                    entity-manager-factory-ref="emf"
                    transaction-manager-ref="transactionManager"/>
    <beans profile="dev">
          <jdbc:embedded-database id="dataSource" type="H2">
             <jdbc:script location="classpath:config/schema.sql"/>
             <jdbc:script location="classpath:config/test-data.sql"/>
       </jdbc:embedded-database>
    </beans>
</beans>
```

如上述配置代码片段所示，使用<beans>标记封装了 dataSource bean，并将 profile 属性的值设置为 dev，表明该数据源仅适用于开发系统。请记住，可以通过将-Dspring.profiles.active = dev 作为系统参数传递给 JVM 来激活配置文件。

13.4.3 Java 配置版本

从引入 Java 配置类开始，XML 正在慢慢失去优势。因此，本书的重点是使用 Java 配置类；而介绍 XML 配置只是为了展示 Spring 配置随着时间发生的演变。前面显示的 XML 配置可以分成两部分：一部分用于覆盖数据源配置(特定于配置文件)，另一部分是事务配置(这对于开发和测试配置来说很常见)。接下来介绍两个类。Java 配置中添加的一项改进是数据库模式的“自动生成”，该过程是通过将 Hibernate 属性 hibernate.hbm2ddl.auto 设置为 create-drop 来完成的。

```
//DataConfig.java
package com.apress.prospring5.ch13.config;

import org.slf4j.Logger;
import org.slf4j.LoggerFactory;
import org.springframework.context.annotation.Bean;
import org.springframework.context.annotation.ComponentScan;
import org.springframework.context.annotation.Configuration;
import org.springframework.context.annotation.Profile;
import org.springframework.jdbc.datasource.embedded.EmbeddedDatabaseBuilder;
import org.springframework.jdbc.datasource.embedded.EmbeddedDatabaseType;

import javax.sql.DataSource;

@Profile("dev")
@Configuration
@ComponentScan(basePackages = {"com.apress.prospring5.ch13.init"} )
public class DataConfig {

    private static Logger logger = LoggerFactory.getLogger(DataConfig.class);

    @Bean
    public DataSource dataSource() {
       try {
          EmbeddedDatabaseBuilder dbBuilder = new EmbeddedDatabaseBuilder();
          return dbBuilder.setType(EmbeddedDatabaseType.H2).build();
```

```
        } catch (Exception e) {
            logger.error("Embedded DataSource bean cannot be created!", e);
            return null;
        }
    }
}

//ServiceConfig.class
package com.apress.prospring5.ch13.config;

import org.springframework.beans.factory.annotation.Autowired;
import org.springframework.context.annotation.Bean;
import org.springframework.context.annotation.ComponentScan;
import org.springframework.context.annotation.Configuration;
import org.springframework.data.jpa.repository.config.EnableJpaRepositories;
import org.springframework.orm.jpa.JpaTransactionManager;
import org.springframework.orm.jpa.JpaVendorAdapter;
import org.springframework.orm.jpa.LocalContainerEntityManagerFactoryBean;
import org.springframework.orm.jpa.vendor.HibernateJpaVendorAdapter;
import org.springframework.transaction.PlatformTransactionManager;

import javax.persistence.EntityManagerFactory;
import javax.sql.DataSource;
import java.util.Properties;

@Configuration
@EnableJpaRepositories(basePackages = {"com.apress.prospring5.ch13.repos"})
@ComponentScan(basePackages = {"com.apress.prospring5.ch13.entities",
  "com.apress.prospring5.ch13.services"})
public class ServiceConfig {

    @Autowired
    DataSource dataSource;

    @Bean
    public Properties hibernateProperties() {
        Properties hibernateProp = new Properties();
            hibernateProp.put("hibernate.dialect",
        "org.hibernate.dialect.H2Dialect");
        hibernateProp.put("hibernate.hbm2ddl.auto", "create-drop");
        hibernateProp.put("hibernate.show_sql", true);
        hibernateProp.put("hibernate.max_fetch_depth", 3);
        hibernateProp.put("hibernate.jdbc.batch_size", 10);
        hibernateProp.put("hibernate.jdbc.fetch_size", 50);
        return hibernateProp;
    }

    @Bean
    public PlatformTransactionManager transactionManager() {
        return new JpaTransactionManager(entityManagerFactory());
    }

    @Bean
    public JpaVendorAdapter jpaVendorAdapter() {
        return new HibernateJpaVendorAdapter();
    }

    @Bean
    public EntityManagerFactory entityManagerFactory() {
        LocalContainerEntityManagerFactoryBean factoryBean =
            new LocalContainerEntityManagerFactoryBean();
        factoryBean.setPackagesToScan("com.apress.prospring5.ch13.entities");
        factoryBean.setDataSource(dataSource);
        factoryBean.setJpaVendorAdapter(new HibernateJpaVendorAdapter());
        factoryBean.setJpaProperties(hibernateProperties());
        factoryBean.setJpaVendorAdapter(jpaVendorAdapter());
        factoryBean.afterPropertiesSet();
        return factoryBean.getNativeEntityManagerFactory();
    }
}
```

13.4.4 实施基础结构类

在实现单个测试用例之前，首先需要实现一些类来支持 Excel 文件中测试数据的填充。此外，为了简化测试用例的开发，还希望引入一个名为@DataSets 的自定义注解，它接收 Excel 文件名作为参数。我们将开发一个自定义测试执行监听器(一个由 Spring 测试框架支持的功能)来检查注解是否存在并相应地加载数据。

实现自定义 TestExecutionListener

在 spring-test 模块中，org.springframework.test.context.TestExecutionListener 接口定义了一个监听器 API，它可以拦截测试用例执行的各个阶段(例如，类被测试之前和之后，方法被测试之前和之后，等等)中的事件。在测试服务层时，将为新引入的@DataSets 注解实现一个自定义监听器。目的是通过对测试用例做简单注解来支持测试数据的填充。例如，要测试 SingerService.findAll()方法，可以使用如下所示的代码片段：

```
@DataSets(setUpDataSet= "/com/apress/prospring5/ch13/SingerServiceImplTest.xls")
@Test
public void testFindAll() throws Exception {
    List<Singer> result = singerService.findAll();
    ...
}
```

将@DataSets 注解应用到测试用例表明在运行测试之前，需要将测试数据从指定的 Excel 文件加载到数据库中。首先，需要定义自定义注解，如下所示：

```
package com.apress.prospring5.ch13;

import java.lang.annotation.ElementType;
import java.lang.annotation.Retention;
import java.lang.annotation.RetentionPolicy;
import java.lang.annotation.Target;

@Retention(RetentionPolicy.RUNTIME)
@Target(ElementType.METHOD)

public @interface DataSets {
    String setUpDataSet() default "";
}
```

自定义注解@DataSets 是一个方法级注解。另外，通过实现以下代码片段所示的 TestExecutionListener 接口，可以开发自定义测试监听器类：

```
package com.apress.prospring5.ch13;

import org.dbunit.IDatabaseTester;
import org.dbunit.dataset.IDataSet;
import org.dbunit.util.fileloader.XlsDataFileLoader;
import org.springframework.test.context.TestContext;
import org.springframework.test.context.TestExecutionListener;

public class ServiceTestExecutionListener implements
      TestExecutionListener {
    private IDatabaseTester databaseTester;

    @Override
    public void afterTestClass(TestContext arg0) throws Exception {
    }

    @Override
    public void afterTestMethod(TestContext arg0) throws Exception {
        if (databaseTester != null) {
            databaseTester.onTearDown();
        }
    }

    @Override
    public void beforeTestClass(TestContext arg0) throws Exception {
    }

    @Override
    public void beforeTestMethod(TestContext testCtx) throws Exception {
        DataSets dataSetAnnotation = testCtx.getTestMethod()
            .getAnnotation(DataSets.class);
        if (dataSetAnnotation == null ) {
            return;
    }

    String dataSetName = dataSetAnnotation.setUpDataSet();

    if (!dataSetName.equals("") ) {
        databaseTester = (IDatabaseTester)
```

```
            testCtx.getApplicationContext().getBean("databaseTester");
        XlsDataFileLoader xlsDataFileLoader = (XlsDataFileLoader)
            testCtx.getApplicationContext().getBean("xlsDataFileLoader");
        IDataSet dataSet = xlsDataFileLoader.load(dataSetName);

        databaseTester.setDataSet(dataSet);
        databaseTester.onSetup();
      }
    }

    @Override
    public void prepareTestInstance(TestContext arg0) throws Exception {
    }
}
```

在实现 TestExecutionListener 接口之后，还需要实现许多方法。但目前，只关注 beforeTestMethod()和 afterTestMethod()方法，这两个方法分别在每个测试方法执行前后完成测试数据的填充和清理。请注意，在每个方法中，Spring 将传入 TestContext 类的一个实例，以便方法可以访问由 Spring 框架启动的底层测试 ApplicationContext。

尤其需要注意 beforeTestMethod()方法。首先，它检查测试方法是否存在@DataSets 注解。如果存在，就从指定的 Excel 文件中加载测试数据。此时，IDatabaseTester 接口(带有实现类 org.dbunit.DataSourceDatabaseTester，该类将在后面讨论)从 TestContext 获得。IDatabaseTester 接口由 DbUnit 提供，并支持基于给定数据库连接或数据源的数据库操作。

其次，从 TestContext 获取 XlsDataFileLoader 类的一个实例。DbUnit 使用 XlsDataFileLoader 类从 Excel 文件中加载数据。该类使用后台的 Apache POI 库来读取 Microsoft Office 格式的文件。然后调用 XlsDataFileLoader.load()方法来加载文件中的数据，并返回 IDataSet 接口的一个实例，它表示所加载的数据集。

最后，调用 IDatabaseTester.setDataSet()方法来设置测试数据，并调用 IDatabaseTester.onSetup()方法来触发数据的填充。

在 afterTestMethod()方法中，调用 IDatabaseTester.onTearDown()方法来清理数据。

实现配置类

继续为测试环境实现配置类。以下代码片段显示了使用 Java Config 样式配置的代码：

```
package com.apress.prospring5.ch13.config;

import javax.sql.DataSource;

import com.apress.prospring5.ch13.init.DBInitializer;
import org.dbunit.DataSourceDatabaseTester;
import org.dbunit.util.fileloader.XlsDataFileLoader;
import org.slf4j.Logger;
import org.slf4j.LoggerFactory;
import org.springframework.context.annotation.*;
import org.springframework.jdbc.datasource.embedded.EmbeddedDatabaseBuilder;
import org.springframework.jdbc.datasource.embedded.EmbeddedDatabaseType;

@Configuration
@ComponentScan(basePackages={"com.apress.prospring5.ch13"},
      excludeFilters = {@ComponentScan.Filter(type = FilterType.ASSIGNABLE_TYPE,
            value = DBInitializer.class)
})
@Profile("test")
public class ServiceTestConfig {
   private static Logger logger = LoggerFactory.getLogger(ServiceTestConfig.class);

   @Bean
   public DataSource dataSource() {
      try {
         EmbeddedDatabaseBuilder dbBuilder = new EmbeddedDatabaseBuilder();
         return dbBuilder.setType(EmbeddedDatabaseType.H2).build();
      } catch (Exception e) {
         logger.error("Embedded DataSource bean cannot be created!", e);
         return null;
      }
   }

   @Bean(name="databaseTester")
   public DataSourceDatabaseTester dataSourceDatabaseTester() {
      DataSourceDatabaseTester databaseTester =
         new DataSourceDatabaseTester(dataSource());
```

```
        return databaseTester;
    }

    @Bean(name="xlsDataFileLoader")
    public XlsDataFileLoader xlsDataFileLoader() {
        return new XlsDataFileLoader();
    }
}
```

ServiceTestConfig 类为服务层测试定义了 ApplicationContext 实现。通过使用@ComponentScan 注解来指示 Spring 扫描想要测试的服务层 bean。

excludeFilters 属性用于确保没有使用实际数据来初始化测试数据库。

@Profile 注解指定该类中定义的 bean 属于测试配置文件。

其次，在该类中声明另一个 dataSource bean，只有在没有任何数据的情况下才对 H2 数据库执行 schema.sql 脚本。用于从 Excel 文件加载测试数据的自定义测试执行监听器使用了 databaseTester 和 xlsDataFileLoader bean。请注意，dataSourceDatabaseTester bean 是使用为测试环境定义的 dataSource bean 构建的。

13.4.5 对服务层进行单元测试

接下来从单元测试查找方法开始，包括 SingerService.findAll()和 SingerService.findByFirstNameAndLastName()方法。首先，需要以 Excel 格式准备测试数据。通常做法是将该文件放入与测试用例类相同的文件夹，且名称相同。因此，此时的文件名是/src/test/java/com/apress/prospring5/ch13/SingerServiceImplTest.xls。

测试数据在工作表中准备。工作表的名称是表名(SINGER)，第一行是表中的列名。从第二行开始，输入名字和姓氏以及出生日期等数据。此外，还指定了 ID 列，但不是一个值，因为 ID 是由数据库填充的。请查看本书源代码中的示例 Excel 文件。

下面的代码片段显示了上述两个查找方法的测试用例的测试类：

```
package com.apress.prospring5.ch13;

import static org.junit.Assert.assertEquals;
import static org.junit.Assert.assertNotNull;
import static org.junit.Assert.assertNull;

import java.util.List;

import javax.persistence.EntityManager;
import javax.persistence.PersistenceContext;
import javax.validation.ConstraintViolationException;

import com.apress.prospring5.ch13.entities.Singer;
import org.junit.Test;
import org.junit.runner.RunWith;
import org.springframework.beans.factory.annotation.Autowired;
import org.springframework.test.context.ActiveProfiles;
import org.springframework.test.context.ContextConfiguration;
import org.springframework.test.context.TestExecutionListeners;
import org.springframework.test.context.junit4.
    AbstractTransactionalJUnit4SpringContextTests;
import org.springframework.test.context.junit4.SpringJUnit4ClassRunner;

@RunWith(SpringJUnit4ClassRunner.class)
@ContextConfiguration(classes = {ServiceTestConfig.class, ServiceConfig.class,
DataConfig.class})
@TestExecutionListeners({ServiceTestExecutionListener.class})
@ActiveProfiles("test")
public class SingerServiceImplTest extends
     AbstractTransactionalJUnit4SpringContextTests {
    @Autowired
    SingerService singerService;

    @PersistenceContext
    private EntityManager em;

    @DataSets(setUpDataSet= "/com/apress/prospring5/ch13/SingerServiceImplTest.xls")
    @Test
    public void testFindAll() throws Exception {
        List<Singer> result = singerService.findAll();
```

```
        assertNotNull(result);
        assertEquals(1, result.size());
    }

    @DataSets(setUpDataSet= "/com/apress/prospring5/ch13/SingerServiceImplTest.xls")
    @Test
    public void testFindByFirstNameAndLastName_1() throws Exception {
        Singer result = singerService.findByFirstNameAndLastName("John", "Mayer");
        assertNotNull(result);
    }
    @DataSets(setUpDataSet= "/com/apress/prospring5/ch13/SingerServiceImplTest.xls")
    @Test
    public void testFindByFirstNameAndLastName_2() throws Exception {
        Singer result = singerService.findByFirstNameAndLastName("BB", "King");
        assertNull(result);
    }
}
```

@RunWith 注解与测试控制器类时使用的@RunWith 注解是相同的。@ContextConfiguration 注解指定应该从 ServiceTestConfig、ServiceConfig 和 DataConfig 类加载 ApplicationContext 配置。虽然并不应该添加 DataConfig 类，但还是添加了，目的是演示即使添加了 Spring 配置文件，实际上也可以工作。@TestExecutionListeners 注解表明应该使用 ServiceTestExecutionListener 类来拦截测试用例执行生命周期。@ActiveProfiles 注解指定要使用的配置文件。此时，应该加载 ServiceTestConfig 类中定义的 dataSource bean，而不是在 datasource-tx-jpa.xml 文件中定义的数据源，因为它属于 dev 配置文件。

此外，该类扩展了 Spring 的 AbstractTransactionalJUnit4SpringContextTests 类，这是 Spring 对 JUnit 提供的支持，同时还包括 Spring 的 DI 和事务管理机制。请注意，在 Spring 的测试环境中，Spring 将在执行完每个测试方法后回滚事务，以便回滚所有数据库更新操作。如果想要控制回滚行为，可以在方法级别使用@Rollback 注解。

针对 findAll()方法有一个测试用例，而针对 testFindByFirstNameAndLastName()方法则有两个测试用例(一个检索结果，而另一个不检索结果)。为所有查找方法都应用了@DataSets 注解(使用了 Excel 中的 Singer 测试数据文件)。另外，SingerService 从 ApplicationContext 自动装配到测试用例中。其余的代码应该比较好理解。在每个测试用例中应用各种断言语句，以确保结果符合预期。

运行测试用例并确保通过。接下来，测试保存操作。主要解测试两种情况。一种是正常情况下有效的 Singer 保存成功，另一种是 Singer 错误而导致异常被抛出。以下代码片段显示了这两个测试用例的附加片段：

```
package com.apress.prospring5.ch13;

import static org.junit.Assert.assertEquals;
import static org.junit.Assert.assertNotNull;
import static org.junit.Assert.assertNull;
import java.util.List;

import javax.persistence.EntityManager;
import javax.persistence.PersistenceContext;
import javax.validation.ConstraintViolationException;
import com.apress.prospring5.ch13.entities.Singer;
import org.junit.Test;
import org.junit.runner.RunWith;
import org.springframework.beans.factory.annotation.Autowired;
import org.springframework.test.context.ActiveProfiles;
import org.springframework.test.context.ContextConfiguration;
import org.springframework.test.context.TestExecutionListeners;
import org.springframework.test.context.junit4.
    AbstractTransactionalJUnit4SpringContextTests;
import org.springframework.test.context.junit4.SpringJUnit4ClassRunner;
@RunWith(SpringJUnit4ClassRunner.class)
@ContextConfiguration(classes = {ServiceTestConfig.class, ServiceConfig.class,
    DataConfig.class})
@TestExecutionListeners({ServiceTestExecutionListener.class})
@ActiveProfiles("test")
public class SingerServiceImplTest extends
        AbstractTransactionalJUnit4SpringContextTests {
    @Autowired
    SingerService singerService;

    @PersistenceContext
    private EntityManager em;

    @Test
    public void testAddSinger() throws Exception {
```

```
        deleteFromTables("SINGER");

        Singer singer = new Singer();
        singer.setFirstName("Stevie");
        singer.setLastName("Vaughan ");

        singerService.save(singer);
        em.flush();

        List<Singer> singers = singerService.findAll();
        assertEquals(1, singers.size());
    }

    @Test(expected=ConstraintViolationException.class)
    public void testAddSingerWithJSR349Error() throws Exception {
        deleteFromTables("SINGER");

        Singer singer = new Singer();
        singerService.save(singer);
        em.flush();

        List<Singer> singers = singerService.findAll();
        assertEquals(0, singers.size());
    }
}
```

在上面的代码清单中，着重看一下 testAddSinger()方法。在该方法中，为了确保 Singer 表中没有数据存在，调用 AbstractTransactionalJUnit4SpringContextTests 类中提供的便捷方法 deleteFromTables()来清理表。请注意，在调用保存操作之后，还需要显式调用 EntityManager.flush()方法来强制 Hibernate 将持久化上下文刷新到数据库，以便 findAll()方法可以正确地从数据库检索信息。

Spring 4.3 中引入了 SpringJUnit4ClassRunner.class 的别名 SpringRunner.class。

在第二个测试方法 testAddSingerWithJSR349Error()方法中，测试存在验证错误的 Singer 对象的保存操作。请注意，在@Test 注解中，传入一个期望的属性，从而指定此测试用例应该抛出具有指定类型的异常，在本例中为 ConstraintViolationException 类。

再次运行测试类并验证结果是否成功。

请注意，目前只介绍了 Spring 的测试框架中最常用的类。Spring 的测试框架提供了许多支持类和注解，从而能够在测试用例生命周期的执行过程中进行精细控制。例如，@BeforeTransaction 和@AfterTransaction 注解允许在 Spring 启动事务之前或者在事务完成之后执行某些逻辑。有关 Spring 的测试框架的各个方面的更详细描述，请参阅 Spring 的参考文档。

13.4.6 丢弃 DbUnit

DbUnit 可能被认为难以使用，因为需要额外的依赖项和配置类。采用 Spring 方式可以更好，不是吗？幸运的是，在 4.0 之后的 Spring 版本中引入了很多有用的注解。其中一个将在以下示例中使用：@Sql。该注解用于注解测试类或测试方法，以配置在集成测试期间针对给定数据库执行的 SQL 脚本和语句。这意味着可以在不使用 DbUnit 的情况下准备测试数据。正因为如此，测试配置也得到了简化，测试类不需要扩展任何东西。

本节的另一个补充内容是使用 JUnit 5(http://junit.org/junit5/)，也被称为 JUnit Jupiter。Spring 自从 4.3 版本以来，甚至在第一个稳定版发布之前，就已经提供对 JUnit 的支持。在编写本书时，当前版本是 5.0.0-M4。JUnit 5 是下一代的 JUnit，目的是在 JVM 上为执行开发人员端的测试创建最新的基础。其中重点关注 Java 8 以及更新的版本，同时支持许多不同类型的测试[1]。

接下来看一看如何修改配置。测试配置文件的数据源配置测试类变得非常简单，因为现在需要的是一个空的数据库。

```
package com.apress.prospring5.ch13.config;

import com.apress.prospring5.ch13.init.DBInitializer;
import org.slf4j.Logger;
import org.slf4j.LoggerFactory;
import org.springframework.context.annotation.*;
```

1 这些内容来自 JUnit 官方网站。

```
import org.springframework.jdbc.datasource.embedded.EmbeddedDatabaseBuilder;
import org.springframework.jdbc.datasource.embedded.EmbeddedDatabaseType;

import javax.sql.DataSource;

@Configuration
@ComponentScan(basePackages={"com.apress.prospring5.ch13"},
    excludeFilters = {@ComponentScan.Filter(type = FilterType.ASSIGNABLE_TYPE,
        value = DBInitializer.class)
  })
@Profile("test")
public class SimpleTestConfig {

    private static Logger logger = LoggerFactory.getLogger(SimpleTestConfig.class);

    @Bean
    public DataSource dataSource() {
        try {
            EmbeddedDatabaseBuilder dbBuilder = new EmbeddedDatabaseBuilder();
            return dbBuilder.setType(EmbeddedDatabaseType.H2).build();
        } catch (Exception e) {
            logger.error("Embedded DataSource bean cannot be created!", e);
            return null;
        }
    }
}
```

测试用例所需的数据和查询将通过 SQL 脚本文件提供。数据由名为 test-data.sql 的文件提供，内容如下所示：

```
insert into singer (first_name, last_name, birth_date,version)
 values ('John', 'Mayer', '1977-10-16', 0);
```

测试数据库的清理将使用 clean-up.sql 脚本来完成。该脚本用于清空数据库，以便一个测试方法所使用的数据不会“污染”另一个测试方法的数据，内容如下所示：

```
delete from singer;
```

测试类将使用一些 JUnit 5 注解来演示如何使用这些注解。每个注解将在代码部分之后进行解释。

```
package com.apress.prospring5.ch13;

import com.apress.prospring5.ch13.config.DataConfig;
import com.apress.prospring5.ch13.config.ServiceConfig;
import com.apress.prospring5.ch13.config.SimpleTestConfig;

import com.apress.prospring5.ch13.entities.Singer;
import com.apress.prospring5.ch13.services.SingerService;
import org.junit.jupiter.api.*;
import org.slf4j.Logger;
import org.slf4j.LoggerFactory;
import org.springframework.beans.factory.annotation.Autowired;
import org.springframework.test.context.ActiveProfiles;
import org.springframework.test.context.jdbc.Sql;
import org.springframework.test.context.jdbc.SqlConfig;
import org.springframework.test.context.jdbc.SqlGroup;
import org.springframework.test.context.junit.jupiter.SpringJUnitConfig;

import java.util.List;

import static org.junit.jupiter.api.Assertions.assertEquals;
import static org.junit.jupiter.api.Assertions.assertNotNull;

@SpringJUnitConfig(classes = {SimpleTestConfig.class, ServiceConfig.class,
DataConfig.class})
@DisplayName("Integration SingerService Test")
@ActiveProfiles("test")
public class SingerServiceTest {

    private static Logger logger =
        LoggerFactory.getLogger(SingerServiceTest.class);

    @Autowired
    SingerService singerService;

    @BeforeAll
    static void setUp() {
        logger.info("--> @BeforeAll -
            executes before executing all test methods in this class");
    }
```

```
    @AfterAll
    static void tearDown(){
        logger.info("--> @AfterAll -
            executes before executing all test methods in this class");
    }

    @BeforeEach
    void init() {
        logger.info("--> @BeforeEach -
            executes before each test method in this class");
    }

    @AfterEach
    void dispose() {
        logger.info("--> @AfterEach -
            executes before each test method in this class");
    }

    @Test
    @DisplayName("should return all singers")
    @SqlGroup({
        @Sql(value = "classpath:db/test-data.sql",
            config = @SqlConfig(encoding = "utf-8", separator = ";",
            commentPrefix = "--"),
            executionPhase = Sql.ExecutionPhase.BEFORE_TEST_METHOD),
        @Sql(value = "classpath:db/clean-up.sql",
            config = @SqlConfig(encoding = "utf-8", separator = ";",
            commentPrefix = "--"),
            executionPhase = Sql.ExecutionPhase.AFTER_TEST_METHOD),
    })
    public void findAll() {
        List<Singer> result = singerService.findAll();
        assertNotNull(result);
        assertEquals(1, result.size());
    }

    @Test
    @DisplayName("should return singer 'John Mayer'")
    @SqlGroup({
        @Sql(value = "classpath:db/test-data.sql",
            config = @SqlConfig(encoding = "utf-8", separator = ";",
            commentPrefix = "--"),
            executionPhase = Sql.ExecutionPhase.BEFORE_TEST_METHOD),
        @Sql(value = "classpath:db/clean-up.sql",
            config = @SqlConfig(encoding = "utf-8", separator = ";",
            commentPrefix = "--"),
            executionPhase = Sql.ExecutionPhase.AFTER_TEST_METHOD),
    })
    public void testFindByFirstNameAndLastNameOne() throws Exception {
        Singer result = singerService.findByFirstNameAndLastName("John", "Mayer");
        assertNotNull(result);
    }
}
```

ApplicationContext 是使用 SpringJUnitJupiterConfig 注解创建的。这是一个组合注解，它将 JUnit Jupiter 的@ExtendWith(SpringExtension.class)与 Spring TestContext 框架的@ContextConfiguration 组合在一起。

@DisplayName 注解是一种典型的 JUnit Jupiter 注解，用于为带注解的测试类或测试方法声明自定义显示值。在支持 JUnit 5 的编辑器中，运行后视图会看起来非常漂亮，如图 13-1 所示。

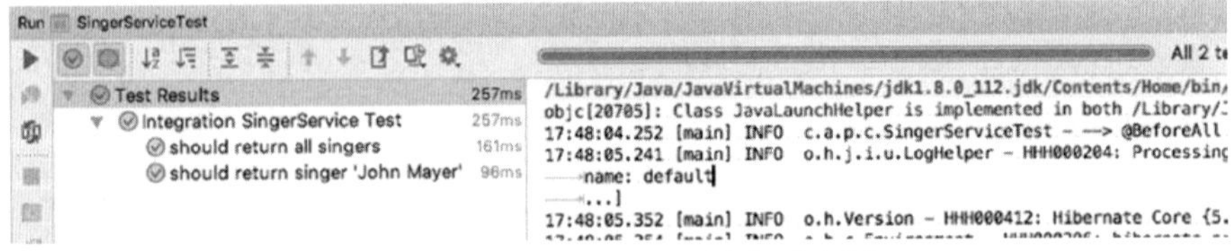

图 13-1 IntelliJ IDEA JUnit 5 测试运行视图

这样做是不是可以更容易看到 SingerService 是否按预期工作？

注解@BeforeAll 和@AfterAll 的名称不言自明，它们用于替代 JUnit 4 的@BeforeClass 和@AfterClass。

对于@BeforeEach 和@AfterEach 也一样，它们用于替换 JUnit 4 的@Before 和@After。@SqlGroup 注解用于对

多个@Sql 注解进行分组。

13.5 实现前端单元测试

另一个特别有趣的测试领域是将前端行为作为整体进行测试，尤其是将 Web 应用程序部署到像 Apache Tomcat 这样的 Web 容器时。

主要原因是，即使测试了应用程序中的每一层，也仍然需要确保视图根据用户的不同行为正确运行。在需要重复测试用例的前端操作时，自动执行前端测试是非常重要的，可以为开发人员和用户节省大量时间。

但是，为前端开发测试用例是一项具有挑战性的任务，特别是对于那些具有大量交互、丰富且基于 Ajax 的组件的 Web 应用程序而言，更是如此。

Selenium 介绍

Selenium 是一个功能强大且全面的工具和框架，用于基于 Web 的自动化前端测试。其主要功能是通过使用 Selenium 来“驱动”浏览器，从而模拟用户与应用程序的交互，并对视图状态进行验证。

Selenium 支持常见的浏览器，包括 Firefox、IE 和 Chrome。在语言方面，支持 Java、C#、PHP、Perl、Ruby 和 Python。此外，在设计 Selenium 时还使用了 Ajax 和富互联网应用程序(RIA)，从而使现代 Web 应用程序的自动化测试成为可能。

如果应用程序有很多前端用户界面并且需要运行大量前端测试，那么 selenium-server 模块会提供内置的网格功能，以支持在一组计算机之间执行前端测试。

Selenium IDE 是一个 Firefox 插件，可以帮助“记录”用户与 Web 应用程序的交互。它还支持重播并将脚本导出为各种格式，这有助于简化测试用例的开发。

从版本 2.0 开始，Selenium 集成了 WebDriver API，从而解决了许多限制，并提供了一个可选且更简单的编程接口。这样就有了一个全面的、面向对象的 API，从而为更多的浏览器提供额外的支持，同时改进了对现代高级 Web 应用程序测试问题的支持。

前端 Web 测试是一个复杂的主题，已超出本书的范围。从前面的简要介绍中，可以知道 Selenium 如何通过跨浏览器兼容性来帮助自动化实现与 Web 应用程序前端的用户交互。有关更多详细信息，请参阅 Selenium 的在线文档(http://seleniumhq.org/docs)。

13.6 小结

在本章中，介绍了如何在基于 Spring 的应用程序中使用常用的框架、库和工具(包括 JUnit、DbUnit 和 Mockito)开发各种单元测试。

首先，本章对企业测试框架进行了详细描述，介绍了在应用程序开发生命周期的每个阶段应执行哪些测试。其次，开发了两种类型的测试，包括逻辑单元测试和集成单元测试。最后，简要介绍了前端测试框架 Selenium。

测试企业级应用程序是一个巨大的话题，如果想更详细地了解 JUnit 库，推荐阅读由 Petar Tahchiev 撰写的 *JUnit in Action*(Manning 出版社于 2011 年出版)一书，其中介绍了 JUnit 4.8。

如果对更多的Spring测试方法感兴趣，可以在 *Pivotal Certified Professional Spring Developer Exam* 一书(www.apress.com/us/book/9781484208120，Apress出版社于 2016 年出版)中找到更多相关信息，其中包括很多的专业章节，里面介绍了更多测试库以及如何使用Spring Boot Test来测试Spring Boot应用程序。

第 14 章

Spring 中的脚本支持

在前面的章节中，介绍了 Spring 框架如何帮助 Java 开发人员创建 JEE 应用程序。通过使用 Spring 框架的 DI 机制及其与每个层的集成(通过 Spring 框架自身模块中的库或通过与第三方库的集成来实现)，可以极大简化实现和维护业务逻辑。

但是，到目前为止所开发的所有逻辑都是用 Java 语言编写的。尽管 Java 是历史上最成功的编程语言之一，但仍因为存在一些缺点而饱受批评，其中包括语言结构以及在大规模并行处理等领域缺乏全面支持。

例如，Java 语言的一个特点是所有变量都是静态类型的。换句话说，在 Java 程序中，声明的每个变量都应该有与它相关联的静态类型(String、int、Object、ArrayList 等)。但是，在某些情况下，动态类型可能更受欢迎，而诸如 JavaScript 之类的动态语言支持动态类型。

为了满足这些需求，已经开发了许多脚本语言。一些最流行的语言包括 JavaScript、Groovy、Scala、Ruby 和 Erlang。几乎所有这些语言都支持动态类型，并且旨在提供 Java 中不可用的功能，以及针对其他特定目的。例如，Scala(www.scala-lang.org)将函数式编程模式与 OO 模式相结合，并支持具有 Actor 和消息传递等概念的更全面且可扩展的并发编程模型。此外，Groovy(http://groovy.codehaus.org)提供了一个简化的编程模型，并支持特定于领域语言(DSL)的实现，使应用程序代码更易于阅读和维护。

这些脚本语言为 Java 开发人员带来的另一个重要概念是闭包(将在本章后面详细讨论)。简而言之，*闭包(closure)*是在封装到对象中的一段代码(或块)。就像 Java 方法一样，闭包是可执行的，可以接收参数并返回对象和值。另外，闭包还是普通的对象，可以在应用程序中通过引用进行传递，就像 Java 中的 POJO 一样。

在本章中，将介绍脚本语言背后的一些主要概念，重点关注 Groovy；你将会看到 Spring 框架如何无缝地使用脚本语言，为基于 Spring 的应用程序提供特定的功能。具体而言，本章包含以下主题：

- **Java 中的脚本支持**：在 JCP 中，JSR-223(用于 Java 平台的脚本)支持 Java 中的脚本语言；自 SE 6 以来，它就可以在 Java 中使用。本章将概述 Java 中的脚本支持。
- **Groovy**：本章将对 Groovy 语言进行详细介绍，该语言是与 Java 一起使用的最流行的脚本语言之一。
- **在 Spring 中使用 Groovy**：Spring 框架为脚本语言提供全面支持。自 3.1 版本以来，Spring 提供对 Groovy、JRuby 和 BeanShell 的开箱即用支持。

我们并不打算把本章作为使用脚本语言的详细参考资料。每种脚本语言都有不错的专业参考书，里面详细讨论了它们的设计和用法。本章的主要目标是介绍 Spring 框架如何支持脚本语言，并举例说明在基于 Spring 的应用程序中使用除 Java 外的脚本语言有什么好处。

14.1 在 Java 中使用脚本支持

从 Java 6 开始，Java Platform API(JSR-223)的脚本就被捆绑到 JDK 中，目的是提供一种标准机制，进而在 JVM 上运行以其他脚本语言编写的逻辑。此外，JDK 6 还捆绑了名为 Mozilla Rhino 的引擎，它可以评估 JavaScript 程序。本节将介绍 JDK 6 中的 JSR-223 支持。

在 JDK 6 中，脚本支持类驻留在 javax.script 包中。首先让我们开发一个简单的程序来检索脚本引擎列表。以下代码片段显示了脚本支持类的内容：

```
package com.apress.prospring5.ch14;

import javax.script.ScriptEngineManager;
```

```
import org.slf4j.Logger;
import org.slf4j.LoggerFactory;

public class ListScriptEngines {
    private static Logger logger =
     LoggerFactory.getLogger(ListS criptEngines.class);

    public static void main(String... args) {
        ScriptEngineManager mgr = new ScriptEngineManager();

        mgr.getEngineFactories().forEach(factory -> {
    String engineName = factory.getEngineName();
    String languageName = factory.getLanguageName();
    String version = factory.getLanguageVersion();
    logger.info("Engine name: " + engineName + " Language: "
    + languageName + " Version: " + version);
        });
    }
}
```

首先，创建ScriptEngineManager类的一个实例，用于发现并维护类路径中的引擎列表(换句话说，实现了javax.script.ScriptEngine 接口的类)。然后，通过调用 ScriptEngineManager.getEngineFactories() 方法检索ScriptEngineFactory接口的列表。ScriptEngineFactory接口用于描述和实例化脚本引擎。从每个ScriptEngineFactory接口可以检索有关脚本语言支持的信息。运行该程序可能会产生不同的输出，具体取决于设置，但在控制台中应该可以看到以下类似的内容：

```
INFO: Engine name: AppleScriptEngine Language: AppleScript Version: 2.5
INFO: Engine name: Oracle Nashorn Language: ECMAScript Version: ECMA - 262 Edition 5.1
```

接下来，编写一个简单的程序来评估基本的 JavaScript 表达式。该程序如下所示：

```
package com.apress.prospring5.ch14.javascript;

import org.slf4j.Logger;
import org.slf4j.LoggerFactory;

import javax.script.ScriptEngine;
import javax.script.ScriptEngineManager;
import javax.script.ScriptException;

public class JavaScriptTest {

    private static Logger logger =
        LoggerFactory.getLogger(JavaScriptTest.class);

    public static void main(String... args) {
        ScriptEngineManager mgr = new ScriptEngineManager();
        ScriptEngine jsEngine = mgr.getEngineByName("JavaScript");
        try {
            jsEngine.eval("print('Hello JavaScript in Java')");
        } catch (ScriptException ex) {
            logger.error("JavaScript expression cannot be evaluated!", ex);
        }
    }
}
```

上述代码首先通过 ScriptEngineManager 类(使用了名称 JavaScript)获取 ScriptEngine 接口的一个实例。然后调用ScriptEngine.eval()方法，并传入一个包含 JavaScript 表达式的 String 参数。请注意，参数也可以是一个 java.io.Reader类，它可以从文件中读取 JavaScript。

运行该程序会产生以下结果：

```
Hello JavaScript in Java
```

通过上面的示例，你应该已经了解了如何在 Java 中运行脚本。但是，使用其他语言转储某些输出是没有任何意义的。在下一节中，将介绍 Groovy，这是一种强大而全面的脚本语言。

14.2 Groovy 介绍

Groovy 在 2003 年由 James Strachan 创建，其主要目的是为 JVM 提供灵活的动态语言，其功能受其他流行脚本语言(包括 Python、Ruby 和 Smalltalk)的启发。Groovy 构建在 Java 基础之上，它扩展了 Java，并弥补了 Java 中的一些缺陷。

在下面的章节中，将讨论 Groovy 背后的一些主要功能和概念，以及它如何通过补充 Java 以解决特定的应用程序需求。请注意，这里提到的许多功能也可用于其他脚本语言(例如 Scala、Erlang、Python 和 Clojure)。

14.2.1 动态类型化

Groovy(以及许多其他脚本语言)和 Java 之间的一个主要区别是支持变量的动态类型化。在 Java 中，所有的属性和变量都应该是静态类型的。换句话说，应该使用 declare 语句来提供声明。但是，Groovy 支持变量的动态类型化。在 Groovy 中，动态类型变量用关键字 def 声明。

接下来通过开发一个简单的 Groovy 脚本来实际看一下。Groovy 类或脚本的文件后缀是 groovy。以下代码片段显示了一个简单的 Groovy 脚本，其中包含动态类型化操作：

```
package com.apress.prospring5.ch14

class Singer {
    def firstName
    def lastName
    def birthDate
    String toString() {
   "($firstName,$lastName,$birthDate)"
    }
}

Singer singer = new Singer(firstName: 'John', lastName: 'Mayer',
    birthDate: new Date(
    (new GregorianCalendar(1977, 9, 16)).getTime().getTime()))

Singer anotherSinger =
    new Singer(firstName: 39, lastName: 'Mayer', birthDate: new Date(
    (new GregorianCalendar(1977, 9, 16)).getTime().getTime()))

println singer
println anotherSinger

println singer.firstName + 39
println anotherSinger.firstName + 39
```

该 Groovy 脚本可以直接在 IDE 中运行，无须编译即可执行(Groovy 提供了一个名为 groovy 的命令行工具，可以直接执行 Groovy 脚本)，也可以首先编译为 Java 字节码，然后像其他 Java 类一样执行。Groovy 脚本不需要 main()方法就可以执行。此外，也不需要与文件名匹配的类声明。

在该例中，定义了一个类 Singer，其属性被设置使用 def 关键字进行动态类型化，共声明了三个属性。然后，使用一个闭包重写了 toString()方法，并返回一个字符串。

接着，构造两个 Singer 对象，并使用 Groovy 提供的简写语法来定义属性。对于第一个 Singer 对象，为 firstName 属性提供一个字符串，为第二个 Singer 对象的 firstName 属性提供一个整数。最后，println 语句(等同于调用 System.out.println()方法)用于打印两个 Singer 对象的相关信息。为了演示 Groovy 如何处理动态类型，定义两条 println 语句来打印 firstName + 39 操作的输出。请注意，在 Groovy 中，当向方法传递参数时，括号是可选的。

运行该程序会产生以下输出：

```
John,Mayer,Sun Oct 16 00:00:00 EET 1977
39,Mayer,Sun Oct 16 00:00:00 EET 1977
John39
78
```

从输出中可以看到，由于 firstName 是使用动态类型定义的，因此在传入 String 或 Integer 作为类型时，该对象会构造成功。另外，在最后两条 println 语句中，加法操作也被正确应用于两个 Singer 对象的 firstName 属性。在第一种情况下，由于 firstName 是一个字符串，因此字符串 39 被附加到字符串的后面；而对于第二种情况，由于 firstName 是一个整数，因此将整数 39 与该整数相加，结果为 78。

Groovy 的动态类型支持为处理应用程序逻辑中的类属性和变量提供了更大的灵活性。

14.2.2 简化的语法

Groovy 还提供了简化的语法，以便可以在 Groovy 中使用较少的代码来实现 Java 中相同的逻辑。一些基本的语法如下所示：

- 语句的末尾不需要分号。
- 在方法中，关键字 return 是可选的。
- 所有方法和类都默认为公共的。因此，除非需要，否则不必为方法声明使用 public 关键字。
- 在一个类中，Groovy 会为声明的属性自动生成 getter/setter 方法。因此，在 Groovy 类中，只需要声明类型和名称(例如，String firstName 或 def firstName)，并且可以通过自动使用 getter/setter 方法访问任何其他 Groovy/Java 类中的属性。另外，还可以简单地访问属性而不需要 get/set 前缀(例如，singer.firstName ='John')。Groovy 将会智能地代为处理。

Groovy 还为 Java Collection API 提供简化的语法和许多有用的方法。以下代码片段显示了一些用于列表操作的常用 Groovy 操作：

```
def list = ['This', 'is', 'John Mayer']
println list

assert list.size() == 3
assert list.class == ArrayList

assert list.reverse() == ['John Mayer', 'is', 'This']

assert list.sort{ it.size() } == ['is', 'This', 'John Mayer']

assert list[0..1] == ['is', 'This']
```

上面的代码只显示了· Groovy 所提供的一小部分功能。有关更详细的说明，请参阅 http://groovy.codehaus.org/JN1015-Collections 上的 Groovy 联机文档。

14.2.3 闭包

Groovy 向 Java 添加的最重要功能之一是支持闭包。闭包允许将一段代码打包为一个对象，并在应用程序内自由传递。闭包是一个强大的功能，可实现智能和动态行为。长期以来，向 Java 语言增加闭包支持的呼声一直很高。现在，JSR-335(Java 编程语言的 lambda 表达式)已被添加到 Java 8 中并由新的 Spring Framework 4 支持，旨在通过在 Java 语言中添加闭包和相关功能来支持多核环境中的编程。

以下代码片段显示了 Groovy 中使用闭包(文件名为 Runner.groovy)的简单示例：

```
def names = ['John', 'Clayton', 'Mayer']

names.each {println 'Hello: ' + it}
```

上述示例首先声明一个列表，然后，使用 each()方法遍历列表中的每一项。each()方法的参数是一个闭包，包含在 Groovy 的大括号中。因此，闭包中的逻辑将被应用于列表中的每一项。闭包内部是 Groovy 用来表示当前上下文中条目的一个特殊变量。所以，闭包会在列表中的每一项前添加字符串“Hello:”并打印出来。运行该脚本会产生以下输出：

```
Hello: John
Hello: Clayton
Hello: Mayer
```

如前所述，可以将闭包声明为变量并在需要时使用。另一个示例如下所示：

```
def map = ['a': 10, 'b': 50]

Closure square = {key, value -> map[key] = value * value}

map.each square

println map
```

在该例中，首先定义了一个映射，然后，声明一个 Closure 类型的变量。闭包接收映射条目的键和值作为参数，并计算值的平方。运行该程序会产生以下输出：

```
[a:100, b:2500]
```

这里只是简单地介绍一下闭包。在下一节中，将使用 Groovy 和 Spring 开发一个简单的规则引擎；封闭也会被使用。有关在 Groovy 中使用闭包的更详细说明，请参阅 http://groovy.codehaus.org/JN2515-Closures 上的联机文档。

14.3 与 Spring 一起使用 Groovy

Groovy 和其他脚本语言为基于 Java 的应用程序带来的主要好处是支持动态行为。通过使用闭包，可以将业务逻辑打包为对象并像任何其他变量一样在应用程序中传递。

Groovy 的另一个主要功能是通过使用简化的语法和闭包来支持开发 DSL。顾名思义，DSL 是针对特定领域的语言，在设计和实现方面具有非常特定的目标。其目的是建立一种不仅可以被开发者理解，也可以被业务分析师和用户理解的语言。大多数时候，涉及的都是商业领域。例如，可以为客户分类、销售费用计算、工资计算等定义 DSL。

在本节中，将演示如何使用 Groovy 实现具有 Groovy DSL 支持的简单规则引擎。该实现参考了 www.pleus.net/articles/grules/grules.pdf 上有关该主题的优秀文章中的示例，并进行了修改。另外，还将介绍 Spring 对可刷新 bean 的支持，讨论如何在无须编译、打包和部署应用程序的情况下即时更新底层规则。

在该例中，实现了一种规则，用于根据年龄将特定歌手划分到不同的类别，其中年龄根据他们的出生日期属性计算得出。

14.3.1 开发 Singer 对象域

如前所述，DSL 针对特定的域，大多数情况下该域指的是某种业务数据。对于将要实现的规则，指的是歌手信息域。

所以，第一步是开发规则适用的域对象模型。本例很简单，只包含一个 Singer 实体类，如下所示。请注意，它是一个 POJO 类，就像前几章中使用的那样。

```
package com.apress.prospring5.ch14;

import org.joda.time.DateTime;

public class Singer {
    private Long id;
    private String firstName;
    private String lastName;
    private DateTime birthDate;
    private String ageCategory;

   ... //getters and setter

    @Override
    public String toString() {
    return "Singer - Id: " + id + ", First name: " + firstName
        + ", Last name: " + lastName + ", Birthday: " + birthDate
        + ", Age category: " + ageCategory;
    }
}
```

此处的 Singer 类包括简单的歌手信息。对于 ageCategory 属性，想要开发一条可用于执行分类的动态规则。该规则将根据 birthDate 属性计算年龄，然后根据规则为 ageCategory 属性赋值(例如，孩子、青少年或成人)。

14.3.2 实现规则引擎

下一步是开发一个简单的规则引擎，在域对象上应用规则。首先，需要定义规则需要包含哪些信息。以下代码片段显示了 Rule 类，它是一个 Groovy 类(文件名为 Rule.groovy)：

```
package com.apress.prospring5.ch14

class Rule {
    private boolean singlehit = true
    private conditions = new ArrayList()
    private actions = new ArrayList()
    private parameters = new ArrayList()
}
```

每条规则都有几个属性。conditions 属性定义了规则引擎应该检查处理的域对象的各种条件。actions 属性定义了在条件匹配时要采取的动作。parameters 属性定义了规则的行为，即不同条件下行为的结果。最后，singlehit 属性定义了规则是否应该在找到匹配的条件时立即结束其执行。

下一步是定义规则执行引擎。以下代码片段显示了 RuleEngine 接口(请注意它是一个 Java 接口)：

```
package com.apress.prospring5.ch14;

public interface RuleEngine {
    void run(Rule rule, Object object);
}
```

该接口只定义了一个方法 run()，它将规则应用于域对象参数。

接下来在 Groovy 中实现规则引擎。以下代码片段显示了 Groovy 类 RuleEngineImpl(文件名为 RuleEngineImpl.groovy)：

```
package com.apress.prospring5.ch14
import org.slf4j.Logger
import org.slf4j.LoggerFactory
import org.springframework.stereotype.Component

@Component("ruleEngine")
class RuleEngineImpl implements RuleEngine {
    Logger logger = LoggerFactory.getLogger(RuleEngineImpl.class);

    void run(Rule rule, Object object) {
    logger.info "Executing rule"

    def exit=false

    rule.parameters.each{ArrayList params ->
        def paramIndex=0
        def success=true
        if(!exit){
            rule.conditions.each{
                logger.info "Condition Param index: " + paramIndex
                success = success && it(object,paramsparamIndex)
                logger.info "Condition success: " + success
                paramIndex++
            }

        if(success && !exit){
            rule.actions.each{
                logger.info "Action Param index: " + paramIndex
                it(object,paramsparamIndex)
                paramIndex++
            }

        if (rule.singlehit){
            exit=true
        }
        }
        }
    }
    }
}
```

首先，RuleEngineImpl 实现了 Java 接口 RuleEngine，并像其他 POJO 一样应用了 Spring 注解。在 run()方法中，将规则中定义的参数传递给一个闭包，以便逐个进行处理。而对于每个参数(此处为一个值列表)，将根据参数列表中的对应项和域对象逐个检查条件(每个条件都是一个闭包)。只有当所有条件都正向匹配时，成功指示器才会变为 true。在这种情况下，规则中定义的动作(每个动作也是一个闭包)将在对象上执行，并使用参数列表中相应的值。最后，如果找到特定参数的匹配项，并且 singlehit 变量为 true，规则执行将停止并立即退出。

为了允许以更灵活的方式检索规则，接下来定义一个 RuleFactory 接口，如下所示。请注意，它是一个 Java 接口。

```
package com.apress.prospring5.ch14;

public interface RuleFactory {
    Rule getAgeCategoryRule();
}
```

由于歌手的年龄类别分类只有一条规则，因此该接口仅定义了一个用于检索规则的方法。

为了使规则引擎对消费者透明，需要开发一个简单的服务层来封装它。以下代码片段分别显示了 SingerService 接口和 SingerServiceImpl 类。请注意，它们都是 Java 实现。

```
//SingerService.java
```

```
package com.apress.prospring5.ch14;

public interface SingerService {
   void applyRule(Singer singer);
}
//SingerServiceImpl.java
package com.apress.prospring5.ch14;

import org.springframework.beans.factory.annotation.Autowired;
import org.springframework.context.ApplicationContext;
import org.springframework.stereotype.Service;

@Service("singerService")
public class SingerServiceImpl implements SingerService {
   @Autowired
   ApplicationContext ctx;

   @Autowired
   private RuleFactory ruleFactory;

   @Autowired
   private RuleEngine ruleEngine;
   public void applyRule(Singer singer) {
      Rule ageCategoryRule = ruleFactory.getAgeCategoryRule();
      ruleEngine.run(ageCategoryRule, singer);
   }
}
```

如你所见，所需的 Spring bean 被自动装配到服务实现类中。在 applyRule()方法中，首先从规则工厂中获取规则，然后应用于 Singer 对象。最终结果是根据规则的定义条件、操作和参数派生出 Singer 对象的 ageCategory 属性。

14.3.3 将规则工厂实现为 Spring 可刷新 bean

现在可以实施规则工厂和年龄分类规则。我们希望能够即时更新规则，并让 Spring 检查所做的更改并应用最新的逻辑。Spring 框架为使用脚本语言编写的 Spring bean(被称为*可刷新 bean*)提供很好的支持。接下来看一看如何将 Groovy 脚本配置为 Spring bean，并指示 Spring 定期刷新 bean。首先看看 Groovy 中规则工厂的实现。为了允许动态刷新，需要将类放入外部文件夹，以便进行修改。可以将外部文件夹命名为 rules。RuleFactoryImpl 类(这是一个 Groovy 类，名为 RuleFactoryImpl.groovy)将被放置到外部文件夹中。以下代码片段显示了该类的内容：

```
package com.apress.prospring5.ch14

import org.joda.time.DateTime
import org.joda.time.Years
import org.springframework.stereotype.Component;

@Component
class RuleFactoryImpl implements RuleFactory {
   Closure age = { birthDate -> return
   Years.yearsBetween(birthDate, new DateTime()).getYears() }
   Rule getAgeCategoryRule() {
   Rule rule = new Rule()

   rule.singlehit=true

   rule.conditions=[ {object, param -> age(object.birthDate) >= param},
      {object, param -> age(object.birthDate) <= param}]

   rule.actions=[{object, param -> object.ageCategory = param}]

   rule.parameters=[
        [0,10,'Kid'],
        [11,20,'Youth'],
        [21,40,'Adult'],
        [41,60,'Matured'],
        [61,80,'Middle-aged'],
        [81,120,'Old']
   ]

   return rule
   }
}
```

该类实现了 RuleFactory 接口，并通过 getAgeCategoryRule()方法来提供规则。在规则中，定义了一个名为 age

的闭包，它将根据 Singer 对象的 birthDate 属性(JodaTime 的 DateTime 类型)来计算年龄。

此处在规则中还定义了两个条件。第一个条件检查歌手的年龄是否大于或等于提供的参数值，第二个条件检查小于或等于参数值的情况。

然后定义了一个动作，将参数中提供的值分配给 Singer 对象的 ageCategory 属性。

最后，用参数定义条件检查和操作的值。例如，第一个参数表示当年龄在 0 到 10 岁之间时将值 Kid 分配给 Singer 对象的 ageCategory 属性，依此类推。因此，对于每个参数，所定义的两个条件将使用前两个值进行年龄范围检查，而最后一个值将用于指定 ageCategory 属性。

下一步是定义 Spring ApplicationContext。以下配置代码片段显示了配置文件(app-context.xml)：

```xml
<?xml version="1.0" encoding="UTF-8"?>
<beans xmlns="http://www.springframework.org/schema/beans"
      xmlns:xsi="http://www.w3.org/2001/XMLSchema-instance"
      xmlns:context="http://www.springframework.org/schema/context"
      xmlns:lang="http://www.springframework.org/schema/lang"
      xsi:schemaLocation="http://www.springframework.org/schema/beans
   http://www.springframework.org/schema/beans/spring-beans.xsd
   http://www.springframework.org/schema/context
   http://www.springframework.org/schema/context/spring-context.xsd
   http://www.springframework.org/schema/lang
   http://www.springframework.org/schema/lang/spring-lang.xsd">
   <context:component-scan base-package="com.apress.prospring5.ch14" />

   <lang:groovy id="ruleFactory" refresh-check-delay="5000"
      script-source="file:rules/RuleFactoryImpl.groovy"/>
</beans>
```

该配置很简单。为了使用脚本语言定义 Spring bean，需要使用 lang-namespace。然后，<lang:groovy>标记用于声明一个带有 Groovy 脚本的 Spring bean。script-source 属性定义了 Spring 将加载的 Groovy 脚本的位置。对于可刷新的 bean，应该提供属性 refresh-check-delay。此时，值为 5000 毫秒，指示 Spring 一旦上次调用所用的时间大于 5 秒，就要检查文件更改。请注意，Spring 并不会每 5 秒就检查一次文件。相反，它只会在相应的 bean 被调用时才检查文件。

14.3.4 测试年龄分类规则

现在准备测试规则。下面显示了测试程序，它是一个 Java 类：

```java
package com.apress.prospring5.ch14;

import org.joda.time.DateTime;
import org.slf4j.Logger;
import org.slf4j.LoggerFactory;
import org.springframework.context.support.GenericXmlApplicationContext;

public class RuleEngineDemo {
   private static Logger logger =
   LoggerFactory.getLogger(RuleEngineTest.class);

   public static void main(String... args) throws Exception {
   GenericXmlApplicationContext ctx = new GenericXmlApplicationContext();
   ctx.load("classpath:spring/app-context.xml");
   ctx.refresh();

   SingerService singerService =
      ctx.getBean("singerService", SingerService.class);

   Singer singer = new Singer();
   singer.setId(1l);
   singer.setFirstName("John");
   singer.setLastName("Mayer");
   singer.setBirthDate(
      new DateTime(1977, 10, 16, 0, 0, 0, 0));
   singerService.applyRule(singer);
   logger.info("Singer: " + singer);

   System.in.read();
   singerService.applyRule(singer);
   logger.info("Singer: " + singer);
```

```
        ctx.close();
    }
}
```

在初始化 Spring 的 GenericXmlApplicationContext 时，首先会创建一个 Singer 对象。然后，获取 SingerService 接口的实例，将规则应用到 Singer 对象上，并将结果输出到控制台。在第二次应用规则之前，程序将被暂停以供用户输入。在暂停期间，可以修改 RuleFactoryImpl.groovy 类，Spring 会刷新 bean，可以看到更改后的规则可以正常运行。

运行测试程序会产生以下输出：

```
00:34:24.814 [main] INFO c.a.p.c.RuleEngineImpl - Executing rule
00:34:24.822 [main] INFO c.a.p.c.RuleEngineImpl - Condition Param index: 0
00:34:24.851 [main] INFO c.a.p.c.RuleEngineImpl - Condition success: true
00:34:24.858 [main] INFO c.a.p.c.RuleEngineImpl - Condition Param index: 1
00:34:24.858 [main] INFO c.a.p.c.RuleEngineImpl - Condition success: false
00:34:24.858 [main] INFO c.a.p.c.RuleEngineImpl - Condition Param index: 0
00:34:24.858 [main] INFO c.a.p.c.RuleEngineImpl - Condition success: true
00:34:24.858 [main] INFO c.a.p.c.RuleEngineImpl - Condition Param index: 1
00:34:24.858 [main] INFO c.a.p.c.RuleEngineImpl - Condition success: false
00:34:24.859 [main] INFO c.a.p.c.RuleEngineImpl - Condition Param index: 0
00:34:24.859 [main] INFO c.a.p.c.RuleEngineImpl - Condition success: true
00:34:24.859 [main] INFO c.a.p.c.RuleEngineImpl - Condition Param index: 1
00:34:24.859 [main] INFO c.a.p.c.RuleEngineImpl - Condition success: true
00:34:24.860 [main] INFO c.a.p.c.RuleEngineImpl - Action Param index: 2
00:34:24.870 [main] INFO c.a.p.c.RuleEngineDemo - Singer: Singer - Id: 1,
    First name: John, Last name: Mayer, Birthday: 1977-10-16T00:00:00.000+03:00,
    Age category: Adult
```

通过输出中的日志记录可以看到，由于歌手的年龄是 39 岁，因此可以看到规则会在第三个参数中找到匹配项(换句话说，即[21,40,'Adult'])。最终的结果是 ageCategory 被设置为 Adult。

现在暂停程序，并且更改 RuleFactoryImpl.groovy 类中的参数。可以在以下代码片段中看到相关修改：

```
rule.parameters=[
    [0,10,'Kid'],
    [11,20,'Youth'],
    [21,30,'Adult'],
    [31,60,'Middle-aged'],
    [61,120,'Old']
]
```

按照上述代码更改并保存文件。紧接着，在控制台区域按 Enter 键，从而对同一个对象第二次应用规则。程序继续运行，并生成以下输出：

```
00:48:50.137 [main] INFO c.a.p.c.RuleEngineImpl - Executing rule
00:48:50.137 [main] INFO c.a.p.c.RuleEngineImpl - Condition Param index: 0
00:48:50.137 [main] INFO c.a.p.c.RuleEngineImpl - Condition success: true
00:48:50.138 [main] INFO c.a.p.c.RuleEngineImpl - Condition Param index: 1
00:48:50.138 [main] INFO c.a.p.c.RuleEngineImpl - Condition success: false
00:48:50.138 [main] INFO c.a.p.c.RuleEngineImpl - Condition Param index: 0
00:48:50.138 [main] INFO c.a.p.c.RuleEngineImpl - Condition success: true
00:48:50.138 [main] INFO c.a.p.c.RuleEngineImpl - Condition Param index: 1
00:48:50.138 [main] INFO c.a.p.c.RuleEngineImpl - Condition success: false
00:48:50.138 [main] INFO c.a.p.c.RuleEngineImpl - Condition Param index: 0
00:48:50.139 [main] INFO c.a.p.c.RuleEngineImpl - Condition success: true
00:48:50.139 [main] INFO c.a.p.c.RuleEngineImpl - Condition Param index: 1
00:48:50.139 [main] INFO c.a.p.c.RuleEngineImpl - Condition success: false
00:48:50.139 [main] INFO c.a.p.c.RuleEngineImpl - Condition Param index: 0
00:48:50.139 [main] INFO c.a.p.c.RuleEngineImpl - Condition success: true
00:48:50.139 [main] INFO c.a.p.c.RuleEngineImpl - Condition Param index: 1
00:48:50.139 [main] INFO c.a.p.c.RuleEngineImpl - Condition success: true
00:48:50.139 [main] INFO c.a.p.c.RuleEngineImpl - Action Param index: 2
00:48:50.139 [main] INFO c.a.p.c.RuleEngineDemo - Singer: Singer - Id: 1,
    First name: John, Last name: Mayer, Birthday: 1977-10-16T00:00:00.000+03:00,
    Age category: Middle-aged
```

在前面的输出中，可以看到规则执行在第四个参数(换句话说，即[31,60，'Middleaged'])处停止。因此，Middle-aged 值被分配给 ageCategory 属性。

为完成此例所参考的文章(http://pleus.net/articles/grules/grules.pdf)还演示了如何将规则参数外部化为 Microsoft Excel 文件，所以用户可以自己准备和更新参数文件。

当然，上述规则很简单，仅仅是为了演示诸如 Groovy 之类的脚本语言如何能够在特定领域(例如，使用带有 DSL 的规则引擎)帮助补充基于 Spring 的 Java EE 应用程序。

你可能会问，“如果可以将规则存储到数据库中，然后让 Spring 的可刷新 bean 功能检测数据库中的更改，那么是不是又进了一步呢？”这样做有助于进一步简化规则的维护，可以为用户(或管理员)提供前端以便动态地将规则更新到数据库中，而不是上传文件。

实际上，就以上方案，Spring 框架中已经存在一个 JIRA 功能(https://jira.springsource.org/browse/ SPR-5106)。请持续关注此功能。同时，提供用户前端来上传规则类也是一个可行的解决方案。当然，此时应该格外小心，在将规则上传到生产环境之前，应该对规则进行彻底测试。

14.3.5 内联动态语言代码

动态语言代码不仅可以从外部源文件执行，还可以直接将这些代码内联到 bean 配置中。虽然这种做法在某些情况下可能有用，例如概念的快速验证等，但从可维护性的角度来看，使用这种做法构建整个应用程序并不是很好。以前面的 Rule 引擎为例，删除 RuleEngineImpl.groovy 文件并将代码移动到内联 bean 定义(在文件 app-context.xml 中)中，如以下代码片段所示：

```
<?xml version="1.0" encoding="UTF-8"?>
<beans xmlns="http://www.springframework.org/schema/beans"
   xmlns:xsi="http://www.w3.org/2001/XMLSchema-instance"
   xmlns:context="http://www.springframework.org/schema/context"
   xmlns:lang="http://www.springframework.org/schema/lang"
   xsi:schemaLocation="http://www.springframework.org/schema/beans
      http://www.springframework.org/schema/beans/spring-beans.xsd
      http://www.springframework.org/schema/context
      http://www.springframework.org/schema/context/spring-context.xsd
      http://www.springframework.org/schema/lang
      http://www.springframework.org/schema/lang/spring-lang.xsd">

   <context:component-scan base-package="com.apress.prospring5.ch14"/>

   <lang:groovy id="ruleFactory" refresh-check-delay="5000">
      <lang:inline-script>
         <![CDATA[

   package com.apress.prospring5.ch14

   import org.joda.time.DateTime
   import org.joda.time.Years
   import org.springframework.stereotype.Component;

   @Component
   class RuleFactoryImpl implements RuleFactory {
      Closure age = { birthDate -> return
         Years.yearsBetween(birthDate, new DateTime()).getYears() }

      Rule getAgeCategoryRule() {

         Rule rule = new Rule()

         rule.singlehit = true

         rule.conditions = [{ object, param -> age(object.birthDate) >= param },
         { object, param -> age(object.birthDate) <= param }]

         rule.actions = [{ object, param -> object.ageCategory = param }]

         rule.parameters = [
             [0, 10, 'Kid'],
             [11, 20, 'Youth'],
             [21, 40, 'Adult'],
             [41, 60, 'Matured'],
             [61, 80, 'Middle-aged'],
             [81, 120, 'Old']
         ]
         return rule
      }
   }
   ]]>
      </lang:inline-script>
   </lang:groovy>
</beans>
```

正如你所看到的，这里添加了 lang:groovy 标记，其 ID 为表示 bean 名称的 ruleFactory 的 ID。然后，使用

lang:inline-script 标记封装来自 RuleFactoryImpl.groovy 的 Groovy 代码。围绕 Groovy 代码的是一个 CDATA 标记，以避免 XML 解析器解析代码。现在，一切就绪，再次运行规则引擎示例。可以看到，除了直接将 Groovy 代码直接内联到 bean 定义中，而不是将它们放在外部文件中之外，工作方式是一样的。使用 RuleFactoryImpl.groovy 中的代码是有意为之，目的是演示当内联大量代码时应用程序会变得多么笨拙。

14.4 小结

在本章中，介绍了如何在 Java 应用程序中使用脚本语言，并演示了 Spring 框架对脚本语言的支持如何帮助为应用程序提供动态行为。

本章首先讨论了 JSR-223(用于 Java 平台的脚本)，它内置于 Java 6 中，支持开箱即用地执行 JavaScript。然后，介绍了在 Java 开发人员社区中非常流行的一种脚本语言 Groovy，并通过与传统的 Java 语言做对比，展示了它的一些主要功能。

最后，本章讨论了 Spring 框架中对脚本语言的支持。通过使用 Groovy 的 DSL 支持设计和实现了一个简单的规则引擎，从而了解该语言的具体运用。还讨论了如何修改规则并让 Spring 框架通过使用可刷新 bean 功能自动获取更改，而无须对应用程序进行编译、打包和部署。另外，本章展示了如何将 Groovy 代码直接内联到配置文件中以定义 bean 的实现代码。

第 15 章

应用程序监控

典型的 JEE 应用程序包含许多层和组件，例如表示层、服务层、持久化层和后端数据源。在开发阶段或将应用程序部署到质量保证(QA)或生产环境之后，通常需要确保应用程序处于健康状态，没有任何潜在问题或瓶颈。

在 Java 应用程序中，各个区域都可能会导致性能问题或服务器资源(如 CPU、内存或 I/O)过载。例如，Java 代码效率低下、内存泄漏(比如，在不释放引用的情况下继续分配新对象的 Java 代码，并且阻止底层 JVM 在垃圾收集过程中释放内存)、错误计算的 JVM 参数、错误计算的线程池参数、过多的数据源配置(例如，允许的并发数据库连接太多)、不正确的数据库设置以及长时间运行的 SQL 查询。

因此，你需要了解应用程序的运行时行为并确定是否存在任何潜在的瓶颈或问题。在 Java 世界中，很多工具可以帮助监控 JEE 应用程序的详细运行时行为。它们中的大多数都建立在 Java 管理扩展(Java Management Extensions，JMX)技术之上。

在本章中，将介绍用于监控基于 Spring 的 JEE 应用程序的常用技术。具体而言，本章包含以下主题：

- **JMX 的 Spring 支持**：将讨论 Spring 对 JMX 的全面支持，并演示如何使用 JMX 工具公开用于监控的 Spring bean。在本章中，将演示如何使用 Java 可执行文件 jvisualvm(https://visualvm.github.io/?Java_VisualVM)作为应用程序监控工具。
- **监控 Hibernate 统计信息**：Hibernate 和许多其他包为使用 JMX 公开操作状态和性能指标提供了支持类和基础结构。本章将演示如何在 Spring 驱动的 JEE 应用程序中启用那些常用组件的 JMX 监控。
- **Spring Boot JMX 支持**：Spring Boot 为 JMX 支持提供了一个启动器库，该库提供了全新的默认配置。

请记住，本章的目的并不是介绍 JMX，而是假定你已经对 JMX 有基本的了解。有关详细信息，请参阅 http://oracle.com/technetwork/java/javase/tech/javamanagement-140525.html 上的 Oracle 在线资源。

15.1 Spring 中的 JMX 支持

在 JMX 中，公开用于 JMX 监控和管理的类被称为*托管 bean*(通常称为 MBean)。Spring 框架支持多种公开 MBean 的机制。本章重点介绍如何将 Spring bean(被开发为简单的 POJO)公开为 MBean，从而进行 JMX 监控。

在下面的章节中，将讨论如何把包含与应用程序相关的统计信息的 bean 公开为用于 JMX 监控的 MBean 的过程。相关主题包括实现 Spring bean、在 Spring ApplicationContext 中将 Spring bean 公开为 MBean 以及使用 VisualVM 监控 MBean。

15.2 将 Spring bean 导出为 JMX

作为示例，本章将使用第 12 章中的 REST 示例。可以查看一下示例应用程序代码所在的章，或者直接跳转到本书的附录 A 部分，里面提供了构建示例所用的源代码。随着 JMX 的加入，我们希望公开数据库中歌手的数量以实现 JMX 监控目的。所以，需要实现相关的接口和类，如下所示：

```
//AppStatistics.java
package com.apress.prospring5.ch15;

public interface AppStatistics {
    int getTotalSingerCount();
}
```

```
//AppStatisticsImpl.java
package com.apress.prospring5.ch15;

import com.apress.prospring5.ch12.services.SingerService;
import org.springframework.beans.factory.annotation.Autowired;
import org.springframework.stereotype.Component;

public class AppStatisticsImpl implements AppStatistics {
    @Autowired
    private SingerService singerService;

    @Override
    public int getTotalSingerCount() {
        return singerService.findAll().size();
    }
}
```

在这个示例中,定义了一个方法来检索数据库中歌手记录的总数。为了将 Spring bean 公开为 JMX,需要在 Spring 的 ApplicationContext 中添加配置。第 12 章介绍了 Spring 安全的 Web 应用程序的配置。现在必须添加两个类型为 MBeanServer 和 MBeanExporter 的基础架构 bean,以便启用对 JMX 管理 bean 的支持。

```
package com.apress.prospring5.ch15.init;
...
import javax.management.MBeanServer;
import org.springframework.jmx.export.MBeanExporter;
import org.springframework.jmx.support.MBeanServerFactoryBean;

@Configuration
@EnableWebMvc
@ComponentScan(basePackages = {"com.apress.prospring5.ch15"})
public class WebConfig implements WebMvcConfigurer {
    //other Web infrastructure specific beans
    ...
    @Bean AppStatistics appStatisticsBean() {
        return new AppStatisticsImpl();
    }

    @Bean
    MBeanExporter jmxExporter() {
        MBeanExporter exporter = new MBeanExporter();
        Map<String, Object> beans = new HashMap<>();
        beans.put("bean:name=ProSpring5SingerApp", appStatisticsBean());
        exporter.setBeans(beans);
        return exporter;
    }
}
```

首先,使用想要公开的统计信息(AppStatisticsImpl)为 POJO 声明 bean。然后,使用实现类 MBeanExporter 声明 jmxExporter bean。

MBeanExporter 类是 Spring 框架对 JMX 提供支持的核心类。它负责使用 JMX MBean 服务器(一台实现了 JDK 的 javax.management.MBeanServer 接口的服务器,通常存在于最常用的 Web 和 JEE 容器中,例如 Tomcat 和 WebSphere)注册 Spring bean。当将 Spring bean 作为 MBean 公开时,Spring 会尝试在服务器上查找正在运行的 MBeanServer 实例,并使用它注册 MBean。例如,在 Tomcat 上会自动创建 MBeanServer 实例,因此不需要其他配置。

在 jmxExporter bean 中,属性 bean 定义了想要公开的 Spring bean。这是一个 Map 对象,可以指定任意数量的 MBean。在本例中,需要公开 appStatisticsBean bean,其中包含想要向管理员显示的歌手应用程序的有关信息。对于 MBean 定义,可以将键用作对应条目值所引用的 Spring bean 的 ObjectName 值(JDK 中的 javax.management. ObjectName 类)。在前面的配置中,appStatisticsBean 在 ObjectName bean:name = Prospring5SingerApp 下被公开。默认情况下,该 bean 的所有公共属性也都作为属性公开,而所有公共方法都作为操作公开。

现在,MBean 可以通过 JMX 进行监控。接下来继续设置 VisualVM 并使用它的 JMX 客户端进行监控。

15.3 使用 Java VisualVM 进行 JMX 监控

VisualVM 是一个有用的工具,可以帮助在各个方面监控 Java 应用程序。它是一个免费工具,位于 JDK 安装文件夹的 bin 文件夹下。也可以从项目网站下载独立版本[1]。在本章中将使用 JDK 安装版本。

VisualVM 使用一个插件系统来支持各种监控功能。为了支持监控 Java 应用程序的 MBean，需要安装 MBean 插件。要安装该插件，请按照下列步骤操作：

(1) 从 VisualVM 的菜单中选择 Tools | Plug-ins。

(2) 单击 Available Plgu-ins 选项卡。

(3) 单击 Check for Newest 按钮。

(4) 选择插件 VisualVM-MBeans，然后单击 Install 按钮。

图 15-1 显示了上述过程。完成安装后，验证 Tomcat 已启动并且示例应用程序正在运行。然后在 VisualVM 的左侧的 Applications 视图中，应该能够看到 Tomcat 进程正在运行。

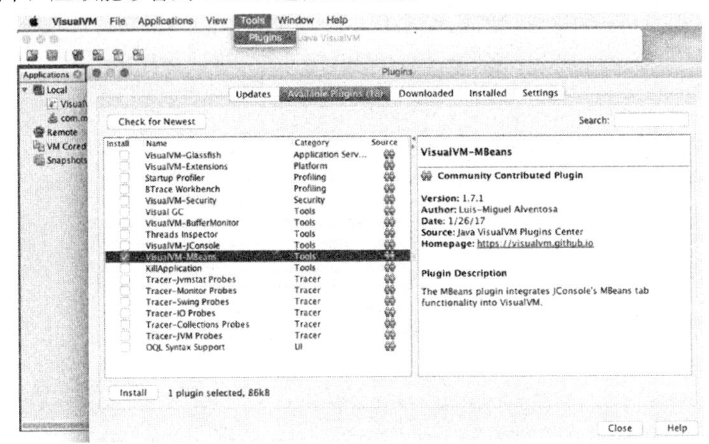

图 15-1　使用 JTA 进行全局事务概览

默认情况下，VisualVM 会扫描在 JDK 平台上运行的 Java 应用程序。双击目标节点将打开监控屏幕。

在安装 VisualVM-MBeans 插件之后，将会看到 MBeans 选项卡。打开该选项卡将显示可用的 MBean。同时还可以看到名为 bean 的节点。展开该节点，将显示已公开的 Prospring5SingerApp MBean。

在右侧，你将看到在该 bean 中实现的方法，并带有属性 TotalSingerCount(由 bean 中的 getTotalSingerCount()方法自动派生)，值为 3，对应于应用程序启动时已添加到数据库中的记录数。在常规应用程序中，这一数字将根据应用程序运行时添加的歌手数量而变化。

图 15-2 显示了公开 Prospring5SingerApp MBean 的 MBeans 选项卡。

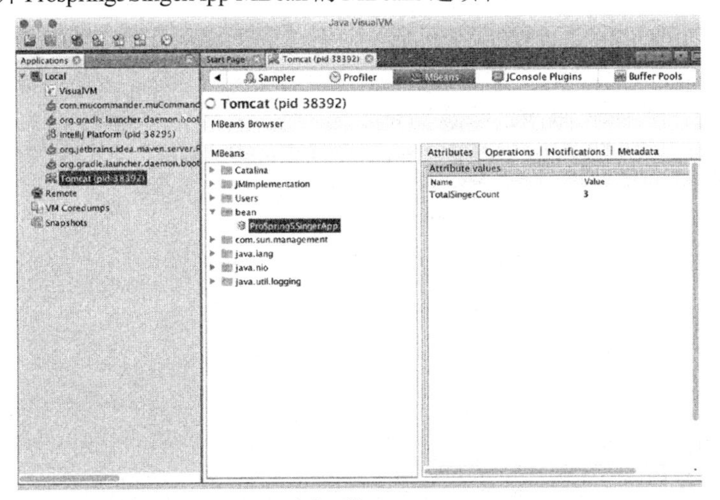

图 15-2　VisualVM 中公开的 Prospring5SingerApp MBean

1　在编写本书时，Java VisualVM 的当前版本是 1.3.9。可以在以下网址找到它：http://visualvm.java.net/download.html。

15.4 监视 Hibernate 统计信息

Hibernate 还可以维护与持久化相关的指标并向 JMX 公开。要启用这些功能，需要在 JPA 配置中添加三个 Hibernate 属性，如下所示：

```
package com.apress.prospring5.ch12.config;

import org.slf4j.Logger;
import org.slf4j.LoggerFactory;
import org.springframework.context.annotation.Bean;
import org.springframework.context.annotation.ComponentScan;
import org.springframework.context.annotation.Configuration;
import org.springframework.data.jpa.repository.config.EnableJpaRepositories;
import org.springframework.jdbc.datasource.embedded.EmbeddedDatabaseBuilder;
import org.springframework.jdbc.datasource.embedded.EmbeddedDatabaseType;
import org.springframework.orm.jpa.JpaTransactionManager;
import org.springframework.orm.jpa.JpaVendorAdapter;
import org.springframework.orm.jpa.LocalContainerEntityManagerFactoryBean;
import org.springframework.orm.jpa.vendor.HibernateJpaVendorAdapter;
import org.springframework.transaction.PlatformTransactionManager;
import javax.persistence.EntityManagerFactory;
import javax.sql.DataSource;
import java.util.Properties;

@Configuration
//using components that were introduced in Chapter 12 project
@EnableJpaRepositories(basePackages = {"com.apress.prospring5.ch12.repos"})
@ComponentScan(basePackages = {"com.apress.prospring5.ch12"} )
public class DataServiceConfig {
    ...

    @Bean
    public Properties hibernateProperties() {
        Properties hibernateProp = new Properties();
        hibernateProp.put("hibernate.dialect", "org.hibernate.dialect.H2Dialect");
        hibernateProp.put("hibernate.hbm2ddl.auto", "create-drop");
        hibernateProp.put("hibernate.show_sql", true);
        hibernateProp.put("hibernate.max_fetch_depth", 3);
        hibernateProp.put("hibernate.jdbc.batch_size", 10);
        hibernateProp.put("hibernate.jdbc.fetch_size", 50);

        hibernateProp.put("hibernate.jmx.enabled", true);
        hibernateProp.put("hibernate.generate_statistics", true);
        hibernateProp.put("hibernate.session_factory_name", "sessionFactory");
    return hibernateProp;
    }
    ...
}
```

属性 hibernate.jmx.enabled 用于启用 Hibernate JMX 行为。

属性 hibernate.generate_statistics 指示 Hibernate 为其 JPA 持久化提供程序生成统计信息，而属性 hibernate.session_factory_name 则定义了 Hibernate 统计信息 MBean 所需的会话工厂名称。

最后，需要将 MBean 添加到 Spring 的 MBeanExporter 配置中。以下配置片段显示了前面在 WebConfig 类中创建的更新的 MBean 配置。用 CustomStatistics 类替代 org.hibernate.jmx.StatisticsService，后者不再是 Hibernate 5 的一部分[1]。

```
package com.apress.prospring5.ch15.init;

...

@Configuration
@EnableWebMvc
@ComponentScan(basePackages = {"com.apress.prospring5.ch15"})
public class WebConfig implements WebMvcConfigurer {
    ...
    // JMX beans
    @Bean AppStatistics appStatisticsBean() {
        return new AppStatisticsImpl();
```

1 如果想丰富代码，那么 https://github.com/manuelbernhardt/hibernate-core/blob/master/hibernate-core/src/main/java/org/hibernate/jmx/StatisticsService.java 上的代码在 GitHub 上仍然可用。

```
    }

    @Bean CustomStatistics statisticsBean(){
        return new CustomStatistics();
    }

    @Autowired
    private EntityManagerFactory entityManagerFactory;

    @Bean SessionFactory sessionFactory(){
        return entityManagerFactory.unwrap(SessionFactory.class);
    }

    @Bean
    MBeanExporter jmxExporter() {
        MBeanExporter exporter = new MBeanExporter();
        Map<String, Object> beans = new HashMap<>();
        beans.put("bean:name=ProSpring5SingerApp", appStatisticsBean());
        beans.put("bean:name=Prospring5SingerApp-hibernate", statisticsBean());
        exporter.setBeans(beans);
        return exporter;
    }
}
```

此处使用 Hibernate 的 org.hibernate.stat.Statistics 实现作为核心组件来声明 statisticsBean()方法。这就是 Hibernate 支持向 JMX 公开统计信息的方式。

现在，Hibernate 统计信息已启用并可通过 JMX 使用。重新加载应用程序并刷新 VisualVM，你将能够看到 Hibernate 统计信息 MBean。单击该节点将在右侧显示详细统计信息。请注意，对于不是 Java 基本类型的信息(例如，列表)，可以单击该字段以将其展开并显示内容。

在 VisualVM 中，还可以看到许多其他指标，例如 EntityNames、SessionOpenCount、SecondCloseCount 和 QueryExecutionMaxTime。这些指标对于理解应用程序中的持久化行为很有用，并可帮助进行故障排除和性能调整。

15.5 使用了 Spring Boot 的 JMX

将以前的应用程序迁移到 Spring Boot 很容易，Spring Boot 提供依赖项并进行自动配置。对于 JMX，不需要启动器依赖项，但可以将 spring-bootstarter-actuator.jar 作为依赖项添加；如果使用特定于 Spring 的插件，则可以在智能编辑器中监控 Spring 应用程序并显示应用程序中的 bean、运行状况和映射。

本章中的示例应用程序是一个没有接口的 Web 应用程序(因为带有接口的 Web 应用程序是第 16 章将要讨论的主题)，同时使用了内存数据库和完整的 JTA 配置。由于这些内容在前面的章节中已经介绍过，因此本章的重点将放在 MBean 上。

接下来通过使用@ManagedResource 来升级 AppStatisticsImpl 类，从而使用 JMX 服务器注册该类的实例。这是非常实用的，因为默认情况下，Spring Boot 将创建一个 MBeanServer，其 bean ID 为 mbeanServer，并公开所有使用 Spring JMX 注解的 bean。

```
package com.apress.prospring5.ch15;

import com.apress.prospring5.ch15.entities.Singer;
import com.apress.prospring5.ch15.services.SingerService;
import org.springframework.beans.factory.annotation.Autowired;
import org.springframework.jmx.export.annotation.ManagedAttribute;
import org.springframework.jmx.export.annotation.ManagedOperation;
import org.springframework.jmx.export.annotation.ManagedResource;
import org.springframework.stereotype.Component;

import java.util.List;

@Component
@ManagedResource(description = "JMX managed resource",
    objectName = "jmxDemo:name=ProSpring5SingerApp")
public class AppStatisticsImpl implements AppStatistics {

    @Autowired
    private SingerService singerService;

    @ManagedAttribute(description = "Number of singers in the application")
    @Override
```

```
    public int getTotalSingerCount() {
        return singerService.findAll().size();
    }

    @ManagedOperation
    public String findJohn() {
        List<Singer> singers = singerService.
            findByFirstNameAndLastName("John", "Mayer");
        if (!singers.isEmpty()) {
            return singers.get(0).getFirstName() + " "
                + singers.get(0).getLastName();
        }
        return "not found";
    }
}
```

默认情况下，Spring Boot 会将管理端点公开为 org.springframework.boot 域下的 JMX MBean。在上面描述的代码片段中，@ManagedResource 注解使用了一个名为 objectName 的属性，其值表示 MBean 的域和名称。为了便于在 VisualVM 中找到显式创建的托管 bean(Spring Boot 提供自己的自动配置 MBean 进行内部监视)，此处使用了域 jmxDemo。

@ManagedAttribute 注解用于将给定的 bean 属性公开为 JMX 属性，而@ManagedOperation 注解用于将给定的方法公开为 JMX 操作。因为上面显示的两个方法的注解不同，所以它们将显示在 VisualVM 的不同选项卡中。调用 getTotalSingerCount()的结果将显示在 Attributes 选项卡中。而在 Operations 选项卡中，两个方法将显示为可单击的按钮。每个注解中作为描述信息提供的字符串可以在 Metadata 选项卡中查看。

图 15-3 所示的 MBeans 选项卡显示了 jmxDemo 域下公开的 Prospring5SingerApp MBean，此外还可以看到 org.springframework.boot 域。

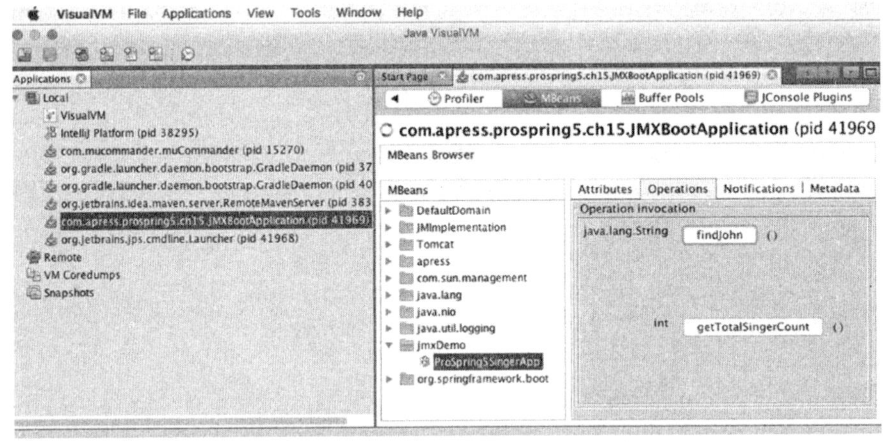

图 15-3　在 VisualVM 中公开的 Prospring5SingerApp MBean

你所需要做的不只这些！还必须启用对 JMX 的支持。这是通过使用@EnableMBeanExport 注解配置类来完成的。该注解允许从 Spring 上下文默认导出所有标准 MBean，以及所有用@ManagedResource 注解的 bean。从根本上讲，该注解告诉 Spring Boot 创建一个名为 mbeanExporter 的 MBeanExporter bean。该 Spring Boot 应用程序的配置类如下所示：

```
package com.apress.prospring5.ch15;

import org.slf4j.Logger;
import org.slf4j.LoggerFactory;
import org.springframework.boot.SpringApplication;
import org.springframework.boot.autoconfigure.SpringBootApplication;
import org.springframework.context.ConfigurableApplicationContext;
import org.springframework.context.annotation.EnableMBeanExport;

import java.io.IOException;

@EnableMBeanExport
@SpringBootApplication(scanBasePackages = {"com.apress.prospring5.ch15"})
public class JMXBootApplication {
```

```
    private static Logger logger = LoggerFactory.
        getLogger(JMXBootApplication.class);
    public static void main(String args) throws IOException {
        ConfigurableApplicationContext ctx =
            SpringApplication.run(JMXBootApplication.class, args);
        assert (ctx != null);
        logger.info("Started ...");
        System.in.read();
        ctx.close();
    }
}
```

现在，可以这么说，以上就是在 Spring Boot 应用程序中使用 JMX 时所需要完成的所有操作。

15.6 小结

在本章中，介绍了关于监控 Spring 驱动的 JEE 应用程序的高级主题。首先讨论了 Spring 对 JXM 的支持，即监控 Java 应用程序的标准。然后讨论了如何实现自定义 MBean 以公开与应用程序相关的信息，以及公开常见组件(比如 Hibernate)的统计信息。最后演示了如何在 Spring Boot 应用程序中使用 JMX 以及 Spring 的特殊功能。

第 16 章

Web 应用程序

在企业级应用程序中，表示层严重影响用户对应用程序的接受程度。表示层是应用程序的前门。它可以让用户执行应用程序提供的业务功能，并呈现应用程序正在维护的信息视图。用户界面的执行方式对应用程序的成功有很大的影响。

由于互联网的爆炸式增长(特别是现在)以及人们使用的各种设备的兴起，开发应用程序的表示层是一项具有挑战性的任务。以下是开发 Web 应用程序时的一些主要注意事项。

- **性能：**性能始终是 Web 应用程序的最高要求。如果用户选择一项功能或者单击一个链接，并且需要执行很长时间(在 Internet 的世界中，三秒钟就像经历了一个世纪！)，那么用户肯定不会对应用程序感到满意。
- **用户友好：**应用程序应易于使用，易于浏览，并带有明确的指令，从而不会让用户混淆。
- **交互性和丰富性：**用户界面应该具有高度的交互性和响应能力。此外，在视觉呈现方面应该丰富，比如图表、界面的仪表板类型等。
- **可访问性：**如今，用户需要通过任何设备从任何地方访问应用程序。在办公室，他们将使用桌面访问应用程序。在路上，则使用各种移动设备(包括笔记本电脑、平板电脑和智能手机)访问应用程序。

虽然开发满足上述要求的 Web 应用程序并不容易，但这些要求被认为是商业用户的强制性要求。幸运的是，目前已经开发出许多新技术和框架来满足这些要求。许多 Web 应用程序框架和库，例如 Spring MVC(Spring Web Flow)、Struts、Tapestry、Java Server Faces(JSF)、Google Web Toolkit(GWT)、jQuery 和 Dojo 等，都提供了工具和丰富的组件库来帮助开发高度互动的 Web 前端。另外，许多框架还提供了针对移动设备(包括智能手机和平板电脑)的工具或相应的部件库。HTML5 和 CSS3 标准的兴起以及大多数网络浏览器和移动设备制造商对这些最新标准的支持，也有助于简化 Web 应用程序的开发，可以随时随地从任何设备访问这些 Web 应用程序。

在 Web 应用程序开发方面，Spring 提供了全面且集约的支持。Spring MVC 模块为 Web 应用程序开发提供了坚实的基础架构和模型-视图-控制器(MVC)框架。在使用 Spring MVC 时，可以使用各种视图技术(例如 JSP 或 Velocity)。另外，Spring MVC 集成了许多常用的 Web 框架和工具包(例如 Struts 和 GWT)，而其他 Spring 项目有助于满足 Web 应用程序的特定需求。例如，当 Spring MVC 与 Spring Web Flow 项目及其 Spring Faces 模块结合使用时，可以为开发具有复杂流程的 Web 应用程序以及 JSF 作为视图技术提供全面的支持。简单地说，在表示层开发方面有很多选择。本章重点介绍 Spring MVC，并讨论如何使用 Spring MVC 提供的强大功能来开发高性能的 Web 应用程序。具体而言，本章包含以下主题：

- **Spring MVC：**讨论 MVC 模式的主要概念并介绍 Spring MVC，重点介绍 Spring MVC 的核心概念，包括其 WebApplicationContext 层次结构以及请求处理生命周期。
- **国际化(i18n)语言环境和主题：**Spring MVC 为常用的 Web 应用程序需求提供全面的支持，包括国际化、语言环境和主题，主要讨论如何使用 Spring MVC 开发满足这些需求的 Web 应用程序。
- **视图和 Ajax 支持：**Spring MVC 支持许多视图技术。在本章中，将重点介绍如何使用 Java Server Pages(JSP)和 Tiles 作为 Web 应用程序的视图部分。在 JSP 之上，将使用 JavaScript 来提供丰富的功能。此外，还可以使用许多杰出和流行的 JavaScript 库，比如 jQuery 和 Dojo。在本章中，将重点讨论如何使用 jQuery，以及用来支持开发高度交互式 Web 应用程序的子项目 jQuery UI 库。
- **分页和文件上传支持：**在演示如何开发本章的示例时，将讨论如何在浏览基于网格的数据时使用 Spring Data JPA 和前端 jQuery 组件提供分页支持。另外，还将介绍如何在 Spring MVC 中实现文件上传。本章并没有继承 Apache Commons File Upload，而是讨论如何使用 Spring MVC 和 Servlet 3.1 容器的内置多部分支持来实现文件上传。

- **安全性**：安全性是 Web 应用程序中的一个重要主题，主要讨论如何使用 Spring Security 来帮助保护应用程序并处理登录和注销。

16.1 实现示例的服务层

在本章的服务层中，仍将使用歌手应用程序作为示例。在本节中，将讨论数据模型以及本章中所使用服务层的实现。

16.1.1 对示例使用数据模型

本章将使用简单的数据模型作为示例，它仅包含用于存储歌手信息的单个 SINGER 表。以下 SQL 代码片段显示了模式创建脚本(schema.sql)：

```
DROP TABLE IF EXISTS SINGER;

CREATE TABLE SINGER (
    ID INT NOT NULL AUTO_INCREMENT
    , FIRST_NAME VARCHAR(60) NOT NULL
    , LAST_NAME VARCHAR(40) NOT NULL
    , BIRTH_DATE DATE
    , DESCRIPTION VARCHAR(2000)
    , PHOTO BLOB
    , VERSION INT NOT NULL DEFAULT 0
    , UNIQUE UQ_SINGER_1 (FIRST_NAME, LAST_NAME)
    , PRIMARY KEY (ID)
);
```

如你所见，SINGER 表仅包含用于存储歌手信息的几个基本字段。值得一提的是使用了二进制大对象(BLOB)数据类型的 PHOTO 列，它用来存储通过文件上传的歌手照片。此时并不会使用上述 SQL 脚本来创建表；相反，Hibernate 将根据以下代码片段中描述的 Singer 实体的配置生成创建表所需的 SQL：

```
package com.apress.prospring5.ch16.entities;

import javax.persistence.*;
import javax.validation.constraints.NotEmpty;
import javax.validation.constraints.Size; import java.io.Serializable;
import java.text.SimpleDateFormat;
import java.util.Date;

import static javax.persistence.GenerationType.IDENTITY;

@Entity
@Table(name = "singer")
public class Singer implements Serializable {
  @Id
  @GeneratedValue(strategy = IDENTITY)
  @Column(name = "ID")
  private Long id;

  @Version
  @Column(name = "VERSION")
  private int version;

  @NotEmpty(message="{validation.firstname.NotEmpty.message}")
  @Size(min=3, max=60, message="{validation.firstname.Size.message}")
  @Column(name = "FIRST_NAME")
  private String firstName;

  @NotEmpty(message="{validation.lastname.NotEmpty.message}")
  @Size(min=1, max=40, message="{validation.lastname.Size.message}")
  @Column(name = "LAST_NAME")
  private String lastName;

  @Temporal(TemporalType.DATE)
  @Column(name = "BIRTH_DATE")
  private Date birthDate;
```

```
    @Column(name = "DESCRIPTION")
    private String description;
    @Basic(fetch= FetchType.LAZY)
    @Lob
    @Column(name = "PHOTO")
    private byte photo;

    public Long getId() {
      return id;
    }

    public int getVersion() {
      return version;
    }

    public String getFirstName() {
      return firstName;
    }

    public String getLastName() {
      return lastName;
    }

    public void setId(Long id) {
      this.id = id;
    }

    public void setVersion(int version) {
      this.version = version;
    }

    public void setFirstName(String firstName) {
      this.firstName = firstName;
    }

    public void setLastName(String lastName) {
      this.lastName = lastName;
    }

    public void setBirthDate(Date birthDate) {
      this.birthDate = birthDate;
    }

    public Date getBirthDate() {
      return birthDate;
    }

    public String getDescription() {
      return description;
    }

    public void setDescription(String description) {
      this.description = description;
    }

    public byte getPhoto() {
      return photo;
    }

    public void setPhoto(byte photo) {
      this.photo = photo;
    }
    @Transient
    public String getBirthDateString() {
      String birthDateString = "";
      if (birthDate != null) {
        SimpleDateFormat sdf = new SimpleDateFormat("yyyy-MM-dd");
        birthDateString = sdf.format(birthDate);
      }
      return birthDateString;
    }

    @Override
    public String toString() {
```

```
    return "Singer - Id: " + id + ", First name: " + firstName
  + ", Last name: " + lastName + ", Birthday: " + birthDate
  + ", Description: " + description;
  }
}
```

如果想要使用填充脚本来填充 SINGER 表，那么可以使用如下所示的脚本：

```
insert into singer first_name, last_name, birth_date values 'John', 'Mayer', '1977-10-16';
insert into singer first_name, last_name, birth_date values 'Eric', 'Clapton', '1945-03-30';
insert into singer first_name, last_name, birth_date values 'John', 'Butler', '1975-04-01';
insert into singer first_name, last_name, birth_date values 'B.B.', 'King', '1925-09-16';
insert into singer first_name, last_name, birth_date values 'Jimi', 'Hendrix', '1942-11-27';
insert into singer first_name, last_name, birth_date values 'Jimmy', 'Page', '1944-01-09';
insert into singer first_name, last_name, birth_date values 'Eddie', 'Van Halen','1955-01-26';
insert into singer first_name, last_name, birth_date values 'Saul Slash', 'Hudson','1965-07-23';
insert into singer first_name, last_name, birth_date values 'Stevie', 'Ray Vaughan','1954-10-03';
insert into singer first_name, last_name, birth_date values 'David', 'Gilmour','1946-03-06';
insert into singer first_name, last_name, birth_date values 'Kirk', 'Hammett', '1992-11-18';
insert into singer first_name, last_name, birth_date values 'Angus', 'Young', '1955-03-31';
insert into singer first_name, last_name, birth_date values 'Dimebag', 'Darrell','1966-08-20';
insert into singer first_name, last_name, birth_date values 'Carlos', 'Santana','1947-07-20';
```

但是在官方的代码示例中，你会发现是通过 DBInitializer 类插入上面所示的数据。在本章中，需要更多测试数据，以便稍后演示分页支持。

16.1.2 实现 DAO 层

实体类、存储库和数据库配置组成被称为 DAO 的应用程序层，负责数据库对象。

前面已经介绍过实体类，并且你可能已经注意到典型的 JPA 注解。但是，这里有两个新的属性。

- 新增了一个名为 birthDateString 的瞬态属性(通过将@Transient 注解应用于 getter 方法)，该属性用于后面示例中的前端显示。
- 对于 photo 属性，使用一个字节数组作为 Java 数据类型，它对应于 RDBMS 中的 BLOB 数据类型。另外，getter 方法用@Lob 和@Basic(fetch = FetchType.LAZY)注解。前一个注解向 JPA 提供程序指明它是一个大对象列，后一个注解表示应该延迟提取该属性以避免加载不需要照片信息的类时影响性能。

本章后面还会介绍一些验证注解。

因为将要使用 Spring Data JPA 的存储库支持，所以需要实现 SingerRepository 接口，如下所示：

```
package com.apress.prospring5.ch16.repo;

import org.springframework.data.repository.PagingAndSortingRepository;

public interface SingerRepository extends
   PagingAndSortingRepository<Singer, Long> {
}
```

该例没有扩展 CrudRepository 接口，而是使用 PagingAndSortingRepository，它是 CrudRepository 的高级扩展，提供了使用分页和排序抽象检索实体的方法。这些方法是非常有用的，因为查询只会返回需要在界面中显示的已排序数据，而不需要进行其他更改。

16.1.3 实现服务层

在本节中，将首先讨论如何通过使用 JPA 2、Spring Data JPA 和 Hibernate 作为持久化服务提供程序来实现 SingerService。然后介绍 Spring 项目中服务层的配置。以下代码片段显示了 SingerService 接口以及想要公开的服务：

```
package com.apress.prospring5.ch16.services;

import java.util.List;

import org.springframework.data.domain.Page;
import org.springframework.data.domain.Pageable;

public interface SingerService {
   List<Singer> findAll();
   Singer findById(Long id);
   Singer save(Singer singer);
   Page<Singer> findAllByPage(Pageable pageable);
}
```

该接口中方法的含义应该是不言自明的。该接口的实现也很简单。由于应用程序很简单，因此不需要对数据进行其他更改，所以 SingerServiceImpl 的作用仅仅是将调用转发到类似的存储库方法。

```
package com.apress.prospring5.ch16.services;

import java.util.List;

import com.apress.prospring5.ch16.repos.SingerRepository;
import com.apress.prospring5.ch16.entitites.Singer;
import org.springframework.beans.factory.annotation.Autowired;
import org.springframework.data.domain.Page;
import org.springframework.data.domain.Pageable;
import org.springframework.stereotype.Service;
import org.springframework.transaction.annotation.Transactional;

import com.google.common.collect.Lists;

@Transactional
@Service("singerService")
public class SingerServiceImpl implements SingerService {
   private SingerRepository singerRepository;

   @Override
   @Transactional(readOnly=true)
   public List<Singer> findAll() {
      return Lists.newArrayList(singerRepository.findAll());
   }

   @Override
   @Transactional(readOnly=true)
   public Singer findById(Long id) {
      return singerRepository.findById(id).get();
   }

   @Override
   public Singer save(Singer singer) {
      return singerRepository.save(singer);
   }

   @Autowired
   public void setSingerRepository(SingerRepository singerRepository) {
      this.singerRepository = singerRepository;
   }

   @Override
   @Transactional(readOnly=true)
   public Page<Singer> findAllByPage(Pageable pageable) {
      return singerRepository.findAll(pageable);
   }
}
```

该实现已基本完成，下一步是在 Web 项目的 Spring 应用程序上下文中配置服务，这将在下一节中讨论。

16.2 配置 SingerService

显然，可以使用两种方法完成配置：XML 配置和 Java 配置。可以在本章的源代码中找到使用 XML 配置的此项目的版本之一。如果对配置的内容感到好奇，可以分析一下，但在本节中，重点将放在 Java 配置类上。要配置 SingerService、数据库访问和事务，可以使用以下类(在第 9 章中已经详细介绍过这些类)：

```
package com.apress.prospring5.ch16.config;

import org.slf4j.Logger;
import org.slf4j.LoggerFactory;
import org.springframework.context.annotation.Bean;
import org.springframework.context.annotation.ComponentScan;
import org.springframework.context.annotation.Configuration;
import org.springframework.data.jpa.repository.config.EnableJpaRepositories;
import org.springframework.jdbc.datasource.embedded.EmbeddedDatabaseBuilder;
import org.springframework.jdbc.datasource.embedded.EmbeddedDatabaseType;
import org.springframework.orm.jpa.JpaTransactionManager;
import org.springframework.orm.jpa.JpaVendorAdapter;
import org.springframework.orm.jpa.LocalContainerEntityManagerFactoryBean;
import org.springframework.orm.jpa.vendor.HibernateJpaVendorAdapter;
```

```
import org.springframework.transaction.PlatformTransactionManager;

import javax.persistence.EntityManagerFactory;
import javax.sql.DataSource;
import java.util.Properties;

@Configuration
@EnableJpaRepositories(basePackages = {"com.apress.prospring5.ch16.repos"})
@ComponentScan(basePackages = {"com.apress.prospring5.ch16"} )
public class DataServiceConfig {

    private static Logger logger = LoggerFactory.getLogger(DataServiceConfig.class);

    @Bean
    public DataSource dataSource() {
        try {
        EmbeddedDatabaseBuilder dbBuilder = new EmbeddedDatabaseBuilder();
        return dbBuilder.setType(EmbeddedDatabaseType.H2).build();
        } catch (Exception e) {
            logger.error("Embedded DataSource bean cannot be created!", e);
            return null;
        }
    }

    @Bean
    public Properties hibernateProperties() {
        Properties hibernateProp = new Properties();
        hibernateProp.put("hibernate.dialect", "org.hibernate.dialect.H2Dialect");
        hibernateProp.put("hibernate.hbm2ddl.auto", "create-drop");
        hibernateProp.put("hibernate.show_sql", true);
        hibernateProp.put("hibernate.max_fetch_depth", 3);
        hibernateProp.put("hibernate.jdbc.batch_size", 10);
        hibernateProp.put("hibernate.jdbc.fetch_size", 50);
        return hibernateProp;
    }

    @Bean
    public PlatformTransactionManager transactionManager() {
        return new JpaTransactionManager(entityManagerFactory());
    }

    @Bean
    public JpaVendorAdapter jpaVendorAdapter() {
        return new HibernateJpaVendorAdapter();
    }

    @Bean
    public EntityManagerFactory entityManagerFactory() {
        LocalContainerEntityManagerFactoryBean factoryBean =
    new LocalContainerEntityManagerFactoryBean();
        factoryBean.setPackagesToScan("com.apress.prospring5.ch16.entities");
        factoryBean.setDataSource(dataSource());
        factoryBean.setJpaVendorAdapter(new HibernateJpaVendorAdapter());
        factoryBean.setJpaProperties(hibernateProperties());
        factoryBean.setJpaVendorAdapter(jpaVendorAdapter());
        factoryBean.afterPropertiesSet();
        return factoryBean.getNativeEntityManagerFactory();
    }
}
```

现在服务层已完成并准备好被公开以供远程客户端使用。

16.3 MVC 和 Spring MVC 介绍

现在，MVC 已经成为 Web 应用程序中一种常用的模式，所以在实现表示层之前，首先介绍一下 MVC 的一些主要概念，以及 Spring MVC 如何在这一方面提供全面的支持。

在下面的章节中，将逐一介绍一些高级概念。首先，对 MVC 进行简要介绍。然后，介绍 Spring MVC 及其 WebApplicationContext 层次结构的高级视图。最后，讨论 Spring MVC 中的请求生命周期。

16.3.1 MVC 介绍

MVC 是实现应用程序表示层的常用模式。MVC 模式的主要原则是为不同组件定义具有明确责任的架构。顾名思义，MVC 模式中有三个参与者。

- **模型(Model)**：模型表示业务数据以及用户上下文中应用程序的"状态"。例如，在电子商务网站中，如果用户在网站上购买商品，模型将包含用户个人信息、购物车数据和订单数据。
- **视图(View)**：以所需格式向用户呈现数据，支持与用户交互，并支持客户端验证、国际化、样式等。
- **控制器(Controller)**：控制器处理前端用户执行操作的请求、与服务层交互、更新模型以及根据执行结果将用户引导至适当的视图。

由于基于 Ajax 的 Web 应用程序的兴起，MVC 模式得到了增强，从而提供了更加灵活且丰富的用户体验。例如，当使用 JavaScript 时，视图可以"监听"用户执行的事件或操作，然后向服务器提交 XMLHttpRequest。在控制器端，将返回原始数据(例如，以 XML 或 JSON 格式)，而不是返回视图，JavaScript 应用程序使用接收到的数据对视图进行"部分"更新。图 16-1 显示了典型 Web 应用程序中的 MVC 模式，可将其视为对传统 MVC 模式的增强。常见的视图请求处理过程如下所示。

(1) **请求**：将请求提交给服务器。在服务器端，大多数框架(例如 Spring MVC 或 Struts)都有一个调度程序(以 servlet 的形式)用来处理请求。

(2) **调用**：调度程序根据 HTTP 请求信息和 Web 应用程序配置将请求分派给适当的控制器。

(3) **服务调用**：控制器与服务层交互。

(4) **填充模型**：控制器使用从服务层获得的信息填充模型。

(5) **创建视图**：根据模型创建视图。

(6) **响应**：控制器将相应的视图返回给用户。

另外，在视图中，将会发生 Ajax 调用。例如，假设用户浏览网格中的数据。当用户单击下一页时，并不是刷新整个页面，而是发生了以下事件：

(a) **请求**：准备 XMLHttpRequest 并提交给服务器。调度程序将请求发送给相应的控制器。

(b) **响应**：控制器与服务层交互，响应数据将被格式化并发送到浏览器。此时并不涉及任何视图。浏览器接收数据并对现有视图进行部分更新。

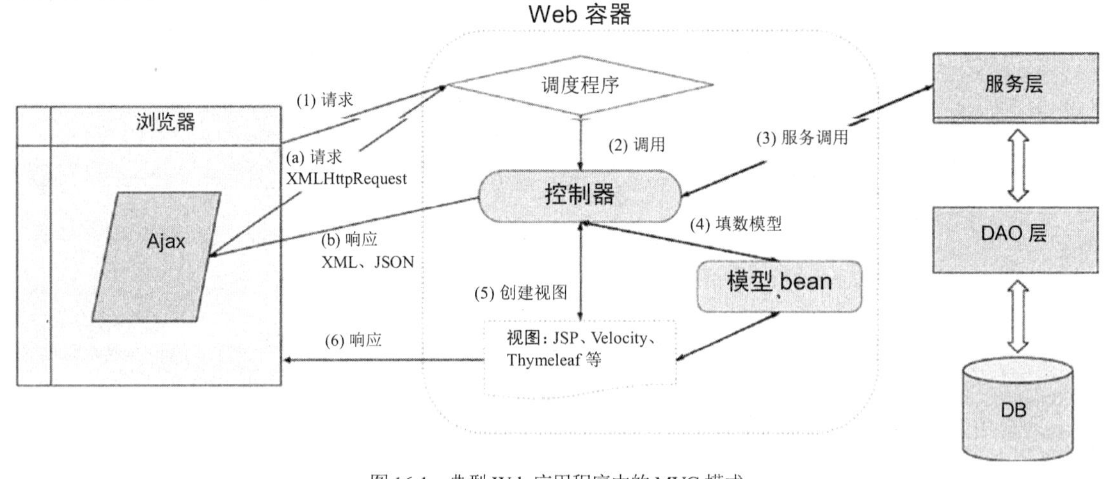

图 16-1 典型 Web 应用程序中的 MVC 模式

16.3.2 Spring MVC 介绍

在 Spring 框架中，Spring MVC 模块提供了对 MVC 模式的全面支持，并支持其他功能(例如主题、国际化、验证以及类型转换和格式化)，从而简化了表示层的实现。

在下面的章节中，将讨论 Spring MVC 的主要概念。主题包括 Spring MVC 的 WebApplicationContext 层次结构、典型的请求处理生命周期和配置。

16.3.3 Spring MVC WebApplicationContext 层次结构

在 Spring MVC 中，DispatcherServlet 是接收请求并将它们分发给相应控制器的中央 servlet。在 Spring MVC 应用程序中，可以有多种用于不同目的(例如，处理用户界面请求和 RESTful-WS 请求)的 DispatcherServlet 实例，并且每个 DispatcherServlet 都有自己的 WebApplicationContext 配置，该配置定义了 servlet 级别的特性，例如支持 servlet 的控制器、处理映射、视图解析、国际化、主题、验证以及类型转换和格式化。

在 servlet 级别的 WebApplicationContext 配置下，Spring MVC 维护着一个根 WebApplicationContext，其中包含应用程序级别的配置，例如后端数据源、安全性以及服务和持久层配置。根 WebApplicationContext 可以供所有 servlet 级别的 WebApplicationContext 使用。

接下来看一个示例。假设在应用程序中有两个 DispatcherServlet 实例。一个支持用户界面(被称为*应用程序 servlet*)，另一个则以 RESTful-WS 的形式向其他应用程序提供服务(被称为 RESTful servlet)。在 Spring MVC 中，将为这两个 DispatcherServlet 实例的 WebApplicationContext 实例以及根 WebApplicationContext 实例定义配置。图 16-2 显示了 Spring MVC 此时维护的 WebApplicationContext 层次结构。

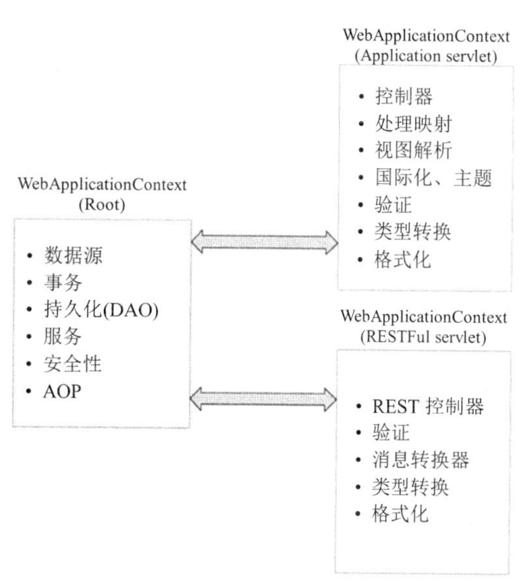

图 16-2 Spring MVC WebApplicationContext 层次结构

16.3.4 Spring MVC 请求生命周期

下面看看 Spring MVC 如何处理请求。图 16-3 显示了在 Spring MVC 中处理请求所涉及的主要组件。这些主要组件及其用途如下所示。

- **过滤器**：过滤器适用于所有请求。下一节将介绍几种常用的过滤器及其用途。
- **调度程序 servlet**：该 servlet 分析请求并将它们分发给相应的控制器进行处理[1]。
- **通用服务**：通用服务将被应用于每个请求，以提供包括国际化、主题、文件上传等支持。它们的配置在 DispatcherServlet 的 WebApplicationContext 中定义。
- **处理映射**：将传入的请求映射到处理程序(即 Spring MVC 控制器类中的方法)。自 Spring 2.5 以来，在大多数情况下，都不需要进行配置，因为 Spring MVC 会自动注册一个 HandlerMapping 实现，该实现根据通过@RequestMapping 注解在控制器类中的类型或方法级别上表示的 HTTP 路径来映射处理程序[2]。
- **处理程序拦截器**：在 Spring MVC 中，可以为处理程序注册拦截器，以便实现通用检查或逻辑。例如，处理程序拦截器可以进行相关检查以确保只能在办公时间内调用处理程序。
- **处理程序异常解析器**：在 Spring MVC 中，HandlerExceptionResolver 接口(在 org.springframework.web.servlet 包中定义)用于处理请求处理期间抛出的意外异常。默认情况下，DispatcherServlet 注册了 DefaultHandlerExceptionResolver 类(来自 org.springframework.web.servlet.mvc.support 包)。该解析器通过设置特定的响应状态码来处理某些标准的 Spring MVC 异常。也可以通过使用@ExceptionHandler 注解控制器方法并传入异常类型作为属性来实现自己的异常处理程序。
- **视图解析**：Spring MVC 的 ViewResolver 接口(来自 org.springframework.web.servlet 包)支持基于控制器返回的逻辑名称的视图解析。有许多实现类支持各种视图解析机制。例如，UrlBasedViewResolver 类支持将逻辑名直接解析为 URL。ContentNegotiatingViewResolver 类可以根据客户端支持的媒体类型(比如 XML、PDF 和 JSON)动态解析视图。还有一些实现可以集成不同的视图技术，例如 FreeMarker(FreeMarkerViewResolver)、Velocity(VelocityViewResolver)和 JasperReports(JasperReportsViewResolver)。

---

1 如果熟悉设计模式，就会意识到 DispatcherServlet 是 Front Controller 设计模式的一种表示。

2 在 Spring 2.5 中，DefaultAnnotationHandlerMapping 是默认实现。而从 Spring 3.1 开始，RequestMappingHandlerMapping 已经成为默认实现，它支持将请求映射到没有注解的处理程序，只要遵守 Spring 关于命名控制器和方法的约定即可。

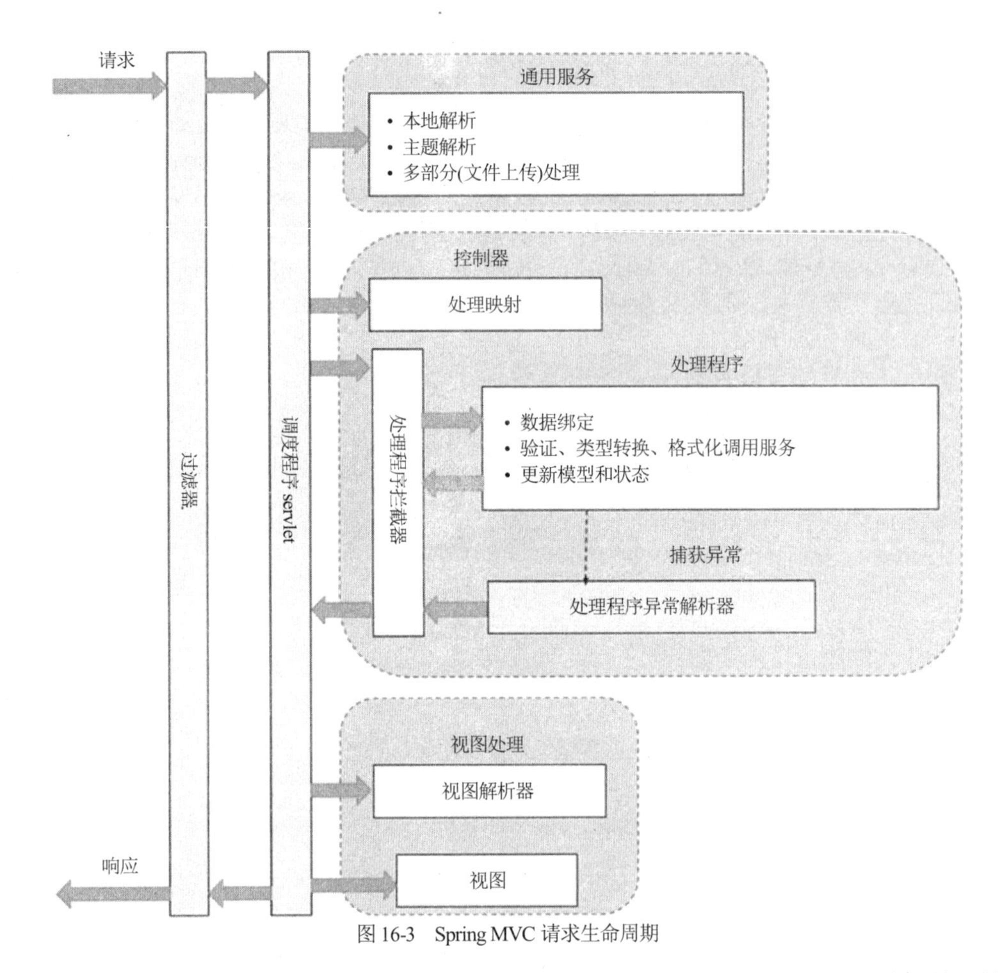

图 16-3　Spring MVC 请求生命周期

上面介绍的只是一些常用的处理程序和解析器。有关完整的描述，请参阅 Spring 框架的参考文档及其 Javadoc。

16.3.5　Spring MVC 配置

要想在 Web 应用程序中启用 Spring MVC，还需要一些初始配置，特别是对于 Web 部署描述符文件 web.xml。自 Spring 3.1 以来，就支持在 Servlet 3.0 Web 容器中进行基于代码的配置。这样就提供了 Web 部署描述符文件 web.xml 中所需 XML 配置的替代方案。

要为 Web 应用程序配置 Spring MVC 支持，需要在 Web 部署描述符中完成以下配置：

- 配置根 WebApplicationContext
- 配置 Spring MVC 所需的 servlet 过滤器
- 在应用程序内配置调度程序 servlet

以下配置类只需要几行代码即可完成上述三种配置：

```
package com.apress.prospring5.ch16.init;

import com.apress.prospring5.ch16.config.DataServiceConfig;
import com.apress.prospring5.ch16.config.SecurityConfig;
import com.apress.prospring5.ch16.config.WebConfig;
import org.springframework.web.filter.CharacterEncodingFilter;
import org.springframework.web.filter.HiddenHttpMethodFilter;
import org.springframework.web.servlet.support.
   AbstractAnnotationConfigDispatcherServletInitializer;
   import javax.servlet.Filter;

   public class WebInitializer extends
      AbstractAnnotationConfigDispatcherServletInitializer {
```

```
        @Override
        protected Class<?>[] getRootConfigClasses() {
            return new Class<?>[]{
            SecurityConfig.class, DataServiceConfig.class
            };
        }

        @Override
        protected Class<?>[] getServletConfigClasses() {
            return new Class<?>[]{
            WebConfig.class
            };
        }

        @Override
        protected String[] getServletMappings() {
            return new String[]{"/"};
        }

        @Override
        protected Filter[] getServletFilters() {
            CharacterEncodingFilter cef = new CharacterEncodingFilter();
            cef.setEncoding("UTF-8");
            cef.setForceEncoding(true);
            return new Filter[]{new HiddenHttpMethodFilter(), cef};
        }
    }
```

要想使用基于代码的配置，必须开发实现了 org.springframework.web.WebApplicationInitializer 接口的类。为了更加实用，在前面的示例中，对 Spring 类 AbstractAnnotationConfigDispatcherServletInitializer(WebApplicationInitializer 的一个实现)进行了扩展，因为它包含一些方法的具体实现，而这些方法是配置使用基于 Java 的 Spring 配置的 Spring Web 应用程序所需要的。

实现了 WebApplicationInitializer 接口的所有类都将被 org.springframework.web.SpringServletContainerInitializer 类(实现了 Servlet 3.0 的 javax.servlet.ServletContainerInitializer 接口)自动检测到，该类可以在任何 Servlet 3.0 容器中自动启动。如前面的示例所示，以下方法被重写以插入自定义配置。

- getRootConfigClasses()：使用此方法返回的配置类创建 AnnotationConfigWebApplicationContext 类型的根应用程序上下文。
- getServletConfigClasses()：使用此方法返回的配置类创建 AnnotationConfigWebApplicationContext 类型的 Web 应用程序上下文。
- getServletMappings()：DispatcherServelt 的映射(上下文)由此方法返回的字符串数组指定。
- getServletFilters()：如方法的名称所示，此方法返回将应用于每个请求的 javax.servlet.Filter 实现的数组。

等一下，如果查看一下前面的示例，你会发现没有在任何地方提到过安全过滤器！这怎么可能？在第 12 章曾经介绍过安全过滤器，但此处忽略了安全过滤器，为什么？答案很简单：可以使用一个专门的 Spring 类完成安全过滤。

```
package com.apress.prospring5.ch16.init;

import org.springframework.security.web.context.
    AbstractSecurityWebApplicationInitializer;
public class SecurityWebApplicationInitializer
    extends AbstractSecurityWebApplicationInitializer {
}
```

通过提供一个扩展了 AbstractSecurityWebApplicationInitializer 的空类，可以告诉 Spring 希望启用 DelegatingFilterProxy，所以 springSecurityFilterChain 将在注册的任何 javax.servlet.Filter 之前使用。

通过使用这种方法，并结合 Spring 的基于 Java 代码的配置，可以实现基于 Spring Web 应用程序的纯 Java 代码配置，而无须在 web.xml 或其他 Spring XML 配置文件中声明任何 Spring 配置。是的，就是这么简单。

返回到过滤器，表 16-1 描述了由 getServletFilters()返回的数组中的每个过滤器。

表 16-1 常用的 Spring MVC servlet 过滤器

过滤器类的全名	描述
org.springframework.web.filter.CharacterEncodingFilter	此过滤器用于指定请求的字符编码
org.springframework.web.filter.HiddenHttpMethodFilter	此过滤器支持除 GET 和 POST 外的 HTTP 方法(例如，PUT)

⚠ 虽然此处不需要(因此没有用在配置中)，但还是应该提到一个过滤器实现：org.springframework.orm.jpa.support.OpenEntityManagerInViewFilter。该实现将 JPA EntityManager 绑定到线程，以便处理整个请求。它适用于 View 模式中的 Open EntityManager，允许在 Web 视图中延迟加载，尽管原始事务已经完成。虽然很实用，但这样做是相当危险的，因为多个请求最终可能会消耗所有数据库允许的开放连接。另外，如果所加载的数据集很大，应用程序可能会冻结。因此，开发人员不愿意使用它，而是通过 Ajax 请求调用特定的处理程序来加载特定于 Web 的视图对象(而不是实体)中的数据。

16.3.6 在 Spring MVC 中创建第一个视图

有了服务层和 Spring MVC 配置，接下来可以开始实现第一个视图。在本节中，将实现一个简单视图来显示最初由 DBInitializer bean 填充的所有歌手。

如前所述，我们将使用 JSPX 来实现视图。JSPX 是格式良好的 XML 格式的 JSP。相比于 JSP，JSPX 具有以下主要优点：

- JSPX 更严格地强制将代码从视图层中分离出来。例如，不能将 Java “scriptlets” 放置在 JSPX 文档中。
- 工具可以执行即时验证(使用 XML 语法)，以便提前发现错误。此时需要使用列出的依赖项来配置项目。

首先，使用以下配置代码片段中列出的依赖项来配置项目：

```
\\pro-spring-15\build.gradle
ext {
  //spring libs
  springVersion = '5.0.0.RC2'
  springSecurityVersion = '5.0.0.M2'
  h2Version = '1.4.194'
  tilesVersion = '3.0.7'

  //persistency libraries
  hibernateVersion = '5.2.10.Final'
  hibernateJpaVersion = '1.0.0.Final'
  hibernateValidatorVersion = '5.4.1.Final'
  ...

  spring = [
    webmvc : "org.springframework:spring-webmvc:$springVersion",
    data : "org.springframework.data:spring-data-jpa:$springDataVersion",
    securityWeb :
   "org.springframework.security:spring-security-web:$springSecurityVersion",
    securityConfig:
   "org.springframework.security:spring-security-config:$springSecurityVersion",
    securityTaglibs:
   "org.springframework.security:spring-security-taglibs:$springSecurityVersion",
    ...
  ]

  hibernate = [
    validator : "org.hibernate:hibernate-validator:$hibernateValidatorVersion",
    em : "org.hibernate:hibernate-entitymanager:$hibernateVersion",
    jpaApi :
  "org.hibernate.javax.persistence:hibernate-jpa-2.1-api:$hibernateJpaVersion",
    ...
  ]

  misc = [
    validation : "javax.validation:validation-api:$javaxValidationVersion",
    castor : "org.codehaus.castor:castor-xml:$castorVersion",
    io : "commons-io:commons-io:2.5",
    tiles : "org.apache.tiles:tiles-jsp:$tilesVersion",
    jstl : "jstl:jstl:1.2",
    ...
  ]

  db = [
  h2 : "com.h2database:h2:$h2Version"
  ]
}
...
\\chapter-16\build.gradle
dependencies {
```

```
    //we exclude common transitive dependencies
    compile spring.contextSupport {
    exclude module: 'spring-context'
    exclude module: 'spring-beans'
    exclude module: 'spring-core'
    }
    compile spring.securityTaglibs {
    exclude module: 'spring-web'
    exclude module: 'spring-context'
    exclude module: 'spring-beans'
    exclude module: 'spring-core'
    }
    compile spring.securityConfig {
    exclude module: 'spring-security-core'
    exclude module: 'spring-context'
    exclude module: 'spring-beans'
    exclude module: 'spring-core'
    }
    Compile misc.slf4jJcl, misc.logback, misc.lang3, hibernate.em,
  hibernate.validator, misc.guava, db.h2, spring.data,
  spring.webmvc, misc.castor, misc.validation, misc.tiles,
  misc.jacksonDatabind, misc.servlet, misc.io,misc.jstl,
  spring.securityTaglibs
}
```

到目前为止，上述配置中列出的大多数库你应该都已经知道了。下面解释一下首次使用的库：

- spring-webmvc 是用于 MVC 支持的 Spring MVC 模块。
- spring-security-web 是 Spring Web 模块，用于添加对 Spring Security 的支持。它是 spring-security-web 的一个直接依赖项，包含在 JSP 页面中使用的安全标记的定义。spring-security-web 包含用于保护 Web 应用程序的 AbstractSecurityWebApplicationInitializer 类以及其他相关的 Spring 组件。
- spring-security-config 是 Spring Security 模块，包含用于在 Spring 应用程序中配置安全性的类。
- tiles-jsp 是 Apache Tiles 模块，包含为 Web 应用程序创建 Java 模板的 Java 类和标记定义[1]。

16.3.7 配置 DispatcherServlet

下一步是配置 DispatcherServlet。这是通过创建一个定义 Spring Web 应用程序所需的所有基础架构 bean 的配置类来完成的。下面显示了基于 Java Config 的类的一个片段，其中包含少量的信息：

```
package com.apress.prospring5.ch16.config;

import org.springframework.context.annotation.Bean;
import org.springframework.context.annotation.ComponentScan;
import org.springframework.context.annotation.Configuration;
import org.springframework.web.servlet.config.annotation.*;
import org.springframework.web.servlet.view.InternalResourceViewResolver;
...

@Configuration
@EnableWebMvc
@ComponentScan(basePackages = {"com.apress.prospring5.ch16"})
public class WebConfig implements WebMvcConfigurer {

  //Declare the static resources.
  @Override
  public void addResourceHandlers(ResourceHandlerRegistry registry) {
    registry.addResourceHandler("/resources/**").addResourceLocations("/")
    .setCachePeriod(31556926);
  }

  @Bean
  InternalResourceViewResolver viewResolver(){
   InternalResourceViewResolver resolver = new InternalResourceViewResolver();
   resolver.setPrefix("/WEB-INF/views"); resolver.setSuffix(".jspx" );
   resolver.setRequestContextAttribute("requestContext"); return resolver;
  }

  //<=> <mvc:view-controller .../>
  @Override
  public void addViewControllers(ViewControllerRegistry registry) {
```

1 可以在 https://tiles.apache.org/上找到更多的内容。

```
    registry.addViewController("/").setViewName("singers/list");
  }

  //<=> <mvc:default-servlet-handler/>
  @Override
  public void configureDefaultServletHandling(
    DefaultServletHandlerConfigurer configurer) {
      configurer.enable();
  }
}
```

接口 WebMvcConfigurer 通过定义回调方法为 Spring MVC 定义了通过@EnableWebMvc 启用的基于 Java 的配置。尽管在 Spring 应用程序中可以有多个基于 Java 的配置类，但只有一个被允许使用@EnableWebMvc 进行注解。在前面的配置中，可以看到有几个方法被重写以自定义配置：

- addResourceHandlers()方法添加了一些处理程序，这些处理程序用于提供来自 Web 应用程序根目录、类路径和其他特定位置的静态资源，比如图像、JavaScript 和 CSS 文件。在这个自定义的实现中，对任何包含资源的 URL 的请求都将绕过所有过滤器而由特殊处理程序进行处理。
- configureDefaultServletHandling()方法启用处理静态资源的处理程序。
- addViewControllers()方法定义了用响应状态码和/或显示响应体的视图预先配置的简单自动化控制器。这些视图没有控制器逻辑，主要用于显示欢迎页面、执行简单的站点 URL 重定向、返回 404 状态等。在前面描述的配置中，使用该方法实现了重定向到 singers/list 视图。
- viewResolver()方法声明了一个类型为 InternalResourceViewResolver 的视图解析器，它将符号视图的名称与/WEB-INF/views 下的* .jspx 模板相匹配。

16.3.8 实现 SingerController

配置完 DispatcherServlet 的 WebApplicationContext 之后，下一步就是实现控制器类。

```
package com.apress.prospring5.ch16.web;
...

@RequestMapping("/singers")
@Controller
public class SingerController {
  private final Logger logger = LoggerFactory.getLogger(SingerController.class);

  private SingerService singerService;
  private MessageSource messageSource;

  @RequestMapping(method = RequestMethod.GET)
  public String list(Model uiModel) {
    logger.info("Listing singers");
    List<Singer> singers = singerService.findAll();
    uiModel.addAttribute("singers", singers);

    logger.info("No. of singers: " + singers.size());

    return "singers/list";
  }

  @Autowired
  public void setSingerService(SingerService singerService) {
    this.singerService = singerService;
  }
}
```

注解@Controller 被应用于该类，表明它是一个 Spring MVC 控制器。类级别的@RequestMapping 注解指示将由控制器处理的根 URL。在这种情况下，所有具有前缀/singers 的 URL 都将被分派给此控制器。在 list()方法上也应用了@RequestMapping 注解，此时该方法被映射到 HTTP GET 方法。这意味着具有 HTTP GET 方法的 URL /singers 将由此方法处理。在 list()方法中，歌手列表被检索并保存到由 Spring MVC 传入到该方法的 Model 接口中。最后，返回名为 singers/list 的逻辑视图。在 DispatcherServlet 配置中，InternalResourceViewResolver 被配置为视图解析器，并且该文件具有前缀/ WEB-INF/views /和后缀.jspx。因此，Spring MVC 将获取文件/WEB-INF/views/singers/list.jspx 作为视图。

16.3.9 实现歌手列表视图

下一步是实现显示歌手信息的视图页面，即文件//src/main/webapp/WEB- INF/views/singers/list.jspx。

```
//src/main/webapp/WEB- INF/views/singers/list.jspx.

<?xml version="1.0" encoding="UTF-8" standalone="no"?>
<div xmlns:jsp="http://java.sun.com/JSP/Page"
  xmlns:c="http://java.sun.com/jsp/jstl/core"
  xmlns:spring="http://www.springframework.org/tags"
  xmlns:fmt="http://java.sun.com/jsp/jstl/fmt" version="2.0">

<h1>Singer Listing</h1>

<c:if test="${not empty singers}">
  <table>
  <thead>
<tr>
<th>First Name</th>
<th>Last Name</th>
<th>Birth Date</th>
</tr>
  </thead>
  <tbody>
  <c:forEach items="${singers}" var="singer">
    <tr>
      <td>${singer.firstName}</td>
      <td>${singer.lastName}</td>
      <td><fmt:formatDate value="${singer.birthDate}"/></td>
    </tr>
  </c:forEach>
      </tbody>
    </table>
  </c:if>
</div>
```

如果曾经使用 JSP 进行过应用程序开发，那么应该对上面所示的代码片段非常熟悉。但由于这是一个 JSPX 页面，所以页面内容被嵌入<div>标记。另外，正在使用的标记库被声明为 XML 名称空间。

首先，<jsp:directive.page>标记定义了应用于整个 JSPX 页面的属性，而<jsp:output>标记控制 JSPX 文档输出的属性。

其次，标记<c:if>检测模型属性 singers 是否为空。因为已经在数据库中填充了一些歌手信息，所以 singers 属性应该包含数据。结果，<c:forEach>标记将在页面内的表格中显示歌手信息。请注意使用<fmt:formatDate>标记来格式化 BirthDate 属性，该属性的类型为 java.utilDate。

16.3.10 测试歌手列表视图

现在准备测试歌手列表视图。首先，构建并部署应用程序；然后，为了测试歌手列表视图，请打开 Web 浏览器并访问 URL http://localhost:8080/singers。此时应该能够看到歌手列表页面。

现在已经有了第一个可以使用的视图。在后续章节中，将使用更多的视图来丰富应用程序，并支持国际化、主题等。

16.4 理解 Spring MVC 项目结构

在深入了解 Web 应用程序各个方面的实现之前，先让我们看一下本章中开发的示例 Web 应用程序的项目结构是什么样子的。

通常，在 Web 应用程序中，需要大量文件来支持各种功能。例如，有很多静态资源文件，例如样式表、JavaScript 文件、图像和组件库，还有支持以各种语言显示界面的文件。当然，也包括 Web 容器将解析和显示的视图页面，以及由模板框架(例如 Apache Tiles)用来提供应用程序一致界面外观的布局和定义文件。

将具有不同用途的文件存储在结构良好的文件夹层次结构中，可以让开发人员更清楚地了解应用程序使用的各种资源，并简化日常维护工作。

表 16-2 描述了本章中开发的示例 Web 应用程序的文件夹结构。请注意，此处介绍的文件夹结构不是强制性的，

但通常在开发人员社区中用于 Web 应用程序开发。

表 16-2 示例 Web 应用程序的文件夹结构及描述

文件夹名称	描述
ckeditor	CKEditor(http://ckeditor.com)是一个 JavaScript 组件库，它在输入表单中提供了一个富文本编辑器。可以使用它对歌手的描述信息进行富文本编辑
jqgrid	jqGrid(http://trirand.com)是一个构建在 jQuery 之上的组件，它提供了各种基于网格的数据显示组件。将使用该库来实现网格以显示歌手信息，以及支持 Ajax 样式的分页
scripts	这是存放所有通用 JavaScript 文件的文件夹。在本章的示例中，将使用 jQuery(http://jquery.org)和 jQuery UI(http://jqueryui.com) JavaScript 库来实现丰富的用户界面。这些脚本都被放置在该文件夹中。内部 JavaScript 库也应放在这里
styles	该文件夹存储样式表文件和用来支持样式的相关图像
WEB-INF/国际化	此文件夹存储支持国际化的文件。application*.properties 文件存储与布局相关的文本(例如页面标题、字段标记和菜单标题)。message*.properties 文件存储各种消息(例如，成功消息、错误消息以及验证消息)。本章示例将支持 English(US)和 Chinese 两种语言
WEB-INF/layouts	该文件夹存储布局视图和定义。这些文件将被 Apache Tiles(http://tiles.apache.org)模板框架使用
WEB-INF/views	该文件夹存储应用程序使用的视图(在本例中为 JSPX 文件)

在后面的章节中，将需要各种文件(例如 CSS 文件、JavaScript 文件和图像)来支持实现，但并不会显示 CSS 和 JavaScript 的源代码。因此，建议下载本章的源代码并将其解压到一个临时文件夹中，以便可以直接将所需的文件复制到项目中。

16.5 实现国际化(i18n)

在开发 Web 应用程序时，在早期阶段启用国际化是一个非常好的习惯，要做的主要工作是将用户界面文本和消息外部化到属性文件中。

虽然刚开始可能并没有国际化要求，但是将语言相关设置外部化是很好的做法，以便日后可以更加容易地支持更多语言。

如果使用 Spring MVC，那么启用国际化是非常简单的。首先，将语言相关的用户界面设置外部化到 /WEB-INF/i18n 文件夹中的各种属性文件中，如表 16-2 中所述。因为此时需要支持 English(US)和 Chinese 语言，所以需要四个文件。application.properties 和 message.properties 文件存储默认语言环境的设置，此时为 English(US)。application_zh_HK.properties 和 message_zh_HK.properties 文件以 Chinese 语言存储相关的设置。

16.5.1 在 DispatcherServlet 配置中配置国际化

完成语言设置之后，下一步就是配置 DispatcherServlet 实例的 WebApplicationContext 以获得国际化支持。以下配置片段描述了基于 Java 的配置类 WebConfig 中的 bean 和方法，对这些 bean 和方法做声明是为了启用和自定义国际化支持。

```
package com.apress.prospring5.ch16.config;

import org.springframework.context.annotation.Bean;
import org.springframework.context.annotation.ComponentScan;
import org.springframework.context.annotation.Configuration;
import org.springframework.context.support.ReloadableResourceBundleMessageSource;
import org.springframework.web.servlet.config.annotation.*;
import org.springframework.web.servlet.i18n.CookieLocaleResolver;
import org.springframework.web.servlet.i18n.LocaleChangeInterceptor;
import org.springframework.web.servlet.mvc.WebContentInterceptor;
import java.util.Locale;
...

@Configuration
```

```
@EnableWebMvc
@ComponentScan(basePackages = {"com.apress.prospring5.ch16"})
public class WebConfig implements WebMvcConfigurer {

  //Declare our static resources.
    @Override
    public void addResourceHandlers(ResourceHandlerRegistry registry) {
  registry.addResourceHandler("/resources/**").addResourceLocations("/")
    .setCachePeriod(31556926);
    }

  //<=> <mvc:default-servlet-handler/>
    @Override
    public void configureDefaultServletHandling(
  DefaultServletHandlerConfigurer configurer) {
  configurer.enable();
    }

    @Bean
    ReloadableResourceBundleMessageSource messageSource() {
  ReloadableResourceBundleMessageSource messageSource =
      new ReloadableResourceBundleMessageSource();
  messageSource.setBasenames(
    "WEB-INF/i18n/messages",
    "WEB-INF/i18n/application");
  messageSource.setDefaultEncoding("UTF-8");
  messageSource.setFallbackToSystemLocale(false);
  return messageSource;
    }
    @Override
    public void addInterceptors(InterceptorRegistry registry) {
  registry.addInterceptor(localeChangeInterceptor());
  ...
    }

    @Bean
    LocaleChangeInterceptor localeChangeInterceptor() {
  return new LocaleChangeInterceptor();
    }

    @Bean
    CookieLocaleResolver localeResolver() {
  CookieLocaleResolver cookieLocaleResolver = new CookieLocaleResolver();
  cookieLocaleResolver.setDefaultLocale(Locale.ENGLISH);
  cookieLocaleResolver.setCookieMaxAge(3600);
  cookieLocaleResolver.setCookieName("locale");
  return cookieLocaleResolver;
    }
  ...
}
```

在上面的配置代码片段中，对资源定义进行了修改，以反映表16-2中展示的新的文件夹结构。addResourceHandlers()方法定义了静态资源文件的位置，这使得Spring MVC能够高效地处理这些文件夹中的文件。在标记中，location属性定义了存放静态资源的文件夹。资源位置(/)表示Web应用程序的根文件夹，即/src/main/webapp。资源处理程序路径/resources/**定义了映射到静态资源的URL；例如，对于URL http://localhost:8080/resources/styles/standard.css，Spring MVC将在/src/main/webapp/styles文件夹中检索文件standard.css。

首先，configureDefaultServletHandling()方法将 DispatcherServlet 映射到 Web 应用程序的根上下文 URL，同时仍允许静态资源请求由容器的默认 servlet 处理。

其次，使用 LocaleChangeInterceptor 类定义了一个 Spring MVC 拦截器，以拦截对 DispatcherServlet 的所有请求。该拦截器支持使用可配置的请求参数进行语言环境切换。在该拦截器的配置中，名为 lang 的 URL 参数是为更改应用程序的语言环境而定义的。

然后，使用 ReloadableResourceBundleMessageSource 类定义了一个 bean。ReloadableResourceBundleMessageSource 类实现了 MessageSource 接口，从定义的文件(本例中为/ WEB-INF/i18n 文件夹中的 message*.properties 和 application *.properties 文件)中加载消息，以支持国际化。请注意属性 fallbackToSystemLocale。该属性指示 Spring MVC 是否在客户端语言环境的特殊资源包未找到时回退到应用程序运行时的系统语言环境。

最后，使用 CookieLocaleResolver 类定义了一个 bean。该类支持从用户浏览器的 cookie 中存储和检索语言环境设置。

16.5.2 为国际化支持而修改歌手列表视图

现在可以更改 JSPX 页面来显示国际化消息。以下 JSPX 片段显示了修改后的歌手列表视图[1]：

```
<?xml version="1.0" encoding="UTF-8" standalone="no"?>
<div xmlns:jsp="http://java.sun.com/JSP/Page"
 xmlns:c="http://java.sun.com/jsp/jstl/core"
 xmlns:fmt="http://java.sun.com/jsp/jstl/fmt"
 xmlns:spring="http://www.springframework.org/tags"
 version="2.0">
 <jsp:directive.page contentType="text/html;charset=UTF-8"/>
 <jsp:output omit-xml-declaration="yes"/>

 <spring:message code="label_singer_list" var="label SingerList"/>
 <spring:message code="label_singer_first_name" var="labelSingerFirstName"/>
 <spring:message code="label_singer_last_name" var="labelSingerLastName"/>
 <spring:message code="label_singer_birth_date" var="labelSingerBirthDate"/>

 <h1>${label SingerList}</h1>

 <c:if test="${not empty singers}">
   <table>
     <thead>
 <tr>
   <th>${labelSingerFirstName}</th>
   <th>${labelSingerLastName}</th>
   <th>${labelSingerBirthDate}</th>
 </tr>
   </thead>
   <tbody>
 <c:forEach items="${singers}" var="singer">
   <tr>
     <td>${singer.firstName}</td>
     <td>${singer.lastName}</td>
     <td><fmt:formatDate value="${singer.birthDate}"/></td>
 </tr>
 </c:forEach>
     </tbody>
   </table>
 </c:if>
</div>
```

如上面的代码片段所示，首先，将 Spring 名称空间添加到页面中。然后，使用<spring:message>标记在相应变量中加载视图所需的消息。最后，将页面标题和标记更改为使用国际化消息。

现在生成并重新部署项目，打开浏览器并指向 URL http://localhost:8080/singers?lang=zh_HK。此时你将看到 Chinese 语言环境下的页面。

由于在 DispatcherServlet 的 WebApplicationContext 中定义了 localeResolver，因此 Spring MVC 会将语言环境设置存储在浏览器的 cookie(名为 locale)中，默认情况下，将为用户会话保留 cookie。如果想要更长时间地保存 cookie，那么可以通过调用 setCookieMaxAge()方法重写 localeResolver bean 定义中从类 org.springframework.web.util.CookieGenerator 继承的属性 cookieMaxAge。

如果想要切换到 English(US)，可以将浏览器中的 URL 更改为 reflect?lang=en_US，此时页面将切换回 English(US)语言环境。虽然没有提供名为 application_en_US.properties 的属性文件，但 Spring MVC 可以回退到使用文件 application.properties，而该文件以默认的 English 语言存储属性。

16.6 使用主题和模板

除了国际化，Web 应用程序还需要适当的界面外观(例如，商业网站需要专业的界面外观，而社交网站需要更加生动的样式)以及一致的布局，以便用户不会在使用 Web 应用程序时感到困惑。

另外，为了提供一致的布局，还需要使用模板框架。在本节中，将使用流行的页面模板框架 Apache Tiles(http://tiles.apache.org)来支持视图模板。Spring MVC 在这方面可以与 Apache Tiles 紧密集成。此外，Spring 还支

1 请记住，目前还没有在其中引入 Tiles 或 JavaScript。

持 Velocity 和 FreeMarker，它们是更通用的模板系统，除了适用于 Web 应用程序之外，也适用于电子邮件模板等。

在下面的章节中，将讨论如何在 Spring MVC 中启用主题支持，以及如何使用 Apache Tiles 定义页面布局。

主题支持

Spring MVC 为主题提供全面的支持，并且可以很容易地在 Web 应用程序中启用主题。例如，在本章的歌手应用程序中，想要创建一个主题并命名为standard。首先，在文件夹/src/main/resources 中，创建一个名为 standard.properties 的文件，内容如下所示：

```
@Override
public void addInterceptors(InterceptorRegistry registry) {
  registry.addInterceptor(localeChangeInterceptor());
  registry.addInterceptor(themeChangeInterceptor());
}

@Bean
ThemeChangeInterceptor themeChangeInterceptor() {
  return new ThemeChangeInterceptor();
}
```

此处，添加新的类型为 ThemeChangeInterceptor 的拦截器，并且拦截每个更改主题的请求。

然后，需要如下 bean 定义：

```
@Bean
ResourceBundleThemeSource themeSource() {
  return new ResourceBundleThemeSource();
}

@Bean
  CookieThemeResolver themeResolver() {
  CookieThemeResolver cookieThemeResolver = new CookieThemeResolver();
  cookieThemeResolver.setDefaultThemeName("standard");
  cookieThemeResolver.setCookieMaxAge(3600);
  cookieThemeResolver.setCookieName("theme");
  return cookieThemeResolver;
}
```

此处定义了两个 bean。第一个 bean 的类型为 ResourceBundleThemeSource，负责加载活动主题的 ResourceBundle bean。例如，如果活动主题为 standard，那么该 bean 将查找文件 standard.properties 作为主题的 ResourceBundle bean。CookieThemeResolver 类型的第二个 bean 用于解析用户的活动主题。属性 defaultThemeName 定义了要使用的默认主题，即标准主题。请注意，顾名思义，CookieThemeResolver 类使用 cookie 来存储用户的主题。此外，还有将 theme 属性存储到用户会话中的 SessionThemeResolver 类。

现在，标准主题已配置好并可以在视图中使用。下面的 JSPX 代码片段显示了具有主题支持的修订后的歌手列表视图(/WEB-INF/views/singers/list.jspx)：

```
<?xml version="1.0" encoding="UTF-8" standalone="no"?>
<div xmlns:jsp="http://java.sun.com/JSP/Page"
    xmlns:c="http://java.sun.com/jsp/jstl/core"
    xmlns:fmt="http://java.sun.com/jsp/jstl/fmt"
    xmlns:spring="http://www.springframework.org/tags"
    version="2.0">
  <jsp:directive.page contentType="text/html;charset=UTF-8"/>
  <jsp:output omit-xml-declaration="yes"/>

  <spring:message code="label_singer_list" var="labelSingerList"/>
  <spring:message code="label_singer_first_name" var="labelSingerFirstName"/>
  <spring:message code="label_singer_last_name" var="labelSingerLastName"/>
  <spring:message code="label_singer_birth_date" var="labelSingerBirthDate"/>

  <head>
    <spring:theme code="styleSheet" var="app_css" />
    <spring:url value="/${app_css}" var="app_css_url" />
    <link rel="stylesheet" type="text/css" media="screen" href="${app_css_url}" />
  </head>
  <h1>${labelSingerList}</h1>

  <c:if test="${not empty singers}">
    <table>
      <thead>
      <tr>
```

```
        <th>${labelSingerFirstName}</th>
        <th>${labelSingerLastName}</th>
        <th>${labelSingerBirthDate}</th>
          </tr>
          </thead>
          <tbody>
          <c:forEach items="${singers}" var="singer">
        <tr>
          <td>${singer.firstName}</td>
          <td>${singer.lastName}</td>
          <td><fmt:formatDate value="${singer.birthDate}"/></td>
        </tr>
            </c:forEach>
          </tbody>
        </table>
      </c:if>
    </div>
```

上述代码在视图中添加了<head>部分，并使用<spring:theme>标记从主题的 ResourceBundle(即样式表文件 standard.css)中检索 styleSheet 属性。最后，将样式表的链接添加到视图中。

在重新生成应用程序并重新部署到服务器之后，打开浏览器并再次指向歌手列表视图的 URL(http://localhost:8080/singers)，此时可以看到歌手列表视图应用了 standard.css 文件中定义的样式。

通过使用 Spring MVC 的主题支持功能，可以轻松添加新主题或更改应用程序中的现有主题。

16.7 使用 Apache Tiles 查看模板

对于使用了 JSP 技术的视图模板，Apache Tiles(http://tiles.apache.org)是最流行的框架。Spring MVC 与 Tiles 紧密集成。为了使用 Tiles 并对数据进行验证，需要将 tilejsp、validation-api 和 hibernate-validator 库添加为依赖项。

在下面的章节中，将讨论如何实现页面模板，包括页面布局设计、定义以及布局中组件的实现。

16.7.1 设计模板布局

首先，需要定义应用程序中所需的模板数量以及每个模板的布局。

在本章的歌手应用程序中，只需要一个模板。布局也相当简单，如图 16-4 所示。如你所见，该模板需要以下页面组件：

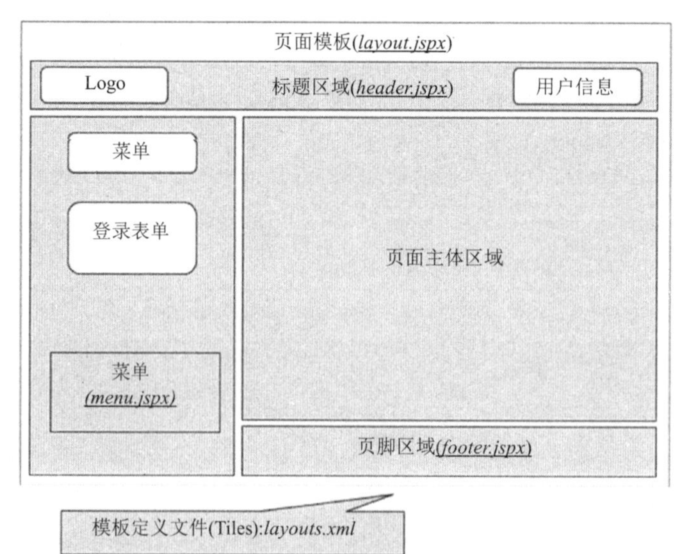

图 16-4 带有布局组件的页面模板

- /WEB-INF/views/header.jspx：此页面提供标题区域。
- /WEB-INF/views/menu.jspx：此页面提供左侧的菜单区域以及本章后面将要实施的登录表单。

- /WEB-INF/views/footer.jspx：此页面提供页脚区域。

我们将使用 Apache Tiles 来定义模板，需要开发页面模板文件以及布局定义文件，如下所示：

- /WEB-INF/layouts/default.jspx：此页面提供特定模板的整体页面布局。
- /WEB-INF/layouts/layouts.xml：该文件存储 Apache Tiles 所需的页面布局定义。

16.7.2 实现页面布局组件

定义布局之后，可以实现页面组件。首先，开发 Apache Tiles 所需的页面模板文件和布局定义文件。

```
<?xml version="1.0" encoding="UTF-8"?>
<!DOCTYPE tiles-definitions PUBLIC
  "-//Apache Software Foundation//DTD Tiles Configuration 2.1//EN"
  "http://tiles.apache.org/dtds/tiles-config_3_0.dtd">
<tiles-definitions>
  <definition name="default" template="/WEB-INF/layouts/default.jspx">
    <put-attribute name="header" value="/WEB-INF/views/header.jspx" />
    <put-attribute name="menu" value="/WEB-INF/views/menu.jspx" />
    <put-attribute name="footer" value="/WEB-INF/views/footer.jspx" />
  </definition>
</tiles-definitions>
```

该文件应该很容易理解。有一个页面模板定义，名为 default。模板代码位于文件 default.jspx 中。在页面中定义了三个组件，分别命名为 header、menu 和 footer。组件的内容将从 value 属性提供的文件中加载。有关 Tiles 定义的详细说明，请参阅项目文档页面(http://tiles.apache.org/)。

以下 JSPX 代码片段显示了 default.jspx 模板文件的内容：

```
<html xmlns:jsp="http://java.sun.com/JSP/Page"
    xmlns:c="http://java.sun.com/jsp/jstl/core"
    xmlns:fn="http://java.sun.com/jsp/jstl/functions"
    xmlns:tiles="http://tiles.apache.org/tags-tiles"
    xmlns:spring="http://www.springframework.org/tags">

<jsp:output doctype-root-element="HTML"
    doctype-system="about:legacy-compat" />

<jsp:directive.page contentType="text/html;charset=UTF-8" />
<jsp:directive.page pageEncoding="UTF-8" />
<head>
  <meta http-equiv="Content-Type" content="text/html; charset=UTF-8" />
  <meta http-equiv="X-UA-Compatible" content="IE=8" />

  <spring:theme code="styleSheet" var="app_css" />
  <spring:url value="/${app_css}" var="app_css_url" />
  <link rel="stylesheet" type="text/css" media="screen" href="${app_css_url}" />

  <!-- Get the user locale from the page context
  (it was set by Spring MVC's locale resolver) -->
  <c:set var="userLocale">
    <c:set var="plocale">${pageContext.response.locale}</c:set>
    <c:out value="${fn:replace(plocale, '_', '-')}" default="en" />
  </c:set>

  <spring:message code="application_name" var="app_name" htmlEscape="false"/>
  <title><spring:message code="welcome_h3" arguments="${app_name}" /></title>
</head>

<body class="tundra spring">
<div id="headerWrapper">
  <tiles:insertAttribute name="header" ignore="true" />
</div>
<div id="wrapper">
  <tiles:insertAttribute name="menu" ignore="true" />
  <div id="main">
    <tiles:insertAttribute name="body"/>
    <tiles:insertAttribute name="footer" ignore="true"/>
  </div>
</div>
</body>
</html>
```

该页面基本上就是一个 JSP 页面，其要点如下所示：

- <spring:theme>标记被放置在模板中，支持模板级别的主题。

- <tiles:insertAttribute>标记用于指示需要从其他文件加载的页面组件，如 layouts.xml 文件中所示。

接下来实现 header、menu 和 footer 组件，具体内容如下所示。header.jspx 文件非常简单，只包含以下文本：

```
<div id="header" xmlns:jsp="http://java.sun.com/JSP/Page"
   xmlns:spring="http://www.springframework.org/tags"
   version="2.0">
  <jsp:directive.page contentType="text/html;charset=UTF-8" />
  <jsp:output omit-xml-declaration="yes" />

  <spring:message code="header_text" var="headerText"/>

  <div id="appname">
    <h1>${headerText}</h1>
  </div>
</div>
```

menu.jspx 文件也很简单，因为该应用程序是用最小的接口设计的，主要的关注点都被放在 Spring 上。

```
<?xml version="1.0" encoding="UTF-8" standalone="no"?>
<div id="menu" xmlns:jsp="http://java.sun.com/JSP/Page"
   xmlns:spring="http://www.springframework.org/tags"
   version="2.0">
  <jsp:directive.page contentType="text/html;charset=UTF-8" />
  <jsp:output omit-xml-declaration="yes" />

  <spring:message code="menu_header_text" var="menuHeaderText"/>
  <spring:message code="menu_add_singer" var="menuAddsinger"/>
  <spring:url value="/singers?form" var="addsingerUrl"/>

  <h3>${menuHeaderText}</h3>
    <a href="${addsingerUrl}"><h3>${menuAddsinger}</h3></a>
</div>
```

footer.jspx 文件包含用于更改界面显示语言的 URL。

```
<?xml version="1.0" encoding="UTF-8" standalone="no"?>
<div id="footer" xmlns:jsp="http://java.sun.com/JSP/Page"
   xmlns:spring="http://www.springframework.org/tags" version="2.0">
  <jsp:directive.page contentType="text/html;charset=UTF-8" />
  <jsp:output omit-xml-declaration="yes" />

  <spring:message code="home_text" var="homeText"/>
  <spring:message code="label_en_US" var="labelEnUs"/>
  <spring:message code="label_zh_HK" var="labelZhHk"/>
  <spring:url value="/singers" var="homeUrl"/>

  <a href="${homeUrl}">${homeText}</a> |
  <a href="${homeUrl}?lang=en_US">${labelEnUs}</a> |
  <a href="${homeUrl}?lang=zh_HK">${labelZhHk}</a>
</div>
```

对于歌手列表视图，可以对其进行修改以适合模板。基本上，只需要删除<head>部分就可以了，因为现在它位于模板页面 default.jspx 中。改进后的版本如下所示：

```
<?xml version="1.0" encoding="UTF-8" standalone="no"?>
<div xmlns:jsp="http://java.sun.com/JSP/Page"
   xmlns:c="http://java.sun.com/jsp/jstl/core"
   xmlns:spring="http://www.springframework.org/tags"
   xmlns:fmt="http://java.sun.com/jsp/jstl/fmt" version="2.0">
  <jsp:directive.page contentType="text/html;charset=UTF-8"/>
  <jsp:output omit-xml-declaration="yes"/>

  <spring:message code="label_singer_list" var="labelSingerList"/>
  <spring:message code="label_singer_first_name" var="labelSingerFirstName"/>
  <spring:message code="label_singer_last_name" var="labelSingerLastName"/>
  <spring:message code="label_singer_birth_date" var="labelSingerBirthDate"/>

  <h1>${labelSingerList}</h1>

  <c:if test="${not empty singers}">
    <table>
      <thead>
      <tr>
  <th>${labelSingerFirstName}</th>
  <th>${labelSingerLastName}</th>
  <th>${labelSingerBirthDate}</th>
    </tr>
    </thead>
```

```
      <tbody>
      <c:forEach items="${singers}" var="singer">
    <tr>
      <td>${singer.firstName}</td>
      <td>${singer.lastName}</td>
      <td><fmt:formatDate value="${singer.birthDate}"/></td>
    </tr>
        </c:forEach>
        </tbody>
      </table>
    </c:if>
  </div>
```

现在，模板、定义和组件已准备就绪；下一步是配置 Spring MVC，以便与 Apache Tiles 集成。

16.8 在 Spring MVC 中配置 Tiles

在 Spring MVC 中配置 Tiles 支持很简单。在 DispatcherServlet 配置(WebConfig 类)中，需要进行相应的修改，以便用 UrlBasedViewResolver 类替换 InternalResourceViewResolver。以下代码片段仅包含配置 Tiles 支持所需的 bean：

```
package com.apress.prospring5.ch16.config;

import org.springframework.context.annotation.Bean;
import org.springframework.context.annotation.ComponentScan;
import org.springframework.context.annotation.Configuration;
import org.springframework.web.servlet.view.UrlBasedViewResolver;
import org.springframework.web.servlet.view.tiles3.TilesConfigurer;
import org.springframework.web.servlet.view.tiles3.TilesView;
...
@Configuration
@EnableWebMvc
@ComponentScan(basePackages = {"com.apress.prospring5.ch16"})
public class WebConfig implements WebMvcConfigurer {

  @Bean
  UrlBasedViewResolver tilesViewResolver() {
    UrlBasedViewResolver tilesViewResolver = new UrlBasedViewResolver();
    tilesViewResolver.setViewClass(TilesView.class);
    return tilesViewResolver;
  }

  @Bean
  TilesConfigurer tilesConfigurer() {
    TilesConfigurer tilesConfigurer = new TilesConfigurer();
    tilesConfigurer.setDefinitions( "/WEB-INF/layouts/layouts.xml",
    "/WEB-INF/views/**/views.xml"
    );
    tilesConfigurer.setCheckRefresh(true);
    return tilesConfigurer;
  }
  ...
}
```

在上面的配置代码片段中，定义了 UrlBasedViewResolver 类的一个 ViewResolver bean，并将属性 viewClass 设置为 TilesView 类，这是 Spring MVC 对 Tiles 提供的支持。最后，定义一个 tilesConfigurer bean，它提供 Tiles 所需的布局配置。

需要准备的最后一个配置文件是/WEB-INF/views/singers/views.xml，该文件定义了歌手应用程序的视图，内容如下所示：

```
<?xml version="1.0" encoding="UTF-8" standalone="no"?>
<!DOCTYPE tiles-definitions PUBLIC "-//Apache Software
   Foundation//DTD Tiles Configuration 3.0//EN"
   "http://tiles.apache.org/dtds/tiles-config_3_0.dtd">
<tiles-definitions>
  <definition extends="default" name="singers/list">
    <put-attribute name="body" value="/WEB-INF/views/singers/list.jspx" />
  </definition>
</tiles-definitions>
```

如上所示，逻辑视图名称被映射到要显示视图的 body 属性。和 SingerController 类一样，list()方法返回逻辑视图名称 singers/list，因此 Tiles 能够将视图名映射到要显示的正确模板和视图主体。

现在可以测试页面。重新生成项目并部署到服务器。再次加载歌手列表视图(http://localhost:8080/singers)，将显示基于模板的视图。

16.9 实现歌手信息视图

接下来继续实现允许用户查看歌手详细信息、创建新歌手或更新现有歌手信息的视图。

在下面的章节中，将讨论 URL 到各种视图的映射，以及视图的实现方式，还会讨论如何在 Spring MVC 中为编辑视图启用 JSR-349 验证支持。

16.9.1 将 URL 映射到视图

首先，需要设计如何将各种 URL 映射到相应的视图。在 Spring MVC 中，最佳实践之一是遵循 REST 风格的 URL 来映射视图。表 16-3 显示了 URL 到视图的映射，以及用来处理操作的控制器方法名称。

表 16-3 URL 到视图的映射

URL	HTTP 方法	控制器方法	描述
/singers	GET	list()	列出歌手的信息
/singers/id	GET	show()	显示单名歌手的信息
/singers/id?form	GET	updateForm()	显示用于更新现有歌手的编辑表单
/singers/id?form	POST	update()	用户更新歌手信息并提交表单。数据将在这里处理
/singers?form	GET	createForm()	显示用于创建新歌手的编辑表单
/singers?form	POST	create()	用户输入歌手信息并提交表单。数据将在这里处理
/singers/photo/id	GET	downloadPhoto()	下载歌手照片

16.9.2 实现显示歌手视图

接下来实现用来显示歌手信息的视图。实现显示歌手视图分为三个步骤：

(1) 实现控制器方法。

(2) 实现显示歌手视图(/views/singers/show.jspx)。

(3) 修改歌手视图的视图定义文件(/views/singers/views.xml)。

以下代码片段显示了用于显示歌手信息的 singerController 类的 show()方法实现：

```
package com.apress.prospring5.ch16.web;

import org.slf4j.Logger;
import org.slf4j.LoggerFactory;
import org.springframework.beans.factory.annotation.Autowired;
import org.springframework.stereotype.Controller;
import org.springframework.ui.Model;
import org.springframework.validation.BindingResult;
import org.springframework.web.bind.annotation.PathVariable;
import org.springframework.web.bind.annotation.RequestMapping;
import org.springframework.web.bind.annotation.RequestMethod;
...

@RequestMapping("/singers")
@Controller
public class SingerController {
  private final Logger logger = LoggerFactory.getLogger(SingerController.class);

  private SingerService singerService;
  private MessageSource messageSource;

  @RequestMapping(value = "/{id}", method = RequestMethod.GET)
  public String show(@PathVariable("id") Long id, Model uiModel) {
    Singer singer = singerService.findById(id);
    uiModel.addAttribute("singer", singer);
```

```
        return "singers/show";
    }

    @Autowired
    public void setSingerService(SingerService singerService) {
        this.singerService = singerService;
    }
    ...
}
```

在 show()方法上应用@RequestMapping 注解，表明该方法将使用 HTTP GET 方法来处理 URL/singers/{id}。而在该方法中，@PathVariable 注解被应用于参数 id，指示 Spring MVC 从 URL 提取 ID 到参数中。然后，歌手将被检索并添加到模型中，并返回逻辑视图名称 singers/show。下一步是实现显示歌手视图/views/singers/show.jspx，如下所示：

```
<?xml version="1.0" encoding="UTF-8" standalone="no"?>
<div xmlns:jsp="http://java.sun.com/JSP/Page"
    xmlns:c="http://java.sun.com/jsp/jstl/core"
    xmlns:spring="http://www.springframework.org/tags"
    xmlns:fmt="http://java.sun.com/jsp/jstl/fmt" version="2.0">
  <jsp:directive.page contentType="text/html;charset=UTF-8"/>
  <jsp:output omit-xml-declaration="yes"/>

  <spring:message code="label_singer_info" var="labelSingerInfo"/>
  <spring:message code="label_singer_first_name" var="labelSingerFirstName"/>
  <spring:message code="label_singer_last_name" var="labelSingerLastName"/>
  <spring:message code="label_singer_birth_date" var="labelSingerBirthDate"/>
  <spring:message code="label_singer_description" var="labelSingerDescription"/>
  <spring:message code="label_singer_update" var="labelSingerUpdate"/>
  <spring:message code="date_format_pattern" var="dateFormatPattern"/>
  <spring:message code="label_singer_photo" var="labelSingerPhoto"/>

  <spring:url value="/singers/photo" var="singerPhotoUrl"/>
  <spring:url value="/singers" var="editSingerUrl"/>

  <h1>${labelSingerInfo}</h1>

  <div id="singerInfo">

    <c:if test="${not empty message}">
      <div id="message" class="${message.type}">${message.message}</div>
    </c:if>

  <table>
    <tr>
    <td>${labelSingerFirstName}</td>
    <td>${singer.firstName}</td>
      </tr>
      <tr>
    <td>${labelSingerLastName}</td>
    <td>${singer.lastName}</td>
      </tr>
      <tr>
    <td>${labelSingerBirthDate}</td>
    <td><fmt:formatDate value="${singer.birthDate}"/></td>
      </tr>
      <tr>
    <td>${labelSingerDescription}</td>
    <td>${singer.description}</td>
      </tr>
      <tr>
    <td>${labelSingerPhoto}</td>
    <td><img src="${singerPhotoUrl}/${singer.id}"></img></td>
      </tr>
    </table>

    <a href="${editSingerUrl}/${singer.id}?form">${labelSingerUpdate}</a>
  </div>
</div>
```

该页面很简单，仅在页面中显示了模型属性 singer。

最后一步是修改视图定义文件/views/singers/views.xml 以映射逻辑视图名称 singers/show。

```
<?xml version="1.0" encoding="UTF-8" standalone="no"?>
<!DOCTYPE tiles-definitions PUBLIC "-//Apache Software
    Foundation//DTD Tiles Configuration 3.0//EN"
```

```
    "http://tiles.apache.org/dtds/tiles-config_3_0.dtd">
  <tiles-definitions>
    <definition extends="default" name="singers/list">
      <put-attribute name="body" value="/WEB-INF/views/singers/list.jspx" />
    </definition>
    <definition extends="default" name="singers/show">
      <put-attribute name="body" value="/WEB-INF/views/singers/show.jspx" />
  </definition>
</tiles-definitions>
```

显示歌手视图已经完成。现在，需要在歌手列表视图(/views/singers/list.jspx)中为每位歌手添加链接到显示歌手视图的锚链接。修改后的文件的内容如下所示：

```
<?xml version="1.0" encoding="UTF-8" standalone="no"?>
<div xmlns:jsp="http://java.sun.com/JSP/Page"
   xmlns:c="http://java.sun.com/jsp/jstl/core"
   xmlns:spring="http://www.springframework.org/tags"
   xmlns:fmt="http://java.sun.com/jsp/jstl/fmt" version="2.0">
  <jsp:directive.page contentType="text/html;charset=UTF-8"/>
  <jsp:output omit-xml-declaration="yes"/>

  <spring:message code="label_singer_list" var="labelSingerList"/>
  <spring:message code="label_singer_first_name" var="labelSingerFirstName"/>
  <spring:message code="label_singer_last_name" var="labelSingerLastName"/>
  <spring:message code="label_singer_birth_date" var="labelSingerBirthDate"/>

  <h1>${labelSingerList}</h1>

  <spring:url value="/singers" var="showSingerUrl"/>

  <c:if test="${not empty singers}">
    <table>
      <thead>
    <tr>
  <th>${labelSingerFirstName}</th>
  <th>${labelSingerLastName}</th>
  <th>${labelSingerBirthDate}</th>
    </tr>
    </thead>
    <tbody>
    <c:forEach items="${singers}" var="singer">
  <tr>
    <td>
    <a href="${showSingerUrl}/${singer.id}">${singer.firstName}</a>
    </td>
    <td>${singer.lastName}</td>
    <td><fmt:formatDate value="${singer.birthDate}"/></td>
  </tr>
      </c:forEach>
      </tbody>
    </table>
  </c:if>
</div>
```

如上所示，使用<spring:url>标记声明了一个 URL 变量，并为 firstName 属性添加了一个锚链接。为了测试显示歌手视图，在重新生成和部署之后，再次打开歌手列表视图。此时，列表应该包含指向显示歌手视图的超链接。单击任何链接都会进入歌手信息视图。

16.9.3 实现编辑歌手视图

接下来实现编辑歌手视图。与显示歌手视图相同；首先将方法 updateForm()和 update()添加到 SingerController 类中。以下代码片段显示了为这两个方法修改后的控制器：

```
package com.apress.prospring5.ch16.web;
...

@RequestMapping("/singers")
@Controller
public class SingerController {
  private final Logger logger = LoggerFactory.getLogger(SingerController.class);

  private SingerService singerService;
    private MessageSource messageSource;

  @RequestMapping(value = "/{id}", params = "form", method = RequestMethod.POST)
```

```
public String update(@Valid Singer singer, BindingResult bindingResult,
  Model uiModel, HttpServletRequest httpServletRequest,
  RedirectAttributes redirectAttributes, Locale locale) {
  logger.info("Updating singer");
  if (bindingResult.hasErrors()) {
    uiModel.addAttribute("message", new Message("error",
messageSource.getMessage("singer_save_fail", new Object[]{}, locale)));
  uiModel.addAttribute("singer", singer);
  return "singers/update";
}
  uiModel.asMap().clear();
  redirectAttributes.addFlashAttribute("message", new Message("success",
messageSource.getMessage("singer_save_success", new Object[]{}, locale)));
  singerService.save(singer);
  return "redirect:/singers/" + UrlUtil.encodeUrlPathSegment(
    singer.getId().toString(), httpServletRequest);
}

@RequestMapping(value = "/{id}", params = "form", method = RequestMethod.GET)
public String updateForm(@PathVariable("id") Long id, Model uiModel) {
  uiModel.addAttribute("singer", singerService.findById(id));
  return "singers/update";
}

@Autowired
public void setSingerService(SingerService singerService) {
  this.singerService = singerService;
}
   @Autowired
  public void setMessageSource(MessageSource messageSource) {
this.messageSource = messageSource;
  }
...
}
```

在上面的配置中，要点如下所示：

- MessageSource 接口被自动装配到控制器，用于检索具有国际化支持的消息。
- 对于 updateForm()方法，首先检索歌手并保存到模型中，然后返回逻辑视图 singers/update，将显示编辑歌手视图。
- 当用户更新歌手信息并单击 Save 按钮时，将触发 update()方法。对该方法需要进行一些解释。首先，Spring MVC 尝试将提交的数据绑定到 Singer 域对象并自动执行类型转换和格式化。如果发现绑定错误(例如，输入的出生日期格式错误)，错误将被保存到 BindingResult 接口(位于 org.springframework.validation 包中)中，同时将错误消息保存到模型中，并重新显示编辑视图。如果绑定成功，则保存数据，并且通过使用 redirect:作为前缀，为显示歌手视图返回逻辑视图名称。请注意，我们希望在重定向后显示成功消息，因此需要使用 RedirectAttributes.addFlashAttribute()方法(包 org.springframework.web.servlet.mvc.support 中的一个接口)在显示歌手视图中显示成功消息。在 Spring MVC 中，flash 属性在重定向之前被临时保存(通常在会话中)，以便在重定向后对请求可用，并立即删除。
- Message 是一个自定义类，存储了从 MessageSource 中检索的消息以及视图在消息区域中显示的消息类型(即成功或错误)。以下是 Message 类的内容：

```
package com.apress.prospring5.ch16.util;

public class Message {
  private String type;
  private String message;

  public Message(String type, String message) {
this.type = type;
this.message = message;
  }

  public String getType() {
return type;
  }

  public String getMessage() {
return message;
  }
}
```

- UrlUtil 是用于编码 URL 以重定向的实用程序类，其内容如下所示：

```
package com.apress.prospring5.ch16.util;

import java.io.UnsupportedEncodingException;

import javax.servlet.http.HttpServletRequest;

import org.springframework.web.util.UriUtils;
import org.springframework.web.util.WebUtils;

public class UrlUtil {
   public static String encodeUrlPathSegment(String pathSegment,
   HttpServletRequest httpServletRequest) {
  String enc = httpServletRequest.getCharacterEncoding();

  if (enc == null) {
  enc = WebUtils.DEFAULT_CHARACTER_ENCODING;
  }

  try {
  pathSegment = UriUtils.encodePathSegment(pathSegment, enc);
  } catch (UnsupportedEncodingException uee) {
  //
  }

  return pathSegment;
    }
}
```

接下来是编辑歌手视图/views/singers/edit.jspx，我们将使用它来更新和创建新歌手。

```
<?xml version="1.0" encoding="UTF-8" standalone="no"?>

<div xmlns:jsp="http://java.sun.com/JSP/Page"
   xmlns:c="http://java.sun.com/jsp/jstl/core"
   xmlns:spring="http://www.springframework.org/tags"
   xmlns:form="http://www.springframework.org/tags/form"
   version="2.0">

  <jsp:directive.page contentType="text/html;charset=UTF-8"/>
  <jsp:output omit-xml-declaration="yes"/>

  <spring:message code="label_singer_new" var="labelSingerNew"/>
  <spring:message code="label_singer_update" var="labelSingerUpdate"/>
  <spring:message code="label_singer_first_name" var="labelSingerFirstName"/>
  <spring:message code="label_singer_last_name" var="labelSingerLastName"/>
  <spring:message code="label_singer_birth_date" var="labelSingerBirthDate"/>
  <spring:message code="label_singer_description" var="labelSingerDescription"/>
  <spring:message code="label_singer_photo" var="labelSingerPhoto"/>

  <spring:eval expression="singer.id == null ? labelSingerNew:labelSingerUpdate"
   var="formTitle"/>

  <h1>${formTitle}</h1>
  <div id="singerUpdate">
  <form:form modelAttribute="singer" id="singerUpdateForm" method="post">

  <c:if test="${not empty message}">
   <div id="message" class="${message.type}">${message.message}</div>
  </c:if>

  <form:label path="firstName">
   ${labelSingerFirstName}*
  </form:label>
  <form:input path="firstName" />
  <div>
   <form:errors path="firstName" cssClass="error" />
  </div>
  <p/>

  <form:label path="lastName">
   ${labelSingerLastName}*
  </form:label>
  <form:input path="lastName" />
  <div>
   <form:errors path="lastName" cssClass="error" />
  </div>
```

```
    <p/>
    <form:label path="birthDate">
      ${labelSingerBirthDate}
    </form:label>
    <form:input path="birthDate" id="birthDate"/>
    <div>
      <form:errors path="birthDate" cssClass="error" />
    </div>
    <p/>

    <form:label path="description">
      ${labelSingerDescription}
    </form:label>
    <form:textarea cols="60" rows="8" path="description"
      id="singerDescription"/>
    <div>
      <form:errors path="description" cssClass="error" />
    </div>
    <p/>

    <label for="file">
      ${labelSingerPhoto}
    </label>
    <input name="file" type="file"/>
    <p/>

    <form:hidden path="version" />
    <button type="submit">Save</button>
    <button type="reset">Reset</button>
    </form:form>
      </div>
  </div>
```

上述模板的要点如下所示：

- <spring:eval>标记使用 Spring 表达式语言来测试歌手 ID 是否为空。如果为空，则表明是一名新歌手；否则，更新歌手信息。显示相应的表单标题。
- 表单中使用各种 Spring MVC <form>标记来显示标签、输入字段和提交表单时绑定失败的错误信息。

接下来，将视图映射添加到视图定义文件/views/singers/views.xml 中。

```
<?xml version="1.0" encoding="UTF-8" standalone="no"?>
<!DOCTYPE tiles-definitions PUBLIC "-//Apache Software
    Foundation//DTD Tiles Configuration 3.0//EN"
  "http://tiles.apache.org/dtds/tiles-config_3_0.dtd">
<tiles-definitions>
    <definition extends="default" name="singers/update">
  <put-attribute name="body" value="/WEB-INF/views/singers/edit.jspx" />
    </definition>
    ...
</tiles-definitions>
```

编辑歌手视图现在已完成。接下来重新生成并部署该项目。单击编辑链接后，将显示编辑歌手视图。更新信息并单击 Save 按钮。如果绑定成功，则会看到成功消息，并显示编辑歌手视图。

16.9.4 实现添加歌手视图

添加歌手视图非常类似于编辑歌手视图。因为将重用 edit.jspx 页面，所以只需要在 SingerController 类和视图定义中添加方法即可。以下所示的 SingerController 类的代码片段描述了为新歌手实现保存操作所需的方法：

```
package com.apress.prospring5.ch16.web;
...

@RequestMapping("/singers")
@Controller
public class SingerController {
  private final Logger logger = LoggerFactory.getLogger(SingerController.class);

  private SingerService singerService;
  private MessageSource messageSource;

  @RequestMapping(method = RequestMethod.POST)
public String create(@Valid Singer singer, BindingResult bindingResult,
  Model uiModel, HttpServletRequest httpServletRequest,
  RedirectAttributes redirectAttributes,Locale locale) {
```

```
logger.info("Creating singer");
if (bindingResult.hasErrors()) {
uiModel.addAttribute("message", new Message("error",
 messageSource.getMessage("singer_save_fail", new Object{}, locale)));
uiModel.addAttribute("singer", singer);
return "singers/create";
}
uiModel.asMap().clear();

redirectAttributes.addFlashAttribute("message", new Message("success",
 messageSource.getMessage("singer_save_success", new Object[]{}, locale)));

logger.info("Singer id: " + singer.getId());
singerService.save(singer);
return "redirect:/singers/";
  }

  @RequestMapping(params = "form", method = RequestMethod.GET)
  public String createForm(Model uiModel) {
Singer singer = new Singer();
uiModel.addAttribute("singer", singer);

return "singers/create";
  }

  @Autowired
  public void setSingerService(SingerService singerService) {
this.singerService = singerService;
  }
...
}
```

接下来，将视图映射添加到视图定义文件/views/singers/views.xml 中。

```
<?xml version="1.0" encoding="UTF-8" standalone="no"?>
<!DOCTYPE tiles-definitions PUBLIC "-//Apache Software
    Foundation//DTD Tiles Configuration 3.0//EN"
  "http://tiles.apache.org/dtds/tiles-config_3_0.dtd">
<tiles-definitions>
    <definition extends="default" name="singers/create">
  <put-attribute name="body" value="/WEB-INF/views/singers/edit.jspx" />
    </definition>
    ...
</tiles-definitions>
```

添加歌手视图现已完成。重新生成并部署项目后，单击菜单区域中的 New Singer 链接，将会显示添加歌手视图，允许输入新歌手的详细信息。

16.9.5 启用 JSR-349(bean 验证)

下面配置 JSR-349(bean 验证)支持来创建和更新歌手动作。首先，将验证约束应用于 Singer 域对象。在本例中，仅为 firstName 和 lastName 属性定义约束。虽然在本章的开始部分已经显示了带有验证注解的 Singer 类，但在本节中将对其进行详细解释。在下面所示的类代码片段中，对感兴趣的字段进行了注解：

```
package com.apress.prospring5.ch16.entities;

import org.hibernate.validator.constraints.NotBlank;

import javax.persistence.*;
import javax.validation.constraints.NotNull;
import javax.validation.constraints.Size;
...
@Entity
@Table(name = "singer")
public class Singer implements Serializable {
...
  @NotBlank(message="{validation.firstname.NotBlank.message}")
    @Size(min=2, max=60, message="{validation.firstname.Size.message}")
    @Column(name = "FIRST_NAME")
    private String firstName;

    @NotBlank(message="{validation.lastname.NotBlank.message}")
    @Size(min=1, max=40, message="{validation.lastname.Size.message}")
  @Column(name = "LAST_NAME")
  private String lastName;
```

```
   ...
}
```

我们为各个字段应用了约束条件。请注意，对于验证消息，通过使用大括号来指定消息密钥。这样就可以从 ResourceBundle 检索验证消息，并支持国际化。

要在 Web 数据绑定过程中启用 JSR-349 验证，需要将@Valid 注解应用于 SingerController 类的 create()和 update()方法的参数。以下代码片段显示了这两个方法的签名：

```
package com.apress.prospring5.ch16.web;
@RequestMapping("/singers")
@Controller
public class SingerController {
...
    @RequestMapping(value = "/{id}", params = "form", method = RequestMethod.POST)
    public String update(@Valid Singer singer, BindingResult bindingResult, ...

    @RequestMapping(method = RequestMethod.POST)
  public String create(@Valid Singer singer, BindingResult bindingResult, ...
...
}
```

此外，还希望 JSR-349 验证消息使用与视图相同的 ResourceBundle 实例。为此，需要在 WebConfig 类的 DispatcherServlet 配置中配置验证器。

```
package com.apress.prospring5.ch16.config;

import com.apress.prospring5.ch16.util.DateFormatter;
import org.springframework.context.annotation.Bean;
import org.springframework.context.annotation.ComponentScan;
import org.springframework.context.annotation.Configuration;
import org.springframework.context.support.ReloadableResourceBundleMessageSource;
import org.springframework.validation.Validator;
import org.springframework.validation.beanvalidation.LocalValidatorFactoryBean;
import org.springframework.web.servlet.config.annotation.*;

@Configuration
@EnableWebMvc
@ComponentScan(basePackages = {"com.apress.prospring5.ch16"})
public class WebConfig implements WebMvcConfigurer {
  @Bean
  public Validator validator() {
    final LocalValidatorFactoryBean validator =
    new LocalValidatorFactoryBean();
  validator.setValidationMessageSource(messageSource());
  return validator;
}
  //<=> <mvc:annotation-driven validator="validator"/>
  @Override
  public Validator getValidator() {
    return validator();
  }

  @Bean
  ReloadableResourceBundleMessageSource messageSource() {
    ReloadableResourceBundleMessageSource messageSource =
      new ReloadableResourceBundleMessageSource();
    messageSource.setBasenames(
      "WEB-INF/i18n/messages",
      "WEB-INF/i18n/application");
    messageSource.setDefaultEncoding("UTF-8");
    messageSource.setFallbackToSystemLocale(false); return messageSource;
  }
  ...
}
```

首先，为了提供 JSR-349 支持而定义一个验证器 bean(使用了类 LocalValidatorFactoryBean)。请注意，此时将 validationMessageSource 属性设置为引用所定义的 messageSource bean，从而指示 JSR-349 验证器通过 messageSource bean 中的代码查找消息。然后实现 getValidator()方法，返回所定义的验证器 bean。

你所要做的就这么多，现在可以测试验证。调出添加歌手视图，然后单击 Save 按钮。返回的页面将会显示验证错误。

切换到 Chinese 语言，并完成同样的操作。这一次，验证消息将以中文显示。

目前，除了删除歌手视图之外，所需的视图基本上已经完成。将删除歌手视图作为练习留给读者来完成。接下来，将开始丰富用户界面。

16.10 使用 jQuery 和 jQuery UI

虽然歌手应用程序的视图运行良好，但用户界面却非常粗糙。例如，对于出生日期字段，如果可以添加日期选择器来输入歌手的出生日期，而不是让用户手动输入日期字符串，那么用户体验将会好得多。

要为 Web 应用程序的用户提供更丰富的界面，一般使用 JavaScript 来实现相关功能，除非使用的富 Internet 应用程序(RIA)技术需要在 Web 浏览器客户端运行特殊运行时(例如，Adobe Flex 需要 Flash，JavaFX 需要 JRE，Microsoft Silverlight 需要 Silverlight)。

但是，使用原始 JavaScript 开发 Web 前端并不容易。语法与 Java 存在很大的不同，此外还需要处理跨浏览器兼容性问题。因此，可以使用很多开源的 JavaScript 库来简化 Web 前端的开发，比如 jQuery 和 Dojo Toolkit。

在接下来的章节中，将讨论如何使用 jQuery 和 jQuery UI 来开发更具响应性和交互性的用户界面。此外，还讨论一些用于特定目的常用 jQuery 插件，比如富文本编辑支持，并讨论一些用于浏览数据的基于网格的组件。

16.10.1 jQuery 和 jQuery UI 介绍

jQuery(http://jquery.org)是用于 Web 前端开发的最流行的 JavaScript 库之一。jQuery 对主要功能提供全面支持，包括用于选择文档中 DOM 元素的强大“选择器”语法、复杂的事件模型和强大的 Ajax 支持。

jQuery UI 库(http://jqueryui.com)构建于 jQuery 之上，它提供了一组丰富的小部件和效果，主要功能包括用于常用用户界面组件(日期选取器、自动完成控件、可折叠控件等)、拖放、效果、动画、主题等的小部件。

jQuery 社区为特定目的开发了大量的 jQuery 插件，本章将重点讨论其中的两个。

本章介绍的只是 jQuery 的一部分内容。有关使用 jQuery 的更多详细信息，推荐阅读 B.M.Harwani 编写的 *jQuery Recipes:A Problem-Solution Approach*(Apress 出版社于 2010 年出版)以及 Bear Bibeault 和 Yehuda Katz 合著的 *jQuery in Action*(Manning 出版社于 2010 年出版)。

16.10.2 在视图中使用 jQuery 和 jQuery UI

为了能够在视图中使用 jQuery 和 jQuery UI 组件，需要包含所需的样式表和 JavaScript 文件。如果已经阅读了本章的 16.4 节“理解 Spring MVC 项目结构”，那么所需的文件应该已经被复制到项目中。以下是需要包含的主要文件：

- /src/main/webapp/scripts/jquery-1.12.4.js：这是核心 jQuery JavaScript 库，本章中使用的版本是 1.12.4。请注意，它是完整的源代码版本。在生产中，应该使用最小化版本(即 jquery-1.12.4.min.js)，该版本经过优化和压缩，从而提高了下载和执行性能。
- /src/main/webapp/scripts/jquery-ui.min.js：这是绑定了主题样式表的 jQuery UI 库，可以从 jQuery UI Themeroller 页面(http://jqueryui.com/ThemeRoller)下载该样式表并进行自定义。此处使用的 jQuery UI 版本是 1.12.1。请注意，这是 JavaScript 的缩小版本。
- /src/main/webapp/styles/custom-theme/jquery-ui.theme.min.css：这是自定义主题的样式表，jQuery UI 将使用它来支持主题。

要包含上面所示的文件，只需要将它们包含在模板页面(即/layouts/default.jspx)中即可。需要将以下代码片段添加到页面中：

```
<html xmlns:jsp="http://java.sun.com/JSP/Page" ..>
...
<head>
  <meta http-equiv="Content-Type" content="text/html; charset=UTF-8" />
<  meta http-equiv="X-UA-Compatible" content="IE=8" />

    <spring:theme code="styleSheet" var="app_css" />
    <spring:url value="/${app_css}" var="app_css_url" />
    <link rel="stylesheet" type="text/css" media="screen" href="${app_css_url}" />
    <spring:url value="/resources/scripts/jquery-1.12.4.js" var="jquery_url" />
    <spring:url value="/resources/scripts/jquery-ui.min.js" var="jquery_ui_url" />
    <spring:url value="/resources/styles/custom-theme/jquery-ui.theme.min.css"
      var="jquery_ui_theme_css" />
    <link rel="stylesheet" type="text/css" media="screen"
```

```
      href="${jquery_ui_theme_css}" />
    <script src="${jquery_url}" type="text/javascript"><jsp:text/></script>
    <script src="${jquery_ui_url}" type="text/javascript"><jsp:text/></script>
  ...
</head>
...
</html>
```

首先，<spring:url>标记用于定义文件的 URL 并将它们存储在变量中。然后，在<head>部分添加对 CSS 和 JavaScript 文件的引用。请注意，这里在<script>标记中使用了<jsp:text />标记。这是因为 JSPX 会自动删除没有正文的标记。所以，文件中的<script ..> </ script>标记在浏览器中将以<script .. />结尾，从而导致页面中的行为不确定。添加<jsp:text />可以确保<script>标记不会在页面中显示，因为这避免了出现意外的问题。

添加完脚本之后，还可以向视图中添加一些更有趣的东西。对于编辑歌手视图，可以让按钮变得更好看一点，并为出生日期字段启用日期选择器组件。以下代码片段显示了对需要添加到/views/singers/edit.jspx 中的按钮和日期字段所做的更改：

```
<?xml version="1.0" encoding="UTF-8" standalone="no"?>

<div xmlns:jsp="http://java.sun.com/JSP/Page"
   xmlns:c="http://java.sun.com/jsp/jstl/core"
   xmlns:spring="http://www.springframework.org/tags"
   xmlns:form="http://www.springframework.org/tags/form"
   version="2.0">

  <script type="text/javascript">
  $(function(){
  $('#birthDate').datepicker({
    dateFormat: 'yy-mm-dd',
    changeYear: true
  });
  });
    </script>
    ...
      <form:form modelAttribute="singer" id="singerUpdateForm" method="post">
      ...
      <button type="submit" class="ui-button ui-widget
      ui-state-default ui-corner-all ui-button-text-only">
      <span class="ui-button-text">Save</span>
    </button>
    <button type="reset" class="ui-button ui-widget
      ui-state-default ui-corner-all ui-button-text-only">
      <span class="ui-button-text">Reset</span>
  </button>
  </form:form>
    </div>
</div>
```

$(function(){}语法指示 jQuery 在文档准备就绪时执行脚本。在该函数中，使用 jQuery UI 的 datepicker()函数对出生日期输入字段(带有 ID birthDate)进行了修饰。然后，向按钮添加了不同的样式类。

现在重新部署应用程序，你将会看到新的按钮样式，并且当单击出生日期字段时，将显示日期选取器组件。

16.10.3 使用 CKEditor 进行富文本编辑

对于歌手信息的描述字段，可以使用 Spring MVC <form:textarea>标记来支持多行输入。启用富文本编辑对于长文本输入(比如用户评论)来说是常见要求。

为了支持该功能，需要使用富文本组件库 CKEditor(http://ckeditor.com)，它是一个常用的富文本 JavaScript 组件，并与 jQuery UI 集成。这些文件位于示例源代码的/src/main/we-bapp/ckeditor 文件夹中。

首先，需要将所需的 JavaScript 文件包含到模板页面 default.jspx 中。以下代码片段显示了需要添加到页面中的内容：

```
<html xmlns:jsp="http://java.sun.com/JSP/Page"
  xmlns:c="http://java.sun.com/jsp/jstl/core"
  xmlns:fn="http://java.sun.com/jsp/jstl/functions"
  xmlns:tiles="http://tiles.apache.org/tags-tiles"
  xmlns:spring="http://www.springframework.org/tags">

<head>
  <!-- CKEditor -->
```

```
  <spring:url value="/resources/ckeditor/ckeditor.js" var="ckeditor_url" />
  <spring:url value="/resources/ckeditor/adapters/jquery.js"
var="ckeditor_jquery_url" />
  <script type="text/javascript" src="${ckeditor_url}"><jsp:text/></script>
  <script type="text/javascript" src="${ckeditor_jquery_url}"><jsp:text/></script>
  ...
  </head>
  ...
</html>
```

上面的 JSPX 代码片段包含两个脚本：核心 CKEditor 脚本和带有 jQuery 的适配器。

下一步是在编辑歌手视图中启用 CKEditor。下面所示的 JSPX 代码片段显示了对页面 edit.jspx 所需要做的更改：

```
<?xml version="1.0" encoding="UTF-8" standalone="no"?>

<div xmlns:jsp="http://java.sun.com/JSP/Page"
    xmlns:c="http://java.sun.com/jsp/jstl/core"
    xmlns:spring="http://www.springframework.org/tags"
    xmlns:form="http://www.springframework.org/tags/form" version="2.0">

  <script type="text/javascript">
$(function(){
$("#singerDescription").ckeditor(
  {
  toolbar : 'Basic',
  uiColor : '#CCCCCC'
  }
);
});
  </script>
  ...
</div>
```

文档准备好后，用 CKEditor 修饰歌手描述字段。重新部署应用程序并转到添加歌手页面，此时描述字段支持富文本编辑。

有关使用和配置 CKEditor 的完整文档，请参阅项目的文档站点(http://docs.cksource.com/)。

16.10.4 使用 jqGrid 实现具有分页支持的数据网格

如果系统中只有少数歌手存在，那么当前的歌手列表视图就没有什么问题了。但是，随着数据增长到数千条甚至更多条记录，性能将成为突出问题。

常见的解决方案是实现用于数据浏览的具有分页支持的数据网格组件，以便用户仅浏览一定数量的记录，从而避免浏览器和 Web 容器之间大量的数据传输。本节将演示如何使用 jqGrid(http://trirand.com/blog)(一种流行的基于 JavaScript 的数据网格组件)实现数据网格。此处使用的版本是 4.6.0。

对于分页支持，将使用 jqGrid 内置的 Ajax 分页支持，它为每个页面发出 XMLHttpRequest 请求并接收 JSON 数据格式的页面数据。要生成 JSON 数据，需要使用 ackson-databind 库。

在下面的章节中，将讨论如何在服务器端和客户端实现分页支持。首先，介绍在歌手列表视图中实现 jqGrid 组件。然后，将讨论如何使用 Spring Data Commons 模块的全面分页支持在服务器端实现分页。

16.10.5 在歌手列表视图中启用 jqGrid

要在视图中启用 jqGrid，首先需要在模板页面 default.jspx 中包含必需的 JavaScript 和样式表文件。

```
<html xmlns:jsp="http://java.sun.com/JSP/Page"
...
  <head>
  ...
    <!-- jqGrid -->
    <spring:url value="/resources/jqgrid/css/ui.jqgrid.css" var="jqgrid_css" />
    <spring:url value="/resources/jqgrid/js/i18n/grid.locale-en.js"
    var="jqgrid_locale_url" />
  <spring:url value="/resources/jqgrid/js/jquery.jqGrid.min.js" var="jqgrid_url" />
    <link rel="stylesheet" type="text/css" media="screen" href="${jqgrid_css}" />
    <script type="text/javascript" src="${jqgrid_locale_url}"><jsp:text/></script>
    <script type="text/javascript" src="${jqgrid_url}"><jsp:text/></script>
    ...
   </head>
  ...
```

```
</html>
```

首先，加载特定于网格的 CSS 文件。然后，需要两个 JavaScript 文件。第一个是语言环境脚本(此处使用英语)，第二个是 jqGrid 核心库文件 jquery.jqGrid.min.js。

下一步是修改歌手列表视图 textitlist.jspx，以便使用 jqGrid。修改后的页面如下所示：

```
<?xml version="1.0" encoding="UTF-8" standalone="no"?>
<div xmlns:jsp="http://java.sun.com/JSP/Page"
   xmlns:c="http://java.sun.com/jsp/jstl/core"
   xmlns:spring="http://www.springframework.org/tags"
   version="2.0">
  <jsp:directive.page contentType="text/html;charset=UTF-8"/>
  <jsp:output omit-xml-declaration="yes"/>
  <spring:message code="label_singer_list" var="labelSingerList"/>
  <spring:message code="label_singer_first_name" var="labelSingerFirstName"/>
  <spring:message code="label_singer_last_name" var="labelSingerLastName"/>
  <spring:message code="label_singer_birth_date" var="labelSingerBirthDate"/>
  <spring:url value="/singers/" var="showSingerUrl"/>

  <script type="text/javascript">
$(function(){
$("#list").jqGrid({
  url:'${showSingerUrl}/listgrid',
  datatype: 'json',
  mtype: 'GET',

  colNames:['${labelSingerFirstName}', '${labelSingerLastName}',
'${labelSingerBirthDate}'],
  colModel :[
{name:'firstName', index:'firstName', width:150},
{name:'lastName', index:'lastName', width:100},
{name:'birthDateString', index:'birthDate', width:100}
  ],
  jsonReader : {
root:"singerData",
page: "currentPage",
total: "totalPages",
records: "totalRecords",
repeatitems: false,
id: "id"
  },
  pager: '#pager',
  rowNum:10, rowList:[10,20,30],
  sortname: 'firstName',
  sortorder: 'asc',
  viewrecords: true,
  gridview: true,
  height: 250,
  width: 500,
  caption: '${labelSingerList}',
  onSelectRow: function(id){
document.location.href ="${showSingerUrl}/" + id;
  }
});
});

</script>
   <c:if test="${not empty message}">
  <div id="message" class="${message.type}">
  ${message.message}
  </div>
   </c:if>

   <h2>${labelSingerList}</h2>

   <div>
  <table id="list"><tr><td/></tr></table>
   </div>
   <div id="pager"></div>
</div>
```

上述代码声明一个 ID 为 list 的<table>标记来显示网格数据。在表格中定义一个 ID 为 pager 的<div>部分，它是 jqGrid 的分页部分。在 JavaScript 中，当文档准备就绪时，指示 jqGrid 将 ID 为 list 的表格装饰成网格并提供详细的配置信息。脚本的一些要点如下：

- url 属性指定发送 XMLHttpRequest 请求的链接，从而获取当前页面的数据。

- datatype 属性指定数据格式，此处为 JSON。jqGrid 也支持 XML 格式。
- mtype 属性定义要使用的 HTTP 方法，即 GET。
- colNames 属性定义网格中所显示数据的列标题，colModel 属性则定义每个数据列的详细信息。
- jsonReader 属性定义服务器返回的 JSON 数据格式。
- pager 属性启用分页支持。
- onSelectRow 属性定义选择一行时要采取的操作。在本例中，将用户引导到歌手 ID 的显示歌手视图。

有关 jqGrid 配置和用法的详细说明，请参阅项目的文档站点(http://trirand.com/jqgridwiki/doku.php?id=wiki:jqgriddocs)。

16.10.6 在服务器端启用分页

在服务器端，实现分页需要几个步骤。首先，使用 Spring Data Commons 模块的分页支持功能。为此，需要修改 SingerRepository 接口来扩展 PagingAndSortingRepository<T, ID extends Serializable>接口而不是 CrudRepository <T, ID extends Serializable>接口。修改后的接口如下所示：

```
package com.apress.prospring5.ch16.repos;

import com.apress.prospring5.ch16.entities.Singer;
import org.springframework.data.repository.PagingAndSortingRepository;

public interface SingerRepository extends
    PagingAndSortingRepository<Singer, Long> {
}
```

下一步是在 SingerService 接口中添加一个新方法来支持分页检索数据。修改后的接口如下所示：

```
package com.apress.prospring5.ch16.services;

import java.util.List;

import com.apress.prospring5.ch16.entities.Singer;
import org.springframework.data.domain.Page;
import org.springframework.data.domain.Pageable;

public interface SingerService {
  List<Singer> findAll();
  Singer findById(Long id);
  Singer save(Singer singer);
  Page<Singer> findAllByPage(Pageable pageable);
}
```

如上所示，添加名为 findAllByPage()的新方法，并将 Pageable 接口的实例作为参数。以下代码片段显示了 SingerServiceImpl 类中 findAllByPage()方法的实现。该方法返回 Page <T>接口的一个实例(属于 Spring Data Commons 并位于 org.springframework.data.domain 包中)。与预期的一样，该服务方法只是调用由 PagingAndSortingRepository<T, ID extends Serializable>接口提供的存储库方法 findAll()。

```
package com.apress.prospring5.ch16.services;

import java.util.List;

import com.apress.prospring5.ch16.repos.SingerRepository;
import com.apress.prospring5.ch16.entities.Singer;
import org.springframework.beans.factory.annotation.Autowired;
import org.springframework.data.domain.Page; import org.springframework.data.domain.
Pageable;
import org.springframework.stereotype.Repository;
import org.springframework.stereotype.Service;
import org.springframework.transaction.annotation.Transactional;

import com.google.common.collect.Lists;

@Repository
@Transactional
@Service("singerService")
public class SingerServiceImpl implements SingerService {
  private SingerRepository singerRepository;
  ...

  @Autowired
```

```
    public void setSingerRepository(SingerRepository singerRepository) {
  this.singerRepository = singerRepository;
    }

    @Override
    @Transactional(readOnly=true)
    public Page<Singer> findAllByPage(Pageable pageable) {
  return singerRepository.findAll(pageable);
    }
}
```

下一步最复杂，就是实现 SingerController 类中的方法，以便从 jqGrid 获取页面数据的 Ajax 请求。以下代码片段显示了实现过程：

```
package com.apress.prospring5.ch16.web;
....

@RequestMapping("/singers")
@Controller
public class SingerController {
  private final Logger logger = LoggerFactory.getLogger(SingerController.class);

  private SingerService singerService;

  @ResponseBody
  @RequestMapping(value = "/listgrid", method = RequestMethod.GET,
    produces="application/json")
  public SingerGrid listGrid(@RequestParam(value = "page",
  required = false) Integer page,
@RequestParam(value = "rows", required = false) Integer rows,
@RequestParam(value = "sidx", required = false) String sortBy,
@RequestParam(value = "sord", required = false) String order) {

logger.info("Listing singers for grid with page: {}, rows: {}",
page, rows);
logger.info("Listing singers for grid with sort: {}, order: {}",
sortBy, order);

//Process order by
Sort sort = null;
String orderBy = sortBy;
if (orderBy != null && orderBy.equals("birthDateString"))
orderBy = "birthDate";

if (orderBy != null && order != null) {
if (order.equals("desc")) {
  sort = new Sort(Sort.Direction.DESC, orderBy);
} else
  sort = new Sort(Sort.Direction.ASC, orderBy);
}

//Constructs page request for current page
//Note: page number for Spring Data JPA starts with 0,
//while jqGrid starts with 1
PageRequest pageRequest = null;
if (sort != null) {
pageRequest = PageRequest.of(page - 1, rows, sort);
} else {
pageRequest = PageRequest.of(page - 1, rows);
}

Page<Singer> singerPage = singerService.findAllByPage(pageRequest);

//Construct the grid data that will return as JSON data
SingerGrid singerGrid = new SingerGrid();

singerGrid.setCurrentPage(singerPage.getNumber() + 1);
singerGrid.setTotalPages(singerPage.getTotalPages());
singerGrid.setTotalRecords(singerPage.getTotalElements());

singerGrid.setSingerData(Lists.newArrayList(singerPage.iterator()));

return singerGrid;
  }

  @Autowired
  public void setSingerService(SingerService singerService) {
this.singerService = singerService;
```

```
    }
    ...
}
```

该方法处理 Ajax 请求，首先从请求中读取参数(页码、每页的记录数、排序方式和排序顺序)(代码示例中的参数名称是 jqGrid 的默认值)，并构造实现了 Pageable 接口的 PageRequest 类的实例。然后调用 SingerService.findAllByPage()方法来获取页面数据。最后，构造 SingerGrid 类的一个实例并以 JSON 格式返回给 jqGrid。以下代码片段显示了 SingerGrid 类：

```
package com.apress.prospring5.ch16.util;

import com.apress.prospring5.ch16.entities.Singer;

import java.util.List;

public class SingerGrid {
  private int totalPages;
  private int currentPage;
  private long totalRecords;
  private List<Singer> singerData;

  public int getTotalPages() {
return totalPages;
  }

  public void setTotalPages(int totalPages) {
this.totalPages = totalPages;
  }
  //other getters and setters
  ...
}
```

现在准备测试新的歌手列表视图。首先确保重新生成并部署项目，然后调用歌手列表视图。此时你应该能看到改进后的歌手列表网格视图。

可以随意使用网格、浏览页面、更改每页的记录数、通过单击列标题更改排序顺序等。同时也支持国际化，可以看到带中文标签的网格。

jqGrid 还支持数据过滤功能。例如，可以根据名字中包含 John 或出生日期在一定日期范围内进行数据过滤。

16.11 处理文件上传

歌手信息具有 BLOB 类型的字段，可用来存储从客户端上传的照片。本节将介绍如何在 Spring MVC 中实现文件上传。

很长一段时间，标准的 servlet 规范不支持文件上传。因此，Spring MVC 需要与其他库(最常见的是 Apache Commons FileUpload 库，详见 http://commons.apache.org/fileupload)一起使用才能实现此目的。Spring MVC 内置了对 Commons FileUpload 的支持。但是，从 Servlet 3.0 开始，文件上传已经成为 Web 容器的一项内置功能。Tomcat 7 支持 Servlet 3.0，Spring 还支持自 3.1 版本以来的 Servlet 3.0 文件上传功能。

在下面的章节中，将讨论如何使用 Spring MVC 和 Servlet 3.0 实现文件上传功能。

16.11.1 配置文件上传支持

在与 Servlet 3.0 兼容的 Web 容器中，配置文件上传分为两个步骤。

首先，在一个基于 Java 的配置类中定义创建 DispatcherServlet 定义所需的一切，并添加一个 StandardServletMultipartResolver 类型的 bean。该类型是基于 Servlet 3.0 javax.servlet.http.Part API 的 MultipartResolver 接口的标准实现。以下代码片段描述了需要添加到 WebConfig 类中的 bean 的声明：

```
package com.apress.prospring5.ch16.config;
...
@Configuration
@EnableWebMvc
@ComponentScan(basePackages = {"com.apress.prospring5.ch16"})
public class WebConfig implements WebMvcConfigurer {
...
  @Bean StandardServletMultipartResolver multipartResolver() {
    return new StandardServletMultipartResolver();
```

```
  }
 ...
}
```

其次是在 Servlet 3.0 环境中启用 MultiParsing，这意味着需要对 WebInitializer 实现进行一些更改。在 AbstractDispatcherServletInitializer 抽象类(该类是 AbstractAnnotationConfigDispatcherServletInitializer 的扩展类)中定义了一个名为 customizeRegistration()的方法。必须实现此方法才能注册 javax.servlet.MultipartConfigElement 的实例。WebInitializer 类的改进版本如下所示：

```
package com.apress.prospring5.ch16.init;

import com.apress.prospring5.ch16.config.DataServiceConfig;
import com.apress.prospring5.ch16.config.SecurityConfig;
import com.apress.prospring5.ch16.config.WebConfig;
import org.springframework.web.filter.CharacterEncodingFilter;
import org.springframework.web.filter.HiddenHttpMethodFilter;
import org.springframework.web.servlet.support.
AbstractAnnotationConfigDispatcherServletInitializer;
import javax.servlet.Filter;
import javax.servlet.MultipartConfigElement;
import javax.servlet.ServletRegistration;
public class WebInitializer extends
    AbstractAnnotationConfigDispatcherServletInitializer {

@Override
protected Class<?>[] getRootConfigClasses() {
  return new Class<?>[]{
SecurityConfig.class, DataServiceConfig.class
  };
}

@Override
protected Class<?>[] getServletConfigClasses() {
return new Class<?>[]{
WebConfig.class
  };
}

@Override
protected String[] getServletMappings() {
  return new String[]{"/"};
}

@Override
protected Filter[] getServletFilters() {
  CharacterEncodingFilter cef = new CharacterEncodingFilter();
  cef.setEncoding("UTF-8");
  cef.setForceEncoding(true);
  return new Filter[]{new HiddenHttpMethodFilter(), cef};
}
  //<=> <multipart-config>

  protected void customizeRegistration(ServletRegistration.Dynamic registration) {
    registration.setMultipartConfig(getMultipartConfigElement());
  }

  @Bean
  private MultipartConfigElement getMultipartConfigElement() {
    return new MultipartConfigElement( null, 5000000, 5000000, 0);
  }
}
```

MultipartConfigElement 的第一个参数是应该存储文件的临时位置。第二个参数是允许上传的最大文件大小，此时大小为 5MB。第三个参数表示请求的大小，此时大小也是 5MB。最后一个参数表示将文件写入磁盘的阈值。

16.11.2 修改视图以支持文件上传

需要修改两个视图来支持文件上传。一个是编辑视图(edit.jspx)，用于支持歌手照片的上传；另一个是用于显示照片的显示视图(show.jspx)。

以下 JSPX 代码片段描述了对 edit.jspx 视图所需要做的更改：

```
<?xml version="1.0" encoding="UTF-8" standalone="no"?>
<div xmlns:jsp="http://java.sun.com/JSP/Page"
...
```

```
<form:form modelAttribute="singer" id="singerUpdateForm" method="post"
  enctype="multipart/form-data">
...
<form:label path="description">
  ${labelSingerDescription}
</form:label>
<form:textarea cols="60" rows="8" path="description" id="singerDescription"/>
<div>
  <form:errors path="description" cssClass="error" />
</div>
<p/>
<label for="file">
  ${labelSingerPhoto}
</label>
<input name="file" type="file"/>
<p/>
...
</form:form>
</div>
```

在上面所示的<form:form>标记中，需要通过指定属性 enctype 来启用多部分文件上传支持。然后将文件上传字段添加到表单中。

还需要修改显示视图以显示歌手的照片。以下 JSPX 代码片段显示了对 show.jspx 视图所需要做的更改：

```
<?xml version="1.0" encoding="UTF-8" standalone="no"?>
<div xmlns:jsp="http://java.sun.com/JSP/Page"
...
 <spring:message code="label_singer_photo" var="labelSingerPhoto"/>
 <spring:url value="/singers/photo" var="singerPhotoUrl"/>
 ...
   <tr>
   <td>${labelSingerDescription}</td>
   <td>${singer.description}</td>
</tr>
<tr>
   <td>${labelSingerPhoto}</td>
   <td><img src="${singerPhotoUrl}/${singer.id}"></img></td>
 </tr>
 ...
</div>
```

在上面的视图模板中，通过指向用于下载照片的 URL，将新行添加到表格中以显示照片。

16.11.3 修改控制器以支持文件上传

最后一步是修改控制器。需要完成两处修改。第一处是修改 create()方法以接收上传文件作为请求参数，第二处是根据提供的歌手 ID 实现照片下载的新方法。以下代码片段显示了修改后的 SingerController 类：

```
package com.apress.prospring5.ch16.web;
...

@RequestMapping("/singers")
@Controller
public class SingerController {
 private final Logger logger = LoggerFactory.getLogger(SingerController.class);

 private SingerService singerService;
  @RequestMapping(method = RequestMethod.POST)
 public String create(@Valid Singer singer, BindingResult bindingResult,
 Model uiModel, HttpServletRequest httpServletRequest,
 RedirectAttributes redirectAttributes,
 Locale locale, @RequestParam(value="file", required=false) Part file) {
logger.info("Creating singer");
if (bindingResult.hasErrors()) {
uiModel.addAttribute("message", new Message("error",

   messageSource.getMessage("singer_save_fail",
 new Object[]{}, locale)));
uiModel.addAttribute("singer", singer);
 return "singers/create";
 }
 uiModel.asMap().clear();

 redirectAttributes.addFlashAttribute("message", new Message("success",
 messageSource.getMessage("singer_save_success",
```

```
        new Object[]{}, locale)));

      logger.info("Singer id: " + singer.getId());

      //Process upload file if (file != null) {
      logger.info("File name: " + file.getName());
      logger.info("File size: " + file.getSize());
      logger.info("File content type: " + file.getContentType()); byte[] fileContent = null;
      try {
        InputStream inputStream = file.getInputStream();
        if (inputStream == null) logger.info("File inputstream is null");
        fileContent = IOUtils.toByteArray(inputStream);
        singer.setPhoto(fileContent);
      } catch (IOException ex) {
        logger.error("Error saving uploaded file");
      }
      singer.setPhoto(fileContent);
      }

      singerService.save(singer);
      return "redirect:/singers/";
        }

        @RequestMapping(value = "/photo/{id}", method = RequestMethod.GET)
        @ResponseBody
        public byte[] downloadPhoto(@PathVariable("id") Long id) {
      Singer singer = singerService.findById(id);

      if (singer.getPhoto() != null) {
      logger.info("Downloading photo for id: {} with size: {}", singer.getId(),
        singer.getPhoto().length);
      }

      return singer.getPhoto();
        }
        ...
    }
```

首先，在 create()方法中，将接口类型 javax.servlet.http.Part 的新请求参数作为参数添加，其值由 Spring MVC 根据请求中的上传内容提供。然后，该方法将获取保存到 Singer 对象的照片字段中的内容。

最后，添加一个名为 downloadPhoto()的新方法来处理文件下载。该方法只是从 Singer 对象中检索照片字段，并直接写入响应流，响应流对应于显示视图中的<img>标记。

要测试文件上传功能，需要重新部署应用程序并添加带有照片的新歌手。完成后，就能够在显示视图中看到照片。

此外，还需要修改用于更改照片的编辑功能，我们将此留作练习。

16.12 用 Spring Security 保护 Web 应用程序

假设现在想要保护歌手应用程序。只有使用有效用户 ID 登录应用程序的用户才能添加新歌手或更新现有歌手。其他用户称为匿名用户，只能查看歌手信息。

Spring Security 是保护基于 Spring 的应用程序的最佳选择。虽然 Spring Security 主要用于表示层，但也可以帮助保护应用程序中的其他层，如服务层。在下面的章节中，将演示如何使用 Spring Security 来保护歌手应用程序。

在第 12 章介绍如何保护 REST 服务时曾经讲过 Spring Security。对于 Web 应用程序来说，可能性是多种多样的，因为还可以使用标记库来保护页面的某些元素。Spring 安全标记库是 spring-security-web 模块的一部分。

16.12.1 配置 Spring 安全性

在前面的章节中曾经讲过，在本章将使用完整的注解配置，即使用基于 Java 的配置类。在 Spring 3.x 之前，通过在 Web 部署描述符文件(web.xml)中配置一个过滤器来实现 Spring Web 应用程序的安全性。该过滤器被命名为 springSecurityFilterChain，可应用于任何请求，除了针对静态组件的请求之外。在 Spring 4.0 中，已经引入了一个 AbstractSecurityWebApplicationInitializer 类，可以通过扩展该类来配置 Spring 安全性。

```
AbstractSecurityWebApplicationInitializer.
```

```
package com.apress.prospring5.ch16.init;

import org.springframework.security.web.context.
 AbstractSecurityWebApplicationInitializer;
public class SecurityWebApplicationInitializer
 extends AbstractSecurityWebApplicationInitializer {
}
```

通过提供一个扩展了 AbstractSecurityWebApplicationInitializer 的空类，就可以告诉 Spring 希望启用 DelegatingFilterProxy，以便在任何其他已注册的 javax.servlet.Filter 之前使用 springSecurityFilterChain。

此外，显然必须使用基于 Java 的配置类配置 Spring Security 上下文，并且必须将该类添加到根 WebApplicationContext 配置中。SecurityConfig 配置类如下所示：

```
package com.apress.prospring5.ch16.config;

import org.springframework.beans.factory.annotation.Autowired;
import org.springframework.context.annotation.Configuration;
import org.springframework.security.config.annotation.
   authentication.builders.AuthenticationManagerBuilder;
import org.springframework.security.config.annotation.
   method.configuration.EnableGlobalMethodSecurity;
import org.springframework.security.config.annotation.web.
   builders.HttpSecurity;
import org.springframework.security.config.annotation.web.
   configuration.EnableWebSecurity;
import org.springframework.security.config.annotation.web.configuration.
   WebSecurityConfigurerAdapter;
import org.springframework.security.web.csrf.CsrfTokenRepository;
import org.springframework.security.web.csrf.HttpSessionCsrfTokenRepository;

@Configuration
@EnableWebSecurity
@EnableGlobalMethodSecurity(prePostEnabled = true)
public class SecurityConfig extends WebSecurityConfigurerAdapter {

  @Autowired
  public void configureGlobal(AuthenticationManagerBuilder auth) {
    try {
      auth.inMemoryAuthentication().withUser("user")
      .password("user").roles("USER");
    } catch (Exception e) {
      e.printStackTrace();
    }
  }

  @Override
  protected void configure(HttpSecurity http) throws Exception {
    http
  .authorizeRequests()
  .antMatchers("/*").permitAll()
  .and()
  .formLogin()
  .usernameParameter("username")
  .passwordParameter("password")
  .loginProcessingUrl("/login")
  .loginPage("/singers")
  .failureUrl("/security/loginfail")
  .defaultSuccessUrl("/singers")
  .permitAll()
  .and()
  .logout()
  .logoutUrl("/logout")
  .logoutSuccessUrl("/singers")
  .and()
  .csrf().disable();
  }
}
```

方法 configure(HttpSecurity http)定义了 HTTP 请求的安全配置。.antMatchers("/*").permitAll()链式调用指定所有用户都允许进入应用程序。你将会看到如何通过使用 Spring Security 的标记库和控制器方法安全性来隐藏视图中的编辑选项，从而保护相关的功能。然后，.formLogin()定义对表单登录的支持，而.and()调用之前的所有调用都是配置登录表单详细信息的方法。正如在布局中所讨论的那样，登录表单将显示在左侧。此外，还通过.logout()调用提供了注销链接。

在 Spring Security 4 中，可以在 Spring 中使用 CSFR 令牌来防止跨站请求伪造[1]。在本例中，为了简单起见，通过调用.csrf().disable()来禁止使用 CSFR 令牌。默认情况下，没有 CSFR 元素配置的配置是无效的，任何登录请求都会进入 403 错误页面，页面内容如下所示：

```
Invalid CSRF Token 'null' was found on the request parameter
'_csrf' or header 'X-CSRF-TOKEN'.
```

configureGlobal(AuthenticationManagerBuilder auth)方法定义了验证机制。在配置中，使用分配的 USER 角色对单个用户进行硬编码。在生产环境中，应该根据数据库、LDAP 或 SSO 机制对用户进行身份验证。

此外，在 Spring 3 之前，默认的登录 URL 为/j_spring_security_check，并且身份验证密钥的默认名称为 j_username 和 j_password。从 Spring 4 开始，默认的登录 URL 为/login，而身份验证密钥的默认名称是 username 和 password。

在前面的配置中，显式设置了用户名、密码和登录 URL，但如果在视图中使用默认名称，则可以跳过配置的以下部分：

```
.usernameParameter("username")
.passwordParameter("password")
.loginProcessingUrl("/login")|
```

尽管 WebInitializer 类的内容在前面已经描述过，但这里再次介绍一下配置，从而强调 SecurityConfig 类的使用位置。

```
package com.apress.prospring5.ch16.init;
...
public class WebInitializer extends
    AbstractAnnotationConfigDispatcherServletInitializer {

  @Override
  protected Class<?>[] getRootConfigClasses() {
    return new Class<?>[]{

    SecurityConfig.class, DataServiceConfig.class
      };
  @Override
  protected Class<?>[] getServletConfigClasses() {
    return new Class<?>[]{
  WebConfig.class
    };
  }
...
}
```

16.12.2 将登录功能添加到应用程序中

要想将登录表单添加到应用程序中，必须更改两个视图。如果用户已登录，以下所示的视图 header.jspx 用来显示用户信息：

```
<div id="header" xmlns:jsp="http://java.sun.com/JSP/Page"
    xmlns:spring="http://www.springframework.org/tags"
    xmlns:sec="http://www.springframework.org/security/tags"
    version="2.0">
  <jsp:directive.page contentType="text/html;charset=UTF-8" />
  <jsp:output omit-xml-declaration="yes" />
  <spring:message code="header_text" var="headerText"/>

  <spring:message code="label_logout" var="labelLogout"/>
  <spring:message code="label_welcome" var="labelWelcome"/>
  <spring:url var="logoutUrl" value="/logout" />

  <div id="appname">
<h1>${headerText}</h1>
  </div>

  <div id="userinfo">
  <sec:authorize access="isAuthenticated()">${labelWelcome}
```

1 这是一种攻击类型，主要是为了在 Web 应用程序中执行未经授权的命令而攻击现有会话。可以在 https://en.wikipedia.org/wiki/cross-site_request_forgery 上阅读更多相关信息。

```
    <sec:authentication property="principal.username" />
    <br/>
    <a href="${logoutUrl}">${labelLogout}</a>
    </sec:authorize>
      </div>
  </div>
```

首先，上述代码为Spring Security标记库添加了带有前缀sec的标记库。然后，添加带有<sec:authorize>标记的<div>部分以检测用户是否已登录。如果已登录(即 isAuthenticated()表达式返回 true)，则显示用户名和注销链接。

要修改的第二个视图是 menu.jspx 文件，在其中添加了登录表单；只有当用户已登录时，New Singer 选项才会显示。

```
<?xml version="1.0" encoding="UTF-8" standalone="no"?>
<div id="menu" xmlns:jsp="http://java.sun.com/JSP/Page"
    xmlns:spring="http://www.springframework.org/tags"
    xmlns:sec="http://www.springframework.org/security/tags"
    version="2.0">
  <jsp:directive.page contentType="text/html;charset=UTF-8" />
  <jsp:output omit-xml-declaration="yes" />
  <spring:message code="menu_header_text" var="menuHeaderText"/>
  <spring:message code="menu_add_singer" var="menuAddSinger"/>
  <spring:url value="/singers?form" var="addSingerUrl"/>

  <spring:message code="label_login" var="labelLogin"/>
  <spring:url var="loginUrl" value="/login" />
 <h3>${menuHeaderText}</h3>
  <sec:authorize access="hasRole('ROLE_USER')">
<a href="${addSingerUrl}"><h3>${menuAddSinger}</h3></a>
  </sec:authorize>

  <sec:authorize access="isAnonymous()">
  <div id="login">
  <form name="loginForm" action="${loginUrl}" method="post">
    <table>
  <caption align="left">Login:</caption>
  <tr>
  <td>User Name:</td>
  <td><input type="text" name="username"/></td>
  </tr>
  <tr>
  <td>Password:</td>
  <td><input type="password" name="password"/></td>
  </tr>
  <tr>
  <td colspan="2" align="center">
    <input name="submit" type="submit" value="Login"/>
    </td>
  </tr>
    </table>
  </form>
  </div>
    </sec:authorize>
 </div>
```

“Add Singer”菜单项仅在用户已登录并被授予 USER 角色(在<sec:authorized>标记中指定)时才会显示。如果用户没有登录(如第二个<sec:authorized>标记所示，当表达式 isAnonymous()返回 true 时)，则会显示登录表单。

重新部署应用程序，将显示登录表单，注意新的歌手链接未显示。

在用户名和密码字段中输入 user 并单击 Login 按钮。用户信息将显示在标题区域。同时，新的歌手链接也显示出来。

此外，还需要对显示视图(show.jspx)进行修改，以便仅为已登录用户显示编辑歌手链接，我们将此留作练习。

当登录信息不正确时，进行相关处理的 URL 是/security/loginfail。所以，需要实现一个控制器来处理这种登录失败的情况。以下代码片段显示了 SecurityController 类：

```
package com.apress.prospring5.ch16.web;

import com.apress.prospring5.ch16.util.Message;
import org.slf4j.Logger;
import org.slf4j.LoggerFactory;
import org.springframework.beans.factory.annotation.Autowired;
import org.springframework.context.MessageSource;
import org.springframework.stereotype.Controller; import org.springframework.ui.Model;
import org.springframework.web.bind.annotation.RequestMapping;
```

```
import java.util.Locale;

@Controller
@RequestMapping("/security")
public class SecurityController {
    private final Logger logger = LoggerFactory.getLogger(SecurityController.class);

    private MessageSource messageSource;

    @RequestMapping("/loginfail")
    public String loginFail(Model uiModel, Locale locale) {
  logger.info("Login failed detected");
  uiModel.addAttribute("message", new Message("error",
    messageSource.getMessage("message_login_fail", new Object{}, locale)));
  return "singers/list";
    }

    @Autowired
    public void setMessageSource(MessageSource messageSource) {
  this.messageSource = messageSource;
    }
}
```

该控制器类将处理所有带有前缀 security 的 URL，方法 loginFail()将处理登录失败的情况。在该方法中，将登录失败消息存储在模型中，然后重定向到首页。现在重新部署应用程序并输入错误的用户信息，首页上将再次显示登录失败消息。

16.12.3 使用注解来保护控制器方法

仅仅隐藏菜单中新的歌手链接是不够的。例如，如果直接在浏览器中输入 http://localhost:8080/users?form URL，即使尚未登录，也仍然可以看到添加歌手页面。原因是没有在 URL 级别对应用程序进行保护。一种保护页面的方法是配置 Spring Security 过滤器链，以便仅截取经过身份验证的用户的 URL。但是，这样做会阻止所有其他用户看到歌手列表视图。

解决此问题的另一种方法是通过使用 Spring Security 的注解支持功能在控制器方法级别应用安全性。通过使用@EnableGlobalMethodSecurity(prePostEnabled=true)注解 SecurityConfig 类来启用方法安全性，并且 prePostEnabled 属性支持在方法上使用前注解和后注解。

现在可以将@PreAuthorize 注解用于想要保护的控制器方法。以下代码片段显示了保护 createForm()方法的一个示例：

```
import org.springframework.security.access.prepost.PreAuthorize;
...

  @PreAuthorize("isAuthenticated()")
  @RequestMapping(params = "form", method = RequestMethod.GET)
  public String createForm(Model uiModel) {
Singer singer = new Singer();
uiModel.addAttribute("singer", singer);

return "singers/create";
  }
```

此时，使用@PreAuthorize 注解来保护 createForm()方法，其中的参数是安全需求的表达式。

现在，可以尝试直接在浏览器中输入新的歌手链接 URL，如果未登录，那么 Spring Security 将会重定向到登录页面，即 SecurityConfig 类中配置的歌手列表视图。

16.13 使用 Spring Boot 创建 Spring Web 应用程序

在本书的前面已经介绍过 Spring Boot，它是快速创建应用程序的实用工具。在本节中，将介绍如何使用 Thymeleaf 创建具有安全性和 Web 网页的完整 Web 应用程序。Thymeleaf 是一个 XML/XHTML/HTML5 模板引擎，可以在 Web 和非 Web 环境中使用。可以非常容易地将它与 Spring 集成。它是 Spring 最适合的模板引擎，因为是 Spring 的创造者和项目负责人 Daniel Fernandez 启动了这个项目，目的是想为 Spring MVC 提供应有的模板引擎[1]。Thymeleaf 是 JSP 或 Tiles 的实用替代方案，SpringSource 团队对其抱有很大期望，所以知道如何配置和使用它可能对你未来的职业生涯很有帮助。

第一个 Thymeleaf 版本于 2011 年 4 月发布。在编写本书时，Thymeleaf 3.0.7 已经发布，其中包括对 Spring 5 的新集成模块的更新。Thymeleaf 有很多扩展(例如，Thymeleaf Spring Security Extension[2]和 Thymeleaf Module for Java 8 Time API 兼容性[3])，由官方 Thymeleaf 团队负责编写和维护。

无论怎样，对 Thymeleaf 的介绍暂时到此为止，接下来开始创建 Spring Boot Web 应用程序。要创建完整的 Spring Web 应用程序，意味着需要 DAO 层和服务层，也就意味着需要特定的启动器库来进行持久化和事务处理。下面显示了一个 Gradle 配置片段，里面描述了创建应用程序所需的库。每个库都会在适当的时候进行详细说明。

```
//pro-spring-15/build.gradle
ext {
  //spring libs
  bootVersion = '2.0.0.M3'

  bootstrapVersion = '3.3.7-1'
  thymeSecurityVersion = '3.0.2.RELEASE'
  jQueryVersion = '3.2.1'
  ...
  spring = [
    ...
    springSecurityTest:
    "org.springframework.security:spring-security-test:$springSecurityVersion"

  ]

  boot = [
    ...
    starterThyme :
  "org.springframework.boot:spring-boot-starter-thymeleaf:$bootVersion",
    starterSecurity :
  "org.springframework.boot:spring-boot-starter-security:$bootVersion"
  ]

web = [
   bootstrap : "org.webjars:bootstrap:$bootstrapVersion",
   jQuery : "org.webjars:jquery:$jQueryVersion",
   thymeSecurity:
 "org.thymeleaf.extras:thymeleaf-extras-springsecurity4:$thymeSecurityVersion"
]
db = [
   ...
   h2 : "com.h2database:h2:$h2Version"
  ]
}

//chapter16/build.gradle
...
apply plugin: 'org.springframework.boot'

dependencies {
  compile boot.starterJpa, boot.starterJta, db.h2, boot.starterWeb,
    boot.starterThyme, boot.starterSecurity,
    web.thymeSecurity, web.bootstrap, web.jQuery
  testCompile boot.starterTest, spring.springSecurityTest
}
```

16.14 设置 DAO 层

Spring Boot JPA 库 spring-boot-starter-data-jpa 包含自动配置的 bean，如果 H2 库位于类路径中，则可用来设置和生成 H2 数据库。开发人员只需要开发一个实体类和一个存储库即可。对于本例，为了尽可能简便，将使用 Singer

1 可以在 http://forum.thymeleaf.org/why-Thymeleaftd3412902.html 上找到关于官方 Thymeleaf 论坛的完整讨论。

2 Thymeleaf Extras Spring Security 库提供了一种方言，允许将 Spring Security(版本 3.x 和 4.x)的多个授权和身份验证方面集成到基于 Thymeleaf 的应用程序中。请参阅 https://github.com/thymeleaf/thymeleaf-extras-springsecurity。

3 这是 Thymeleaf Extras 模块而不是 Thymeleaf 核心的一部分(因此遵循自己的版本模式)，但 Thymeleaf 团队完全支持它。请参阅 https://github.com/thymeleaf/thymeleaf-extras-java8time。

类的简单版本。

```
package com.apress.prospring5.ch16.entities;

import org.hibernate.validator.constraints.NotBlank;

import javax.persistence.*;
import javax.validation.constraints.NotNull;
import javax.validation.constraints.Size;
import java.io.Serializable;
import java.util.Date;

import static javax.persistence.GenerationType.IDENTITY;

@Entity
@Table(name = "singer")
public class Singer implements Serializable {

  @Id
  @GeneratedValue(strategy = IDENTITY)
  @Column(name = "ID")
  private Long id;

  @Version
  @Column(name = "VERSION")
  private int version;

  @NotBlank(message = "{validation.firstname.NotBlank.message}")
  @Size(min = 2, max = 60, message = "{validation.firstname.Size.message}")
  @Column(name = "FIRST_NAME")
  private String firstName;

  @NotBlank(message = "{validation.lastname.NotBlank.message}")
  @Size(min = 1, max = 40, message = "{validation.lastname.Size.message}")
  @Column(name = "LAST_NAME")
  private String lastName;

  @NotNull
  @Temporal(TemporalType.DATE)
  @Column(name = "BIRTH_DATE")
  private Date birthDate;

  @Column(name = "DESCRIPTION")
  private String description;

  //setters and getters
  ...
}
```

另外，为了简单，将使用 CrudRepository 实例的最简单扩展。

```
package com.apress.prospring5.ch16.repos;

import com.apress.prospring5.ch16.entities.Singer;
import org.springframework.data.repository.CrudRepository;

public interface SingerRepository extends CrudRepository<Singer, Long> {

}
```

16.14.1 设置服务层

服务层也很简单；它由 SingerServiceImpl 类和 DBInitializer 类组成，用于初始化数据库并用 Singer 记录进行填充。下面的代码片段描述了 SingerServiceImpl 类。初始化类在前面的章节中已经介绍过了，所以在此不再赘述。

```
package com.apress.prospring5.ch16.services;

import com.apress.prospring5.ch16.entities.Singer;
import com.apress.prospring5.ch16.repos.SingerRepository;
import com.google.common.collect.Lists;
import org.springframework.beans.factory.annotation.Autowired;
import org.springframework.stereotype.Service;

import java.util.List;

@Service
public class SingerServiceImpl implements SingerService {
```

```
    private SingerRepository singerRepository;

    @Override
    public List<Singer> findAll() {
      return Lists.newArrayList(singerRepository.findAll());
    }

    @Override
    public Singer findById(Long id) {
      return singerRepository.findById(id).get();
    }

    @Override
    public Singer save(Singer singer) {
      return singerRepository.save(singer);
    }

    @Autowired
    public void setSingerRepository(SingerRepository singerRepository) {
      this.singerRepository = singerRepository;
    }
  }
```

16.14.2 设置 Web 层

Web 层仅包含最简单的 SingerController 版本。@RequestMapping 注解被替换为不再需要指定 HTTP 方法的等效注解。该类如下所示：

```
package com.apress.prospring5.ch16.web;

import com.apress.prospring5.ch16.entities.Singer;
import com.apress.prospring5.ch16.services.SingerService;
import org.slf4j.Logger;
import org.slf4j.LoggerFactory;
import org.springframework.beans.factory.annotation.Autowired;
import org.springframework.stereotype.Controller;
import org.springframework.ui.Model;
import org.springframework.web.bind.annotation.*;

import javax.validation.Valid;
import java.util.List;

@Controller
@RequestMapping(value = "/singers")
public class SingerController {

  private final Logger logger = LoggerFactory.getLogger(SingerController.class);
  @Autowired SingerService singerService;

  @GetMapping
  public String list(Model uiModel) {
    logger.info("Listing singers");
    List<Singer> singers = singerService.findAll();
    uiModel.addAttribute("singers", singers);
    logger.info("No. of singers: " + singers.size());
    return "singers";
  }

  @GetMapping(value = "/{id}")
  public String show(@PathVariable("id") Long id, Model uiModel) {
    Singer singer = singerService.findById(id);
    uiModel.addAttribute("singer", singer);
    return "show";
  }

  @GetMapping(value = "/edit/{id}")
  public String updateForm(@PathVariable Long id, Model model) {
    model.addAttribute("singer", singerService.findById(id));
    return "update";
  }

  @GetMapping(value = "/new")
  public String createForm(Model uiModel) {
    Singer singer = new Singer();
    uiModel.addAttribute("singer", singer);
```

```
    return "update";
  }

  @PostMapping
  public String saveSinger(@Valid Singer singer) {
    singerService.save(singer);
    return "redirect:/singers/" + singer.getId();
  }
}
```

updateForm 和 createForm 方法返回相同的视图。唯一的区别是，updateForm 方法接收一个现有 Singer 实例的 ID 作为参数，该 Singer 实例被用作 update 视图的模型对象。update 视图包含一个按钮，单击该按钮将调用 createSinger 方法。如果 Singer 实例有一个 ID，则在数据库中更新该对象；否则，创建一个新的 Singer 实例并将其保存到数据库中。本例中暂不包括验证以及验证失败后进行相关处理的内容，我们将这作为练习留给读者来完成。

Spring Boot Web 启动器库 spring-boot-starter-web 可以发现和创建控制器 bean。

16.14.3 设置 Spring 安全性

Spring Boot提供了一个名为spring-boot-starter-security的Spring Security启动器库。如果此库位于类路径中，Spring Boot 会自动使用基本身份验证来保护所有 HTTP 端点。但是默认的安全设置可以进一步自定义。在本节中，假设只能使用根(/)上下文(http:\localhost:8080/)访问应用程序的首页。自定义配置如下所示：

```
package com.apress.prospring5.ch16;

import org.springframework.beans.factory.annotation.Autowired;
import org.springframework.context.annotation.Configuration;
import org.springframework.security.config.annotation.authentication.builders.
  AuthenticationManagerBuilder;
import org.springframework.security.config.annotation.web.builders.HttpSecurity;
import org.springframework.security.config.annotation.web.configuration.
  EnableWebSecurity;
import org.springframework.security.config.annotation.web.configuration.
  WebSecurityConfigurerAdapter;

@Configuration
@EnableWebSecurity
public class WebSecurityConfig extends WebSecurityConfigurerAdapter {
   @Override
   protected void configure(HttpSecurity http) throws Exception {
  http
      .authorizeRequests()
      .antMatchers("/", "/home").permitAll()
      .and()
      .authorizeRequests().antMatchers("/singers/**").authenticated()
      .and()
      .formLogin()
      .loginPage("/login")
      .permitAll()
      .and()
      .logout()
      .permitAll();
   }

   @Autowired
   public void configureGlobal(AuthenticationManagerBuilder auth) throws Exception {
  auth
    .inMemoryAuthentication()
    .withUser("user").password("user").roles("USER");
  }
}
```

16.15 创建 Thymeleaf 视图

在跳转到使用 Thymeleaf 模板引擎创建视图之前，先介绍一下 Spring Boot Web 应用程序的基本结构。Spring Boot Web 应用程序使用 resources 目录作为 Web 资源目录，因此不需要 webapp 目录。如果 resources 目录的内容按照 SpringBoot 的默认结构要求进行组织，则不需要编写大量配置，因为 Spring Boot 启动器库提供了预配置 bean。要使用带有默认配置的 Thymeleaf 模板引擎，Spring Boot Thymeleaf 启动器库 spring-bootstarter-thymeleaf 必须位于项目的类路径中。

图 16-5 显示了 spring-boot-starter-thymeleaf 库的可传递依赖项，以及你可能感兴趣的其他初始库。

正如你所看到的，Spring Boot Thymeleaf 启动器版本 2.0.0.M3 随 Thymeleaf 3.0.7 一起发布，它是编写本书时最新发布的版本。

请注意，spring-boot-starter-web 以嵌入式 Tomcat 服务器为依赖项，它将用于运行应用程序。

现在已经拥有类路径中的所有库，接下来让我们分析一下 resources 目录的结构，如图 16-6 所示。默认情况下，Spring Boot 将 Thymeleaf 引擎配置为从模板目录中读取模板文件。如果单独使用 Spring MVC 而没有使用 Spring Boot，则需要通过定义三个 bean 来显式配置 Thymeleaf 引擎：SpringResourceTemplateResolver 类型的 bean、SpringTemplateEngine 类型的 bean 以及 ThymeleafViewResolver 类型的 bean[1]。但如果使用了 Spring Boot，那么开发人员所要做的就是开始创建模板并将它们放到/resources/templates 目录中。

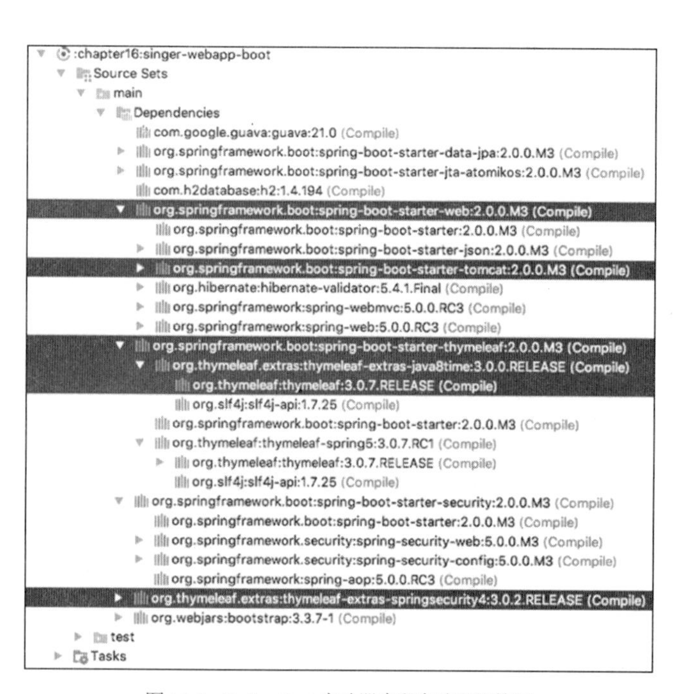

图 16-5　Spring Boot 启动器库和启动器依赖项

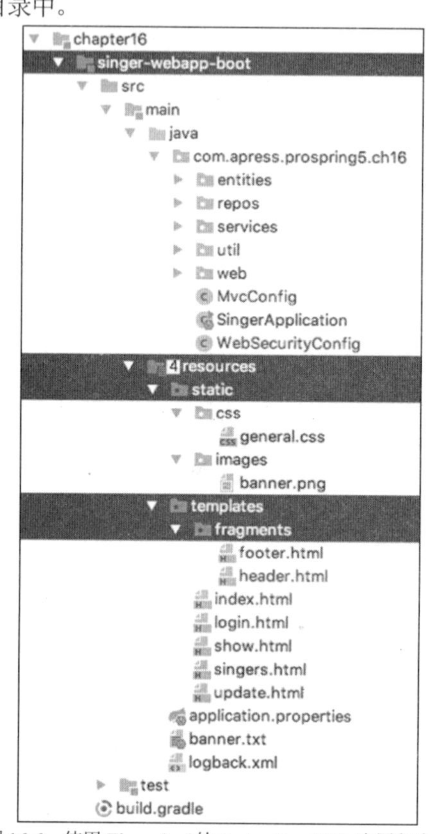

图 16-6　使用 Thymeleaf 的 Spring Boot Web 应用程序的 resources 目录的默认结构

创建 Thymeleaf 模板的操作非常简单，语法只是简单的 HTML。接下来从一个简单的例子开始。

```
<!DOCTYPE html>
<html xmlns:th="http://www.thymeleaf.org">
<head lang="en">

  <title>Spring Boot Thymeleaf Sample</title>
  <meta http-equiv="Content-Type" content="text/html; charset=UTF-8"/>
</head>
<body>
<h1>Hello World</h1>

<h2>Just another simple text here</h2>
</body>
</html>
```

1 Thymeleaf 官方网站提供了关于如何使用 Thymeleaf 和 Spring 的很好的教程。请参阅 http://thymeleaf.org/doc/tutorials/3.0/thymeleafspring.html#spring-mvc-configuration。

代码行<html xmlns:th ="http://www.thymeleaf.org">是 Thymeleaf 名称空间声明。该声明很重要，如果没有它，模板将只是一个简单的静态 HTML 文件。作为模板引擎，Thymeleaf 允许定义和自定义如何处理模板的细节。在引擎内部，模板由一个或多个处理器处理，它识别以 Thymeleaf 标准方言编写的某些元素。大多数标准方言由属性处理器组成；这些处理器可以让浏览器正确显示模板，即使没有处理文件，也是如此，因为未知属性将被忽略。接下来看一个表单输入文本字段的示例：

```
<input type="text" name="singerName" value="John Mayer" />
```

上面声明了一个静态的 HTML 组件，浏览器知道如何解释它。如果想要使用 Spring 表单元素编写相同的元素，可以编写如下代码：

```
<form:inputText name="singerName" value="${singer.name}" />
```

此时浏览器将无法显示上面的声明，因此，如果使用 Apache Tiles 编写模板，那么每次想要在浏览器中查看表单时，都必须编译项目。而 Thymeleaf 通过使用普通的 HTML 元素解决了这个问题，这些元素可以使用 Thymeleaf 属性进行配置。下面的代码行显示了如何使用 Thymeleaf 编写前面所示的元素：

```
<input type="text" name="singerName" value="John Mayer"th:value="${singer.name}" />
```

通过使用上面的声明，浏览器可以显示该元素，并且还可以为该元素实际设置一个值，从而可以看到组件是如何在页面中组合在一起的。该值将被模板处理过程中$ {singer.name}的评估结果取代。

正如前面所述，当为 Web 应用程序编写接口时，可能需要隔离公共部分并将它们包含在其他模板中，以避免代码重复。例如，页眉、页脚和菜单是应用程序所有页面的公共部分。这些公共部分可以使用称为*片段(fragment)*的特殊模板来表示。默认情况下，这些部分模板应该在/templates/fragments 目录下定义，但如果情况需要，也可以使用不同的目录名称[1]。下面显示了定义片段的语法，其中介绍了 footer.html 示例：

```
<!DOCTYPE html>
<html xmlns:th="http://www.thymeleaf.org">
<head>
   <meta http-equiv="Content-Type" content="text/html; charset=UTF-8"/>
   <title>
  ProSpring5 Singer Boot Application
   </title>
  </head>
<body>
<div th:fragment="footer" th:align="center">
  <p>Copyright (c) 2017 by Iuliana Cosmina and Apress. All rights reserved.</p>
</div>
</body>
</html>
```

请注意 th:fragment 属性，它声明了片段的名称，稍后将通过插入或替换某些元素，从而在其他模板中使用该片段。下面看一下使用片段的语法，其中显示了一个简单的 index.html 模板：

```
<!DOCTYPE html>
<html xmlns:th="http://www.thymeleaf.org">
head>
   <meta http-equiv="Content-Type" content="text/html; charset=UTF-8"/>
   <title>
  ProSpring5 Singer Boot Application
   </title>
</head>

<body>
<div th:replace="~{fragments/header :: header}">Header</div>
<div class="container">
  ...
</div>
</body>
</html>
```

有三种使用片段的方法：

- th:replace 用指定的片段替换主机标记。
- th:insert 插入指定的片段作为主机标记的主体。
- th:include 插入片段的内容[2]。

1 一个 Web 应用程序可以有多个主题，并且每个主题都可以由一组片段表示。
2 从版本 3.0 开始，不建议使用该方法。

如果想了解更多关于 Thymeleaf 模板的信息，可以在官方网站 http://thymeleaf.org/documentation.html 上找到最好的资源。

需要介绍的最后一个目录是 static。static 目录包含应用程序模板的静态资源，比如 CSS 文件和图像。想要在 Thymeleaf 模板中包含自定义 CSS 类，可以使用以下语法：

```
<!DOCTYPE html>
<html xmlns:th="http://www.thymeleaf.org">
<head>
  <meta http-equiv="Content-Type" content="text/html; charset=UTF-8"/>

  <link href="../../public/css/general.css"
th:href="@{css/general.css}" rel="stylesheet" media="screen"/>

  <title>
    ProSpring5 Singer Boot Application
      </title>
  </head>
  <body>
  ...
</body>
</html>
```

href 属性包含相对于模板的路径，当在浏览器中打开模板时需要使用该属性，以便加载 general.css 样式。而在处理模板时则使用 th:href 属性，并将{css/general.css}解析为应用程序的 URL。

16.16 使用 Thymeleaf 扩展

在本节介绍的应用程序中，使用两个 Thymeleaf 扩展。首先使用 thymeleaf-extrasspringsecurity4 扩展，因为正在使用 Thymeleaf 模板构建安全的应用程序。例如，当用户未登录时，应该显示 Log In 选项；而当用户登录时，则应该显示 Log Out 选项。

该扩展是一个 Thymeleaf Extras 模块，虽然它不是 Thymeleaf 核心的一部分，但得到了 Thymeleaf 团队的支持。虽然数字 4 出现在名称中，但需要的是对一些特殊安全实用工具对象的支持。因此，它与 Spring Security 5.0.x 兼容，Spring Security 5.0.x 是 Spring Boot 启动器安全库的可传递依赖项。接下来看一个扩展后的 header.html 文件，其中包含一个导航菜单，里面包含针对授权用户和未授权用户的不同菜单项：

```
<!DOCTYPE html>
<html xmlns:th="http://www.thymeleaf.org">
<head>
    <meta http-equiv="Content-Type" content="text/html; charset=UTF-8"/>
    <link href="../../public/css/bootstrap.min.css"
  th:href="@{/webjars/bootstrap/3.3.7-1/css/bootstrap.min.css}"
  rel="stylesheet" media="screen"/>
    <script src="http://cdn.jsdelivr.net/webjars/jquery/3.2.1/jquery.min.js"
  th:src="@{/webjars/jquery/3.2.1/jquery.min.js}"></script>
    <title>
  ProSpring5 Singer Boot Application
    </title>
  </head>
  <body>
  <div class="container">
    <div th:fragment="header">
  <nav class="navbar navbar-inverse">
  <div class="container-fluid">
    <div class="navbar-header">
  <a class="navbar-brand" href="#" th:href="@{/}">ProSpring 5</a>
    </div>
    <ul class="nav navbar-nav">
  <li><a href="#" th:href="@{/singers}">Singers</a></li>
  <li><a href="#" th:href="@{/singers/new}">Add Singer</a></li>
    </ul>
    <ul class="nav navbar-nav navbar-right" >
  <li th:if="${#authorization.expression('!isAuthenticated()')}">
  <a href="/login" th:href="@{/login}">
    <span class="glyphicon glyphicon-log-in"></span> Log in
  </a>
  </li>
  <li th:if="${#authorization.expression('isAuthenticated()')}">
  <a href="/logout" th:href="@{#}" onclick="$('#form').submit();">
    <span class="glyphicon glyphicon-log-out"></span> Logout
```

```
    </a>
    <form style="visibility: hidden" id="form" method="post" action="#"
      th:action="@{/logout}"></form>
    </li>
      </ul>
    </div>
    </nav>
      </div>
  </div>
  </body>
  </html>
```

该模块提供了一种名为 org.thymeleaf.extras.springsecurity4.dialect.SpringSecurityDialect 的新方言，其中包括示例中使用的#authorization 对象，该对象是一个表达式实用工具对象，里面包含根据表达式、URL 以及访问控制列表检查授权的相关方法[1]。该方言由 Spring Boot 配置。

应用程序中所使用的第二个 Thymeleaf 扩展是 thymeleaf-extras-java8time。该依赖项必须被显式添加到项目的类路径中，并为 Java 8 Time API 提供支持。Thymeleaf 官方团队也完全支持该扩展。它在表达式评估过程中将#temporals 对象(以及诸如#dates 之类的其他对象)添加到 ApplicationContext 作为实用工具对象处理器。这意味着可以评估对象图导航语言(OGNL)和 Spring 表达式语言(SpringEL)中的表达式。下面看一下显示歌手细节所需的模板，名为 show.html 模板。

```
  <!DOCTYPE html>
  <html xmlns:th="http://www.thymeleaf.org">

  <head>
    <meta http-equiv="Content-Type" content="text/html; charset=UTF-8"/>
    <link href="../../public/css/bootstrap.min.css"
        th:href="@{/webjars/bootstrap/3.3.7-1/css/bootstrap.min.css}"
        rel="stylesheet" media="screen"/>
    <script src="http://cdn.jsdelivr.net/webjars/jquery/3.2.1/jquery.min.js"
      th:src="@{/webjars/jquery/3.2.1/jquery.min.js}"></script>

    <title>
      ProSpring5 Singer Boot Application
    </title>
  </head>
    <body>
    <div th:replace="~{fragments/header :: header}">Header</div>
    <div class="container">

      <h1>Singer Details</h1>

      <div>
        <form class="form-horizontal">
          <div class="form-group">
            <label class="col-sm-2 control-label">First Name:</label>
            <div class="col-sm-10">
            <p class="form-control-static" th:text="${singer.firstName}">
                Singer First Name
              </p>
                </div>
        </div>
        <div class="form-group">
          <label class="col-sm-2 control-label">Last Name:</label>
          <div class="col-sm-10">
            <p class="form-control-static" th:text="${singer.lastName}">
            Singer Last Name
            </p>
        </div>
      </div>
      <div class="form-group">
        <label class="col-sm-2 control-label">Description:</label>
        <div class="col-sm-10">
          <p class="form-control-static" th:text="${singer.description}">
            Singer Description
          </p>
        </div>
      </div>
      <div class="form-group">
        <label class="col-sm-2 control-label">BirthDate:</label>
```

1 由于此扩展不在本书的讨论范围，因此如果你对它感兴趣，可以访问官方网站 https://github.com/thymeleaf/thymeleaf-extras-springsecurity 以获取更多信息。

```
    <div class="col-sm-10">
      <p class="form-control-static"
        th:text="${#dates.format(singer.birthDate, 'dd-MMM-yyyy')}">
          Singer BirthDate
        </p>
    </div>
  </div>
 </form>
</div>
<div th:insert="~{fragments/footer :: footer}">
      (c) 2017 Iuliana Cosmina & Apress</div>
</div>
</body>
</html>
```

在上面的示例中，使用#dates 实用工具对象对 java.util.Date 类型的 singer.birthDate 进行了格式化。此时，不需要日期格式化程序对象，并且接收的日期模式在模板中被硬编码。

这看起来是不是很实用？

假设定义了如以下代码片段所示的 DateFormatter 类：

```
package com.apress.prospring5.ch16.util;

import org.springframework.format.Formatter;

import java.text.ParseException;
import java.text.SimpleDateFormat;
import java.util.Date;
import java.util.Locale;

public class DateFormatter implements Formatter<Date> {
   public static final SimpleDateFormat formatter =
      new SimpleDateFormat("yyyy-MM-dd");

    @Override
    public Date parse(String s, Locale locale) throws ParseException {
       return formatter.parse(s);
    }

    @Override
    public String print(Date date, Locale locale) {
       return formatter.format(date);
    }
}
```

接下来使用最小化的 Web 配置类在应用程序类路径中配置此格式化程序。

```
package com.apress.prospring5.ch16;
import com.apress.prospring5.ch16.util.DateFormatter;
import org.springframework.context.annotation.Bean;
import org.springframework.context.annotation.Configuration;
import org.springframework.format.FormatterRegistry;
import org.springframework.web.servlet.config.annotation.ViewControllerRegistry;
import org.springframework.web.servlet.config.annotation.WebMvcConfigurer;

@Configuration
public class MvcConfig implements WebMvcConfigurer {

    @Override
    public void addViewControllers(ViewControllerRegistry registry) {
       registry.addViewController("/index").setViewName("index");
       registry.addViewController("/").setViewName("index");
       registry.addViewController("/login").setViewName("login");
    }

    @Override
    public void addFormatters(FormatterRegistry formatterRegistry) {
       formatterRegistry.addFormatter(dateFormatter());
    }

    @Bean
    public DateFormatter dateFormatter() {
       return new DateFormatter();
    }
}
```

一旦在应用程序中注册了日期格式化程序，就可以通过双括号语法在 Thymeleaf 模板中使用，代码如下所示：

```
<p class="form-control-static"
        th:text="${#dates.format(singer.birthDate, 'dd-MMM-yyyy')}">
            Singer BirthDate
</p>
```

上述代码可以重写为如下所示的代码：

```
<p class="form-control-static" th:text="${{singer.birthDate}}">
    Singer BirthDate
</p>
```

因此，日期模式的硬编码被删除，并且日期格式化程序类可以在外部进行修改，而不需要对 Thymeleaf 模板进行任何更改。这样做更实用，不是吗？

使用 webjars

在前面的 Thymeleaf 模板示例中，你可能已经注意到<head>元素中一些奇怪的链接。

```
<!DOCTYPE html>
<html xmlns:th="http://www.thymeleaf.org">
<head>
    <meta http-equiv="Content-Type" content="text/html; charset=UTF-8"/>
    <link href="../../public/css/bootstrap.min.css"
  th:href="@{/webjars/bootstrap/3.3.7-1/css/bootstrap.min.css}"
  rel="stylesheet" media="screen"/>
    <script src="http://cdn.jsdelivr.net/webjars/jquery/3.2.1/jquery.min.js"
  th:src="@{/webjars/jquery/3.2.1/jquery.min.js}"></script>
    <title>
  ProSpring5 Singer Boot Application
    </title>
</head>
<body>
...
</body>
</html>
```

自从 Bootstrap[1]出现以后，创建一个看起来很棒的 HTML 网页变得更加简单。Bootstrap 是用于在 Web 上开发响应式的、移动优先项目的、最流行的 HTML、CSS 和 JS 框架。要想在模板中使用 Bootstrap，只需要在模板中链接到 Bootstrap CSS 文件即可。很长一段时间，CSS 样式和 JavaScript 代码都是 Web 应用程序的一部分，并被开发人员手动复制到一个特殊的目录中。但最近，创建 Web 网页的最常用框架(比如 jQuery 和 Bootstrap)可以以不同方式使用，方法是将它们作为依赖项添加到应用程序中，并打包为 Java 归档文件(称为 *webjars*[2])。webjar 被部署到 Maven 中央存储库中，它们一旦被声明为项目的依赖项，就会被构建工具(Maven、Gradle 等)自动下载并添加到应用程序中。

在上面的示例中，jQuery 和 Bootstrap webjars 是使用 th:href 属性声明的。将它们链接到模板文件中可确保将所有 Bootstrap 类都应用于模板中的元素(例如 class="container-fluid")，并且可以在处理模板并将应用程序部署到服务器时访问所有 jQuery 函数(onclick="$('#form').submit();)。

完成所有这些设置后，如果运行下面所示的 SingerApplication 类并访问 http://localhost:8080/，则应该看到 index.html 处理后的模板。

```
package com.apress.prospring5.ch16;

import org.slf4j.Logger;
import org.slf4j.LoggerFactory;
import org.springframework.boot.SpringApplication;
import org.springframework.boot.autoconfigure.SpringBootApplication;
import org.springframework.context.ConfigurableApplicationContext;

@SpringBootApplication
public class SingerApplication {

    private static Logger logger =
        LoggerFactory.getLogger(SingerApplication.class);

    public static void main(String... args) throws Exception {
        ConfigurableApplicationContext ctx =
```

1 官方网站为 http://getbootstrap.com/。

2 参见 https://www.webjars.org/.

```
            SpringApplication.run(SingerApplication.class, args);
        assert (ctx != null);
        logger.info("Application started...");

        System.in.read();
        ctx.close();
    }
}
```

请注意，除了@SpringBootApplication 之外，不需要其他注解。只需要在类路径中添加 JPA 和 Web Spring Boot 启动器库，就会自动拾取实体、控制器和存储库。现在，如果在浏览器中访问 RUL http://localhost:8080/，你应该会看到应用程序的首页。如果一切顺利并且 Bootstrap webjars 被处理正确，则会看到如图 16-7 所示的页面。

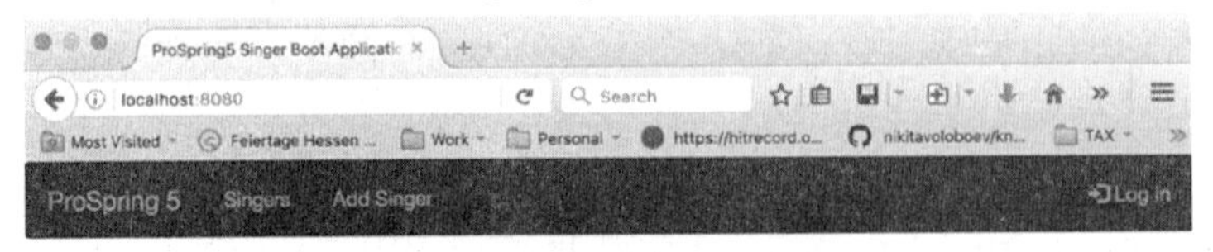

图 16-7　本节中构建的 Spring Boot 应用程序的首页

16.17　小结

在本章中，介绍了许多与使用 Spring MVC 进行 Web 开发相关的主题。首先，我们讨论了 MVC 模式的高级概念，然后介绍了 Spring MVC 的架构，包括其 WebApplicationContext 层次结构、请求处理生命周期和配置。

接下来，我们学习了如何使用 Spring MVC 开发歌手管理示例应用程序，并将 JSPX 作为视图技术。在开发示例的过程中，详细阐述了不同的领域。主要的主题包括国际化、主题以及使用 Apache Tiles 的模板支持功能。此外，我们学习了如何使用 jQuery、jQuery UI 和其他 JavaScript 库来丰富界面，示例包括日期选择器、富文本编辑器和具有分页支持的数据网格。随后还讨论了如何使用 Spring Security 来保护 Web 应用程序。

我们最后学习了与 Servlet 3.0 兼容的 Web 容器的一些功能，例如基于代码的配置，而不是使用 web.xml 文件；还演示了如何在 Servlet 3.0 环境中处理文件上传。

由于 Spring Boot 是 Spring 团队引以为傲的功能，因此必须介绍一下如何使用它构建一个完整的 Web 应用程序。当然，还应该引入最适合 Spring 的模板引擎 Thymeleaf。

在下一章中，将介绍 Spring 通过引入 WebSocket 为 Web 应用程序开发带来的更多功能。

第 17 章

WebSocket

传统上，Web 应用程序使用标准的请求/响应 HTTP 功能来提供客户端和服务器之间的通信。随着网络的发展，需要更多的交互功能，其中一些需要从服务器上推/拉或实时更新。随着时间的推移，已经实施了各种方法，例如连续轮询(continuous polling)、长轮询(long polling)和 Comet。每种方法都有各自的优缺点，WebSocket 协议试图从这些需求和不足中吸取经验，为构建交互式应用程序创建更简单、更强大的方法。HTML5 WebSocket 规范定义了一个使网页能够使用 WebSocket 协议与远程主机进行双向通信的 API。

本章涵盖 WebSocket 协议的高级概述以及 Spring 框架所能提供的主要功能。具体而言，本章包含以下主题：

- **WebSocket 介绍**：对 WebSocket 协议进行一般介绍。这些内容并不打算作为 WebSocket 协议的详细参考资料，而是作为高级概述[1]。
- **与 Spring 一起使用 WebSocket**：深入介绍一些使用 WebSocket 和 Spring 框架的细节；具体来说，介绍如何使用 Spring 的 WebSocket API、如何利用 SockJS 作为支持非 WebSocket 的浏览器的后备选项，以及如何使用 SockJS/WebSocket 通过 STOMP(简单(或流式)的面向文本的消息协议)发送消息。

17.1 WebSocket 介绍

WebSocket 是作为 HTML5 计划的一部分而开发的一种规范，可以实现在客户端和服务器之间发送消息的全双工单套接字连接。过去，需要实时更新功能的 Web 应用程序会通过打开多个连接或使用长轮询来定期轮询服务器端组件以获取更新数据。

使用 WebSocket 进行双向通信可以避免对客户端(例如 Web 浏览器)和 HTTP 服务器之间的双向通信执行 HTTP 轮询。WebSocket 协议旨在取代所有现有的使用 HTTP 作为传输的双向通信方法。WebSocket 的单套接字模型产生了一个更简单的解决方案，能避免每个客户端需要多个连接并减少开销。例如，不需要为每条消息发送一个 HTTP 头。

WebSocket 在初始握手期间使用 HTTP，这反过来又允许在标准 HTTP(80)和 HTTPS(443)端口上使用它。WebSocket 规范定义了 ws://和 wss://方案来指示不安全和安全的连接。WebSocket 协议有两部分：首先是客户端和服务器之间的握手，然后是数据传输。WebSocket 连接是在客户端和服务器之间的初始握手期间，通过在相同的底层 TCP/IP 连接上发出从 HTTP 到 WebSocket 协议的升级请求来建立的。在通信的数据传输部分，客户端和服务器都可以同时向对方发送消息，正如你所想象的那样，这为应用程序添加更强大的实时通信功能打开了大门。

17.2 与 Spring 一起使用 WebSocket

从版本 4.0 开始，Spring 框架支持 WebSocket 样式的消息传递以及 STOMP 作为应用程序级别的子协议。在框架内，可以在 spring-websocket 模块中找到对 WebSocket 的支持，该模块与 JSR-356(Java WebSocket)兼容[2]。

应用程序开发人员还必须认识到，尽管 WebSocket 带来了新的令人振奋的机会，但并非所有的 Web 浏览器都支持该协议。鉴于此，应用程序必须继续为用户工作，并利用某些后备技术尽可能模拟预期的功能。为了处理这种情

---

1 有关协议的详细信息，请参阅 http://tools.ietf.org/html/rfc6455 或 https://www.websocket.org/上的 RFC-6455。
2 请参阅 www.oracle.com/technetwork/articles/java/jsr356-1937161.html。

况，Spring 通过 SockJS 协议提供了透明的后备选项，本章的后面将介绍这些选项。

与基于 REST 的应用程序(在这些应用程序中，服务由不同的 URL 表示)不同，WebSocket 使用单个 URL 来建立初始握手，并在同一连接上执行数据流。这种类型的消息传递功能更像传统的消息传递系统。从 Spring Framework 4 开始，基于消息的核心接口(比如 Message)已经从 Spring Integration 项目迁移到名为 spring-messaging 的新模块中，以支持 WebSocket 样式的消息应用程序。

当提到使用 STOMP 作为应用程序级子协议时，其实就是在讨论通过 WebSocket 传输的协议。WebSocket 本身是一个简单的将字节转换为消息的低级协议。应用程序需要了解通过线路发送的内容，而这恰恰是诸如 STOMP 之类的子协议发挥作用的地方。在初始握手期间，客户端和服务器可以使用 Sec-WebSocket-Protocol 标头来定义要使用的子协议。虽然 Spring 框架提供对 STOMP 的支持，但 WebSocket 并没有强制任何特定的东西。

现在我们已经理解了 WebSocket 是什么以及 Spring 提供的支持，接下来的问题是应该在什么地方使用这种技术？鉴于 WebSocket 的单一套接字特性及其提供连续双向数据流的能力，WebSocket 非常适用于需要高频率消息传递和低延迟通信的应用程序。可使用 WebSocket 的应用程序包括游戏、实时群组协作工具、消息传递系统、时间敏感的定价信息(如财务更新)等。当打算使用 WebSocket 来设计应用程序时，必须考虑消息的频率和延迟要求。这将有助于确定是使用 WebSocket 还是使用诸如 HTTP 长轮询之类的其他技术。

17.3 使用 WebSocket API

正如本章前面提到的，WebSocket 只是将字节转换为消息并在客户端和服务器之间传输它们。这些消息仍然需要应用程序本身能够理解，而这恰恰是 STOMP 等子协议发挥作用的地方。如果想直接使用较低级别的 WebSocket API，则 Spring 框架提供了一个可以与之交互的 API。当使用 Spring 的 WebSocket API 时，通常需要实现 WebSocketHandler 接口或使用便利的子类(比如用于处理二进制消息的 BinaryWebSocketHandler、用于处理 SockJS 消息的 SockJsWebSocketHandler 或使用基于 String 消息的 TextWebSocketHandler)。在本例中，为了简单起见，将使用 TextWebSocketHandler 通过 WebSocket 传递 String 消息。下面先来看看如何使用 Spring WebSocket API 来接收和处理低级别的 WebSocket 消息。

如果喜欢，本章中的每个示例也可以通过 Java 配置进行配置。但在我们看来，XML 名称空间以更加简洁的方式表示了配置的方方面面，所以本章将使用这种方法。有关 Java 配置的更多信息，请阅读参考手册。接下来开始添加所需的依赖项。以下 Gradle 配置代码片段列出了这些库：

```
//pro-spring-15/build.gradle
ext {
    springVersion = '5.0.0.RC3'
    twsVersion = '9.0.0.M22'
   ...

    spring = [
      ...
       context          : "org.springframework:spring-context:$springVersion",
       webmvc           : "org.springframework:spring-webmvc:$springVersion",
       webSocket        : "org.springframework:spring-websocket:$springVersion",
       messaging        : "org.springframework:spring-messaging:$springVersion"

   ]
...
    web = [
      ...
       jacksonDatabind : "com.fasterxml.jackson.core:jackson-databind:$jacksonVersion",
       tomcatWsApi      : "org.apache.tomcat:tomcat-websocket-api:$twsVersion",
       tomcatWsEmbed    : "org.apache.tomcat.embed:tomcat-embed-websocket:$twsVersion",
       httpclient       : "org.apache.httpcomponents:httpclient:$httpclientVersion",
       websocket        : "javax.websocket:javax.websocket-api:1.1"
]
}
...
//pro-spring-15/chapter17/build.gradle
compile (web.tomcatWsApi) {
       exclude module: 'tomcat-embed-core'
    }
    compile (web.tomcatWsEmbed) {
       exclude module: 'tomcat-embed-core'
    }
```

```
    compile spring.context, spring.webSocket, spring.messaging,
        spring.webmvc, web.websocket, misc.slf4jJcl,
        misc.logback, misc.lang3, web.jacksonDatabind
```

在下面的配置代码片段中，可以看到需要配置的 WEB-INF/web.xml 文件的内容，因此可以使用带有标准 Spring MVC 调度程序 servlet 的 WebSocket：

```
<web-app xmlns="http://java.sun.com/xml/ns/javaee"
        xmlns:xsi="http://www.w3.org/2001/XMLSchema-instance"
        xsi:schemaLocation="http://java.sun.com/xml/ns/javaee
        http://java.sun.com/xml/ns/javaee/web-app_3_0.xsd"
        version="3.0">

    <display-name>Spring WebSocket API Sample</display-name>

    <servlet>
        <servlet-name>websocket</servlet-name>
        <servlet-class>
            org.springframework.web.servlet.DispatcherServlet
        </servlet-class>
        <init-param>
            <param-name>contextConfigLocation</param-name>
            <param-value>/WEB-INF/spring/root-context.xml</param-value>
        </init-param>
        <load-on-startup>1</load-on-startup>
    </servlet>

    <servlet-mapping>
        <servlet-name>websocket</servlet-name>
        <url-pattern>/*</url-pattern>
    </servlet-mapping>
</web-app>
```

首先使用 Spring 的 DispatcherServlet 创建 servlet 定义，并提供一个配置文件(/WEB-INF/spring/root-context.xml)。然后提供 servlet 映射，指示所有请求都应该通过 DispatcherServlet。

现在创建根上下文文件，其中包含 WebSocket 配置，如下所示：

```
<beans xmlns="http://www.springframework.org/schema/beans"
    xmlns:xsi="http://www.w3.org/2001/XMLSchema-instance"
    xmlns:websocket="http://www.springframework.org/schema/websocket"
    xmlns:mvc="http://www.springframework.org/schema/mvc"
    xsi:schemaLocation="
        http://www.springframework.org/schema/beans
        http://www.springframework.org/schema/beans/spring-beans.xsd
        http://www.springframework.org/schema/websocket
        http://www.springframework.org/schema/websocket/spring-websocket.xsd
        http://www.springframework.org/schema/mvc
        http://www.springframework.org/schema/mvc/spring-mvc.xsd">
    <websocket:handlers>
        <websocket:mapping path="/echoHandler" handler="echoHandler"/>
    </websocket:handlers>

    <mvc:default-servlet-handler/>

    <mvc:view-controller path= "/" view-name="/static/index.html" />

    <bean id="echoHandler"
        class="com.apress.prospring5.ch17.EchoHandler"/>
</beans>
```

首先，配置名为 index.html 的静态资源。该文件包含用于与后端 WebSocket 服务进行通信的静态 HTML 和 JavaScript。然后，通过使用 websocket 名称空间，配置处理程序以及相应的 bean 来处理请求。在本例中，定义一个处理程序映射，在/echoHandler 处接收请求，并使用 ID 为 echoHandler 的 bean 接收消息，最后通过将提供的消息回送给客户端来进行响应。

你可能对上面所示的配置非常熟悉，因为在本书中并没有过多地使用 XML 配置。所以，接下来切换到 Java 配置类。下面所示是 Spring MVC 的配置：

```
package com.apress.prospring5.ch17.config;

import org.springframework.context.annotation.ComponentScan;
import org.springframework.context.annotation.Configuration;
import org.springframework.web.servlet.config.annotation.*;

@Configuration
```

```
@EnableWebMvc
@ComponentScan(basePackages = {"com.apress.prospring5.ch17"})
public class WebConfig implements WebMvcConfigurer {

    //<=> <mvc:default-servlet-handler/>
    @Override
    public void configureDefaultServletHandling(

        DefaultServletHandlerConfigurer configurer) {
          configurer.enable();
    }

    //<=> <mvc:view-controller .../>
    @Override
    public void addViewControllers(ViewControllerRegistry registry) {
        registry.addViewController("/").setViewName("/static/index.html");
    }
}
```

继续通过使用替换 web.xml 的类来配置 DispatcherServlet。

```
package com.apress.prospring5.ch17.config;

import org.springframework.web.servlet.support.
    AbstractAnnotationConfigDispatcherServletInitializer;

public class WebInitializer extends
    AbstractAnnotationConfigDispatcherServletInitializer {

    @Override
    protected Class<?>[] getRootConfigClasses() {
        return new Class<?>[]{
            WebSocketConfig.class
        };
    }

    @Override
    protected Class<?>[] getServletConfigClasses() {
        return new Class<?>[]{
            WebConfig.class
        };
    }

    @Override
    protected String[] getServletMappings() {
        return new String[]{"/"};
    }
}
```

WebConfig 类包含 Spring MVC 应用程序的基础结构，因为我们希望在使用 Java 配置时尊重关注点分离的原则，所以需要一个不同的配置类来支持 WebSocket 通信。该类必须实现定义了回调方法的 WebSocketConfigurer 接口来配置 WebSocket 请求处理。

```
package com.apress.prospring5.ch17.config;

import com.apress.prospring5.ch17.EchoHandler;
import org.springframework.context.annotation.Bean;
import org.springframework.context.annotation.Configuration;
import org.springframework.web.socket.config.annotation.EnableWebSocket;
import org.springframework.web.socket.config.annotation.WebSocketConfigurer;
import org.springframework.web.socket.config.annotation.WebSocketHandlerRegistry;

@Configuration
@EnableWebSocket
public class WebSocketConfig implements WebSocketConfigurer {

    @Override
    public void registerWebSocketHandlers(WebSocketHandlerRegistry registry) {
        registry.addHandler(echoHandler(), "/echoHandler");
    }

    @Bean
    public EchoHandler echoHandler() {
        return new EchoHandler();
    }
}
```

需要将@EnableWebSocket 注解添加到@Configuration 类，以便配置处理 WebSocket 请求。

接下来准备实现 TextWebSocketHandler 的一个子类(src/main/java/com/apress/prospring5/ch17/EchoHandler.java)，从而以一种方便的方式处理基于 String 的消息，如下所示：

```
package com.apress.prospring5.ch17;

import org.springframework.web.socket.TextMessage;
import org.springframework.web.socket.WebSocketSession;
import org.springframework.web.socket.handler.TextWebSocketHandler;

import java.io.IOException;

public class EchoHandler extends TextWebSocketHandler {
    @Override
    public void handleTextMessage(WebSocketSession session,
        TextMessage textMessage) throws IOException {
        session.sendMessage(new TextMessage(textMessage.getPayload()));
    }
}
```

正如你所看到的，这是一个基本的处理程序，它接收提供的消息并简单地将其回送给客户端。所接收的 WebSocket 消息的内容包含在 getPayload()方法中。

以上就是在后端需要完成的所有工作。鉴于 EchoHandler 是一个典型的 Spring bean，因此可以在普通的 Spring 应用程序中做任何事情(比如注入服务)，从而执行处理程序可能需要完成的任何功能。

接下来创建一个简单的可以与后端 WebSocket 服务进行交互的前端客户端。前端是一个带有一点 JavaScript 代码的简单 HTML 页面，它使用浏览器的 API 来创建 WebSocket 连接；此外还包含一些 jQuery 来处理按钮单击事件和数据显示。前端应用程序能够建立连接、断开连接、发送消息并向屏幕显示状态更新。以下代码片段显示了前端客户端页面(src/main/webapp/static/index.html)的代码：

```
<html>
<head>
    <meta charset="UTF-8">
    <title>WebSocket Tester</title>
    <script language="javascript" type="text/javascript"
        src="http://code.jquery.com/jquery-2.1.1.min.js"></script>
    <script language="javascript" type="text/javascript">
        var ping;
        var websocket;

        jQuery(function ($) {
            function writePing(message) {
                $('#pingOutput').append(message + '\n');
            }

            function writeStatus(message) {
                $("#statusOutput").val($("#statusOutput").val() + message + '\n');
            }

            function writeMessage(message) {
                $('#messageOutput').append(message + '\n')
            }

            $('#connect')
                .click(function doConnect() {
                    websocket = new WebSocket($("#target").val());

                    websocket.onopen = function (evt) {
                        writeStatus("CONNECTED");

                    var ping = setInterval(function () {
                        if (websocket != "undefined") {
                            websocket.send("ping");
                        }
                    }, 3000);
                };

                websocket.onclose = function (evt) {
                    writeStatus("DISCONNECTED");
                };

                websocket.onmessage = function (evt) {
                    if (evt.data === "ping") {
                        writePing(evt.data);
                    } else {
```

```
                    writeMessage('ECHO: ' + evt.data);
                }
            };
                websocket.onerror = function (evt) {
                    onError(writeStatus('ERROR:' + evt.data))
                };
            });

        $('#disconnect')
            .click(function () {
                if (typeof websocket != 'undefined') {
                    websocket.close();
                    websocket = undefined;
                } else {
                    alert("Not connected.");
                }
            });

        $('#send')
            .click(function () {
                if (typeof websocket != 'undefined') {
                    websocket.send($('#message').val());
                } else {
                    alert("Not connected.");
                }
            });
    });
    </script>
</head>

<body>
    <h2>WebSocket Tester</h2> Target:
    <input id="target" size="40"
       value="ws://localhost:8080/websocket-api/echoHandler"/>
    <br/>
    <button id="connect">Connect</button>
    <button id="disconnect">Disconnect</button>
    <br/>
    <br/>Message:
    <input id="message" value=""/>
    <button id="send">Send</button>
    <br/>
    <p>Status output:</p>
    <pre><textarea id="statusOutput" rows="10" cols="50"></textarea></pre>
    <p>Message output:</p>
    <pre><textarea id="messageOutput" rows="10" cols="50"></textarea></pre>
    <p>Ping output:</p>
    <pre><textarea id="pingOutput" rows="10" cols="50"></textarea></pre>
</body>
</html>
```

以上代码片段提供了一个 UI，它允许回调 WebSocket API 并查看屏幕上显示的实时结果。

生成项目并将其部署到 Web 容器中。然后导航到 http://localhost:8080/websocket-api/index.html 以显示 UI。单击 Connect 按钮后，将会在 Status Output 文本区域中看到 CONNECTED 消息，并且每隔三秒会在 Ping Output 文本区域中显示 ping 消息。继续在 Message 文本框中输入消息，然后单击 Send 按钮。此消息将被发送到后端 WebSocket 服务并显示在“Message Output”文本区域中。发送完消息后，可以随时单击 Disconnect 按钮，此时将在 Status Output 文本区域中看到 DISCONNECTED 消息。在重新连接到 WebSocket 服务之前，将无法再发送任何消息或断开连接。虽然该例在底层 WebSocket API 的基础上使用了 Spring 抽象，但仍然可以清楚地看到该技术可能给应用程序带来令人兴奋的各种可能性。接下来，看看当浏览器不支持 WebSocket 并且需要一个后备选项时应该如何处理。可以使用 http://websocket.org/echo.html 等网站测试浏览器的兼容性。

使用 SockJS

由于并不是所有的浏览器都支持 WebSocket，同时应用程序始终需要为最终用户正常运行，因此 Spring 框架提供了一个使用 SockJS 的后备选项。通过使用 SockJS，可以在无须更改应用程序代码的情况下在运行时提供尽可能接近 WebSocket 的行为。SockJS 协议通过 JavaScript 库在客户端使用。Spring 框架的 spring-websocket 模块包含相关的 SockJS 服务器端组件。当使用 SockJS 提供无缝后备选项时，客户端将通过/info 路径向服务器发送 GET 请求，以从服务器获取传输信息。SockJS 将首先尝试使用 WebSocket，然后是 HTTP 流，最后将使用 HTTP 长轮询作为最终手段。要详细了解 SockJS 及其各种项目，请参阅 https://github.com/sockjs。

通过 websocket 名称空间支持来启用 SockJS 是非常简单的，只需要在<websocket:handlers>块中添加一条指令即可。接下来用原始的 WebSocket API 构建一个类似的应用程序，但此处使用 SockJS。src/main/webapp/WEB-INF/spring/root-context.xml 文件如下所示：

```
<beans ...>

    <websocket:handlers>
        <websocket:mapping path="/echoHandler" handler="echoHandler"/>
        <websocket:sockjs/>
    </websocket:handlers>

    <mvc:default-servlet-handler/>

    <mvc:view-controller path= "/" view-name="/static/index.html" />

    <bean id="echoHandler" class="com.apress.prospring5.ch17.EchoHandler"/>
</beans>
```

注意在文件中添加了<websocket:sockjs>标记。基本上，添加了该标记就可以启用 SockJS。可以重用 WebSocket API 示例中的 EchoHandler 类，因为所提供的功能是相同的。

<websocket:sockjs />名称空间标记还提供了其他属性来设置配置选项，例如处理会话 cookie(默认启用)、自定义客户端库加载位置(在编写本书时，默认为 https://d1fxtkz8shb9d2.cloudfront.net/sockjs-0.3.4.min.js)、heartbeat 配置、消息大小限制等。这些选项应根据应用程序需求和传输类型进行适当的审查和配置。在 web.xml 文件中不需要添加过多的代码来反映 SockJS servlet，如下所示：

```
<web-app xmlns="http://java.sun.com/xml/ns/javaee"
        xmlns:xsi="http://www.w3.org/2001/XMLSchema-instance"
        xsi:schemaLocation="http://java.sun.com/xml/ns/javaee
        http://java.sun.com/xml/ns/javaee/web-app_3_0.xsd"
        version="3.0">

    <display-name>Spring SockJS API Sample</display-name>

    <servlet>
        <servlet-name>sockjs</servlet-name>
        <servlet-class>
            org.springframework.web.servlet.DispatcherServlet
        </servlet-class>
         <init-param>
            <param-name>contextConfigLocation</param-name>
            <param-value>/WEB-INF/spring/root-context.xml</param-value>
        </init-param>
        <load-on-startup>1</load-on-startup>
        <async-supported>true</async-supported>
    </servlet>

    <servlet-mapping>
        <servlet-name>sockjs</servlet-name>
        <url-pattern>/*</url-pattern>
    </servlet-mapping>
</web-app>
```

正如你所猜测的那样，还可以使用 Java 配置。使用 SockJS 支持 WebSocket 通信需要修改两个地方。首先，需要支持异步消息传递，在前面的配置中是通过使用<async-supported> true </ asyncsupported>来启用的。此任务是通过使用另一个注解 EnableAsync 注解 Java 配置类(已用@Configuration 注解的类)来完成的。如果查看一下官方的 Spring Javadoc，就会发现该注解启用了 Spring 的异步方法执行功能，从而为整个 Spring 应用程序上下文启用注解驱动的异步处理功能[1]。

```
package com.apress.prospring5.ch17.config;

import org.springframework.context.annotation.ComponentScan;
import org.springframework.context.annotation.Configuration;
import org.springframework.scheduling.annotation.EnableAsync;
import org.springframework.web.servlet.config.annotation.DefaultServletHandlerConfigurer;
import org.springframework.web.servlet.config.annotation.EnableWebMvc;
import org.springframework.web.servlet.config.annotation.ViewControllerRegistry;
import org.springframework.web.servlet.config.annotation.WebMvcConfigurer;
```

1 Spring Javadoc 的官方网站为 https://docs.spring.io/spring/docs/current/javadocapi/org/springframework/scheduling/annotation/EnableAsync.html。

```
@Configuration
@EnableWebMvc
@EnableAsync
@ComponentScan(basePackages = {"com.apress.prospring5.ch17"})
public class WebConfig implements WebMvcConfigurer {

    @Override
    public void configureDefaultServletHandling(
       DefaultServletHandlerConfigurer configurer) {
          configurer.enable();
    }

    @Override
    public void addViewControllers(ViewControllerRegistry registry) {
       registry.addViewController("/").setViewName("/static/index.html");
    }
}
```

其次，必须更改 WebSocketConfig，以便为处理程序启用 SockJS 支持。

```
package com.apress.prospring5.ch17.config;

import com.apress.prospring5.ch17.EchoHandler;
import org.springframework.context.annotation.Bean;
import org.springframework.context.annotation.Configuration;
import org.springframework.messaging.simp.config.MessageBrokerRegistry;
import org.springframework.web.socket.config.annotation.*;

@Configuration
@EnableWebSocket
public class WebSocketConfig implements WebSocketConfigurer {

    @Override
    public void registerWebSocketHandlers(WebSocketHandlerRegistry registry) {
       registry.addHandler(echoHandler(),
         "/echoHandler").withSockJS();
    }

    @Bean
    public EchoHandler echoHandler() {
       return new EchoHandler();
    }
}
```

接下来，需要像在 WebSocket API 示例中那样创建一个 HTML 页面，但此时使用 SockJS 负责传输协商。最显著的差异是直接使用 SockJS 库而不是使用 WebSocket，并使用典型的 http://方案而不是 ws://连接到端点。简单的 HTML 客户端代码如下所示：

```
<html>
<head>
    <meta charset="UTF-8">
    <title>SockJS Tester</title>
    <script language="javascript" type="text/javascript"
    src="https://d1fxtkz8shb9d2.cloudfront.net/sockjs-0.3.4.min.js">
    </script>
    <script language="javascript" type="text/javascript"
    src="http://code.jquery.com/jquery-2.1.1.min.js">
    </script>
    <script language="javascript" type="text/javascript">
       var ping;
       var sockjs;

       jQuery(function ($) {
          function writePing(message) {
             $('#pingOutput').append(message + '\n');
          }

          function writeStatus(message) {
             $("#statusOutput").val($("#statusOutput").val() + message + '\n');
          }

          function writeMessage(message) {
             $('#messageOutput').append(message + '\n')
          }

          $('#connect')
                .click(function doConnect() {
                    sockjs = new SockJS($("#target").val());
```

```
                sockjs.onopen = function (evt) {
                   writeStatus("CONNECTED");

                   var ping = setInterval(function () {
                      if (sockjs != "undefined") {
                         sockjs.send("ping");
                      }
                   }, 3000);
                };

                sockjs.onclose = function (evt) {
                   writeStatus("DISCONNECTED");
                };

                sockjs.onmessage = function (evt) {
                   if (evt.data === "ping") {
                      writePing(evt.data);
                   } else {
                      writeMessage('ECHO: ' + evt.data);
                   }
                };

                sockjs.onerror = function (evt) {
                   onError(writeStatus('ERROR:' + evt.data))
                };
             });

          $('#disconnect')
                .click(function () {
                   if(typeof sockjs != 'undefined') {
                      sockjs.close();
                      sockjs = undefined;
                   } else {
                      alert("Not connected.");
                   }
                });
          $('#send')
              .click(function () {
                 if(typeof sockjs != 'undefined') {
                    sockjs.send($('#message').val());
                 } else {
                    alert("Not connected.");
                 }
              });
       });
    </script>
</head>
<body>
<h2>SockJS Tester</h2>
    Target:
    <input id="target" size="40"
       value="http://localhost:8080/sockjs/echoHandler"/>
    <br/>
    <button id="connect">Connect</button>
    <button id="disconnect">Disconnect</button>
    <br/>
    <br/>Message:
    <input id="message" value=""/>
    <button id="send">Send</button>
    <br/>
    <p>Status output:</p>
    <pre><textarea id="statusOutput" rows="10" cols="50"></textarea></pre>
    <p>Message output:</p>
    <pre><textarea id="messageOutput" rows="10" cols="50"></textarea></pre>
    <p>Ping output:</p>
    <pre><textarea id="pingOutput" rows="10" cols="50"></textarea></pre>
</body>
</html>
```

实现新的 SockJS 代码之后，生成项目并将其部署到容器，导航到位于 http://localhost:8080/sockjs/index.html 的 UI，该 UI 具有 WebSocket 示例的所有特征和功能。要想测试 SockJS 后备功能，需要尝试在浏览器中禁用 WebSocket 支持。例如，在 Firefox 中，导航到 about:config 页面，然后搜索 network.websocket.enabled。将此设置切换为 false，然后重新加载示例 UI 并连接。使用诸如 Live HTTP Headers 之类的工具可以检查从浏览器到服务器的通信流量，以便进行验证。验证完行为之后，将 Firefox 设置 network.websocket.enabled 切换回 true，重新加载页面并连接。通过

Live HTTP Headers 查看流量将会显示 WebSocket 握手。在这个简单的示例中，所有操作都应该与使用 WebSocket API 是一样的。

17.4 使用 STOMP 发送消息

在使用 WebSocket 时，通常会使用 STOMP 等子协议作为客户端和服务器之间的通用格式，以便客户端和服务器都知道应该发生什么并作出相应的反应。Spring 框架支持 STOMP，我们在示例中将使用该协议。

STOMP 是一种简单的基于帧的消息传递协议(建立在 HTTP 基础之上)，可用于任何可靠的双向流网络协议，比如 WebSocket。STOMP 有标准的协议格式；在浏览器中可以利用 JavaScript 客户端支持来发送和接收消息，并可选择插入支持 STOMP 的传统消息代理(比如 RabbitMQ 和 ActiveMQ)。Spring 框架支持一个简单的代理，可用来处理订阅请求以及向内存中链接的客户端广播消息。在本例中，将使用这个简单的代理，并将全功能的代理设置作为练习[1]。

有关 STOMP 协议的完整说明，请参阅 http://stomp.github.io/stomp-specification-1.2.html。

在 STOMP 示例中，将创建一个简单的股票行情应用程序，该应用程序显示一些预定义的股票代码、它们的当前价格以及价格变化时的时间戳。新的股票代码和开盘价格也可以通过用户界面添加。任何连接的客户端(即选项卡中的其他浏览器或其他网络上的全新客户端)将看到与订阅消息广播相同的数据。每一秒，每只股票的价格都会更新为新的随机价格并更新时间戳。

为了确保客户能够使用该股票行情应用程序，即使他们的浏览器不支持 WebSocket，也会使用 SockJS 来透明地处理任何传输交换。在深入研究代码之前，值得注意的是，由 spring-messaging 库提供的 STOMP 消息支持。

接下来，首先创建 Stock 域对象，该对象包含股票的信息，比如股票代码和价格，如下所示：

```
package com.apress.prospring5.ch17;

import java.util.Date;
import java.io.Serializable;
import java.text.DateFormat;
import java.text.SimpleDateFormat;

public class Stock implements Serializable {
    private static final long serialVersionUID = 1L;
    private static final String DATE_FORMAT = "MMM dd yyyy HH:mm:ss";

    private String code;
    private double price;
    private Date date = new Date();
    private DateFormat dateFormat =
       new SimpleDateFormat(DATE_FORMAT);

    public Stock() { }

    public Stock(String code, double price) {

    this.code = code;
    this.price = price;
    }
    //setters and getters
   ...
}
```

现在需要添加一个 MVC 控制器来处理传入的请求，如下所示：

```
package com.apress.prospring5.ch17;

import org.springframework.beans.factory.annotation.Autowired;
import org.springframework.messaging.handler.annotation.MessageMapping;
import org.springframework.messaging.simp.SimpMessagingTemplate;
import org.springframework.scheduling.TaskScheduler;
import org.springframework.stereotype.Controller;
```

1 更多有关信息，请参阅 http://docs.spring.io/spring/docs/current/spring-framework-reference/html/websocket.html#websocket-stomp-handle-broker-relay-configure。

```java
import javax.annotation.PostConstruct;
import java.util.ArrayList;
import java.util.Date;
import java.util.List;
import java.util.Random;

@Controller
public class StockController {
    private TaskScheduler taskScheduler;
    private SimpMessagingTemplate simpMessagingTemplate;
    private List<Stock> stocks = new ArrayList<Stock>();
    private Random random = new Random(System.currentTimeMillis());

    public StockController() {
       stocks.add(new Stock("VMW", 1.00d));
       stocks.add(new Stock("EMC", 1.00d));
       stocks.add(new Stock("GOOG", 1.00d));
       stocks.add(new Stock("IBM", 1.00d));
    }

    @MessageMapping("/addStock")
    public void addStock(Stock stock) throws Exception {
       stocks.add(stock);
       broadcastUpdatedPrices();
    }

    @Autowired
    public void setSimpMessagingTemplate(
       SimpMessagingTemplate simpMessagingTemplate) {
       this.simpMessagingTemplate = simpMessagingTemplate;
    }

    @Autowired
    public void setTaskScheduler(TaskScheduler taskScheduler) {
       this.taskScheduler = taskScheduler;
    }

    private void broadcastUpdatedPrices() {
       for(Stock stock : stocks) {
          stock.setPrice(stock.getPrice() +
             (getUpdatedStockPrice() * stock.getPrice()));
          stock.setDate(new Date());
       }

       simpMessagingTemplate.convertAndSend("/topic/price", stocks);
    }

    private double getUpdatedStockPrice() {
       double priceChange = random.nextDouble() * 5.0;

       if (random.nextInt(2) == 1) {
          priceChange = -priceChange;
       }

       return priceChange/100.0;
    }

    @PostConstruct
    private void broadcastTimePeriodically() {
       taskScheduler.scheduleAtFixedRate(new Runnable() {
          @Override
          public void run() {
             broadcastUpdatedPrices();
          }
       }, 1000);
    }
}
```

此时，该控制器做了以下几件事情。首先，在列表中添加了一些预定义的股票代码和它们的开盘价格以供演示。然后定义了一个 addStock()方法，该方法接收一个 Stock 对象并将其添加到股票列表中，最后将股票广播给所有订阅者。在广播股票时，遍历所有已添加的股票、更新每只股票的价格并通过使用连接的 SimpMessagingTemplate 发送给所有/topic/price 的订阅者。还可以使用 TaskExecutor 实例向所有订阅的客户端连续广播更新的股票价格列表，每秒一次。

完成控制器之后，接下来创建 HTML UI 以显示给客户端(src/main/webapp/static/in-dex.html)，如以下 HTML 代码片段所示：

```
<html>
<head>
    <title>Stock Ticker</title>
    <script src="https://d1fxtkz8shb9d2.cloudfront.net/sockjs-0.3.4.min.js"/>
    <script src="http://cdnjs.cloudflare.com/ajax/libs/stomp.js/2.3.2/stomp.min.js"/>
    <script src="http://code.jquery.com/jquery-2.1.1.min.js"/>
    <script>
        var stomp = Stomp.over(new SockJS("/stomp/ws"));

        function displayStockPrice(frame) {
            var prices = JSON.parse(frame.body);

            $('#price').empty();

            for (var i in prices) {
                var price = pricesi;

                $('#price').append(
                    $('<tr>').append(
                        $('<td>').html(price.code),
                        $('<td>').html(price.price.toFixed(2)),
                        $('<td>').html(price.dateFormatted)
                    )
                );
            }
        }
        var connectCallback = function () {
            stomp.subscribe('/topic/price', displayStockPrice);
        };

        var errorCallback = function (error) {
            alert(error.headers.message);
        };

        stomp.connect("guest", "guest", connectCallback, errorCallback);

        $(document).ready(function () {
            $('.addStockButton').click(function (e) {
                e.preventDefault();

                var jsonstr = JSON.stringify({ 'code': $('.addStock .code').val(),
                    'price': Number($('.addStock .price').val()) });

                stomp.send("/app/addStock", {}, jsonstr);

                return false;
            });
        });
    </script>
</head>
<body>
<h1><b>Stock Ticker</b></h1>
<table border="1">
    <thead>
    <tr>
        <th>Code</th>
        <th>Price</th>
        <th>Time</th>
    </tr>
    </thead>
    <tbody id="price"></tbody>
</table>
<p class="addStock">
    Code: <input class="code"/><br/>
    Price: <input class="price"/><br/>
    <button class="addStockButton">Add Stock</button>
</p>
</body>
</html>
```

与前面的示例类似，使用一些HTML标记并混合JavaScript来更新显示[1]。使用jQuery来更新HTML数据，SockJS

1 将 HTML 和 JavaScript 混合使用的原因虽然不符合常识性编程规则，但可以尽可能简化 Spring MVC 配置。

提供传输选择，STOMP JavaScript 库 stomp.js 用于与服务器进行通信。通过 STOMP 消息发送的数据以 JSON 格式编码，可以从事件中提取数据。通过 STOMP 连接，订阅/topic/price 以接收股票价格更新。

现在，在 root-context.xml(src/main/webapp/WEB-INF/spring/root-context.xml)中配置内置的 STOMP 代理。

```
<beans ...">

    <mvc:annotation-driven />

    <mvc:default-servlet-handler/>

    <mvc:view-controller path= "/" view-name="/static/index.html" />

    <context:component-scan base-package="com.apress.prospring5.ch17" />

    <websocket:message-broker application-destination-prefix="/app">
        <websocket:stomp-endpoint path="/ws">
            <websocket:sockjs/>
        </websocket:stomp-endpoint>
        <websocket:simple-broker prefix="/topic"/>
    </websocket:message-broker>

    <bean id="taskExecutor"
        class="org.springframework.core.task.SimpleAsyncTaskExecutor"/>
</beans>
```

在大多数情况下，上述配置应该看起来很熟悉。在本例中，使用 WebSocket 名称空间来配置 message-broker、定义 STOMP 端点并启用 SockJS。此外，还配置了订阅者用来接收消息的前缀。所配置的 TaskExecutor 用于在控制器类中定义的时间间隔内提供股票价格。当使用名称空间支持时，会自动创建 SimpMessagingTemplate，并可注入到 bean 中。

现在，剩下要做的就是配置 web.xml 文件(src/main/webapp/WEB-INF/web.xml)，如下面的配置代码片段所示：

```
<web-app xmlns="http://java.sun.com/xml/ns/javaee"
       xmlns:xsi="http://www.w3.org/2001/XMLSchema-instance"
       xsi:schemaLocation="http://java.sun.com/xml/ns/javaee
       http://java.sun.com/xml/ns/javaee/web-app_3_0.xsd"
       version="3.0">

    <display-name>Spring STOMP Sample</display-name>

    <servlet>
        <servlet-name>stomp</servlet-name>
        <servlet-class>
            org.springframework.web.servlet.DispatcherServlet
        </servlet-class>
        <init-param>
            <param-name>contextConfigLocation</param-name>
            <param-value>/WEB-INF/spring/root-context.xml</param-value>
        </init-param>
        <load-on-startup>1</load-on-startup>
        <async-supported>true</async-supported>
    </servlet>
    <servlet-mapping>
        <servlet-name>stomp</servlet-name>
        <url-pattern>/*</url-pattern>
    </servlet-mapping>
</web-app>
```

介绍完 XML 配置之后，接下来介绍使用 Java 配置类的非传统配置。要想启用 Spring 异步调用和任务执行，必须使用@EnableAsync 注解 WebConfig，并且必须声明 org.springframework.core.task.TaskExecutor 类型的 bean。

```
package com.apress.prospring5.ch17.config;

import org.springframework.context.annotation.Bean;
import org.springframework.context.annotation.ComponentScan;
import org.springframework.context.annotation.Configuration;
import org.springframework.core.task.SimpleAsyncTaskExecutor;
import org.springframework.core.task.TaskExecutor;
import org.springframework.scheduling.annotation.EnableAsync;
import org.springframework.web.servlet.config.annotation.DefaultServletHandlerConfigurer;
import org.springframework.web.servlet.config.annotation.EnableWebMvc;
import org.springframework.web.servlet.config.annotation.ViewControllerRegistry;
import org.springframework.web.servlet.config.annotation.WebMvcConfigurer;

@Configuration
```

```
@EnableWebMvc
@EnableAsync
@ComponentScan(basePackages = {"com.apress.prospring5.ch17"})
public class WebConfig implements WebMvcConfigurer {

    //<=> <mvc:default-servlet-handler/>
    @Override
    public void configureDefaultServletHandling(
        DefaultServletHandlerConfigurer configurer) {
            configurer.enable();
    }

    //<=> <mvc:view-controller .../>
    @Override
    public void addViewControllers(ViewControllerRegistry registry) {
        registry.addViewController("/").setViewName("/static/index.html");
    }

    @Bean TaskExecutor taskExecutor() {
        return new SimpleAsyncTaskExecutor();
    }
}
```

需要在 WebSocketConfig 类中进行一些重要的更改。首先，该类必须实现 org.springframework.web.s 或扩展 AbstractWebSocketMessageBrokerConfigurer 抽象类，从而有助于确定实际要实现的方法。其次，该类还要定义从 WebSoC 客户端使用简单消息协议(比如 STOMP)配置消息处理的方法。最后，还必须使用名为@EnableWebSocketMessageBroker 的注解来注解类，该注解将使用高级消息子协议在 WebSocket 上启用代理支持的消息传递。

```
package com.apress.prospring5.ch17.config;

import org.springframework.context.annotation.Configuration;
import org.springframework.messaging.simp.config.MessageBrokerRegistry;
import org.springframework.web.socket.config.annotation.AbstractWebSocketMessageBrokerConf
import org.springframework.web.socket.config.annotation.EnableWebSocketMessageBroker;
import org.springframework.web.socket.config.annotation.StompEndpointRegistry;

@Configuration
@EnableWebSocketMessageBroker
public class WebSocketConfig extends
    AbstractWebSocketMessageBrokerConfigurer {

        //<=> <websocket:stomp-endpoint ... />
        @Override
        public void registerStompEndpoints(StompEndpointRegistry registry) {
            registry.addEndpoint("/ws").withSockJS();
        }

        //<=> websocket:message-broker../>
        @Override
        public void configureMessageBroker(MessageBrokerRegistry config) {
            config.setApplicationDestinationPrefixes("/app");
            config.enableSimpleBroker("/topic");
        }
}
```

在上面的配置中，重写的方法等价于前面的 XML 元素，用于配置 STOMP 端点和消息代理。

17.5 小结

在本章中，介绍了 WebSocket 的一般概念；讨论了 Spring 框架对低级 WebSocket API 的支持，以及如何使用 SockJS 作为后备选项来根据客户端浏览器选择适当的传输；最后介绍了作为 WebSocket 子协议的 STOMP，被用于在客户端和服务器之间传递消息。本章针对所有示例，介绍了 XML 和 Java 配置类，但业务的整体趋势是完全放弃 XML。

在下一章中，将讨论可以混合到应用程序中以提供更强大功能的 Spring 子项目。

第18章

Spring 项目：批处理、集成和 XD 等

本章将对 Spring 项目组合中的一些项目进行高级概述，特别是 Spring Batch、Integration、XD 以及 Spring Framework 5 中新增的一些功能。本章并不打算详细介绍每个项目，而只是提供足够的信息和示例来帮助你入门。虽然 Spring 项目组合包含的项目比本章中讨论的更多，但本章介绍的项目都被广泛使用，有些还是新的或即将推出的项目。可以在 http://spring.io/projects 上查看 Spring 项目的完整列表。本章包含以下主题。

- **Spring Batch**：介绍 Spring 批处理框架的核心概念，包括为开发人员提供的内容，以及 Spring Batch 3.0 对新的 JSR-352 的支持。
- **Spring Integration**：集成模式用于许多企业级应用程序，Spring Integration 为实现这些模式提供强大的框架。我们以批处理示例为基础来演示如何使用 Spring Integration 作为启动批处理作业的工作流的一部分。
- **Spring XD**：Spring XD 将许多现有的 Spring 项目联系在一起，为大数据应用程序提供统一且可扩展的系统。Spring XD 是一个专注于数据摄取、实时分析、批处理和数据导出的分布式系统。本章演示如何在 shell 界面中借助简单的 DSL，通过使用 Spring XD 来实现批处理和集成示例中的应用程序。
- **Spring 5 中引入的新功能**：关于 Spring Framework 5 的新功能有很多讨论，但随着官方发布日期的临近，最终的功能列表已经确定。除了诸如将完整的代码库迁移到 Java 8[1]之类的底层功能之外，还包括通过集成 Commons Logging 桥接模块[2](被命名为 spring-jcl 而不是标准的 Commons Logging，并且自动检测 Log4j 2.x、SLF4J 和 JUL，而无需任何额外的桥梁)来解决日志混乱的问题，添加*候选组件索引(candidate component index)*作为类路径扫描的替代方法，以及其他一些显著的改进[3]。本章主要介绍以下三个改进：
  - **功能性 Web 框架**：spring-webflux 是 spring-mvc 的补充，并且建立在反应式基础之上。由于 Reactive Streams API 是 Java 9 的官方组成部分，因此在实现了 Reactive Streams API 规范的 Project Reactor(http://projectreactor.io/)基础之上构建 Spring Framework 5 流支持。
  - **与 Java 9 的完全互操作性**：Spring Framework RC3 版本于 2017 年 7 月发布，并针对最新的 JDK 9 候选版本进行了全面测试。Java 9 引入了不少有趣的功能，其中包括：Jigsaw 项目/ Java 模块化，支持 HTTP 2 协议和 WebSocket 握手的新的 HTTP 客户端，改进的进程 API，用于 try-with-resources、钻石操作符(diamond operator)和接口私有方法等功能的改进语法，响应式编程的发布-订阅框架以及一组新的 API。Oracle 网站上提供了正式的更改和功能列表[4]，但基本上只有两个与 Spring 相关的功能：JDK 的模块化功能和响应式框架。
  - **完全支持 JUnit 5 Jupiter**[5]：JUnit 5 的 Jupiter 编程和扩展模型在 Spring Framework 5 中得到完全支持，包括支持 Spring TestContext 框架中并行测试的执行。

由于每个主题都可以用一整章或一本书来介绍，因此详细介绍每个项目及其提供的每种功能是不可能的。本章

---

1 Spring 5 的目的是与 Java 9 完全兼容，但由于 Java 9 的发布时间晚了大约 18 个月，因此 Spring 团队决定继续使用 Java 8。Java 9 于 2017 年 9 月发布，并且发布的 5.x 版本可以与 Spring 5 完全整合。

2 你可能会记得，这样设置项目是为了避免 Commons Logging，因为该模块中存在问题的运行时发现算法；可以通过 https://docs.spring.io/spring/docs/current/spring-framework-reference/htmlsingle/#overview-avoiding-commons-logging 在 Spring 的官方参考页面上阅读这篇文章。

3 可以在 https://github.com/spring-projects/spring-framework/wiki/What's-New-in-the-Spring-Framework#whats-new-in-spring-framework-5x 上查看完整列表。

4 请参阅 https://docs.oracle.com/javase/9/whatsnew/toc.htm。

5 到官方网站为 http ://junit.org/junit5/docs/current/user-guide/。

的目的是通过介绍和基本示例让你感兴趣，从而促使你对这些主题进行进一步的探索。

18.1 Spring Batch

Spring Batch 是一个批处理框架，是 Spring 项目组合的一部分。它轻巧、灵活，旨在让开发人员具备以最小努力创建强大的批处理应用程序的能力。Spring Batch 为各种技术提供许多现成的组件，大多数情况下，甚至可以通过单独使用提供的组件来构建批处理应用程序。

典型的批处理应用程序包括日常发票的生成、工资单系统以及 ETL(提取、转换、加载)流程。虽然这些都是人们可能想到的典型示例，但 Spring Batch 还可以用于任何需要无人值守运行的进程，而不止用于以上应用。与所有其他 Spring 项目一样，Spring Batch 构建在核心 Spring 框架上，并且可以完全访问其所有功能。

在较高层次上，批处理作业包含一个或多个步骤。每个步骤都能够执行单个工作单元(由 tasklet 实现表示)或参与所谓的*面向块的处理*(*chunk-oriented processing*)。当进行面向块的处理时，首先使用 ItemReader 读取某种形式的数据，然后使用可选的 ItemProcessor 对数据进行任何所需的转换，最后使用 ItemWriter 将数据写出。另外，还可以完成各种配置属性的设置，例如配置块大小(每个事务处理的数据量)、启用多线程执行、跳过限制等。可以在步骤级别以及作业级别使用监听器，以接收在批处理作业生命周期中发生的各种事件的通知。例如，在某个步骤开始之前、在某个步骤结束时、在面向块的处理过程中等。

尽管大多数作业能够以单线程、单进程的方式完美运行，但 Spring Batch 还提供了用于作业缩放和并行处理的选项。目前，Spring Batch 提供了以下可伸缩性选项：

- **多线程步骤**：这是实现多线程步骤的最简单方法。只需要将选择的 TaskExecutor 实例添加到步骤配置中，面向块的处理设置中的每个项目块都将在各自的执行线程中处理。
- **并行步骤**：比如，需要在作业开始时读入两个包含不同数据的大文件。起初可以创建两个步骤，一个步骤会在另一个步骤之后执行。但如果这两个数据文件加载不相互依赖，那么为什么不同时处理它们？在这种情况下，Spring Batch 允许定义包含流元素的分割，并封装这些并行执行的任务。
- **远程分块**：该可伸缩性选项允许将工作远程分发给多个远程工作人员，并通过 AMQP 或 JMS 等持久中间件进行通信。当应用程序的瓶颈是块数据的写入和可选处理而不是数据读取时，通常使用远程分块。数据块通过中间件发送给从属节点进行拾取和处理，然后传送给主节点，告知数据块处理的状态。
- **分区**：当想要处理一系列数据时，通常使用该可伸缩性选项，针对每个范围使用一个线程。典型的应用场景是使用具有数字标识符列的数据填充的数据库表。通过分区，可以用一定数量的记录在单独的线程中对要处理的数据进行“分区”。Spring Batch 为开发人员提供了这种功能，以便与分区方案挂钩，如下面的示例所示。分区可以在本地线程中完成，也可以远程工作(类似于远程分块可伸缩性选项)。

批处理的一个基本而常见的用例是读取某种文件，通常是采用分隔格式的平面文件(例如用逗号分隔)，然后需要将其加载到数据库中，并且在写入数据库之前选择性地处理每条记录。接下来看看如何在 Spring Batch 中实现这个用例。首先，需要添加所需的依赖项，如下面的 Gradle 配置所示：

```
//pro-spring-15/build.gradle
ext {
   //spring libs
   ...
   springBatchVersion = '4.0.0.M3'
   ...

   spring = [
      context    : "org.springframework:spring-context:$springVersion",
      jdbc       : "org.springframework:spring-jdbc:$springVersion",
      batchCore  : "org.springframework.batch:spring-batch-core:$springBatchVersion"
      ...
   ]
   misc = [
      io : "commons-io:commons-io:2.5",
      ...
   ]

   db = [
      ...
      dbcp2 : "org.apache.commons:commons-dbcp2:$dbcpVersion",
```

```
        h2 : "com.h2database:h2:$h2Version",

        //needed for the Batch JSR-352 module
        hsqldb: "org.hsqldb:hsqldb:2.4.0"
    dbcp : "commons-dbcp:commons-dbcp:1.4",
    ]
}
...
//pro-spring-15/chapter18/build.gradle
dependencies {
        if (!project.name.contains("boot")) {
            compile(spring.jdbc) {
                //exclude these as batchCore will bring them
                //on as transitive dependencies
                exclude module: 'spring-core'
                exclude module: 'spring-beans'
                exclude module: 'spring-tx'
            }
            compile spring.batchCore, db.dbcp2, db.h2, misc.io,
                    misc.slf4jJcl, misc.logback
        }
}
```

在上面的配置中，可以看到需要添加到 Spring Batch 项目(而不是 Spring Boot 项目)的核心依赖项。这也是使用 f(!project.name.contains("boot"))条件的原因，可以防止显式设置版本的库与本章中 Spring Boot 项目的依赖项产生混合。

添加完依赖项之后，下面研究一下具体的代码。首先，根据要读取文件中的数据创建一个表示歌手的域对象，如下所示：

```
package com.apress.prospring5.ch18;

public class Singer {
    private String firstName;
    private String lastName;
    private String song;
    ... //setters & getters

    @Override
    public String toString() {
        return "firstName: " + firstName + ", lastName: "
            + lastName + ", song: " + song;
    }
}
```

接下来，创建 ItemProcessor 的一个实现，用于将 Singer 对象表示的每位歌手的名字、姓氏和歌曲转换为大写形式，如下所示：

```
package com.apress.prospring5.ch18;

import org.slf4j.Logger;
import org.slf4j.LoggerFactory;
import org.springframework.batch.item.ItemProcessor;
import org.springframework.stereotype.Component;

@Component("itemProcessor")
public class SingerItemProcessor implements
    ItemProcessor<Singer, Singer> {
    private static Logger logger =
        LoggerFactory.getLogger(SingerItemProcessor.class);

    @Override
    public Singer process(Singer singer) throws Exception {
        String firstName = singer.getFirstName().toUpperCase();
        String lastName = singer.getLastName().toUpperCase();
        String song = singer.getSong().toUpperCase();

        Singer transformedSinger = new Singer();
        transformedSinger.setFirstName(firstName);
        transformedSinger.setLastName(lastName);
        transformedSinger.setSong(song);

        logger.info("Transformed singer: " + singer + " Into: " +
            transformedSinger);
```

```
        return transformedSinger;
    }
}
```

请注意，在面向块的处理场景中不需要 ItemProcessor；只需要 ItemReader 和 ItemWriter。此处使用 ItemProcessor 作为示例来演示如何在写入之前对数据进行转换。

接下来，创建一个步骤级别的 StepExecutionListener 实现，并告知在该步骤完成后写入的记录条数，如以下代码片段所示：

```
package com.apress.prospring5.ch18;

import org.slf4j.Logger;
import org.slf4j.LoggerFactory;
import org.springframework.batch.core.ExitStatus;
import org.springframework.batch.core.StepExecution;
import org.springframework.batch.core.listener.StepExecutionListenerSupport;
import org.springframework.stereotype.Component;

@Component
public class StepExecutionStatsListener extends
    StepExecutionListenerSupport {
    public static Logger logger = LoggerFactory.
        getLogger(StepExecutionStatsListener.class);
    @Override
    public ExitStatus afterStep(StepExecution stepExecution) {
        logger.info("--> Wrote: " + stepExecution.getWriteCount()
            + " items in step: " + stepExecution.getStepName());
        return null;
    }
}
```

如果需要，StepExecutionListener 还允许修改返回的 ExitStatus 值；或者简单地返回 null 以保持它不变。到目前为止，已经装配了核心组件，但在介绍配置和调用代码之前，先看看数据模型和数据本身。该作业的数据模型很简单(src/main/resources/support/singer.sql)，下面所示的是包含测试数据的 src/main/resources/support/test-data.sql 文件：

```
-- singer.sql
DROP TABLE singer IF EXISTS;

CREATE TABLE singer (
    singer_id BIGINT IDENTITY NOT NULL PRIMARY KEY,
    first_name VARCHAR(20),
    last_name VARCHAR(20),
    song VARCHAR(100)
);

-- test-data.sql
John,Mayer,Helpless
Eric,Clapton,Change The World
John,Butler,Ocean
BB,King,Chains And Things
```

现在需要创建 Spring Batch 配置文件、定义作业并设置嵌入式数据库和相关的作业组件。由于 XML 配置比较麻烦，并且在本书的前一版中已经介绍过，因此本章将只关注 Java 配置类。按照常识性编程规则，分别为批处理配置和数据源配置创建单独的配置类，从而实现两者的分离。DataSourceConfig 配置类如下所示：

```
package com.apress.prospring5.ch18.config;

import org.slf4j.Logger;
import org.slf4j.LoggerFactory;
import org.springframework.context.annotation.Bean;
import org.springframework.context.annotation.Configuration;
import org.springframework.jdbc.datasource.embedded.EmbeddedDatabaseBuilder;
import org.springframework.jdbc.datasource.embedded.EmbeddedDatabaseType;
import javax.sql.DataSource;

@Configuration
public class DataSourceConfig {

    private static Logger logger = LoggerFactory.getLogger(DataSourceConfig.class);

    @Bean
    public DataSource dataSource() {
        try {
            EmbeddedDatabaseBuilder dbBuilder = new EmbeddedDatabaseBuilder();
```

```
            return dbBuilder.setType(EmbeddedDatabaseType.H2)
                .addScripts("classpath:/org/springframework/batch/core/schema-h2.sql",
                    "classpath:support/singer.sql" ).build();
        } catch (Exception e) {
            logger.error("Embedded DataSource bean cannot be created!", e);
            return null;
        }
    }
}
```

你对上述配置应该很熟悉了，因此我们只解释 schema-h2.sql 文件。该文件是 spring-batch-core 库的一部分，并且包含创建 Spring Batch 实用程序表所需的 DML 语句。

DataSourceConfig 类将被导入到 BatchConfig 类中，如下所示：

```
package com.apress.prospring5.ch18.config;

import com.apress.prospring5.ch18.Singer;
import com.apress.prospring5.ch18.StepExecutionStatsListener;
import org.springframework.batch.core.Job;
import org.springframework.batch.core.Step;
import org.springframework.batch.core.configuration.annotation.EnableBatchProcessing;
import org.springframework.batch.core.configuration.annotation.JobBuilderFactory;
import org.springframework.batch.core.configuration.annotation.StepBuilderFactory;
import org.springframework.batch.item.ItemProcessor;
import org.springframework.batch.item.ItemReader;
import org.springframework.batch.item.ItemWriter;
import org.springframework.batch.item.database.BeanPropertyItemSqlParameterSourceProvider;
import org.springframework.batch.item.database.JdbcBatchItemWriter;
import org.springframework.batch.item.file.FlatFileItemReader;
import org.springframework.batch.item.file.mapping.BeanWrapperFieldSetMapper;
import org.springframework.batch.item.file.mapping.DefaultLineMapper;
import org.springframework.batch.item.file.transform.DelimitedLineTokenizer;
import org.springframework.beans.factory.annotation.Autowired;
import org.springframework.beans.factory.annotation.Qualifier;
import org.springframework.context.annotation.Bean;
import org.springframework.context.annotation.ComponentScan;
import org.springframework.context.annotation.Configuration;
import org.springframework.context.annotation.Import;
import org.springframework.core.io.ResourceLoader;
import javax.sql.DataSource;

@Configuration
@EnableBatchProcessing
@Import(DataSourceConfig.class)
@ComponentScan("com.apress.prospring5.ch18")
public class BatchConfig {

    @Autowired
    private JobBuilderFactory jobs;

    @Autowired
    private StepBuilderFactory steps;

    @Autowired DataSource dataSource;

    @Autowired ResourceLoader resourceLoader;
    @Autowired StepExecutionStatsListener executionStatsListener;

    @Bean
    public Job job(@Qualifier("step1") Step step1) {
        return jobs.get("singerJob").start(step1).build();
    }

    @Bean
    protected Step step1(ItemReader<Singer> reader,
        ItemProcessor<Singer,Singer> itemProcessor,
        ItemWriter<Singer> writer) {
        return steps.get("step1").listener(executionStatsListener)
            .<Singer, Singer>chunk(10)
            .reader(reader)
            .processor(itemProcessor)
            .writer(writer)
            .build();
    }

    @Bean
    public ItemReader itemReader() {
```

```
        FlatFileItemReader itemReader = new FlatFileItemReader();
        itemReader.setResource(resourceLoader.getResource(
            "classpath:support/test-data.csv"));
        DefaultLineMapper lineMapper = new DefaultLineMapper();

        DelimitedLineTokenizer tokenizer = new DelimitedLineTokenizer();
        tokenizer.setNames("firstName","lastName","song");
        tokenizer.setDelimiter(",");
        lineMapper.setLineTokenizer(tokenizer);

        BeanWrapperFieldSetMapper<Singer> fieldSetMapper =
            new BeanWrapperFieldSetMapper<>();
        fieldSetMapper.setTargetType(Singer.class);
        lineMapper.setFieldSetMapper(fieldSetMapper);
        itemReader.setLineMapper(lineMapper);
        return itemReader;
    }
    @Bean
    public ItemWriter itemWriter() {
        JdbcBatchItemWriter<Singer> itemWriter = new JdbcBatchItemWriter<>();
        itemWriter.setItemSqlParameterSourceProvider(
            new BeanPropertyItemSqlParameterSourceProvider<>());
        itemWriter.setSql("INSERT INTO singer (first_name, last_name, song)
            VALUES (:firstName, :lastName, :song)");
        itemWriter.setDataSource(dataSource);
        return itemWriter;
    }
}
```

BatchConfig 虽然看起来代码很多，但它并没有 XML 配置那么庞大。接下来解释一下里面定义的每个 bean。

- @EnableBatchProcessing 注解的工作方式与所有的@Enable* Spring 注解类似。它为构建批处理作业提供了基本配置。如果使用该注解来注解配置类，会发生以下情况：
  - 创建 org.springframework.batch.core.scope.StepScope 的一个实例。该作用域内的对象将使用 Spring 容器作为对象工厂，所以每个执行步骤中这种 bean 的实例只有一个。
  - 生成一组可用于自动装配的特定的批处理基础结构 bean：jobRepository(JobRepository 类型)、jobLauncher(jobLauncher 类型)、jobBuilders(类型为 JobBuilderFactory) 和 stepBuilders(类型为 StepBuilderFactory)。这意味着不必显式地声明它们(比如在 XML 中)。
- job bean 是通过调用 JobBuilderFactory.get()创建的名为 singerJob 的批处理作业。
- step1 bean 是通过调用 StepBuilderFactory.get()创建的，并且被配置用于面向块的处理。Spring 容器会自动注入在上下文中找到的 ItemReader、ItemProcessor 和 ItemWriter 实例。但是，StepExecutionStatsListener bean 必须显式设置。
- database bean 被用在 ItemWriter 的声明中，因为该 bean 用于将 Singer 实例写入嵌入式数据库。

最后，需要一个驱动程序来启动作业，如下所示：

```
package com.apress.prospring5.ch18;

import com.apress.prospring5.ch18.config.BatchConfig;
import org.springframework.batch.core.Job;
import org.springframework.batch.core.JobParameters;
import org.springframework.batch.core.JobParametersBuilder;
import org.springframework.batch.core.launch.JobLauncher;
import org.springframework.context.annotation.AnnotationConfigApplicationContext;
import org.springframework.context.support.GenericApplicationContext;
import java.util.Date;

public class SingerJobDemo {

    public static void main(String... args) throws Exception {
        GenericApplicationContext ctx =
            new AnnotationConfigApplicationContext(BatchConfig.class);
        Job job = ctx.getBean(Job.class);
        JobLauncher jobLauncher = ctx.getBean(JobLauncher.class);
        JobParameters jobParameters = new JobParametersBuilder()
            .addDate("date", new Date())
            .toJobParameters();
        jobLauncher.run(job, jobParameters);

        System.in.read();
```

```
        ctx.close();
    }
}
```

你应该对上述代码非常熟悉了，主要是创建上下文、获得一些 bean 并且调用它们的方法。有个对象你可能会注意到，那就是 JobParameters 对象。该对象封装用于区分不同作业实例的参数。作业识别对于确定作业的最后状态(如果有的话)是非常重要的，当然也起到其他作用，例如重启作业。在本例中，只使用当前日期作为 Job 参数。JobParameters 对象可以是多种类型，并且这些参数可以作为引用数据在作业中访问。接下来准备测试新的作业。编译代码并运行 SingerJobDemo 类。你会在屏幕上看到一些日志语句，你感兴趣的内容如下：

```
o.s.b.c.l.s.SimpleJobLauncher - Job: [SimpleJob: [name=singerJob]] launched with the
   following parameters: [{date=1501418591075}]
o.s.b.c.j.SimpleStepHandler - Executing step: [step1]
c.a.p.c.SingerItemProcessor - Transformed singer: firstName: John, lastName: Mayer,
   song: Helpless Into: firstName: JOHN, lastName: MAYER, song: HELPLESS
c.a.p.c.SingerItemProcessor - Transformed singer: firstName: Eric, lastName: Clapton,
   song: Change The World Into: firstName: ERIC, lastName: CLAPTON, song: CHANGE THE WORLD
c.a.p.c.SingerItemProcessor - Transformed singer: firstName: John, lastName: Butler,
   song: Ocean Into: firstName: JOHN, lastName: BUTLER, song: OCEAN
c.a.p.c.SingerItemProcessor - Transformed singer: firstName: BB, lastName: King,
   song: Chains And Things Into: firstName: BB, lastName: KING, song: CHAINS AND THINGS
c.a.p.c.StepExecutionStatsListener - --> Wrote: 4 items in step: step1
o.s.b.c.l.s.SimpleJobLauncher - Job: [SimpleJob: [name=singerJob]] completed with the
   following parameters: [{date=1501418591075}] and the following status: [COMPLETED]
```

以上就是所有代码。现在，已经构建了一个简单的批处理作业，首先从 CSV 文件中读取数据，并通过 ItemProcessor 转换数据，从而将歌手的姓名和歌曲更改为大写形式，然后将结果写入数据库。此外，还使用 StepListener 输出在步骤中写入的项目数量。有关 Spring Batch 的更多信息，请参阅 http://projects.spring.io/spring-batch/上的项目页面。

18.2 JSR-352

JSR-352(针对 Java 平台的批处理应用程序)深受 Spring Batch 的影响。如果选择使用 JSR-352 来完成作业，那么你会注意到两者之间存在很多相似之处。对于 Spring Batch 用户，在使用 JSR-352 时应该会感到非常顺手。在大多数情况下，Spring Batch 和 JSR-352 具有相似的构造，从 Spring Batch 3.0 开始，Spring Batch 就完全支持 JSR。与 Spring Batch 一样，JSR-352 作业也是通过一种 XML 模式(称为作业说明语言(Job Specification Language，JSL))配置的。因为 JSR-352 只定义了一个规范和一个 API，所以无法像使用 Spring Batch 时那样使用现成的基础架构组件。严格遵守 JSR-352 API 意味着需要实现 JSR-352 接口并自行编写所有基础结构组件(如 ItemReader 和 ItemWriter)。

在本例中，将对前面的批处理示例进行转换，以利用 JSR-352 JSL，但并不是重新编写自己的基础结构组件，而是利用相同的 ItemReader、ItemProcessor 和 ItemWriter 以及利用 Spring 进行依赖注入等。实现 100%符合 JSR-352 规范的作业将作为练习留给读者完成。

如上所述，在本例中，将重用 Spring Batch 示例中的大部分代码，并稍作修改，具体修改内容稍后介绍。如果还没有运行前面的 Spring Batch 示例，那么可以先运行，然后继续在本节中应用这些更改。

对于本例，H2 数据库被 HSQLDB 替换，DBCP 2 被替换为 DBCP，因为更新的版本不支持。此外，JSR 需要在 src/main/resources/META-INF/batch-jobs /下声明 singerJob.xml 配置文件，并且在启动作业时，所需的全部文件名不带.xml 扩展名。所以，JSR-352 必须使用 XML。示例中的 XML 配置文件如下所示：

```
<?xml version="1.0" encoding="UTF-8"?>

<beans xmlns="http://www.springframework.org/schema/beans"
       xmlns:xsi="http://www.w3.org/2001/XMLSchema-instance"
       xmlns:jdbc="http://www.springframework.org/schema/jdbc"
       xsi:schemaLocation="
       http://www.springframework.org/schema/jdbc
       http://www.springframework.org/schema/jdbc/spring-jdbc.xsd
       http://www.springframework.org/schema/beans
       http://www.springframework.org/schema/beans/spring-beans.xsd
       http://xmlns.jcp.org/xml/ns/javaee
       http://xmlns.jcp.org/xml/ns/javaee/jobXML_1_0.xsd">

    <job id="singerJob" xmlns="http://xmlns.jcp.org/xml/ns/javaee" version="1.0">
       <step id="step1">
```

```
            <listeners>
                <listener ref="stepExecutionStatsListener"/>
            </listeners>
            <chunk item-count="10">
                <reader ref="itemReader"/>
                <processor ref="itemProcessor"/>
                <writer ref="itemWriter"/>
            </chunk>
            <fail on="FAILED"/>
            <end on="*"/>
        </step>
    </job>
    <jdbc:embedded-database id="dataSource" type="HSQL">
        <jdbc:script location="classpath:support/singer.sql"/>
    </jdbc:embedded-database>

    <!-- no transaction manager needed -->
    <bean id="stepExecutionStatsListener" ../>
    <bean id="itemReader" ../>
    <bean id="itemProcessor" ../>
    <bean id="itemWriter" ../>
</beans>
```

上面未显示的 bean 具有与使用 Spring Batch 时相同的定义。可以在本书附带的源代码中找到完整的 XML 定义。下面的代码片段显示了如何使用 Spring Batch XML 配置前面的作业:

```
<?xml version="1.0" encoding="UTF-8"?>

<beans xmlns="http://www.springframework.org/schema/beans"
       xmlns:xsi="http://www.w3.org/2001/XMLSchema-instance"
       xmlns:batch="http://www.springframework.org/schema/batch"
       xmlns:jdbc="http://www.springframework.org/schema/jdbc"
       xmlns:p="http://www.springframework.org/schema/p"
       xsi:schemaLocation="
       http://www.springframework.org/schema/batch
       http://www.springframework.org/schema/batch/spring-batch.xsd
       http://www.springframework.org/schema/jdbc
       http://www.springframework.org/schema/jdbc/spring-jdbc.xsd
       http://www.springframework.org/schema/beans
       http://www.springframework.org/schema/beans/spring-beans.xsd">

    <batch:job id="singerJob">
        <batch:step id="step1">
            <batch:tasklet>
                <batch:chunk reader="itemReader"
                processor="itemProcessor"
                writer="itemWriter"
                          commit-interval="10"/>
                <batch:listeners>
                    <batch:listener ref="stepExecutionStatsListener"/>
                </batch:listeners>
            </batch:tasklet>
            <batch:fail on="FAILED"/>
            <batch:end on="*"/>
        </batch:step>
    </batch:job>

    <jdbc:embedded-database id="dataSource" type="H2">
    <jdbc:script location="classpath:/org/springframework/batch/core/schema-h2.sql"/>
        <jdbc:script location="classpath:support/singer.sql"/>
    </jdbc:embedded-database>
    <batch:job-repository id="jobRepository"/>

    <bean id="jobLauncher"

        class="org.springframework.batch.core.launch.support.SimpleJobLauncher"
            p:jobRepository-ref="jobRepository"/>
    <bean id="transactionManager"

      class="org.springframework.jdbc.datasource.DataSourceTransactionManager"
            p:dataSource-ref="dataSource"/>
    <bean id="stepExecutionStatsListener" ../>
    <bean id="itemReader" ../>
    <bean id="itemProcessor" ../>
    <bean id="itemWriter" ../>
</beans>
```

除了作业定义使用 JSR-352 JSL 之外，这两个配置看起来几乎相同，可以删除一些 bean(transactionManager、jobRepository 和 jobLauncher)，因为它们已经以某种方式提供给我们。此外，你还会注意到使用 jobXML_1.0.xsd 的其他架构定义。对该架构的支持是通过 JSR-352 API JAR 获得的，并在使用诸如 Gradle 等构建工具时自动将其作为传递依赖项进行处理。如果需要手动获取依赖项，请参阅本节末尾列出的项目页面。需要修改的另一部分是 SingerJobDemo 类，因为将使用 JSR-352 特定的代码来调用作业，如下所示：

```
package com.apress.prospring5.ch18;

import org.springframework.batch.core.Job;
import org.springframework.batch.core.JobParameters;
import org.springframework.batch.core.JobParametersBuilder;
import org.springframework.batch.core.launch.JobLauncher;
import org.springframework.context.ApplicationContext;
import org.springframework.context.support.ClassPathXmlApplicationContext;

import java.util.Date;

public class SingerJobDemo {
   public static void main(String... args) throws Exception {
      ApplicationContext applicationContext
            = new ClassPathXmlApplicationContext("/spring/singerJob.xml");

      Job job = applicationContext.getBean(Job.class);
      JobLauncher jobLauncher = applicationContext.getBean(JobLauncher.class);

      JobParameters jobParameters = new JobParametersBuilder()
            .addDate("date", new Date())
            .toJobParameters();

      jobLauncher.run(job, jobParameters);
   }
}
```

与其他示例有所不同，这里直接使用 ApplicationContext 和 bean。在创建 JSR-352 作业时，使用 JsrJobOperator 来启动和控制作业。并没有使用 JobParameters 对象向作业提供参数，而是使用 Properties 对象。所使用的 Properties 对象是一个标准的 java.util.Properties 类，并且应该使用 String 键和值来创建作业参数。你可能会注意到的另一处有趣的变化是 waitForJob()方法。JSR-352 默认以异步方式启动所有作业。因此，如下所示，在独立程序中，在程序终止之前需要等待作业处于可接收的状态。如果代码运行在某个容器(比如某种应用服务器)中，则可能不需要此代码。现在编译并运行 SingerJobDemo 类，将生成以下相关的日志语句：

```
o.s.b.c.r.s.JobRepositoryFactoryBean - No database type set,
   using meta data indicating: HSQL
o.s.b.c.j.c.x.JsrXmlApplicationContext - Refreshing org.springframework.batch.core.jsr.
   configuration.xml.JsrXmlApplicationContext@48c76607
o.s.b.c.j.SimpleStepHandler - Executing step: [step1]
c.a.p.c.SingerItemProcessor - Transformed singer: firstName: John, lastName: Mayer,
   song: Helpless Into: firstName: JOHN, lastName: MAYER, song: HELPLESS
c.a.p.c.SingerItemProcessor - Transformed singer: firstName: Eric, lastName: Clapton,
   song: Change The World Into: firstName: ERIC, lastName: CLAPTON, song: CHANGE THE WORLD
c.a.p.c.SingerItemProcessor - Transformed singer: firstName: John, lastName: Butler,
   song: Ocean Into: firstName: JOHN, lastName: BUTLER, song: OCEAN
c.a.p.c.SingerItemProcessor - Transformed singer: firstName: BB, lastName: King,
   song: Chains And Things Into: firstName: BB, lastName: KING, song: CHAINS AND THINGS
c.a.p.c.StepExecutionStatsListener - --> Wrote: 4 items in step: step1
o.s.b.c.j.c.x.JsrXmlApplicationContext - Closing ... JsrXmlApplicationContext
```

日志输出看起来差不多，但现在使用 JSR-352 来定义和运行作业，并且使用 Spring 的依赖注入功能以及 Spring Batch 的基础设施组件，而不是自己编写。

有关 JSR-352 的更多信息，请参阅其项目页面，网址为 https://jcp.org/en/jsr/detail?id=352。

18.3 Spring Boot Batch

正如预期的那样，Spring Boot 为 Spring Batch 提供一个特殊的启动器库，从而可以更好地完成配置。好的方面是，当在类路径中使用 Spring Batch 的 Spring Boot 启动器库时，不需要过多地处理依赖项。不好的方面是，当涉及配置时，涉及的 Spring Batch 的细节不会减少太多。

接下来尝试修改前面运行的批处理示例，并将 StepExecutionStatsListener 替换为 JobExecutionStatsListener 类；该

类扩展了 JobExecutionListenerSupport 类，后者是 JobExecutionListener 接口的空的抽象实现，用于在作业生命周期内的特定时间点提供回调。在本例中，JobExecutionStatsListener 类将查询数据库以检查歌手条目是否确实保存在数据库中。

下面的代码描述了类 JobExecutionStatsListener：

```
package com.apress.prospring5.ch18;

import org.slf4j.Logger;
import org.slf4j.LoggerFactory;
import org.springframework.batch.core.BatchStatus;
import org.springframework.batch.core.ExitStatus;
import org.springframework.batch.core.JobExecution;
import org.springframework.batch.core.StepExecution;
import org.springframework.batch.core.listener.JobExecutionListenerSupport;
import org.springframework.batch.core.listener.StepExecutionListenerSupport;
import org.springframework.beans.factory.annotation.Autowired;
import org.springframework.jdbc.core.JdbcTemplate;
import org.springframework.jdbc.core.RowMapper;
import org.springframework.stereotype.Component;

import java.sql.ResultSet;
import java.sql.SQLException;
import java.util.List;

@Component
public class JobExecutionStatsListener extends JobExecutionListenerSupport {

   public static Logger logger = LoggerFactory.
      getLogger(JobExecutionStatsListener.class);

   private final JdbcTemplate jdbcTemplate;

   @Autowired
   public JobExecutionStatsListener(JdbcTemplate jdbcTemplate) {
      this.jdbcTemplate = jdbcTemplate;
   }
   @Override
   public void afterJob(JobExecution jobExecution) {
      if(jobExecution.getStatus() == BatchStatus.COMPLETED) {
         logger.info(" --> Singers were saved to the database. Printing results ...");
         jdbcTemplate.query("SELECT first_name, last_name, song FROM SINGER",
            (rs, row) -> new Singer(rs.getString(1),
               rs.getString(2), rs.getString(3))).forEach(
            singer -> logger.info(singer.toString())
         );
      }
   }
}
```

正如你所看到的，在上面的示例中 lambda 表达式被大量使用，此时可以很容易地知道在 afterJob()回调方法的主体中完成了哪些操作。

接下来看看使用 BatchConfig 类能够做什么。

```
package com.apress.prospring5.ch18;

import org.springframework.batch.core.Job;
import org.springframework.batch.core.Step;
import org.springframework.batch.core.configuration.annotation.EnableBatchProcessing;
import org.springframework.batch.core.configuration.annotation.JobBuilderFactory;
import org.springframework.batch.core.configuration.annotation.StepBuilderFactory;
import org.springframework.batch.item.ItemProcessor;
import org.springframework.batch.item.ItemReader;
import org.springframework.batch.item.ItemWriter;
import org.springframework.batch.item.database.BeanPropertyItemSqlParameterSourceProvider;
import org.springframework.batch.item.database.JdbcBatchItemWriter;
import org.springframework.batch.item.file.FlatFileItemReader;
import org.springframework.batch.item.file.mapping.BeanWrapperFieldSetMapper;
import org.springframework.batch.item.file.mapping.DefaultLineMapper;
import org.springframework.batch.item.file.transform.DelimitedLineTokenizer;
import org.springframework.beans.factory.annotation.Autowired;
import org.springframework.beans.factory.annotation.Qualifier;
import org.springframework.context.annotation.Bean;
import org.springframework.context.annotation.ComponentScan;
import org.springframework.context.annotation.Configuration;
import org.springframework.core.io.ClassPathResource;
```

```
import org.springframework.core.io.ResourceLoader;

import javax.sql.DataSource;

@Configuration
@EnableBatchProcessing
public class BatchConfig {

    @Autowired
    private JobBuilderFactory jobs;
    @Autowired
    private StepBuilderFactory steps;

    @Autowired DataSource dataSource;

    @Autowired SingerItemProcessor itemProcessor;

    @Bean
    public Job job(JobExecutionStatsListener listener) {
        return jobs.get("singerJob")
                .listener(listener)
                .flow(step1())
                .end()
                .build();
    }

    @Bean
    protected Step step1() {
        return steps.get("step1")
            .<Singer, Singer>chunk(10)
            .reader(itemReader())
            .processor(itemProcessor)
            .writer(itemWriter())
            .build();
    }

    //adding lambda expressions
    @Bean
    public ItemReader itemReader() {
        FlatFileItemReader<Singer> itemReader = new FlatFileItemReader<>();
        itemReader.setResource(new ClassPathResource("support/test-data.csv"));
        itemReader.setLineMapper(new DefaultLineMapper<Singer>() {{
            setLineTokenizer(new DelimitedLineTokenizer() {{
                setNames(new String { "firstName", "lastName", "song" });
            }});
            setFieldSetMapper(new BeanWrapperFieldSetMapper<Singer>() {{
                setTargetType(Singer.class);
            }});
        }});
        return itemReader;
    }

    @Bean
    public ItemWriter itemWriter() {
    ... //same as before
    }
}
```

除了在 itemReader bean 的声明中大量使用 lambda 表达式之外，更改最多的 bean 是 job bean。执行步骤不再由基于限定符的 Spring 容器自动执行，而是调用 flow()方法来创建一个新的作业生成器，它将执行一个或一系列步骤。

这与 Spring Boot 非常相似，剩下需要做的就是通过执行典型的 Application 类来启动应用程序。

```
package com.apress.prospring5.ch18;

import org.slf4j.Logger;
import org.slf4j.LoggerFactory;
import org.springframework.boot.SpringApplication;
import org.springframework.boot.autoconfigure.SpringBootApplication;
import org.springframework.context.ConfigurableApplicationContext;

@SpringBootApplication
public class Application {

    private static Logger logger = LoggerFactory
        .getLogger(Application.class);

    public static void main(String... args) throws Exception {
```

```
        ConfigurableApplicationContext ctx =
            SpringApplication.run(Application.class, args);
        assert (ctx != null);
        logger.info("Application started...");

        System.in.read();
        ctx.close();
    }
}
```

最后，可以检查日志以获取 JobExecutionStatsListener 类打印的预期结果，如下所示：

```
o.s.b.c.l.s.SimpleJobLauncher - Job: [FlowJob: [name=singerJob]] launched with
   the following parameters: [{}]
o.s.b.c.j.SimpleStepHandler - Executing step: [step1]
c.a.p.c.SingerItemProcessor - Transformed singer: firstName: John, lastName: Mayer,
   song: Helpless Into: firstName: JOHN, lastName: MAYER, song: HELPLESS
c.a.p.c.SingerItemProcessor - Transformed singer: firstName: Eric, lastName: Clapton,
   song: Change The World Into: firstName: ERIC, lastName: CLAPTON, song: CHANGE THE WORLD
c.a.p.c.SingerItemProcessor - Transformed singer: firstName: John, lastName: Butler,
   song: Ocean Into: firstName: JOHN, lastName: BUTLER, song: OCEAN
c.a.p.c.SingerItemProcessor - Transformed singer: firstName: BB, lastName: King,
   song: Chains And Things Into: firstName: BB, lastName: KING, song: CHAINS AND THINGS
c.a.p.c.JobExecutionStatsListener - --> Singers were saved to the database. Printing results ...
c.a.p.c.JobExecutionStatsListener - firstName: JOHN, lastName: MAYER, song: HELPLESS
c.a.p.c.JobExecutionStatsListener - firstName: ERIC, lastName: CLAPTON, song: CHANGE THE WORLD
c.a.p.c.JobExecutionStatsListener - firstName: JOHN, lastName: BUTLER, song: OCEAN
c.a.p.c.JobExecutionStatsListener - firstName: BB, lastName: KING, song: CHAINS AND THINGS
o.s.b.c.l.s.SimpleJobLauncher - Job: [FlowJob: [name=singerJob]] completed with the
   following parameters: [{}] and the following status: [COMPLETED]
```

18.4 Spring Integration

Spring Integration 项目提供了众所周知的企业集成模式(Enterprise Integration Pattern，EIP)的开箱即用型实现。Spring Integration 侧重于消息驱动架构。它为集成解决方案、异步功能和松耦合组件提供了一个简单的模型，并且它是为可扩展性和可测试性而设计的。

在它的核心中，Message 包装器在框架中起着核心作用。Java 对象的这个通用包装器与框架使用的元数据(更具体地说,是有效载荷和标头)相结合，用于确定如何处理该对象。

Message 通道位于管道和过滤器架构中，生产者将消息发送到该通道并由消费者从该通道接收消息。另一方面，Message 端点表示管道和过滤器架构中的过滤器，它们将应用程序代码连接到消息传递框架。Spring Integration 提供的一些 Message 端点包括 Transformers、Filters、Routers 和 Splitters，每个端点都有自己的角色和职责。

Spring Integration 还提供大量的集成端点(在编写本书时大约有 20 个)，可以在 http://docs.spring.io/spring-integration/reference/htmlsingle/#endpoint-summary 上的“Endpoint Quick Reference”部分找到它。这些端点提供连接各种资源的能力，例如 AMQP、文件、HTTP、JMX、Syslog 和 Twitter。除了 Spring Integration 提供的开箱即用功能之外，另一个名为 Spring Integration Extensions 的项目是一个基于社区的贡献模型，位于 https://github.com/spring-projects/spring-batch-extensions，它包含更多集成的可能性，包括亚马逊 Web 服务(AWS)、Apache Kafka、短消息点对点(SMPP)和 Voldemort。在开箱即用组件和扩展项目组件之间，Spring Integration 提供大量现成的组件，这意味着大大降低自己编写组件的可能性。

下面的示例建立在前面的批处理示例基础之上，将演示如何使用 Spring Integration 在给定的时间间隔内监视目录。当文件到达时，检测到文件并启动批处理作业进行处理。

在本例中，将再次使用本章开头初始的“纯”Spring Batch 项目。在继续之前，请务必再看一下该项目并运行，因为本节仅介绍新类和配置修改。

显而易见，必须将一些新的依赖项添加到项目中。

```
//pro-spring-15/build.gradle
ext {
    //spring libs
    ...
    springBatchVersion = '4.0.0.M3'
    springIntegrationVersion = '5.0.0.M6'
    springBatchIntegrationVersion = '4.0.0.M3'
    ...
```

```
    spring = [
        context           : "org.springframework:spring-context:$springVersion",
        jdbc              : "org.springframework:spring-jdbc:$springVersion",
        batchCore         : "org.springframework.batch:spring-batch-core:$springBatchVersion"
        batchIntegration :
"org.springframework.batch:spring-batch-integration:$springBatchIntegrationVersion",
        integrationFile :
"org.springframework.integration:spring-integration-file:$springIntegrationVersion"
        ...
    ]

    misc = [
            io            : "commons-io:commons-io:2.5",
            ...
    ]

    db = [
            ...
            dbcp2         : "org.apache.commons:commons-dbcp2:$dbcpVersion",
            h2            : "com.h2database:h2:$h2Version",

            //needed for the Batch JSR-352 module
            hsqldb: "org.hsqldb:hsqldb:2.4.0"
    dbcp : "commons-dbcp:commons-dbcp:1.4",
    ]
}
...

//pro-spring-15/chapter18/build.gradle
dependencies {
      if (!project.name.contains("boot")) {
         compile(spring.jdbc) {
            //exclude these as batchCore will bring them
            //on as transitive dependencies
            exclude module: 'spring-core'
            exclude module: 'spring-beans'
            exclude module: 'spring-tx'
         }
      compile spring.batchCore, db.dbcp2, db.h2, misc.io,
         spring.batchIntegration, spring.integrationFile,
         misc.slf4jJcl, misc.logback
    }
}
```

添加新的依赖项之后，接下来创建一个充当 Spring Integration 转换器的类。该 Transformer 实例将从入站通道接收一个 Message 实例(该实例代表找到的文件)并启动批处理作业，如以下代码片段所示：

```
package com.apress.prospring5.ch18;

import org.springframework.batch.core.Job;
import org.springframework.batch.core.JobParametersBuilder;
import org.springframework.batch.integration.launch.JobLaunchRequest;
import org.springframework.messaging.Message;

import java.io.File;

public class MessageToJobLauncher {
   private Job job;
   private String fileNameKey;

   public MessageToJobLauncher(Job job, String fileNameKey) {
      this.job = job;
      this.fileNameKey = fileNameKey;
   }

   public JobLaunchRequest toRequest(Message<File> message) {
      JobParametersBuilder jobParametersBuilder = new JobParametersBuilder();
      jobParametersBuilder.addString(fileNameKey, message.getPayload().
         getAbsolutePath());
      return new JobLaunchRequest(job, jobParametersBuilder.
         toJobParameters());
   }
}
```

现在，修改 BatchConfig 类以支持批量集成。真正需要做的是修改 itemReader bean，在每次需要处理新文件时创建该 bean。这意味着必须注入路径位置到该 bean 中，也就是说，该 bean 不再具有单例作用域。

```
package com.apress.prospring5.ch18.config;
...
@Configuration
@EnableBatchProcessing
@Import(DataSourceConfig.class)
@ComponentScan("com.apress.prospring5.ch18")
public class BatchConfig {
    ... //autowired beans

    @Bean
    public Job singerJob() {
        return jobs.get("singerJob").start(step1()).build();
    }

    @Bean
    protected Step step1() {
        ...//no change
    }

    @Bean
    @StepScope

public FlatFileItemReader itemReader(
    @Value("file://#{jobParameters['file.name']}") String filePath) {
    FlatFileItemReader itemReader = new FlatFileItemReader();
    itemReader.setResource(resourceLoader.getResource(filePath));
    DefaultLineMapper lineMapper = new DefaultLineMapper();
    DelimitedLineTokenizer tokenizer = new DelimitedLineTokenizer();
    tokenizer.setNames("firstName", "lastName", "song");
    tokenizer.setDelimiter(",");
    lineMapper.setLineTokenizer(tokenizer);
    BeanWrapperFieldSetMapper<Singer> fieldSetMapper =
        new BeanWrapperFieldSetMapper<>();
        fieldSetMapper.setTargetType(Singer.class);
        lineMapper.setFieldSetMapper(fieldSetMapper);
        itemReader.setLineMapper(lineMapper);
        return itemReader;
    }

    @Bean
    public ItemWriter<Singer> itemWriter() {
        ...//no change
    }
}
```

@StepScope 是一个便捷的注解，等同于@Scope(value ="step",proxyMode =TARGET_CLASS)，可以让 itemReader bean 具有 proxy 作用域并将作用域命名为 step，从而可以清楚地知道该 bean 是如何使用的。

现在已经完成了批处理配置，接下来添加集成典型配置。我们将使用 XML 配置文件来完成此工作，因为它确实更实用，如下所示：

```
<?xml version="1.0" encoding="UTF-8"?>

<beans xmlns="http://www.springframework.org/schema/beans"
       xmlns:xsi="http://www.w3.org/2001/XMLSchema-instance"
       xmlns:batch-int="http://www.springframework.org/schema/batch-integration"
       xmlns:int="http://www.springframework.org/schema/integration"
       xmlns:int-file="http://www.springframework.org/schema/integration/file"
       xmlns:context="http://www.springframework.org/schema/context"
       xsi:schemaLocation="http://www.springframework.org/schema/batch-integration
http://www.springframework.org/schema/batch-integration/spring-batch-integration.xsd
       http://www.springframework.org/schema/beans
       http://www.springframework.org/schema/beans/spring-beans.xsd
       http://www.springframework.org/schema/integration
       http://www.springframework.org/schema/integration/spring-integration.xsd
       http://www.springframework.org/schema/integration/file
http://www.springframework.org/schema/integration/file/spring-integration-file.xsd
       http://www.springframework.org/schema/context
       http://www.springframework.org/schema/context/spring-context.xsd">

    <bean name="/BatchConfig" class="com.apress.prospring5.ch18.config.BatchConfig"/>

    <context:annotation-config/>

    <int:channel id="inbound"/>
    <int:channel id="outbound"/>
    <int:channel id="loggingChannel"/>
```

```
    <int-file:inbound-channel-adapter id="inboundFileChannelAdapter" channel="inbound"
          directory="file:/tmp/" filename-pattern="*.csv">
       <int:poller fixed-rate="1000"/>
    </int-file:inbound-channel-adapter>

    <int:transformer input-channel="inbound"
             output-channel="outbound">
       <bean class="com.apress.prospring5.ch18.MessageToJobLauncher">
          <constructor-arg ref="singerJob"/>
          <constructor-arg value="file.name"/>
       </bean>
    </int:transformer>

    <batch-int:job-launching-gateway request-channel="outbound"
       reply-channel="loggingChannel"/>

    <int:logging-channel-adapter channel="loggingChannel"/>
</beans>
```

在配置中添加的主要是以 int:和 batch-int:名称空间为前缀的部分。首先创建一些命名通道来传递数据。然后配置一个 inbound-channel-adapter，它专门用于在给定的一秒钟时间间隔内查看指定的目录。接下来配置 Transformer bean，它把接收的文件作为包装在 Message 中的标准 java.io.File 对象。再接下来配置 job-launching-gateway，它接收来自 Transformer 的 job-launch 请求，并实际调用批处理作业。最后但同样重要的是创建一个 logging-channel-adapter，它将在作业完成后打印信息通知。通过配置的 Channel 属性可以看到，每个组件都通过 Channel 实例使用或生成消息，又或者两者兼具。下面创建一个加载该配置文件的简单驱动程序类。所有驱动程序类都使用该配置加载应用程序上下文，并在连续轮询指定的目录时持续运行，直到终止进程。

```
package com.apress.prospring5.ch18;

import org.slf4j.Logger;
import org.slf4j.LoggerFactory;
import org.springframework.context.support.GenericXmlApplicationContext;

public class FileWatcherDemo {

   private static Logger logger =
      LoggerFactory.getLogger(FileWatcherDemo.class);

   public static void main(String... args) throws Exception {
      GenericXmlApplicationContext ctx
         = new GenericXmlApplicationContext(
               "classpath:spring/integration-config.xml");
      assert (ctx != null);
      logger.info("Application started...");
      System.in.read();
      ctx.close();
   }
}
```

现在编译代码并运行 FileWatcherDemo 类。当应用程序启动时，你可能会看到某些日志消息被打印到屏幕上，但并没有发生任何其他事情。这是因为 Spring Integration 文件适配器正在等待在轮询时间间隔内被放置在所配置位置的文件，并且只有在检测到该位置存在文件时才会发生事情。在 src/main/resources/support/目录中，你会发现四个 CSV 文件，名为 singer1.csv 到 singer4.csv。它们中的每一个都包含一行数据，其中包括歌手的姓名及其一首歌曲。请将这些文件逐个复制到 tmp 目录中，然后观察控制台，看看会发生什么。

```
o.s.i.f.FileReadingMessageSource - Created message: [GenericMessage [
   payload=/tmp/singers1.csv, headers={file_originalFile=/tmp/singers1.csv,
   id=ca0ec15e-b9b6-5dc2-2fc6-44fdb4b433f4, file_name=singers1.csv,
   file_relativePath=singers1.csv, timestamp=1501442201624}]]
o.s.b.c.l.s.SimpleJobLauncher - Job: [SimpleJob: [name=singerJob]] launched with
   the following parameters: [{file.name=/tmp/singers1.csv}]
o.s.b.c.j.SimpleStepHandler - Executing step: [step1]
c.a.p.c.SingerItemProcessor - Transformed singer: firstName: John, lastName: Mayer,
   song: Helpless Into: firstName: JOHN, lastName: MAYER, song: HELPLESS
INFO c.a.p.c.StepExecutionStatsListener - --> Wrote: 1 items in step: step1
o.s.b.c.l.s.SimpleJobLauncher - Job: [SimpleJob: [name=singerJob]] completed with
   the following parameters: [{file.name=/tmp/singers1.csv}] and the following
   status: [COMPLETED]
o.s.i.h.LoggingHandler - JobExecution: id=1, version=2,
   startTime=Sun Jul 30 22:16:41 EEST 2017, endTime=Sun Jul 30 22:16:41 EEST 2017,
   lastUpdated=Sun Jul 30 22:16:41 EEST 2017, status=COMPLETED,
```

```
exitStatus=exitCode=COMPLETED;exitDescription=, job=[JobInstance: id=1,
version=0, Job=[singerJob]], jobParameters=[{file.name=/tmp/singers1.csv}]
```

从日志语句中可以看到，Spring Integration 检测到文件并创建 Message，然后调用转换 CSV 文件内容的批处理作业，最后将内容写入内存数据库。虽然这是一个非常简单的示例，但它演示了如何使用 Spring Integration 在各种类型的应用程序之间构建复杂且解耦的工作流程。

有关 Spring Integration 的更多信息，请参阅 http://projects.spring.io/spring-integration/上的项目页面。

18.5 Spring XD

Spring XD 是一个可扩展的运行时服务，专为分布式数据采集、实时分析、批处理和数据导出而设计。Spring XD 基于许多现有的 Spring 组合项目，主要是 Spring Framework、Spring Batch 和 Spring Integration。Spring XD 的目标是提供一种统一的方式，用以将许多系统集成到一个统一的大数据解决方案中，从而帮助减少许多常见用例的复杂性。

Spring XD 可以以单一独立模式运行(通常用于开发和测试目的)，也可以以完全分布式模式运行，从而能够提供具有高可用性的主节点以及任意数量的工作节点。Spring XD 能够通过 shell 界面(利用 Spring shell)以及图形化 Web 界面来管理服务。通过这些界面，可以定义如何组装各种组件，以便通过 shell 应用程序使用 DSL 类型语法或将数据输入到 Web 界面来完成数据处理需求。

Spring XD DSL 基于几个概念，特别是流、模块、源、处理器、接收器、作业和 tap。通过简洁的语法组合这些组件，可以很容易创建流来连接各种技术，以提取数据、处理数据并最终将数据输出到外部源，甚至运行批处理作业以进行进一步处理。接下来快速了解一下这些概念：

- 流定义了数据如何从源流向接收器，并可能通过任意数量的处理器。DSL 用于定义流；例如，从源到接收器的基本定义可能看起来如下所示：http |file。
- 模块封装了由流组成的可重用工作单元。模块根据其角色按类型分类。在编写本书时，Spring XD 包含源、处理器、接收器和作业类型的模块。
- Spring XD 中的源轮询外部资源或由某种事件触发。源仅向下游组件提供输出，并且流中的第一个模块必须是源。
- 处理器在本质上类似于你在 Spring Batch 中看到的处理器。处理器的作用是接收输入，对提供的对象执行转换或业务逻辑，并返回输出。
- 在源的另一端，接收器接收输入源并将数据输出到目标资源。接收器是流中的最后一站。
- 作业是定义了 Spring Batch 作业的模块。这些作业的定义方式与本章开头描述的相同，并被部署到 Spring XD 中以提供批处理功能。
- 顾名思义，tap 可以对流中的数据进行监听，并允许在单独的数据流中处理窃听到的数据。tap 的概念与 WireTap 企业集成模式相似。

正如期望的那样，Spring XD 提供了许多开箱即用的源、处理器、接收器、作业和 tap。但作为一名开发人员，不应该仅限于这些内容，还应该可以自由构建自己的模块和组件。有关如何构建自己的模块和组件的更多详细信息，请参阅下面这些定制关注点的参考手册。

- **模块**：http://docs.spring.io/spring-xd/docs/current/reference/html/#_creating_a_module
- **源**：http://docs.spring.io/spring-xd/docs/current/reference/html/#creating-a-source-module
- **处理器**：http://docs.spring.io/spring-xd/docs/current/reference/html/#creating-a-processor-module
- **接收器**：http://docs.spring.io/spring-xd/docs/current/reference/html/#creating-a-sink-module
- **作业**：http://docs.spring.io/spring-xd/docs/current/reference/html/#creating-a-job-module

在本例中，将演示如何使用 Spring XD 的现成组件来复制 Spring Batch 和 Spring Integration 示例中创建的内容，所有操作都是通过使用了 XD shell 和 DSL 的简单命令行配置来完成的。

在开始之前，必须先安装 Spring XD。请参阅 http://docs.spring.io/spring-xd/docs/current/reference/html/#getting-started 上的用户手册中的“Getting Started”一节内容，其中详细介绍了在计算机上安装 Spring XD 的各种方法。选择哪种安装方法取决于个人偏好，不会对示例产生任何影响。安装完 Spring XD 之后，按照文档中所述，以单节点模式启动运行时。

为了在 Spring XD 中复制 Spring Batch 和 Spring Integration 示例中创建的内容，只需要执行一些基本任务。首先，需要创建一个要导入的 CSV 文件，如下所示，并将其放在/tmp/singers.csv 中：

```
John,Mayer,Helpless
Eric,Clapton,Change The World
John,Butler,Ocean
BB,King,Chains And Things
```

在 Spring XD shell 控制台中，输入以下命令：

```
job create singerjob --definition "filejdbc --resources=file:///tmp/singers.csv
   --names=firstname,lastname,song --tableName=singer
   --initializeDatabase=true" --deploy
```

在控制台中按 Enter 键之后，你应该看到一条消息，表明“已成功创建和部署作业 singerjob”。如果没有看到这条消息，请检查启动单节点 Spring XD 容器的终端中的控制台输出以获取更多详细信息。

至此已创建新的作业定义，但尚未发生任何事情，因为尚未启动。在 shell 中，输入以下命令以启动作业：

```
job launch singerjob
```

现在，shell 应该会响应一条消息，指出已成功提交 singerjob 的启动请求。

通过拆分提供的 DSL，Spring XD 知道我们想要创建一个批处理作业，该作业通过 job create 语句和 filejdbc 源，从文件读取数据并通过 JDBC 输出到数据库。此外，还通过使用 tableName 参数自动创建表，从 names 参数中获取列名，以及从 resources 参数(在该参数中提供 CSV 文件的文件路径)中读取数据。

如果要检查导入的数据，可以使用喜欢的数据库工具连接到安装时使用的数据库(嵌入的或真实的 RDBMS)，然后从 Singer 表中选择记录进行验证。如果看不到数据，请检查单节点容器正在其中运行的控制台中的日志语句。

此处在 shell 中键入了两条命令，但没有编写任何代码或复杂的配置。然而，我们已经以最小的努力将 CSV 文件的内容导入到了数据库中。具体是通过使用 Spring XD 的预构建批处理作业来完成这一任务，首先用简单的命令行 DSL 语法定义作业，然后从 shell 中启动作业。

如你所见，Spring XD 提供了许多开箱即用的功能，开发人员无须创建常见的用例场景。因为在之前的示例中完成了一些将人的名字和姓氏改为大写形式的转换，所以将这部分作为练习，以便可以进一步探索 Spring XD！

有关 Spring XD 的更多信息，请参阅 http://projects.spring.io/spring-xd/上的项目页面。

18.6 Spring 框架的五个最显著的功能

在编写本书时，Spring Framework RC3 版本已经发布。Spring 5 是对核心框架的一次重要修订，附带了一个用 Java 8 重写的代码库，同时随着 RC3 版本的发布，Spring 开始适应 Java 9。该版本包含如下几个主要功能：

- 附带了基于 Reactor 3.1 的 spring-webflux 模块，支持 RxJava 1.3 和 2.1，也称为 Reactive Web Framework。它是对 spring-webmvc 的反应式补充，提供了一个能在 Tomcat、Jetty 或 Undertow 上运行异步 API 的 Web 编程模型。
- Kotlin 支持是通过一个用于 bean 注册和功能性 Web 端点的完全无安全性的 API 提供的。
- 与 Java EE 8 API 的集成包括对 Servlet 4.0、Bean Validation 2.0、JPA 2.2 和 JSON Binding API(作为 Spring MVC 中 Jackson/Gson 的替代方案)的支持。
- 完全支持 JUnit 5 Jupiter，从而允许开发人员在 JUnit 5 中编写测试和扩展，并与 Spring TestContext Framework 并行执行它们。
- Java 9 的互操作性是 Spring 团队期望 Spring 5 具备的功能。但由于 Oracle 将 Java 9 的发布日期推迟了几个月，因此 Spring 框架最终用 Java 8 开发并发布，并使用 Project Reactor 来提供反应式编程支持。但是完全支持 Java 9 的承诺依然有效，并且开始用 RC3 来实现。
- 还有更多功能可用[1]。

以下部分仅介绍上述列表中的前三项功能。

1 为了跟踪 Spring 项目的发布和内容，建议查看官方的 Spring 博客 https://spring.io/blog。

18.6.1 功能性 Web 框架

如前所述，功能性 Web 框架(spring-webflux 模块)是对 spring-webmvc 的反应式补充，它提供了一个针对异步 API 设计的 Web 编程模型。该框架是根据反应式编程原则构建的。*反应式编程(reactive programming)*可以用最简单的方式解释为“用反应流编程”。流是反应式编程模型的核心，它们可用来支持任何异步处理。简而言之，反应式库可以将任何东西(变量、用户输入、缓存、数据结构等)作为流来使用，从而支持特定于流的操作。比如，对流进行过滤以创建另一个流、合并流、将值从一个流映射到另一个流等。反应式编程中的反应式部分意味着流将成为一个“可观察”的对象，由组件进行观察并根据发射的对象对其做出反应。流可以发射三种类型的对象：值、错误或“完成”信号。

要将普通应用程序转换为反应式应用程序，第一个逻辑步骤是修改组件以生成和处理数据流。共有两种类型的数据流。

- reactor.core.publisher.Flux[1]：这是一个[0..n]元素流。创建 Flux 的最简单方法如下：

```
Flux simple = Flux.just("1", "2", "3");
//or from an existing list: List<Singer> Flux<Singer>
fromList = Flux.fromIterable(list);
```

- reactor.core.publisher.Mono[2]：这是一个[0..1]元素流。创建 Mono 的最简单方法如下：

```
Mono simple = Mono.just("1");
//or from an existing object of type Singer
Mono<Singer> fromObject = Mono.justOrEmpty(singer);
```

这两个类都是 org.reactivestreams.Publisher <T>的实现，你可能会怀疑此时是否需要 Mono 实现。答案是肯定的，实际原因是：根据流的发射值的操作类型，知道基数总是有用的。例如，设想有一个反应式库类，那么 findOne()方法返回 Flux 是否合理？

以上简短介绍应该能够让你掌握反应式编程的基础知识，并且能够理解 spring-webflux 所带来的功能[3]。

在本节中，将使用第 16 章中介绍的使用了 Thymeleaf 的 Spring Boot Web 应用程序，并将其转换为使用 Spring WebFlux。第一步是将 spring-boot-starterwebflux 添加为依赖项，并删除不必要的依赖项。在下面的代码片段中，可以看到所需的库，这些库在父项目中配置，并在 webflow-boot 模块中使用：

```
//pro-spring-15/build.gradle
ext {
   //spring libs
   bootVersion = '2.0.0.M3'
   springDataVersion = '2.0.0.M3'
   junit5Version = '5.0.0-M4'
...
boot = [
...
      springBootPlugin:
         "org.springframework.boot:spring-boot-gradle-plugin:$bootVersion",
      starterWeb :
         "org.springframework.boot:spring-boot-starter-web:$bootVersion",
      starterTest :
         "org.springframework.boot:spring-boot-starter-test:$bootVersion",
      starterJpa :
         "org.springframework.boot:spring-boot-starter-data-jpa:$bootVersion",
      starterJta :
         "org.springframework.boot:spring-boot-starter-jta-atomikos:$bootVersion",
      starterWebFlux :
         "org.springframework.boot:spring-boot-starter-webflux:$bootVersion"
   ]
   testing = [
   ...
      junit5 : "org.junit.jupiter:junit-jupiter-engine:$junit5Version"
   ]

   db = [
   ...
      h2 : "com.h2database:h2:$h2Version"
```

1 可以在 https://projectreactor.io/docs/core/release/reference/#flux 上找到详细的解释。

2 可以在 https://projectreactor.io/docs/core/release/reference/#mono 上找到详细的解释。

3 如果需要了解关于反应式编程模型和反应流编程的更多资源，可以参考 https://projectreactor.io/docs 上的参考文档、代码示例以及 Project Reactor(https://projectreactor.io/docs)的 Javadoc，或者参考 https://gist.github.com/staltz/868e7e9bc2a7b8c1f754 上关于反应式编程的详细介绍。

```
    ]
}

//chapter18/webflux-module/build.gradle
buildscript {
   repositories {
      ...
   }

   dependencies {
      classpath boot.springBootPlugin
   }
}

apply plugin: 'org.springframework.boot'

dependencies {
   compile boot.starterWebFlux, boot.starterWeb,
      boot.starterJpa, boot.starterJta, db.h2
   testCompile boot.starterTest, testing.junitJupiter, testing.junit5
}
```

spring-boot-starter-webflux 模块依赖于一些将被自动添加到应用程序中的反应式库，如 IntelliJ IDEA 中的 Gradle 视图所示，图 18-1 显示了应用程序的依赖项。请注意 reactive-streams 库。该库包含反应流基本接口、标准规范和四个接口：Publisher、Subscriber、Subscription 和 Processor。可以在 www.reactive-streams.org 上了解更多关于该库的内容，因为流 API 并不是本节的重点。本节在应用程序中使用的流实现是由 reactor-core 库提供的，可以在 https://projectreactor.io/上了解有关该库的更多信息。

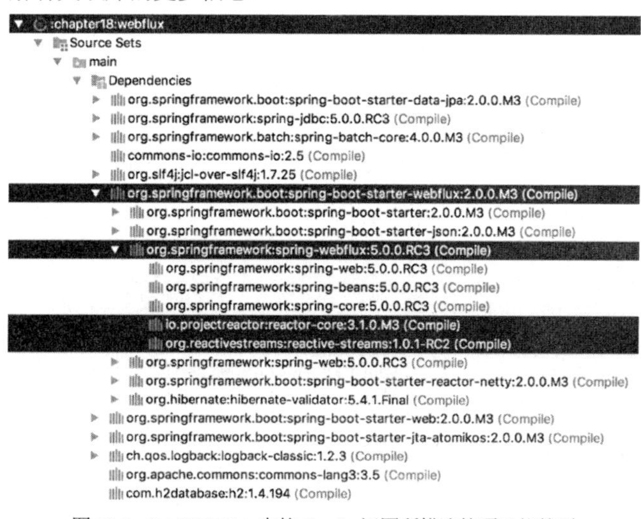

图 18-1　IntelliJ IDEA 中的 Gradle 视图所描述的项目依赖项

为了简单起见，安全层将被删除，界面也被删除了。我们将使用 REST 以及一组测试类来测试应用程序。此外，为了简化 Singer 类，所有的验证器注解也被删除了，因为它们并不是本节的重点。

```
package com.apress.prospring5.ch18.entities;

import javax.persistence.*;
import java.io.Serializable;
import java.text.SimpleDateFormat;
import java.util.Date;

import static javax.persistence.GenerationType.IDENTITY;
@Entity
@Table(name = "singer")

   public class Singer implements Serializable {

      @Id
      @GeneratedValue(strategy = IDENTITY)
      @Column(name = "ID")
      private Long id;
      @Version
```

```
        @Column(name = "VERSION")
        private int version;
        @Column(name = "FIRST_NAME")
        private String firstName;
        @Column(name = "LAST_NAME")
        private String lastName;
        @Temporal(TemporalType.DATE)
        @Column(name = "BIRTH_DATE")
        private Date birthDate;
           public Singer() {
    }

    public Singer(String firstName,
        String lastName, Date birthDate) {
        this.firstName = firstName;
        this.lastName = lastName;
        this.birthDate = birthDate;
    }
    //setters and getters
    ...
}
```

存储库实现是 CrudRepository<Singer,Long>接口的一个空白扩展，在此不再描述。此处的新奇之处在于，该存储库将由反应式存储库的实现使用。下面的代码显示了 ReactiveSingerRepo 接口的内容，其中使用 Mono 和 Flux 来处理 Singer 对象：

```
package com.apress.prospring5.ch18.repos;

import com.apress.prospring5.ch18.entities.Singer;
import reactor.core.publisher.Flux;
import reactor.core.publisher.Mono;

public interface ReactiveSingerRepo {

    Mono<Singer> findById(Long id);

    Flux<Singer> findAll();

    Mono<Void> save(Mono<Singer> singer);
}
```

该接口的实现如下所示：

```
package com.apress.prospring5.ch18.repos;

import com.apress.prospring5.ch18.entities.Singer;
import org.springframework.beans.factory.annotation.Autowired;
import org.springframework.stereotype.Service;
import reactor.core.publisher.Flux;
import reactor.core.publisher.Mono;

@Service
public class ReactiveSingerRepoImpl implements ReactiveSingerRepo {

    @Autowired
    SingerRepository singerRepository;

    @Override public Mono<Singer> findById(Long id) {
        return Mono.justOrEmpty(singerRepository.findById(id));
    }

    @Override public Flux<Singer> findAll() {
        return Flux.fromIterable(singerRepository.findAll());
    }

    @Override public Mono<Void> save(Mono<Singer> singerMono) {
        return singerMono.doOnNext(singer ->
            singerRepository.save(singer)
        ).thenEmpty((Mono.empty()));
    }
}
```

接下来分别讨论该类中的每个方法：

- findById()方法返回一个简单的流，如果对 singerRepository.findById(id)的调用返回 1，则发射一个 Singer 对象；否则，会发出 onComplete 信号。这意味着返回的将是一个不包含任何内容的 Mono<Singer>对象。

- findAll()方法返回一个包含 textitsingerRepository.findAll()方法返回的所有 Singer 对象的流。
- save() 方法接收一个包含 Singer 实例的 Mono<Singer>实例类型的参数。声明 java.util.function.Consumer<Singer>的一个实例，以便当 Singer 对象被成功发射时，由 singerRepository 将其保存到数据库中。该方法返回一个空的 Mono 实例，因为没有任何东西可以返回。

现在已经有了反应式存储库，还需要一个反应式处理程序。因为典型的 Spring @Controller 只会使用流并返回视图中的内容，所以需要使用不同的类。该类被称为 SingerHandler，如下所示：

```
package com.apress.prospring5.ch18.web;

import com.apress.prospring5.ch18.entities.Singer;
import com.apress.prospring5.ch18.repos.ReactiveSingerRepo;
import org.springframework.beans.factory.annotation.Autowired;
import org.springframework.stereotype.Component;
import org.springframework.web.reactive.function.server.ServerRequest;
import org.springframework.web.reactive.function.server.ServerResponse;
import reactor.core.publisher.Flux;
import reactor.core.publisher.Mono;

import static org.springframework.http.MediaType.APPLICATION_JSON;
import static org.springframework.web.reactive.function.BodyInserters.fromObject;

@Component
public class SingerHandler {

    @Autowired ReactiveSingerRepo reactiveSingerRepo;

    public Mono<ServerResponse> list(ServerRequest request) {
        Flux<Singer> singers = reactiveSingerRepo.findAll();
        return ServerResponse.ok().contentType(APPLICATION_JSON)
            .body(singers, Singer.class);
    }

    public Mono<ServerResponse> show(ServerRequest request) {
        Mono<Singer> singerMono = reactiveSingerRepo.findById(Long
            .valueOf(request.pathVariable("id")));
        Mono<ServerResponse> notFound = ServerResponse.notFound().build();
        return singerMono
                .flatMap(singer -> ServerResponse.ok().contentType(APPLICATION_JSON)
                .body(fromObject(singer)))
                .switchIfEmpty(notFound);
    }

    public Mono<ServerResponse> save(ServerRequest request) {
        Mono<Singer> data = request.bodyToMono(Singer.class);
        reactiveSingerRepo.save(data);
        return ServerResponse.ok().build(reactiveSingerRepo.save(data));
    }
}
```

请求由一个处理函数处理，该函数将 ServerRequest 作为参数并返回一个 Mono<ServerResponse>。这两个接口都是不可更改的，并提供对底层 HTTP 消息的访问。两者都是完全反应式的；ServerRequest 将主体公开为 Flux 或 Mono，而 Mono<ServerResponse>接收任何反应式流作为主体。

ServerRequest 还提供对其他 HTTP 相关数据的访问，比如正在处理的 URI、标题和路径变量。bodyToMono()或等效的 bodyToFlux()方法提供对主体的访问。ServerResponse 提供对 HTTP 响应的访问。由于是不可更改的，因此必须使用生成器类和 HTTP 响应状态来创建。标题和主体则通过调用各种方法来设置。代码片段中的所有示例都会创建具有 200(OK)状态、JSON 内容类型和主体的响应。

但功能性 Web 框架从何而来呢？lambda 是这一切的源头。如果看一下前面的代码片段，就会注意到可以很容易地将 list 和 crete 编写为函数。下面显示了两个 lambda 函数：

```
package com.apress.prospring5.ch18.web;

import com.apress.prospring5.ch18.entities.Singer;
import com.apress.prospring5.ch18.repos.ReactiveSingerRepo;
import org.springframework.beans.factory.annotation.Autowired;
import org.springframework.stereotype.Component;
import org.springframework.web.reactive.function.server.HandlerFunction;
import org.springframework.web.reactive.function.server.ServerRequest;
import org.springframework.web.reactive.function.server.ServerResponse;
import reactor.core.publisher.Flux;
```

```
import reactor.core.publisher.Mono;

import static org.springframework.http.MediaType.APPLICATION_JSON;
import static org.springframework.web.reactive.function.BodyInserters.fromObject;

@Component
public class SingerHandler {

   @Autowired ReactiveSingerRepo reactiveSingerRepo;

   public HandlerFunction<ServerResponse> list =
          serverRequest -> ServerResponse.ok().contentType(APPLICATION_JSON)
          .body(reactiveSingerRepo.findAll(), Singer.class);

   public Mono<ServerResponse> show(ServerRequest request) {
      Mono<Singer> singerMono = reactiveSingerRepo.findById(
              Long.valueOf(request.pathVariable("id")));
   Mono<ServerResponse> notFound = ServerResponse.notFound().build();
   return singerMono
            .flatMap(singer -> ServerResponse.ok()
            .contentType(APPLICATION_JSON).body(fromObject(singer)))
            .switchIfEmpty(notFound);
   }

   public HandlerFunction<ServerResponse> save =
      serverRequest -> ServerResponse.ok()

      .build(reactiveSingerRepo.save(serverRequest.bodyToMono(Singer.class)));
}
```

HandlerFunction<ServerResponse>接口本质上是一个 Function<Request, Response<T>>，它没有任何副作用，因为它直接返回响应，而不是将其作为参数。这使得这种类型的函数非常实用，因为它们可以更容易测试、编写和优化。

代码非常简洁，对吧？但是请记住，当谈到 lambda 时，过度地使用它们会付出降低可读性和可维护性的代价。另外，你现在可能在想：*好的，反应式流和 lambda 表达式很酷，但映射在哪里呢*？容器应该如何知道哪个函数被映射到 HTTP 请求？为了解答这些问题，需要一些新的类。

下面所示的是一个 Spring 应用程序类，必须运行该类以启动应用程序，稍后对其进行仔细分析：

```
package com.apress.prospring5.ch18;

import com.apress.prospring5.ch18.web.SingerHandler;
import org.slf4j.Logger;
import org.slf4j.LoggerFactory;
import org.springframework.beans.factory.annotation.Autowired;
import org.springframework.boot.SpringApplication;
import org.springframework.boot.autoconfigure.SpringBootApplication;
import org.springframework.boot.web.servlet.ServletRegistrationBean;
import org.springframework.context.ConfigurableApplicationContext;
import org.springframework.context.annotation.Bean;

import org.springframework.http.server.reactive.HttpHandler;
import org.springframework.http.server.reactive.ServletHttpHandlerAdapter;

import org.springframework.web.reactive.function.server.RouterFunction;

import org.springframework.web.reactive.function.server.ServerResponse;
import org.springframework.web.server.WebHandler;
import org.springframework.web.server.adapter.WebHttpHandlerBuilder;

import static org.springframework.web.reactive.function.BodyInserters.fromObject;
import static org.springframework.web.reactive.function.server.RequestPredicates.GET;
import static org.springframework.web.reactive.function.server.RequestPredicates.POST;
import static org.springframework.web.reactive.function.server.RouterFunctions.route;
import static org.springframework.web.reactive.function.server.RouterFunctions.
toHttpHandler;
import static org.springframework.web.reactive.function.server.ServerResponse.ok;

@SpringBootApplication
public class SingerApplication {

   private static Logger logger = LoggerFactory.getLogger(SingerApplication.class);

   @Autowired
   SingerHandler singerHandler;

   private RouterFunction<ServerResponse> routingFunction() {
```

```
        return route(GET("/test"), serverRequest -> ok().body(fromObject("works!")))
            .andRoute(GET("/singers"), singerHandler.list)
            .andRoute(GET("/singers/{id}"), singerHandler::show)
            .andRoute(POST("/singers"), singerHandler.save)
            .filter((request, next) -> {
                logger.info("Before handler invocation: " + request.path());
                return next.handle(request);
            });
    }

    @Bean
    public ServletRegistrationBean servletRegistrationBean() throws Exception {
        HttpHandler httpHandler = RouterFunctions.toHttpHandler(routingFunction());
        ServletRegistrationBean registrationBean = new ServletRegistrationBean<>
                (new ServletHttpHandlerAdapter(httpHandler), "/");
        registrationBean.setLoadOnStartup(1);
        registrationBean.setAsyncSupported(true);
        return registrationBean;
    }

    public static void main(String... args) throws Exception {
        ConfigurableApplicationContext ctx =
                SpringApplication.run(SingerApplication.class, args);
        assert (ctx != null);
        logger.info("Application started...");

        System.in.read();
        ctx.close();
    }
}
```

新的功能性 Web 框架通过使用 RouterFunction 公开了请求处理函数，而 RouterFunctions.toHttpHandler 实用方法使用函数创建了一个 HttpHandler。该 HttpHandler 用于创建一个 ServletRegistrationBean，它是一个用于在 Servlet 3.0+容器中注册 Servlet 的 ServletContextInitializer。

RouterFunction 对请求 URI 进行评估并检查是否存在匹配的处理函数；如果不存在，则返回一个空的结果。它的行为类似于@RequestMapping 注解，但是 RouterFunction 的优势在于表示路由不仅限于使用注解值可以定义的内容，并且不会分散在类中。所有代码都在一个地方，可以非常轻松地重写或替换。编写 RouterFunction 的语法非常灵活；在前面的代码示例中，采用的是最实用的语法[1]。

对应用程序进行初始测试的最简单方法是运行 SingerApplication 并在浏览器中访问 RouterFunction 中定义的 URI。例如，如果访问 http://localhost:8080/test，则应该看到文本“works!”。

到目前为止，还没有提供流的示例，每个值都在发射时被处理。接下来创建一个 REST 控制器，该控制器直接返回歌手的 Flux 对象，并每隔两秒发射一次。

```
package com.apress.prospring5.ch18;

import com.apress.prospring5.ch18.entities.Singer;
import com.apress.prospring5.ch18.repos.ReactiveSingerRepo;
import org.slf4j.Logger;
import org.slf4j.LoggerFactory;
import org.springframework.beans.factory.annotation.Autowired;
import org.springframework.boot.SpringApplication;
import org.springframework.boot.autoconfigure.SpringBootApplication;
import org.springframework.context.ConfigurableApplicationContext;
import org.springframework.http.MediaType;
import org.springframework.web.bind.annotation.GetMapping;
import org.springframework.web.bind.annotation.RestController;
import reactor.core.publisher.Flux;
import reactor.util.function.Tuple2;

import java.time.Duration;

@SpringBootApplication
@RestController
public class ReactiveApplication {

    private static Logger logger = LoggerFactory.getLogger(ReactiveApplication.class);
    @Autowired ReactiveSingerRepo reactiveSingerRepo;
```

1 如果想深入了解相关内容，官方的 Spring 博客上有一篇关于它的博客文章：https://spring.io/blog/2016/09/22/new-in-spring-5-functional-web-framework。

```
    @GetMapping(value = "/all", produces = MediaType.TEXT_EVENT_STREAM_VALUE)
    public Flux<Singer> oneByOne() {
        Flux<Singer> singers = reactiveSingerRepo.findAll();
        Flux<Long> periodFlux = Flux.interval(Duration.ofSeconds(2));

        return Flux.zip(singers, periodFlux).map(Tuple2::getT1);
    }

    @GetMapping(value = "/one/{id}")
    public Mono<Singer> one(@PathVariable Long id) {
        return reactiveSingerRepo.findById(id);
    }

    public static void main(String... args) throws Exception {
        ConfigurableApplicationContext ctx =
            SpringApplication.run(ReactiveApplication.class, args);
        assert (ctx != null);
        logger.info("Application started...");
        System.in.read();
        ctx.close();
    }
}
```

MediaType.TEXT_EVENT_STREAM_VALUE 表示特殊类型的纯文本响应。这样可以确保服务器使用 text/event-stream Content-Type 创建一个响应，并遵循服务器发送的事件格式。响应应该包含“data:”行，后面跟着消息。在本例中，应该为 Singer 对象的文本表示，最后用两个“\ n”字符结束流。该格式仅适用于一条消息，意味着正在发射一个对象。如果发射多个对象，则生成多个“data:”行。以“data:”开头的两个或更多个连续行将被视为单个数据，意味着只有一个消息事件被触发。每行应该以一个“\ n”结尾(最后一行以两个“\n”结尾)。

oneByOne()方法组合了两个流。一个流是由反应式存储库返回的 Flux 实现，包含歌手实例。另一个流是一个包含间隔秒数(由 java.time.Duration 类生成)的流。通过使用 zip()方法将两个流组合在一起，这样就可以等待所有流发射一个元素，并将这些元素组合成一个输出值(由提供的组合器生成)。该操作将一直进行下去，直到任何源完成为止。此时，组合器就是 map(Tuple2 :: getT1)语句，这里表示为方法引用；也可以写成 map(t -> t.getT1())，这样就更容易地知道流中的一个元素被映射到另一个元素。

接下来测试一下两个新的处理函数。可以使用两种方法进行测试：使用浏览器以及使用 curl 命令。在浏览器中，如果尝试访问 localhost:8080/，将会获得一个空白页面，浏览器会询问是否将其全部保存到文件中。如果回答是，将会显示一个弹出窗口，显示正在下载的文件。当文件下载完成时，文件的内容将在所选的编辑器中显示。如果打开文件，你会看到其中的内容。如果显示了不可见字符，你应该能够看到行结束符“\ n”。图 18-2 显示了已下载文件的内容。但是，除了缓慢的下载时间之外，我们还没有清楚地看到每两秒钟显示一位歌手。此时，只能使用基于 UNIX 的系统中可用的 curl 命令。对于 Windows，可以尝试 PowerShell 中的 Invoke-RestMethod。如果使用上面示例中映射的 URI 调用 curl，则会看到如下所示的内容。curl http://localhost:8080/all 命令的输出需要一段时间才能显示出来，并且每两秒钟打印一行。

```
$ curl http://localhost:8080/one/1
{"id":1,"version":0,"firstName":"John","lastName":"Mayer","birthDate":"1977-10-15"}
$ curl http://localhost:8080/all
data:{"id":1,"version":0,"firstName":"John","lastName":"Mayer","birthDate":"1977-10-15"}

data:{"id":2,"version":0,"firstName":"Eric","lastName":"Clapton","birthDate":"1945-03-29"}

data:{"id":3,"version":0,"firstName":"John","lastName":"Butler","birthDate":"1975-03-31"}

data:{"id":4,"version":0,"firstName":"B.B.","lastName":"King","birthDate":"1925-10-15"}

data:{"id":5,"version":0,"firstName":"Jimi","lastName":"Hendrix","birthDate":"1942-12-26"}

data:{"id":6,"version":0,"firstName":"Jimmy","lastName":"Page","birthDate":"1944-02-08"}

data:{"id":7,"version":0,"firstName":"Eddie","lastName":"Van Halen","birthDate":"1955-02-25"}

data:{"id":8,"version":0,"firstName":"Saul Slash","lastName":"Hudson","birthDate":"1965-08-22"}

data:{"id":9,"version":0,"firstName":"Stevie","lastName":"Ray Vaughan","birthDate":"1954-11-02"}

data:{"id":10,"version":0,"firstName":"David","lastName":"Gilmour","birthDate":"1946-04-05"}

data:{"id":11,"version":0,"firstName":"Kirk","lastName":"Hammett","birthDate":"1992-12-17"}
```

```
data:{"id":12,"version":0,"firstName":"Angus","lastName":"Young","birthDate":"1955-04-30"}

data:{"id":13,"version":0,"firstName":"Dimebag","lastName":"Darrell","birthDate":"1966-09-19"}

data:{"id":14,"version":0,"firstName":"Carlos","lastName":"Santana","birthDate":"1947-08-19"}

$
```

```
all-1 — Downloads
data:{"id":1,"version":0,"firstName":"John","lastName":"Mayer","birthDate":"1977-10-15"}

data:{"id":2,"version":0,"firstName":"Eric","lastName":"Clapton","birthDate":"1945-03-29"}

data:{"id":3,"version":0,"firstName":"John","lastName":"Butler","birthDate":"1975-03-31"}

data:{"id":4,"version":0,"firstName":"B.B.","lastName":"King","birthDate":"1925-10-15"}

data:{"id":5,"version":0,"firstName":"Jimi","lastName":"Hendrix","birthDate":"1942-12-26"}

data:{"id":6,"version":0,"firstName":"Jimmy","lastName":"Page","birthDate":"1944-02-08"}

data:{"id":7,"version":0,"firstName":"Eddie","lastName":"Van Halen","birthDate":"1955-02-25"}

data:{"id":8,"version":0,"firstName":"Saul (Slash)","lastName":"Hudson","birthDate":"1965-08-22"}

data:{"id":9,"version":0,"firstName":"Stevie","lastName":"Ray Vaughan","birthDate":"1954-11-02"}

data:{"id":10,"version":0,"firstName":"David","lastName":"Gilmour","birthDate":"1946-04-05"}

data:{"id":11,"version":0,"firstName":"Kirk","lastName":"Hammett","birthDate":"1992-12-17"}

data:{"id":12,"version":0,"firstName":"Angus","lastName":"Young","birthDate":"1955-04-30"}

data:{"id":13,"version":0,"firstName":"Dimebag","lastName":"Darrell","birthDate":"1966-09-19"}

data:{"id":14,"version":0,"firstName":"Carlos","lastName":"Santana","birthDate":"1947-08-19"}
```

图 18-2　使用 Content-Type text/event-stream 的响应

对于 http://localhost:8080/one/1 URI，数据会自动转换为文本，因为这是默认的接收标头，仅返回一个条目。而对于 http://localhost:8080/all，则会将请求挂起，并且每条消息在到达时都会转换为文本。请注意语法的不同；对于具有多条消息的流，使用“data:”前缀来发送每个元素。

访问针对 http://localhost:8080/all 的 HTTP 请求所返回的流的另一种方法是使用反应式 Web 客户端。Spring 为此提供了一个接口。接下来演示如何使用该接口。因为需要共享上下文，所以在同一个 Spring Boot Application 类中声明客户端，并且如果查看一下控制台，你会看到客户端在启动应用程序后立即运行。

```
package com.apress.prospring5.ch18;
...
@SpringBootApplication
@RestController
public class ReactiveApplication {
...

    @Bean WebClient client() {
        return WebClient.create("http://localhost:8080");
    }

    @Bean CommandLineRunner clr(WebClient client) {
        return args -> {
            client.get().uri("/all")
                    .accept(MediaType.TEXT_EVENT_STREAM)
                    .exchange()
                    .flatMapMany(cr -> cr.bodyToFlux(Singer.class))
                    .subscribe(System.out::println);
        };
    }
}
```

WebClient bean 是 REST 模板的一种反应式、非阻塞的替代方案，在未来的 Spring 版本中，它将取代 AsyncRestTemplate。在 Spring 5 中，它已被标记为弃用。

Spring WebFlux 中有许多新组件可用于编写反应式 Web 应用程序；如果在此详细介绍这些组件，恐怕本书的篇幅会翻一倍。只需要参照 Spring 的官方博客 https://spring.io/blog 上的指导，即可快速更新所有引入的新组件并了解它们的使用方法。接下来介绍另一种选择。

18.6.2 Java 9 互操作性

由于 Java 9 的发布日期不断推迟，Spring 5 必须基于 Java 8 发布，但开发团队仍继续努力，通过使用 JDK 的早期版本以适应 Java 9。在编写本书时，Java 9 的大部分新组件都已稳定下来，并且已于 2017 年 9 月发布。因此，本书将根据早期可访问的版本介绍与 Java 9 的互操作性。下面进入正题。

18.6.3 JDK 模块化

JDK 模块化被认为是 Java 9 中最大的改进之处。JDK 模块化具有可扩展性的好处，因为现在 Java 可以部署在较小的设备上。模块化功能被称为 Project Jigsaw，原本计划成为 Java 8 的一部分，但直到 Java 9 才实现，因为它被认为不够稳定。因此，Java 9 引入了模块概念，模块是一个软件单元，在名为 module-info.java 的文件中配置，该文件位于项目的源代码目录中，并包含以下信息。

- **应遵循包命名约定的模块名称**：按照约定，模块应与根包的名称相同。
- **一组导出的包**：这些包被视为公共 API 并可供其他模块使用。如果某个类位于导出的包中，就可以在模块之外访问和使用该类。
- **所需的一组包**：这些是一个模块所依赖的其他模块。在这些模块导出的包中，所有公共类型都可以访问，并且可以由相关模块使用。

上述信息意味着可访问性不再由四个 Java 分类器表示：public、private、default 和 protected。由于一个模块的配置决定了另一个模块可以从它的依赖模块中使用什么内容，因此应该完成以下内容：

- 在模块级别公开，对读取此模块的所有模块公开(exports)。
- 在模块级别公开，对所选择的模块列表(即过滤性访问，export to)公开。
- 在模块中公开，并对模块内部的其他任何类公开。
- 模块内部私有，典型的私有访问。
- 模块内部的<default>，典型的默认访问。
- 模块内部受保护，典型的受保护访问。

将模型应用于 JDK，显而易见，很多周期性和不直观的依赖项被删除。此外还涉及一些清理操作，因为有些包超出 JSE 的范围[1]。

该主题对于 Spring 互操作性非常重要，因为 JDK 的组件可能无法直接访问。直到目前，本书中并没有真正涉及 Java 9，因为项目是使用 Java 8 构建的，Java 9 不够稳定，无法使用。但现在时机已成熟。因此，在更改 JDK 后，必须将 Gradle 安装升级到最新的 4.2 里程碑版本，以支持 Java 9。之后，项目的第一个版本将失败，并显示以下消息：

```
$ gradle clean build -x test
> Task :chapter03:collections:compileJava
/workspace/pro-spring-15/chapter03/collections/src/main/java/com/apress/prospring5/ch3/
annotated/CollectionInjection.java:13: error: package javax.annotation is not visible
import javax.annotation.Resource;
            ^
    (package javax.annotation is declared in module java.xml.ws.annotation,
    which is not in the module graph)
Note: Some input files use unchecked or unsafe operations.
Note: Recompile with -Xlint:unchecked for details.
1 error

FAILURE: Build failed with an exception.
```

此处发生的情况是，javax.annotation 包不是其中一个导出对象，因此其中的任何注解都不能再使用。该包以及许多其他相关的包不再可访问的原因是，它们包含了更适合成为 JEE 一部分的企业级组件。这些包是名为 java.se.ee 的模块的一部分，根据之前提到的内容，需要将以下模块添加到文件中，此时所有内容都应该可以正常工作：

```
module collections.main {
    requires java.se.ee;
}
```

1 请参阅 https://blog.codecentric.de/en/2015/11/first-steps-with-java9-jigsaw-part-1/上的内容。

但事实并非如此，因为如果打开 java.se.ee module-info.java，就可以找到以下内容：

```
@java.lang.Deprecated(since = "9", forRemoval = true)
module java.se.ee {
   requires transitive java.xml.bind;
   requires transitive java.activation;
   requires transitive java.corba;
   requires transitive java.se;
   requires transitive java.transaction;
   requires transitive java.xml.ws;
   requires transitive java.xml.ws.annotation;
}
```

如你所见，上述代码中并没有包含 exports java.xml.ws.annotation;(这是包含 javax.annotation 包的模块)这行代码。那么应该如何解决呢？当然，可以添加包含该包的 JEE 依赖项。可以在 Gradle 配置文件中进行如下定义：

```
misc = [
   ...
   jsr250       : "javax:javaee-endorsed-api:7.0"
   ...
]
```

当然，必须将 misc.jsr250 添加到使用该包的所有注解中。下面再试一次，你可能会看到以下类似内容：

```
$ gradle clean build -x test
# all good until here
...
:chapter05:aspectj-aspects:compileAspect
[ant:iajc] java.nio.file.NoSuchFileException:
   /Library/Java/JavaVirtualMachines/jdk-9.jdk/Contents/Home/jrt-fs.jar
[ant:iajc] at java.base/sun.nio.fs.UnixException.translateToIOExceptionUnixException.java:92
[ant:iajc] at java.base/sun.nio.fs.UnixException.rethrowAsIOExceptionUnixException.java:111
...
[ant:iajc] /workspace/pro-spring-15/chapter05/aspectj-aspects/src/main/java/com/apress/
   prospring5/ch5/MessageWriter.java:5 [error] Implicit super constructor Object is undefined.
   Must explicitly invoke another constructor
[ant:iajc] public MessageWriter {
[ant:iajc] ^^^^^^^^^^
[ant:iajc] /workspace/pro-spring-15/chapter05/aspectj-aspects/src/main/java/com/apress/
prospring5/ch5/MessageWriter.java:9 [error] System cannot be resolved
[ant:iajc] System.out.println"foobar!";
[ant:iajc]
[ant:iajc] /workspace/pro-spring-15/chapter05/aspectj-aspects/src/main/java/com/apress/
prospring5/ch5/MessageWriter.java:13 [error] System cannot be resolved
[ant:iajc] System.out.println"foo";
[ant:iajc]
[ant:iajc]
[ant:iajc] 11 errors
:chapter05:aspectj-aspects:compileAspect FAILED
276 actionable tasks: 186 executed, 90 up-to-date
```

此处发生的事情是，aspectj-aspects 不能被编译，因为切面无法被识别。这是因为根据 Gradle aspects 插件，没有编译所需的 JAR。该插件显然已过时，并且与新的 JDK 内部结构不匹配，因此需要进行一处小的修改：需要将 jrt-fs.jar 文件从$JAVA_HOME/libs 目录复制到插件所查找的位置。

当再次运行时，虽然会产生大量的弃用警告信息，但至少生成成功。现在可以开始为所有模块添加 module-info.java 文件。接下来看看 chapter02/hello-world 项目，并定义 module-info.java，因为它是一个小项目，所以应该很简单。

```
//pro-spring-15/chapter02/hello-world/src/main/java/module-info.java
module com.apress.prospring5.ch2 {
   requires spring.context;
   requires logback.classic;
   exports com.apress.prospring5.ch2;
}
```

就是这样。为了简化模块名称，Spring 团队决定打破惯例；必须添加 requires org.springframework.context，而不是 requires spring.context。[1]

Project Jigsaw 中添加的模块化功能不仅仅是分割 JDK，而且还限制对某些包和组件的访问(反射对非导出模块无效)；该功能在编译时检测循环依赖项，提高可读性，检测仅版本不同的重复模块依赖项，以及消除应用程序的冲突或混乱行为。由于以上这些开发优势，本书专门用一节来介绍 Java 9 的这一新功能。

1 请查看一下导致这一决定的相关讨论：https://jira.spring.io/browse/SPR-14579。

18.6.4 使用 Java 9 和 Spring WebFlux 进行反应式编程

在前几节，介绍了反应式模型以及标准 API。在本节中，将不使用 reactive-streams.jar，因为新版本中引入了新的模块。接下来介绍 java.base 模块，从该模块导出了包 java.util.concurrent，除此之外，它还包含其他四个功能接口，它们在 java.util.concurrent.Flow 最终类中定义相同的用途。

- Flow.Processor，相当于 org.reactivestreams.Processor<T>
- Flow.Publisher，相当于 org.reactivestreams.Publisher<T>
- Flow.Subscriber，相当于 org.reactivestreams.Subscriber<T>
- Flow.Subscription，相当于 org.reactivestreams.Subscription<T>

这些接口具有与 Reactive Streams 项目定义的相同的功能，用于支持创建发布-订阅应用程序，即反应式应用程序。JDK 9 提供了一个简单的 Publisher 实现，可用于简单的用例，也可以根据需要进行扩展。SubmissionPublisher<T> 是 Flow.Publisher<>接口的实现类，它同时也实现了 AutoCloseable，可以在 try-with-resources 模块中使用。

RxJava[1]是为支持数据/事件序列而设计的 JVM 的反应式实现之一，并添加了允许流组合和过滤的运算符，这些运算符通过隐藏低级别关注点来实现线程同步和线程安全。在编写本书时，RxJava 有两个版本：RxJava 1.x 和 RxJava 2.x。RxJava 1.x 是 ReactiveX(Reactive Extensions)[2]的一个实现，它是一个用可观察流进行异步编程的 API。RxJava 1.x 将被弃用，因为它在上一节介绍的 Reactive Streams API 的基础上进行了重写。虽然 Spring 5 可以使用这两个版本，但是由于 RxJava 1.x 将在 2018 年弃用，所以我们只关注 RxJava 2.x。

在使用 Project Reactor 的 Spring 反应式编程示例中，我们实现了反应式存储库和反应式客户端。接下来将通过使用 RxJava 2.x 来完成相同的功能。定义一个名为 Rx2SingerRepo 的反应式存储库界面：

```
package com.apress.prospring5.ch18.repos;
import com.apress.prospring5.ch18.entities.Singer;
import io.reactivex.Flowable;
import io.reactivex.Single;

public interface Rx2SingerRepo {
   Single<Singer> findById(Long id);
   Flowable<Singer> findAll();
   Single<Void> save(Single<Singer> singer);
}
```

仔细观察上面的代码片段，你可能已经注意到：Flowable 类是元素流的实现，而 Single 类是[0..1]元素流的实现。创建它们非常简单，此外你还会注意到 Single 类在创建空流时的细微 API 差异。

```
Flowable simple = Flowable.just("1", "2", "3");
//or from an existing list: List<Singer>
Flowable<Singer> fromList = Flowable.fromIterable(list);

Single simple = Single.just("1");
//or from an existing object of type Singer
Single<Singer> fromObject = Single.just(null);
```

下面看看如何使用 RxJava 2.x 实现反应式存储库：

```
package com.apress.prospring5.ch18.repos;

import com.apress.prospring5.ch18.entities.Singer;
import io.reactivex.Flowable;
import io.reactivex.Single;
import org.springframework.beans.factory.annotation.Autowired;
import org.springframework.stereotype.Component;

@Component
public class Rx2SingerRepoImpl implements Rx2SingerRepo {

   @Autowired
   SingerRepository singerRepository;

   @Override public Single<Singer> findById(Long id) {
      return Single.just(singerRepository.findById(id).get());
   }
```

1 参见 https://github.com/ReactiveX/RxJava。
2 参见 https://reactivex.io/。

```
    @Override public Flowable<Singer> findAll() {
       return Flowable.fromIterable(singerRepository.findAll());
    }

    @Override public Single<Void> save(Single<Singer> singerSingle) {
          singerSingle.doOnSuccess(singer -> singerRepository.save(singer));
       return Single.just(null);
    }
}
```

正如你所看到的，语法是相似的，但要考虑 API 中的细微差别。如果要重写映射方法，如下所示：

```
package com.apress.prospring5.ch18;
...
@SpringBootApplication
@RestController
public class Rx2ReactiveApplication {

   private static Logger logger = LoggerFactory.getLogger(Rx2ReactiveApplication.class);

   @Autowired Rx2SingerRepo rx2SingerRepo;

   @GetMapping(value = "/all", produces = MediaType.TEXT_EVENT_STREAM_VALUE)
   public Flowable<Singer> all() {
      Flowable<Singer> singers = rx2SingerRepo.findAll();
      Flowable<Long> periodFlowable = Flowable.interval(2, TimeUnit.SECONDS);
      return singers.zipWith(periodFlowable, (singer, aLong) -> {
         Thread.sleep(aLong);
         return singer;
      });
   }

   @GetMapping(value = "/one/{id}")
   public Single<Singer> one(@PathVariable Long id) {
      return rx2SingerRepo.findById(id);
   }
...
}
```

用 RxJava 2.x 实现压缩功能似乎更复杂一点，不是吗？

在测试映射时，仍然可以使用 Spring 反应式 WebClient，它非常酷。也可以从 JDK 9 中获取新的 HTTP 客户端进行测试运行。

```
import jdk.incubator.http.HttpClient;
import jdk.incubator.http.HttpRequest;
import jdk.incubator.http.HttpResponse;
...

URI oneURI = new URI("http://localhost:8080/one/1");
HttpClient client = HttpClient
             .newBuilder()
          .build();
HttpRequest httpRequest = HttpRequest.newBuilder().GET().build();
HttpResponse httpResponse = client.send(httpRequest,
   HttpResponse.BodyHandler.asString());

System.out.println(httpResponse.statusCode());
System.out.println(httpResponse.body());
```

确保在 module-info.java 文件的 jdk.incubator.httpclient 模块中添加一个依赖项。

以上就是 Java 9 互操作性的所有内容。与 Spring 框架相关的最大变化是模块化和 Spring 已经支持的新反应式 API RxJava 2.x。Apress 出版的另一本书 *Java 9 Revealed*[1]中详细介绍了其他新的 Java 9 功能。

18.6.5 Spring 支持 JUnit 5 Jupiter

JUnit 5 在第 13 章中简要提到过，接下来进行深入介绍。如果想要了解 JUnit 5 的最新功能，可以在 http://junit.org/junit5/docs/current/user-guide/#overview 上的官方文档中找到最简单的答案(如图 18-3 所示)。

1 可以从 www.apress.com/la/book/9781484225912 订购。

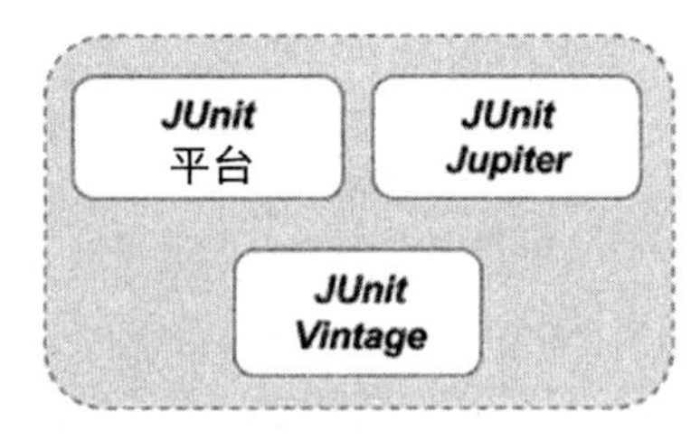

图 18-3 JUnit 5

JUnit 平台是在 JVM 上启动测试框架的基础。它带有一个 Console Launcher，用于从命令行启动平台并为 Gradle 和 Maven 构建插件。该启动器可以用来发现、过滤和执行测试；因此，Gradle 测试任务不需要 Surefire 或自定义设置。此外，Spock、Cucumber 和 FitNesse 等第三方库可通过提供自定义的 TestEngine 插入到 JUnit 平台的启动基础架构中。

JUnit Jupiter 是新的编程模型(基于 org.junit.jupiter.api 包中一组新的 JUnit 5 注解)和扩展模型(带有@ExtendWith 注解(取代了 JUnit 4 的 Runner @Rule 和@ClassRule)的 Extension API)的组合，用于在 JUnit 5 中编写测试和扩展。Jupiter 子项目提供了一个 TestEngine，用于在平台上运行基于 Jupiter 的测试。

正如你所猜到的，JUnit Vintage 提供了一个 TestEngine，用于在平台上运行基于 JUnit 3 和 JUnit 4 的测试。

接下来，通过测试 SingerHandler 类看看具体的测试过程。下面从一个负面的测试方法开始；具体来说，尝试获取一位不存在的歌手。

```
package com.apress.prospring5.ch18;

import com.apress.prospring5.ch18.entities.Singer;
import org.junit.jupiter.api.BeforeAll;
import org.junit.jupiter.api.Test;
import org.slf4j.Logger;
import org.slf4j.LoggerFactory;
import org.springframework.http.HttpMethod;
import org.springframework.http.HttpStatus;
import org.springframework.http.client.reactive.ReactorClientHttpConnector;
import org.springframework.web.reactive.function.BodyInserters;
import org.springframework.web.reactive.function.client.ClientRequest;
import org.springframework.web.reactive.function.client.ClientResponse;
import org.springframework.web.reactive.function.client.ExchangeFunction;
import org.springframework.web.reactive.function.client.ExchangeFunctions;
import reactor.core.publisher.Flux;
import reactor.core.publisher.Mono;

import java.net.URI;
import java.util.Date;
import java.util.GregorianCalendar;
import java.util.List;

import static org.junit.jupiter.api.Assertions.*;
public class SingerHandlerTest {

   private static Logger logger = LoggerFactory.getLogger(SingerHandlerTest.class);

   public static final String HOST = "localhost";

   public static final int PORT = 8080;

   private static ExchangeFunction exchange;

   @BeforeAll
   public static void init(){
      exchange = ExchangeFunctions.create(new ReactorClientHttpConnector());
   }

   @Test
   public void noSinger(){
      //get singer
      URI uri = URI.create(String.format("http://%s:%d/singers/99", HOST, PORT));
      logger.debug("GET REQ: "+ uri.toString());
      ClientRequest request = ClientRequest.method(HttpMethod.GET, uri).build();
      Mono<Singer> singerMono = exchange.exchange(request)
            .flatMap(response -> response.bodyToMono(Singer.class));
```

```
        Singer singer = singerMono.block();
        assertNull(singer);
    }
...
}
```

ExchangeFunction 是一个功能性接口，用于为延迟的 ClientResponse 交换 ClientRequest，并可用作 WebClient 的替代方案。可以将这种实现请求发送到正在运行的服务器，并可以分析响应。在前面的示例中，应该注意以下几点：

- @BeforeAll 注解类似于 JUnit 中的@BeforeClass；因此，在当前类中，使用了@BeforeAll 注解的方法必须在使用了@Test(或类似 RepeatedTest 的派生注解)注解的方法之前执行。该注解的特殊之处在于，如果在类级别使用@TestInstance(Lifecycle.PER_CLASS)注解，那么可以把它设置在非静态方法上。
- org.junit.jupiter.api 包中的@Test 注解等同于 JUnit 中的相同注解；唯一的区别是该注解没有声明任何属性，因为 JUnit Jupiter 中的测试扩展将会基于它们自己的专用注解进行操作。
- 类 org.junit.jupiter.api.Assertions 中的 assertNull 语句类似于 JUnit 中的相同注解。对于 Assertions 类中的所有静态函数，都有类似的实现和一些额外的实现，它们很适合与 Java 8 lambda 一起使用。

下面看一个更详细的例子。此处将测试对 Singer 实例的编辑。首先，通过使用分组的断言检查 FirstName 和 LastName 来确定是否确实检索到了正确的歌手。

```
package com.apress.prospring5.ch18;

import com.apress.prospring5.ch18.entities.Singer;
import org.junit.jupiter.api.BeforeAll;
import org.junit.jupiter.api.Test;
import org.slf4j.Logger;
import org.slf4j.LoggerFactory;
import org.springframework.http.HttpMethod;
import org.springframework.http.HttpStatus;
import org.springframework.http.client.reactive.ReactorClientHttpConnector;
import org.springframework.web.reactive.function.BodyInserters;
import org.springframework.web.reactive.function.client.ClientRequest;
import org.springframework.web.reactive.function.client.ClientResponse;
import org.springframework.web.reactive.function.client.ExchangeFunction;
import org.springframework.web.reactive.function.client.ExchangeFunctions;
import reactor.core.publisher.Flux;
import reactor.core.publisher.Mono;
import java.net.URI;
import java.util.Date;
import java.util.GregorianCalendar;
import java.util.List;

import static org.junit.jupiter.api.Assertions.*;

public class SingerHandlerTest {

    private static Logger logger = LoggerFactory.getLogger(SingerHandlerTest.class);

    public static final String HOST = "localhost";

    public static final int PORT = 8080;

    private static ExchangeFunction exchange;

    @BeforeAll
    public static void init(){
        exchange = ExchangeFunctions.create(new ReactorClientHttpConnector());
    }
    @Test
    public void editSinger() {
        //get singer
        URI uri = URI.create(String.format("http://%s:%d/singers/1", HOST, PORT));
        logger.debug("GET REQ: "+ uri.toString());
        ClientRequest request = ClientRequest.method(HttpMethod.GET, uri).build();

        Mono<Singer> singerMono = exchange.exchange(request)
            .flatMap(response -> response.bodyToMono(Singer.class));
        Singer singer = singerMono.block();
        assertNotNull(singer);

        assertAll("singer",
            () -> assertEquals("John", singer.getFirstName()),
            () -> assertEquals("Mayer", singer.getLastName()));
```

```
        logger.info("singer:" + singer.toString());

        //edit singer
        singer.setFirstName("John Clayton");
        uri = URI.create(String.format("http://%s:%d/singers", HOST, PORT));
        logger.debug("UPDATE REQ: "+ uri.toString());
        request = ClientRequest.method(HttpMethod.POST, uri)
            .body(BodyInserters.fromObject(singer)).build();

        Mono<ClientResponse> response = exchange.exchange(request);
        assertEquals(HttpStatus.OK, response.block().statusCode());
        logger.info("Update Response status: " + response.block().statusCode());
    }
}
```

此外，还进行了其他的测试。在本例中，可以通过 assertNotNull 断言来确定 assertEquals 语句的执行情况。如果返回错误的歌手，则会执行 assertEquals 语句并且失败[1]。下面所示的代码声明了依赖于 assertNotNull 执行的 assertEquals 执行：

```
assertAll("singer", () -> {
        assertNotNull(singer);
    assertAll("singer",
            () -> assertEquals("John", singer.getFirstName()),
            () -> assertEquals("Mayer", singer.getLastName()));
});
```

接下来将停止使用 JUnit 特定的测试组件，转而使用更多新的 Spring 测试功能。除 WebClient 接口外，对于反应式编程模型，spring-test 还包含 WebTestClient，用于集成对 Spring WebFlux 的测试支持。新的 WebTestClient 与 MockMvc 类似，不需要运行服务器。但这并不意味着它不能与现有的应用程序一起使用。在以下示例中，将 WebTestClient 与从 SingerApplication 类运行的应用程序一起使用：

```
package com.apress.prospring5.ch18;

import com.apress.prospring5.ch18.entities.Singer;
import org.junit.jupiter.api.Assertions;
import org.junit.jupiter.api.BeforeAll;
import org.junit.jupiter.api.Test;
import org.springframework.http.MediaType;
import org.springframework.test.web.reactive.server.FluxExchangeResult;
import org.springframework.test.web.reactive.server.WebTestClient;
import reactor.core.publisher.Mono;

import java.util.Date;
import java.util.GregorianCalendar;

import static org.junit.jupiter.api.Assertions.*;

public class AnotherSingerHandlerTest {

    private static WebTestClient client;

    @BeforeAll
    public static void init() {
        client = WebTestClient
        .bindToServer()
        .baseUrl("http://localhost:8080")
        .build();
    }

    @Test
    public void getSingerNotFound() throws Exception {
        client.get().uri("/singers/99").exchange().expectStatus().isNotFound()
            .expectBody().isEmpty();
    }

    @Test
    public void getSingerFound() throws Exception {
        client.get().uri("/singers/1").exchange().expectStatus().isOk()
            .expectBody(Singer.class).consumeWith(seer -> {
        Singer singer = seer.getResponseBody();
        assertAll("singer", () ->
```

1 可以试着将 URL 更改为 http://%s:%d/singers/2。

```
        {
            assertNotNull(singer);
            assertAll("singer",
                () -> assertEquals("John", singer.getFirstName()),
                () -> assertEquals("Mayer", singer.getLastName()));
            });
        });
    }

    @Test
    public void getAll() throws Exception {
        client.get().uri("/singers").accept(MediaType.TEXT_EVENT_STREAM)
            .exchange()
            .expectStatus().isOk()
            .expectHeader().contentType(MediaType.APPLICATION_JSON)
            .expectBodyList(Singer.class).consumeWith(Assertions::assertNotNull);
    }

    @Test
    public void create() throws Exception {
        Singer singer = new Singer();
        singer.setFirstName("Ed");
        singer.setLastName("Sheeran");
        singer.setBirthDate(new Date(
                (new GregorianCalendar(1991, 2, 17)).getTime().getTime()));
        client.post().uri("/singers").body(Mono.just(singer), Singer.class)
            .exchange().expectStatus().isOk();
    }
}
```

WebTestClient 是测试 WebFlux 服务器端点的主要组件。它具有与 WebClient 类似的 API，并将大部分工作委托给主要侧重于提供测试环境的内部 WebClient。如果想要在实际运行的服务器上运行集成测试，就必须调用 bindToServer()方法。上面的示例很简单。WebTestClient 被绑定到实际应用程序正在运行的地址；因此，它不需要定义自己的映射或路由功能。但是在某些情况下，如有必要，也可以这样做，因为还有更多可用的配置方法。以下代码片段中的测试方法通过调用 bindToRouterFunction(function)来使用自定义的 RouterFunction：

```
@Test
public void testCustomRouting(){
    RouterFunction function = RouterFunctions.route(
        RequestPredicates.GET("/test"),
        request -> ServerResponse.ok().build()
    );

    WebTestClient
        .bindToRouterFunction(function)
        .build().get().uri("/test")
        .exchange()
        .expectStatus().isOk()
        .expectBody().isEmpty();
}
```

另一个值得一提的功能是当使用 Spring JUnit 5 时，集成测试可以并行运行。现在，可以暂时离开 Singer 应用程序，以便保持代码示例的指向性。为了演示如何以并行方式运行测试，将使用另一个简单的 Spring Boot 应用程序，该应用程序仅声明了一个 FluxGenerator 类型的 bean，顾名思义，它生成 Flux 实例。下面所示的代码描述了应用程序的配置/入口点以及简单的 bean 类型 FluxGenerator：

```
//FluxGenerator.java
package com.apress.prospring5.ch18;

import org.springframework.stereotype.Component;
import reactor.core.publisher.Flux;

@Component
public class FluxGenerator {

    public Flux<String> generate(String... args){
        return Flux.just(args);
    }
}

//Application.java
package com.apress.prospring5.ch18;

import org.slf4j.Logger;
```

```
import org.slf4j.LoggerFactory;
import org.springframework.boot.autoconfigure.SpringBootApplication;
import org.springframework.boot.builder.SpringApplicationBuilder;
import org.springframework.context.ConfigurableApplicationContext;

@SpringBootApplication
public class Application {

   private static final Logger logger = LoggerFactory.getLogger(Application.class);
   public static void main(String args) throws Exception {
      ConfigurableApplicationContext ctx =
         new SpringApplicationBuilder(Application.class)
            .run(args);
      assert (ctx != null);
      logger.info("Application started...");

      System.in.read();
      ctx.close();
   }
}
```

接下来，将创建两个测试类——IntegrationOneTest 和 IntegrationTwoTest，它们将分别声明两个测试方法；这两个方法不会测试任何东西，只是使用 FluxGenerator bean 来获取 Flux 实例并打印其内容。下面的代码描述了这两个类，正如你所看到的，值的集合是不同的，所以可以在命令行中跟踪执行并确保它们并行完成。

```
//IntegrationOneTest.java
package com.apress.prospring5.ch18.test;

import com.apress.prospring5.ch18.FluxGenerator;
import org.junit.Test;
import org.junit.runner.RunWith;
import org.slf4j.Logger;
import org.slf4j.LoggerFactory;
import org.springframework.beans.factory.annotation.Autowired;
import org.springframework.boot.test.context.SpringBootTest;
import org.springframework.test.context.junit4.SpringRunner;

@RunWith(SpringRunner.class)
@SpringBootTest
public class IntegrationOneTest {

   private final Logger logger = LoggerFactory.getLogger(IntegrationOneTest.class);
   @Autowired FluxGenerator fluxGenerator;

   @Test
   public void test1One() {
      fluxGenerator.generate("1", "2", "3").collectList().block()
      .forEach(s -> executeSlow(2000, s) );
   }

   @Test
   public void test2One() {
      fluxGenerator.generate("11", "22", "33").collectList().block()
      .forEach(s -> executeSlow(1000, s));
   }

   private void executeSlow(int duration, String s) {
      try {
      Thread.sleep(duration);
         logger.info(s);
         } catch (InterruptedException e) {
      }
   }
}

//IntegrationTwoTest.java
package com.apress.prospring5.ch18.test;
... //same imports as above
@RunWith(SpringRunner.class)
@SpringBootTest
public class IntegrationTwoTest {
   private final Logger logger =
      LoggerFactory.getLogger(IntegrationTwoTest.class);

   @Autowired FluxGenerator fluxGenerator;

   @Test
```

```
    public void test1One() {
        fluxGenerator.generate(2, "a", "b", "c").collectList().block()
            .forEach(logger::info);
    }

    @Test
    public void test2One() {
        fluxGenerator.generate(3, "aa", "bb", "cc").collectList().block()
            .forEach(logger::info);
    }
}
```

下面的代码定义了执行上述测试的类。它包含两个测试方法：一个并行执行测试；另一个以线性方式一个接一个执行测试。

```
package com.apress.prospring5.ch18.test;

import org.junit.experimental.ParallelComputer;
import org.junit.jupiter.api.Test;
import org.junit.runner.Computer;
import org.junit.runner.JUnitCore;

public class ParallelTests {
    @Test
    void executeTwoInParallel() {
        final Class<?> classes = {
            IntegrationOneTest.class, IntegrationTwoTest.class
        };

        JUnitCore.runClasses(new ParallelComputer(true, true), classes);
    }

    @Test
    void executeTwoLinear() {
        final Class<?> classes = {
            IntegrationOneTest.class, IntegrationTwoTest.class
        };

        JUnitCore.runClasses(new Computer(), classes);
    }
}
```

JUnitCore 是用于运行测试的装饰，它支持 JUnit 4 和 JUnit 3.8 测试以及混合测试。它接收 Computer 实例或 ParallelComputer 实例的扩展作为参数。Computer 类用于正常执行测试，以线性方式依次执行。ParallelComputer 实例允许并行运行测试，其构造函数接收两个布尔参数。第一个用于类的并行执行，第二个用于方法的并行执行。在前面的示例中，将它们都设置为“true”，从而告知 JUnit 运行程序以并行方式执行类以及类中的方法。这就是为什么 IntegrationOneTest 与对应的 IntegrationTwoTest 类略有不同的原因。

如果运行 executeTwoInParallel()方法，你应该能在控制台中看到类似如下输出的日志：

```
...
17:29:30.460 [pool-2-thread-2] INFO c.a.p.c.t.IntegrationTwoTest - aa
17:29:30.460 [pool-2-thread-2] INFO c.a.p.c.t.IntegrationTwoTest - bb
17:29:30.460 [pool-2-thread-1] INFO c.a.p.c.t.IntegrationTwoTest - a
17:29:30.460 [pool-2-thread-2] INFO c.a.p.c.t.IntegrationTwoTest - cc
17:29:30.460 [pool-2-thread-1] INFO c.a.p.c.t.IntegrationTwoTest - b
17:29:30.460 [pool-2-thread-1] INFO c.a.p.c.t.IntegrationTwoTest - c
17:29:31.463 [pool-1-thread-2] INFO c.a.p.c.t.IntegrationOneTest - 11
17:29:32.461 [pool-1-thread-1] INFO c.a.p.c.t.IntegrationOneTest - 1
17:29:32.468 [pool-1-thread-2] INFO c.a.p.c.t.IntegrationOneTest - 22
17:29:33.472 [pool-1-thread-2] INFO c.a.p.c.t.IntegrationOneTest - 33
17:29:34.466 [pool-1-thread-1] INFO c.a.p.c.t.IntegrationOneTest - 2
17:29:36.471 [pool-1-thread-1] INFO c.a.p.c.t.IntegrationOneTest - 3
...
```

最后介绍的主题是 JUnit 5 Extension API。其核心是 Extension 接口，它只是组件的一个标记接口，可以使用@ExtendWith 或通过 Java 的 ServiceLoader 机制自动注册。在 Spring 5 中，SpringExtension 已被添加到 spring-test 模块中，该模块实现了许多@ExtendWith 的 Jupiter 接口衍生物，从而将 Spring TestContext Framework 集成到 JUnit 5 的 Jupiter 编程模型中。要使用此扩展，只需要使用@ExtendWith(SpringExtension.class)、@SpringJUnitConfig 或@SpringJUnitWebConfig 注解基于 JUnit Jupiter 的测试类即可。

接下来看一个简单的示例。首先创建一个名为 TestConfig 的类，它声明了一个 FluxGenerator 类型的 bean，然后创建一个测试该 bean 的测试类。这两个类如下所示：

```
//TestConfig.java
package com.apress.prospring5.ch18.test.config;

import com.apress.prospring5.ch18.FluxGenerator;
import org.springframework.context.annotation.Bean;
import org.springframework.context.annotation.Configuration;
@Configuration
public class TestConfig {

   @Bean FluxGenerator generator(){
      return new FluxGenerator();
   }
}

package com.apress.prospring5.ch18.test;

import com.apress.prospring5.ch18.Application;
import com.apress.prospring5.ch18.FluxGenerator;
import com.apress.prospring5.ch18.test.config.TestConfig;
import org.junit.jupiter.api.Test;
import org.junit.jupiter.api.extension.ExtendWith;
import org.springframework.beans.factory.annotation.Autowired;
import org.springframework.test.context.ContextConfiguration;
import org.springframework.test.context.junit.jupiter.SpringExtension;

import java.util.List;
import static org.junit.jupiter.api.Assertions.assertEquals;

//JUnit5IntegrationTest.java
@ExtendWith(SpringExtension.class)
@ContextConfiguration(classes = TestConfig.class)
public class JUnit5IntegrationTest {

   @Autowired FluxGenerator fluxGenerator;

   @Test
   public void testGenerator() {
      List<String> list = fluxGenerator.generate("2", "3")
         .collectList().block();
      assertEquals(2, list.size());
   }
}
```

在 Spring JUnit 5 支持方面还有很多工作要做。因此，本节介绍的源代码在未来的 Spring 版本中可能会有一些修改。该领域是非常新的，甚至还没有在 Spring 文档中介绍过[1]，可能会在未来某个时候更新。

18.7 小结

在本章中，提供了对 Spring 组合中几个项目的高级概述。我们介绍了 Spring Batch、JSR-352、Integration、XD、WebFlux 以及 Spring 对 JUnit 5 的支持，其中每个项目都提供自己的独特功能，目的是简化开发人员的特定任务。其中一些项目是新的，有些已经被证明是稳定和可靠的，能为其他框架奠定理想的基础。建议大家更深入地了解这些项目，因为它们可以大大简化 Java 项目。

1 Spring 参考文档的测试章节还没有提到 JUnit 5，具体请参阅 https://docs.spring.io/spring/docs/current/spring-framework-reference/htmlsingle/#testing。

附录 A

设置开发环境

本附录将帮助读者设置开发环境，你将在该环境下编写、编译和执行本书中讨论的 Spring 应用程序。

A.1 pro-spring-15 项目介绍

pro-spring-15 项目是一个三层 Gradle 项目，如图 A-1 所示。pro-spring-15[1] 是根项目，是名为 chapter02~chapter18 的项目的父项目，而每个 chapter**项目又是嵌套在其下的模块项目的父项目。pro-spring-15 项目的这种设计方式有助于以代码为导向，并容易将每章与各自相关的源代码匹配起来。

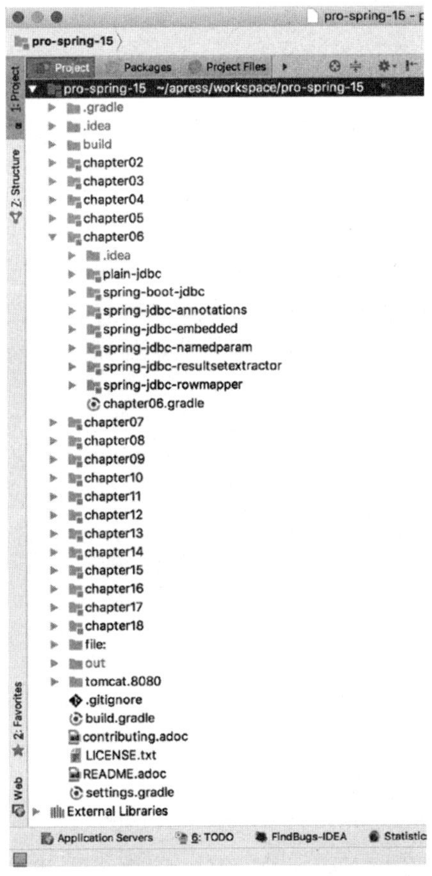

图 A-1 IntelliJ IDEA 中的 pro-spring-15 项目结构

1 在编写本附录时，该项目仍处于初始开发状态。在开发完成后，它被重新命名为 pro-spring-5，并托管在 Apress 官方存储库中，详见 https://github.com/Apress/pro-spring-5。除项目名称外，以上所有内容仍然有效。

A.2 了解 Gradle 配置

pro-spring-15 项目定义了一组可供子模块使用的库，它在通常名为 build.gradle 的文件中完成 Gradle 配置。第二级上的所有模块(换句话说，即 chapter**项目)都有一个名为[module_name].gradle 的 Gradle 配置文件(例如 chapter02.gradle)，因此可以为该章项目中常见的设置快速找到配置文件。此外，pro-spring-15/settings.gradle 中还有一个 closure 元素，用于验证在构建时所有章节模块都具有其配置文件。

```
rootProject.children.each { project ->
    project.buildFileName = "${project.name}.gradle"
    assert project.projectDir.isDirectory()
    assert project.buildFile.exists()
    assert project.buildFile.isFile()
}
```

如果 Gradle 构建文件的命名与模块不同，那么在执行任何 Gradle 任务时都会引发错误。该错误类似于以下代码片段中描述的错误，其中 chapter02.gradle 文件被重命名为 chapter02_2.gradle:

```
$ gradle clean
FAILURE: Build failed with an exception.

* Where:
Settings file '/workspace/pro-spring-15/settings.gradle' line: 328

* What went wrong:
A problem occurred evaluating settings 'pro-spring-15'.
> assert project.buildFile.exists
         |       |          |
         |       |          false
         |       /workspace/pro-spring-15/chapter02/chapter02.gradle
         :chapter02
* Try:
Run with --stacktrace option to get the stack trace. Run with --info or --debug
      option to get more log output.

BUILD FAILED in 0s
```

这是一种开发选择；这样一来，章节模块的配置文件在编辑器中就可见了。另外，如果要修改章节模块的配置文件，只需要使用唯一名称就可以轻松地在 IntelliJ IDEA 中找到该文件。

在 pro-spring-15 项目中，第三级中的每个模块(换句话说，即 chapter**项目的子项目)都有一个名为 build.gradle 的 Gradle 配置文件。这也是一种开发选择；在 Gradle 的这个特定配置中，在该级别拥有唯一命名的 Gradle 配置文件是不可能的。

对于多模块项目，另一种方法是在主 build.gradle 文件中使用特定的闭包来自定义每个模块的配置。但是，本着良好开发实践的精神，还是决定将模块的配置尽可能分离，并与模块内容保持在同一位置。

针对每个正在使用的软件版本，pro-spring-15/build.gradle 配置文件都包含一个变量。这些变量用于自定义以特定用途或技术命名的数组中的依赖声明字符串。该文件还包含一个公共存储库列表，从中可以下载依赖项。

在以下配置中，可以看到项目中使用的持久化库的版本和数组：

```
ext {
    ...
    //persistency libraries
    hibernateVersion = '5.2.10.Final'
    hibernateJpaVersion = '1.0.0.Final'
    hibernateValidatorVersion = '5.4.1.Final' //6.0.0.Beta2
    atomikosVersion = '4.0.4'

    hibernate = [
        validator   : "org.hibernate:hibernate-validator:$hibernateValidatorVersion",
        jpaModelGen : "org.hibernate:hibernate-jpamodelgen:$hibernateVersion",
        ehcache     : "org.hibernate:hibernate-ehcache:$hibernateVersion",
        em          : "org.hibernate:hibernate-entitymanager:$hibernateVersion",
        envers      : "org.hibernate:hibernate-envers:$hibernateVersion",
        jpaApi      : "org.hibernate.javax.persistence:hibernate-jpa-2.1-api:
           $hibernateJpaVersion",
        querydslapt : "com.mysema.querydsl:querydsl-apt:2.7.1",
        tx          : "com.atomikos:transactions-hibernate4:$atomikosVersion"
    ]
}
...
```

```
subprojects {
   version '5.0-SNAPSHOT'

   repositories {
      mavenLocal()
      mavenCentral()
      maven { url "http://repo.spring.io/release" }
      maven { url "http://repo.spring.io/snapshot" }
      maven { url "https://repo.spring.io/libs-snapshot" }
      maven { url "http://repo.spring.io/milestone" }
      maven { url "https://repo.spring.io/libs-milestone" }
   }
}

tasks.withType(JavaCompile) {
   options.encoding = "UTF-8"
}
```

在层次结构的第二级项目(换句话说，即 chapter**项目)中，可以找到 chapter**.gradle 配置文件。这些依赖项通过数组名称以及与之关联的名称引用。此外，这些文件还包含特定于该章的项目组名称、其他插件和额外的 Gradle 任务。下面显示了 chapter08.gradle 文件：

```
subprojects {
   group 'com.apress.prospring5.ch08'
   apply plugin: 'java'

   /*Task that copies all the dependencies under build/libs */
   task copyDependenciestype: Copy {
      from configurations.compile
      into 'build/libs'
   }

   dependencies {
      if !project.name.contains"boot" {
         compile spring.contextSupport {
            exclude module: 'spring-context'
            exclude module: 'spring-beans'
            exclude module: 'spring-core'
         }
         compile spring.orm, spring.context, misc.slf4jJcl,
           misc.logback, db.h2, misc.lang3, hibernate.em
      }
      testCompile testing.junit
   }
}
```

此处的 if(!project.name.contains("boot"))是必需的，因为在项目 chapter08 下嵌套了一个或多个 Spring Boot 项目。由于 Spring Boot 为这些项目附带了一组固定的依赖项和版本，因此不希望此文件中定义的配置被继承，因为这可能导致冲突或不可预知的行为。

对于第三级项目，可以通过添加自己的依赖项以及声明清单文件规范、插件和任务来进一步自定义从 chapter** 继承的配置。下面显示的是一个名为 chapter12\spring-invoker\build.gradle 的简单文件(如果想了解更多自定义的 Gradle 配置，请查看 chapter08\jpa-criteria\build.gradle 文件)：

```
apply plugin: 'war'

dependencies {
   compile project(':chapter12:base-remote')
   compile spring.webmvc, web.servlet
   testCompile spring.test
}

war {
   archiveName = 'remoting.war'
   manifest {
      attributes("Created-By" : "Iuliana Cosmina",
         "Specification-Title": "Pro Spring 5",
         "Class-Path" : configurations.compile.collect { it.getName() }.join(' '))
   }
}
```

A.3 构建和故障排除

在克隆(或下载)源代码之后，需要在 IntelliJ IDEA 编辑器中导入项目。为此，请按照下列步骤操作：

(1) 从 IntelliJ IDEA 菜单中依次选择 File | New | Project from Existing Sources，如图 A-2 所示。

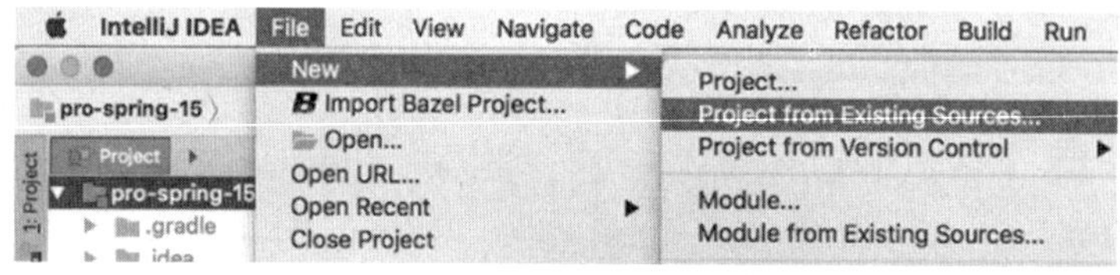

图 A-2 IntelliJ IDEA 中的项目导入菜单选项

选择适当的选项后，会出现一个弹出窗口，要求确定项目的位置，如图 A-3 所示。

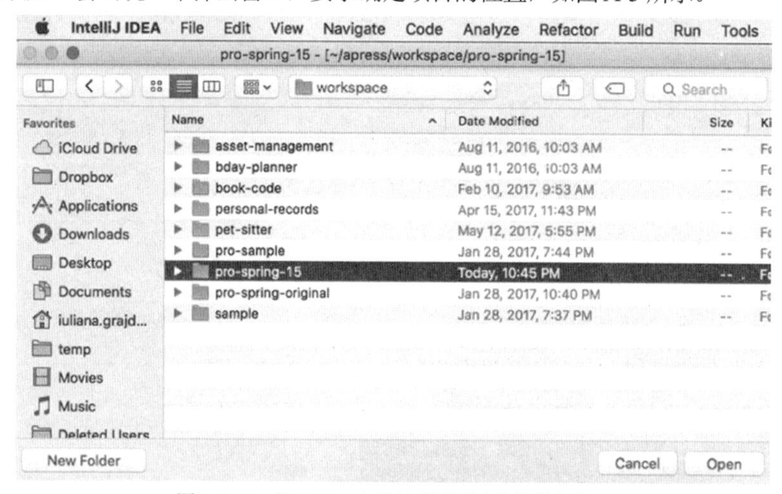

图 A-3 IntelliJ IDEA 中选择项目根目录的弹出窗口

(2) 选择 pro-spring-15 目录。弹出窗口会询问项目类型。IntelliJ IDEA 可以根据所选的源创建自己的项目类型，并使用内部 Java 构建器来构建，但此时该选项没有用，因为 pro-spring-15 是一个 Gradle 项目。

(3) 选中 Import project from external model 单选按钮并从下方的列表框中选择 Gradle，如图 A-4 所示。

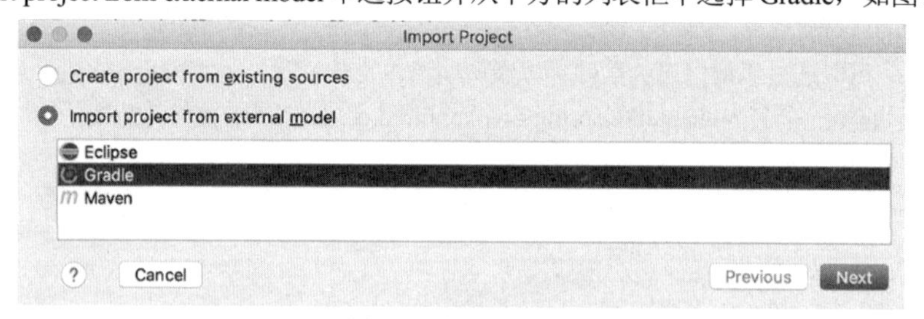

图 A-4 选择项目类型

(4) 出现最后一个弹出窗口并询问 build.gradle 文件和 Gradle 可执行文件的位置。这些选项默认已设置好，如图 A-5 所示。如果在系统上安装了 Gradle，则可能需要使用它。

在开始后续工作之前，应该构建项目。可单击 Refresh 按钮，通过 IntelliJ IDEA 来完成，如图 A-6 中的(1)标记所示。单击该按钮将使 IntelliJ IDEA 扫描项目的配置并解析依赖项，包括下载缺失的库以及对项目进行内部的轻量化构建(足以消除由于缺少依赖项而导致的编译时错误)。

Gradle compileJava 任务(在图 A-6 中用(3)标记)执行完整的项目构建，也可以通过执行 Gradle build 命令从命令行完成相同的操作，如下所示：

```
.../workspace/pro-spring-15 $ gradle build
```

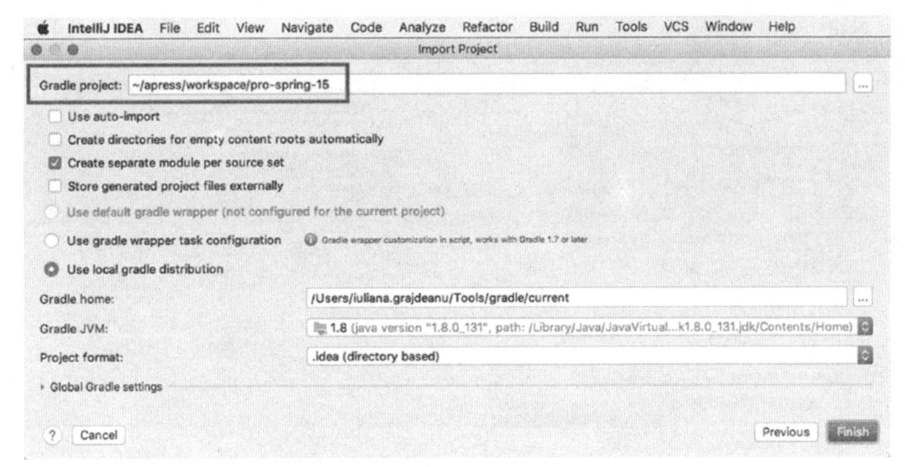

图 A-5　用于在 IntelliJ IDEA 中导出项目的弹出窗口

也可以在 IntelliJ IDEA 中通过双击构建任务来完成此操作，如图 A-6 中的标记(2)所示。

随后，将会在每个模块上执行以下一组任务：

```
:chapter02:hello-world:compileJava
:chapter02:hello-world:processResources
:chapter02:hello-world:classes
:chapter02:hello-world:jar
:chapter02:hello-world:assemble
:chapter02:hello-world:compileTestJava
:chapter02:hello-world:compileTestResources
...
```

在前面的示例中，这些任务仅用于模块 chapter02。Gradle 构建任务将执行它所依赖的所有任务。正如你所看到的，它不会运行 clean 任务，所以需要确保在构建项目时多次手动运行该任务，以确保使用最新版本的类。

由于此项目包含旨在尝试失败的测试，因此执行此任务将失败。也可以只执行任务 clean clean 和 compileJava。另一种选择是执行 Gradle 构建任务，但通过使用“-x”参数跳过测试：

```
.../workspace/pro-spring-15 $ gradle build -x test
```

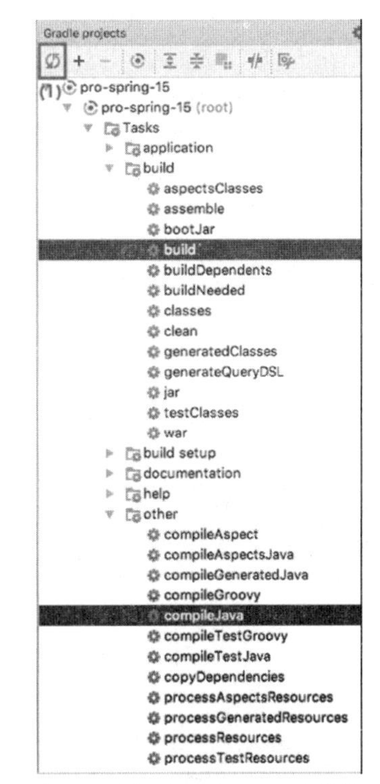

图 A-6　用于在 IntelliJ IDEA 中进行项目导入的弹出窗口

A.4　部署到 Apache Tomcat 上

在 pro-spring-15 项目下有几个 Web 应用程序。其中大多数是在嵌入式服务器上运行的 Spring Boot 应用程序，但使用 Apache Tomcat 服务器等外部容器有一定的优势。以调试模式启动服务器并使用断点来调试应用程序要容易得多，但这只是优势之一。外部容器可以一次运行多个应用程序，而无须停止服务器。嵌入式服务器对测试、快速开发和教育都非常有用，例如 Spring Boot；但在生产中，应用程序服务器是首选。

如果有兴趣使用外部服务器，则必须执行以下操作。首先从官方站点下载最新版本的 Apache Tomcat[1]。可以获得版本 8 或 9；它们都可以与本书的资源一起使用。将下载的文件解压到系统的某个位置。然后配置 IntelliJ IDEA 启动程序以启动服务器并部署选定的应用程序。这很容易做到，按照以下步骤操作：

(1) 从可运行的配置菜单中，选择 Edit Configuration，如图 A-7 中的标记(1)所示。这会显示一个弹出窗口并列出一组启动器。单击加号(+)并选择 Tomcat Server 选项。此时菜单将展开；选择 Local，如图 A-7 中的标记(2)所示，

1　Apache Tomcat 官方网站位于 http://tomcat.apache.org/。

因为正在使用计算机上安装的服务器。

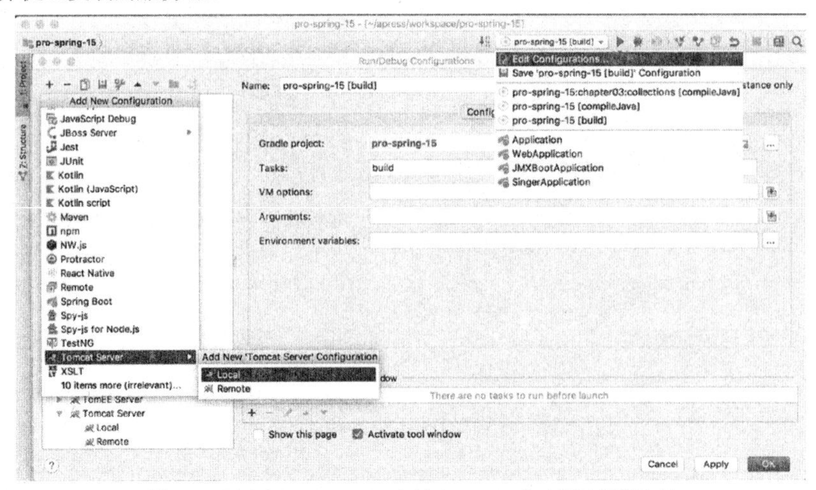

图 A-7　用于在 IntelliJ IDEA 中创建 Tomcat 启动器的菜单选项

(2) 如图 A-8 所示的弹出窗口将会出现，并会请求一些信息。

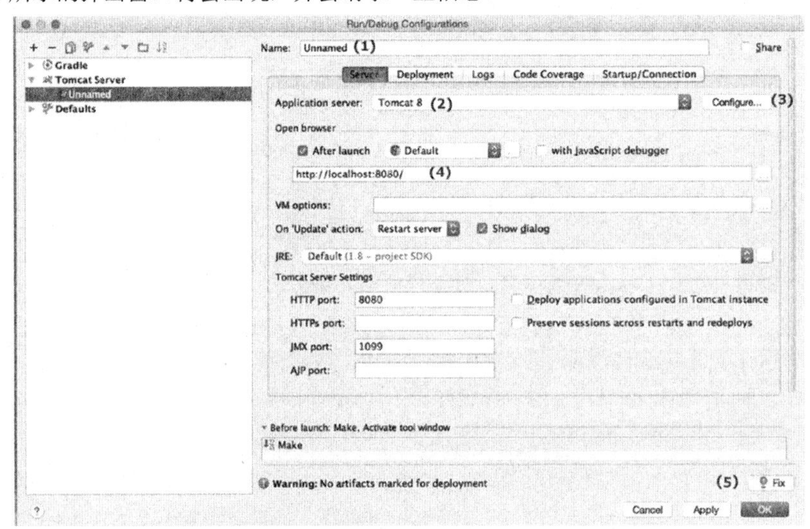

图 A-8　用来在 IntelliJ IDEA 中创建 Tomcat 启动器的弹出窗口

在图 A-8 中，你会看到一些项目被编号。下面是这些数字的含义：

(1) 是启动器名称。可以是任何名称，但最好是正在部署的应用程序的名称。

(2) 是 Tomcat 实例名称。

(3) 是打开弹出窗口以插入 Tomcat 实例位置的按钮，如图 A-9 所示。

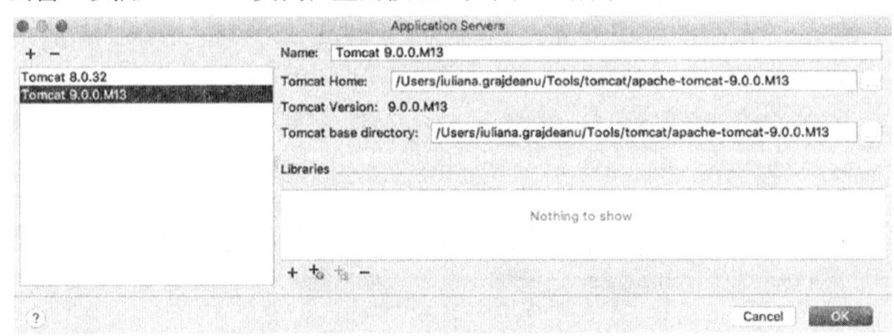

图 A-9　在 IntelliJ IDEA 中配置 Tomcat 实例

(4) 是可以访问 Tomcat 服务器的 URL。

(5) 是 Fix 按钮，用于选择要在 Tomcat 实例上部署的工件。如果没有要部署到 Tomcat 的 Web 归档文件集，该按钮将显示为红色的灯泡图标。

(6) 单击 Fix 按钮并选择一个工件。IntelliJ IDEA 将检测所有可用的工件(如图 A-10 所示)并将它们显示在列表中。如果打算以调试模式打开服务器并在代码中使用断点，请选择名称中后缀为(exploded)的工件；这样一来，IntelliJ IDEA 将管理 WAR 的内容，并且可以将浏览器中的操作与代码中的断点连接起来。

图 A-10　IntelliJ IDEA 中可部署工件的列表

(7) 单击 OK 按钮完成配置。可通过在 Application Context 字段中插入新值来指定不同的应用程序上下文。选择不同的应用程序上下文将告诉 Tomcat 在给定名称下部署应用程序。应用程序将通过下面的 URL 访问：

```
http://localhost:8080/[app_context_name]/
```

只要 IntelliJ IDEA 提供对应的插件，其他应用程序服务器就能够以类似的方式使用。启动器配置可以重复，并且多个 Tomcat 实例可以同时启动，只要它们在不同的端口上运行，这样就可以加快测试和实现之间的比较。IntelliJ IDEA 非常灵活和实用，这就是为什么推荐使用它来完成本书中练习的原因。Gradle 项目也可以导入到支持 Gradle 的 Eclipse 和其他 Java 编辑器中。